D0975939

NUMERICAL
METHODS

電子計算機數字方法

NUMERICAL METHODS

GERMUND DAHLQUIST

Department of Computer Sciences
Royal Institute of Technology, Stockholm

ÅKE BJÖRCK

Department of Mathematics
Linköping University, Sweden

Translated by

NED ANDERSON

Department of Computer Sciences
Royal Institute of Technology, Stockholm

DAHLQUIST, GERMUND.
Numerical Methods.

"An extended and updated translation of":
Numeriska metoder, by Å. Björck and G. Dahlquist.
Bibliography: p.
1. Electronic data processing—Numerical analysis.
I. Björck, Åke, joint author. II. Björck, Åke,
Numeriska metoder. III. Title.

著作權執照號碼：台內著字第七七三〇號

This edition was translated from Numeriska Metoder by Åke Björck
and Germund Dahlquist, published by CWK Gleerup, Bokfoerlag,
P. O. Box 1205, Lund, Sweden, © 1969.

Original American edition published by
Prentice-Hall, Inc., Englewood Cliffs,
New Jersey, U.S.A.

中華民國六十九年 四月十六日第一版
局版臺業字第〇六七五號

發行人：孫國仁
住址：臺北市峨眉街107號
發行所：虹橋書店
發行所地址：臺北市峨眉街107號
印刷所：合興彩色印刷有限公司
印刷所地址：台北市大理街130巷2弄1號

To the memory

of

George E. Forsythe

CONTENTS

Both section and page numbers are given for the topics below. The reader is recommended to use section numbers (at the top of the pages of the text proper) when looking up a given topic, since a better idea of the structure of the book will be obtained.

In general, review questions and/or problems follow after each one-decimal sub-section.

PREFACE

This book is an extended and updated translation of a textbook published in Swedish by the CWK Gleerup Co. in 1969. Prerequisites for most of the book are sophomore courses in mathematics (in particular calculus and linear algebra) as well as some knowledge of a problem-oriented programming language. The latter can be studied in parallel with the first three chapters of the book. Parts of the book are more difficult. We hope that it will be easy for teachers who will use the book as a text to give instructions as to what depth the various parts are to be studied in a particular course.

We have tried to select methods which are important to large-scale computing as well as techniques for small-scale computing with simple tools. It is helpful if the reader has access to a time-sharing system for getting acquainted with some of the algorithms, though in most cases a desk calculator or even a slide rule and mathematical tables can be useful enough.

We hope that the book will be useful as a handbook for computation in science and technology. The last chapter contains an extensive bibliography and a list of published algorithms.

The general ideas and concepts of scientific computation are introduced in the first chapter, while the second chapter is devoted to error analysis. There and everywhere else we try to stress those aspects which are of importance for the design of algorithms. In contrast to the general survey style of the first two chapters, the rest of the book is mostly concerned with the treatment of various classes of problems.

The most important difference between this book and the Swedish edition is the increased number of exercises. We have had the use of material from a Swedish collection of problems by our colleagues Nils Hagander and Yngve Sundblad. Their generosity is gratefully acknowledged. Section 7.7 on functions of several variables is new, and so is Section 10.5 on non-linear optimization. The related sections on non-linear equations is thoroughly revised.

We also have a new treatment of methods for partial differential equations, other than difference methods. In addition to this there has been an overall modernization, particularly in the chapter on linear algebra. A chapter on nomography and some peripheral details have been excluded.

We have benefited greatly from stimulating discussions with colleagues. In particular we mention Peter Pohl, Håkan Ramsin, and the group at TRU (Swedish board for television and radio in education).

We are especially grateful to our colleague Ned Anderson, who has not only translated the Swedish text but improved the presentation in many places, and to Maja Anderson for typing most of the manuscript. Christina Landing also typed a large part of the manuscript. It is a pleasure to express our gratitude for excellent cooperation and invaluable help.

We also thank the American and Swedish publishers for excellent cooperation. It was the encouragement of the late Professor George E. Forsythe that stimulated us to translate the book.

Last but not least we would like to thank our families for help, encouragement, and patience when the work intruded greatly on their leisure time.

GERMUND DAHLQUIST

ÅKE BJÖRCK

CONVENTIONS

The book consists of thirteen chapters numbered in a way that is clear from the Contents. The reference "see 4.3.3" indicates Section 4.3.3 in Chapter 4. The reference (4.3.5) indicates a formula which can be found in one of the sections whose first decimal is 3. [The formula (4.3.5) is in Section 4.3.3.] A name followed by digits enclosed in brackets, e.g. Fröberg [7] refers to a book in the Bibliography, Chapter 13.

Besides the generally accepted mathematical abbreviations and notations (see e.g., James and James [38], pp. 467–471), the following notations are used in the book:

$\log x$	logarithm to base 10
$\ln x$	natural logarithm
e^x and $\exp(x)$	both denote the exponential function
$\sinh x$, $\cosh x$	hyperbolic sine, hyperbolic cosine
$\operatorname{sgn} x$	sign of x, see 3.1.2
$fl\,(x + y)$	floating-point operations, see 2.3.5
$\operatorname{tab}(f)$	tabulation operator, see 4.1.1
$E_n(f)$	see 4.3.1
$\|\|\cdot\|\|_p$, $\|\|\cdot\|\|_{p,w}$	norms and seminorms, see 4.1.3 and 5.5.2
$\{x_i\}_{i=0}^n$	denotes the set $\{x_0, x_1, \ldots, x_n\}$
$[a, b]$	closed interval $(a \leq x \leq b)$
(a, b)	open interval $(a < x < b)$, or scalar product of the functions a and b
$\operatorname{int}(a, b, \ldots, w)$	the least interval which contains $a, b, \ldots, w$
$k \leq i, j \leq n$	means $k \leq i \leq n$ and $k \leq j \leq n$

The notations $\approx$, $\ll$, $<$, O, o are defined in 2.1.1. Vectors and matrices are printed in boldface italics. $\mathbf{A}^T$ and $\mathbf{x}^T$ denote the transpose of the matrix $\mathbf{A}$ and the vector $\mathbf{x}$ respectively. $(\boldsymbol{A}, \boldsymbol{B})$ means a partitioned matrix (see Sec. 5.2). The rest of the notation for linear algebra can be found in Section 5.2. Notations for differences and difference operators, e.g., $\Delta^2 y_n$, $f[x_0, x_1, x_2]$, $\delta^2 y$ are defined in Chapter 7.

1 SOME GENERAL PRINCIPLES OF NUMERICAL CALCULATION

1.1. INTRODUCTION

Mathematics is used in one form or another within most of the areas of science and industry. There has always been a close interaction between mathematics on the one hand and science and technology on the other. During the present century, advanced mathematical models and methods have been used more and more even within other areas—for example, in medicine, economics, and social science.

Very often, applications lead to mathematical problems which in their complete form cannot be conveniently solved with exact formulas. One often restricts oneself then to special cases or simplified models which can be exactly analyzed. In most cases, one thereby reduces the problem to a linear problem —for example, a linear differential equation. Such an approach can be very effective, and leads quite often to concepts and points of view which can at least qualitatively be used even in the unreduced problem.

But occasionally such an approach does not suffice. One can instead treat a less simplified problem with the use of a large amount of numerical calculation. The amount of work depends on the demand for accuracy. With computers, which have been developed during the past twenty-five years, the possibilities of using numerical methods have increased enormously. The points of view which one has taken toward them have also changed.

To develop a numerical method means, in most cases, that one applies a small number of general and relatively simple ideas. One combines these ideas in an inventive way with one another and with such knowledge of the given problem as one can obtain in other ways—for example, with the methods of mathematical analysis. Some knowledge of the background of the problem is also of value; among other things, one should take into account the order of magnitude of certain numerical data of the problem.

In this book we shall illustrate the use of the general ideas behind numerical methods on some problems which often occur as subproblems or computational details of larger problems, though as a rule they occur in a less pure form and on a larger scale than they do here. When we present and analyze numerical methods, we use to some degree the same approach which was described first above: we study in detail special cases and simplified situations, with the aim of uncovering more generally applicable concepts and points of view which can be a guide in more difficult problems.

In this chapter we shall throw some light upon some important ideas and problems in abbreviated form. A more systematic treatment comes in the chapters following.

1.2. SOME COMMON IDEAS AND CONCEPTS IN NUMERICAL METHODS

One of the most frequently recurring ideas in many contexts is **iteration** (from the Latin *iteratio*, "repetition") or **successive approximation.** Taken generally, iteration means the repetition of a pattern of action or process. Iteration in this sense occurs, for example, in the repeated application of a numerical process—perhaps very complicated and itself containing many instances of the use of iteration in the somewhat narrower sense to be described below—in order to successively improve previous results. To illustrate a more specific use of the idea of iteration, we consider the problem of solving an equation of the form

$$x = F(x). \tag{1.2.1}$$

Here F is a differentiable function whose value we can compute for any given value of the real variable x (within a certain interval). Using the method of iteration, one starts with an initial approximation x_0, and computes the sequence

$$x_1 = F(x_0), \qquad x_2 = F(x_1), \qquad x_3 = F(x_2), \ \ldots \tag{1.2.2}$$

Each computation of the type

$$x_{n+1} = F(x_n)$$

is called an iteration. If the sequence $\{x_n\}$ converges to a limiting value α, then we have $\lim F(x_n) = F(\alpha)$, so $x = \alpha$ satisfies the equation $x = F(x)$. As n grows, we would like the numbers x_n to be better and better estimates of the desired root. One stops the iterations when sufficient accuracy has been attained.

A geometric interpretation is shown in Fig. 1.2.1. A root of Eq. (1.2.1) is given by the abscissa (and ordinate) of an intersection point of the curve $y = F(x)$ and the line $y = x$. Using iteration and starting from $(x_0, F(x_0))$ we obtain $x_1 = F(x_0)$ and the point x_1 on the x-axis is obtained by first drawing a horizontal line from the point $(x_0, F(x_0)) = (x_0, x_1)$ until it intersects the line $y = x$ in the point (x_1, x_1). From there we draw a vertical line to $(x_1, F(x_1)) = (x_1, x_2)$ and so on. In Fig. 1.2.1 it is obvious that $\{x_n\}$ converges monotonely to α. Figure 1.2.2 shows a case where F is a decreasing function. There we also have convergence but not monotone convergence: the successive iterates x_n are alternately to the right and to the left the of root α.

But there are also divergent cases, exemplified by Figs. 1.2.3 and 1.2.4. One can see geometrically that the quantity which determines the rate of convergence (or divergence) is the slope of the curve $y = F(x)$ in the neighborhood of the root. Indeed, from the mean value theorem we have

$$\frac{x_{n+1} - x_n}{x_n - x_{n-1}} = \frac{F(x_n) - F(x_{n-1})}{x_n - x_{n-1}} = F'(\xi_n),$$

where ξ_n lies between x_{n-1} and x_n. Thus convergence is faster the smaller

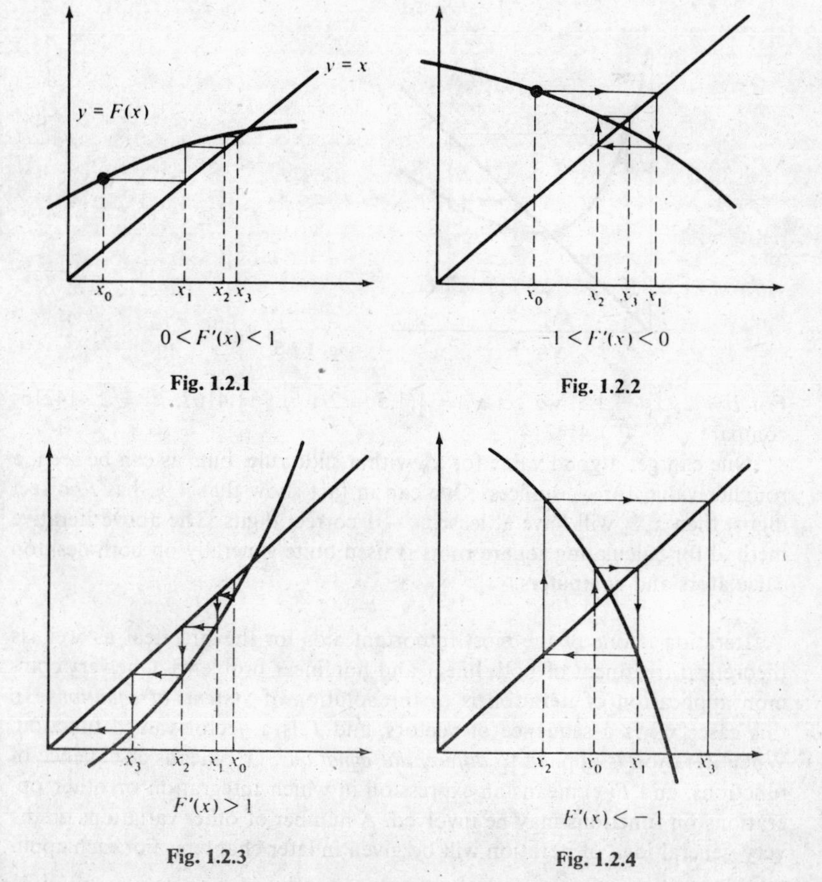

Fig. 1.2.1

Fig. 1.2.2

Fig. 1.2.3

Fig. 1.2.4

$|F'(x)|$ is in a neighborhood of the root. Convergence is assured if $|F'(x)| < 1$ for all x in a neighborhood of the root containing x_0 and x_1. But if $|F'(\alpha)| > 1$, x_n converges to α only in very exceptional cases, no matter how close to α one chooses $x_0 (x_0 \neq \alpha)$.

Example 1.2.1. *A Fast Method for Calculating Square Roots*

The equation $x^2 = c$ can be written in the form $x = F(x)$, where

$$F(x) = \frac{1}{2}\left(x + \frac{c}{x}\right), \qquad c > 0$$

(Fig. 1.2.5). The limiting value is $\alpha = c^{1/2}$ and $F'(\alpha) = 0$. (Show this!) Thus we set

$$x_{n+1} = \frac{1}{2}\left(x_n + \frac{c}{x_n}\right).$$

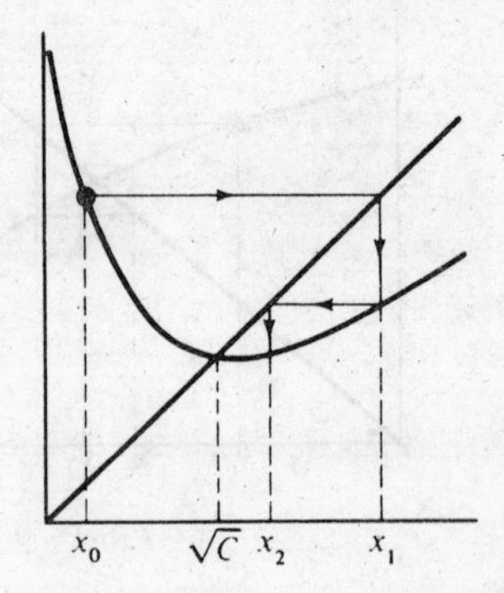

Fig. 1.2.5

For $c = 2$, $x_0 = 1.5$, we get $x_1 = \frac{1}{2}(1.5 + 2/1.5) = 1.4167$, $x_2 = 1.414216$; compare $\sqrt{2} = 1.414214 \ldots$.

One can get a good value for x_0 with a slide rule, but, as can be seen, a rougher value for x_0 suffices. One can in fact show that if x_n has t correct digits, then x_{n+1} will have at least $2t - 1$ correct digits. The above iterative method for calculating square roots is used quite generally on both desktop calculators and computers.

Iteration is one of the most important aids for the practical as well as theoretical treatment of both linear and nonlinear problems. One very common application of iteration is to the solution of *systems of equations*. In this case $\{x_n\}$ is a sequence of vectors, and F is a vector-valued function. When iteration is applied to *differential equations*, $\{x_n\}$ means a sequence of functions, and $F(x)$ means an expression in which integration or other operations on functions may be involved. A number of other variations on the very general idea of iteration will be given in later chapters. For each application it is of course necessary to find a suitable way to put the equations in a form similar to Eq. (1.2.1) and to choose a suitable initial approximation. One has a certain amount of choice in these matters which should be used in order to reduce the number of iterations one will have to make.

Example 1.2.2

The equation $x^2 = 2$ can also be written, among other ways, in the form $x = 2/x$. In the previous example we saw that the form $x = \frac{1}{2}(x + 2/x)$ gave rapid convergence when iteration was applied. On the other hand, the formula $x_{n+1} = 2/x_n$ gives a sequence which goes back and forth between x_0 (for even n) and $2/x_0$ (for n odd)—the sequence does not converge.

Another often recurring idea is that **one locally** (that is, in a small neighborhood) **approximates a complicated function with a linear function**. We shall illustrate the use of this idea in the solution of the equation $f(x) = 0$. Geometrically, this means that we are seeking the intersection point between the x-axis and the curve $y = f(x)$ (Fig. 1.2.6). Assume that we have an approximating value x_0 to the root. We then approximate the curve with its **tangent** at the point $(x_0, f(x_0))$. Let x_1 be the abscissa of the point of intersection between the x-axis and the tangent. Normally x_1 will be a much better approximation to the root than x_0. In most cases, x_1 will have nearly twice as many correct digits as x_0, but if x_0 is a very poor initial approximation, then it is possible that x_1 will be worse than x_0.

A combination of the ideas of iteration and local linear approximation gives rise to a much used and, ordinarily, rapidly convergent process which is called **Newton-Raphson's method** (Fig. 1.2.7). In this iterative method x_{n+1} is defined as the abscissa of the point of intersection between the x-axis and the tangent to the curve $y = f(x)$ in the point $(x_n, f(x_n))$.

The approximation of the curve $y = f(x)$ with its tangent at the point $(x_0, f(x_0))$ is equivalent to replacing the function with the first-degree terms in its Taylor series about $x = x_0$. The corresponding approximation for functions of many variables also has important uses.

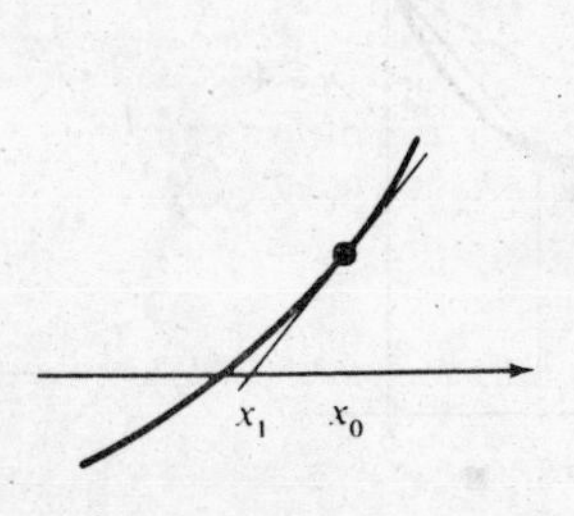

Fig. 1.2.6

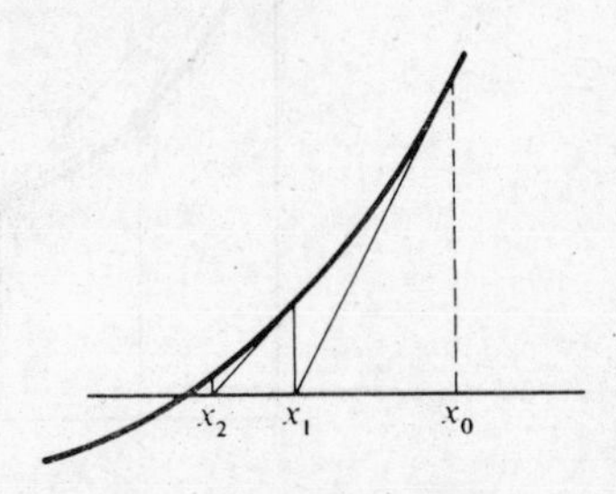

Fig. 1.2.7

Another way (instead of drawing the tangent) to approximate a curve locally is to choose two neighboring points on the curve and to approximate the curve with the **secant** which joins the two points (Fig. 1.2.8). In a later chapter, we shall discuss more closely the **secant method** for the solution of equations, which is based on the above approximation.

The same secant approximation is useful in many other contexts. It is, for instance, generally used when one "reads between the lines" in a table of numerical values. In this case the secant approximation is called **linear interpolation.**

When the secant approximation is used in the approximate calculation of a definite integral,

$$I = \int_a^b y(x)\,dx,$$

numerical integration (Fig. 1.2.9), it is called the **trapezoidal rule.** With this method, the area between the curve $y = y(x)$ and the x-axis is approximated with sum $T(h)$ of the areas of a series of parallel trapezoids. Using the notation of Fig. 1.2.9 we have

$$T(h) = \tfrac{1}{2}h \sum_{i=0}^{n-1} (y_i + y_{i+1}), \qquad (nh = b - a)$$

(in the figure, $n = 4$). We shall show in a later chapter that the error $(T(h) - I)$ in the above approximation is very nearly proportional to h^2 when

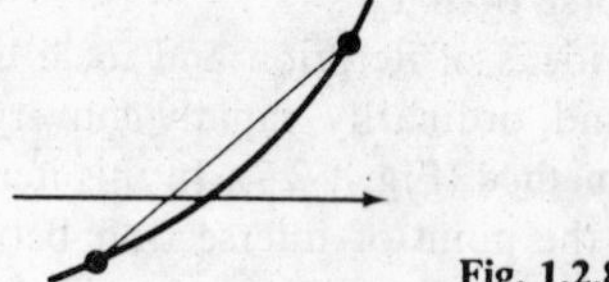

Fig. 1.2.8

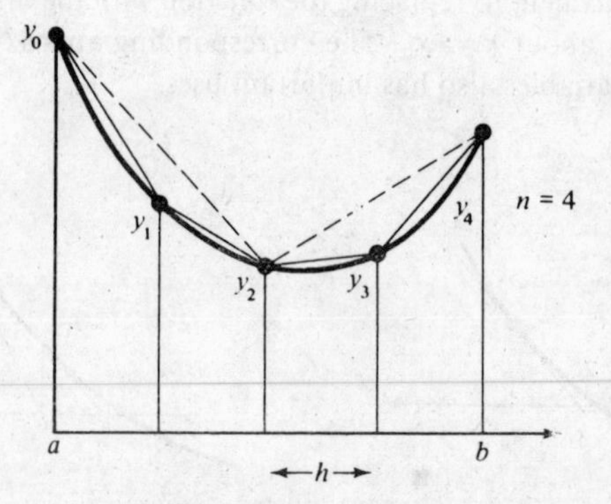

Fig. 1.2.9

h is small. One can then, in principle, attain arbitrarily high accuracy by choosing h sufficiently small; except that the computational work involved (the number of points where $y(x)$ must be computed) is inversely proportional to h. Thus the computational work grows rapidly as one demands higher accuracy (smaller h).

Numerical integration is a fairly common problem because in fact it is quite seldom that the "primitive" function can be analytically calculated in a finite expression containing only elementary functions. It is not possible, for example, for such simple functions as $\exp(x^2)$ or $(\sin x)/x$. In order to obtain **higher accuracy** with significantly less work than the trapezoidal rule requires, one can use one of the following two important ideas:

(a) **Local approximation** of the integrand with a polynomial of higher degree (or with a function of some other class, for which one knows the primitive function).
(b) Computation with the trapezoidal rule for several values of h and then *extrapolation to $h = 0$*, so-called **Richardson extrapolation** or **the deferred approach to the limit**, with the use of general results concerning the dependence of the error upon h.

The technical details for the various ways of approximating a function with a polynomial, among others Taylor expansions, interpolation, and the method of least squares, are treated in later chapters.

The extrapolation idea can easily be applied to numerical integration with the trapezoidal rule. As was mentioned previously, the trapezoidal approximation to $I = \int_a^b y(x)\,dx$ has an error approximately proportional to the square of the step size. Thus, using two step sizes, h and $2h$, one has:

$$
\begin{aligned}
&(T(h) - I) \approx kh^2 \\
&(T(2h) - I) \approx k(2h)^2 \\
&\therefore \quad 4(T(h) - I) \approx T(2h) - I \\
&3I \approx 4T(h) - T(2h) \\
&I \approx T(h) + \tfrac{1}{3}(T(h) - T(2h)).
\end{aligned}
$$

Thus, by adding the corrective term, $\frac{1}{3}(T(h) - T(2h))$, to $T(h)$, one should get an estimate of I which is much better than $T(h)$. In Chap. 7 we shall see that the improvement is in most cases quite striking. That chapter also contains a further development of the extrapolation idea, **Romberg's method**.

Example 1.2.3

Compute

$$\int_{10}^{12} f(x)\,dx$$

for $f(x) = x^3$ and $f(x) = x^4$ by the trapezoidal method. Extrapolate and compare with the exact results.

$f(x)$	x^3	x^4
$f(10)$	1,000	10,000
$f(11)$	1,331	14,641
$f(12)$	1,728	20,736
$T(2)$	2,728	30,736
$T(1)$	2,695	30,009
Extrapolation	2,684	29,766.67
Exact result	2,684	29,766.4

We have seen above that some knowledge of the behavior of the error can, together with the idea of extrapolation, lead to a powerful method for improving results. Such a line of reasoning is useful not only for the common problem of numerical integration, but also in many other types of problems.

Approximate solution of *differential equations* is a very important problem which now, since the development of computers, one has the possibility of treating to a much larger extent than previously. Nearly all the areas of science and technology contain mathematical models which lead to systems of ordinary or partial differential equations. Let us consider a case of just one ordinary differential equation,

$$\frac{dy}{dx} = f(x, y)$$

with initial condition $y(0) = p$. The differential equation indicates, at each point (x, y), the direction of the tangent to the solution curve which passes through the point in question. The direction of the tangent changes continuously from point to point, but the simplest approximation (which was pro-

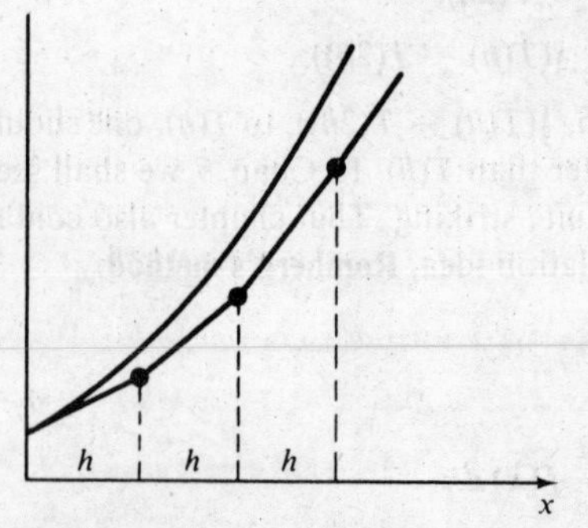

Fig. 1.2.10

posed as early as the 18th century, by Euler) is that one studies the solution for only certain values of $x = 0, h, 2h, 3h, \ldots$ (h is called the "step" or "step length") and assumes that dy/dx is constant between the points. In this way, the solution curve is approximated by a polygon segment (Fig. 1.2.10) which joins the points $(0, y_0)$, (h, y_1), $(2h, y_2)$, ... where

$$y_0 = p, \qquad \frac{y_{n+1} - y_n}{h} = f(nh, y_n). \tag{1.2.3}$$

Thus we have a simple *recursion formula*, **(Euler's method)**:

$$y_0 = p, \qquad y_{n+1} = y_n + hf(nh, y_n), \qquad n = 0, 1, 2, 3, \ldots. \tag{1.2.4}$$

During the computation, each y_n occurs first on the left-hand side, then *recurs* later on the right-hand side of an equation: hence the name **recursion formula**. (One could also call Eq. (1.2.4) an iteration formula, but one usually reserves the word "iteration" for the special case where a recursion formula is used solely as a means of calculating the limiting value.)

The only disadvantage of the above method is that the step length must be quite short if reasonable accuracy is desired. In order to improve the method one can—just as in the case of numerical integration—choose either the use of local approximation with a polynomial of higher degree or the use of extrapolation to $h = 0$. As we shall see in Chap. 8, a combination of these two possibilities gives good results.

In Eq. (1.2.3) the derivative $y'(nh)$ is replaced by a **difference quotient** $(y_{n+1} - y_n)/h$. The approximation of derivatives with difference quotients is also one of the most frequently encountered devices in the construction of numerical methods, among other things in the numerical treatment of more complicated differential equations. Observe, though, that $(y_{n+1} - y_n)/h$, which is the slope of the secant line between (nh, y_n) and $((n + 1)h, y_{n+1})$, in reality is a better approximation for the derivative at the midpoint of the interval $[nh, (n + 1)h]$ than at its left end point (see Fig. 1.2.11). The value of the

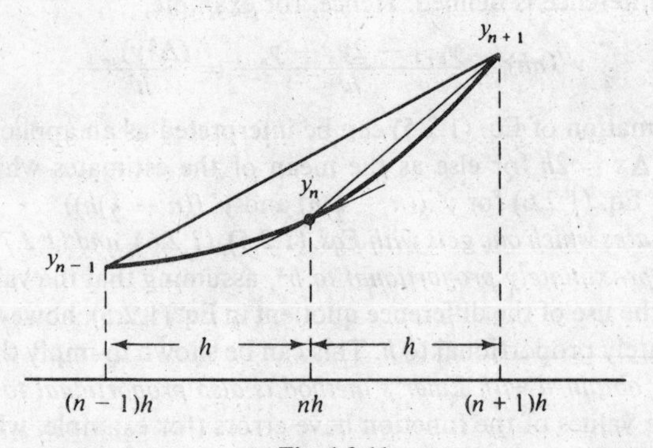

Fig. 1.2.11

derivative at the point $x = nh$ is more accurately approximated by a centered difference quotient (see Fig. 1.2.11),

$$y'(nh) \approx \frac{y_{n+1} - y_{n-1}}{2h}. \tag{1.2.5}$$

The above approximation is in most situations preferable to the one mentioned previously. There are, however, situations where the first mentioned suffices, but where the centered difference quotient is entirely unusable, for reasons which have to do with how errors are propagated to later stages in the calculation. We shall not discuss this more closely here, but mention it only to intimate some of the surprising and fascinating mathematical questions which can arise in the study of numerical methods.

Higher derivatives are approximated with **higher differences,** that is, differences of differences, another central concept in numerical calculation. We define:

$$(\Delta y)_n = y_{n+1} - y_n$$
$$(\Delta^2 y)_n = (\Delta(\Delta y))_n = (y_{n+2} - y_{n+1}) - (y_{n+1} - y_n) = y_{n+2} - 2y_{n+1} + y_n$$
$$(\Delta^3 y)_n = (\Delta(\Delta^2 y))_n = \ldots = y_{n+3} - 3y_{n+2} + 3y_{n+1} - y_n$$

etc.

For simplicity one often omits the parentheses and writes, for example, $\Delta^2 y_5$ instead of $(\Delta^2 y)_5$ The coefficients that appear here in the expressions for the higher differences are, by the way, the binomial coefficients. In addition, if we denote the step length by Δx instead of by h, we get the following formulas, which are easily remembered:

$$\frac{dy}{dx} \approx \frac{\Delta y}{\Delta x}, \quad \frac{d^2 y}{dx^2} \approx \frac{\Delta^2 y}{(\Delta x)^2}, \quad \text{etc.} \tag{1.2.6}$$

Here we mean the value of the derivative for an x which lies *right between* the largest and the smallest x where the corresponding value of y is needed, so that the difference is defined. Hence, for example,

$$y''(nh) \approx \frac{y_{n+1} - 2y_n + y_{n-1}}{h^2} = \frac{(\Delta^2 y)_{n-1}}{h^2}. \tag{1.2.7}$$

The approximation of Eq. (1.2.5) can be interpreted as an application of Eq. (1.2.6) with $\Delta x = 2h$ (or else as the mean of the estimates which one gets according to Eq. (1.2.6) for $y'((n + \frac{1}{2})h)$ and $y'((n - \frac{1}{2})h)$).

The estimates which one gets with Eqs. (1.2.5), (1.2.6), and (1.2.7) have errors which are approximately proportional to h^2, assuming that the values of y are exact. With the use of the difference quotient in Eq. (1.2.3), however, the error is approximately proportional to h. This can be shown to imply that *the error in the results obtained with Euler's method is also proportional to h (not h^2).*

When the values of the function have errors (for example, when they are rounded numbers), the difference quotients become more and more uncertain the less h is. Thus if one wishes to compute the derivatives of a function given by a table, one should as a rule use a step length which is greater than the table step.

Example 1.2.4

For $y(x) = \cos(x)$ one has, using a six-figure table:

x	y	Δy	$\Delta^2 y$
0.59	0.830941		
		−5605	
0.60	0.825336		−83
		−5688	
0.61	0.819648		

Using Eq. (1.2.5) one gets $y'(0.60) \approx (0.819648 - 0.830941)/0.02 = -0.56465$. Using Eq. (1.2.7) one gets $y''(0.60) \approx -83 \cdot 10^{-6}/(0.01)^2 = -0.83$. The correct values are, with six decimals, $y'(0.60) = -0.564642$, $y''(0.60) = -0.825336$. The arrangement of the numbers in the example is called a **difference scheme**.

In many of the applications in which differential equations occur, x is the time variable and the differential equation expresses the rule or law of nature which directs the changes in the given system. The method of calculation of Eq. (1.2.4) means, then, that one **simulates** the passing of time. Simulations analogous to the above-mentioned are often used whenever the changes in the system are described with a mathematically much more complicated type of equation than an ordinary differential equation. In recent years, computers have been used to simulate automobile traffic flow and battles between tank units, among other things. Using simulation, one often bypasses the conventional mathematical formulation of the problem, as, for example, a system of differential equations. Instead, one proceeds directly from a verbal or graphical description of the system to a computer program. The technique of simulation is also of great value in studying the influence of random factors on a complicated system. In this connection one uses so-called **random numbers**; the values of certain variables are determined by a process comparable to dice throwing.

We have now seen a variety of ideas and concepts which can be used in the development of numerical methods. A small warning is perhaps warranted here: it is not certain that the methods will work as well in practice as one might expect. This is because approximations and the restriction of numbers to a certain number of decimals or digits introduce errors which are propagated to later stages of a calculation. The manner in which errors are propagated is decisive for the practical usefulness of a given numerical method.

We shall examine such questions first in Example 1.3.3 and, above all, in Chap. 2. Later chapters will treat **propagation of errors** in connection with various typical problems. The risk that error propagation may upstage the desired result of a numerical process should not, however, dissuade one from the use of numerical methods. It is often wise, though, to experiment with a proposed method on a simplified problem before using it in a larger context. As a rule, a mixture of careful experiment and analysis leads to the best results.

REVIEW QUESTIONS

1. Make a list of the concepts and ideas which have been introduced. Review their use in the various types of problems mentioned.
2. Discuss the convergence condition and the rate of convergence of the method of iteration for solving $x = F(x)$.
3. What is the trapezoidal rule? What is said about the dependence of its error on the step length h?

PROBLEMS

(In these problems, six-place tables should be used.)

1. Calculate $\sqrt{10}$ to five decimal places using the method in Example 1.2.1. Begin with $x_0 = 3$ and check the final result in a table of square roots.
2. Calculate $\int_0^{1/2} e^x \, dx$
   (a) to six decimals, by determining the primitive function.
   (b) with the trapezoidal rule, step length $h = \frac{1}{4}$.
   (c) using extrapolation to $h = 0$ on the results which one gets with $h = \frac{1}{2}, \frac{1}{4}$.
   (d) Compute the ratio between the error in the result in (c) to that of (b).
3. What is the relationship between x_{n+1} and x_n in the application of Newton's method to the equation $f(x) = 0$? (Set up the equation for the tangent to the curve $y = f(x)$ in the point $(x_n, f(x_n))$.) What formula does one get when $f(x) = x^2 - c$? Have you seen this before?
4. Integrate numerically the differential equation $dy/dx = y$, with initial condition $y(0) = 1$, to $x = 0.4$. Use Euler's method;
   (a) with step length $h = 0.2$.
   (b) with $h = 0.1$.
   (c) Extrapolate to $h = 0$, using the fact that the error is approximately proportional to the step length (not to the square of the step length). Compare the result with the exact solution to the differential equation. What is the ratio between the errors in the results in (b) and (c), respectively.
   (d) How many steps would one have needed in order to attain, without using extrapolation, the same accuracy as was obtained in (c)?
5. In Example 1.2.4 we computed $y''(0.6)$ for $y = \cos(x)$, with step length

$h = 0.01$. Make similar calculations using $h = 0.1$, $h = 0.05$, and $h = 0.001$. Which value of h gives the best result (using a six-place table)? Discuss qualitatively the influences of both the rounding errors in the table values and the error in the approximation of a derivative with a difference quotient on the result for various values of h.

6. Show that $F'(\alpha) = 0$ in Example 1.2.1.

1.3. NUMERICAL PROBLEMS AND ALGORITHMS

1.3.1. Definitions

By a **numerical problem** we mean a clear and unambiguous description of the *functional connection* between **input data**—that is, the "independent variables" in the problem—and **output data**—that is, the desired results. Input and output data consist of a finite number of real quantities. (Since a complex number is a pair of real numbers, complex input and output data is included in this definition.) Input and output data are thus representable by finite dimensional vectors. The functional connection can be expressed in either explicit or implicit form.

By an **algorithm** for a given numerical problem we mean a *complete description of well-defined operations* through which each permissible input data vector is transformed into an output data vector. By "operations" we mean here arithmetic and logical operations which a computer can perform, together with references to previously defined algorithms. (The concept "algorithm" can be analogously defined for problems completely different from "numerical problems," with other types of input data and fundamental operations—for example, inflection, merging of words, and other transformations of words in a given language.)

For a given numerical problem one can consider many differing algorithms. These can give approximate answers which have widely varying accuracy.

Example 1.3.1

To determine the largest real root of the equation

$$x^3 + a_2x^2 + a_1x + a_0 = 0,$$

with real coefficients a_0, a_1, a_2, is a numerical problem. The input data vector is (a_0, a_1, a_2). The output data is the root x; it is an implicitly defined function of the input data. An algorithm for this problem can be based on Newton-Raphson's method, supplemented with rules for how the initial approximation should be chosen and how the iteration process is to be terminated. One could also use other iterative methods, or even algorithms based upon Cardan's exact solution of the cubic equation. Cardan's solution uses square roots and cube roots, so one needs to assume that algorithms for the computation of these functions have been specified previously.

One often begins the construction of an algorithm for a given problem by breaking down the problem into subproblems in such a way that the output data from one subproblem is the input data to the next subproblem. Thus the distinction between problem and algorithm is not always so easy to make. The essential point is that, in the formulation of the problem, one is only concerned with the initial state and the final state. In an algorithm, however, one should clearly define each step along the way, from start to finish.

Example 1.3.2

The problem of solving the differential equation

$$\frac{d^2y}{dx^2} = x^2 + y^2$$

with boundary conditions $y(0) = 0$, $y(5) = 1$, is, according to the definition stated above, not a "numerical problem." This is because the output data is the *function* y, which cannot, in any conspicuous way, be specified by a finite number of parameters. The above problem is a *mathematical problem*, which *can be approximated with a numerical problem* if one specifies the output data to be values approximating $y(x)$ for $x = h, 2h, 3h, \ldots, 5 - h$, and one approximates the derivative with a difference quotient according to Eq. (1.2.7). In this way, one gets a system of nonlinear equations with $5/h - 1$ unknowns. We shall not go further here into how the domain of variation of the unknowns must be restricted in order to show that the problem has a unique solution. This can be done, however, and one can also give a number of algorithms for solving the system, some good and some bad with respect to the number of calculations made and the accuracy obtained.

1.3.2. Recursive Formulas; Horner's Rule

One of the most important and interesting parts of the preparation of a problem for a computer is to *find a recursive description of the task*. Sometimes an enormous amount of computation can be described by a small set of recursive formulas. Euler's method for the step-by-step solution of ordinary differential equations (p. 9) is an example. Other examples will be given in this section and in Sec. 1.3.3. See also the problems at the end of the chapter.

A common computational task is the evaluation of a polynomial, at a given point z where, say,

$$p(z) = a_0z^3 + a_1z^2 + a_2z + a_3.$$

This can be reformulated as

$$p(z) = ((a_0z + a_1)\cdot z + a_2)\cdot z + a_3.$$

For computation by hand, the following scheme, **Horner's rule**, illustrates the algorithm indicated by the above reformulation:

$$\begin{array}{cccc} a_0 & a_1 & a_2 & a_3 \\ & z\cdot b_0 & z\cdot b_1 & z\cdot b_2 \\ \hline b_0 & b_1 & b_2 & b_3, \end{array} \qquad p(z) = b_3.$$

Example 1.3.3

Compute $p(8)$, where $p(x) = 2x^3 + x + 7$.

$$\begin{array}{cccc} 2 & 0 & 1 & 7 \\ & 16 & 128 & 1{,}032 \\ \hline 2 & 16 & 129 & 1{,}039 \end{array} \qquad p(8) = 1{,}039.$$

Horner's rule for evaluating a polynomial of degree n,

$$p(x) = a_0x^n + a_1x^{n-1} + \cdots + a_{n-1}x + a_n,$$

at a point z, is described by the recursive formula:

$$b_0 = a_0, \qquad b_i = a_i + z\cdot b_{i-1} \qquad (i = 1, 2, \ldots, n), \quad b_n = p(z). \qquad (1.3.1)$$

If the intermediate b_i are of no interest, then, in most programming languages, the algorithm can be described without subscripts for the b_i, such as in the flowchart in Fig. 1.3.1 and the corresponding Algol-fragment:

```
b := a[0];
for i := 1 step 1 until n do
    b := a[i] + z * b;
```

(The symbol := is read "is given the value of.")

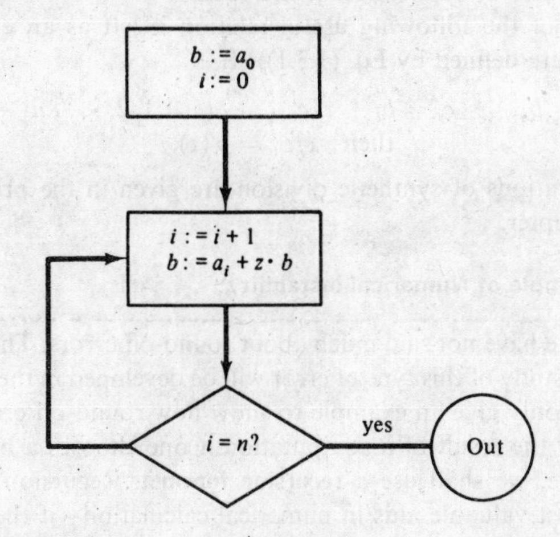

Fig. 1.3.1

Sometimes, however, the b_i are of intrinsic interest because of the following result, often called **synthetic division**:

THEOREM 1.3.1

$$\frac{p(x) - p(z)}{x - z} = \sum_{i=0}^{n-1} b_i x^{n-1-i},$$

where the b_i are defined by Eq. (1.3.1).

Proof. Denote the right-hand side by $g(x)$. Then

$$\begin{aligned}(x - z)g(x) &= \sum_{i=0}^{n-1} b_i x^{n-i} - \sum_{i=0}^{n-1} b_i z x^{n-1-i} \\ &= \sum_{i=0}^{n-1} b_i x^{n-i} - \sum_{i=1}^{n} b_{i-1} z x^{n-i} \\ &= \sum_{i=1}^{n} (b_i - b_{i-1} z) x^{n-i} + b_0 x^n - b_n x^0.\end{aligned}$$

Hence, by Eq. (1.3.1),

$$\begin{aligned}(x - z)g(x) &= \sum_{i=1}^{n} a_i x^{n-i} + a_0 x^n - p(z) \\ &= p(x) - p(z),\end{aligned}$$

and the theorem is proved.

Synthetic division is used, for instance, in the solution of algebraic equations, when already-computed roots are sucessively eliminated. Then, after each elimination, one can deal with an equation of lower degree. This process is called *deflation*. In Chap. 6, however, it is shown that some care is necessary in the numerical application of this idea.

The proof of the following useful relation is left as an exercise to the reader (the b_i are defined by Eq. (1.3.1)): If

$$c_0 = b_0, \qquad c_i = b_i + z \cdot c_{i-1} \qquad (i = 1, 2, \ldots, n-1),$$
$$\text{then} \quad c_{n-1} = p'(z). \tag{1.3.2}$$

Further applications of synthetic division are given in the problems at the end of this chapter.

1.3.3. An Example of Numerical Instability

Thus far, we have not said much about round-off errors. The terminology necessary to a study of this type of error will be developed in the next chapter. Now we shall only give an example to show how round-off errors can completely destroy the result of a computation if one chooses a bad algorithm. In the example, we shall use a **recursion formula**. Recursion formulas are among the most valuable aids in numerical calculation—if they are used in the right way. As intimated in Sec. 1.3.2, one can specify very extensive calculations in relatively short computer programs with the help of such formulas.

Example 1.3.4

Compute for $n = 0, 1, \ldots, 8$

$$y_n = \int_0^1 \frac{x^n}{x+5}\,dx.$$

Use the recursion formula,

$$y_n + 5y_{n-1} = \frac{1}{n},$$

which follows from

$$y_n + 5y_{n-1} = \int_0^1 \frac{x^n + 5x^{n-1}}{x+5}\,dx = \int_0^1 \frac{x^{n-1}(x+5)}{x+5}\,dx = \int_0^1 x^{n-1}\,dx = \frac{1}{n}.$$

We use three decimals throughout the example.

Algorithm 1. Compute

$$y_0 = \int_0^1 \frac{dx}{x+5} = [\ln(x+5)]_0^1 = \ln 6 - \ln 5 \approx 0.182,$$

$$y_1 = 1 - 5y_0 = 1 - 0.910 \approx 0.0090,$$

$$y_2 = \tfrac{1}{2} - 5y_1 \approx 0.050,$$

$$y_3 = \tfrac{1}{3} - 5y_2 \approx 0.083, \qquad \text{strange that } y_3 > y_2!$$

$$y_4 = \tfrac{1}{4} - 5y_3 \approx -0.165, \qquad \text{obviously absurd that } y_4 < 0!$$

The reason for the absurd result is that the round-off error ϵ in y_0, whose magnitude can be as high as $5 \cdot 10^{-4}$ is *multiplied* by -5 in the calculation of y_1, which then has an error of -5ϵ. That error produces an error in y_2 of 25ϵ, etc. Thus the error in y_4 is 625ϵ, the value of which can be as large as $625 \cdot 5 \cdot 10^{-4} = 0.3125$. On top of this comes the round-off error committed in the various steps of the calculation, which, however, in this case can be shown to be relatively unimportant.

If one uses more decimal places of accuracy throughout the calculation, the absurd results will show up at a later stage. The above algorithm is an example of a disagreeable phenomenon, called **numerical instability**. We shall now see that one can avoid numerical instability by choosing a more suitable algorithm.

Algorithm 2. We use the recursion formula in the other direction,

$$y_{n-1} = \frac{1}{5n} - \frac{y_n}{5}.$$

Now the error will be *divided* by -5 in each step. But we need a starting value. We can see directly from the definition that y_n decreases as n increases. One can also surmise that y_n decreases slowly when n is large (the reader is recommended to motivate this). Thus we try setting $y_{10} \approx y_9$, and it follows that:

$$y_9 + 5y_9 \approx \tfrac{1}{10}, \qquad \text{or } y_9 \approx \tfrac{1}{60} \approx 0.017$$

$$\text{(show that } 0 < y_{10} < \tfrac{1}{60} < y_9)$$

$$y_8 = \frac{1}{45} - \frac{y_9}{5} \approx 0.019,$$

$$y_7 = \frac{1}{40} - \frac{y_8}{5} \approx 0.021,$$

$$y_6 \approx 0.025,$$

$$y_5 \approx 0.028,$$

$$y_4 \approx 0.034,$$

$$y_3 \approx 0.043,$$

$$y_2 \approx 0.058,$$

$$y_1 \approx 0.088,$$

$$y_0 \approx 0.182. \qquad \text{(Correct!)}$$

Algorithm 3. The same as Algorithm 2 except that one takes as starting value $y_{10} = 0$. One then gets $y_9 = 0.020$, $y_8 = 0.018$, and the rest of the y_n have the same values as in Algorithm 2. The difference in the values for y_{10} in the two algorithms is 0.017. The subsequent values $y_9, y_8, \ldots, y_0$ in the two algorithms are quite close because the error is divided by -5 in each step. A closer analysis is given in Example 2.2.12; the results obtained with Algorithm 2 have errors which are less than 10^{-3} for $n \leq 8$.

The reader is warned, however, not to draw erroneous conclusions from the above example. The use of a recursion formula "backwards" is not a universal recipe! Compare Problem 10 at the end of this section!

In this book, we mean by the term **numerical method** a procedure which is often useful, either to approximate a mathematical problem with a numerical problem or to solve a numerical problem (or at least to reduce the numerical problem to a simpler problem). The transformation of a differential equation problem to a system of nonlinear equations (as in Example 1.3.2) is a numerical method—even without instructions as to how to solve the system of equations. When, as in Example 1.3.4, we specify a recursion formula for a sequence of integrals, we are also giving a numerical method—even without instruction as to how the recursion formula is to be used. Thus we require that a numerical method be more generally applicable than an algorithm, and set lesser emphasis on the completeness of the computational details. Newton-Raphson's method is, for example, a numerical method for determining a root of an almost arbitrary equation, but in order to get an algorithm one must add conditions for starting and stopping the iteration process; these should be designed with regard to the type of equation and the context in which the equation occurs.

REVIEW QUESTIONS

1. Explain the concepts *numerical problem*, *algorithm*, and *numerical method.*
2. Give a concise explanation why Algorithm 1 of Example 1.3.4 didn't work, and why the other two algorithms did work.

PROBLEMS

1. Use Horner's scheme to compute $p(2)$ where

$$p(x) = 2 - 3x^2 + 2x^3 + x^4.$$

2. Count the number of multiplications and additions required for the calculation of $p(z)$ (see Sec. 1.3.2) by Horner's rule. Compare with the work needed when the powers of x are calculated by $x^i = x \cdot x^{i-1}$ and subsequently multiplied by a_{n-i}.

3. (a) Prove formula (1.3.2) in Sec. 1.3.2.
   (b) If

$$g_0(x, z) = p(x), \qquad g_{k+1}(x, z) = \frac{g_k(x, z) - g_k(z, z)}{x - z} \qquad (k = 0, 1, \ldots, n),$$

   show that $g_k(z, z) = \dfrac{p^{(k)}(z)}{k!}$.

   (If you find this difficult, study Problem 4, below, first.)

4. From the computational scheme given below, one can read off that the polynomial $x^4 + 2x^3 - 3x^2 + 2$, after the substitution $y = x - 2$, is transformed to $y^4 + 10y^3 + 33y^2 + 44y + 22$. Investigate and give a theoretical explanation for how the scheme is constructed.

$$\begin{array}{rrrrr}
1 & 2 & -3 & 0 & 2 \\
 & 2 & 8 & 10 & 20 \\
\hline
1 & 4 & 5 & 10 & 22 \\
 & 2 & 12 & 34 & \\
\hline
1 & 6 & 17 & 44 & \\
 & 2 & 16 & & \\
\hline
1 & 8 & 33 & & \\
 & 2 & & & \\
\hline
1 & 10 & & &
\end{array}$$

5. Write a program for the computation of a scalar product

$$S = \sum_{i=1}^{n} a_i b_i.$$

6. Given the continued fraction

$$f = b_0 + \frac{a_1}{b_1 + a_2/(b_2 + a_3/\cdots + a_n/b_n)}.$$

(a) Show that f can be computed using the algorithm

$$d_n = b_n, \qquad d_{n-i-1} = b_{n-i-1} + \frac{a_{n-i}}{d_{n-i}}, \qquad i = 0, 1, \ldots, n-1,$$

where $f = d_0$.

(b) Write a program which reads in $n, b_0, \ldots, b_n, a_1, \ldots, a_n$, performs the calculation, and prints the value of f.

7. Write a program which reads in a sequence of equidistant function values $f_0, f_1, \ldots, f_n$ to the variables $y[0], y[1], \ldots, y[n]$, $(n \leq 20)$, and then computes the differences $f_0, \Delta f_0, \ldots, \Delta^n f_0$ and stores them where the f_i were stored earlier. (Thus all the values of the function except f_0 are destroyed.) Differences were defined in Sec. 1.2. The program should not use any memory space (variables) other than the $y[i]$.

8. The coefficients of two polynomials f and g,

$$f(x) = \sum_{i=1}^{m} a_i x^{i-1}, \qquad g(x) = \sum_{j=1}^{n} b_j x^{j-1},$$

are given. Derive recursive formulas and write a program for the computation of the coefficients of the product of the polynomials.

9. Let x, y be nonnegative integers, with $y \neq 0$. The division x/y yields the quotient q and remainder r. Show that if x and y have a common factor, then that number is a factor of r as well. Use this remark to design an algorithm for the determination of the greatest common factor of x and y (Euclid's algorithm). Write a program which uses this algorithm and prints out the reduction of a fraction to lowest terms.

10. Derive a recursion formula for calculating the integrals

$$y_n = \int_0^1 \frac{x^n}{4x + 1}\, dx.$$

Give one algorithm that works well and another that works poorly (both based on the recursion formula).

2 HOW TO OBTAIN AND ESTIMATE ACCURACY IN NUMERICAL CALCULATIONS

2.1. BASIC CONCEPTS IN ERROR ESTIMATION

2.1.1. Introduction

Approximation is a central concept in almost all the uses of mathematics. One must often be satisfied with approximate values of the quantities with which one works. Another type of approximation occurs when one ignores some quantities which are small compared to other quantities. Such approximations are often necessary to insure that the mathematical and numerical treatment of a problem does not become hopelessly complicated.

We shall now introduce some notations, useful in practice, though their definitions are not exact in a mathematical sense:

$a \ll b$ (or $b \gg a$) is read: "a is much smaller than b" (or "b is much greater than a"). What is meant by "much smaller" (or "much greater") depends on the context—among other things, on the desired precision. In a given instance it can be sufficient that $a < \frac{1}{2}b$; in another instance perhaps $a < b/100$ is necessary.

$a \approx b$ is read: "a is approximately equal to b" and means the same as $|a - b| \ll c$, where c is chosen appropriate to the context. We *cannot generally* say, for example, that $10^{-6} \approx 0$.

$a \lesssim b$ (or $b \gtrsim a$) is read: "a is less than or approximately equal to b" and means the same as "$a < b$ or $a \approx b$."

Occasionally we shall have use for the following more precisely defined mathematical concepts:

$f(x) = O(g(x))$ when $x \longrightarrow a$, which means that $|f(x)/g(x)|$ is bounded as $x \longrightarrow a$ (a can be finite, $+\infty$, or $-\infty$).

$f(x) = o(g(x))$ when $x \to a$, which means that $\lim_{x \to a} f(x)/g(x) = 0$.

2.1.2. Sources of Error

Numerical results are influenced by many types of errors. Some sources of error are difficult to influence; others can be reduced or even eliminated by, for example, rewriting formulas or making other changes in the computational sequence.

A. Errors in Given Input Data. Input data can be the result of measurements which have been influenced by systematic errors or by temporary disturbances. Round-off errors occur, for example, whenever an irrational number is shortened ("rounded off") to a fixed number of decimals. Round-off errors can also occur when a decimal fraction is converted to the form used in the computer.

B. Round-off Errors During the Computations. If the calculating device which one is using cannot handle numbers which have more than, say, s digits, then the exact product of two s-digit numbers (which contains $2s$ or $2s - 1$ digits) cannot be used in the subsequent calculations; the product must be rounded off. The effect of such roundings can be quite noticeable in an extensive calculation, or in an algorithm which is numerically unstable (defined in Example 1.3.3).

C. Truncation Errors. These are errors committed when a limiting process is truncated (broken off) before one has come to the limiting value. Truncation occurs, for example, when an infinite series is broken off after a finite number of terms, or when a derivative is approximated with a difference quotient (although in this case the term **discretization error** is better). Another example is when a nonlinear function is approximated with a linear function. Observe the distinction between truncation error and round-off error.

D. Simplifications in the Mathematical Model. In most of the applications of mathematics, one makes **idealizations**. In a mechanical problem, for example, one might assume that a string in a pendulum has zero mass. In many other types of problems it is advantageous to consider a given body to be homogeneously filled with matter, instead of being built up of atoms. For a calculation in economics, one might assume that the rate of interest is constant over a given period of time. The effects of such sources of error are usually more difficult to estimate than the types named in *A*, *B*, and *C*.

E. "Human" Errors and Machine Errors. In all numerical work, one must expect that clerical errors, errors in hand calculation, and misunderstandings will occur. One should even be aware that printed tables, etc., may contain errors. When one uses computers, one can expect errors in the program itself, errors in the punched cards, operator errors, and machine errors.

Errors which are purely machine errors are responsible for only a very small part of the strange results which (occasionally with great publicity) are produced by computers year after year. Most of the errors depend on the so-called human factor. As a rule, the effect of this type of error source cannot be analyzed with the help of the theoretical considerations of this chapter! We take up these sources of error in order to emphasize that both the person who carries out a calculation and the person who guides the work of others can plan so that such sources of error are not damaging. One can reduce the risk for such errors by suitable adjustments in working conditions and routines. Stress and tiredness are common causes of such errors.

One should also carefully consider what kind of checks can be made, either in the final result or in certain stages of the work, to prevent the necessity of redoing a whole project for the sake of a small error in an early stage. One can often discover whether calculated values are of the wrong order of magnitude or are not sufficiently regular (see difference checks, Chap. 7). Occasionally one can check the credibility of several results at the same time by checking that certain relations are true. In linear problems, one often has the possibility of sum checks. In physical problems, one can check, for example, to see whether energy is conserved, although because of the error sources A–D one cannot expect that it will be exactly conserved. In some situations, it can be best to treat a problem in two independent ways, although one can usually (as intimated above) check a result with less work than this.

2.1.3. Absolute and Relative Errors

Let $\tilde{a}$ be an approximate value for a quantity whose exact value is a. We define:

The **absolute error** in $\tilde{a}$ is $\tilde{a} - a$.

The **relative error** in $\tilde{a}$ is $(\tilde{a} - a)/a$ if $a \neq 0$. The relative error is often given as a percentage—for example, 3 percent relative error means that the relative error is 0.03.

In some books the error is defined with opposite sign to that which we use here. It makes almost no difference which convention one uses, as long as one is consistent. Using our definition, then, $a - \tilde{a}$ is the *correction* which should be added to $\tilde{a}$ to get rid of the error, $\tilde{a} - a$. The correction and the error have, then, the same magnitude but different signs.

It is important to make a distinction between the error, which can be positive or negative, and a positive bound for the magnitude of the error, an **error bound.** We shall have reason to compute error bounds in many situations.

In the above definition of absolute error, a and $\tilde{a}$ need not be real numbers: they can also be vectors or matrices. (If we let $\|\cdot\|$ denote a vector norm (see Sec. 5.5.2), then the magnitude of the absolute and relative errors for the vector $\tilde{a}$ are defined by $\|\tilde{a} - a\|$ and $\|\tilde{a} - a\|/\|a\|$ respectively.)

The notation $a = \tilde{a} \pm \epsilon$ means, in this book, $|\tilde{a} - a| \leq \epsilon$. For example, $a = 0.5876 \pm 0.0014$ means $0.5862 \leq a \leq 0.5890$. In many applications, the same notation as above denotes the "standard error" (see Sec. 2.2.2) or some other measure of deviation of a statistical nature.

2.1.4. Rounding and Chopping

When one gives the *number of digits* in a numerical value one should not include zeros in the beginning of the number, as these zeros only help to denote where the decimal point should be. If one is counting the *number of decimals*, one should of course include leading zeros to the right of the decimal point.

Example

The number 0.00147 is given with three digits but has five decimals. The number 12.34 is given with four digits but has two decimals.

If the magnitude of the error in $\tilde{a}$ does not exceed $\frac{1}{2}\cdot 10^{-t}$, then $\tilde{a}$ is said to have **t correct decimals.** The *digits* in $\tilde{a}$ which occupy positions where the unit is greater than or equal to 10^{-t} are called, then, **significant digits** (any initial zeros are not counted).

Example

0.001234 ± 0.000004 has five correct decimals and three significant digits, while 0.001234 ± 0.000006 has four correct decimals and two significant digits.

The number of correct decimals gives one an idea of the magnitude of the absolute error, while the number of significant digits gives a rough idea of the magnitude of the relative error.

There are **two ways of rounding off** numbers to a given number (t) of decimals. In **chopping**, one simply leaves off all the decimals to the right of the tth. That way of abridging a number is *not recommended* since the error has, systematically, the opposite sign of the number itself. Also, the magnitude of the error can be as large as 10^{-t}. A surprising number of computers use chopping on the results of every arithmetical operation. This usually does not do so much harm, because the number of digits used in the operations is generally far greater than the number of significant digits in the data.

In **rounding** (sometimes called "correct rounding"), one chooses, among the numbers which can be expressed with t decimals, a number which is closest to the given number. Thus if the part of the number which stands to the right of the tth decimal is less than $\frac{1}{2}\cdot 10^{-t}$ in magnitude, then one should leave the tth decimal unchanged. If it is greater than $\frac{1}{2}\cdot 10^{-t}$, then one raises the tth decimal by 1. In the boundary case, when that which stands to the right of the tth decimal is exactly $\frac{1}{2}\cdot 10^{-t}$, one should raise the tth decimal if it is odd or leave it unchanged if it is even. In this way, the error is positive or negative about equally often. Most computers which perform rounding always, in the boundary case mentioned above, raise the number by $\frac{1}{2}\cdot 10^{-t}$ (or the corresponding operation in a base other than 10), because this is easier to realize technically. Whichever convention one chooses in the boundary case, the error in rounding will always lie in the interval $[-\frac{1}{2}\cdot 10^{-t}, \frac{1}{2}\cdot 10^{-t}]$.

Example

Shortening to three decimals.

0.2397	rounds to	0.240	(is chopped to	0.239),
−0.2397	rounds to	−0.240	(is chopped to	−0.239),
0.23750	rounds to	0.238	(is chopped to	0.237),
0.23650	rounds to	0.236	(is chopped to	0.236),
0.23652	rounds to	0.237	(is chopped to	0.236).

Observe that when one rounds off a numerical value one produces an error; thus it is occasionally wise to give more decimals than those which are correct. Take, for example, $a = 0.1237 \pm 0.0004$, which has three correct decimals according to the definition given previously. If one rounds to three decimals, one gets 0.124; here the third decimal is not correct, since the least possible value for a is 0.1233.

One consequence of these rounding conventions is that numerical results which are not followed by any error estimations should often, though not always, be considered as having an uncertainty of $\frac{1}{2}$ unit in the last decimal place.

In presenting numerical results, it is a good habit, if one does not want to go to the difficulty of presenting an error estimate with each result, to give explanatory remarks such as:

"All the digits given are thought to be significant."
"The data has an uncertainty of at most 3 units in the last digit."
"For an ideal two-atomed gas, $c_P/c_V = 1.4$ (exactly)."

REVIEW QUESTIONS

1. Clarify (with examples) the various types of error sources which occur in numerical work.
2. Define absolute error, relative error. What is meant by an *error bound*?

PROBLEM

Give π to four decimals using (a) chopping, (b) rounding.

2.2. PROPAGATION OF ERRORS

2.2.1. Simple Examples of Error Analysis

Example 2.2.1

If $x_1 = 2.31 \pm 0.02$ and $x_2 = 1.42 \pm 0.03$, what is the error bound for $x_1 - x_2$?

The greatest possible value of x_1 is 2.33, and the least possible value for x_2 is 1.39. Thus the greatest possible value for $x_1 - x_2$ is $2.33 - 1.39 = 0.94$. Similarly, the least possible value for $x_1 - x_2$ is $2.29 - 1.45 = 0.84$. Hence, $0.84 \leq x_1 - x_2 \leq 0.94$; that is,

$$x_1 - x_2 = 0.89 \pm 0.05.$$

More generally, if $x_1 = \tilde{x}_1 \pm \epsilon_1$, $x_2 = \tilde{x}_2 \pm \epsilon_2$, we have

$$\tilde{x}_1 - \epsilon_1 - (\tilde{x}_2 + \epsilon_2) \leq x_1 - x_2 \leq \tilde{x}_1 + \epsilon_1 - (\tilde{x}_2 - \epsilon_2)$$
$$\tilde{x}_1 - \tilde{x}_2 - (\epsilon_1 + \epsilon_2) \leq x_1 - x_2 \leq \tilde{x}_1 - \tilde{x}_2 + (\epsilon_1 + \epsilon_2)$$
$$x_1 - x_2 = \tilde{x}_1 - \tilde{x}_2 \pm (\epsilon_1 + \epsilon_2).$$

A similar calculation gives

$$x_1 + x_2 = \tilde{x}_1 + \tilde{x}_2 \pm (\epsilon_1 + \epsilon_2).$$

Using induction, one can show, for an arbitrary number of terms:

THEOREM 2.2.1

In **addition** *and* **subtraction**, *the bounds for the* **absolute** *error in the result are given by the sum of the bounds for the absolute errors of the operands.*

The error bound given in the above proposition can, for various reasons, be a coarse overestimate of the real error; we shall see this in Examples 2.2.8 and 2.2.11.

According to the definition of relative error (Sec. 2.1.3), we have the following relation between an exact value x, its estimate $\tilde{x}$, and the estimate's *real relative error* r:

$$\tilde{x} = x + xr = x(1 + r).$$

If $\tilde{x}_1$, $\tilde{x}_2$ have relative errors r_1 and r_2, respectively, then

$$\tilde{x}_1\tilde{x}_2 = x_1(1 + r_1)x_2(1 + r_2) = x_1x_2(1 + r_1)(1 + r_2).$$

Thus, the relative error in $\bar{x}_1\bar{x}_2$ is

$$(1 + r_1)(1 + r_2) - 1 = r_1 + r_2 + r_1r_2 \approx r_1 + r_2, \quad \text{if } |r_1| \ll 1, |r_2| \ll 1. \tag{2.2.1}$$

For example, if $r_1 = 0.02$, $r_2 = -0.01$, then the relative error in the product is $0.0098 \approx 0.01$. In the same way, one finds for the relative error in the quotient, x_1/x_2, that:

$$\frac{1 + r_1}{1 + r_2} - 1 = \frac{r_1 - r_2}{1 + r_2} \approx r_1 - r_2, \quad \text{if } |r_1| \ll 1, |r_2| \ll 1. \tag{2.2.2}$$

If the *bounds* (see Sec. 2.1.3) for the relative errors in x_1, x_2 are ρ_1 and ρ_2, respectively, then $\rho_1 + \rho_2$ is the best upper bound for $|r_1 + r_2|$ as well as $|r_1 - r_2|$. Thus:

THEOREM 2.2.2

In **multiplication** *and* **division**, *the bounds for the* **relative** *errors in the operands are added.* (As one can see from the derivation above, this theorem is only approximately valid; see Examples 2.2.5, 2.4.3.)

We note here that the approximations used in Eqs. (2.2.1) and (2.2.2) are very useful in many other situations.

Error analysis is more than just a means for judging the reliability of calculated results; it has an even more important function as a means for **planning a given calculation**—for example, in the choosing of an algorithm (see Example 1.3.3)—and in making certain **decisions** during a calculation. Examples of such decisions are the choice of step length during a numerical integration or, especially in hand calculation, the choice of the number of digits to be used in the various parts of a computation. Increased accuracy is often bought at the price of more time-consuming or more troublesome calculations.

Example 2.2.2

The following type of situation occurs quite often. Assume that a product or a quotient is to be calculated, $y = x_1x_2$ or $y = x_1/x_2$. The quantity x_1 is already known to a certain amount of accuracy and x_2 is to be calculated. How accurately should one compute x_2 (assuming that the work to calculate x_2 grows the more quickly as one demands higher accuracy)?

Since the limit for the relative error in y is equal to the sum of the bounds for the relative errors in x_1 and x_2, there is no use making the relative error in x_2 very much less than the relative error in x_1. To take along *one* more significant digit in x_2 than in x_1 can, however, often be wise.

The error-propagation formulas are also of great interest in the **planning and analysis of scientific experiments**. One can shed some light on certain *questions analogous to the previous example*—for example, to what degree it is advisable to obtain a new apparatus to improve the measurements of a given variable when the measurements of other variables are subject to error as well.

Cancellation. One very common reason for poor accuracy in the result of a calculation is that one has somewhere carried out a subtraction in which the difference between the operands is considerably less than either of the operands. This is called **cancellation of terms**. From Theorem 2.2.1 we see, if we denote the error in x_1 and x_2 by Δx_1 and Δx_2, respectively, that:

$$y = x_1 - x_2 \Rightarrow |\Delta y| \leq |\Delta x_1| + |\Delta x_2| \Rightarrow \left|\frac{\Delta y}{y}\right| \leq \frac{|\Delta x_1| + |\Delta x_2|}{|x_1 - x_2|}.$$

This shows that *there can be very poor relative accuracy in the difference between two nearly equal numbers*. For example, if $x_1 = 0.5764 \pm \frac{1}{2}\cdot 10^{-4}$ and $x_2 = 0.5763 \pm \frac{1}{2}\cdot 10^{-4}$, then $x_1 - x_2 = 0.0001 \pm 0.0001$—the error bound is just as large as the estimate of the result.

One should try to avoid cancellation by appropriate **rewriting of formulas**, or by other changes in the algorithm.

Example 2.2.3

The quadratic equation $x^2 - 56x + 1 = 0$ has the roots

$$x_1 = 28 - \sqrt{783} \approx 28 - 27.982 = 0.018 \pm \tfrac{1}{2}\cdot 10^{-3},$$
$$x_2 = 28 + \sqrt{783} = 55.982 \pm \tfrac{1}{2}\cdot 10^{-3}.$$

In spite of the fact that the square root is given to five digits, we get only two significant digits in x_1, while the relative error in x_2 is less than 10^{-5}. *It is worthwhile to notice that the subtraction itself, in the calculation of x_1, has been carried out exactly. The subtraction only gives an indication of the unhappy consequence of a loss of information in the past* due to the rounding of one of the operands.

Since $x_1 x_2 = 1$, one can instead use $x_1 = 1/55.982 = 0.01786288$ with a relative error of less than 10^{-5}. Thus $x_1 = 0.0178629 \pm 0.0000002$; the same estimate of the square root gave x_1 to five significant digits instead of two because we also used information from the quadratic equation which was the source of the numerical data.

More generally, if $|\delta| \ll x$, then one should rewrite

$$\sqrt{x+\delta} - \sqrt{x} = \frac{x + \delta - x}{\sqrt{x+\delta} + \sqrt{x}} = \frac{\delta}{\sqrt{x+\delta} + \sqrt{x}}.$$

There are other exact ways of rewriting formulas which are as useful as the above; for example,

$$\cos(x + \delta) - \cos x = -2\sin(\tfrac{1}{2}\delta)\sin(x + \tfrac{1}{2}\delta).$$

If one cannot find an exact way of rewriting a given expression of the form $f(x + \delta) - f(x)$, it is often advantageous to use one or more terms in the Taylor series

$$f(x + \delta) - f(x) = f'(x)\delta + \tfrac{1}{2}f''(x)\delta^2 + \cdots.$$

In Example 2.2.3 we got a warning that cancellation would occur, since x_1 was found as the difference between two nearly equal numbers each of which was, relatively, much larger than the difference itself. In practice, one does not always get such a warning, for two reasons: first, in using a computer one has no direct contact with the individual steps of a calculation; secondly, cancellation can be spread over a great number of operations, as in the following example.

Example 2.2.4

Set $y_0 = 28$ and define y_n, for $n = 1, 2, \ldots, 100$, by the recursion formula:

$$y_n = y_{n-1} - \tfrac{1}{100}\sqrt{783}.$$

As previously, we use the approximate value 27.982 for the square root. We then compute with five decimals in each operation in order to make the effect of further round-off errors negligible. We get the same bad value for y_{100} that we got for x_1 in the previous example. Of all the subtractions, only the last would lead one to suspect cancellation, $y_{100} = 0.29782 - 0.27982 = 0.01800$, but this result in itself gives one no reason to suspect that only two digits are significant. (With four significant digits, the result is 0.01786.)

2.2.2. The General Formula for Error Propagation; Maximum Error and Standard Error

Consider a function of one variable, $y(x)$, and suppose we want to estimate a bound for the magnitude of the error $\Delta y = y(\tilde{x}) - y(x^{(0)})$ in $y(x^{(0)})$, where $\tilde{x}$ is an approximate value for $x^{(0)}$. A natural way (see Fig. 2.2.1) to approximate Δy is with the differential of y. In this way, one locally approxi-

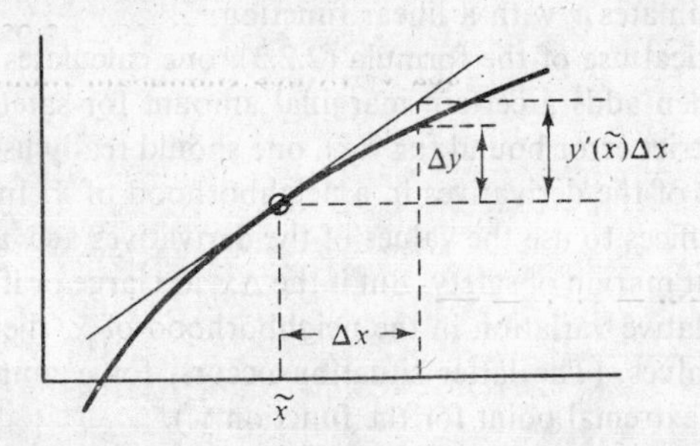

Fig. 2.2.1

mates $y(x)$ with a linear function. The quantity $|y'(x)|$ can be interpreted as a measure of **the sensitivity of $y(x)$ for disturbances in the argument** x.

Suppose that one knows approximate values $\tilde{x}_1, \tilde{x}_2, \ldots, \tilde{x}_n$ for certain variables $x_1, x_2, \ldots, x_n$, whose exact values are $x_1^{(0)}, x_2^{(0)}, \ldots, x_n^{(0)}$. Let y be a function of the variables $x_1, x_2, \ldots, x_n$. We shall now estimate the error in the approximate value $y(\tilde{x}_1, \tilde{x}_2, \ldots, \tilde{x}_n)$. This general problem is of great interest in experimental science and other areas where empirical data are put into mathematical models.

Introduce the vector notation

$$\tilde{x} = (\tilde{x}_1, \tilde{x}_2, \ldots, \tilde{x}_n), \qquad x_0 = (x_1^{(0)}, x_2^{(0)}, \ldots, x_n^{(0)}),$$

and set

$$\Delta x_i = \tilde{x}_i - x_i^{(0)}, \qquad \Delta y = y(\tilde{x}) - y(x_0).$$

THEOREM 2.2.3 **General Error-Propagation Formula**

$$\Delta y \approx \sum_{i=1}^{n} \frac{\partial y}{\partial x_i}(\tilde{x})\cdot\Delta x_i, \qquad \textit{hence} \quad |\Delta y| \lesssim \sum_{i=1}^{n} \left|\frac{\partial y}{\partial x_i}(\tilde{x})\right|\cdot|\Delta x_i|. \tag{2.2.3}$$

The formula is derived as follows: one moves from x_0 to $\tilde{x}$ in n steps, by changing one coordinate at a time. Thus for $i = 1, 2, \ldots, n-1$, set

$$x_i = (\tilde{x}_1, \ldots, \tilde{x}_i, x_{i+1}^{(0)}, \ldots, x_n^{(0)}), \qquad x_n = \tilde{x}.$$

In moving from x_{i-1} to x_i, the ith coordinate is changed from $x_i^{(0)}$ to $\tilde{x}_i$, while the rest of the coordinates are left unchanged. Then, by the mean-value theorem, we have, for some point ξ_i between x_{i-1} and x_i, that

$$y(x_i) - y(x_{i-1}) = \frac{\partial y}{\partial x_i}(\xi_i)\cdot(\tilde{x}_i - \tilde{x}_i^{(0)}) \approx \frac{\partial y}{\partial x_i}(\tilde{x})\cdot\Delta x_i.$$

The theorem now follows, since

$$\Delta y = y(x_n) - y(x_0) = \sum_{i=1}^{n} (y(x_i) - y(x_{i-1})).$$

As one can see from Theorem 2.2.3, Δy was approximated with the *total differential* of y. This means that, in a small neighborhood of $\tilde{x}$, containing x_0, one approximates y with a linear function.

In the practical use of the formula (2.2.3), one calculates $|\partial y/\partial x_i|$ at the point $\tilde{x}$, and then adds a certain marginal amount for safety. However, in order to get a strict error bound for $y(\tilde{x})$, one should really use the maximum absolute values of the derivatives in a neighborhood of $\tilde{x}$. In most practical situations, it suffices to use the values of the derivatives at $\tilde{x}$ and then to add a 5 to 10 percent margin of safety. But if the Δx_i are large or if the derivatives have a large relative variation in the neighborhood of $\tilde{x}$, then one must use the maximal values. (The latter situation occurs, for example, in a neighborhood of an extremal point for the function y.)

Example 2.2.5

The relative error in a quantity is approximately equal to the absolute error in its natural logarithm, since

$$\Delta(\ln y) \approx d(\ln y) = \frac{dy}{y} \approx \frac{\Delta y}{y}.$$

Using an exact formula, one has $\Delta(\ln y) = \ln(y + \Delta y) - \ln(y) = \ln(1 + \Delta y/y)$. A comparison follows:

$\frac{\Delta y}{y} =$ 0.001	0.01	0.1	0.2	−0.2	−0.1
$\Delta(\ln y) =$ 0.00100	0.00995	0.095	0.182	−0.223	−0.105

For logarithms to base 10 ("common logarithms") one has

$$\Delta(\log_{10} y) = \log_{10} e \cdot \Delta(\ln y) \approx 0.434 \frac{\Delta y}{y}.$$

Example 2.2.6

Compute $y = x_1^2 - x_2$, for $x_1 = 1.03 \pm 0.01$, $x_2 = 0.45 \pm 0.01$.

$$\left|\frac{\partial y}{\partial x_1}\right| = |2x_1| \le 2.1, \qquad \left|\frac{\partial y}{\partial x_2}\right| = |-1| = 1,$$

$$|\Delta y| \le 2.1 \cdot 0.01 + 1 \cdot 0.01 = 0.031.$$

We find $y = 1.061 - 0.450 \pm 0.032 = 0.611 \pm 0.032$; the error bound has been raised 0.001 because of the rounding in the calculation of x_1^2. If one rounded the results to two decimals, one would write $y = 0.61 \pm 0.04$.

We shall now give a generalization of Theorem 2.2.2:

THEOREM 2.2.4

If $y = x_1^{m_1} x_2^{m_2} x_3^{m_3} \ldots x_n^{m_n}$, *and* $x_i \neq 0$, *then*

$$\left|\frac{\Delta y}{y}\right| \lesssim \sum_{i=1}^{n} |m_i| \left|\frac{\Delta x_i}{x_i}\right|.$$

Proof. $\ln y = m_1 \ln x_1 + m_2 \ln x_2 + \ldots + m_n \ln x_n$. Differentiate! We get

$$\frac{1}{y}\frac{\partial y}{\partial x_i} = \frac{m_i}{x_i},$$

and the theorem is proved.

It is very seldom, in practice, that one is asked to give mathematically guaranteed error bounds, but it occasionally happens. More often, it is satisfactory to give an estimate of the *order of magnitude* of the anticipated error. The bound for $|\Delta y|$ obtained, for example, with Theorem 2.2.3 covers the worst possible cases, where the sources of error Δx_i contribute with the same sign and magnitudes equal to the error bounds for the individual variables. For this reason, the bound for $|\Delta y|$ thus obtained is called the **maximal error**. *In practice, the trouble with formula* (2.2.3) *is that it gives bounds which are too coarse.*

As a complement to the maximal error, which is often much too pessimistic when the number of variables is large, one uses the **standard error**. The standard error of an estimate of a given quantity is the same as the *standard deviation of its sampling distribution*. The theory of standard error is based on probability theory and will not be treated in detail here. However, we give without proof the following theorem, which a reader familiar with probability theory can derive using the first of the formulas (2.2.3):

THEOREM 2.2.5

Assume that the errors $\Delta x_1, \Delta x_2, \ldots, \Delta x_n$ are independent random variables with mean zero and standard deviations $\epsilon_1, \epsilon_2, \ldots, \epsilon_n$. Then the standard error ϵ for $y = y(x_1, x_2, \ldots, x_n)$ is given by the formula:

$$\epsilon \approx \left(\sum_{i=1}^{n} \left(\frac{\partial y}{\partial x_i}\right)^2 \epsilon_i^2\right)^{1/2}$$

Example 2.2.7

For the problem in Example 2.2.6 we get

$$\epsilon^2 \approx (2.1)^2 \cdot 10^{-4} + 1 \cdot 10^{-4} = 5.41 \cdot 10^{-4}.$$

Thus the standard error $\epsilon \approx 0.024$.

Example 2.2.8. *Maximum Error and Standard Error for Sums*

From the derivation of Theorem 2.2.1 it follows that:

$$y = x_1 + x_2 + \ldots + x_n \Rightarrow \Delta y = \Delta x_1 + \Delta x_2 + \ldots + \Delta x_n.$$

If each x_i has error bound δ, then the maximal error for y is $n\delta$. Thus, *the maximal error grows proportionally to n.* If n is large—for example, $n = 1{,}000$—then it is in fact highly improbable that the real error will be anywhere near $n\delta$, since that bound is attained only when every Δx_i has the same sign and the same maximal magnitude. Observe, though that if one is adding positive numbers, each of which has been abridged to t decimals by "chopping," then each Δx_i has the same sign and a magnitude which is on the average $\frac{1}{2}\delta$, where $\delta = 10^{-t}$. Thus, using the above method (not to be recommended) for abridging the x_i, the real error is often about 500δ.

If the numbers are rounded (instead of chopped), and *if one can assume that the errors in the various terms are stochastically independent* with standard deviation ϵ, then the standard error in y becomes (using Theorem 2.2.5)

$$(\epsilon^2 + \epsilon^2 + \epsilon^2 + \ldots + \epsilon^2)^{1/2} = \epsilon \cdot \sqrt{n}.$$

Thus the growth of the standard error of the sum of n terms is only proportional to $\sqrt{n}$.

If $n \gg 1$, then the error in y is, under the assumptions made above, approximately **normally distributed** with standard deviation $\sigma = \epsilon \cdot \sqrt{n}$. This distribution is illustrated in Fig. 2.2.2; the curve shown there is also called the Gauss curve. The assumption that the error is normally distributed with standard deviation σ means, for example, that the statement "the magnitude of the error is less than 2σ" (see the shaded area of Fig. 2.2.2) is false in about only 5 percent of all cases (the clear areas under the curve). More generally: *the assertion that the magnitude of the error is less than* σ, 2σ, *and* 3σ, *respectively, is false in about (respectively) 32 percent, 5 percent, and 0.27 percent of all cases.*

The assumption that the errors are normally distributed is justified in many computational situations and scientific experiments where the error can be considered to have arisen from the addition of a large number of *independent* error sources of about the same order of magnitude.

One can show that if the individual terms have a uniform probability distribution in the interval $[-\frac{1}{2}\delta, \frac{1}{2}\delta]$, then the standard deviation of an individual term is $\delta/\sqrt{12}$. In only 5 percent of all cases, then, is the error in the sum of 1,000 terms greater than $2 \cdot \delta \cdot \sqrt{1000/12} \approx 18\delta$. This example shows that rounding can be far superior to chopping when a statistical interpretation (especially, the assumption of independence) can be given to the *principal sources of error*. Observe that, in the above, we have only considered the propagation of the errors which were present in the original data, and have ignored the effect of possible round-off errors in the additions themselves.

In science and technology, one generally is careful to discriminate between **systematic error** and **random error**. A systematic error can, for example, be produced by insufficiencies in the construction of an instrument; such an error is the same in each trial. Random errors depend on variations in the experimental environment which cannot be controlled; then the formula

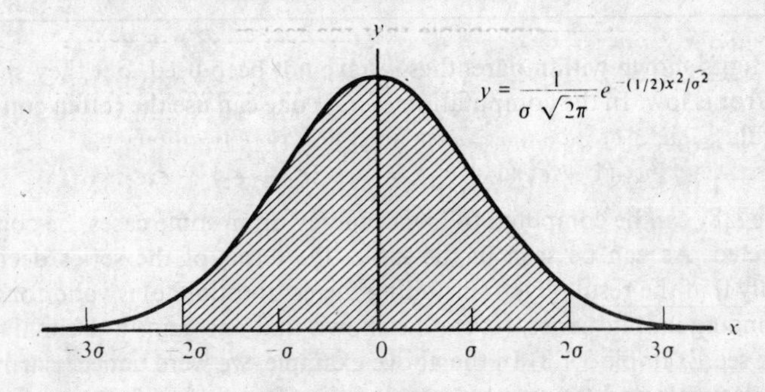

Fig. 2.2.2

for standard error is used. For systematic errors, however, the formula for maximal error (2.2.3) should be used.

2.2.3. On the Practical Application of Error Estimation

In hand calculation or calculation with a desk calculator, it is much easier to work with few digits than with many. On most computers, however, no simplification is gained if an operand contains fewer digits. (Many computers do not even make any shortcuts when one of the operands is zero.) An essential question, then, in hand calculation, is: how many decimals or how many digits should one use? On a computer, one often works either with a constant number of digits (computation with "floating" decimal or binary point) or else with a constant number of decimal or binary digits in the fractional part of a number (computation with "fixed" decimal or binary point). In hand computation, one can (and should) have a more flexible adaptation to the given situation.

Example 2.2.9

Compute $\sum_{n=1}^{6} n^{1/2}\pi^{-2n}$ with an error which is less than 10^{-5}. It is appropriate to compute the terms to six *decimals* (where the last decimal can have error 1). One can adapt the number of *digits* in $n^{1/2}$ and π^{-2n} to the number of digits which is needed in each term (compare Example 2.2.3).

n	π^{-2n}	$n^{1/2}$	Term
1	0.101321(2)	1	0.101321(2) π^{-2}
2	0.010266	1.4142(1)	0.014518
3	0.001040	1.732(0)	0.001801(3)
4	0.0001053	2	0.000211
5	0.000011	2.2	0.000024
6	0.0000011	2.4	0.000003
			$\sum = 0.117878 \pm 5\cdot 10^{-6} \approx 0.11788$

The digits shown within parentheses have not been used, but they show that the error is low. In the computation of π^{-2n} one can use the (often convenient) formula

$$(1+\epsilon_1)(1+\epsilon_2) = 1 + (\epsilon_1+\epsilon_2) + \epsilon_1\epsilon_2,$$

where $\epsilon_1\epsilon_2$ can be computed to low accuracy or, in some cases, be completely neglected. As can be seen in the table, the terms of the series decrease so rapidly that the result given (including the error estimate) is valid for the corresponding infinite series. (It can be shown that one gains a decimal for each term; see Example 3.1.3) In the above example, we were unnecessarily stingy with decimals and terms; error analysis is often easier if one works with a slight surplus of digits and terms.

In the application of mathematics, there often occur situations where one "sees" (or is in some other way convinced) that a certain quantity can be ignored, but one cannot, without difficulty, give a strict proof of the fact. Of course, if the proof looks easy, then it is worthwhile, for the added security, to try to prove the fact. On the other hand, numerical methods are often used just because the problem is too difficult for analytic treatment. It can then be wise to drop the demand for a proof. Mathematical clarity is a good resource, but the concomitant demand for strictness can have a crippling effect. Small gaps in the argumentation are natural elements in the mathematical treatment of a difficult problem, but they should be openly presented. The more important and general the problem to be treated is, the more relevant becomes the demand for mathematical stringency. Thus some research in numerical analysis is concerned with developing methods which enable a computer to give strict error estimates for computed results in a wide range of typical problems.

A decision to forgo a proof should, though, be based on experience. One should know that the situation, in all its essential elements, is analogous to a simpler situation where a proof can be given. One should also keep a certain margin of safety; for example, one might use more terms or more digits than one anticipates is necessary.

The criteria which one adopts should not be too coarse; this comment is especially relevant in the case of slowly converging processes. For example, it is far from always true that the magnitude of the remainder of a series is less than the most recently used term, or that decimals which remain unchanged in several consecutive terms of a sequence are correct.

Example 2.2.10

In the series

$$\sum_{n=1}^{\infty} \frac{1}{n(n+1)}$$

the thousandth term is less than 10^{-6}, but the remainder is one thousand times as large, since

$$\sum_{n=1,001}^{\infty} \frac{1}{n(n+1)} = \sum_{1,001}^{\infty} \left(\frac{1}{n} - \frac{1}{n+1}\right)$$
$$= \frac{1}{1,001} - \frac{1}{1,002} + \frac{1}{1,002} - \frac{1}{1,003} + \cdots = \frac{1}{1,001}.$$

The sum of the 999, 1,000, 1,001, . . . , 1,008 first terms is, to six decimals,

$$0.999000, 0.999001, 0.999002, \ldots, 0.999009.$$

The fifth decimal is unchanged in ten consecutive partial sums. In spite of this, the error is one unit in the third decimal.

Actually, one should, when one decides how many digits to take along in a quantity to be used later in the computation, give consideration to the entire context of the computations. It can in fact occur that the errors in many operands depend upon each other in such a way that they cancel each other. Such **cancellation of error** (a completely different phenomenon from the previously mentioned (p. 28) "cancellation of terms") is most common in larger problems, but will be illustrated here with a simple example.

Example 2.2.11

Compute to full accuracy $y = z_1 + z_2$, where

$$z_1 = (x^2 + 1)^{1/2}, \qquad z_2 = 200 - x, \qquad x = 100 \pm 1.$$

One finds

$$\frac{dz_1}{dx} = x(x^2 + 1)^{-1/2}, \qquad \frac{dz_2}{dx} = -1, \qquad z_1 = 100 \pm 1, \qquad z_2 = 100 \pm 1.$$

It is tempting to be satisfied with the result $y = 200 \pm 2$. But the errors in z_1 and z_2 cancel each other, since

$$\frac{\Delta y}{\Delta x} \approx \frac{dy}{dx} = \frac{dz_1}{dx} + \frac{dz_2}{dx} = x(x^2 + 1)^{-1/2} - 1$$

$$= \frac{x - \sqrt{x^2 + 1}}{\sqrt{x^2 + 1}} = \frac{x^2 - (x^2 + 1)}{\sqrt{x^2 + 1}(x + \sqrt{x^2 + 1})} \approx \frac{-1}{2x^2}.$$

Thus $|\Delta y| \lesssim \frac{1}{2} \cdot 10^{-4}$. z_1, and z_2 should then be computed to four decimals even though the last integer digits are uncertain! The result:

$$y = 200.0050 \pm \tfrac{1}{2} \cdot 10^{-4}.$$

In the above example one could (and should) avoid the calculation of the square root to many digits. By rewriting the expression for y,

$$y = (\sqrt{x^2 + 1} - x) + 200 = \frac{x^2 + 1 - x^2}{\sqrt{x^2 + 1} + x} + 200$$

$$= \frac{1}{x + \sqrt{x^2 + 1}} + 200,$$

one obtains the result even with low accuracy in the square root. In larger problems, such a cancellation of error can occur even though one cannot easily give a way to rewrite the expressions involved. The authors have seen examples where the final result was a sum of seven terms, and it was obtained correctly to eight decimals even though the terms (which were complicated functions of the solutions to a system of nonlinear equations with fourteen unknowns) had errors already in the second decimal!

2.2.4. The Use of Experimental Perturbations

In larger calculational problems, the relations between input data and output data are so complicated that it is diffcult to directly apply the general formulas for the propagation of error. One should then investigate the sensitivity of the output data for errors in the input data by means of an *experimental perturbational calculation:* one performs the calculations many times with perturbed input data and studies the relation between the changes (perturbations) in the input data and the changes in the output data.

Important data, such as the step length in a numerical integration or the parameter which determines when an iterative process is to be broken off, should be varied with all the other data left unchanged. If one can easily *vary the precision of the machine* in the arithmetic operations one can get an idea of the influence of *round-off errors.* It is generally not necessary to make a perturbational calculation for each and every input data component; one can instead *perturb many input data simultaneously*—for example, by using random numbers.

Such a perturbational calculation often gives not only an error estimate, but also greater insight into the problem. Occasionally, it can be difficult to interpret the perturbational data correctly, since the disturbances in the output data depend not only on the mathematical problem but also on the choice of numerical method and the details in the design of the algorithm. The round-off errors during the computation are not the same for the perturbed and the unperturbed problem. Thus if the output data reacts more sensitively than one had anticipated, it can be difficult to immediately point out the source of the error. It can then be profitable to plan a series of perturbation experiments with the help of which one can separate the effects of the various sources of error. If the dominant source of error is the method or the algorithm, then one should try another method or another algorithm.

It is beyond the scope of this book to give further comments on the planning of such experiments; imagination and the general insights regarding error analysis which this chapter is meant to give play a large role. Even in the special literature, the discussion of the planning of such numerical experiments is surprisingly meager.

2.2.5. Automatic Control of Accuracy

Efforts have been made to design the computational unit of a computer so that it gives, in every arithmetic operation, only those digits of the result which are judged to be significant (possibly with a fixed number of extra digits), so-called *unnormalized floating arithmetic*. This method reveals poor construction in algorithms, but in many other cases it gives a significant and unnecessary loss of accuracy. The mechanization of the rules which a knowledgeable and routined person uses for control of accuracy in hand calculation is not as free from problems as one might expect. As a complement to arithmetical operations of conventional type, the above type of arithmetic is of some interest. At present, it has only been used in a preliminary and experimental way on computers, and it is doubtful that its use will be much wider.

Another effort to automate the control of accuracy (even truncation error) is **interval analysis**, developed by R. E. Moore. Interval analysis is partly an automatization of calculation with maximal error bounds. The ordinary computational unit of the computer is used. The kernel of the method consists of programs for the four operations of arithmetic, with intervals as input data and output data. Let $I_1 = [a_1, \ b_1]$, and $I_2 = [a_2, \ b_2]$. Then, for example, $I_1 - I_2$ is defined as the shortest interval which contains all the numbers of the form $x_1 - x_2$, where $x_1 \in I_1$, $x_2 \in I_2$. Thus $I_1 - I_2 = [a_1 - b_2, b_1 - a_2]$. If rounding off is needed, then the lower bound is always lowered and the upper bound is always raised. Interval analysis has been used successfully in, among other things, the numerical treatment of certain ordinary differential equations and giving strict error bounds for the solutions to systems of linear equations. Interval analysis often requires a refined design in the given algorithm in order to prevent the bounds for the intervals from becoming unacceptably coarse.

For both interval analysis and unnormalized floating arithmetic, the error bounds can be unnecessarily large when *cancellation between errors* occurs, as in Example 2.2.11.

Example 2.2.12

We shall treat the Example 1.3.3, Algorithm 2, using interval analysis. There it was pointed out that $0 < y_{10} < \frac{1}{60}$. Thus

$$y_{10} \in I_{10} = [0, 0.0167],$$

$$y_9 \in I_9 = \frac{1}{50} - \frac{I_{10}}{5} \subset [0.0200 - 0.0034, 0.0200] = [0.0166, 0.0200],$$

$$y_8 \in I_8 = \frac{1}{45} - \frac{I_9}{5} \subset [0.0222 - 0.0040, 0.0223 - 0.0033]$$

$$= [0.0182, 0.0190], \text{ etc.}$$

Example 2.2.13

The recursion formulas

$$x_{n+1} = 2^{-1/2}(x_n - y_n)$$
$$y_{n+1} = 2^{-1/2}(x_n + y_n),$$

mean a series of 45-degree rotations in the xy-plane (see Fig. 2.4.5). By a two-dimensional interval one means a rectangle whose sides are parallel to the coordinate axes.

If (x_0, y_0) is given as some interval $|x_0 - 1| \leq \epsilon, |y_0| \leq \epsilon$ (see the dashed square, in the leftmost portion of Fig. 2.4.5), then (x_n, y_n) will, with *exact* performance of the transformations, also be a square with side 2ϵ, for all n (see the other squares in Fig. 2.4.5). If the computations are made using interval analysis, rectangles with sides parallel to the coordinate axes will, in each step, be circumscribed about the exact image of the interval one had in the previous step. Thus the interval is multiplied by $\sqrt{2}$ in each step. After forty steps, for example, the interval has been multiplied by more than 10^6 (since $2^{20} > 10^6$).

If one used, for example, polar coordinates instead of rectangular, there would not have been any difficulties of the above type. This illustrates that one occasionally, with the use of interval analysis, needs to consider the formulation of the algorithm in order to avoid unrealistic overestimates of the error.

REVIEW QUESTIONS

1. In the commentary to Theorem 2.2.1 (on the error bounds for addition and subtraction), it is mentioned that the bounds given there can "for various reasons" be a coarse overestimate of the real error. Give (preferably with examples) two such reasons.

2. Explain (and give an example of) what is meant by "cancellation of terms."

3. Explain the terms "maximum error" and "standard error." What statistical assumption is made about the terms of a sum in calculating the standard error in the sum due to round-off?

PROBLEMS

1. Compute, to three correct decimals:
(a) $1.3134 \cdot \pi$; (b) $0.3761 \cdot e$; (c) $\pi \cdot e$.
(Analyze first how many decimals must be taken along in each of the terms.)

2. Determine the maximum error and the relative error in the following results where $x = 2.00$, $y = 3.00$, and $z = 4.00$ are correctly rounded.

(a) $f = 3x + y - z$; (b) $f = x \cdot \frac{y}{z}$; (c) $f = x \cdot \sin \frac{y}{40}$.

3. (a) Determine the maximum error for $y = x_1 x_2^2/\sqrt{x_3}$ where $x_1 = 2.0 \pm 0.1$, $x_2 = 3.0 \pm 0.2$, $x_3 = 1.0 \pm 0.1$. Which of the variables contributes most to the error in y?
(b) Compute the standard error using the same data as in (a), assuming that the error estimates for the x_i indicate standard deviations.

4.
$$\begin{cases} 3x + ay = 10 \\ 5x + by = 20 \end{cases}$$
$$a = 2.100 \pm 5 \cdot 10^{-4}$$
$$b = 3.300 \pm 5 \cdot 10^{-4}.$$

How accurately can $x + y$ be determined?

5. One wishes to compute the expression

$$f = (\sqrt{2} - 1)^6, \qquad (1)$$

using the approximate value 1.4 for $\sqrt{2}$. One can then choose between substituting the approximate value in (1) or else in one of the following equivalent expressions:

$$\frac{1}{(\sqrt{2}+1)^6} \tag{2}$$

$$(3-2\sqrt{2})^3 \tag{3}$$

$$\frac{1}{(3+2\sqrt{2})^3} \tag{4}$$

$$99-70\sqrt{2} \tag{5}$$

$$\frac{1}{99+70\sqrt{2}} \tag{6}$$

Which alternative gives the best result?

6. The expression $\ln(x-\sqrt{x^2-1})$ is to be computed for $x = 30$. If the square root is obtained from a six-place table, how large does the error in the logarithm become? Give an expression which is mathematically equivalent to the above expression but better numerically, and compute the error in the logarithm.

7. One is making observations of a satellite in order to determine its speed. At the first observation, $R = 30{,}000 \pm 10$ km. Five seconds later, the distance has increased by $r = 125.0 \pm 0.5$ km and the change in angle was $\varphi = 0.00750 \pm 0.00002$ radians. What is the speed of the satellite, assuming that it moves in a straight line and with constant speed in the given time interval?

8. One has measured two sides and the included angle of a triangle to be $a = 100.0 \pm 0.1$, $b = 101.0 \pm 0.1$, and the angle $c = 1.00° \pm 0.01°$.
   (a) How accurately is it possible to give the third side c?
   (b) How accurately does one get c if one uses the cosine theorem and the value $\cos 1° = 0.9998$ from a four-place table?
   (c) Derive a formula with which it is possible to compute c to full accuracy with the help of a four-place table.

   Note: As a *convention* for hand computation, we define *full accuracy* to mean

$$|R_C + R_T| \leq 0.2\,|R_X|,$$

   where
   R_X denotes a bound for the absolute error due to uncertainty in the input data; it can occasionally be separated into R_{XX}, uncertainty in the argument, and R_{XF}, uncertainty in the values of the function;
   R_C denotes a bound for the absolute error from round-off errors made during the calculation; and
   R_T denotes truncation error (defined in Sec. 2.1.2).
   (In this problem, $R_T = 0$.)

9. In the statistical treatment of data, one often needs to compute the quantities

$$\bar{x} = n^{-1}\sum_{i=1}^{n} x_i, \qquad s^2 = n^{-1}\sum_{i=1}^{n}(x_i - \bar{x})^2.$$

(*Note:* If the numbers x_i are the results of statistically independent measurements of a quantity with expected value m, then $\bar{x}$ is an estimate of m, whose standard error is estimated by $(n-1)^{-1/2}s$.)

(a) Show that $s^2 = n^{-1}\sum_{i=1}^{n} x_i^2 - \bar{x}^2$.

(b) Let α be an arbitrary number and set $x_i' = x_i - \alpha$. (α is sometimes called a provisional mean. In practice, it should be chosen to be of the same order of magnitude as the x_i.) Show that

$$\bar{x} = \alpha + n^{-1}\sum_{i=1}^{n} x_i', \qquad s^2 = n^{-1}\sum_{i=1}^{n}(x_i')^2 - (\bar{x} - \alpha)^2.$$

(c) In sixteen measurements of a quantity x one got the following results:

i	x_i	i	x_i	i	x_i	i	x_i
1	546.85	5	546.81	9	546.96	13	546.84
2	546.79	6	546.82	10	546.94	14	546.86
3	546.82	7	546.88	11	546.84	15	546.84
4	546.78	8	546.89	12	546.82	16	546.84

Take $\alpha = 546.85$. Make a table with one column for $10^2 x_i'$ and one column for $(10^2 x_i')^2$, where the x_i' are defined according to (b). Compute $\bar{x}$ and s^2, s^2 to two significant digits.

(d) In the computations in (c), one never needed more than three digits. If one uses the formulas in (a), how many digits must one have in x_i^2 in order to get two significant digits in s^2? If one uses five digits throughout the computations, why is the cancellation with the use of the formulas in (a) more fatal than the cancellation in the subtraction $x_i' = x_i - \alpha$? (One can even get negative values for s^2!)

(e) Discuss how important (or unimportant) the choice of a good provisional mean is. Take into consideration quick estimates made without a calculating machine, work with a desk calculator, and calculation with a computer!

(f) What advantage do formulas of type (a) or (b) have on a computer as opposed to direct calculation of s according to definition, assuming that the amount of data is so great that it cannot be stored in the central memory?

(g) Set $\bar{y} = n^{-1}\sum_{i=1}^{n} y_i$, $\quad y_i' = y_i - \beta$. Derive a formula for computing $n^{-1}\sum_{i=1}^{n}(x_i - \bar{x})(y_i - \bar{y})$, of the same type as the formula in (b).

10. Continue the calculations in Example 2.2.12.

11. One has an algorithm for computing the integral

$$I(a, b) = \int_0^1 \frac{e^{-bx}}{a + x^2}\,dx.$$

The physical quantities x and y have been measured to be

$$x = 0.400 \pm 0.003, \qquad y = 0.340 \pm 0.005.$$

Using the algorithm for various values of a and b one obtained:

a	b	I
0.39	0.34	1.425032
0.40	0.32	1.408845
0.40	0.34	1.398464
0.40	0.36	1.388198
0.41	0.34	1.372950

How large is the uncertainty in $I(x, y)$?

12. Show that if the errors in the quantities x_i, $i = 1, 2, \ldots, n$ are independent random variables with standard deviation ϵ, then the mean

$$\bar{x} = \frac{x_1 + x_2 + \ldots + x_n}{n}$$

has standard error $\epsilon/\sqrt{n}$.

2.3. NUMBER SYSTEMS; FLOATING AND FIXED REPRESENTATION

2.3.1. The Position System

In order to represent numbers, we use in daily life a **position system** with base 10 (the decimal system). Thus for the numbers we use ten different characters, and the magnitude with which the digit a contributes to a number's value depends on the digit's position in the number in such a way that, if the digit stands n steps to the left of the decimal point, the value contributed is $a \cdot 10^{n-1}$, or if it stands n steps to the right of the decimal point, the value contributed is $a \cdot 10^{-n}$. The sequence of digits 4711.303 means, then,

$$4 \cdot 10^3 + 7 \cdot 10^2 + 1 \cdot 10^1 + 1 \cdot 10^0 + 3 \cdot 10^{-1} + 0 \cdot 10^{-2} + 3 \cdot 10^{-3}.$$

Every real number has a unique representation in the above way, except for the possibility of infinite sequences of nines—for example, the infinite decimal fraction 0.3199999 . . . represents the same number as 0.32.

One can very well consider other position systems with base different from 10. Any natural number $B \geq 2$ can be used as base. One can show that every positive real number has (with exceptions analogous to the nines-sequences mentioned above) a unique representation of the form

$$a_n B^n + a_{n-1} B^{n-1} + \ldots + a_1 B + a_0 + a_{-1} B^{-1} + a_{-2} B^{-2} + \ldots,$$

where the coefficients a_i are the "digits" in the system with base B—that is, positive integers a_i such that $0 \leq a_i \leq B - 1$.

One of the greatest advantages of the position system is that one can give simple, general rules for the arithmetical operations. The smaller the base is, the simpler these rules become. For this reason, among others, most computers operate in base 2, the **binary number system**. In this system, the addition and multiplication tables take the following simple form:

$$0 + 0 = 0; \quad 0 + 1 = 1 + 0 = 1; \quad 1 + 1 = 10;$$
$$0 \cdot 0 = 0; \quad 0 \cdot 1 = 1 \cdot 0 \quad = 0; \quad 1 \cdot 1 = 1.$$

In the binary system, the number seventeen, for example, becomes 10001, since

$$1 \cdot 2^4 + 0 \cdot 2^3 + 0 \cdot 2^2 + 0 \cdot 2^1 + 1 \cdot 2^0 = \text{sixteen} + \text{one} = \text{seventeen}.$$

Put another way $(10001)_2 = (17)_{10}$; i.e., we use an index to denote the base of the number system (in decimal representation). The numbers become longer written in the binary system; large integers become about 3.3 times as long, since N binary digits suffice to represent integers less than $2^N = 10^{N \cdot \log 2} \approx 10^{N/3.3}$.

Occasionally one groups together the binary digits in subsequences of three or four, which is equivalent to using 2^3 and 2^4, respectively, as base. These systems are called the **octal** and **hexadecimal** number systems, respectively. The octal system uses the digits from 0 to 7; in the hexadecimal system, the digits 0 through 9 and the letters A, B, C, D, E, F ("ten" through "fifteen") are used.

Example

$$(17)_{10} = (10001)_2 = (21)_8 = (11)_{16},$$
$$(13.25)_{10} = (1101.01)_2 = (15.2)_8 = (D.4)_{16},$$
$$(0.1)_{10} = (0.19999 \ldots)_{16}.$$

The "points" used to separate the integer and fractional part of a number (corresponding to the decimal point) are called the binary point, octal point, and hexadecimal point. The digits in the binary system are called bits (= **b**inary dig**its**). One very seldom converts (translates) from one number system to another manually; even a binary computer is fed decimal data and prints results in decimal form—it performs the conversions itself. There are also tables for conversion between the various number systems.

We are so accustomed to the position system that we forget that it is built upon an ingenious idea. The reader can puzzle over how the rules for arithmetical operations would look if one used roman numerals, a number system without the position principle described above.

2.3.2. Floating and Fixed Representation

A computer is usually equipped with two types of arithmetic operations, calculation with **fixed point** and with **floating point**. "Point" means the decimal point if the base is 10, or the binary point if the base is 2, etc. For simplicity of expression in the discussion that follows, we shall treat only the decimal case. A few of the most important results, however, will be formulated for a general base.

Computation with **floating** decimal point means that one works with a constant number of **digits;** computation with **fixed** decimal point means that one works with a constant number of **decimals**. Commonly, in the output data (because of various sources of error) some digits will be insignificant, or, respectively, some decimals will be incorrect.

2.3.3. Floating Decimal Point

By **normalized** *floating-decimal representation* of a number a, we mean a representation in the form:

$$a = m \cdot 10^q, \qquad 0.1 \leq |m| < 1, \qquad q \text{ an integer.}$$

Such a representation is possible for all numbers a, and unique if $a \neq 0$. The variable m is called the fractional part or **mantissa** and q is called the **exponent**.

In a computer, the number of digits for q and m is limited. This means that only a finite set of numbers can be represented in the machine. The numbers in this set (for a given q and m, and usually a base different from 10) are called **floating-point numbers**. The limited number of digits in the exponent implies that a is limited to an interval which is called the machine's *floating-point variable* **range**. With, for example, three digits in the exponent (plus one bit for the sign of the exponent), the range is

$$10^{-1,000} \leq |a| < 10^{999}$$

For $a = 0$ one sets $m = 0$, and it is also practical to make q as small as possible; in the above example one would then take $q = -999$. Minor deviations from the conventions mentioned here occur on certain machines, but such deviations are of little importance for the questions which are studied here.

In the computer, a is represented by the floating number

$$\bar{a} = \bar{m} \cdot 10^q$$

where $\bar{m}$ means the mantissa m, rounded off to t decimals. The **precision** of the machine is said, then, to be t decimal digits. (One should be careful to distinguish between the precision of the machine and the accuracy of the input or output data of a problem.) Then, according to Sec. 2.1.4, we have

$$|\bar{m} - m| \leq \begin{cases} \frac{1}{2} \cdot 10^{-t} & \text{for rounding,} \\ 10^{-t} & \text{for chopping.} \end{cases}$$

(There is one exception. If $|m|$ after rounding should be raised to 1, then $|\bar{m}|$ is set equal to 0.1 and q is raised by 1.)

The magnitude of the relative error in $\bar{a}$ is, then, since $m \geq 0.1$, at the most equal to

$$\frac{\frac{1}{2}10^{-t}\cdot 10^{q}}{m\cdot 10^{q}} \le \frac{1}{2}\cdot 10^{1-t} \quad \text{for rounding,}$$

and twice as large for chopping. By analogous reasoning for an arbitrary base B we get the following theorem:

THEOREM 2.3.1

Suppose that the floating numbers in a machine have base B and a mantissa with t digits. (The binary digit which gives the sign of the number is not counted.) Then, every real number in the floating-point range of the machine can be represented with a relative error which does not exceed the **machine unit (round-off unit)** *u which is defined by:*

$$u = \begin{cases} \frac{1}{2}\cdot B^{1-t} & \text{if rounding is used,} \\ B^{1-t} & \text{if chopping is used.} \end{cases}$$

In most machines, u lies between 10^{-15} and 10^{-6}. The quantity u is, in many contexts, a natural unit for relative changes and relative errors. We shall occasionally use terminology of the type "the quantity is perturbed by a few u."

Input and output data seldom are so large in magnitude that the range of the machine is not sufficient. In the rare cases where the above does not hold, one can, for example, use double (long) precision or else work with logarithms or some other transformation of the data. One should, however, keep in mind the risk that intermediate results in a calculation can produce *exponent spill*, i.e., an exponent which is too large (exponent overflow) or too small (underflow) for the machine representation. Different machines take different actions in such situations, as well as for division by zero. One should acquaint oneself with the conventions used by the particular machine one is working with and write one's programs so that no irritating occurrences of the above phenomena go overlooked.

Too small an exponent is usually, but not always, unprovoking. If the machine does not signal that a result has an exponent which is too small, but simply sets the result equal to zero, then one can—if there is a risk of annoying consequences—put in a test to see whether the quantity is zero and then have the program either print a message and stop the run or continue at some other appropriate point in the program. (Most machines give a signal when a division by zero occurs, and the program is then often terminated.) Occasionally, but not often, "unexplainable errors" in output data are the product of an "underflow" somewhere in the computations.

Example 2.3.1

Even simple programs can quickly give exponent spill. If

$$x_0 = 2, \qquad x_{n+1} = x_n^2,$$

then already $x_{13} = 2^{8,192}$ is larger than most machines permit. One should also be careful in computation with factorials, for example $200! > 10^{374}$.

2.3.4. Fixed Decimal Point

Computation with fixed decimal point means that all real numbers are shortened to t decimals. Since the length of a computer word is usually constant (say, s digits), then only numbers in the interval $I = [-10^{s-t}, 10^{s-t}]$ are permitted. This restriction is much narrower than the corresponding one with floating representation with division of the word into mantissa and exponent. Some common conventions in fixed point are $t = s$ (fraction convention) or $t = 0$ (integer convention). Occasionally, the limitation on I causes difficulty, since even if $x \in I$, $y \in I$, we can have, for example, that $x + y$ or x/y lies outside of I.

If one writes programs for a fixed-point machine, then one must see to it that all numbers (even uninteresting intermediate results) remain within I. This can be attained by multiplying the variables by appropriate factors, so-called **scale factors**, and then transforming the equations accordingly. One can interpret this as changing the units of measurement of the variables of the problem. This complicates preparatory work, since there is a risk of another type: if one chooses the scale factors carelessly, certain intermediate results can have many leading zeros. The number of significant digits then becomes low; this can lead to poor accuracy in the final results. On the other hand, addition in fixed point is usually much faster than addition in floating point. One can say that in floating-point calculation the machine itself chooses a scale factor in each operation.

Most of the algebraic programming languages (Algol, Fortran, etc.) assume that elementary arithmetical operations shall be performable so long as the values of the variables are not unusually large. Practically, this means that floating representation is assumed. As a consequence, fixed point is seldom used, except when a program will be used so much (and in unmodified form) that the savings in computer time become greater than the additional cost from the increase in programming. In certain problems, the choice of scale factors is trivial; in others, it may be impracticable.

In practice, hexadecimal or binary representation is more common than decimal, which we have talked about here in order to simplify the mode of expression. The reader will surely have no difficulty in modifying the discussion for the binary or hexadecimal cases. See also the problems at the end of this section.

2.3.5. Round-off Errors in Computation with Floating Arithmetic Operations

Even if the operands in an arithmetic operation are exactly represented in the machine, it is not certain that the *exact* result of the operation is a floating-point number. For example, the exact product of two floating-point t-digit numbers has $2t$ or $2t - 1$ digits.

Let x and y be two floating-point machine numbers. Denote by

$$fl(x + y), \quad fl(x - y), \quad fl(xy), \quad fl\left(\frac{x}{y}\right)$$

the results of floating addition, subtraction, multiplication, and division, which the machine stores in memory (after rounding or chopping). In many computers, these results are equal to the rounded or chopped value of the exact result of the operation. We assume that such is the case, and thus, according to Theorem 2.3.1, we have,

$$|fl(x \text{ op } y) - x \text{ op } y| \leq |x \text{ op } y| \cdot u, \tag{2.3.1}$$

where "op" stands for one of the four symbols for elementary operations; $+, -, \cdot, /$. *The operations $fl\,(x \text{ op } y)$ have, to some degree, other properties than the exact arithmetic operations.*

Example 2.3.2

Associativity does not, in general, hold for floating addition. Consider floating addition using seven decimals in the mantissa,

$$a = 0.1234567 \cdot 10^0, \qquad b = 0.4711325 \cdot 10^4, \qquad c = -b.$$

The following schema indicates how floating addition is performed.

$$fl(b + c) = 0, \qquad fl(a + fl(b + c)) = a = 0.1234567 \cdot 10^0$$

$a =$	0.0000123	$4567 \cdot 10^4$	The digits to the right of the vertical line are thrown away.
$+b =$	0.4711325	$\cdot 10^4$	
$fl(a + b) =$	0.4711448	$\cdot 10^4$	
$c =$	-0.4711325	$\cdot 10^4$	

$$fl(fl(a + b) + c) = 0.0000123 \cdot 10^4 = 0.1230000 \cdot 10^0 \neq fl(a + fl(b + c)).$$

Example 2.3.3

Using a hexadecimal machine (floating representation, $t = 6$, with chopping, $u = 16^{-5} \approx 10^{-6}$) one computes

$$\sum_{n=1}^{10,000} n^{-2} \approx 1.644834$$

in two different orders. Using the natural ordering $((1 + \frac{1}{4}) + \frac{1}{9}) + \ldots$, the error was $1{,}317 \cdot 10^{-6}$, but summing in the opposite order, the error was only $2 \cdot 10^{-6}$. This was not unexpected. Each operation is an addition, where the sum s is increased by n^{-2}; thus, in each operation, one commits an error of about $s \cdot u$, and all of these errors are added. Using the first summation order, we have $1 \leq s \leq 2$ in every step, but using the other order of summation, we have $s < 10^{-2}$ in 9,900 of the 10,000 additions.

If the terms in a sum (of positive terms) are of varying orders of magnitude, one should add the terms in increasing order if one needs high accuracy in the result. It is seldom that one needs to think of this in practice, but important instances in which this idea can be applied do occur; see Chap. 8.

The previous example is mainly of interest as a simple illustration of the important thesis: *mathematically equivalent algorithms are not always numerically equivalent.* By **mathematical equivalence** of two algorithms we mean here that the algorithms give exactly the same results from the same input data, if the computations are made without round-off error ("with infinitely many decimals"). The one algorithm can then as a rule be derived from the other purely formally using the rules of algebra for real numbers (or else with the help of other mathematical identities). Two algorithms are **numerically equivalent** if their respective results, using the same input data, do not differ by more than what the problem's exact output data would be changed by if the input data were perturbed by a few u (u was defined in Theorem 2.3.1).

Example 2.3.4. *Increasing the Precision of Addition in Floating-Point Arithmetic*

In Example 2.3.3, when adding in the natural order, about the last *9,000* terms give *no* contribution to the sum because of the summation order, the fact that a computer has a fixed finite word length, and the way floating-point addition is performed (recall $u \approx 10^{-6}$). There the remedy was to sum the terms in the opposite order (smallest to largest). Often, however, one has no a priori information about the relative magnitudes of the terms to be summed. If there are a large number of such terms, then it can be time-consuming to sort them into order of increasing magnitude before adding. Also, the number of extra memory cells (equal to the number of terms in the sum) required for the sorting may not be available.

Consider the "usual" program segment for computing a sum $\sum_{i=1}^{N} X_i$:

```
      S = 0.0
      DO 33 I = 1,N
   33 S = S + X(I)
```

Let S_i denote the partial sum (value of the variable S) after adding the term X_i. Since for each i the additional error is bounded by $|S_i|\cdot u$, the total error is bounded approximately by $N\cdot(\sum_{i=1}^{N}|X_i|)\cdot u$ (see also the more exact expression given in the table in Sec. 2.4.1). This may be unsatisfactory for N large. We would like a way to compute sums with floating-point arithmetic which gives a small error bound independent of N.

To illustrate the basic idea which can be used, take $A = 0.1234567\cdot 10^0$, $B = 0.4711325\cdot 10^4$. Here $fl(A + B) = 0.4711448\cdot 10^4$ (denote this by S). Suppose we form

$$C = fl(fl(B - fl(A + B)) + A)$$

corresponding to the Fortran statements

```
S = A + B
C = (B - S) + A
```

Then

$$C = (-.1230000 \cdot 10^0) + (.1234567 \cdot 10^0)$$
$$= 0004567 \cdot 10^0 = 4567000 \cdot 10^{-3}.$$

Thus the variable C has picked up the information that was lost in the operation

```
S = A + B.
```

W. Kahan, at the 1971 IFIP Congress, gave the following program for computing $\sum_{J=1}^{N} X(J, \ldots)$ which implements the above idea and also contains some other sophistications to account for the differences in performance of floating-point operations in existing computers:

```
    S = 0.
    C = 0.
    DO 999  J=1, N
    Y = C + X(J, . . .)
    T = S + Y
    F = 0.
    IF(sign (Y) equals sign (S)) F = (0.46*T - T) + T
    C = ((S-F)-(T-F)) + Y
999 S = T
    S = S + C
```

Kahan states that "on all current North American machines with floating point hardware," this program keeps $|\xi_j|$ less than $5u + O(Nu^2)$ in the expression for the *computed* sum of N terms,

$$S_N = \sum_{j=1}^{N} (1 + \xi_j) X_j.$$

(This formulation is a typical example. of backward error analysis—see Section 2.4.)
Thus the above program gives a very small error bound in the sum, and the error bound is (essentially) independent of N.

We do not go into more detail here. For a comprehensive analysis of floating-point operations, and references to the literature, see Knuth [13].†

REVIEW QUESTIONS

1. What are the binary, octal, and hexadecimal number systems?

†Bibliographical references are listed in Chapter 13, starting in Sec. 13.2.

2. What is meant by "floating-point representation"?

3. How large can the maximum relative error be in the floating-point representation of a number?

4. Give examples to show that some of the axioms for arithmetic with real numbers need not always hold for floating-point arithmetic.

PROBLEMS

1. (a) Which numbers can be expressed with a finite number of binary digits to the right of the binary point?
   (b) Give a number which can be expressed with a finite decimal fraction, but which in binary notation requires an infinite number of binary digits!

2. The (fictional) space probe *Venus 17* sent back pictures of a Venusian rock-carving, assumed to constitute an addition. Which number system do the inhabitants of Venus count in? How many fingers can one presume that they have on each hand? (Almost every culture on earth uses one of the numbers 5, 10, or 20 as base for its number system.)

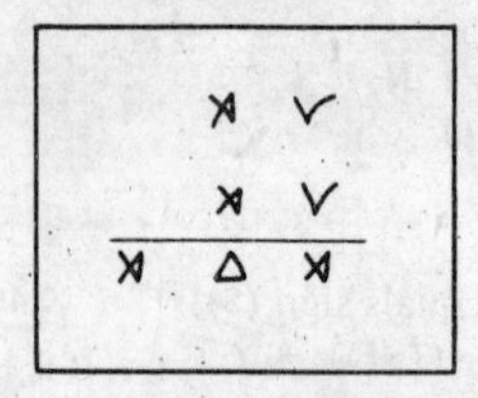

3. Give in decimal representation: (a) $(10000)_2$; (b) $(100)_8$; (c) $(64)_{16}$; (d) $(FF)_{16}$; (e) $(0.11)_8$; (f) $(0.0631463146314\ldots)_8$; (g) the largest positive integer which can be written with thirty-one binary digits (answer with one significant digit).

4. In the fictional computer EDP 4711, a memory cell has two binary positions for storing signs (of the mantissa and of the exponent) and eleven decimal positions for the exponent. The number π is, for example, stored in the following way:

+	3	1	4	1	5	9	2	6	5	3	6

+	0	0	1

With this representation, numbers in the interval $(-10^{999}, 10^{999})$ are stored to eleven significant digits. Show how the following numbers are stored: (a) 2.7182818285; (b) -1073741824; (c) 0.577216; (d) $-123\cdot10^{-45}$.

5. How large (maximally) can the relative error be when a number is stored using the representation in Problem 4?

In the CDC 3200, a floating-point number is stored using forty-eight binary positions, in the following way:

1	11	36
s.	exp.	mantissa

A given number is written in the form $p = \pm p' 2^{p''}$, where the mantissa p' must satisfy $\frac{1}{2} \leq p' < 1$. The mantissa is stored as a (correctly rounded) binary fraction. The exponent, which must satisfy $|p''| < 2^{10}$, is represented by the binary integer $p'' + 2^{10}$ if $p'' \geq 0$, otherwise with $p'' + (2^{10} - 1)$. The sign bit is set equal to zero. If $p < 0$, then the ones-complement of the representation obtained above is taken—that is, in all forty-eight positions the ones are changed to zeros and the zeros to ones. (Thus, in particular, the sign bit is set to 1 if $p < 0$.) If $p = 0$, the number is stored as all zeros.

Show how the following numbers would be stored:

(a) 1.0; (b) −0.0625; (c) 250.25; (d) 0.1.

(e) Give (in decimal notation) the largest and smallest positive numbers which can be stored.

(f) What is the maximal relative error with which a number is stored?

7. In IBM system 360, every memory cell consists of thirty-two binary positions. A number p is stored in floating hexadecimal form in the following way:

One sets

$$p = p' \cdot 16^{(p''-64)},$$

where $p'' \geq 0$ and $\frac{1}{16} \leq |p'| < 1$; and one stores the mantissa p' with sign and the modified exponent p''.

In single-precision representation, one memory cell is used; p'' is allocated 7 bits (pos. 2–8); p' is given 1 bit for sign (pos. 1) and 24 bits for p' (pos. 9–32).

In double-precision representation, two memory cells are used; p'' is again allocated 7 bits (pos. 2–8, cell 1), and p' is allocated 1 bit for sign (pos. 1, cell 1), and 56 bits (pos. 9–32 in cell 1 and all of cell 2) thereafter.

How are the following numbers represented in single precision?

(a) 1.0; (b) 0.5; (c) −15.0

(d) Give (decimally) the largest and smallest positive numbers which can be stored.

(e) What is the maximal relative error with which a number can be stored in single and double precision, respectively? (Give the answers in decimal notation.)

2.4. BACKWARD ERROR ANALYSIS; CONDITION NUMBERS

2.4.1. Backward Error Analysis

Let us use the notation of Sec. 2.3.5 and let "op" denote one of the operations $+, -, \cdot, /$. By Eq. (2.3.1), it follows that there exists a number δ, $|\delta| \leq u$ (where δ depends on x and y) such that

$$fl(x \text{ op } y) = (x \text{ op } y)(1 + \delta), \quad (|\delta| \leq u). \tag{2.4.1}$$

This means, for example, that in multiplication, $fl(xy)$ is the exact product of x and $y(1 + \delta)$ for some δ, $|\delta| \leq u$. In the same way, the results using the three other operations can be interpreted as the result of exact operations where the operands have been perturbed by a relative amount which does not exceed u. To use **backward error analysis** means that one applies the above interpretation step by step backwards in an algorithm. By means of backward error analysis it has been shown, even for many quite complicated algorithms, that the output data which the algorithms produce (under the influence of round-off error) is the exact output data of a problem of the same type in which the input data has been changed (relatively) by a few u. That change, measured with u as a unit, is called the **condition number of the algorithm**. Clearly, the condition number has a small value for a good algorithm and a large value for a poor algorithm.

Using backward error analysis, one transfers the problem of estimating the effects of round-off errors during the computation back to the problem of estimating the effect of disturbances in input data. In this way, the general error-propagation formula (see Sec. 2.2.2) can then be used in the estimation of the effect of the disturbances. A graphical illustration of backward error analysis is given in Fig. 2.4.4.

Example 2.4.1

By repeatedly using formula (2.4.1) in the case of multiplication, one can show that $fl(x_1 x_2 \ldots x_n)$ is exactly equal to a product of the form

$$x_1x_2(1+\delta_2)x_3(1+\delta_3)\ldots x_n(1+\delta_n),$$

where

$$|\delta_i| \leq u, \quad i = 2, 3, \ldots, n.$$

From this, it follows that

$$|fl(x_1x_2\ldots x_n) - x_1x_2\ldots x_n| = \epsilon\,|x_1x_2\ldots x_n|,$$

where

$$\epsilon = |(1+\delta_2)(1+\delta_3)\ldots(1+\delta_n) - 1|.$$

Thus

$$\epsilon \leq (1+u)^{n-1} - 1.$$

For convenience, let $j = n - 1$ and make the (realistic) assumption that $(n-1)\cdot u < 0.1$. Then:

$$\ln(1+u)^j = j\ln(1+u) < ju, \quad (u < 1)$$

so

$$(1+u)^j < e^{ju}$$

$$(1+u)^j - 1 < (ju) + \frac{(ju)^2}{2!} + \frac{(ju)^3}{3!} + \cdots$$

$$(1+u)^j - 1 < ju\left(1 + \left(\frac{ju}{2}\right) + \left(\frac{ju}{2}\right)^2 + \left(\frac{ju}{2}\right)^3 + \cdots\right)$$

$$(1+u)^j - 1 < ju\left(1 + \frac{0.05}{1-0.05}\right) < 1.06\,ju$$

Thus $\epsilon < 1.06(n-1)u$, for $(n-1)u < 0.1$.

Example 2.4.2

Recall the notation (introduced in Problem 8 at the end of Sec. 2.2):

R_X = a bound for the part of the *absolute* error in a result which depends on the error in the input data.

R_C = a bound for the *absolute* error in a result which depends on round-off errors during the calculations.

If the magnitude of the relative error in the indata $(x_i, y_i, a_i, x), i = 0, 1, \ldots, n$ does not exceed r, then, using backward error analysis analogously to Example 2.4.1, one finds the following results for some commonly occurring expressions (assuming that $2nu \leq 0.1$, $nr \leq 0.1$):

	$\frac{R_X}{1.06r}$	$\frac{R_C}{1.06u}$
$\sum_{i=1}^{n} x_i$	$\sum \lvert x_i \rvert$	$\sum \lvert (n+1-i)x_i \rvert$
$\prod_{i=1}^{n} x_i$	$n \prod \lvert x_i \rvert$	$(n-1) \prod \lvert x_i \rvert$
$\sum_{i=1}^{n} x_i y_i$	$2 \sum \lvert x_i y_i \rvert$	$\sum \lvert (n+2-i)x_i y_i \rvert$
$\sum_{i=0}^{n} a_i x^i = F(x)$	$\lvert xF'(x) \rvert + \sum \lvert a_i x^i \rvert$	$\sum \lvert (2i+1)a_i x^i \rvert$

Here, the sum, the product, and the scalar product are assumed to be computed in the natural order, i.e., for the sum $(\ldots(((x_1 + x_2) + x_3) + x_4) + \ldots + x_n)$, and analogously for the products; the polynomial is assumed to be computed by Horner's rule, see Sec. 1.3.2. The reader is of course recommended to verify the results given above (Problem 1 at the end of this section).

2.4.2. Condition Numbers for Problems and Algorithms

There can be many reasons for poor accuracy in output data. For example, the algorithm can be poorly constructed. But poor accuracy can also depend on the problem itself; that is, the output data can be very sensitive to disturbances in the input data—independently of the choice of algorithm. In the first case, we say that the algorithm is ill-conditioned; in the second case, the problem is said to be **ill-conditioned**. One also says that the algorithm is **numerically unstable** or that the problem is (mathematically) unstable.

Consider a numerical problem P, with given (possibly error-afflicted) input data, and an algorithm A, which is used on a computer with round-off unit u (floating point). Since floating point is used, it is natural to look at relative errors.

The **condition number C_A for the algorithm** A was defined in Sec. 2.4.1. We shall consider the output data produced by algorithm A as the exact output data for the problem with the same structure as P, except that the input data has been changed. The condition number of the algorithm indicates a sufficient size for such a relative change using u as a unit of measurement.

The **condition number C_P for the problem** P with given input data is the largest relative change, measured with u as a unit, that the exact output data of the problem can have, if there is a relative disturbance in the input data of size u.

If r is a bound for the relative error in the input data, we have then

$$C_P(r + C_A u)$$

as *a bound for the relative error in the output data.*

The condition numbers for P and A depend on the input data. A problem or an algorithm can be well-conditioned (have a small condition number) for certain input data, and ill-conditioned (have a large condition number) for other input data. Furthermore, C_P and C_A are independent of each other; C_P can be small even if C_A is large, and vice versa. In principle, C_P can be estimated using the general error-propagation formula.

Example 2.4.3

Let P be the problem of computing the value $f(x)$ for the function f at the point x. Then, from the definition of C_P, we have that C_P is equal to the largest of the two values that

$$\left|\frac{(f(x+\Delta x) - f(x))/f(x)}{\Delta x/x}\right|$$

takes for $\Delta x = ux$ and $\Delta x = -ux$, respectively. Thus we find that (practically):

$$C_P = \left|\frac{f'(x)/f(x)}{1/x}\right|.$$

Example 2.4.4

Compute x from the system

$$x + \alpha y = 1$$

$$\alpha x + y = 0.$$

The solution is $x = (1 - \alpha^2)^{-1}$. The matrix is singular for $\alpha = 1$. The condition number for the problem of computing x, when α is the only input data value which has been rounded, is

$$C_P = \frac{x'(\alpha)/x(\alpha)}{1/\alpha} = \frac{2\alpha^2}{1 - \alpha^2}.$$

Thus, the problem is **ill-conditioned** when $\alpha^2 \approx 1$, but very **well-conditioned** when $\alpha^2 \ll 1$. The condition number is thus related to how near the matrix of coefficients lies to a singular matrix. The problem of computing $y = -\alpha(1 - \alpha^2)^{-1}$ is also **ill-conditioned** when $\alpha^2 \approx 1$.

On the other hand, the **problem** of computing

$$z = x + y = \frac{1}{1 - \alpha^2} - \frac{\alpha}{1 - \alpha^2} = \frac{1}{1 + \alpha} \tag{2.4.2}$$

is **well-conditioned** even when $\alpha \approx 1$. (The condition number is then about 0.5.) But there would be a fatal cancellation if, when $\alpha \approx 1$, one first solved for x and y from the system of equations and then got z by adding x and y. For example, for $\alpha = 0.9900$ the condition number for solving for x and y is about 100. With floating-decimal arithmetic (using four digits), one gets

$$z = x + y = 0.5025 \cdot 10^2 - 0.4975 \cdot 10^2 = 0.5000,$$

while the correct answer, using Eq. (2.4.2), is $z = 0.5025$. In this case, it is the **algorithm** for computing z, with the use of rounded values for x and y, which is **ill-conditioned,** not the problem. In order to get four digits in z with that algorithm, one would need six digits in x and y. In comparison with what can happen in larger problems, however, this is a very mild degree of poor conditioning.

If the relative accuracy in the output data is unacceptably poor, and if this is because C_A is large, then one should first investigate whether a simple **rewriting of the computational sequence** can improve things (eliminate cancellations, change the order of summation, etc.). If this does not work, then one can investigate whether it is sufficient to use **higher-precision arithmetic** in a small part of the computation—that is, use more digits than normal. This is possible on most computers, but computational speed is reduced and there can also be a great deal of extra programming involved.

Notice, in addition, that C_A is related to the *relative accuracy* in a result which A gives. If one does not need high relative accuracy in a quantity with a small absolute magnitude, then it is not necessary to reject the algorithm, even if C_A is large. (Condition numbers can, on the other hand, be defined with respect to absolute accuracy.)

If the poor accuracy is caused by a large C_P, then one might at first think that nothing can be done about the situation; the difficulty lies, of course, with P. But *the difficulty can depend on the form one has chosen to represent the input and output data of the problem.*

Example 2.4.5

The polynomial P,

$$P(x) = (x - 10)^4 + 0.200(x - 10)^3 + 0.0500(x - 10)^2 \\ - 0.00500(x - 10) + 0.00100,$$

is identical with a polynomial Q which, if the coefficients are rounded to six digits, becomes

$$\tilde{Q}(x) = x^4 - 39.8000x^3 + 594.050x^2 - 3941.00x + 9805.05.$$

One finds that $P(10.11) = 0.0015 \pm 10^{-4}$, where only three digits are needed in the computation, while $\tilde{Q}(10.11) = -0.0481 \pm \frac{1}{2} \cdot 10^{-4}$, in spite of the fact that eight digits were used in the computation. The rounding to six digits of the coefficients of Q has thus caused an error in the polynomial's value at $x = 10.11$; the erroneous value is more than 30 times larger than the correct value $P(10.11)$. When the coefficients of Q are input data, the problem of computing the value of the polynomial for $x \approx 10$ is far more ill-conditioned than when the coefficients of P are input data.

2.4.3. Geometrical Illustrations of Error Analysis

The problem of error propagation can to some degree be illustrated geometrically. A numerical problem P means a mapping of the space X of possible input data onto the space Y of the output data. The dimensions of these spaces are usually quite large. In Fig. 2.4.1 we satisfy ourselves with two di-

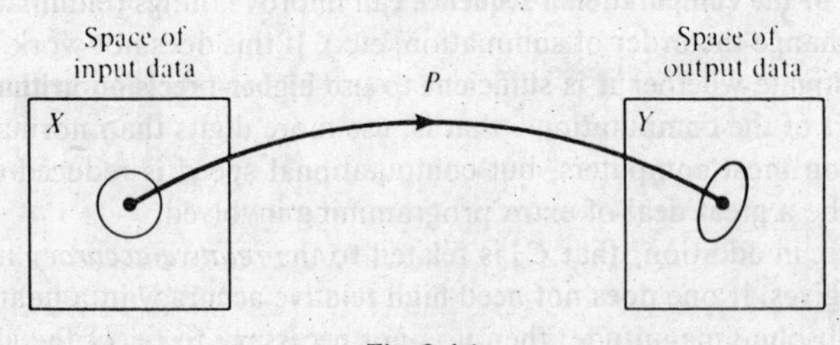

Fig. 2.4.1

mensions, and, since we are considering relative changes, we can consider that the coordinate axes are logarithmically scaled. A small circle of radius r is mapped onto an ellipse whose major axis is about $C_p r$.

In the construction of an algorithm for a given problem, one often breaks down the problem into a chain of subproblems, $P_1, P_2, \ldots, P_k$ for which algorithms $A_1, A_2, \ldots, A_k$ are known, in such a way that the output data from P_{i-1} is the input data to P_i (Fig. 2.4.2). Different ways of decomposing the problem give different algorithms, as a rule with different condition numbers. It is dangerous if the last part of such a chain—for example, the last link—is ill-conditioned. On the other hand, it need not be dangerous if the first link of such a decomposition of the problem is ill-conditioned, if the problem itself is well-conditioned.

In Fig. 2.4.3 we see two examples of a decomposition of the problem P into two subproblems. From X to X'' there is a strong contraction which is followed by an expansion about equally strong in the mapping from X'' to Y. The round-off errors which are made in X'' when the intermediate results are stored have as a consequence that one arrives somewhere in the surrounding circle, which is then transformed to a very large region in Y. The situation in Problem 9 in Sec. 2.2 corresponds approximately to this picture—but with other proportions.

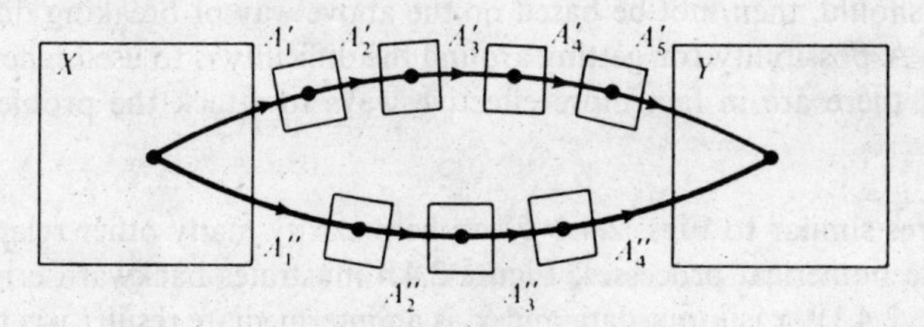

Fig. 2.4.2

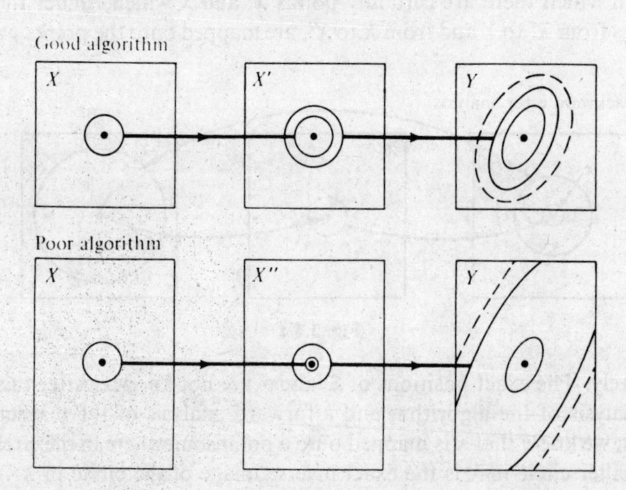

Fig. 2.4.3

Example 2.4.6

One can show that the problem P of determining the largest eigenvalue of a real symmetric matrix is always well-conditioned, $C_P \approx 1$. It seems natural to break down this problem into the two subproblems:

P_1: to compute the coefficients of the characteristic equation of the matrix.

P_2: to solve the characteristic equation.

But there is a matrix of order 40 where the condition number for P_2 is 10^{14}—in spite of the fact that the origin lies exactly between the largest and smallest eigenvalues, so that one cannot blame the high condition number on a difficulty of the same type as that encountered in Example 2.4.5. This is an extreme case of the situation which the lower picture in Fig. 2.4.3 is intended to illustrate.

A general method for solving the eigenvalue problem for symmetric matrices should, then, not be based on the above way of breaking down the problem. A possibility for getting around the difficulty is to use higher precision. But there are in fact more effective ways to attack the problem; see Chap. 5.

Pictures similar to Figs. 2.4.1–5 can help clarify many other relations in composite numerical processes. Figure 2.4.4 illustrates backward error analysis (Sec. 2.4.1); x is input data and x' is an intermediate result; y is the output data obtained by the numerical computation, which were influenced by rounding errors. The results of the backward analysis are the circles in X' and X, in which there are certainly points $\hat{x}'$ and $\hat{x}$ which, under the exact mappings from X' to Y and from X to X', are mapped onto the points y and $\hat{x}'$,

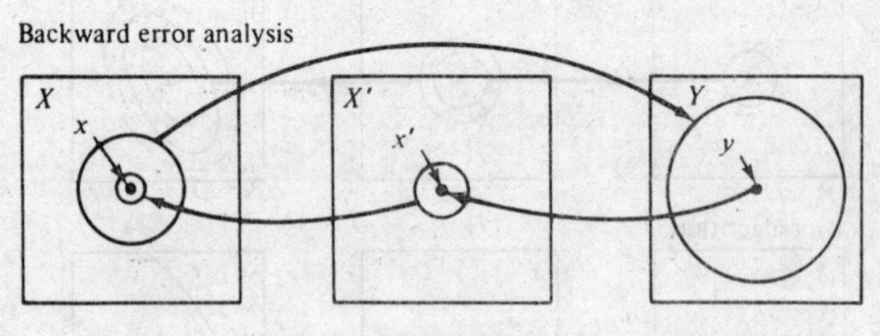

Fig. 2.4.4

respectively. The exact positions of $\hat{x}'$ and $\hat{x}$ are not known. After this backward analysis of the algorithm and a forward analysis of the mathematical problem, we know that x is mapped onto a point somewhere in the circle in Y. (The smaller circle in X is the exact inverse image of the circle in X'.) Backward error analysis is often more adequate than forward error analysis.

Finally, Fig. 2.4.5 gives an illustration of one difficulty which can occur with the use of interval analysis (see Example 2.2.13).

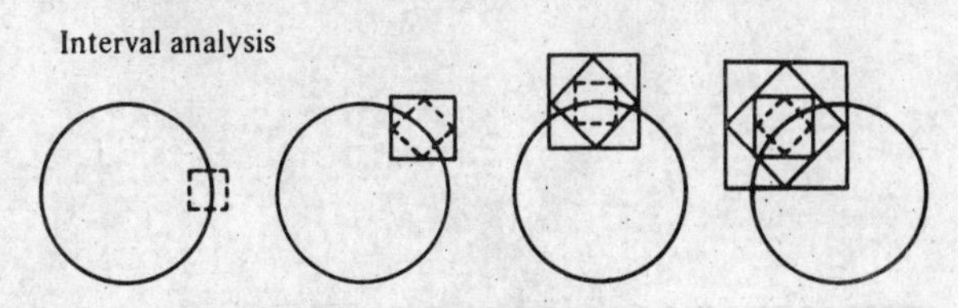

Fig. 2.4.5

REVIEW QUESTIONS

1. Define condition number for:
   (a) a numerical problem;
   (b) an algorithm.
2. Give examples of well-conditioned and ill-conditioned problems.
3. Give an example of an ill-conditioned algorithm for a well-conditioned problem, and vice versa.
4. What is backward error analysis? Give an example of its use.
5. A problem with condition number C_P is treated using an algorithm which has condition number C_A, on a computer with round-off unit (machine unit) u. Show that if the input data has a relative error whose absolute value is less than r, then the absolute value of the relative error in the output data is less than $C_P(r + C_A u)$.

PROBLEM

Verify some of the results given in the table in Example 2.4.2.

3 NUMERICAL USES OF SERIES

3.1. ELEMENTARY USES OF SERIES

3.1.1. Simple Examples

Series expansions are a very important aid in numerical calculations, especially for quick estimates made in hand calculation—for example, in evaluating functions, integrals, or derivatives. Solutions to differential equations can often be expressed in terms of series expansions. Since the advent of computers, it has, however, become more common to treat differential equations directly, using difference approximations (see Chap. 8) instead of series expansions. But in connection with the development of automatic methods for formula manipulation, one can anticipate renewed interest for series methods. These methods have some advantages, especially in multidimensional problems.

Example 3.1.1

Compute to four correct decimals

$$I = \int_0^1 \exp(-t^2)\, dt.$$

The primitive function cannot be expressed in a finite expression in terms of elementary functions. By making the substitution $x = -t^2$ in the familiar expansion

$$e^x = \sum_{n=0}^{\infty} \frac{x^n}{n!}$$

and integrating term by term, one gets

$$I = \sum_{n=0}^{\infty} \int_0^1 \frac{(-1)^n t^{2n}}{n!}\, dt = \sum_{n=0}^{\infty} \frac{(-1)^n}{(2n+1)n!} \approx 0.7468 \quad \text{(with seven terms).}$$

The series converges quickly. The eighth term is already less than $2 \cdot 10^{-5}$. For an alternating series where the absolute value of the terms decreases monotonely to zero, the first neglected term gives a strict error estimate. (This will be proved later; see Theorem 3.1.4.)

Notice that the series expansion for the integrand was not derived by successively differentiating the integrand and substitution into the Taylor expansion for the integrand, but by a simple operation with a known series, the substitution $x = -t^2$.

Example 3.1.2

Compute, to five decimals, $y(0.5)$, where $y(x)$ is the solution to the differential equation

$$y'' = -xy,$$

with initial conditions $y(0) = 1$, $y'(0) = 0$. The solution cannot be simply expressed in terms of elementary functions. We shall use the **method of undetermined coefficients**. Thus we try substituting a series of the form:

$$y(x) = \sum_{n=0}^{\infty} c_n x^n = c_0 + c_1 x + c_2 x^2 + \ldots.$$

Differentiate twice:

$$\begin{aligned} y''(x) &= \sum n(n-1)c_n x^{n-2} \\ &= 2c_2 + 6c_3 x + 12c_4 x^2 + \ldots + (m+2)(m+1)c_{m+2}x^m + \ldots \\ -xy(x) &= -c_0 x - c_1 x^2 - c_2 x^3 - \ldots - c_{m-1}x^m - \ldots. \end{aligned}$$

Equate the coefficients of x^m in these series:

$$c_2 = 0, \quad (m+1)(m+2)c_{m+2} = -c_{m-1} \quad \text{for} \quad m \geq 1.$$

From the initial conditions, it follows that $c_0 = 1$, $c_1 = 0$. Thus

$$c_3 = -\frac{c_0}{6} = -\frac{1}{6}, \qquad c_6 = -\frac{c_3}{30} = \frac{1}{180},$$

$$c_9 = -\frac{c_6}{72} = -\frac{1}{12{,}960}, \ldots$$

$$c_n = 0, \quad \text{if } n \text{ is not a multiple of 3.}$$

Thus

$$y(x) = 1 - \frac{x^3}{6} + \frac{x^6}{180} - \frac{x^9}{12{,}960} + \ldots$$

$$y(0.5) = 0.97925.$$

The x^9-term is ignored, since it is less than $2 \cdot 10^{-7}$. In this example, also, the first neglected term gives a strict bound for the error (i.e., for the remaining terms), since the absolute value of the terms decreases, and the terms alternate in sign.

Since the calculation was based on a trial substitution, one should, strictly speaking, prove that the series obtained defines a function which satisfies the given problem. Clearly, the series converges at least for $|x| < 1$, since the coefficients are bounded. (In fact, the series converges for all x.) Since a power series can be differentiated term by term in the interior of its interval of convergence, the proof presents no difficulty. Notice, in addition, that the finite series obtained for $y(x)$ by breaking off after the x^9-term is the exact solution to the following differential equation:

$$y'' = -xy - \frac{x^{10}}{12{,}960}, \qquad y(0) = 1, \qquad y'(0) = 0,$$

where the "perturbational term," $-x^{10}/12{,}960$, has magnitude less than 10^{-7} for $|x| \leq 0.5$. A similar "backward analysis" can (even in analogous, more complicated cases) be extended to give a strict error estimate, but that would lead us outside the scope of this book.

3.1.2. Estimating the Remainder

In practice, one is seldom seriously concerned about a strict error bound when the computed terms decrease rapidly and it is "obvious" that the terms will continue to decrease equally quickly. One can then break off the series and use either the last included term or a coarse estimate of the **first neglected term** as an estimate of the remainder. However, it is natural to feel a need for more stringency if one does not yet have much computational experience. For this reason, we shall formulate a few theorems, with which one can often transform the feeling that "the remainder is negligible" to a mathematical proof. There are, in addition, actually numerically useful *divergent* series; see Sec. 3.2.5. When one uses such series, estimates of the remainder are clearly essential.

Assume that we want to compute a quantity S, which can be expressed in a series expansion, $a_0 + a_1 + a_2 + \ldots$, and set

$$S_n = \sum_{j=0}^{n} a_j, \qquad R_n = S - S_n.$$

We call $a_{n+1} + a_{n+2} + a_{n+3} + \ldots$ the **tail** of the series; a_n is the "last included term" and a_{n+1} is the "first neglected term."

The tail of a convergent series can often be compared to a series with a known sum (for example, a geometric series), or with an integral which can be computed directly.

THEOREM 3.1.1. *Comparison with a Geometric Series*

If $|a_{j+1}| \le k|a_j|$, *where* $k < 1$, *for all* $j \ge n$, then

$$|R_n| \le \frac{|a_{n+1}|}{1-k} \le \frac{k|a_n|}{1-k}.$$

Proof. By induction, one finds that

$$|a_j| \le k^{j-1-n}|a_{n+1}| \quad \text{for all} \quad j \ge n+1, \quad \text{since}$$

$$|a_j| \le k^{j-1-n}|a_{n+1}| \Rightarrow |a_{j+1}| \le k|a_j| \le k^{j-n}|a_{n+1}|.$$

Thus

$$|R_n| \le \sum_{j=n+1}^{\infty} |a_j| \le \sum_{n+1}^{\infty} k^{j-1-n}|a_{n+1}| = \frac{|a_{n+1}|}{1-k} \le \frac{|a_n|k}{1-k},$$

according to the formula for the sum of an infinite geometric series.

Example 3.1.3

$a_j = j^{1/2}\pi^{-2j}$. According to Example 2.2.9, $|a_6| < 3 \cdot 10^{-6}$.

$$\left|\frac{a_{j+1}}{a_j}\right| \le \frac{(j+1)^{1/2}}{j^{1/2}} \frac{\pi^{-2j-2}}{\pi^{-2j}} \le \left(1 + \frac{1}{6}\right)^{1/2} \pi^{-2} < 0.11, \quad j \ge 6.$$

Thus

$$|R_6| < 3 \cdot 10^{-6} \frac{0.11}{1 - 0.11} < 4 \cdot 10^{-7}$$

Example 3.1.4

If $k < \frac{1}{2}$, then it is strictly valid that the absolute value of the remainder is less than the last included term, since we have then $k/(1-k) < 1$.

THEOREM 3.1.2. *Comparison with an Integral*

If $|a_j| \le f(j)$ *for all* $j \ge n$, *where* $f(x)$ *is a decreasing function for* $x \ge n$, *then*

$$|R_n| \le \int_n^{\infty} f(x)\,dx.$$

Proof: See Fig. 3.1.1.

Example 3.1.5

$a_j = (j^3 + 1)^{-1}$. Choose $f(x) = x^{-3}$. Then

$$R_n \le \int_n^{\infty} x^{-3}\,dx = \frac{n^{-2}}{2}.$$

In addition, this bound gives an asymptotically correct estimate of the remainder (when $n \longrightarrow \infty$), which shows that R_n is here significantly larger than the first neglected term.

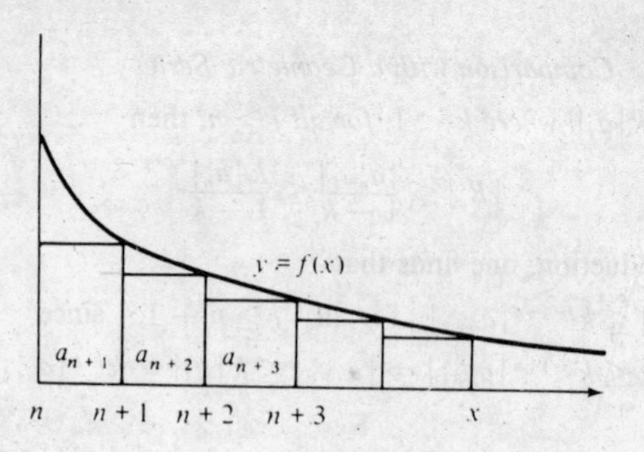

Fig. 3.1.1

DEFINITION

A series is **alternating** for $j \geq n$ if, for all $j \geq n$, a_j and a_{j+1} have opposite signs.

DEFINITION

sgn x, read "signum of x," is defined by:

$$\operatorname{sgn} x = \begin{cases} +1 & \text{if} \quad x > 0, \\ 0 & \text{if} \quad x = 0, \\ -1 & \text{if} \quad x < 0. \end{cases}$$

THEOREM 3.1.3

If for a certain n it holds that R_n and R_{n+1} have opposite signs, then S lies between S_n and S_{n+1}. Furthermore, it holds that

$$S = \tfrac{1}{2}(S_n + S_{n+1}) \pm \tfrac{1}{2}|a_{n+1}|, \tag{3.1.1}$$

and we also have the weaker results:

$$|R_n| \leq |a_{n+1}|, \qquad |R_{n+1}| \leq |a_{n+1}|, \qquad \operatorname{sgn} R_n = \operatorname{sgn} a_{n+1}.$$

(No assumptions are made about sgn R_j for $j > n + 1$ or $j < n$.)

Proof. The fact that R_{n+1} and R_n have opposite signs means, quite simply, that one of S_{n+1} or S_n is too large, and the other is too small—i.e., that S lies between S_{n+1} and S_n. Since $a_{n+1} = S_{n+1} - S_n$, one has, for positive values of a_{n+1}, the situation shown in Fig. 3.1.2. From this figure (and an analogous one for a_{n+1} negative) the remaining assertions of the theorem clearly follow.

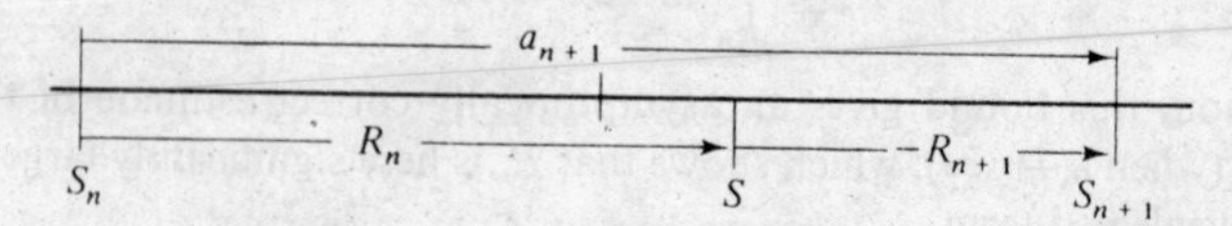

Fig. 3.1.2

We shall see that, in many Taylor series, the remainder has the same sign as the first neglected term. If the series is alternating, then the above theorem can be used. An important consequence is:

THEOREM 3.1.4

For an alternating series, the absolute values of whose terms approach zero monotonically, the remainder has the same sign as the first neglected term, and the absolute value of the remainder does not exceed the absolute value of the first neglected term.

Proof. That the theorem is true is perhaps clear from Fig. 3.1.3 (and an analogous figure for the case $a_{n+1} < 0$). The figure shows how S_j depends on j when the premises of the theorem are fulfilled. A formal proof is left to the reader; see Problem 8 at the end of this section. For the use of the theorem, see Examples 3.1.1 and 3.1.2.

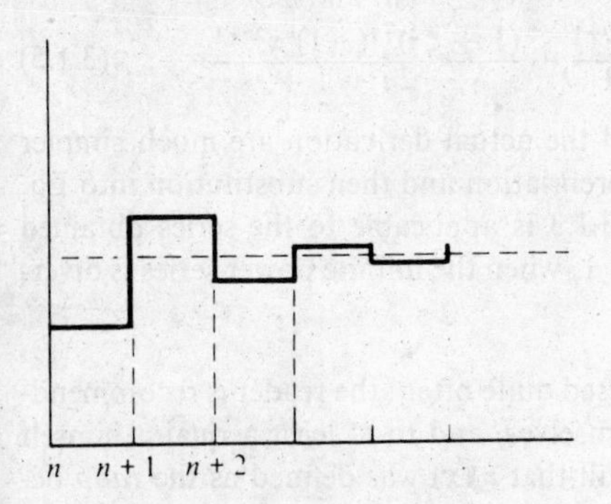

Fig. 3.1.3

3.1.3. Power Series

Taylor's formula can be written as

$$f(a+x) = f(a) + xf'(a) + \frac{x^2 f''(a)}{2!} + \dots + \frac{x^{n-1}f^{(n-1)}(a)}{(n-1)!} + \frac{r_n(x)x^n f^{(n)}(a)}{n!} \quad (\text{if} \quad f^{(n)}(a) \neq 0), \tag{3.1.2}$$

where $r_n(x)$ *is the ratio between the remainder and the first neglected term*. According to Lagrange's formula for the remainder, we have

$$r_n(x) = \frac{f^{(n)}(\xi)}{f^{(n)}(a)}, \tag{3.1.3}$$

where ξ lies between a and $a + x$. The usefulness of Lagrange's remainder term is limited because of the difficulty of computing high-order derivatives. But, according to Theorem 3.1.3, it is occasionally sufficient (with alternating series) to know the sign of the remainder in order to be able to give a strict error estimate.

For certain functions, one can quite easily compute the remainder term and the coefficients without using substitution in Eq. (3.1.2) (see Example 3.1.1). For $f(x) = (1 - x)^{-1}$, one has the important *geometric series:*

$$\frac{1}{1-x} = 1 + x + x^2 + x^3 + \dots + x^{n-1} + \frac{x^n}{1-x} \quad (\text{if} \quad x \neq 1). \tag{3.1.4}$$

Example 3.1.6

Set $x = -t^2$ in the geometric series, and integrate:

$$\int_0^x (1 + t^2)^{-1}\, dt = \sum_{j=0}^{n-1} \int_0^x (-t^2)^j\, dt + \int_0^x (-t^2)^n (1 + t^2)^{-1}\, dt.$$

Use the mean-value theorem of integral calculus on the last term. We get, for some $\xi \in [0, x]$,

$$\arctan x = \sum_{j=0}^{n-1} \frac{(-1)^j x^{2j+1}}{2j+1} + \frac{(1 + \xi^2)^{-1}(-1)^n x^{2n+1}}{2n+1}, \tag{3.1.5}$$

where both the remainder term and the actual derivation are much simpler than what one would get using differentiation and then substitution into Eq. (3.1.2). Notice also that Theorem 3.1.3 is applicable to the series obtained above for all x and n, even for $|x| > 1$, when the infinite power series is divergent.

The formulas in Table 3.1.1 are used quite often; the reader is recommended to memorize the expansions themselves, and to at least acquaint himself with the expressions for $r_n(x)$. (Recall that $r_n(x)$ was defined as the *ratio* be-

Table 3.1.1

Function	Expansion	$r_n(x)$
$(1 - x)^{-1}$	$1 + x + x^2 + x^3 + \dots$ if $\lvert x\rvert < 1$	$(1 - x)^{-1}$ if $x \neq 1$
$(1 + x)^k$	$1 + \binom{k}{1}x + \binom{k}{2}x^2 + \dots$ if $\lvert x\rvert < 1$	$(1 + \xi)^{k-n}$ if $x > -1$
$\ln(1 + x)$	$x - \frac{x^2}{2} + \frac{x^3}{3} - \frac{x^4}{4} + \dots$ if $\lvert x\rvert < 1$	$(1 + \xi)^{-1}$ if $x > -1$
$\exp(x)$	$1 + x + \frac{x^2}{2!} + \frac{x^3}{3!} + \dots$ all x	$\exp(\xi)$, all x
$\sin x$	$x - \frac{x^3}{3!} + \frac{x^5}{5!} - \frac{x^7}{7!} + \dots$ all x	$\cos\xi$, all x, n odd
$\cos x$	$1 - \frac{x^2}{2!} + \frac{x^4}{4!} - \frac{x^6}{6!} + \dots$ all x	$\cos\xi$, all x, n even
$\frac{1}{2}\ln\left[\frac{(1+x)}{(1-x)}\right]$	$x + \frac{x^3}{3} + \frac{x^5}{5} + \dots$ if $\lvert x\rvert < 1$	$\frac{1}{1-\xi^2}$, $\lvert x\rvert < 1$, n even
$\arctan(x)$	$x - \frac{x^3}{3} + \frac{x^5}{5} - \dots$ if $\lvert x\rvert < 1$	$\frac{1}{1+\xi^2}$, all x

tween the remainder and the first neglected term.) In the table, ξ means a number between 0 and x.

Series expansions for many other functions can be found for example in Abramowitz and Stegun, *Handbook of Mathematical Functions* [27], or in several other of the references listed in Chap. 13.

The following two examples give a hint as to how one can arrange the computation of coefficients in series expansions for composite functions—even for formula manipulation using a computer.

Example 3.1.7

Derive a recursion formula for the coefficients in the Taylor series about the origin of the function

$$f(x) = \frac{x}{2}\frac{e^x + 1}{e^x - 1}.$$

f is an *even* function—i.e., $f(x) = f(-x)$ for all x, since

$$f(-x) = -\frac{x}{2}\frac{e^{-x} + 1}{e^{-x} - 1} = -\frac{x}{2}\frac{1 + e^x}{1 - e^x} = f(x).$$

Hence the coefficients of odd powers of x in the expansion are zero. We set, then,

$$f(x) = c_0 + c_2x^2 + c_4x^4 + \dots.$$

From $(e^x - 1)f(x) = \frac{1}{2}x(e^x + 1)$ it follows that

$$\left(x + \frac{x^2}{2!} + \frac{x^3}{3!} + \dots\right)(c_0 + c_2x^2 + c_4x^4 + \dots)$$
$$= \frac{x}{2}\left(2 + x + \frac{x^2}{2!} + \frac{x^3}{3!} + \dots\right). \tag{3.1.6}$$

Comparing the coefficients of x^{2n+1} on both sides, we get $c_0 = 1$,

$$\frac{c_0}{(2n+1)!} + \frac{c_2}{(2n-1)!} + \frac{c_4}{(2n-3)!} + \dots + \frac{c_{2n}}{1!} = \frac{1}{2}\frac{1}{(2n)!} \quad (\text{for } n \geq 1).$$

$$n = 1 \Rightarrow c_2 = \frac{1}{4} - \frac{c_0}{6} = \frac{1}{12},$$

$$n = 2 \Rightarrow c_4 = \frac{1}{48} - \frac{c_2}{6} - \frac{c_0}{120} = -\frac{1}{720}, \quad \text{etc.}$$

(Check that the coefficients of x^2, x^4, x^6 on both sides of (3.1.6) also agree.)

Example 3.1.8

If the power series for $f(x)$ is known, then one can find the expansion of $y(x) = \exp(f(x))$ by differentiating this last relation—that is,

$$y'(x) = f'(x)\exp(f(x)) = f'(x)y(x)$$

—and then treating the differential equation $y'(x) = f'(x)y(x)$ analogously to the differential equation in Example 3.1.2.

Example 3.1.9

In Table 3.1.2 we have collected some useful approximate formulas. They consist of the first term of the respective function's Taylor series. A bound is also given, within which the formula represents the function with an

Table 3.1.2

Function	Formula	10^{-3}-bound	Next Term
$(1+x)^a$	$1+ax$		$\binom{a}{2}x^2$
$(1+x)^2$	$1+2x$	0.031	x^2
$(1+x)^3$	$1+3x$	0.018	$3x^2$
$\frac{1}{(1+x)}$	$1-x$	0.031	x^2
$\sqrt{1+x}$	$1+\frac{1}{2}x$	0.089	$-\frac{x^2}{8}$
$(1+x)^{-1/2}$	$1-\frac{1}{2}x$	0.050	$\frac{3x^2}{8}$
e^x	$1+x$	0.044	$\frac{x^2}{2}$
$\ln(1+x)$	x	0.044	$-\frac{x^2}{2}$
$\sin x$	x	0.18	$-\frac{x^3}{6}$
$\cos x$	1	0.044	$-\frac{x^2}{2}$
$\cos x$	$1-\frac{1}{2}x^2$	0.39	$\frac{x^4}{24}$
$\tan x$	x	0.14	$\frac{x^3}{3}$
$\arcsin x$	x	0.18	$\frac{x^3}{6}$
$\arctan x$	x	0.14	$-\frac{x^3}{3}$
$(1+x)(1+y)$	$1+x+y$	$\lvert xy\rvert < 10^{-3}$	

absolute error which is less than 10^{-3}. The next term in the series expansion is also given. It gives (with less than 15 percent relative uncertainty) a good estimate of the size of the error for values of x which lie within 3 times the given 10^{-3}-bound.

Example 3.1.10

The following calculations indicate how the formulas in Table 3.1.2 can be used to give quick estimates in hand computation:

$\cos 0.2 = 0.980.$

$$\sqrt{404} = 20\cdot\sqrt{1.01} = 20\left(1.00500 \pm \frac{10^{-4}}{8}\right) = 20.100.$$

$$\frac{1}{396} = \frac{1}{400}(1-0.01)^{-1} = 2.5\cdot 10^{-3}\left(1.010 \pm \frac{10^{-3}}{9.9}\right) = 2.525\cdot 10^{-3}.$$

$$\cot 0.090 = \frac{1}{\tan 0.090} = \frac{1}{0.090 \pm 10^{-3}/3} = 11.1.$$

$$\log_{10} 204 = \log_{10} 200 + \log_{10} 1.02 = 2.301 + 0.434\cdot 0.020$$
$$= 2.310, \text{ using that } \log_{10} x = 0.434 \ln x, \qquad \log_{10} 2 = 0.301.$$

Example 3.1.11

Compute $1.001^{1,000}$ to six significant digits; that is, compute $y(0.001)$ where $y(x) = (1+x)^{1/x}$. (Notice that as $x \to 0$ we get the indefinite expression 1^∞.)

The binomial series converges too slowly when the exponent is 1,000. It is much better to take the logarithm,

$$\ln y(x) = \frac{1}{x}\ln(1+x) = \frac{1}{x}\left(x - \frac{x^2}{2} + \frac{x^3}{3} - \dots\right)$$
$$= 1 - \frac{x}{2} + \frac{x^2}{3} - \dots = 0.999500 \pm \frac{1}{3}\cdot 10^{-6} \quad \text{for} \quad x = 0.001.$$

$y(0.001) = \exp(0.999500) = 2.71692$ (from a table). The reader is recommended to make an error analysis of the above computation! Using a six-place logarithm table, one would have gotten $\log_{10}\ y(0.001) = 1000\cdot\log_{10}\ 1.001 = 0.434$; hence (from a log table) $y(0.001) = 2.71644$, barely four significant digits.

Example 3.1.12

Compute, with at most 1 percent error, $f'''(10)$, where $f(x) = (x^3+1)^{-1/2}$. x is large, but x^{-1} is fairly small. Expand in powers of x^{-1}:

$$f(x) = (x^3+1)^{-1/2} = x^{-3/2}(1+x^{-3})^{-1/2}$$
$$= x^{-1.5}\left(1 - 0.5\cdot x^{-3} + \frac{0.5\cdot 1.5}{2}x^{-6} - \dots\right)$$
$$= x^{-1.5} - \frac{1}{2}x^{-4.5} + \frac{3}{8}x^{-7.5} - \dots.$$

Differentiate three times!

$$f'''(x) = -x^{-4.5}\left(\frac{105}{8} - \frac{1,287}{16}x^{-3} + \dots\right).$$

For $x = 10$ the second term is less than 1 percent of the first; the terms after the second decrease quickly and are negligible. (The reader is recommended to carry through the calculation in more detail to convince himself of this.) One can show that the magnitude of each term is less than $8\cdot x^{-3}$ of the previous term. Hence we get $f'''(10) = -4.14\cdot 10^{-4}$ to the desired accuracy.

Example 3.1.13

On a binary computer with machine unit $u = 2^{-36} \approx 1.6 \cdot 10^{-11}$, one wishes to compute $\sinh x$ with good relative accuracy, for all $x \in [-5, 5]$. Assume that e^x is computed with a relative error of less than $5u$ in the given interval (by a library subroutine). The formula $\frac{1}{2}(e^x - e^{-x})$ for $\sinh x$ is sufficiently accurate except when $|x|$ is small; then cancellation occurs. Hence, for $|x| \ll 1$, e^x and e^{-x} can have **absolute** errors of order of magnitude $5u$. Then the absolute error in $\frac{1}{2}(e^x - e^{-x})$ can also have error of the same order of magnitude, which means that the relative error can be more than 100 percent. ($|5u/\sinh x| \approx |5u/x|$, which becomes 500 percent for $x \approx 10^{-11}$.)

If one instead uses, when $|x|$ is small, two terms in the series expansion for $\sinh x$,

$$\frac{1}{2}(e^x - e^{-x}) = \frac{1}{2}\left(1 + x + \frac{x^2}{2!} + \frac{x^3}{3!} + \dots\right)$$
$$- \frac{1}{2}\left(1 - x + \frac{x^2}{2!} - \frac{x^3}{3!} + \dots\right) = x + \frac{x^3}{6} + \frac{x^5}{120} + \dots,$$

one gets an absolute truncation error which is about $x^5/120$, and a round-off error of the order of $2ux$. Thus the formula $x + x^3/6$ is better than $\frac{1}{2}(e^x - e^{-x})$ if

$$\frac{|x|^5}{120} + 2u|x| < 5u.$$

If $2u|x| \ll 5u$, we have $|x|^5 < 600u \approx 300 \cdot 2^{-35}$, or $|x| < 300^{1/5} \cdot 2^{-7} \approx 0.0243$ (which shows that $2u|x|$ really could be ignored in this rough calculation). Thus, if for $|x| \gtrsim 0.0243$, one switches from $x + x^3/6$ to $\frac{1}{2}(e^x - e^{-x})$ —a switch quite easy to program—one will have a relative error which nowhere exceeds $5u/0.0243 \approx 4 \cdot 10^{-9}$. If one needs higher accuracy, one can take more terms in the series for $\sinh x$, so that the switch can occur at a larger value of $|x|$.

For large values of $|x|$, however, one must expect a relative error of order of magnitude $|xu|$ because of round-off error in the argument x.

REVIEW QUESTIONS

1. Formulate three general theorems which can be used to estimate the remainder term in numerical series.
2. Give the Taylor series expansions for $\ln(1 + x)$, e^x, $\sin x$, $\cos x$, $(1 + x)^k$, $(1 - x)^{-1}$, $\ln[(1 + x)/(1 - x)]$.

PROBLEMS

1. In how large a neighborhood of $x = 0$ does one get, respectively, four and six correct decimals using the following approximations?

(a) $\sin x \approx x$;
(b) $\cos x \approx 1 - x^2/2$;
(c) $1/\sqrt{1 - x^2} \approx 1 + x^2/2$.

2. Derive the following approximate formula (for x small) and estimate the error when $x = 0.1$ and $x = 0.01$.

$$\frac{e\sqrt{\cos x}}{\sqrt{1 + x^2}} \approx e\left(1 - \frac{3x^2}{4}\right).$$

3. Compute to six significant digits:
(a) 1.01^{100}; (b) $10 - \sqrt[3]{999}$.

4. Use the series

$$\frac{1}{2}\ln\frac{1 + y}{1 - y} = y + \frac{y^3}{3} + \frac{y^5}{5} + \cdots.$$

to compute ln 1.2 with an error of less than 10^{-7}. A strict error estimate should be made.

How many terms would be needed if the expression for $\ln(1 + y)$ with $y = 0.2$ were used instead?

5. Compute to five correct decimals:

$$\int_0^{0.1} (1 - 0.1 \sin t)^{1/2}\, dt.$$

6. Compute some of the first coefficients in an expansion of the form

$$y = c_1 x + c_2 x^2 + c_3 x^3 + \cdots$$

which satisfies the algebraic equation

$$y^3 + y = x.$$

Can one say right off that certain coefficients are zero?

7. Compute to four significant digits:

$$\int_{10}^{\infty} (x^3 + x)^{-1/2}\, dx.$$

8. Give a formal proof of Theorem 3.1.4.

3.2. ACCELERATION OF CONVERGENCE

3.2.1. Slowly Converging Alternating Series

Example 3.2.1

From Eq. (3.1.5) it follows that for $x = 1$ $(n \longrightarrow \infty)$,

$$\frac{\pi}{4} = \arctan 1 = 1 - \frac{1}{3} + \frac{1}{5} - \frac{1}{7} + \frac{1}{9} - \cdots$$

$$= \sum_{j=0}^{\infty} (-1)^j (2j + 1)^{-1}.$$

This series converges very slowly. Even after 500 terms there still occur changes in the third decimal.

One can, however, accelerate the convergence in many different ways, for example by **repeated averaging** of the partial sums. Let S_n be the sum of the first $n + 1$ terms. The columns to the right of the S_n-column in the scheme given below are formed by building averages: each number in such a column is the mean of the two numbers which stand to the upper left and lower left of the number itself. In other words, each number is the mean of its "northwest" and "southwest" neighbor. Only the digits which are different from those in the previous column are written out.

	n	S_n	M_1	M_2	M_3	M_4	M_5	M_6
$n_0 =$	5	0.744 012						
			782 474					
	6	0.820 935		5038				
			787 602		340			
	7	0.754 268		5641		387		
			783 680		434		396	
	8	0.813 092		5228		405		785398
			786 776		376		400	
	9	0.760 460		5523		395		
			784 270		414			
	10	0.808 079		5305				
			786 340					
	11	0.764 601						

Notice that the values in each column oscillate. One can show in general that for alternating series, *if the absolute value of the* jth *term (considered as a function of* j*) has a* kth *derivative which approaches zero monotonically for* $j > n_0$, *then every other value in column* M_k *is larger than the sum, and every other value is smaller.* The above premise is satisfied here, since if $f(j) = (2j + 1)^{-1}$, then $f^{(k)}(j) = c_k(2j + 1)^{-1-k}$ which approaches zero monotonically. From column M_5, then, it follows that

$$0.785396 \leq \frac{\pi}{4} \leq 0.785400$$

(if round-off error is ignored). To take account of round-off error, we set $\pi/4 = 0.785398 \pm 0.000003$. The correct value is $\pi/4 = 0.785398163$.

Using twelve terms and repeated averaging one obtained in fact six correct decimals; using ordinary summation, 500,000 terms would have been needed for the same accuracy. The above method can be shown to be identical to a special case of **Euler's transformation**, compare Problem 3, Sec. 7.6. The formula described in that problem has certain computational advantages, but the scheme with repeated averaging is more easily remembered.

3.2.2. Slowly Converging Series with Positive Terms

Example 3.2.2

Compute $\sum_{j=1}^{\infty} j^{-2}$. The sum of the first nine terms is 1.5398. It immediately occurs to one to compare the tail of the series with the integral of x^{-2} from 10 to ∞. We approximate the integral according to the trapezoidal rule; see Sec. 1.2 and Fig. 3.2.1:

$$\int_{10}^{\infty} x^{-2}\,dx \approx T_1 + T_2 + T_3 + \cdots = \tfrac{1}{2}(10^{-2} + 11^{-2}) + \tfrac{1}{2}(11^{-2} + 12^{-2}) + \tfrac{1}{2}(12^{-2} + 13^{-2}) + \cdots = \sum_{10}^{\infty} j^{-2} - \tfrac{1}{2}10^{-2}.$$

$$\therefore \sum_{j=10}^{\infty} j^{-2} \approx [-x^{-1}]_{10}^{\infty} + \tfrac{1}{2}10^{-2} = 0.1050.$$

$$\sum_{j=1}^{\infty} j^{-2} = \sum_{j=1}^{9} j^{-2} + \sum_{j=10}^{\infty} j^{-2} \approx 1.5398 + 0.1050 = 1.6448.$$

The answer to four decimals is $\pi^2/6 = 1.6449$. We would have needed about 10,000 terms to get the same accuracy by direct addition of the terms!

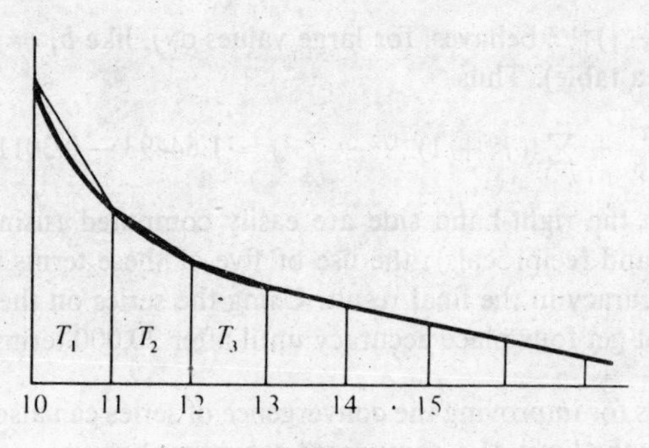

Fig. 3.2.1

The approximation of a sum by an integral by means of the trapezoidal rule is not a coincidental trick, but a very useful method. A further systematic development of the method (using corrections for the truncation error of the trapezoidal rule) leads to **Euler-Maclaurin's summation formula**. This formula often gives startlingly high accuracy with only simple calculations:

$$\sum_{j=k}^{\infty} f(j) = \int_{k}^{\infty} f(x)\,dx + \frac{1}{2}f(k) - \frac{f'(k)}{12} + \frac{f'''(k)}{720} - \frac{f^{v}(k)}{30{,}240} + \cdots \tag{3.2.1}$$

The complete formula, together with an error estimate, is given in Sec. 7.4.4. In many important applications the series will be alternating and the first neglected term often gives a bound for the error (see Theorem 7.4.2 for the necessary premises).

3.2.3. Other Simple Ways to Accelerate Convergence

If the terms in the series $\sum a_j$ behave, for large j, like the terms of a series $\sum b_j$ whose sum is known ($\sum b_j = s$)—that is, if

$$\lim_{j \to \infty} \frac{a_j}{b_j} = 1$$

—then one can write

$$\sum_{j=1}^{\infty} a_j = s + \sum_{j=1}^{\infty} (a_j - b_j),$$

where the series on the right-hand side converges more quickly than the given series. We call this *making use of a* **simple comparison problem**. The same idea is used in many other contexts—for example, in the computation of integrals where the integrand has a singularity. (Compare also Theorems 3.1.1 and 3.1.2.)

Example 3.2.3

$a_j = (j^4 + 1)^{-1/2}$ behaves, for large values of j, like $b_j = j^{-2}$, whose sum is $\pi^2/6$ (from a table). Thus

$$\sum_{j=1}^{\infty} a_j = \frac{\pi^2}{6} + \sum_{j=1}^{\infty} ((j^4 + 1)^{-1/2} - j^{-2}) = 1.64493 - 0.30119 = 1.3437.$$

The terms on the right-hand side are easily computed (using a table with square roots and reciprocals); the use of five of these terms is sufficient for four-place accuracy in the final result. Using the series on the left-hand side, one would not get four-place accuracy until after 20,000 terms.

Procedures for improving the convergence of series can also be used when one is seeking the limit of a convergent sequence, because

$$\lim_{n \to \infty} x_n = a \iff a = x_j + \sum_{p=1}^{\infty} (x_{p+j} - x_{p+j-1}).$$

If the differences $x_{p+j} - x_{p+j-1}$, $p = 1, 2, 3, \ldots$ form a geometric series, then

$$a = x_j + (x_j - x_{j-1}) \sum_{p=1}^{\infty} k^p = x_j + \frac{(x_j - x_{j-1})k}{1 - k}.$$

Since $k = (x_j - x_{j-1})/(x_{j-1} - x_{j-2})$, we obtain $a = x'_j$, where

$$x'_j = x_j - \frac{(x_j - x_{j-1})^2}{x_j - 2x_{j-1} + x_{j-2}}.$$

The application of this formula to the convergence acceleration of a sequence, in which the differences approximately form a geometric series, is called **Aitken extrapolation** *or exponential extrapolation.* In fact, it can be shown (see [6, p. 73]) that

$$\lim_{j\to\infty} x_j = a, \quad \lim_{j\to\infty} \frac{x_{j+1} - x_j}{x_j - x_{j-1}} = k, \quad |k| < 1 \Longrightarrow \lim_{j\to\infty} \frac{x'_j - a}{x_j - a} = 0.$$

In other words: under the assumption stated, the sequence $\{x'_j\}$ converges faster to a than does the sequence $\{x_j\}$. This assumption can often be verified for sequences arising from iterative processes and for many other applications.

Example 3.2.4

Using the iteration formula

$$x_0 = 0.8, \qquad x_{j+1} = 1 - \tfrac{1}{2}x_j^2,$$

one gets

$$x_1 = 0.68, \qquad x_2 = 0.7688, \qquad x'_2 = 0.7688 - \frac{(0.0888)^2}{0.2088} = 0.73103.$$

One can show that $\lim_{j\to\infty} x_j = \sqrt{3} - 1 = 0.73205$ to five decimals. The error in x'_2 is only about 3 percent of the error in x_2.

The Aitken extrapolation is similar in spirit to the *Richardson extrapolation* mentioned in Sec. 1.2. In that method general results on how the error of an approximative method (e.g., the trapezoidal rule for integration) depends on the step size h make it possible to obtain an improved estimate of the limiting value for $h \to 0$, when results are known for two (or more) different values of the step size. We shall return to this in Sec. 7.2.

3.2.4. Ill-Conditioned Series

Slow convergence is not the only numerical difficulty which occurs in connection with infinite series. There are also series where the size of certain of the terms are many orders of magnitude larger than the sum of the series. Then there occurs a complicated cancellation between the terms of the series. Small relative errors in the computation of the large terms can then produce large errors in the result.

Example 3.2.5

The Taylor series for e^{-x},

$$e^{-x} = \sum_{n=0}^{\infty} \frac{(-1)^n x^n}{n!},$$

converges for all x. The ratio between a given term and the previous term is

$$-\frac{x^{n+1}}{x^n}\frac{n!}{(n+1)!} = \frac{-x}{n+1}.$$

Thus the magnitude of the terms grows, as long as $n + 1 < x$. For $x \approx 20$, then, the largest term is about $20^{20}/(20!) \approx 4 \cdot 10^7$ [log $(x!)$ can be found in a table], but the sum is only $e^{-20} \approx 2 \cdot 10^{-9}$. Hence a relative error of 10^{-16} in the largest term can lead to an error of more than 100 percent in the result. The convergence of the series is quite slow to begin with, but not worse than what one is forced to accept in many other situations. The remainder term after 100 terms is $20^{100}/(100!) \approx 7 \cdot 10^{-20}$. Thus one could, after 100 terms, have about ten significant digits in the result—but this would require the use of thirty digits in the calculations. Hence this series can be considered as quite difficult from a numerical point of view, in spite of the fact that it is convergent.

Fortunately, the exponential function has many convenient properties with the help of which one can find excellent ways to compute its value when the magnitude of the argument is large. But there are analogous cases where it is not so easy to give an effective alternative to the use of a power series to compute the value of a function. In such cases, one should seek a method of attacking the problem which avoids the use of the series expansion in question.

3.2.5. Numerical Use of Divergent Series

In the previous sections, we saw that the fact that a series is convergent is no guarantee that it is numerically useful. In this section, we shall see examples of the reverse situation: a divergent series can be of use in numerical computations. This sounds strange, but it refers to series where the size of the terms decreases rapidly at first and then grows later (see Fig. 3.2.2), and where one can compare the magnitude of the remainder with the absolute value of the first neglected term. Such series are sometimes called **semiconvergent**.

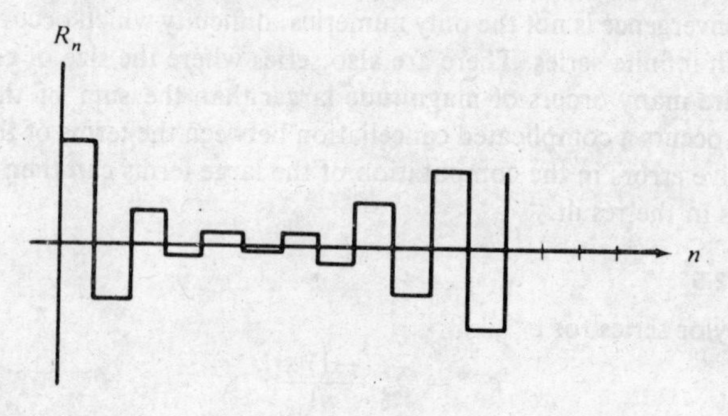

Fig. 3.2.2

Example 3.2.6

Give a semiconvergent series for the computation of

$$f(x) = e^x \int_x^\infty e^{-t}t^{-1}\,dt = \int_0^\infty e^{-u}(u+x)^{-1}\,du$$

for large positive values of x. (The second integral was obtained from the first by the substitution $t = u + x$.) The expression $(u + x)^{-1}$ should first be expanded in a *geometric series with remainder term, valid even for* $u/x > 1$:

$$(u+x)^{-1} = x^{-1}\left(1+\frac{u}{x}\right)^{-1} = x^{-1}\left(1 - \frac{u}{x} + \left(\frac{u}{x}\right)^2 - \cdots \right.$$
$$\left. + (-1)^{n-1}\left(\frac{u}{x}\right)^{n-1} + (-1)^n\left(\frac{u}{x}\right)^n\left(1+\frac{u}{x}\right)^{-1}\right)$$
$$= x^{-1}\sum_{j=0}^{n-1}(-1)^j x^{-j}u^j + (-1)^n(u+x)^{-1}\left(\frac{u}{x}\right)^n.$$

We can write

$$f(x) = S_n(x) + R_n(x),$$
$$S_n(x) = x^{-1}\sum_{j=0}^{n-1}(-1)^j x^{-j}\int_0^\infty u^j e^{-u}\,du = \frac{1}{x} - \frac{1!}{x^2} + \frac{2!}{x^3} - \cdots$$
$$+ (-1)^{n-1}\frac{(n-1)!}{x^n},$$
$$R_n(x) = (-1)^n\int_0^\infty (u+x)^{-1}\left(\frac{u}{x}\right)^n e^{-u}\,du.$$

The terms in $S_n(x)$ behave as in Fig. 3.2.2. The ratio between the last term in $S_{n+1}(x)$ and the last term in $S_n(x)$ is

$$-\frac{n!}{x^{n+1}}\frac{x^n}{(n-1)!} = \frac{-n}{x}, \tag{3.2.3}$$

and since the absolute value of that ratio for fixed x is unbounded as $n \longrightarrow \infty$, the sequence $\{S_n(x)\}_{n=1}^\infty$ diverges for every positive x. But since $\operatorname{sgn} R_n(x) = (-1)^n$ for $x > 0$, it follows from Theorem 3.1.3 that

$$f(x) = \frac{1}{2}[S_n(x) + S_{n+1}(x)] \pm \frac{1}{2}\frac{n!}{x^{n+1}}.$$

The idea now is to choose n so that the estimate of the remainder is as small as possible. According to Eq. (3.2.3), this happens when n is equal to the integer part of x. For $x = 5$ we choose $n = 5$,

$$S_5(5) = 0.2 - 0.04 + 0.016 - 0.0096 + 0.00768 = 0.17408,$$
$$S_6(5) = S_5(5) - 0.00768 = 0.16640,$$
$$f(5) = 0.17024 \pm 0.00384.$$

The correct value is 0.17042. (The actual error is thus only 5 percent of the error bound.)

For larger values of x the accuracy attainable increases. One can show that the bound for the relative error using the above computational scheme decreases approximately as $(\pi \cdot x/2)^{1/2}e^{-x}$. Thus the above divergent series gives extremely good accuracy for large values of x, if one stops at the smallest term.

One can derive the same series expansion as the above by repeated partial integration—see Problem 10 at the end of this section. Repeated partial integration is often a good way to derive numerically useful divergent series.

The series in Example 3.2.6 is an expansion in negative powers of x, with the property that for all n, the remainder (when $x \longrightarrow \infty$) approaches zero faster than the last included term. Such an expansion is said to **represent** $f(x)$ **asymptotically** as $x \longrightarrow \infty$. Such an **asymptotic series** can be either convergent or divergent.

An expansion in *positive* powers of $x - a$,

$$f(x) \sim \sum_{v=0}^{n-1} c_v(x-a)^v + R_n(x),$$

represents $f(x)$ asymptotically when $x \longrightarrow a$ if

$$\lim_{x \to a} (x-a)^{-(n-1)} R_n(x) = 0.$$

Asymptotic expansions of the error in positive powers of a step length h are of great importance in the more advanced study of numerical methods. Such expansions form the basis of simple and effective extrapolation methods for improving numerical results.

In many other branches of applied mathematics, divergent asymptotic series are an important aid, though they are often needlessly surrounded with an air of mysticism.

REVIEW QUESTIONS

1. Describe three procedures for improving the speed of convergence of certain series. Give examples of their use.
2. Define what is meant when one says that the series
$$\sum_{n=0}^{\infty} a_n x^n$$
(a) converges to a function $f(x)$ in the interval $[0, R]$;
(b) represents a function $f(x)$ asymptotically as $x \longrightarrow 0$.

PROBLEMS

1. (a) Compute ln 2 to 3 decimals by using seven terms of the series
$$1 - \tfrac{1}{2} + \tfrac{1}{3} - \tfrac{1}{4} + \cdots$$
and building averages.
(b) How many terms would one need in order to get three correct decimals with direct addition of terms?
2. (a) Write an Algol function procedure
$$\text{AVESUM}\ (a, k, n, eps)$$

which computes $\sum_{k=0}^{\infty}(-1)^k a_k$ by using repeated averages. The algorithm should be stopped when two elements in the same column differ by less than *eps* or when n terms have been used.

(b) Make a call to the procedure which computes $\sum_{k=2}^{\infty}(-1)^k/\ln k$ with an error $<10^{-6}$.

3. (a) Compute

$$\sum_{n=1}^{\infty} n^{-3/2} = 2.612375$$

to two correct decimals by using the method in Example 3.2.2. Four terms suffice.

(b) How many terms would have been needed for two correct decimals if the terms had been added directly? (Estimate the error term with an integral.)

(c) Improve the result by using more terms on the right-hand side of Eq. (3.2.1).

4. Compute $\sum_{k=1}^{\infty} k/(k^4+1)$ with an error of less than 10^{-4}. (*Hint:* sum the first three terms directly and then expand the summand in negative powers of k and use Euler-Maclaurin's summation formula.)

5. Compute to three decimals

$$\sum_{n=1}^{\infty} \frac{1}{n^2+1}$$

using

(a) Euler-Maclaurin's formula;

(b) the comparison technique of Example 3.2.3.

6. When the current through a galvanometer changes suddenly, its indicator begins to oscillate toward a new stationary value s. The relation between the successive turning points $v_0, v_1, v_2, \ldots$ is

$$v_n - s \approx A\cdot(-k)^n \quad \text{where} \quad 0 < k < 1.$$

Determine, from the following series of measurements, Aitken extrapolated values v'_2, v'_3, v'_4 which are all approximations to s (compute with one decimal):

$$v_0 = 659; \quad v_1 = 236; \quad v_2 = 463;$$

$$v_3 = 340; \quad v_4 = 406.$$

7. The sequence $y_{n+1} = (1-N)y_n + 1$ converges to $1/N$ if $0 < N < 2$.

(a) Let $N = \frac{1}{2}$ and $y_0 = 1$. Compute the elements y_1, y_2, y_3, y_4 of the sequence and the values y'_2, y'_3, y'_4 obtained with Aitken extrapolation.

(b) Show that the sequence $\{y_j\}$ fulfills the assumptions for Aitken extrapolation.

8. A function $g(t)$ has the form

$$g(t) = c - kt + \sum_{n=1}^{\infty} a_n e^{-\lambda_n t}$$

where c, k, a_n and λ_n, $n = 1, 2, \ldots$ are unknown constants $0 < \lambda_1 < \lambda_2 < \lambda_3 < \cdots$. $g(t)$ is known numerically for $t_\nu = \nu\cdot h$, $\nu = 0, 1, 2, 3, 4$. Formulate a way to determine k by Aitken extrapolation and verify that the conditions for Aitken extrapolation are satisfied. Determine k using the following values for $g(t_v)$, such that one gets two values for k which do not differ by more than $2 \cdot 10^{-2}$. The numerical values for $g(t_v)$ are:

$$g(t_0) = 2.14789; \quad g(t_1) = 1.82207; \quad g(t_2) = 1.59763;$$
$$g(t_3) = 1.40680; \quad g(t_4) = 1.22784; \quad h = 0.1.$$

9. In order to compute the Bessel function $J_0(x)$ for $x = 20$ to six correct decimals one has decided to use the power-series expansion

$$J_0(x) = \sum_{n=0}^{\infty} (-1)^n \frac{(x/2)^{2n}}{(n!)^2},$$

which converges for all x.
(a) How many terms must one compute?
(b) How large is the largest term?
(c) How many digits must one use during the calculations?

10. Derive the expansion of Example 3.2.6 by repeated partial integration.

11. Derive three terms of a semiconvergent series for

$$2 \exp(x^2) \int_x^{\infty} \exp(-t^2)\, dt, \quad (x \longrightarrow \infty).$$

12. The formula for Aitken extrapolation is sometimes given in the form

$$x_n - \frac{(x_{n+1} - x_n)^2}{x_{n+2} - 2x_{n+1} + x_n} \quad \text{or} \quad x_n - \frac{(\Delta x_n)^2}{\Delta^2 x_n}.$$

Show that this is equivalent to x'_{n+2} in the notation of Eq. (3.2.2).

4 APPROXIMATION OF FUNCTIONS

4.1. BASIC CONCEPTS IN APPROXIMATION

4.1.1. Introduction

One fundamental problem which occurs in many variants, is to approximate a function f by a member f^* of a class of functions which is easy to work with mathematically (for example, polynomials, rational functions, or trigonometric polynomials), where each particular function in the class is specified by the numerical values of a number of parameters. In this chapter we shall mainly confine ourselves, in the discussion of the above problem, to functions of *one* variable on a closed interval.

There are two types of shortcomings to take into account: shortcomings in the input data and shortcomings in the particular model (class of functions, form) which one intends to adapt to the input data. For ease in discussion we shall call these shortcomings the **measurement error** and the **error in the model,** respectively.

Example 4.1.1

The points in Fig. 4.1.1 show, for $n = 1, 2, 3, 4, 5$, the time t for the nth passage of a swinging pendulum through its point of equilibrium. The conditions of the experiment were such that a relation of the form $t = a + bn$ can be assumed to be valid to very high accuracy. Random errors in measurement are the dominant cause of the deviation from linearity which is shown in the figure. This deviation causes the values of the parameters a and b to be uncertain. We have five points and only two parameters to determine; the

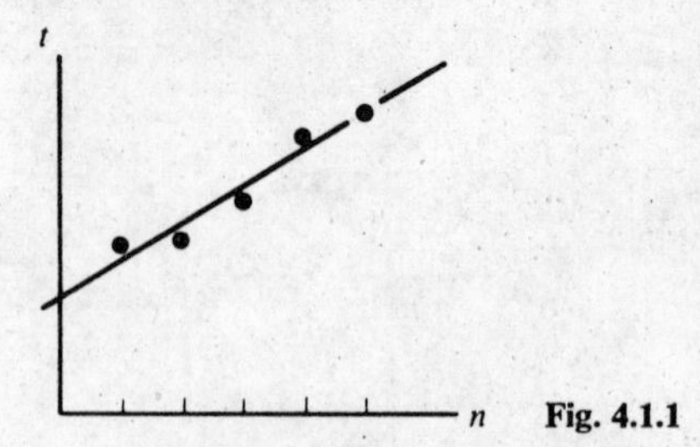

Fig. 4.1.1

problem is said to be **overdetermined**. In Sec. 4.2, we shall see how such overdetermined problems can be treated by the method of least squares.

Example 4.1.2

One wishes to use a small number of arithmetic operations to compute

$$f(x) = \int_0^x \exp(-t^2)\,dt,$$

for arbitrary $x \in [-1, 1]$ with a relative error of less than 10^{-4}. One can then (see Example 3.1.1) approximate the function by the polynomial of degree 13, obtained from taking seven terms of the Maclaurin series for $\exp(-t^2)$ and integrating term by term,

$$f^*(x) = \int_0^x \sum_{j=0}^{6} \frac{(-1)^j t^{2j}}{j!}\,dt = \sum_{j=0}^{6} \frac{(-1)^j x^{2j+1}}{(2j+1)j!}.$$

Here there are no errors in measurement, but the "model" of approximating the function with a polynomial is insufficient. The function is demonstrably not a polynomial. There is always a truncation error which one can, however, make as small as one wants by choosing the degree of the polynomial sufficiently large (i.e., by taking along more terms in the Maclaurin series for

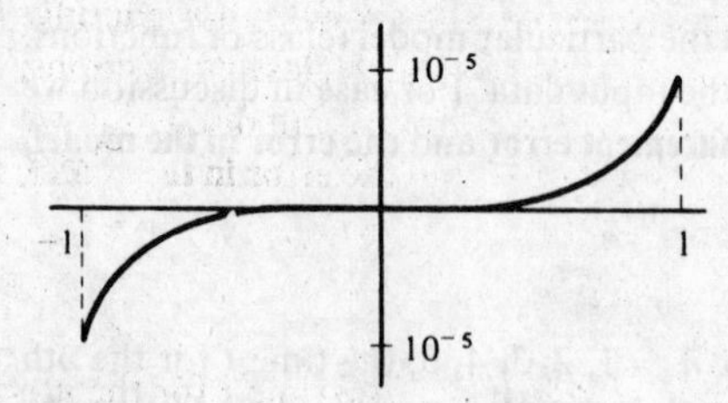

Fig. 4.1.2

$\exp(-t^2)$). Figure 4.1.2 below shows the graph of the **error function** $e(x) = f^*(x) - f(x)$. In general, whenever a function f^* is proposed as an approximation to another function f, we shall refer to the graph of the error function $f^* - f$ as the error curve

Later we shall see that if one uses, instead of the Taylor series, other methods for approximation by polynomials described later in this chapter, one can in fact obtain the desired accuracy using a seventh-degree polynomial.

The above examples showed two isolated situations. In practice, both the input data and the model are as a rule insufficient. One can consider approximation as a special case of a more general and quite important problem: to fit a mathematical *model* to given data and other known facts. One can also see approximation problems as analogous to the task of a communications engineer, to filter away the **noise** from the **signal**. The questions of this chapter are connected with both **mathematical statistics** and the mathematical discipline **approximation theory**. A few very useful results from these areas will be given later in the chapter. The proofs of some of these results are quite intricate; in such cases references are given to standard works in the above fields for the proof.

We shall mainly be concerned with the problem of **linear approximation**; i.e., a function f is to be approximated using a function f^* that can be expressed as a **linear** combination,

$$f^*(x) = c_0\varphi_0(x) + c_1\varphi_1(x) + \dots + c_n\varphi_n(x), \tag{4.1.1}$$

of $n + 1$ functions, $\varphi_0, \varphi_1, \dots, \varphi_n$, which are chosen in advance; $c_0, c_1, \dots, c_n$ are parameters whose value is to be determined. For example, if $\varphi_i(x) = x^i$, the class of possible f^* is the class of polynomials of degree n. The set $\{1, x, x^2, \dots, x^n\}$ is said to be a **basis** of the set of all polynomials of degree n. (There are other possible bases for the set of nth-degree polynomials which may be better suited for particular numerical applications; see Sec. 4.3.)

The function f can be given in different ways. A common situation is that f is known in the form of a table $f(x_0), f(x_1), \dots, f(x_m)$ of its values on a **net** $G = \{x_i\}_{i=0}^m$. The table of values of f on the net can be considered as a column vector, which we denote by tab (f),

$$\text{tab}\,(f) = [f(x_0), f(x_1), \dots, f(x_m)]^T.$$

(^{T} denotes transpose; i.e., the column vector is written as the transpose of a row vector.) We *assume* throughout that the $m + 1$ points of the net are *distinct*.

If the function was originally given by a curve or by a complicated formula, then such a table can be obtained—using a large value for m.

Suppose we now want to determine the $n + 1$ parameters $c_0, c_1, \dots, c_n$ so that $f^*(x_i) = f(x_i)$ holds exactly or as close as possible at all the $m + 1$ net points. This leads to a linear equation system with $n + 1$ unknowns and $m + 1$ equations:

$$\begin{aligned}
\varphi_0(x_0)c_0 + \varphi_1(x_0)c_1 + \dots + \varphi_n(x_0)c_n &= f(x_0) \\
\varphi_0(x_1)c_0 + \varphi_1(x_1)c_1 + \dots + \varphi_n(x_1)c_n &= f(x_1) \\
\dots \\
\varphi_0(x_m)c_0 + \varphi_1(x_m)c_1 + \dots + \varphi_n(x_m)c_n &= f(x_m).
\end{aligned} \tag{4.1.2}$$

Or, in "tab" form:

$$[\operatorname{tab}(\varphi_0)\,|\,\operatorname{tab}(\varphi_1)\,|\ldots|\,\operatorname{tab}(\varphi_n)]\begin{bmatrix} c_0 \\ c_1 \\ \cdot \\ \cdot \\ \cdot \\ c_n \end{bmatrix} = \operatorname{tab}(f). \tag{4.1.3}$$

If $m = n$, then the above system of equations normally has exactly one solution. In this case, the above way of determining f^* is called **interpolation** or collocation. (Linear interpolation is the special case where $n = 1$, $\varphi_0(x) = 1$, $\varphi_1(x) = x$.) The condition which must be satisfied in order to give a unique solution is that the vectors $\{\operatorname{tab}(\varphi_j)\}_{j=0}^{n}$ must be linearly independent. That condition can also be expressed by saying that the functions φ_j must be linearly independent on the net.

If $m > n$, then only in exceptional cases can one get $f^*(x) = f(x)$ at all the net points. The system has more equations than unknowns. One says that it is **overdetermined**. In such a case one must be content that the equations are satisfied only approximately. See Example 4.1.1, where a straight line could not be made to pass through the five points ($n = 1, m = 4$).

Overdetermination is used to attain *two different types of* **smoothing**:

(a) to reduce the effect of random errors in the values of the function;
(b) to give the curve a smoother shape between the net points (even when the function values are perfect).

An important method for the treatment of overdetermined linear equation systems is the **method of least squares**. Its application to the approximation of functions leads to relatively simple computation, as we shall see in Sec. 4.2.3, and in many applications it can be motivated by statistical arguments (see Sec. 4.5.5). More general overdetermined systems will be treated in Sec. 5.7.

4.1.2. The Idea of a Function Space

In the previous section a table of function values was considered as a vector. This is one instance of a very useful way of thinking about functions, that is, *looking at a function as a vector*. There are many other ways of representing functions as vectors:

Example 4.1.3

Consider the class of all nth-degree polynomials. An individual polynomial

$$p(x) = c_0 + c_1 x + \ldots + c_n x^n$$

is uniquely determined by $n + 1$ coefficients $(c_0, c_1, \ldots, c_n)$, which may be looked upon as a coordinate vector in a "function space"—the $(n+1)$-dimensional space of all polynomials of degree n.

Note that a function in this class can be uniquely specified by other sets of $n+1$ data or "coordinates"—e.g., the coefficients in the Taylor series expansion around some point, or by its values on a net of $n+1$ points; see also Sec. 4.3.3.

Example 4.1.4

The set of tables of a function on a net of 20 points,

$$[f(x_0), f(x_1), \ldots, f(x_{19})],$$

defines a space of 20 dimensions. The set of tables of quadratic polynomials, $f(x) = c_0 + c_1 x + c_2 x^2$, is a three-dimensional linear subspace of this space.

A net with a finite number of points is not sufficient for the specification of an arbitrary differentiable function. One therefore says that the space of differentiable functions is **infinite-dimensional**. However, any given requirement of accuracy for representing a function whose first derivative is bounded in magnitude can be met by using the values of the function on a finite grid (together with linear interpolation). Therefore, the conceptual difficulties of infinite-dimensional function spaces play a small part in practical numerical analysis. Finite-dimensional function spaces are thus useful models for the visualization of computation with functions; see Sec. 4.1.4.

4.1.3. Norms and Seminorms

The geometrical concept of the length of a vector has many natural applications in connection with function spaces and approximation. Just as in ordinary ("Euclidean") space of two or three dimensions we have a way of measuring the distance between two vectors v and w (or what is the same, the length of $v - w$), we would like to use length to measure the goodness of an approximation (the "length" of a certain error function). In this connection the word *norm* is often used (a formal definition will be given later).

Example 4.1.5

On the equidistant net defined by $x_i = i/n$, $(0 \leq i \leq n)$ the square root of $N(f)$, where

$$N(f) = \frac{1}{n} \sum_{i=0}^{n} |f(x_i)|^2$$

defines a norm for the $(n+1)$-dimensional space of function vectors tab(f). One can easily see that as n becomes large—i.e., as we take a finer and finer net—$N(f)$ approaches the square of the Euclidean norm (L_2-norm) for a continuous function f; i.e.,

$$\lim_{n\to\infty} N(f) = \int_0^1 \left| f(x) \right|^2 dx.$$

The following results can be shown (see, e.g., Cheney [36]) to hold for the Euclidean norm, and in fact they are taken as axioms which any general norm must satisfy:

(1) $\|f\| \geq 0$ for all vectors f in the space
(2) $\|\alpha f\| = |\alpha| \cdot \|f\|$ for all real α,
(3) $\|f + g\| \leq \|f\| + \|g\|$, the **triangle inequality**, analogous to the well-known inequality for the ordinary concept of length in two and three dimensions, for all f and g in the space,
(4) $\|f\| = 0$ if and only if $f = 0$.

In the particular case of a space of continuous functions on the interval $[a, b]$, the condition $f = 0$ means that $f(x) = 0$ for all $x \in [a, b]$.

DEFINITION

A real-valued function is called a **norm** on a vector space if it is defined everywhere on the space and satisfies the conditions (1)–(4) above.

The most common norms for a continuous function f on a closed interval $[a, b]$ are:

maximum norm: $$\|f\|_\infty = \max_{x \in [a,b]} |f(x)|, \tag{4.1.4}$$

Euclidean (or L_2) norm: $$\|f\|_2 = \left(\int_a^b |f(x)|^2\, dx\right)^{1/2}, \tag{4.1.5}$$

weighted Euclidean norm: $$\|f\|_{2,w} = \left(\int_a^b |f(x)|^2 w(x)\, dx\right)^{1/2}, \tag{4.1.6}$$

where the function w is called a weight function. It is assumed to be continuous and strictly positive on the open interval (a, b). Integrable singularities at the end points of the interval are permitted; an important example (see Sec. 4.4.1) is $w(x) = (1 - x^2)^{-1/2}$ on the interval $[-1, 1]$.

The Euclidean norm and the maximum norm are special cases or limiting cases ($p = 2$ and $p \rightarrow \infty$, respectively) of a family of norms, the L_p-*norms*,

$$\|f\|_p = \left(\int_a^b |f(x)|^p\, dx\right)^{1/p}, \quad p \geq 1.$$

(One can prove that all of these norms satisfy conditions (1)–(4) above, though the proofs are difficult for general p.)

On a given net G, the Euclidean (vector) norm of tab (f) is given by

$$\|\text{tab}\,(f)\| = \left(\sum_{j=0}^{n} |f(x_j)|^2\right)^{1/2}. \tag{4.1.7}$$

It is tempting to consider this as a norm for the function f. Strictly speaking, this is incorrect, since condition (4) is not satisfied. (A function can be zero on the net without being zero everywhere between the net points.) The above expression is called a **seminorm** for f with respect to the interval $[x_0, x_n]$. We shall use the same notation $\|\cdot\|$ for norms and seminorms, and when necessary use subscripts to distinguish between them; e.g., $\|\cdot\|_{2,G}$ is the Euclidean seminorm, while $\|\cdot\|_2$ is the ordinary Euclidean norm.

Many approximation methods are based on the principle of minimizing some norm or seminorm of the error function $y = f^* - f$, where f^* is to approximate f and has some predetermined form. Indeed, even if f^* has been computed by some other method—e.g., Taylor expansion—it is of interest to measure how good the approximation is with some norm. *It is quite important that the norm be chosen with regard to what f^* is to be used for.* The maximum norm for the error on the interval $[a, b]$ is not so relevant if f^* is to be used for computing approximate values for $f(x)$ for x far outside of $[a, b]$. It is important that the norm (seminorm) take account of the values of the function in a point set which is representative with respect to the *use* of f^*. The choice between maximum norm and Euclidean norm or the choice of a weight function is, as a rule, less critical.

4.1.4. Approximation of Functions as a Geometric Problem in Function Space

We mentioned in the previous section that many approximation methods are based on seeking a function $f^* = c_0\varphi_0 + c_1\varphi_1 + \ldots, + c_n\varphi_n$, for which some given norm of the error function $\|f^* - f\|$ is as small as possible. We have seen that functions can be looked upon as vectors—e.g., a table of function values on a net can be considered as a vector in a space which has as many dimensions as there are net points. The more dimensions, the more accurate a description of the function we have. Thus we can see a class of functions—e.g., the class of all continuous functions on a closed interval $[a, b]$—as a linear space, a *function space*. The difficulty lies in taking the step to an infinite number of dimensions. However, the reader perhaps recalls that the axioms for linear spaces make no requirement that the number of dimensions be finite.

The set of linear combinations of $n + 1$ given, linearly independent functions $\varphi_0, \varphi_1, \ldots, \varphi_n$ build an $(n + 1)$-dimensional linear subspace. ($\varphi_0, \varphi_1, \ldots, \varphi_n$ are linearly independent if $\sum_{j=0}^{n} c_j\varphi_j = 0$ implies $c_j = 0$ for all j.) The approximation problem then means that, in the linear subspace spanned by $\varphi_0, \varphi_1, \ldots, \varphi_n$, we seek the vector which lies the shortest distance from the "vector" f.

The solution to the approximation problem, when the Euclidean norm is used, is simply a generalization of the well-known geometrical fact from two and three dimensions: the shortest distance from a point to a linear subspace is the length of the vector (between the point and a point in the subspace) which is perpendicular to the subspace. Thus, in the problem represented in

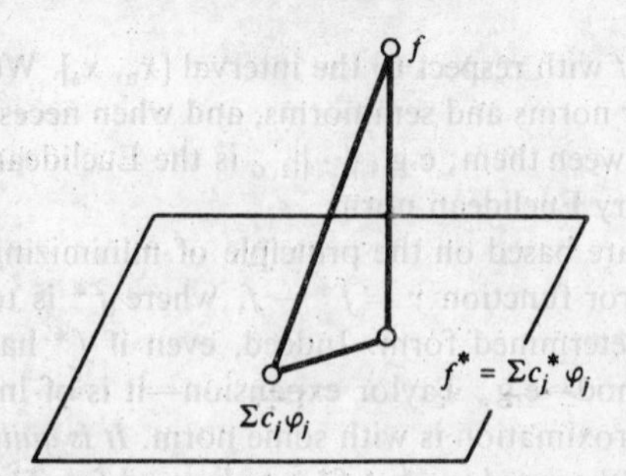

Fig. 4.1.3

Fig. 4.1.3, this would lead to the requirement that the error vector, $f - f^*$, be perpendicular to the subspace spanned by $\varphi_0, \varphi_1, \ldots, \varphi_n$. As we shall see in Sec. 4.2, this *geometrical* observation will enable us to *numerically* solve the least-squares approximation problem; see Theorem 4.2.5.

Similar geometrical considerations used in connection with the idea of a function space make it easier to find (or remember) essential results. The idea of a function space is fundamental not only for numerical analysis but for all of modern mathematical analysis and its applications.

REVIEW QUESTIONS

1. (a) Which two types of "deficiencies" should one take into account when fitting a function to empirical data?
   (b) What is meant by overdetermination, and why does one use it?
2. State the axioms which any norm must satisfy. Define maximum norm and Euclidean norm.

PROBLEMS

1. Compute $\|f\|_\infty$ and $\|f\|_2$ for the function $f(x) = (1 + x)^{-1}$ on the interval [0, 1].
2. (a) Show that $\|f - g\| \geq \|f\| - \|g\|$ for all norms. (Use the axioms mentioned in Sec. 4.1.3.)
   (b) Show that if $\{c_j\}_0^n$ is a set of real numbers and if $\{f_j\}_0^n$ is a set of vectors, then $\| \sum c_j f_j \| \leq \sum |c_j| \cdot \|f_j\|$

4.2. THE APPROXIMATION OF FUNCTIONS BY THE METHOD OF LEAST SQUARES

4.2.1. Statement of the Problems

Let f be a continuous function which in the interval (a, b) is to be approximated by a linear combination,

$$f^*(x) = c_0\varphi_0(x) + c_1\varphi_1(x) + \ldots + c_n\varphi_n(x), \tag{4.2.1}$$

of $n+1$ given functions, $\varphi_0, \varphi_1, \varphi_2, \ldots \varphi_n$. We shall now determine the coefficients $c_0, c_1, \ldots, c_n$ such that a weighted Euclidean norm or seminorm of the error function $f^* - f$ becomes as small as possible—that is, such that

$$\| f^* - f \|^2 = \int_a^b | f^*(x) - f(x) |^2 w(x)\, dx \quad \text{(continuous case)}$$

$$\| f^* - f \|^2 = \sum_{i=0}^{m} | f^*(x_i) - f(x_i) |^2 w_i \quad \text{(discrete case)}$$

becomes as small as possible. This is called the **least-squares approximation problem**.

4.2.2. Orthogonal Systems

We now introduce a formalism related to the geometrical ideas mentioned in Sec. 4.1.4, which is convenient in the study of least-squares approximation.

DEFINITION

The **inner product** or *scalar product* of two real-valued continuous functions f and g is denoted (f, g) and is defined by the relation

$$(f, g) = \begin{cases} \displaystyle\int_a^b f(x)g(x)w(x)\, dx, & \text{(continuous case)} \\ \displaystyle\sum_{i=0}^{m} f(x_i)g(x_i)w_i & \text{(discrete case).} \end{cases}$$

In the discrete case, if all the weights $w_i = 1$, then (f, g) is the same as the vector inner product of tab (f) and tab (g). (The notation tab (f) was defined on page 83.

One can make computations using the more general definition of (f, g) given above in the same way that one does with scalar products in linear algebra. The following fundamental rules are easy to derive (f, g, φ are functions; c_1, c_2 are real numbers):

$$(f, g) = (g, f) \quad \text{(commutativity)}$$

$$(c_1 f + c_2 g, \varphi) = c_1(f, \varphi) + c_2(g, \varphi) \quad \text{(linearity)}$$

$$(f, f) \geq 0 \quad \text{(positivity).}$$

From the rule of linearity it follows (e.g., by induction) that

$$\left(\sum_{j=0}^{n} c_j \varphi_j, \varphi_k\right) = \sum_{j=0}^{n} c_j (\varphi_j, \varphi_k). \tag{4.2.2}$$

THEOREM 4.2.1

$(f, f) = \| f \|^2$, where $\| \cdot \|$ denotes the weighted Euclidean norm or seminorm in the continuous and discrete cases, respectively.

The proof follows immediately from the definitions.

Notice also that if a sequence of functions $\varphi_0, \varphi_1, \ldots, \varphi_n$ is **linearly independent** (defined in Sec. 4.1.4), then $\| \sum_{j=0}^{n} c_j \varphi_j \| = 0$ is true only if $c_j = 0$ for all j. In the discrete case this means that the vectors $\{\text{tab}\,(\varphi_j)\}_{j=0}^{n}$ are linearly independent. Thus in the discrete case, there cannot be more independent functions than there are net points. If a set of continuous functions is linearly independent on a net, then they are also linearly independent on the interval which contains the net; the converse, however, does not always hold.

Example 4.2.1

Set $\prod_{i=0}^{m} (x - x_i) \equiv x^{m+1} - \sum_{j=0}^{m} b_j x^j$. On the net $\{x_i\}_0^m$, x^{m+1} is a linear combination of $1, x, x^2, \ldots, x^m$, since

$$x_i^{m+1} = \sum_{j=0}^{m} b_j x_i^j \quad \text{for} \quad i = 0, 1, 2, \ldots, m.$$

On the other hand, the sequence of functions $1, x, x^2, \ldots, x^m$ is linearly independent on the net, since otherwise there would exist a polynomial of degree m, not identically zero, which would have all of the $m + 1$ net points as zeros. This is of course impossible. For every polynomial $p_m \neq 0$ of degree m, then, it is valid (even for the seminorm on a net of at least $m + 1$ points) that $\| p_m \| \neq 0$.

DEFINITION

Two functions f and g are said to be **orthogonal** if $(f, g) = 0$. A finite or infinite sequence of functions $\varphi_0, \varphi_1, \varphi_2, \ldots, \varphi_n$ build an **orthogonal system,** if $(\varphi_i, \varphi_j) = 0$ for $i \neq j$ and $\| \varphi_i \| \neq 0$ for all i. If, in addition, $\| \varphi_i \| = 1$ for all i, then the sequence is called an **orthonormal system.**

THEOREM 4.2.2. *The Pythagorean Theorem for Functions*

If $(f, g) = 0$, then

$$\| f + g \|^2 = \| f \|^2 + \| g \|^2.$$

Proof.

$$\begin{aligned} \| f + g \|^2 &= (f + g, f + g) = (f, f) + (g, f) + (f, g) + (g, g) \\ &= \| f \|^2 + 0 + 0 + \| g \|^2. \end{aligned}$$

With the help of an inductive proof, the above can easily be generalized to:

THEOREM 4.2.3

If $\varphi_0, \varphi_1, \varphi_2, \ldots, \varphi_n$ are an orthogonal system, then

$$\| \sum_{j=0}^{n} c_j \varphi_j \|^2 = \sum_{j=0}^{n} c_j^2 \| \varphi_j \|^2.$$

From this it follows that the functions in an orthogonal system are linearly independent.

The following theorem gives an important example of an orthogonal system.

THEOREM 4.2.4

The sequence of functions $\varphi_j(x) = \cos jx$, $j = 0, 1, 2, \ldots, m$, *build an orthogonal system, with the inner product*

$$(f, g) = \int_0^{\pi} f(x)g(x)\,dx \qquad \text{(continuous case, } m = \infty\text{)},$$

where

$$\|\varphi_j\|^2 = \tfrac{1}{2}\pi \quad \text{for } j > 0, \qquad \|\varphi_0\|^2 = \pi,$$

and also with the inner product

$$(f, g) = \sum_{i=0}^{m} f(x_i)g(x_i), \quad \textit{where } x_i = \frac{2i+1}{m+1}\frac{\pi}{2} \qquad \text{(discrete case)},$$

whereby

$$\|\varphi_j\|^2 = \tfrac{1}{2}(m+1) \quad \text{for} \quad 1 \le j \le m, \qquad \|\varphi_0\|^2 = m+1.$$

Proof. In the continuous case, if $j \neq k$, $j \geq 0$, $k \geq 0$, it holds that

$$(\varphi_j, \varphi_k) = \int_0^{\pi} \cos jx \cos kx\,dx = \int_0^{\pi} \frac{1}{2}(\cos(j-k)x + \cos(j+k)x)\,dx$$
$$= \frac{1}{2}\left(\frac{\sin(j-k)\pi}{j-k} + \frac{\sin(j+k)\pi}{j+k}\right) = 0,$$

whereby orthogonality is proved for the continuous case.

In the discrete case, set

$$h = \frac{\pi}{m+1}, \qquad x_\mu = \frac{1}{2}h + \mu h,$$

$$(\varphi_j, \varphi_k) = \sum_{\mu=0}^{m} \cos jx_\mu \cos kx_\mu = \tfrac{1}{2}\sum_{\mu=0}^{m} (\cos(j-k)x_\mu + \cos(j+k)x_\mu).$$

Using notation from complex numbers we have, then,

$$(\varphi_j, \varphi_k) = \tfrac{1}{2}\,\mathrm{Re}\left(\sum_{\mu=0}^{m} e^{i(j-k)[(1/2)h+\mu h]} + \sum_{\mu=0}^{m} e^{i(j+k)[(1/2)h+\mu h]}\right). \tag{4.2.3}$$

The sums in Eq. (4.2.3) are geometric series with ratios

$$\exp(i(j-k)h) \quad \text{and} \quad \exp(i(j+k)h),$$

respectively. If $j \neq k$, $0 \leq j \leq m$, $0 \leq k \leq m$, then the ratios are never equal to 1, since

$$0 < |(j \pm k)h| \leq 2m\frac{\pi}{m+1} < 2\pi.$$

The first sum in Eq. (4.2.3) is, then, using the formula for the sum of a geometric series,

$$e^{i(j-k)(1/2)h}\frac{e^{i(j-k)(m+1)h} - 1}{e^{i(j-k)h} - 1} = \frac{e^{i(j-k)\pi} - 1}{e^{i(j-k)(1/2)h} - e^{-i(j-k)(1/2)h}} = \frac{(-1)^{j-k} - 1}{2i\sin(j-k)\frac{1}{2}h}.$$

The real part of the last expression is clearly zero. An analogous computation shows that the real part of the other sum in Eq. (4.2.3) is also zero. Thus the orthogonality property holds in the discrete case also. It is left to the reader to show that the expressions for $\|\varphi_j\|^2$ given in the theorem are correct.

Orthogonal systems give rise to extraordinary formal simplifications in many situations. There are many other examples of orthogonal systems than those given above—see, for example, Theorem 4.4.2. One can in fact derive an orthogonal system from any system of linearly independent functions, by a process analogous to Gram-Schmidt orthogonalization (Sec.5.7.2). In addition, orthogonal systems occur in a natural way in connection with eigenvalue problems for differential equations, which are quite common in mathematical physics.

4.2.3. Solution of the Approximation Problem
(see Fig. 4.1.3, p. 88)

THEOREM 4.2.5

When $\varphi_0, \varphi_1, \ldots, \varphi_n$ are linearly independent, the least-squares approximation problem has a unique solution,

$$f^* = \sum_{j=0}^{n} c_j^* \varphi_j, \tag{4.2.4}$$

where the coefficients c_j^* satisfy the so-called normal equations (Eq. (4.2.6)).

The solution is characterized by the orthogonality property that $f^* - f$ is orthogonal to all φ_j, $(j = 0, 1, \ldots, n)$.

An important special case is when $\varphi_0, \varphi_1, \ldots, \varphi_n$ form an orthogonal system; then the coefficients are computed more simply by the formula

$$c_j^* = \frac{(f, \varphi_j)}{(\varphi_j, \varphi_j)}, \quad (j = 0, 1, \ldots, n). \tag{4.2.5}$$

The coefficients c_j^* are called the **orthogonal coefficients** *(or occasionally, Fourier coefficients).*

Proof. (In all the sums, j varies from 0 to n.) Let $(c_0, c_1, \ldots, c_n)$ be a sequence of coefficients with $c_j \neq c_j^*$ for at least one j. Then

$$\sum c_j \varphi_j - f = \sum (c_j - c_j^*)\varphi_j + (f^* - f).$$

If $f^* - f$ is orthogonal to all the φ_j, then it is also orthogonal to the linear combination $\sum (c_j - c_j^*)\varphi_j$. According to the Pythagorean theorem (Theorem 4.2.2) we have, then,

$$\|\sum c_j \varphi_j - f\|^2 = \|\sum (c_j - c_j^*)\varphi_j\|^2 + \|f^* - f\|^2 > \|f^* - f\|^2.$$

Thus *if* f^*—f is orthogonal to all φ_k, then f^* is a solution to the approximation problem. It remains, then, to show that the orthogonality conditions

$$(\sum c_j^* \varphi_j - f, \varphi_k) = 0, \quad (k = 0, 1, 2, \ldots, n)$$

can be fulfilled. The above conditions are, according to Eq. (4.2.2), equivalent to the system of equations

$$(\varphi_0, \varphi_0)c_0^* + (\varphi_0, \varphi_1)c_1^* + \dots + (\varphi_0, \varphi_n)c_n^* = (\varphi_0, f)$$
$$(\varphi_1, \varphi_0)c_0^* + (\varphi_1, \varphi_1)c_1^* + \dots + (\varphi_1, \varphi_n)c_n^* = (\varphi_1, f)$$
$$\dots$$
$$(\varphi_n, \varphi_0)c_0^* + (\varphi_n, \varphi_1)c_1^* + \dots + (\varphi_n, \varphi_n)c_n^* = (\varphi_n, f).$$

Or, more concisely

$$\sum_{j=0}^{n} (\varphi_j, \varphi_k)c_j^* = (f, \varphi_k), \quad (k = 0, 1, 2, \dots, n). \qquad (4.2.6)$$

This linear system of equations for the determination of the coefficients c_j^* is called the **normal equations.**

If $\{\varphi_j\}_{j=0}^n$ build an orthogonal system, then the normal equations can be solved immediately, since in each equation all the terms with $j \neq k$ are zero. The formula (4.2.5) follows, then, from the equations

$$(\varphi_k, \varphi_k)c_k^* = (f, \varphi_k).$$

Suppose now that we know only that $\{\varphi_j\}_{j=0}^n$ are linearly independent. We shall now show that the solution to the normal equations exists and is unique. This follows unless the homogenous system,

$$\sum_{j=0}^{n} (\varphi_j, \varphi_k)c_j = 0, \quad (k = 0, 1, \dots, n),$$

has a nontrivial solution $(c_0, c_1, \dots, c_n)$ (i.e., with at least one $c_i \neq 0$). But this would lead to

$$\|\sum_{j=0}^{n} c_j\varphi_j\|^2 = (\sum_{j=0}^{n} c_j\varphi_j, \sum_{k=0}^{n} c_k\varphi_k) = \sum_{k=0}^{n}\sum_{j=0}^{n} (\varphi_j, \varphi_k)c_jc_k = \sum_{k=0}^{n} 0\cdot c_k = 0,$$

which contradicts that the φ_j were linearly independent. Thus the theorem is proved.

Example 4.2.2

Use the method of least squares to fit a function of the form $f^*(x) = c_0 + c_1x$ to the following measured data. All the weights are equal to one.

x	1	3	4	6	7
$f(x)$	-2.1	-0.9	-0.6	0.6	0.9

Using the notation of the theorem, we have $\varphi_0(x) = 1, \varphi_1(x) = x$. The normal equations (4.2.6) are, then:

$$(\varphi_0, \varphi_0)c_0 + (\varphi_1, \varphi_0)c_1 = (f, \varphi_0)$$
$$(\varphi_0, \varphi_1)c_0 + (\varphi_1, \varphi_1)c_1 = (f, \varphi_1).$$

The necessary inner products are scalar products of the vectors

$$\text{tab}(\varphi_0) = \begin{bmatrix} 1 \\ 1 \\ 1 \\ 1 \\ 1 \end{bmatrix}, \quad \text{tab}(\varphi_1) = \begin{bmatrix} 1 \\ 3 \\ 4 \\ 6 \\ 7 \end{bmatrix}, \quad \text{tab}(f) = \begin{bmatrix} -2.1 \\ -0.9 \\ -0.6 \\ 0.6 \\ 0.9 \end{bmatrix}.$$

Thus, for example, $(f, \varphi_1) = -2.1 \cdot 1 - 0.9 \cdot 3 - 0.6 \cdot 4 + 0.6 \cdot 6 + 0.9 \cdot 7 = 2.7$. One gets

$$5c_0 + 21c_1 = -2.1$$
$$21c_0 + 111c_1 = 2.7$$

with the result

$$c_0 = -2.542, \qquad c_1 = 0.5053.$$

Check:

$$\begin{aligned} \text{tab}(f^* - f) &= \text{tab}(c_0 + c_1 x - f) \\ &= (0.063, -0.126, 0.079, -0.110, 0.095)^{\mathsf{T}}. \end{aligned}$$

Notice that the sum of the components of $\text{tab}(f^* - f)$ is 0.001. Without round-off error, the sum would have been 0. (Why?)

Example 4.2.3

Consider the case $n = 0$, $\varphi_0(x) = 1$. Set

$$(f, g) = \sum_{i=0}^{m} w_i f(x_i) g(x_i).$$

The "normal equations" (4.2.6) now reduce to the single equation

$$(\varphi_0, \varphi_0)c_0 = (f, \varphi_0);$$

hence

$$c_0 = \frac{\sum_{i=0}^{m} w_i f(x_i)}{\sum_{i=0}^{m} w_i}.$$

c_0 is said to be a *weighted mean* of the values of the function. If all the weights are equal, then one gets

$$c_0 = \frac{\sum_{i=0}^{m} f(x_i)}{(m+1)};$$

that is, the computation of a mean is a special case of the method of least squares.

Notice that, in the case where $\{\varphi_j\}_{j=0}^{n}$ form an orthogonal system, the Fourier coefficient c_j^* is independent of n (see formula (4.2.5)). This has the advantage that one can increase the total number of parameters without recalculating any previous ones. Orthogonal systems are advantageous not only because they simplify calculations; using them, one can often avoid numerical difficulties with round-off error which can come up when one solves the normal equations for a nonorthogonal set of basis functions.

For every continuous function f one can associate with it an infinite series,

$$f \sim \sum_{j=0}^{\infty} c_j^* \varphi_j, \quad c_j^* = \frac{(f, \varphi_j)}{(\varphi_j, \varphi_j)}.$$

Such a series is called an **orthogonal expansion**. For certain orthogonal systems this series converges (with very mild requirements on the function f; see Chap. 9).

THEOREM 4.2.6

If f^* is defined by formulas (4.2.4), (4.2.5), then

$$\| f^* - f \|^2 = \| f \|^2 - \| f^* \|^2 = \| f \|^2 - \sum_{j=0}^{n} (c_j^*)^2 \| \varphi_j \|^2.$$

Proof. Since $f^* - f$ is, according to Theorem 4.2.5, orthogonal to all φ_j, $0 \leq j \leq n$, then $f^* - f$ is orthogonal to f^*. The theorem then follows directly from the Pythagorean theorem and Theorem 4.2.3.

COROLLARY

$$\sum_{j=0}^{n} (c_j^*)^2 \| \varphi_j \|^2 \leq \| f \|^2 \qquad \textbf{(Bessel's inequality)}.$$

The series $\sum_{j=0}^{\infty} (c_j^*)^2 \| \varphi_j \|^2$ is convergent. If $\| f^* - f \| \rightarrow 0$ as $n \rightarrow \infty$, then the sum of the latter series is equal to $\| f \|^2$ **(Parseval's formula)**.

THEOREM 4.2.7

If $\{\varphi_j\}_{j=0}^{m}$ are linearly independent on the net $\{x_i\}_{i=0}^{m}$, then the interpolation problem of determining the coefficients $\{c_j\}_{j=0}^{m}$ such that

$$\sum_{j=0}^{m} c_j \varphi_j(x_i) = f(x_i), \quad (i = 0, 1, 2, \ldots, m), \tag{4.2.7}$$

has exactly one solution. Interpolation is a special case ($n = m$) of the method of least squares. If $\{\varphi_j\}$ is an orthogonal system, then the coefficients c_j are equal to the orthogonal coefficients in Eq. (4.2.5).

Proof. The system of equations (4.2.7) has a unique solution, since its columns are the vectors tab (φ_j), $(j = 0, 1, \ldots, m)$, which are linearly independent. For the solution of the interpolation problem it holds that $\| \sum c_j \varphi_j - f \| = 0$; that is, the error function has the least possible seminorm. The remainder of the theorem follows then from Theorem 4.2.5.

In the next section, we shall give a convenient way for computing the interpolation polynomial and introduce some further results and notation concerning polynomials.

REVIEW QUESTIONS

1. Prove the Pythagorean theorem for functions.
2. Define and give examples of orthogonal systems of functions.
3. Derive the normal equations, and solve them in the case when $\{\varphi_i\}_0^n$ form an orthogonal system. Give a geometric interpretation.
4. Formulate and prove Bessel's inequality and Parseval's identity, and interpret them geometrically.

PROBLEMS

1. Determine straight lines which approximate the curve $y = e^x$ such that
   (a) the Euclidean seminorm of the error function on the net $(-1, -0.5, 0, 0.5, 1)$ is as small as possible;
   (b) the Euclidean norm of the error function on the interval $[-1, 1]$ is as small as possible;
   (c) the line is tangent to $y = e^x$ at the point $(0, 1)$ (i.e., Taylor approximation at the midpoint of the interval).

   Compute the errors at $x = 1$. Use three decimals in the calculations.
2. Determine, for $f(x) = \pi^2 - x^2$, the "cosine polynomial" $f^*(x) = \sum_{j=0}^n c_j \cos jx$, which makes $\|f^* - f\|_2$ on the interval $[0, \pi]$ as small as possible.
3. The water level in the North Sea is mainly determined by the so-called M_2-tide, whose period is about 12 hours and thus has the form

$$H(t) = h_0 + a_1 \sin \frac{2\pi t}{12} + a_2 \cos \frac{2\pi t}{12}, \quad t \text{ in hours.}$$

   One has made the following measurements:

t	0	2	4	6	8	10	hours
$H(t)$	1.0	1.6	1.4	0.6	0.2	0.8	meters

   Fit $H(t)$ to the series of measurements using the method of least squares. Note that the method of least squares cannot be applied to the equivalent expression

$$H(t) = h_0 + A \sin \frac{2\pi(t - t_0)}{12},$$

   since the parameters do not occur linearly in this form.
4. (a) Let the scalar product be defined by

$$(f, g) = \int_0^1 f(x)g(x)\,dx.$$

   Calculate the matrix of the normal equations, when

$$\varphi_j(x) = x^j, \quad j = 0, 1, 2, \ldots, n$$

(b) Do the same for the scalar product

$$(f, g) = \int_{-1}^{1} f(x)g(x)\,dx.$$

Show how the normal equations can be easily decomposed into two systems, with approximately $(n + 1)/2$ equations in each.

5. Verify the formulas for $\|\varphi_j\|^2$ given in Theorem 4.2.4.

4.3. POLYNOMIALS

4.3.1. Basic Terminology: The Weierstrass Approximation Theorem

By an (algebraic) polynomial of degree n we mean a function of the form:

$$p(x) = a_n x^n + a_{n-1}x^{n-1} + \ldots + a_1 x + a_0.$$

a_n is called the *leading coefficient*. If $a_n \neq 0$, then the polynomial is called a *genuine* nth-degree polynomial. The class of nth-degree polynomials contains all polynomials of lower degree as a special case. A constant is a polynomial of degree zero.

We shall study the use of polynomials to approximate functions in a closed, bounded interval $[a, b]$.

DEFINITION

Let f be a given continuous function on the interval $[a, b]$. The lower bound (infimum) of all possible values of $\|f - p\|_\infty$, where p varies over the set of all nth-degree polynomials, is denoted by $E_n(f)$.

THEOREM 4.3.1. *Weierstrass Approximation Theorem*

For every continuous function f defined on a closed, bounded interval it holds that

$$\lim_{n\to\infty} E_n(f) = 0.$$

For the proof of this fundamental theorem we refer the reader to a standard text on approximation—for example, Rice [46]. The smoother f is, the quicker $E_n(f)$ decreases. $E_n(f)$ is of course different for different intervals, and the narrower the interval, the less $E_n(f)$ becomes.

In many cases $E_n(f)$ decreases so slowly toward zero (as n grows) that it is impractical to attempt to approximate f with only one polynomial in the entire interval $[a, b]$.

The most immediately available methods for constructing polynomial approximations often give approximations whose maximal errors are significantly larger than $E_n(f)$, and one cannot be sure that the error goes to zero as $n \longrightarrow \infty$, even if f is quite regular (i.e., smooth, or many times differentiable).

Example 4.3.1

The function $f(x) = (1 + x^2)^{-1}$ has continuous derivatives of arbitrary order on every bounded interval—for example, on the interval $[-10, 10]$—but the Taylor series,

$$(1 + x^2)^{-1} = 1 - x^2 + x^4 - x^6 + \dots,$$

whose partial sums give polynomial approximations of higher and higher degree, converges only in the subinterval $[-1, 1]$.

Even equidistant interpolation can give rise to convergence difficulties, when the number of interpolation points becomes large, as we shall see in Sec. 4.3.4.

4.3.2. Triangle Families of Polynomials

There are many ways of specifying polynomials. One way is to give the coefficients in an expansion of the form

$$\sum_{j=0}^{n} a_j x^j,$$

but this is not the only possibility. One can, for example, give the coefficients in an expansion of the form (for some $c \neq 0$):

$$\sum_{j=0}^{n} b_j (x - c)^j. \tag{4.3.1}$$

Mathematically, it makes no difference whether one writes the polynomial in the first way or the second, but computationally, working with *rounded values of the coefficients*, the choice of representation of the polynomial can make a great difference. If one were to use the form of Eq. (4.3.1) for polynomial approximation on the interval $[a, b]$, then one should choose $c = \frac{1}{2}(a + b)$, i.e., choose c to be the midpoint of the interval.

There are in fact other representations which are often even more advantageous. A sequence of polynomials $\varphi_0, \varphi_1, \varphi_2, \dots$ (finite or infinite)

$$\begin{aligned}
\varphi_0(x) &= a_{00} \\
\varphi_1(x) &= a_{10} + a_{11}x \\
\varphi_2(x) &= a_{20} + a_{21}x + a_{22}x^2 \\
&\dots \\
\varphi_n(x) &= a_{n0} + a_{n1}x + a_{n2}x^2 + \dots + a_{nn}x^n \\
&\dots
\end{aligned}$$

(where $a_{ii} \neq 0$ for all i) is defined to be a **triangle family** of polynomials.

Powers of x can be expressed recursively and uniquely as linear combinations of $\varphi_0, \varphi_1, \varphi_2, \dots$; e.g.,

$$x^0 = \frac{\varphi_0(x)}{a_{00}}, \qquad x^1 = \frac{\varphi_1(x) - a_{10}}{a_{11}} = \frac{\varphi_1(x) - a_{10}\varphi_0(x)/a_{00}}{a_{11}}, \dots$$

Set

$$
\begin{aligned}
x^0 &= b_{00}\varphi_0 \\
x^1 &= b_{10}\varphi_0 + b_{11}\varphi_1 \\
x^2 &= b_{20}\varphi_0 + b_{21}\varphi_1 + b_{22}\varphi_2 \\
&\dots \\
x^n &= b_{n0}\varphi_0 + b_{n1}\varphi_1 + b_{n2}\varphi_2 + \dots + b_{nn}\varphi_n \\
&\dots
\end{aligned}
$$

(If we complement the two triangular schemes of coefficients by setting $a_{ij} = b_{ij} = 0$ for $i < j$, then we get two matrices $\{a_{ij}\}$, $\{b_{ij}\}$ which are each other's inverse.) From these equations it follows that *every polynomial of degree n can be written in a unique way in the form*

$$p(x) = c_0\varphi_0(x) + c_1\varphi_1(x) + \dots + c_n\varphi_n(x).$$

4.3.3. A Triangle Family and Its Application to Interpolation

Suppose that $x_0, x_1, \dots, x_m$ are $m + 1$ given constants (with $x_i \neq x_j$) for $i \neq j$). The formulas

$$
\begin{aligned}
&\varphi_0(x) = 1, \\
&\varphi_i(x) = (x - x_0)(x - x_1)\dots(x - x_{i-1}) \quad (i = 1, 2, \dots, m + 1) \qquad (4.3.3)
\end{aligned}
$$

define a triangle family. (The leading coefficient is of course 1 for all φ_i.) The representation

$$
\begin{aligned}
p(x) = c_0 + c_1(x - x_0) + c_2(x - x_0)(x - x_1) + \dots \\
+ c_m(x - x_0)(x - x_1)\dots(x - x_{m-1}) \qquad (4.3.4)
\end{aligned}
$$

is very convenient, especially for **interpolation problems.**

THEOREM 4.3.2

The interpolation problem of determining an mth-degree polynomial which agrees with the values of a given function on a net of $m + 1$ points always has a unique solution, which can be easily expressed in the form of Eq. (4.3.4).

Proof. We shall show that the coefficients $c_0, c_1, \dots, c_m$ in Eq. (4.3.4) can be uniquely determined in such a way that $p(x)$ takes on arbitrary predetermined values on the net $\{x_i\}_{i=0}^m$. Since $\varphi_i(x_j) = 0$ when $i > j$, we have, from Eq. (4.3.3),

$$
\begin{aligned}
p(x_0) &= c_0 \\
p(x_1) &= c_0 + c_1(x_1 - x_0) \\
p(x_2) &= c_0 + c_1(x_2 - x_0) + c_2(x_2 - x_0)(x_2 - x_1) \\
&\text{etc.}
\end{aligned}
$$

One can compute $c_0, c_1, c_2, \dots$ recursively from the above triangular system of equations. Thus the interpolation problem has exactly one solution.

DEFINITION

By int $(a, b, c, \ldots, w)$ we mean the smallest interval which contains the points $a, b, c, \ldots, w$.

THEOREM 4.3.3. *The Remainder Term in Interpolation*

Let f be a given function with continuous derivatives of order at least $m + 1$. Denote by p the mth-degree polynomial for which $p(x_i) = f(x_i)$, $i = 0, 1, \ldots, m$. Then

$$f(x) - p(x) = \frac{f^{(m+1)}(\xi)}{(m+1)!}(x - x_0)(x - x_1)\ldots(x - x_m),$$

for some point $\xi \in \text{int}(x, x_0, x_1, \ldots, x_m)$.

Notice the similarity to the remainder term in Taylor's formula (which in fact is a kind of limiting case, when all the points x_i coincide). Notice also that the right-hand side is zero at all the net points—as it should be.

Proof. Introduce a new variable z, and set

$$\Phi(z) = (z - x_0)(z - x_1)\ldots(z - x_m),$$
$$G(z) = f(z) - p(z) - R(x)\Phi(z),$$

where $R(x)$ is determined so that $G(x) = 0$. Then we have that

$$G(z) = 0, \quad \text{for} \quad z = x, x_0, x_1, \ldots, x_m.$$

From repeated use of Rolle's theorem it follows that there exists a point $\xi \in \text{int}(x, x_0, \ldots, x_m)$, where $G^{(m+1)}(\xi) = 0$. But $p^{(m+1)}(z) = 0$ and $\Phi^{(m+1)}(z) = (m + 1)!$ for all z. Thus,

$$G^{(m+1)}(z) = f^{(m+1)}(z) - R(x)(m + 1)!$$

If we now put $z = \xi$, we get

$$R(x) = \frac{f^{(m+1)}(\xi)}{(m+1)!}.$$

Put this into the definition of $G(z)$, and set $z = x$. Since $G(x) = 0$, it follows that

$$f(x) - p(x) - \frac{\Phi(x)f^{(m+1)}(\xi)}{(m+1)!} = 0,$$

and the theorem follows.

If a polynomial is given in the form

$$p(x) = c_0 + c_1(x - x_0) + c_2(x - x_0)(x - x_1) + \ldots$$
$$+ c_n(x - x_0)(x - x_1)\ldots(x - x_{n-1}),$$

then $p(\alpha)$ is easily computed (for a given numerical value α) using the recursion formula:

$$b_n = c_n, \quad b_{n-i} = (\alpha - x_{n-i})b_{n-i+1} + c_{n-i} \quad \text{for} \quad i = 1, 2, \ldots, n,$$
$$p(\alpha) = b_0. \tag{4.3.5}$$

The reader should show this, as it is a very useful exercise in how one often rewrites mathematical expressions in a form very suitable for automatic computation. Notice in addition that the special (limiting) case where all the $x_i = 0$ just gives us Horner's rule (see Chap. 1) for computing

$$\sum_{j=0}^{n} c_j \alpha^j.$$

4.3.4. Equidistant Interpolation and the Runge Phenomenon

Having shown that the interpolation problem has a unique solution, we now turn to a problem that can come up in equidistant polynomial interpolation when the number of interpolation points is fairly large; this problem is called Runge's phenomenon and we illustrate it in the following example:

Example 4.3.2

The function f, whose graph is the continuous curve shown in Fig. 4.3.1, is approximated in two different ways by a polynomial of degree 10 in the interval $[-1, 1]$.

The dashed curve has been determined by interpolation on the **equidistant** net with eleven points ($m = 10$)

$$x_i = -1 + \frac{2i}{m}, \quad (i = 0, 1, 2, \ldots, m), \tag{4.3.6}$$

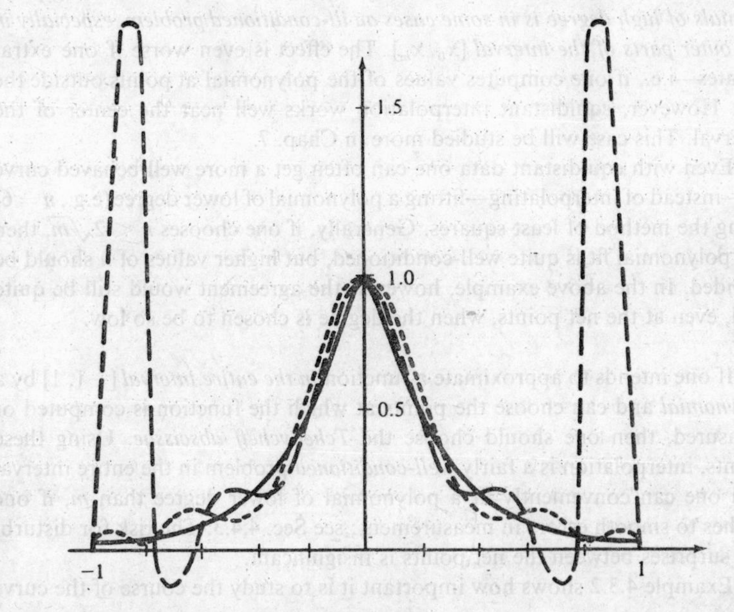

Fig. 4.3.1

The graph of the polynomial so obtained has—unlike the graph of f—a disturbing course between the net points. The agreement with f near the ends of the interval is especially bad, while near the center of the interval, $[-\frac{1}{5}, \frac{1}{5}]$, the agreement is fairly good. Such behavior is typical of *equidistant interpolation with polynomials of high degree*, and can be explained theoretically (**Runge's phenomenon**).

The *dotted* curve has been determined by interpolation in the so-called **Tchebycheff abscissae**,

$$x_i = \cos \frac{2i+1}{m+1}\frac{\pi}{2}, \quad i = 0, 1, 2, \ldots, m, \quad (m = 10). \tag{4.3.7}$$

This procedure is studied more closely in Sec. 4.4.2. The agreement with f is now much better than with equidistant interpolation, but still not good. The function is not at all suited for approximation by one polynomial over the entire interval. Here one would get a much better result using approximation by rational functions (somewhat of a trick, since the curve shown is the graph of $f(x) = 1/(1 + 25x^2)$), or with piecewise polynomials—see Sec. 4.6.

Notice that the difference between the values of the two polynomials is much smaller at the net points themselves ($x_i = -1 + 2i/10$) than in certain points between the net points, especially in the outer parts of the interval. This intimates that the values which one gets by equidistant interpolation with a polynomial of high degree can be very sensitive to disturbances in the given values of the function. Put another way: *equidistant interpolation using polynomials of high degree is in some cases an ill-conditioned problem, especially in the outer parts of the interval* $[x_0, x_m]$. The effect is even worse if one extrapolates—i.e., if one computes values of the polynomial at points outside the net. However, equidistant interpolation works well near the center of the interval. This case will be studied more in Chap. 7.

Even with equidistant data one can often get a more well-behaved curve by—instead of interpolating—fitting a polynomial of lower degree (e.g., $n = 6$) using the method of least squares. Generally, if one chooses $n < 2\sqrt{m}$, then the polynomial fit is quite well-conditioned, but higher values of n should be avoided. In the above example, however, the agreement would still be quite bad, even at the net points, when the degree is chosen to be so low.

If one intends to approximate a function *in the entire interval* $[-1, 1]$ by a *polynomial* and can choose the points at which the function is computed or measured, then one should choose the *Tchebycheff abscissae*. Using these points, interpolation is a fairly *well-conditioned* problem in the entire interval and one can conveniently fit a polynomial of lower degree than m, if one wishes to smooth errors in measurement; see Sec. 4.4.3. The risk for disturbing surprises between the net points is insignificant.

Example 4.3.2 shows how important it is to study the course of the curve

$y = f^*(x)$ between those points which are used in the calculations before one accepts the approximation. When one uses procedures for approximation for which one does not have a complete theoretical analysis, one should make an *experimental perturbational calculation*. In the above case such a calculation would very probably reveal that the interpolation polynomial reacts quite strongly if the values of the function are disturbed by small amounts, say $\pm 10^{-3}$, where the sign is chosen randomly (for example, using random numbers generated by the computer—see Chap. 11). This would give a basis for rejecting the unpleasing dashed curve in the example, even if one knew nothing more about the function than its values at the equidistant net points.

In many applications, the maximum norm over the interval in which the function is to be used is the most natural choice of norm—at least when the truncation error is of greater importance than the errors in measurement. In recent years, a great deal of work has been aimed at methods for producing algorithms which determine the best approximating nth-degree polynomial f^*, measured in the maximum norm, to a given function f. These methods are quite laborious and the theory for them is advanced; they are not treated in this book. However, we shall give some results which show that other procedures give results which are nearly optimal—even when measured in the maximum norm. Among these other procedures, the method of least squares (discussed in Sec. 4.2) and its corresponding continuous formulation (the minimization of a Euclidean norm of the error function) is of great interest, because of its simplicity, even when one is not concerned about smoothing statistically independent errors in measurement. In the next section we shall see other easily used and effective methods, expansion of the function in a series of orthogonal polynomials, and Tchebycheff interpolation.

REVIEW QUESTIONS

1. Define $E_n(f)$, and state the Weierstrass approximation theorem.
2. Give an exact formulation of and prove the theorem which says that the interpolation problem for polynomials has precisely one solution.
3. (a) Give the remainder term in interpolation.
   (b) Derive it.
4. Describe the Runge phenomenon.

PROBLEMS

1. The Hermite polynomials are defined by the recurrence relation

$$H_{n+1}(x) = 2xH_n(x) - 2nH_{n-1}(x)$$
$$H_0(x) = 1, \qquad H_1(x) = 2x.$$

(a) Show that they form a triangle family and calculate H_n for $n \leq 4$.
(b) Express, conversely, $1, x, x^2, x^3, x^4$ in terms of the Hermite polynomials.

2. Verify formula (4.3.5).
3. Write a Fortran subroutine INTERP(X,Y,C,N) which computes coefficients c_i, $i = 1, \ldots, N+1$ for the interpolation polynomial $p(x) = c_1 + c_2(x - x_1) + c_3(x - x_1)(x - x_2) + \cdots + c_{N+1}(x - x_1)(x - x_2) \ldots (x - x_N)$, which interpolates to the points (x_i, y_i), $i = 1, \ldots, N+1$.
4. Write a Fortran program which calls the subroutine INTERP to interpolate $\exp(x)$ at the points $x_i = -1 + 2(i-1)/m$, $i = 1, 2, \ldots, m+1$. Use a finer net (say, $x_i = -1 + 2(i-1)/(4m)$, $i = 1, 2, \ldots, 4m+1$) and a recursion formula like Eq. (4.3.5) to estimate the maximum error of the interpolation polynomial.

4.4. ORTHOGONAL POLYNOMIALS AND APPLICATIONS

4.4.1. Tchebycheff Polynomials

The Tchebycheff polynomials are perhaps the most important example of a triangle family of orthogonal polynomials. Their properties can be derived by rather simple methods; for other families of orthogonal polynomials we shall develop a general theory in Sec 4.4.3.

Consider the (easily verified) formula

$$\cos(n+1)\varphi + \cos(n-1)\varphi = 2\cos\varphi\cos n\varphi, \quad (n \geq 1). \tag{4.4.1}$$

This formula enables one to express $\cos(n\varphi)$ as a polynomial in $\cos\varphi$; e.g.,

$$\cos 2\varphi = 2\cos^2\varphi - 1,$$
$$\cos 3\varphi = 2\cos\varphi\cos 2\varphi - \cos\varphi = 4\cos^3\varphi - 3\cos\varphi,$$
$$\cos 4\varphi = 2\cos\varphi\cos 3\varphi - \cos 2\varphi = 8\cos^4\varphi - 8\cos^2\varphi + 1,$$
$$\ldots$$

If we now set

$$x = \cos\varphi, \qquad \text{thus} \quad \varphi = \arccos x,$$

then we obtain a triangle family of polynomials, the Tchebycheff polynomials, $(-1 \leq x \leq 1, n = 0, 1, 2, \ldots)$ defined by the formula

$$T_n(x) = \cos(n \arccos x). \tag{4.4.2}$$

Thus, from the previous formulas for $\cos(n\varphi)$, we get, e.g.,

$$T_0(x) = 1, \quad T_1(x) = x, \quad T_2(x) = 2x^2 - 1,$$
$$T_3(x) = 4x^3 - 3x, \quad T_4(x) = 8x^4 - 8x^2 + 1.$$

The Tchebycheff polynomials have many useful properties:

1. **Recursion formula,**

$$T_0(x) = 1, \qquad T_1(x) = x = xT_0(x),$$
$$T_{n+1}(x) = 2xT_n(x) - T_{n-1}(x), \quad n \geq 1. \tag{4.4.3}$$

This follows directly from Eq. (4.4.1).

2. **The leading coefficient** is 2^{n-1} for $n \geq 1$ and 1 for $n = 0$.

3. **Symmetry property,** $T_n(-x) = (-1)^n T_n(x)$.

Properties 2 and 3 can be obtained by induction with the help of the recursion formula.

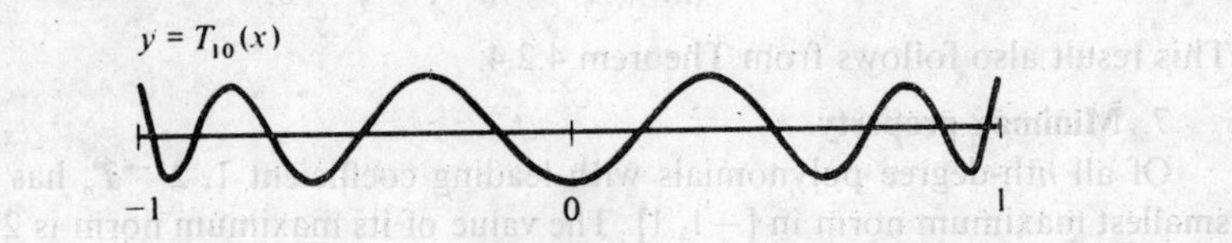

Fig. 4.4.1

4. $T_n(x)$ **has** n **zeros in** $[-1, 1]$ given by

$$x_k = \cos\left(\frac{2k+1}{n}\frac{\pi}{2}\right), \quad \text{the **Tchebycheff abscissae**,}$$
$$(k = 0, 1, 2, \ldots, n-1), \tag{4.4.4}$$

and $n + 1$ **extrema,**

$$x'_k = \cos\frac{k\pi}{n}, \qquad T_n(x'_k) = (-1)^k, \quad (k = 0, 1, 2, \ldots, n). \tag{4.4.5}$$

These results follow directly from the fact that $\cos(n\varphi) = 0$ for

$$\varphi = \frac{2k+1}{n}\frac{\pi}{2}$$

and that $|\cos n\varphi|$ has maxima for $\varphi = k\pi/n$.

5. **Orthogonality property, continuous case.**

Set

$$(f, g) = \int_{-1}^{1} f(x)g(x)(1 - x^2)^{-1/2}\, dx.$$

Then

$$(T_i, T_j) = \begin{cases} 0 & \text{if} \quad i \neq j, \\ \frac{1}{2}\pi & \text{if} \quad i = j \neq 0, \\ \pi & \text{if} \quad i = j = 0. \end{cases} \tag{4.4.6}$$

Proof. Set $x = \cos\varphi$.

$$(T_i, T_j) = \int_{-1}^{1} T_i(x)T_j(x)(1 - x^2)^{-1/2}\, dx = \int_0^{\pi} \cos i\varphi \cos j\varphi \, d\varphi.$$

The result follows then from Theorem 4.2.4.

6. **Orthogonality property, discrete case.**
Set

$$(f, g) = \sum_{k=0}^{m} f(x_k)g(x_k), \tag{4.4.7}$$

where $\{x_k\}$ are the zeros of $T_{m+1}(x)$. Then for $0 \leq i \leq m, 0 \leq j \leq m$, we have

$$(T_i, T_j) = \begin{cases} 0 & \text{if} \quad i \neq j \\ \frac{1}{2}(m+1) & \text{if} \quad i = j \neq 0 \\ m+1 & \text{if} \quad i = j = 0. \end{cases}$$

This result also follows from Theorem 4.2.4.

7. **Minimax property.**
Of all nth-degree polynomials with leading coefficient 1, $2^{1-n}T_n$ has the smallest maximum norm in $[-1, 1]$. The value of its maximum norm is 2^{1-n}.

Proof. Indirect proof: Suppose that there were a polynomial $p_n(x)$, with leading coefficient 1, such that $|p_n(x)| < 2^{1-n}$ for all x in $[-1, 1]$. Let x'_k, $k = 0, 1, 2, \ldots, n$ be the abscissae of the extrema of T_n (see property 4). Then we would have (see Fig. 4.4.1)

$$p_n(x'_0) < 2^{1-n}T_n(x'_0),$$
$$p_n(x'_1) > 2^{1-n}T_n(x'_1),$$
$$p_n(x'_2) < 2^{1-n}T_n(x'_2), \text{ etc., up to } x'_n.$$

From this, it follows that the polynomial

$$p_n(x) - 2^{1-n}T_n(x)$$

changes sign in each of the n intervals (x'_{k+1}, x'_k), $k = 0, 1, 2, \ldots, n-1$. This is impossible, since the polynomial is of degree $n - 1$ (p_n and $2^{1-n}T_n$ have the same leading coefficient). Thus we have proved the minimax property.

Expansions in terms of Tchebycheff polynomials are an important aid in the study of functions on the *interval* $[-1, 1]$. If one is working in terms of a parameter t which varies in the interval $[a, b]$, then one should make the substitution,

$$t = \tfrac{1}{2}(a + b) + \tfrac{1}{2}(b - a)x, \qquad t \in [a, b] \longleftrightarrow x \in [-1, 1], \tag{4.4.8}$$

4.4.2. Tchebycheff Interpolation and Smoothing

The remainder term in interpolation using the values of the function f at the points x_i, $i = 0, 1, 2, \ldots, m$ is, according to Theorem 4.3.3 equal to

$$\frac{(x - x_0)(x - x_1)\ldots(x - x_m)f^{(m+1)}(\xi)}{(m+1)!} \tag{4.4.9}$$

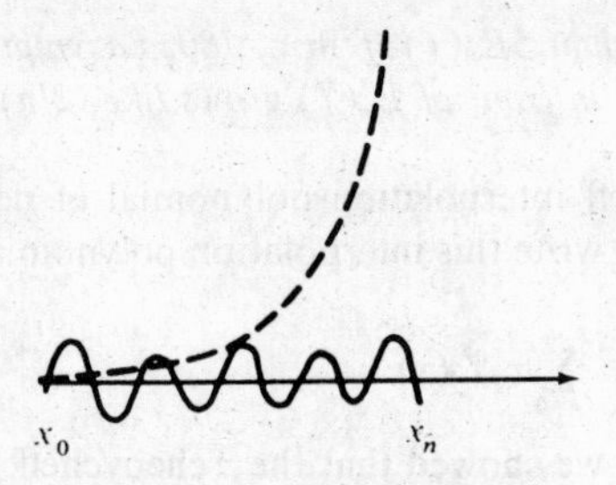

Fig. 4.4.2

Here ξ depends on x, but one can say that the error curve behaves for the most part like a polynomial curve $y = a(x - x_0)(x - x_1) \ldots (x - x_m)$; see the continuously drawn curve of Fig. 4.4.2. A similar oscillating curve is also typical for error curves arising from least-squares approximation. The zeros of the error are then about the same as the zeros for the first neglected term in the orthogonal expansion. This contrasts sharply with the error curve for Taylor approximation, whose usual behavior is described approximately by the formula $y = a(x - x_0)^{m+1}$ (the dashed curve in Fig. 4.4.2). The above observations are most relevant when $E_n(f)$ decreases quickly as n becomes large.

It is natural to ask *what the optimal placing of* $x_0, x_1, \ldots, x_m$ *should be in order to minimize the maximum magnitude of* $q(x) = (x - x_0)(x - x_1) \ldots (x - x_m)$ *in the interval where the formula is to be used.* If the interval is $[a, b]$, then one can first make the substitution (4.4.8). *For the interval* $[-1, 1]$, the answer is given directly by the minimax property of the Tchebycheff polynomials; i.e., *choose* $q = T_{m+1}/2^m$. *Thus*

$$x_k = \cos\left(\frac{2k+1}{m+1}\frac{\pi}{2}\right), \quad (k = 0, 1, \ldots, m). \tag{4.4.10}$$

We call interpolation using these abscissae *Tchebycheff interpolation*. We have already seen an example of the use of this type of interpolation (Sec. 4.3.4). *Near the boundary, Tchebycheff interpolation is much better than equidistant interpolation; this is partly because* $q(x)$ *is so different in the two cases.* (One must also keep in mind that ξ in (4.4.9) does not have the same value in the two cases.)

The following theorem by M.J.D. Powell, which we give without proof, gives a different way of expressing the effectivity of Tchebycheff interpolation. It shows that *whenever* it is reasonable to use polynomial approximation, then Tchebycheff interpolation gives an excellent result and should be used if one has the possibility of freely choosing the points where the function values are to be used.

THEOREM 4.4.1

For an arbitrary continuous function f on the interval $[-1, 1]$, *the magnitude of the error with the use of Tchebycheff interpolation of the mth degree is less* *than $4E_m(f)$ if $m \leq 20$, and less than $5E_m(f)$ if $m \leq 100$. (Asymptotically, for large values of m the coefficient in front of $E_m(f)$ grows like $(2/\pi)$* ln *m.)*

Let $p(x)$ denote the Tchebycheff interpolation polynomial of degree m. For many reasons, it is practical to write this interpolation polynomial in the form

$$p(x) = \sum_{j=0}^{m} c_j T_j(x) \tag{4.4.11}$$

(see Theorem 4.2.7). In Sec. 4.4.1, we showed that the Tchebycheff polynomials were orthogonal with respect to the scalar product,

$$(f, g) = \sum_{k=0}^{m} f(x_k)g(x_k), \quad x_k = \cos\left(\frac{2k+1}{m+1} \cdot \frac{\pi}{2}\right).$$

Thus, using Eq. (4.2.5),

$$c_i = \frac{(f, T_i)}{\| T_i \|^2} = \frac{\sum_{k=0}^{m} f(x_k)T_i(x_k)}{\| T_i \|^2}, \tag{4.4.12}$$

where (see Sec. 4.4.1, property 6)

$$\| T_i \|^2 = \tfrac{1}{2}(m+1) \quad \text{for} \quad i > 0, \qquad \| T_0 \|^2 = m + 1.$$

The recursion formula (4.4.3) can be used for calculating the orthogonal coefficients according to Eq. (4.4.12). For computing $p(x)$ with Eq. (4.4.11), one can use Clenshaw's recursion formula; see the next section (note that $\alpha_k = 2$ for $k > 0$ but $\alpha_0 = 1$). Occasionally one is interested in the partial sums of Eq. (4.4.11). For example, in order to smooth errors in measurement it can be advantageous to break off the summation before one has come to the last term. In such cases, it is better to use Eq. (4.4.3) also for the calculation of the values of the function according to Eq. (4.4.11).

If the values of the function are afflicted with statistically independent errors in measurement with standard deviation σ, then the series can be broken off when for the first time

$$\| f - \sum_{j=0}^{n} c_j T_j \| < \sigma m^{1/2}.$$

(See Sec. 4.5.5.)

4.4.3. General Theory of Orthogonal Polynomials

By a family of orthogonal polynomials we mean a triangle family of polynomials which is an orthogonal system with respect to some given weight function. In the previous section we saw an example, the Tchebycheff polynomials, which were orthogonal with respect to the weight function $(1 - x^2)^{-1/2}$. **Expansions of functions in terms of orthogonal polynomials** are very useful. They are **easy to manipulate**, have **good convergence properties**, and they give a **well-conditioned representation** of a function (with exception of weight distributions on certain nets; see Sec. 4.4.4). The theory of orthogonal polynomials also constitutes the background for numerical methods for many problems which at first sight seem to have little connection with least-squares approximation (e.g., numerical integration, continued fractions, and the algebraic eigenvalue problem).

In this section we shall prove some results from the general theory. The next section will then treat some additional examples of orthogonal polynomials, the *Legendre polynomials* (which arise from a uniform weight distribution on the interval $[-1, 1]$) and the *Gram polynomials* (which arise from a uniform weight distribution on an equidistant net).

THEOREM 4.4.2

For every weight distribution there is an associated orthogonal system $\varphi_0, \varphi_1, \varphi_2, \ldots$ *which is a triangle family of polynomials. The family is uniquely determined apart from the fact that the leading coefficients* $A_0, A_1, A_2, \ldots$ *can be given arbitrary nonzero values. For a weight distribution on a net with* $m+1$ *points, the family ends with* $\varphi_m(x)$. *(*$\varphi_{m+1}(x)$ *becomes zero at each net point.) In the continuous case, the family has infinitely many members.*

For $n \geq 0$ *the orthogonal polynomials satisfy a three-term recursion formula,*

$$\varphi_{n+1}(x) = \alpha_n(x - \beta_n)\varphi_n(x) - \gamma_n\varphi_{n-1}(x),$$
$$\varphi_{-1}(x) = 0, \qquad \varphi_0(x) = A_0. \tag{4.4.13}$$

$\alpha_n = A_{n+1}/A_n$. β_n *and* γ_n *are given in Eqs. (4.4.16) and (4.4.17) respectively.*

Note: If the weight distribution is symmetric about $x = \beta$, then $\beta_n = \beta$ for all n.

Proof. By induction: Suppose that the φ_j have been constructed for $0 \leq j \leq n$, $\varphi_j \neq 0$ $(n \geq 0)$. We now seek a polynomial of degree $n+1$ which

(a) has leading coefficient $A_{n+1} = \alpha_n A_n$;

(b) is orthogonal to $\varphi_0, \varphi_1, \varphi_2, \ldots, \varphi_n$.

Since $\{\varphi_j\}_0^n$ is a triangle family, every nth-degree polynomial can be expressed as a linear combination of these polynomials. Every polynomial which fulfills condition (a) can therefore be written in the form

$$\varphi_{n+1} = \alpha_n x\varphi_n - \sum_{i=0}^{n} c_{ni}\varphi_i. \tag{4.4.14}$$

Condition (b) is fulfilled if and only if

$$\alpha_n(x\varphi_n, \varphi_j) - \sum_{i=0}^{n} c_{ni}(\varphi_i, \varphi_j) = 0, \quad (j = 0, 1, 2, \ldots, n).$$

But $(\varphi_i, \varphi_j) = 0$ for $i \neq j$. Thus

$$c_{nj}\|\varphi_j\|^2 = \alpha_n(x\varphi_n, \varphi_j). \tag{4.4.15}$$

This determines the coefficients c_{nj} uniquely. From the definition of inner product in Sec. 4.2.2, it follows that $(x\varphi_n, \varphi_j) = (\varphi_n, x\varphi_j)$. But $x\varphi_j$ is a polynomial of degree $j+1$. Thus if $j+1 < n$, then it is orthogonal to φ_n. So $c_{nj} = 0$ for $j < n-1$. From Eq. (4.4.14) it follows, then, that

$$\varphi_{n+1} = \alpha_n x\varphi_n - c_{nn}\varphi_n - c_{n,n-1}\varphi_{n-1},$$

which has the same form as the original assertion of the theorem if (using Eq. (4.4.15)) we set

$$\beta_n = \frac{c_{nn}}{\alpha_n} = \frac{(x\varphi_n, \varphi_n)}{\|\varphi_n\|^2}, \tag{4.4.16}$$

$$\gamma_n = c_{n,n-1} = \frac{\alpha_n(\varphi_n, x\varphi_{n-1})}{\|\varphi_{n-1}\|^2},$$

(For $n = 0$, γ_n is not needed, since $\varphi_{-1} = 0$.) The division in Eq. (4.4.16) can always be performed, as long as $n \leq m$ (see Example 4.2.1). In the continuous case, no reservation need be made.

The expression for γ_n can be written in another way. If we scalar-multiply Eq. (4.4.14) by φ_{n+1}, we get

$$(\varphi_{n+1}, \varphi_{n+1}) = \alpha_n(\varphi_{n+1}, x\varphi_n) - \sum_{i=0}^{n} c_{ni}(\varphi_{n+1}, \varphi_i) = \alpha_n(\varphi_{n+1}, x\varphi_n).$$

Thus

$$(\varphi_{n+1}, x\varphi_n) = \frac{\|\varphi_{n+1}\|^2}{\alpha_n}.$$

If we then diminish all indices by 1, we get:

$$(\varphi_n, x\varphi_{n-1}) = \frac{\|\varphi_n\|^2}{\alpha_{n-1}}, \quad (n \geq 1).$$

Put this in the expression for γ_n, thus

$$\gamma_n = \frac{\alpha_n \|\varphi_n\|^2}{\alpha_{n-1}\|\varphi_{n-1}\|^2}. \tag{4.4.17}$$

The proof leads to a unique construction of φ_{n+1}. In the discrete case, with the net $\{x_i\}_{i=0}^m$, this holds only as long as $n \leq m$. For $n = m$, the constructed polynomial must be equal to

$$A_{m+1}(x - x_0)(x - x_1)\ldots(x - x_m),$$

because this polynomial is zero at all the net points, and thus orthogonal to all functions. Condition (b) is thus fulfilled and obviously also condition (a). From this it follows that $\|\varphi_{m+1}\| = 0$; thus the computation of β_n cannot be carried out for $n = m + 1$. The unique construction stops, then, at $n = m$. This is natural, since there cannot be more than $m + 1$ orthogonal (or even linearly independent) functions on a net with $m + 1$ points (see Example 4.2.1). The theorem is thus proved.

In the computation of the coefficients in an expansion of the form

$$p(x) = c_0\varphi_0(x) + c_1\varphi_1(x) + \dots + c_n\varphi_n(x), \tag{4.4.18}$$

using the orthogonal coefficient formula $c_j = (p, \varphi_j)/\|\varphi_j\|^2$, one can make use of the recursion formula (4.4.13). In the discrete case, it is advisable to compute recursively the vectors tab (φ_j), $j = 0, 1, 2, \dots, n$.

The easiest way to compute values of a function defined by an orthogonal expansion is to use **Clenshaw's recursion formula.** Use the notation of formula (4.4.13) and set

$$y_{n+2} = y_{n+1} = 0.$$

Then, for $k = n, n-1, \dots, 1, 0$, compute

$$y_k = \alpha_k(x - \beta_k)y_{k+1} - \gamma_{k+1}y_{k+2} + c_k. \tag{4.4.19}$$

Then $p(x) = A_0 y_0$. The proof is left as a somewhat difficult exercise to the reader.

Expansions of functions in terms of orthogonal polynomials have, in the continuous case, good convergence properties. Let $\hat{p}_n$ denote the polynomial of degree n for which $\|f - \hat{p}_n\|_\infty = E_n(f)$, and set

$$p_n = \sum_{j=0}^{n} c_j\varphi_j,$$

where c_j is the jth Fourier coefficient of f. If we use the weighted Euclidean norm, $\hat{p}_n$ is of course not a better approximation than p_n. In fact,

$$\begin{aligned}\|f - p_n\|_{2,w}^2 &= \int_a^b |f(x) - p_n(x)|^2 w(x)\,dx \\ &\leq \int_a^b |f(x) - \hat{p}_n(x)|^2 w(x)\,dx \leq E_n(f)^2 \int_a^b w(x)\,dx.\end{aligned} \tag{4.4.20}$$

This can be interpreted as saying that *a kind of weighted mean of* $|f(x) - p_n(x)|$ *is less than or equal to* $E_n(f)$—this is about as good a result as one could demand. The error curve has an oscillatory behavior (Fig. 4.4.3). In small sub-

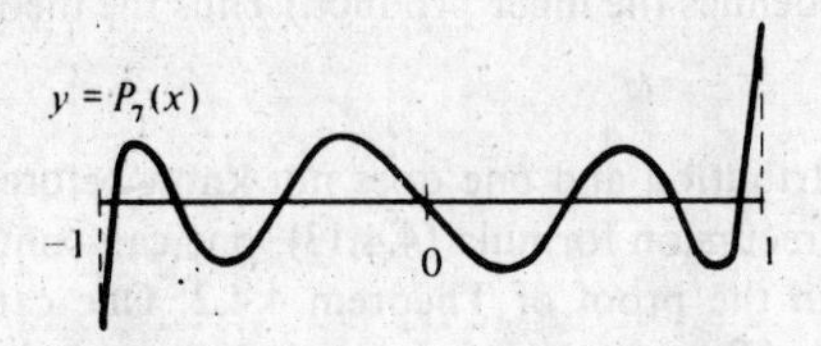

Fig. 4.4.3

intervals, $|f(x) - p_n(x)|$ can be significantly greater than $E_n(f)$. This usually happens near the ends of the interval or in subintervals where $w(x)$ is relatively small. From Eq. (4.4.20) and the Weierstrass approximation theorem (Theorem 4.3.1), it follows that

$$\lim_{n\to\infty} \|f - p_n\|_{2,w} = 0$$

for every continuous function f. From Eq. (4.4.20) and Theorem 4.2.6 with corollary, one gets, after some calculation,

$$\sum_{j=n+1}^{\infty} c_j^2 \|\varphi_j\|^2 = \|f - p_n\|_{2,w}^2 \leq E_n(f)^2 \int_a^b w(x)\,dx,$$

which gives one an idea of how quickly the terms in the orthogonal expansion decrease.

In the discrete case, one has no convergence problem in a mathematical sense. The orthogonal expansion has only $m + 1$ terms, and its sum is equal to the polynomial which agrees with $f(x)$ on the net (see Theorem 4.2.7), the interpolation polynomial. When m is large, however, one has use for the fact that Eq. (4.4.18)—assuming that $E_n(f)$ decreases quickly—provides a good representation for f even for small values of n. With some nets (for example, equidistant nets) one should choose n less than $2\sqrt{m}$, so that the approximation polynomial will not have a very large oscillatory behavior between the net points.

THEOREM 4.4.3

The nth-degree polynomial in a family of orthogonal polynomials associated with weight function w on the interval [a, b] has n simple zeros, all of which lie in the interior of [a, b].

Proof. Indirect proof: Suppose that $\varphi_n(x)$ has k sign changes $0 \leq k < n$ in the interior of the interval—say, at $t_1, t_2, \ldots, t_k$. Then

$$\varphi_n(x) \prod_{i=0}^{k} (x - t_i) \qquad (\text{or} \quad \varphi_n(x), \quad \text{if} \quad k = 0)$$

has constant sign in the interval. But this contradicts the fact that

$$(\varphi_n, p) = \int_a^b \varphi_n(x) p(x) w(x)\,dx = 0$$

for all polynomials p of degree less than n. (In the discrete case one has the same conclusion for the sum which defines the inner product.) Thus the theorem follows.

If one has a discrete weight distribution and one does not know beforehand the coefficients $\{\beta_j, \gamma_j\}$ in the recursion formula (4.4.13), one can compute them by imitating the steps in the proof of Theorem 4.4.2. One can recursively compute the coefficients $\{\beta_j, \gamma_j\}$, the vectors tab (φ_j), and the orthogonal coefficients $\{c_j\}$ for a given function f, for all $j \leq n$, with a total of about $4mn$ operations (one "operation" = a multiplication or division together with an addition), assuming that the net is symmetric and that the leading coefficient of the polynomials is set equal to 1. If there are differing weights, then about mn additional operations are needed; similarly, mn additional operations are required if the net is not symmetric.

If the orthogonal coefficients are determined simultaneously for several functions on the same net, then only about mn additional operations per function are required. (In the above, we assume $m \gg 1$, $n \gg 1$.)

The above way of solving the problem is not only more economical than the classical method of using the normal equations (which requires about $n(n+5)m/2 + n^3/6$ operations); it also has the very important advantage of avoiding the difficulties with ill-conditioned systems of equations which one has even for moderate n when the normal equations are used to determine the coefficients in a polynomial $\sum_{j=0}^{n} k_j x^j$, when the function values are given on an equidistant net. We shall return to this problem in Sec. 4.5.1 and in Chap. 5.

4.4.4. Legendre Polynomials and Gram Polynomials

The Legendre polynomials are defined by the formula

$$P_0(x) = 1, \qquad P_n(x) = \frac{1}{2^n n!} \frac{d^n}{dx^n}[(x^2-1)^n], \quad (n = 1, 2, \ldots). \qquad (4.4.21)$$

Since $(x^2-1)^n$ is a polynomial of degree $2n$, $P_n(x)$ is a polynomial of degree n. The leading coefficient A_n is the same as that for the polynomial

$$\frac{1}{2^n n!} \frac{d^n}{dx^n}(x^{2n}) = \frac{1}{2^n n!} 2n(2n-1)(2n-2)\ldots(n+1)x^n.$$

Thus

$$\alpha_n = \frac{A_{n+1}}{A_n} = \frac{2n+1}{n+1}.$$

Properties.

1. **Orthogonality property**

If

$$(f, g) = \int_{-1}^{1} f(x)g(x)\,dx,$$

then

$$(P_n, P_j) = \begin{cases} 0 & \text{if} \quad n \neq j \\ \dfrac{2}{2n+1} & \text{if} \quad n = j. \end{cases} \qquad (4.4.22)$$

Proof (sketch): Suppose $j \leq n$ (this is no restriction). Set $f = P_n$ and $g = P_j$ using Eq. (4.4.21), and insert this in Eq. (4.4.22). Then use integration by parts (differentiating P_j and integrating P_n) $j+1$ times (or j times if $n = j$). Use the fact that the kth derivative of $(x^2-1)^n$ is zero at $x = -1$ and $x = 1$, when $k < n$. For $n = j$ one also needs the formula

$$\int_{-1}^{1} (1-x^2)^n\,dx = 2\int_0^{\pi/2} \sin^{2n+1} t\,dt = 2\frac{2n}{2n+1}\frac{2n-2}{2n-1}\frac{2n-4}{2n-3}\cdots\frac{2}{3}.$$

2. **Symmetry property**

$$P_n(-x) = (-1)^n P_n(x).$$

3. **Recursion formula**

$$P_{n+1}(x) = \frac{2n+1}{n+1} x P_n(x) - \frac{n}{n+1} P_{n-1}(x),$$
$$P_0(x) = 1, \qquad P_1(x) = x, \qquad P_2(x) = \tfrac{1}{2}(3x^2 - 1). \tag{4.4.23}$$

Proof (*sketch*). This follows from the general three-term recursion formula, Theorem 4.4.2. Since the weight distribution in Eq. (4.4.22) is symmetric about the origin, we have $\beta_n = 0$. Substitute into formula (4.2.5) the expressions for α_n and (P_n, P_n) given above (together with the corresponding expressions for α_{n-1} and (P_{n-1}, P_{n-1})) and simplify!

4. $$|P_n(x)| \leq 1 \quad \text{for} \quad x \in [-1, 1].$$

There seems to be no easy proof for this result; see [9], p. 219.

As is apparent from the orthogonality property above, the Legendre polynomials are quite useful for approximation with the Euclidean norm and weight distribution $w(x) = 1$.

When the method of least squares is to be used on equidistant data, the **Gram polynomials** $\{P_{n,m}\}_{n=0}^{m}$ are of interest. These polynomials are orthogonal with respect to the inner product

$$(f, g) = \sum_{i=0}^{m} f(x_i)g(x_i), \quad x_i = -1 + \frac{2i}{m} \tag{4.4.24}$$

$$(P_{n,m}, P_{j,m}) = \begin{cases} 0 & \text{if} \quad n \neq j \\ 1 & \text{if} \quad n = j. \end{cases} \tag{4.4.25}$$

The recursion formula is

$$P_{n+1,m}(x) = \alpha_{n,m} x P_{n,m}(x) - \gamma_{n,m} P_{n-1,m}(x),$$

where

$$\alpha_{n,m} = \frac{m}{n+1}\left(\frac{4(n+1)^2 - 1}{(m+1)^2 - (n+1)^2}\right)^{1/2}, \qquad \gamma_{n,m} = \frac{\alpha_{n,m}}{\alpha_{n-1,m}}$$
$$P_{0,m}(x) = (m+1)^{-1/2}, \qquad P_{-1,m}(x) = 0. \tag{4.4.26}$$

When $n \ll m^{1/2}$, these polynomials are very similar to the Legendre polynomials, but when $n \gg m^{1/2}$, they have very large oscillations between the net points, and a large maximum norm in $[-1, 1]$. Related to this is the fact that when fitting a polynomial to *equidistant* data, one should never choose n larger than about $2m^{1/2}$.

REVIEW QUESTIONS

1. What are orthogonal polynomials? Give an example of a family of orthogonal polynomials together with the three-term recursion formula which its members satisfy.

2. (a) Give the general form of the three-term recursion formula satisfied by orthogonal polynomials.
   (b) Prove Theorem 4.4.1 (on the existence of orthogonal polynomials and a three-term formula for them).

3. Formulate and prove the theorem concerning the zeros of orthogonal polynomials.

4. (a) Give some reasons for using orthogonal polynomials in polynomial approximation with the method of least squares.
   (b) Give some arguments against the assertion that orthogonal polynomials are difficult to work with.

5. (a) Define Tchebycheff polynomials, and derive their recursion formula.
   (b) State the orthogonality properties of the Tchebycheff polynomials and prove orthogonality in the continuous case.
   (c) Prove the minimum property of the Tchebycheff polynomials.

6. What are Tchebycheff interpolation and Tchebycheff smoothing, and why are they useful?

7. Define the Legendre polynomials, and derive their orthogonality property.

PROBLEMS

1. What first-degree polynomial does one get with Tchebycheff interpolation to $f(x) = 1/(3 + x)$, $x \in [-1, 1]$? What is the maximum norm of the error?

2. Determine the orthogonal polynomials $\varphi_n(x)$, $n = 0, 1, 2, 3$, with leading coefficient 1, for the weight function $w(x) = 1 + x^2$, $-1 \leq x \leq 1$.

   Hint: Either use the method of undetermined coefficients taking advantage of symmetry, and express the fact that φ_n should be orthogonal to x^m, $m < n$, *or* set $\varphi_{n+1}(x) = x\varphi_n(x) - \gamma_n\varphi_{n-1}(x)$ for $n = 1, 2$, and compute γ_n using Eq. (4.4.17).

3. Set $\mu_j = \int_a^b x^j w(x)\,dx$, (the jth **moment of the weight distribution** w). Show that the system of equations

$$\begin{bmatrix} \mu_0 & \mu_1 & \mu_2 & \cdots & \mu_{n-1} \\ \mu_1 & \mu_2 & \mu_3 & \cdots & \mu_n \\ \mu_2 & \mu_3 & \mu_4 & \cdots & \mu_{n+1} \\ \cdots & \cdots & \cdots & \cdots & \cdots \\ \mu_{n-1} & \mu_n & \mu_{n+1} & \cdots & \mu_{2n-2} \end{bmatrix} \begin{bmatrix} c_0 \\ c_1 \\ c_2 \\ \cdots \\ c_{n-1} \end{bmatrix} = -\begin{bmatrix} \mu_n \\ \mu_{n+1} \\ \mu_{n+2} \\ \cdots \\ \mu_{2n-1} \end{bmatrix}$$

has as solution the coefficients of a polynomial $x^n + \sum_{j=0}^{n-1} c_j x^j$, which is a member of the family of orthogonal polynomials associated with the weight function w. (A matrix of this type is called a *Hankel matrix*.)

4. Give a family of orthogonal polynomials associated with the following weight distributions:
(a) $w(x) = (x(1-x))^{-1/2}, \quad x \in [0, 1];$
(b) $w(x) = 1, \quad x \in [0, 1].$

5. (a) Show that if $x = \cos \varphi$, then

$$U_n(x) = \frac{\sin (n+1)\varphi}{\sin \varphi}$$

is an nth-degree polynomial which satisfies the same recursion formula as the Tchebycheff polynomials. What are the initial conditions U_0, U_1?
(b) Show that $\{U_n\}$ is an orthogonal system for the weight distribution $w(x) = (1-x^2)^{1/2}$, $x \in [-1, 1]$.
(Notice that $w(x) = \sin \varphi$. $\{U_n\}$ are called the Tchebycheff polynomials of the second kind.)

6. Derive Clenshaw's recursion formula, Eq. (4.4.19).

7. Make an algorithm (in the form of a Fortran or Algol program) for:
(a) generating the orthogonal polynomials (with leading coefficient 1) associated with a weight distribution on a net $\{x_i\}_{i=0}^m$ where all the weights are equal to 1;
(b) computing the orthogonal coefficients $\{c_j\}_{j=0}^n$ for a function f for which the table $\{f(x_i)\}_{i=0}^m$ is known.

8. (a) **Laguerre polynomials.** Show that the polynomials

$$L_n(x) = e^x \frac{d^n}{dx^n}(x^n e^{-x})$$

are orthogonal with respect to the weight distribution

$$w(x) = e^{-x}, \quad 0 \leq x < \infty.$$

Calculate L_n for $n \leq 3$.
(b) **Hermite polynomials.** Show that the polynomials

$$H_n(x) = (-1)^n \exp (x^2) \frac{d^n}{dx^n} (\exp (-x^2))$$

are orthogonal with respect to the weight distribution

$$w(x) = \exp (-x^2), \quad -\infty < x < \infty.$$

Verify that these polynomials are identical to those mentioned in Problem 1 of Sec. 4.3 (at least for $n \leq 4$).

9. Show the following **minimum property** of orthogonal polynomials: Among all nth-degree polynomials p_n with leading coefficient 1, the smallest value of

$$\|p_n\|^2 = \int_a^b p_n^2(x) w(x)\, dx, \quad w(x) \geq 0$$

is obtained for $p_n = \varphi_n/A_n$, where φ_n is the orthogonal polynomial (with leading coefficient A_n) associated with the weight distribution $w(x)$.

Hint: Determine the best approximation to x^n in the above norm *or* consider the expansion $p_n = \varphi_n/A_n + \sum_{j=0}^{n-1} c_j\varphi_j$.

4.5. COMPLEMENTARY OBSERVATIONS ON POLYNOMIAL APPROXIMATION

4.5.1. Summary of the Use of Polynomials

The following comments are the more relevant the higher the degree of the polynomial one is using. With the use of polynomials of degree 4 (or lower), they can for the most part be ignored.

One should ask oneself the following questions:

1. *Is the given function suited for approximation by polynomials?*
2. *Does the input data define a polynomial satisfactorily?*
3. *Have I chosen suitable parameters to define the polynomial?*
4. *Have I chosen a good algorithm?*

A discussion of each of these four questions follows:

1. The two basic concepts in connection with fitting a function to data are the choice of **form** and the choice of **norm**. The choice of norm was discussed in Sec. 4.1.3 and will not be discussed further here; recall that the choice of norm should be made chiefly with regard to what the approximation is to be used for (and also with regard to computational convenience).

The choice of form, however, is often more a matter of intuition. "Form" refers to the class of functions to be adapted to the data. Here one must distinguish between a situation where one has strong reason to believe in a certain form (as in the pendulum example, 4.1.1), and the situation where one only wants to have a form which is easy to work with, a concise approximate description of the function (see Example 4.1.2). In the latter case, it is natural to try to fit a **polynomial** to the function. *But there are many functions which are not at all suited for approximation by a single polynomial in the entire interval which is of interest.* Functions whose graphs have sharp rises surrounded by weakly curved stretches (see Sec. 4.3.1) are one example. Neither is a polynomial a good choice for form if the function (or some of its lower derivatives) has a singularity or some other type of discontinuity in the interval (including the end points).

Many functions are suitably approximated piecewise, using many different simple functions in the various parts of the interval. An example is linear interpolation—that is, approximation of a curve by secants, which was illustrated in Fig. 1.2.9. The technique of representing functions piecewise by a polynomial has, during the last twenty-five years, gotten a solid foundation in the theory of *spline functions;* see Sec. 4.6.

Another aspect of the choice of form is **transformation** of variables. It is possible, for instance, that $f(\log x)$, $\log(f(x))$, $f(1/x)$, or $xf(x)$ are better suited for approximation by a polynomial than is $f(x)$, to take just a few examples.

It is generally very important to give some thought to the *choice of variables* when one intends to use numerical methods. A change of variable—often seen to be quite obvious after the fact—can lead to a drastic decrease in the computational work. This is in many numerical problems related to the fact that one uses relations which are based upon approximating a function locally by a polynomial.

If the problem is to reproduce a curve, then it is not certain that it should be described by a relation of the form $y = f(x)$. It can be better to use, for example, polar coordinates and to seek a relation of the form $r = f(\varphi)$ between these coordinates. This applies especially to closed curves. In Fig. 4.5.1, A is an appropriate and B an inappropriate choice for the pole (origin) of a polar coordinate system. (Why?)

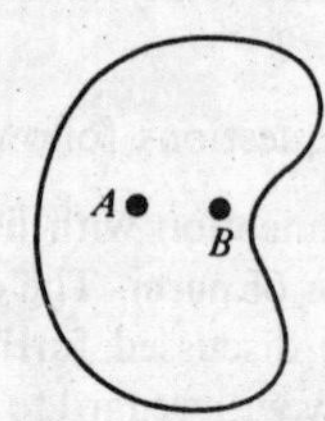

Fig. 4.5.1

In choosing the form one should also *consider what the approximation is to be used for*. For the computation of values of a function, **rational functions** (ratios of two polynomials) are as a rule more accurate and almost as easy to work with as a polynomial with the same number of parameters. For other uses—for example, in the computation of integrals where $f(x)$ appears in the integrand—rational functions are usually less convenient to work with.

One difficulty with rational functions is that the parameters appear nonlinearly. It is not a linear-approximation problem, in the sense of Sec. 4.1.1. Interpolation by rational functions can, however, be formulated as a set of linear simultaneous equations (see Problem 7 at the end of this section), but for least-squares problems we have to refer to the methods of Sec. 10.5.4.

Spline functions (Sec. 4.6) are often useful if one wants the approximating function to approximate both a function and its derivative.

Another important and easily used class of functions is the **trigonometric polynomials;** they are more closely studied in Chap. 9.

The general fitting of **exponentials**—e.g.,

$$f(t) = c_1 + c_2 e^{c_3 t} + c_4 e^{c_5 t},$$

where all five coefficients are undetermined—is, however, a fairly difficult nonlinear-approximation problem; see Example 10.5.1 and Sec. 10.5.4.

2. Mathematically, a polynomial of degree m is uniquely determined by $m + 1$ function values. But if the degree of a polynomial is high, the polynomial can react quite strongly to disturbances in the input data. Hence, when $m + 1$ *equidistant* function values are given, one should not choose the degree to be much larger than $2m^{1/2}$. The sensitivity to perturbations is especially great in the outer parts of the interval if one chooses the degree too large. If the abscissae have a distribution similar to that of the zeros of the Tchebycheff polynomial $T_{m+1}(x)$, then there is usually no risk in choosing $n = m$.

Often a function is specified with data other than function values; in this case it may be that the degree should be chosen much lower than the total number of input data. *Experimental perturbational calculations* (see Sec. 2.2.4, and the last part of Sec. 4.3.4) are a general aid for investigating how output data react to disturbances in the input data.

3. This question has not been considered much previously. Let us take an example. The forms

$$\sum_{j=0}^{10} a_j x^j, \qquad \sum_{j=0}^{10} b_j \left(\frac{9 + x}{10}\right)^j, \qquad \sum_{j=0}^{10} c_j T_j(x),$$

for the representation of the same polynomial in $[-1, 1]$ are mathematically equivalent. However, from a numerical viewpoint—when the coefficients have a limited number of digits—these representations have very different properties. Of the three formulas $\{c_j\}_0^{10}$ clearly gives the best representation, and $\{b_j\}_0^{10}$ is clearly worst.

We shall not analyze the above example completely here. The essential point is that there is an approximate linear dependence between the members of the family $\{((9 + x)/10)^j\}_0^{10}$ and to some extent also between the members of the family $\{x^j\}_0^{10}$. (This assertion will be motivated below.) This gives rise to numerical difficulties, especially in the *computation* of the coefficients $\{b_j\}$ and $\{a_j\}$ respectively. As a rule, an ill-conditioned system of equations must be solved, particularly if the function being treated is defined by means of numerical function values. There can also be difficulties with the *use* of the corresponding polynomial representations for computing values of the function; see Example 2.4.5.

The above difficulties are similar to what can happen in determining position in a plane when one uses a coordinate system whose two axes are nearly parallel and whose directions are known only to limited accuracy.

The functions x^j, $((9 + x)/10)^j$, and $T_j(x)$ all have a maximum norm of 1 on $[-1, 1]$, so the following comparison is a fair one: Using property 2 (see Sec. 4.4.1) of the Tchebycheff polynomials, we see that $x^{10} = 2^{-9}T_{10}(x) +$ (a polynomial of lower degree). Thus x^{10} can be expressed as a linear combination of the other members of the family $\{x^j\}_0^{10}$ with an error whose magnitude is less than $2^{-9} \approx 2 \cdot 10^{-3}$! In the same way, $((9 + x)/10)^{10} = (x/10)^{10}$ + (a polynomial of lower degree) $= 10^{-10}2^{-9}T_{10}(x)$ + (a polynomial of lower degree). Thus $((9 + x)/10)^{10}$ can be expressed as a linear combination of the rest of the members of the family with an error whose magnitude is less than $2 \cdot 10^{-13}$! On the other hand, $T_{10}(x)$ cannot be expressed as a linear combination of the other members of the family with an error which is less than 1. Such a result would contradict the minimax property of the Tchebycheff polynomials (Sec. 4.4.1, property 7).

4. Concerning the fourth question, see Secs. 4.4.2 and 4.4.3 and the discussion on overdetermined linear equation systems in Chap. 5.

4.5.2. Some Inequalities for $E_n(f)$ with Applications to the Computation of Linear Functionals

THEOREM 4.5.1

If $|f^{n+1}(t)| \leq M$ *for all* $t \in [a, b]$, *then*

$$E_n(f) \leq \frac{2M}{(n+1)!}\left(\frac{b-a}{4}\right)^{n+1} \quad \text{on} \quad [a, b].$$

Proof. Let $p(t)$ be the polynomial which interpolates $f(t)$ in the points

$$t_k = \frac{1}{2}(b + a) + \frac{1}{2}(b - a) \cos\left(\frac{2k + 1}{n + 1}\frac{\pi}{2}\right).$$

We shall see that $|f(t) - p(t)|$ does not exceed the bound for $E_n(f)$ given in the theorem. Using the error term in interpolation (Theorem 4.3.2) we have

$$|f(t) - p(t)| \leq \max \frac{|f^{(n+1)}|}{(n+1)!} \cdot \prod_{k=0}^{n} |t - t_k|.$$

Set $t = \frac{1}{2}(b + a) + \frac{1}{2}(b - a)x$. Then $x \in [-1, 1]$, and

$$|f(t) - p(t)| \leq \frac{M}{(n+1)!} \prod_{k=0}^{n} \left|\frac{1}{2}(b - a)\left(x - \cos\frac{2k+1}{n+1}\frac{\pi}{2}\right)\right|.$$

But

$$\cos\left(\frac{2k+1}{n+1}\frac{\pi}{2}\right)$$

are the zeros of the Tchebycheff polynomial $T_{n+1}(x)$, whose leading coefficient is 2^n. Thus

$$|f(t) - p(t)| \leq \frac{M}{(n+1)!}\left(\frac{1}{2}(b - a)\right)^{n+1}\left|\frac{T_{n+1}(x)}{2^n}\right|$$

$$\leq \frac{M}{(n+1)!}\left(\frac{1}{2}(b - a)\right)^{n+1} 2^{-n},$$

from which the theorem follows.

Example 4.5.1

According to the above theorem, the function e^t on the interval [0, 1] can be approximated by a fifth-degree polynomial with an error whose magnitude does not exceed

$$\frac{2e}{6!}\left(\frac{1}{4}\right)^6 \approx 2\cdot 10^{-6}.$$

This accuracy is attained (according to the proof of the theorem) by Tchebycheff interpolation, after a linear transformation of variable using Eq. (4.4.8) which maps [0, 1] onto [−1, 1]. If one instead uses a Taylor series about $t = 0$, then the remainder is $e^\theta/(6!) \geq 1.3\cdot 10^{-3}$; with a Taylor series about $t = \frac{1}{2}$ the remainder is $e^\theta/(\frac{1}{2})^6/(6!) \geq 2\cdot 10^{-5}$, significantly less accurate than Tchebycheff interpolation. Using economization of power series (see Sec. 4.5.4), one gets about the same accuracy as with Tchebycheff interpolation.

Another *upper bound for* $E_n(f)$ is given in a *theorem of Jackson* (see, e.g., Cheney [36]): if $|f^{(k)}(x)| \leq M$ for $x \in [-1, 1]$, then

$$E_n(f) \leq \frac{M(\pi/2)^k}{(n+1)n\cdot\ldots\cdot(n-k+2)}, \quad \text{if} \quad n \geq k.$$

Thus $E_n(f) = O(n^{-k})$ as $n \to \infty$, if f has continuous derivatives up to the kth order. The more regular f is, the quicker $E_n(f)$ decreases. A *theorem of Bernstein* says that if f is analytic in an ellipse in the complex plane whose foci are -1 and 1 and the sum of whose semiaxes is R, and if $|f(z)| \leq M$ on the boundary of the ellipse, then $E_n(f)$ decreases at least exponentially. More exactly (see [55]),

$$E_n(f) \leq \frac{R^{-n}2M}{R-1}.$$

Let $f(x)$ be a real function and consider the expressions

$$3f(0) - 2f(1), \qquad \int_0^1 e^{-x}f(x)\,dx, \qquad f(0) + f'(0).$$

The values of these expressions depend on the choice of the function f. The two first expressions have a meaning for all continuous functions, the third requires differentiability. Such expressions are called **functionals**—*real-valued functions with a function as argument*. Denote any one of the above three expressions by $L(f)$ or Lf. Clearly it is true that if α is a real constant, then

$$L(\alpha f) = \alpha L(f), \qquad L(f+g) = Lf + Lg.$$

A functional which fulfills the above two conditions is called *linear*. $\|f\|$ and $E_n(f)$ are functionals, but they are not linear.

If one has defined a norm for functions, then the **norm of a linear functional**

$\|L\|$ is defined by

$$\|L\| = \sup_{\|f\|=1} |Lf|$$

From this it follows that

$$|Lf| \leq \|L\| \cdot \|f\|.$$

Example 4.5.2

If we choose $\|f\|$ to be the maximum norm on $[-1, 1]$, then in the space C^1 of functions with continuous first derivative we have

$$L(f) = \sum_{i=0}^{m} a_i f(x_i) \Rightarrow \|L\| = \sum_{i=0}^{m} |a_i|,$$

$$L(f) = 3f(0) - 2f(1) \Rightarrow \|L\| = 3 + 2 = 5,$$

$$L(f) = \int_0^1 e^{-x} f(x)\, dx \Rightarrow \|L\| = \int_0^1 e^{-x}\, dx = 1 - e^{-1},$$

$$L(f) = f(0) + f'(0) \Rightarrow \|L\| = \infty.$$

See Problem 1 at the end of this section for the derivation of the above results.

Many formulas for numerical computation of derivatives or integrals are based on linear expressions which give exact results for all polynomials of a certain degree. Thus it is of interest to study linear functionals which are zero for all nth-degree polynomials. The following theorem often gives a convenient estimate of the remainder term in such formulas:

THEOREM 4.5.2

If L is a linear functional for which $L(p) = 0$ for all nth-degree polynomials p, then (for all f)

$$|L(f)| \leq \|L\| \cdot E_n(f) \qquad (\textit{with the maximum norm}).$$

Proof. Let $\hat{p}$ be an nth-degree polynomial for which $\|f - \hat{p}\| = E_n(f)$. Then

$$|L(f)| = |L(f) - L(\hat{p})| = |L(f - \hat{p})| \leq \|L\| \cdot \|f - \hat{p}\| = \|L\| \cdot E_n(f).$$

This theorem has two different types of applications:

1. to give *error estimates for approximation formulas* when $E_n(f)$ can be estimated;
2. to give a simple *lower bound for* $E_n(f)$ when $L(f)$ can be easily computed.

Example 4.5.3

Set

$$L(f) = \tfrac{1}{2}h[f(0) + f(h)] - \int_0^h f(x)\, dx.$$

$L(f)$ means, then, the error in computing an integral with the trapezoidal rule in one step. Thus $L(p)$ is zero for all first-degree polynomials. Take $\|f\|$ to be the maximum norm on $[0, h]$. Then

$$\|L\| \leq \tfrac{1}{2}h(1 + 1) + \int_0^h 1\,dx = 2h.$$

From Theorems 4.5.2 and 4.5.1, it follows that

$$|L(f)| \leq 2hE_1(f) \leq 2h\frac{2\|f''\|}{2!}\left(\frac{h}{4}\right)^2 = \frac{\|f''\|h^3}{8}.$$

The general method for error estimation used above does not always give the best possible results, but as a rule it gives no gross overestimates. In the previous example one can show that $\|f''\|h^3/12$ is the best possible estimate.

Example 4.5.4

For all first-degree polynomials on $[-1, 1]$, the functional

$$L(f) = f(1) - 2f(0) + f(-1)$$

is zero. If the maximum norm is used, then $\|L\| = 1 + 2 + 1 = 4$. Thus

$$E_1(f) \geq \frac{|L(f)|}{\|L\|} = \frac{|f(1) - 2f(0) + f(-1)|}{4}.$$

From this result, it follows, for example, that the curve $y = e^x$ cannot be approximated by a straight line in $[-1, 1]$ with an error which is less than $(e - 2 + e^{-1})/4 \approx 0.271$.

The following generalization of the result of Example 4.5.4 can be derived and used in a similar way, in the interval $[-1, 1]$:

$$E_n(f) \geq 2^{-n-1}\left|\sum_{k=0}^{n+1}(-1)^k\binom{n+1}{k}f\left(-1 + \frac{2k}{n+1}\right)\right|. \tag{4.5.1}$$

The expression on the right-hand side is a difference of order $n + 1$ which (according to Theorem 7.1.3) is zero for all polynomials of degree n. There is another inequality which is usually sharper but less convenient to use than (4.5.1):

$$E_n(f) \geq \frac{1}{n+1}\left|\sum_{k=0}^{n+1}(-1)^k a_k f\left(\cos\frac{k\pi}{n+1}\right)\right|, \tag{4.5.2}$$

where

$$a_k = \begin{cases} \tfrac{1}{2} & \text{for} \quad k = 0, n + 1 \\ 1 & \text{otherwise.} \end{cases}$$

With the help of Theorem 4.4.3 one can also get lower bounds for $E_n(f)$ which can be used to show the impracticality of approximating a given function by an nth-degree polynomial in the entire interval $[-1, 1]$. If one wants to study the corresponding question in the interval $[a, b]$, then one should use the substitution (4.4.8) to get the "standard interval" $[-1, 1]$.

4.5.3. Approximation in the Maximum Norm

From the minimax property of Tchebycheff polynomials (Sec. 4.4.1, property 7) it follows that the best approximation in the maximum norm to the function $f(x) = x^n$ on $[-1, 1]$ by a polynomial of lower degree is given by the polynomial

$$f_n^*(x) = x^n - 2^{1-n}T_n(x). \tag{4.5.3}$$

Because of property 3 (Sec. 4.4.1), this polynomial has in fact degree $n - 2$. The error function $-2^{1-n}T_n(x)$ assumes its extrema 2^{1-n} in a sequence of $n + 1$ points, $x_i = \cos(i\pi/n)$. The sign of the error alternates at these points. (See Sec. 4.4.1, property 4.) Thus

$$E_{n-1}(f) = E_{n-2}(f) = 2^{1-n} \quad \text{for} \quad f(x) = x^n.$$

The above property can be generalized. We shall now give (without proof) a theorem which is the basis of many algorithms. (Notice that we replace n by $n + 1$ below.) The proof is given in [12] and [43].

THEOREM 4.5.3

Let f be a continuous function on $[a, b]$ and let $\hat{p}$ be the nth-degree polynomial which best approximates f in the maximum norm. Then $\hat{p}$ is characterized by the fact that there exist at least $n + 2$ points,

$$a \le \xi_0 < \xi_1 < \xi_2 < \ldots < \xi_{n+1} \le b,$$

where the error function $r = \hat{p} - f$ takes on its maximal magnitude with alternating signs; i. e.,

$$r(\xi_{i+1}) = -r(\xi_i), \quad (i = 0, 1, 2, \ldots, n), \qquad |r(\xi_i)| = \|r\|_\infty.$$

This characterization constitutes both a necessary and sufficient condition. If $\operatorname{sgn} f^{(n+1)}(x)$ *is constant in $[a, b]$ then $\xi_0 = a$, $\xi_{n+1} = b$.*

Example 4.5.5. *Approximation in the Maximum Norm of a Convex Function by a Straight Line (see Fig. 4.5.2)*

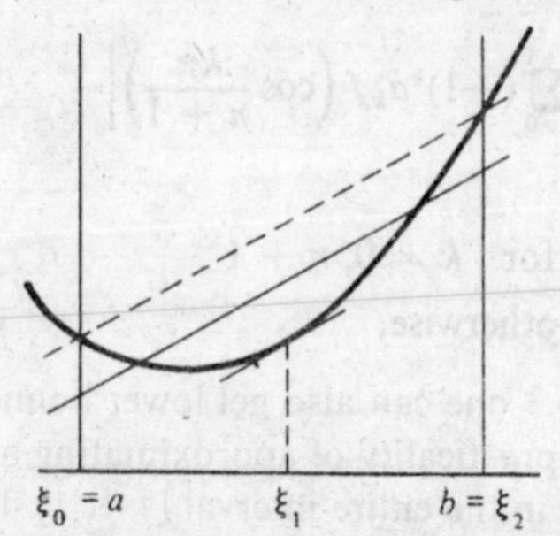

Fig. 4.5.2

Notice that linear interpolation (dotted line) using $y(a)$ and $y(b)$ gives a maximum error which is exactly twice as large as $E_1(f)$.

Approximation in the maximum norm is often called Tchebycheff approximation. We avoid this terminology, since Tchebycheff's name is also naturally associated with two entirely different approximation methods: interpolation in the Tchebycheff abscissae, and expansion in a series of Tchebycheff polynomials.

4.5.4. Economization of Power Series; Standard Functions

Suppose that one has used some simple method—e.g., Taylor expansion—to get a polynomial approximation to a given function. Then one can often get a polynomial approximation of lower degree, which still has an acceptable error, by successively replacing the highest powers x^n with $f_n^*(x)$ according to Eq. (4.5.3) (if the approximation is desired in $[-1, 1]$; if one has some other interval, then the substitution (4.4.8) should be made first). This process is called *economization of power series.*

Example 4.5.6

If the series expansion

$$\cos x = 1 - \frac{x^2}{2} + \frac{x^4}{24} - \cdots$$

is truncated after the x^2-term, then the maximum error becomes 0.042 in $[-1, 1]$. If one truncates after the x^4-term, the maximum error becomes 0.0014. Since $T_4(x) = 8x^4 - 8x^2 + 1$, it holds that

$$\frac{x^4}{24} \approx \frac{x^2}{24} - \frac{1}{192}$$

with an error which does not exceed $\max |T_4(x)|/192 = 0.0052$. Thus the approximation

$$\cos x \approx (1 - \tfrac{1}{192}) - x^2(\tfrac{1}{2} - \tfrac{1}{24}) = 0.99479 - 0.45833\, x^2$$

has an error whose magnitude in $[-1, 1]$ does not exceed $0.0052 + 0.0014 < 0.007$. This is less than one-sixth of the maximum error which was obtained when the power series was truncated after the x^2-term. Since $T_2(x) = 2x^2 - 1$, replacing x^2 by $\frac{1}{2}$ would increase the error by $\frac{1}{2}\cdot\frac{11}{24} < 0.23$.

For a function whose Taylor series converges slowly, there can be a significant reduction in degree. Note another thing in the above example: cos (0) is not approximated by 1 but by 0.99479. Even if an error of 0.007 is acceptable for some values of x, it is possible that this error is unacceptable for just that particular value of x, $(x = 0)$, in many situations. Exact values in important special cases and exact relations between functions can be lost in approximation. During the first ten years of computer technology similar things occurred in standard programs, but we hope that that belongs to the past. By reformulating the approximation problem, one can often take account of similar side conditions (e.g., where one wants certain exact relations to be preserved). In the above example, one could have asked for a polynomial approximation to $(1 - \cos x)/x^2$.

In computers, one often uses special properties of a given function in order to reduce the problem of computing the value of a function for an arbitrary value of the argument to the problem of computing its value for another argument. This is done in such a way that the new argument lies in an interval which is so small that a polynomial of low degree suffices there. If, for example, $\cos x$ is to be computed, then the machine can first compute a suitable comparison angle $\tilde{x}$ in the first quadrant from which one can easily compute $\cos \tilde{x}$. (In fact, this is what one does even when using tables!) Sometimes, however, it is more appropriate to approximate using a rational function instead of a polynomial. One common device in programs which compute standard functions is the use of different rational function approximations over different parts of a certain interval. But we shall not go into this subject more here; the interested reader is referred to Hart, Cheney, et. al. [41] and to Rice (ed.) [22].

4.5.5. Some Statistical Aspects of the Method of Least Squares

We mentioned previously that one of the motivations for the method of least squares was that it effectively reduces the influence of random errors in measurement.

Suppose that the values of a function have been measured in the points $x_0, x_1, \ldots, x_m$. Let $f(x_p)$ be the measured value, and let $\bar{f}(x_p)$ be the "true" (unknown) function value, which is assumed to be the same as the *expected value* of the measured value. Thus *no systematic errors* are assumed to be present. Suppose further, that *the errors in measurement at the various points are statistically independent*. Let the symbol E denote expected value; then we have

$$E\{f(x_p)\} = \bar{f}(x_p), \tag{4.5.4}$$

$$E\{[f(x_p) - \bar{f}(x_p)][f(x_q) - \bar{f}(x_q)]\} = \begin{cases} 0 & \text{if} \quad q \neq p \\ \sigma_p^2 & \text{if} \quad q = p. \end{cases} \tag{4.5.5}$$

The problem is to use the measured data to estimate the coefficients in the series

$$f(x) = \sum_{j=0}^{n} c_j \varphi_j(x),$$

where $\varphi_0, \varphi_1, \ldots, \varphi_n$ are known functions, $n \leq m$. Let c_j^* denote the result of an estimate of c_j. One can think of many different methods for computing c_j^*, all of which are more or less influenced by the measurement errors. It is natural (but not necessary) to demand that the following conditions be satisfied by any given method for computing an estimate c_j^*:

1. *No systematic errors in the estimates* (no bias): $E\{c_j^*\} = c_j$.
2. *Linearity:* The quantities c_j^* should be linear functions of the measurement data.

According to the **Gauss-Markov theorem,** the estimates which one gets by minimizing the sum

$$\sum_{p=0}^{m} w_p(f(x_p) - \sum_{j=0}^{n} c_j\varphi_j(x_p))^2, \quad \text{where} \quad w_p = \frac{1}{\sigma_p^2},$$

have a smaller variance than the values which one gets by any other estimation method which fulfills the conditions 1 and 2 above. This minimum property holds for estimates not just of the coefficients c_j, but also for every linear function of the coefficients—for example, the estimate

$$f_n^*(\alpha) = \sum_{j=0}^{n} c_j^*\varphi_j(\alpha) \tag{4.5.6.}$$

of the value $f(\alpha)$ at an arbitrary point α.

A derivation and a more general formulation of this theorem is given in Todd [24], Chap. 14.

Suppose now that $\sigma_p = \sigma$ for all p and that the functions $\{\varphi_j\}_{j=0}^n$ form an *orthogonal system* with respect to the inner product

$$(f, g) = \sum_{p=0}^{m} f(x_p)g(x_p), \quad \|f\|^2 = (f, f).$$

According to Eq. (4.2.5), then, with the use of the method of least squares,

$$c_j^* = \frac{(f, \varphi_j)}{\|\varphi_j\|^2}, \quad \bar{c}_j = \frac{(\bar{f}, \varphi_j)}{\|\varphi_j\|^2}.$$

We shall now see that *the estimates c_j^* and c_k^* are uncorrelated if $j \neq k$:*

$$E\{(c_j^* - \bar{c}_j)(c_k^* - \bar{c}_k)\} = E\left\{\frac{(f - \bar{f}, \varphi_j)}{\|\varphi_j\|^2}\frac{(f - \bar{f}, \varphi_k)}{\|\varphi_k\|^2}\right\}$$

$$= \frac{E\{\sum_{p=0}^{m}[f(x_p) - \bar{f}(x_p)]\varphi_j(x_p)\cdot\sum_{q=0}^{m}[f(x_q) - \bar{f}(x_q)]\varphi_k(x_q)\}}{(\|\varphi_j\|\cdot\|\varphi_k\|)^2}$$

$$= \sum_{p=0}^{m}\sum_{q=0}^{m}\frac{E\{[f(x_p) - \bar{f}(x_p)][f(x_q) - \bar{f}(x_q)]\}\varphi_j(x_p)\varphi_k(x_q)}{(\|\varphi_j\|\cdot\|\varphi_k\|)^2}.$$

Because of Eq. (4.5.5), only those terms for which $p = q$ give a contribution. Thus

$$E\{(c_j^* - \bar{c}_j)(c_k^* - \bar{c}_k)\} = \sum_{p=0}^{m} \frac{\sigma^2 \varphi_j(x_p)\varphi_k(x_p)}{(\|\varphi_j\| \cdot \|\varphi_k\|)^2} = \frac{\sigma^2(\varphi_j, \varphi_k)}{(\|\varphi_j\| \cdot \|\varphi_k\|)^2}$$

$$= \begin{cases} 0, & \text{if } j \neq k. \\ \left(\dfrac{\sigma}{\|\varphi_j\|}\right)^2 & \text{if } j = k \end{cases} \qquad (4.5.7)$$

As a by-product we get that *the variance of the estimate* c_j^* *is* $(\sigma/\|\varphi_j\|)^2$.

From this it follows that (if D^2 denotes "variance of"),

$$D^2\{f_n^*(\alpha)\} = D^2\{\sum_{j=0}^{n} c_j^* \varphi_j(\alpha)\} = \sum_{j=0}^{n} D^2\{c_j^*\} |\varphi_j(\alpha)|^2 = \sigma^2 \sum_{j=0}^{n} \frac{|\varphi_j(\alpha)|^2}{\|\varphi_j\|^2}.$$

As an average, *taken over the net of measurement points*, the variance of the smoothed function values is $\sigma^2(n+1)/(m+1)$, since

$$\frac{1}{m+1} \sum_{i=0}^{m} D^2\{f_n^*(x_i)\} = \sigma^2 \frac{1}{m+1} \sum_{j=0}^{n} \sum_{i=0}^{m} \frac{|\varphi_j(x_i)|^2}{\|\varphi_j\|^2} = \sigma^2 \frac{n+1}{m+1}.$$

Between the net points, however, the variance can in many cases be significantly larger. For example, when fitting a polynomial to measurements in *equidistant points*, the orthogonal polynomials (Gram polynomials—see Sec. 4.4.3) can be much larger between the net points if $j \gg m^{1/2}$. Set

$$\sigma_I^2 = \sigma^2 \sum_{j=0}^{n} \frac{1}{2} \int_{-1}^{1} \frac{|\varphi_j(\alpha)|^2 d\alpha}{\|\varphi_j\|^2}.$$

Thus σ_I^2 is an average variance for $f_n^*(\alpha)$, *taken over the entire interval* $[-1, 1]$. The following values for the ratio k between σ_I^2 and $\sigma^2\ (n+1)/(m+1)$ when $m = 41$ were obtained by H. Björk.

n	5	10	15	20	25	30	35
k	1.0	1.1	1.7	26	$7 \cdot 10^3$	$1.7 \cdot 10^7$	$8 \cdot 10^{11}$

These results are related to the previously made recommendation that one should choose $n < 2m^{1/2}$ when fitting a polynomial to equidistant data. This recommendation seems to contradict the Gauss-Markov theorem, but in fact it means that one just gives up the requirement that there be no systematic errors. Still it is remarkable that this can lead to such a drastic reduction of the variance of the estimates f_n^*.

If the measurement points are the *Tchebycheff abscissae*, then no difficulties arise in fitting a polynomial to the data. The orthogonal polynomials (Tchebycheff polynomials—see Sec. 4.4.1, 4.4.2) have in this case a magnitude between the net points which is not much larger than their magnitude at the net points. The average variance for $f_n^*(\alpha)$ becomes the same on the interval $[-1, 1]$ as on the net of measurement points, $\sigma^2\ (n+1)/(m+1)$.

The choice of n, when m is given, is a question of compromising between taking into account the systematic error (i.e., the truncation error, which decreases as n increases), and taking into account the random errors (which grow as n increases). In the Tchebycheff case, $|c_j|$ decreases quickly with j if f is a sufficiently smooth function, while the part of c_j^* which comes from errors in measurement varies randomly with magnitude of about $\sigma \cdot (2/(m+1))^{1/2}$, using Eq. (4.5.7) and property 6 of the Tchebycheff polynomials. The expansion should then be broken off when the coefficients begin to "behave randomly." The coefficients in an expansion in terms of Tchebycheff polynomials can hence be used for filtering away the "noise" from the signal, even when σ is initially unknown.

Example 4.5.7

Fifty-one equidistant values of a certain analytic function were rounded to four decimals. In Fig. 4.5.3, a semilog diagram (produced with a computer's line printer) is given which shows how $|c_j^*|$ varies in an expansion in terms of Tchebycheff polynomials of the above data. For $j > 20$ (approximately) the contribution due to noise dominates the contribution due to signal. Thus it is sufficient to break off the series at $n = 20$.

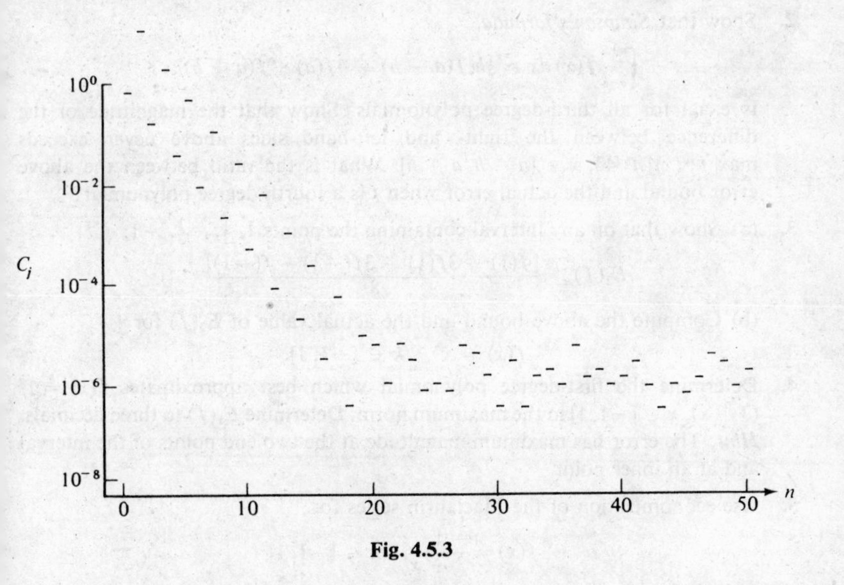

Fig. 4.5.3

REVIEW QUESTIONS

1. Give examples to show that the use of different sets of parameters to represent the same polynomial can be of varying suitability in numerical applications.

2. How quickly does $E_n(f)$ decrease (at least) as $n \longrightarrow \infty$, if f has a continuous derivative of order k?

3. What is a functional? Give examples of linear and nonlinear functionals. What is meant by the norm of a linear functional?

4. One of the formulas in the text is
$$|L(f)| \leq \|L\| \cdot E_n(f).$$
What assumption is made about the functional L? Derive the formula and give two differing examples of its use.

5. What is the characteristic property of the nth-degree polynomial which best approximates a given continuous function in the maximum norm on a closed bounded interval?

6. State the Gauss-Markov theorem.

PROBLEMS

1. In Example 4.5.2, give in each of the four cases a function $f \in C^1[-1, 1]$ such that $\|f\|_\infty = 1$ and $\|L\| = |Lf|$; i.e., show that the value given for $\|L\|$ is attained for these functions.

2. Show that *Simpson's formula*,
$$\int_{a-h}^{a+h} f(x)\,dx \approx \tfrac{1}{3}h(f(a-h) + 4f(a) + f(a+h)),$$
is exact for all third-degree polynomials. Show that the magnitude of the difference between the right- and left-hand sides above never exceeds $\max|f^{\text{iv}}(x)|h^5/48$, $x \in [a-h, a+h]$. What is the ratio between the above error bound and the actual error when f is a fourth-degree polynomial?

3. (a) Show that on any interval containing the points $1, \frac{1}{3}, -\frac{1}{3}, -1$,
$$E_2(f) \geq \frac{|f(1) - 3f(\frac{1}{3}) + 3f(-\frac{1}{3}) - f(-1)|}{8}.$$
(b) Compute the above bound and the actual value of $E_2(f)$ for
$$f(x) = x^3, \quad x \in [-1, 1].$$

4. Determine the first-degree polynomial which best approximates $f(x) = 1/(3 + x)$, $x \in [-1, 1]$ in the maximum norm. Determine $E_1(f)$ to three decimals. *Hint:* The error has maximum magnitude at the two end points of the interval and at an inner point.

5. Use economization of the Maclaurin series for
$$f(x) = \frac{1}{3 + x}, \quad x \in [-1, 1]$$
to determine a first-degree approximation to f. Compute the value of the error at $x = -1$.

6. Use economization of power series to compute a fourth-degree polynomial which approximates
$$f(x) = \int_0^x \frac{e^t - 1}{t}\,dt, \quad x \in [-1, 1],$$
with an error whose magnitude is less than $2 \cdot 10^{-4}$.

7. Interpolate a rational function of the form $(x - a)/(bx + c)$ through the points $(0, -2)$, $(1, -\frac{1}{3})$, $(-1, 3)$.

8. Compare, for large values of n, the bounds for $E_n(f)$ obtained by Theorem 4.5.1 and by the Bernstein and Jackson theorems quoted in Sec. 4.5.2, in the case

$$f(x) = e^x, \quad x \in [-1, 1].$$

Apply the bounds to the case $n = 13$.
Hint: Apply, in the former case, Stirling's formula, $n! \approx (n/e)^n\sqrt{2\pi n}$.

9. Prove the inequality in (4.5.2).

4.6. SPLINE FUNCTIONS

In engineering design, shipbuilders and others often use a **spline**, an elastic ruler which can be bent so that it passes through a given set of points (x_i, y_i), $i = 0, 1, \ldots, m$. Let $y = f(x)$ be the equation for the curve which is defined by the spline. Under certain assumptions (according to elasticity theory), $f(x)$ can be approximately described as being built up of different third-degree polynomials (cubic polynomials) in such a way that $f(x)$ and its two first derivatives are everywhere continuous. The third derivative, however, can have discontinuities at the points x_i. Such a function is called a **cubic spline function**, and the points x_i, $i = 0, 1, \ldots, m$ are called **nodes**.

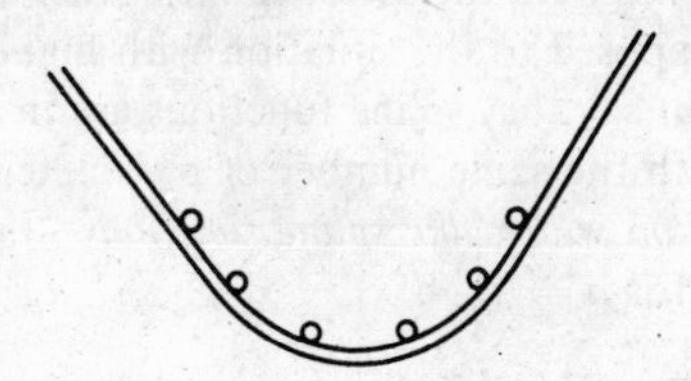

Fig. 4.6.1

The concept of spline function can be generalized in the following way.

DEFINITION

Let $a = x_0 < x_1 < \ldots < x_n = b$ be a subdivision of the interval $[a, b]$. A spline function of degree p with nodes at the points x_i, $i = 0, 1, \ldots, m$ is a function s with the properties:

(a) On each subinterval $[x_i, x_{i+1}]$, $i = 0, 1, \ldots, m - 1$, $s(x)$ is a polynomial of degree p.

(b) $s(x)$ and its first $(p - 1)$ derivatives are continuous on $[a, b]$.

Note that from the definition it follows that if $s_1(x)$ and $s_2(x)$ are spline functions of the same degree, so is $c_1s_1(x) + c_2s_2(x)$. Thus the set of spline functions forms a linear space.

Example 4.6.1

In the simplest case, $p = 1$, $s(x)$ is a *piecewise linear function*. Thus th first degree spline which interpolates given values y_i at the points x_i, $i = 0, 1, \ldots, n$ is uniquely determined. Introduce the "triangle-shaped" func tions (see Fig. 8.6.7 in Sec. 8.6.5)

$$\psi_i(x) = \begin{cases} 0 & , x \notin [x_{i-1}, x_{i+1}] \\ (x - x_{i-1})/(x_i - x_{i-1}), & x \in [x_{i-1}, x_i] \\ (x_{i+1} - x)/(x_{i+1} - x_i), & x \in [x_i, x_{i+1}], \end{cases}$$

and notice that

$$\psi_i(x_j) = \begin{cases} 0 & j \neq i \\ 1 & j = i. \end{cases}$$

Then the interpolating spline can be written as

$$s(x) = \sum_{i=0}^{m} y_i \psi_i(x).$$

An arbitrary first degree spline with nodes on the net $\{x_i\}_{i=0}^{m}$ obviously has unique representation in this form, i.e., the set of functions $\{\psi_i(x)\}_{i=0}^{m}$ form basis for the linear space of first degree splines.

During recent years, spline functions have been more and more used i numerical calculations. They were introduced by Schoenberg in 1946. Splin functions can be used in many contexts where a polynomial approximatio over an entire interval is insufficient. With the use of splines, one has no rea son to fear equidistant data (as opposed to the situation with higher-degre polynomials). However, one *cannot* say that spline functions are in *all* situa tions better than a polynomial with the same number of parameters.

We shall now study *interpolation with cubic spline functions*—i.e., deter mine spline functions f such that for $a = x_0 < x_1 < x_2 \ldots < x_m = b$,

$$f(x_i) = y_i, \quad (0 \leq i \leq m). \tag{4.6.1}$$

This problem always has infinitely many solutions.

THEOREM 4.6.1

Set

$$x_i - x_{i-1} = h_i, \qquad \frac{y_i - y_{i-1}}{h_i} = d_i,$$

$$\frac{x - x_{i-1}}{h_i} = t \quad \text{for} \quad x \in [x_{i-1}, x_i].$$

Every cubic spline function with nodes on the net $\{x_i\}_{i=0}^{n}$ *which fulfills the condi tion (4.6.1) is for* $x \in [x_{i-1}, x_i)$ *equal to a third-degree polynomial of the forn*

$$q_i(x) = ty_i + (1 - t)y_{i-1} + h_i t(1 - t)[(k_{i-1} - d_i)(1 - t) - (k_i - d_i)t],$$
$$(i = 1, 2, 3, \ldots, m), \tag{4.6.2}$$

where $k_0, k_1, \ldots, k_m$ satisfy the linear tridiagonal system of equations,

$$h_{i+1}k_{i-1} + 2(h_i + h_{i+1})k_i + h_i k_{i+1} = 3(h_i d_{i+1} + h_{i+1} d_i) \quad (i = 1, 2, 3, \ldots, m-1). \tag{4.6.3}$$

Since the above system has $m + 1$ unknowns but only $m - 1$ equations, two additional conditions are needed in order to uniquely determine the spline function.

The derivation is (briefly) as follows. The reader can verify the formulas below by substitution and differentiation:

$$\begin{aligned} &x = x_{i-1} \longleftrightarrow t = 0, \qquad x \to x_i \longleftrightarrow t \to 1, \qquad \frac{dt}{dx} = \frac{1}{h_i} \\ &q_i(x_{i-1}) = y_{i-1}, \qquad q_i(x_i) = y_i, \\ &q_i'(x_{i-1}) = k_{i-1}, \qquad q_i'(x_i) = k_i. \end{aligned} \tag{4.6.4}$$

Thus the parameter k_i is the derivative of the spline function at the point x_i. If we replace i by $i + 1$ everywhere in the above, we get

$$q_{i+1}'(x_i) = k_i, \qquad \text{thus} \qquad q_{i+1}'(x_i) = q_i'(x_i).$$

The system of equations (4.6.3) expresses that

$$q_{i+1}''(x_i) = q_i''(x_i), \quad (i = 1, 2, \ldots, m-1).$$

The continuity of both the first and second derivatives is thereby assured.

The two additional conditions mentioned in the theorem can be given in different ways. The most common way is based on the fact that the physical spline is straight outside the interval $[a, b]$—i.e., $f''(x) = 0$ for $x \leq a$ or $x \geq b$. Thus

$$q_1''(a) = q_m''(b) = 0.$$

This leads, after some computation, to the equations

$$\begin{aligned} 2k_0 + k_1 &= 3d_1 \\ k_{m-1} + 2k_m &= 3d_m. \end{aligned} \tag{4.6.5}$$

When Eq. (4.6.5) is applied for approximating a function whose second derivative is not zero at the boundary points ($x = a$ and $x = b$), one still gets a good fit near the center of the interval (error of order h^4 if all $h_i = h$), but in the outermost parts of the interval the error is commonly larger. By formulating the additional conditions in a different way one can get the (asymptotic) order of magnitude of the error to be uniformly $O(h^4)$. For the central parts of the interval, however, replacing Eq. (4.6.5) by some other conditions has little effect. See Problem 4 at the end of this section.

Equations (4.6.3) and (4.6.5) give rise to a *linear tridiagonal system of equations* for determining the derivatives k_i. *The total number of operations required to solve a system of equations of this particular type grows only linearly with m* (see Example 5.4.3, p. 166); hence, the work required to determine

a spline function is quite reasonable, even when the number of points i
large.,

Example 4.6.2

In a case with six equidistant points, the system of equations becomes (i
matrix form, after dividing Eq. (4.6.3) by h):

$$\begin{bmatrix} 2 & 1 & 0 & 0 & 0 & 0 \\ 1 & 4 & 1 & 0 & 0 & 0 \\ 0 & 1 & 4 & 1 & 0 & 0 \\ 0 & 0 & 1 & 4 & 1 & 0 \\ 0 & 0 & 0 & 1 & 4 & 1 \\ 0 & 0 & 0 & 0 & 1 & 2 \end{bmatrix} \begin{bmatrix} k_0 \\ k_1 \\ k_2 \\ k_3 \\ k_4 \\ k_5 \end{bmatrix} = 3 \begin{bmatrix} d_1 \\ d_2 + d_1 \\ d_3 + d_2 \\ d_4 + d_3 \\ d_5 + d_4 \\ d_5 \end{bmatrix}.$$

The designation "tridiagonal" is easily understood.

Example 4.6.3

For the curve in Example 4.1.4, the maximum norm of the error is 0.02
in interpolation with cubic spline functions at the eleven equidistant point
$x_i = -1 + 0.2i$, $0 \leq i \leq 10$. This good result contrasts sharply with th
unpleasant experience of interpolation with a tenth-degree polynomial i
Fig. 4.1.3.

Example 4.6.4

Spline functions can also be used for **surface fitting** in the following way
Consider in two dimensions the rectangular net consisting of points (x_i, y_j)
$i = 0, 1, \ldots, m$, $j = 0, 1, \ldots, n$. We can define a first degree spline func
tion on this net to be a function of the form

$$s(x, y) = \sum_{i=0}^{m} \sum_{j=0}^{n} y_{ij}\psi_i(x)\psi_j(y),$$

where we have used the basis functions in Example 4.6.1. Two-dimensiona
splines of higher degree can be similarly defined (see Problem 6).

Approximate values for derivatives, integrals, etc. of $f(x)$ can be obtainec
fairly easily by differentiating or integrating Eq. (4.6.2). Spline functions car
also be used successfully in the numerical treatment of boundary-value problems for differential equations. The concept of spline function can be generalized in many ways; for example, the continuity requirement in (b) can be
relaxed. For more complete information, the reader is referred to Ahlberg,
Nilson, and Walsh [35] and Schultz [59a].

REVIEW QUESTIONS

1. What is meant by a "cubic spline function"? Give an example where such a function is better suited than a polynomial for approximation over the whole interval.
2. About how many operations are required to interpolate a cubic spline function to $m + 1$ given values of a function ($m \gg 1$)?

PROBLEMS

1. Give the details of the derivation of Theorem 4.6.1.
2. (a) Show that the function

$$x_+^3 = \begin{cases} x^3 & x \geq 0 \\ 0 & x < 0 \end{cases}$$

is a cubic spline.

(b) Show that a cubic spline on the net $\{x_i\}_{i=0}^m$ has a unique representation

$$s(x) = p(x) + \sum_{i=1}^{m-1} c_i(x - x_i)_+^3,$$

where $p(x)$ is a third degree polynomial. (This representation is ill-conditioned and should *not* be used for numerical purposes.)

3. Show that the formula

$$\int_{x_0}^{x_m} f(x)\,dx = \frac{1}{2}\sum_{i=1}^{m} h_i(y_{i-1} + y_i) + \sum_{i=1}^{m} \frac{(k_{i-1} - k_i)h_i^2}{12},$$

is exact for cubic spline functions. (Notation according to Sec. 4.6.) How is the formula simplified if all $h_i = h$?
Hint: Integrate Eq. (4.6.2) from x_{i-1} to x_i.

4. One can get a more accurate alternative to the conditions in Eq. (4.6.5) if $f(x)$ is also known at $x = x_{1/2} = \frac{1}{2}(x_0 + x_1)$ and at $x = x_{m-1/2} = \frac{1}{2}(x_{m-1} + x_m)$, if one requires that $q'''(x)$ be continuous at these points. Show that this gives the equations

$$k_1 - k_0 = \frac{4(y_0 - 2y_{1/2} + y_1)}{h_1},$$

$$k_m - k_{m-1} = \frac{4(y_{m-1} - 2y_{m-1/2} + y_m)}{h_m}.$$

5. Cubic spline functions are sometimes used for the representation of periodic functions. Then $y_0 = y_m$, $d_0 = d_m$, $k_0 = k_m$. Set up the linear system for the determination of $k_1, k_2, \ldots, k_m$ in the equidistant case.

6. (a) Let $x_k = x_0 + kh$, $k = 0, 1, \ldots, m$ be given equidistant points. Determine a cubic spline $B_i(x)$ which satisfies the conditions

$$B_i(x_j) = \begin{cases} 0 & j < i - 1 \text{ or } j > i + 1 \\ 1 & j = i, \end{cases}$$

and

$$B_i'(x) = B_i''(x) = 0 \quad \text{for } x = x_{i-2} \text{ and } x = x_{i+2}.$$

Show that $B_i(x) = 0$ when $|x - x_i| \geq 2h$, and that $B_i(x) > 0$ for $|x - x_i| < 2h$.

(b) The functions $B_i(x)$ are called *B-splines* (*B*ell-shaped). Show that they form a set of basis functions, i.e., show that every cubic spline with nodes on the net $\{x_i\}_{i=0}^{m}$ has a unique representation (for $x_0 \leq x \leq x_m$) as

$$s(x) = \sum_{i=-1}^{m+1} a_i B_i(x).$$

(c) Describe how B-splines can be used for cubic spline interpolation in two dimensions on a rectangular net (x_i, y_j). (See Example 4.6.4.)

5 NUMERICAL LINEAR ALGEBRA

5.1. INTRODUCTION

Linear operators are the simplest type of mathematical operator and are natural models for an applied mathematician. Thus linear problems are very common in all applications, and it is important to be able to solve them quickly and accurately.

Let A be a given square matrix. The two fundamental problems of linear algebra are:

I. **solving the linear system of equations** $Ax = b$;

II. **solving the eigenvalue problem**—i.e., determining the eigenvalues λ_k and eigenvectors x_k such that $Ax_k = \lambda_k x_k$, $k = 1, 2, \ldots, n$.

It has been estimated that the solution of a linear system of equations enters in at some stage in about 75 percent of all scientific problems. Common sources are interpolation and approximation with linear families of functions and solution of differential equations with difference methods. Even in the solution of nonlinear systems of equations, one often approximates the nonlinear system with a sequence of linear equations (e.g., in Newton's method).

Eigenvalue problems stem from such physical problems as vibration problems, resonance (e.g., the flutter of airplane wings), criticality of reactors, and many others. Even statistical problems—e.g., factor analysis—can give rise to eigenvalue problems.

Linear algebra is a good illustration of the difference between classical mathematics and numerical analysis. Even though the theory has been known for centuries, decisive steps in the numerical treatment of problems I and II have first been made during the last few decades. The *explicit determinant formulas* for the inverse matrix and for the solution of a linear system of equations (Cramer's rule) are *very uneconomical* in numerical calculations, with the exception of 2×2 or 3×3 matrices or possibly matrices with very special structure.

Computer programs for the algorithms to be described here can be found in various periodicals and books. Programs of high quality are to be found in the periodical *Numerische Mathematik*, which publishes programs later to be given out in a handbook series. A volume on linear algebra edited by Wilkinson and Reinsch [86], has already been published. Since the treatment of small, critical details in algorithms can influence the accuracy of the result in a way that the neophyte can hardly be expected to see, the reader is recommended to make use of these publications (see also Sec. 13.5 for a list of references).

5.2. BASIC CONCEPTS OF LINEAR ALGEBRA

5.2.1. Fundamental Definitions

In this section we briefly review the basic definitions in linear algebra.

A matrix $\boldsymbol{A}$ is a collection of $m \times n$ real or complex numbers ordered in a scheme

$$\boldsymbol{A} = \begin{bmatrix} a_{11} & a_{12} & \dots & a_{1n} \\ a_{21} & a_{22} & \dots & a_{2n} \\ \dots \\ a_{m1} & a_{m2} & \dots & a_{mn} \end{bmatrix}.$$

The notation $(m \times n)$ means that the matrix has m rows and n columns. We shorten this by writing $\boldsymbol{A} = (a_{ij})$, $i = 1, 2, \dots, m$, $j = 1, 2, \dots, n$. If $m = n$, then the matrix is said to be square. A *column vector* $\boldsymbol{x} = (x_i)$, $i = 1, 2, \dots, m$ is a matrix which consists of just one column. We shall write matrices and vectors in boldface type, $\boldsymbol{A}, \boldsymbol{B}, \dots$, and $\boldsymbol{a}, \boldsymbol{b}, \dots$, and we shall often use Greek letters $\alpha, \beta, \dots$ to denote scalars.

One important class of matrices is the *diagonal matrices*

$$\boldsymbol{D} = \begin{bmatrix} d_1 & 0 & \dots & 0 \\ 0 & d_2 & \dots & 0 \\ \dots \\ 0 & 0 & \dots & d_n \end{bmatrix} = \text{diag}\,(d_1, d_2, \dots, d_n).$$

In particular, the unit matrix, or identity matrix $\boldsymbol{I}_n$, is defined by $\boldsymbol{I}_n = \text{diag}\,(1, 1, \dots, 1)$; i.e., $\boldsymbol{I}_n = (\delta_{ij})$, where δ_{ij} is the *Kronecker delta*, $\delta_{ij} = 0$ for $i \neq j$, $\delta_{ij} = 1$ for $i = j$.

Two matrices are said to be equal, $A = B$, if $a_{ij} = b_{ij}$, $i = 1, 2, \ldots, m$, $j = 1, 2, \ldots, n$. The product of a matrix A and a scalar α is a matrix $\alpha A = (\alpha a_{ij})$. The sum of two matrices, $C = A + B$ is a matrix with elements $c_{ij} = a_{ij} + b_{ij}$. The product of two matrices A $(m \times p)$ and B $(p \times n)$ is a matrix C $(m \times n)$ with elements

$$c_{ij} = \sum_{k=1}^{p} a_{ik} b_{kj}.$$

Matrix multiplication satisfies the rules

$$A(BC) = (AB)C, \qquad A(B + C) = AB + AC.\dagger$$

However, in general $AB \neq BA$, i.e., multiplication is not commutative.

The *transpose*, A^T, of a matrix A is the matrix whose rows are the columns of A. If we set $A^T = B = (b_{ij})$, then we have $b_{ij} = a_{ji}$. A column vector x can be written as $x = (x_1, x_2, \ldots, x_n)^T$—i.e., as the transpose of a row vector. For the transpose of a product, the rule $(AB)^T = B^T A^T$ holds. If each element of a matrix is replaced by its complex conjugate, then we get the complex conjugated matrix $\bar{A}$. *The matrix* $\bar{A}^T$ *is denoted by* A^H.

By a *triangular matrix, we mean a matrix of the form*

$$L = \begin{bmatrix} l_{11} & 0 & \ldots & 0 \\ l_{21} & l_{22} & \ldots & 0 \\ \ldots & & & \\ l_{n1} & l_{n2} & \ldots & l_{nn} \end{bmatrix} \quad \text{or} \quad R = \begin{bmatrix} r_{11} & r_{12} & \ldots & r_{1n} \\ 0 & r_{22} & \ldots & r_{2n} \\ \ldots & & & \\ 0 & 0 & \ldots & r_{nn} \end{bmatrix}.$$

L is said to be left-triangular or lower triangular, R right-triangular or upper triangular. Sums, products, and inverses of triangular matrices are again triangular matrices of the same type.

The *determinant* of a square matrix A is denoted by $\det(A)$. The following three rules are sufficient for computing $\det(A)$ for an arbitrary matrix A.

i. The value of the determinant is unchanged if one adds a row (column) multiplied by a number α to another row (column).
ii. The determinant of a triangular matrix is equal to the product of the elements in the main diagonal,

$$\det(R) = r_{11} r_{22} \ldots r_{nn}, \qquad \det(L) = l_{11} l_{22} \ldots l_{nn}.$$

iii. If two rows (columns) are interchanged, the value of the determinant is multiplied by (-1).

The following rules are also valid:

$$\det(A) = \det(A^T) \quad \text{and} \quad \det(AB) = \det(A)\det(B).$$

If $\det(A) \neq 0$, then A is said to be *nonsingular*. For every nonsingular matrix

†The number of arithmetic operations required to compute, respectively, the left- and right-hand sides of these equations can be very different.

A there exists an inverse matrix A^{-1} with the property $A^{-1}A = AA^{-1} = I$. For the inverse of a product, we have $(AB)^{-1} = B^{-1}A^{-1}$.

A symmetric matrix is a matrix which is equal to its transpose, $A = A^T$. The product of two symmetric matrices A and B is symmetric only if $AB = BA$. An orthogonal matrix Q is an $m \times n$ matrix such that $Q^TQ = I$. If $m = n$, then it follows that $Q^T = Q^{-1}$ and thus also that $QQ^T = I$. *Real* symmetric and orthogonal matrices play an important part in linear algebra. In the complex case, the corresponding classes are the Hermitian matrices, for which $A = A^H$, and the unitary matrices, for which $U^HU = I$. A real symmetric (Hermitian) matrix is called positive-definite if for the corresponding quadratic form, x^TAx, we have $x^TAx > 0$ for all real $x \neq 0$ ($x^HAx > 0$ for all $x \neq 0$).

5.2.2. Partitioned Matrices

It is often suitable to partition a matrix into blocks; e.g.,

$$\begin{bmatrix} 7 & 5 & 8 & 4 \\ 2 & 3 & 9 & 5 \\ 1 & 8 & 4 & 2 \end{bmatrix} = \left[\begin{array}{c|ccc} 7 & 5 & 8 & 4 \\ \hline 2 & 3 & 9 & 5 \\ 1 & 8 & 4 & 2 \end{array}\right] = \left[\begin{array}{cc|cc} 7 & 5 & 8 & 4 \\ \hline 2 & 3 & 9 & 5 \\ 1 & 8 & 4 & 2 \end{array}\right].$$

In this way, a matrix can be thought of as being built up of matrices of lower order

$$A = \begin{bmatrix} A_{11} & A_{12} & \cdots & A_{1n} \\ A_{21} & A_{22} & \cdots & A_{2n} \\ \cdots \\ A_{m1} & A_{m2} & \cdots & A_{mn} \end{bmatrix},$$

where A_{ij} is a $(p_i \times q_j)$ matrix. Such a matrix $A = (A_{ij})$ is called a *partitioned* matrix. Usually we are only interested in the case where the matrices in the diagonal, A_{ii}, are square. Then $m = n, p_i = q_i, i = 1, 2, \ldots, n$. For matrices partitioned in this way, addition and multiplication can be performed formally in the same way as if the blocks were scalars. For the product $C = AB$, we have $C = (C_{ij})$, where

$$C_{ij} = \sum_{k=1}^{n} A_{ik}B_{kj}.$$

A *block-diagonal* matrix is a matrix which can be written in partitioned form, as $A = \text{diag}\,(A_{11}, A_{22}, \ldots, A_{nn})$, where the A_{ii} are square matrices. One can define *block-triangular* matrices analogously. For such a (say, right-block-triangular) matrix it holds that $\det(R) = \det(R_{11}) \det(R_{22}) \cdots \det(R_{nn})$, and similarly for a left-block-triangular matrix.

In many cases it is desirable to avoid complex arithmetic. One possible way is to replace complex matrices and vectors by real ones with twice the order. Suppose, for example, that $A = B + iC$, $x = y + iz$, with real B, C,

, and z. Then A is associated with $\tilde{A}$ and x with $\tilde{x}$, where

$$\tilde{A} = \begin{pmatrix} B & -C \\ C & B \end{pmatrix}, \qquad \tilde{x} = \begin{pmatrix} y \\ z \end{pmatrix}.$$

It is easy to verify such rules of computation as $\widetilde{(Ax)} = \tilde{A}\tilde{x}$, $\widetilde{(AB)} = \tilde{A}\tilde{B}$, $\widetilde{A^{-1}}) = (\tilde{A})^{-1}$, etc.

5.2.3. Linear Vector Spaces

An ordered set of n real (or complex) numbers $(x_1, x_2, \ldots, x_n)$ define a *vector*. The set of all such vectors forms a *vector space* $\boldsymbol{R}_n$ (or $\boldsymbol{C}_n$) of dimension n. If the numbers x_i are interpreted as rectangular coordinates, then the *inner product* (scalar product) of two vectors x and y is defined (in the complex case) by

$$(x, y) = \bar{x}_1 y_1 + \bar{x}_2 y_2 + \ldots + \bar{x}_n y_n.$$

If x is considered as a column vector, then the inner product can be interpreted as a special case of matrix multiplication, and $(x, y) = x^H y$, which in the real case simplifies to $(x, y) = x^T y$.

The vector $y = c_1 x_1 + c_2 x_2 + \ldots + c_k x_k$ is said to be a *linear combination* of the vectors $x_1, x_2, \ldots, x_k$. These vectors are said to be *linearly independent* if $c_1 x_1 + c_2 x_2 + \ldots + c_k x_k = 0$ only if $c_1 = c_2 = \ldots = c_k = 0$. Otherwise, they are said to be *linearly dependent*. The set of vectors given by all possible linear combinations of $x_1, x_2, \ldots, x_k$ forms a *linear subspace* $\boldsymbol{R}$ in $\boldsymbol{R}_n$. We say that the vectors $x_1, x_2, \ldots, x_k$ *span* the space $\boldsymbol{R}$.

The maximum number of linearly independent vectors in $\boldsymbol{R}_n$ is n. Any set of n linearly independent vectors $y_1, y_2, \ldots, y_n$ form a *basis* for $\boldsymbol{R}_n$. Every vector x in $\boldsymbol{R}_n$ can then be written

$$x = \alpha_1 y_1 + \alpha_2 y_2 + \ldots + \alpha_n y_n,$$

where $\alpha_1, \alpha_2, \ldots, \alpha_n$ are called the coordinates of x with respect to the basis $y_1, y_2, \ldots, y_n$. The simplest example of a basis in $\boldsymbol{R}_n$ is given by the n columns in the unit matrix $I = (e_1, e_2, \ldots, e_n)$. Clearly, we have $x_1 e_1 + x_2 e_2 + \ldots + x_n e_n$.

A matrix A can be thought of as being built up of its column vectors or of its row vectors

$$A = (a_{\cdot 1}, a_{\cdot 2}, \ldots, a_{\cdot n}), \qquad A = (a_{1\cdot}^T, a_{2\cdot}^T, \ldots, a_{m\cdot}^T)^T.$$

The maximum number of linearly independent column vectors in A is equal to the maximum number of linearly independent row vectors in A. This number r is called the *rank* of A, and we write rank $(A) = r$. It follows immediately that $r \leq \min(m, n)$. In particular, if $r = m = n$, then A is nonsingular.

A linear system of equations with m equations and n unknowns

$$\left.\begin{aligned} a_{11}x_1 + a_{12}x_2 + \ldots + a_{1n}x_n &= b_1 \\ \ldots \\ a_{m1}x_1 + a_{m2}x_2 + \ldots + a_{mn}x_n &= b_m \end{aligned}\right\}$$

can be written in matrix notation as $\boldsymbol{Ax} = \boldsymbol{b}$. If $\boldsymbol{b} = \boldsymbol{0}$, then the system is said to be homogeneous. A homogeneous system of equations always has the trivial solution $\boldsymbol{x} = \boldsymbol{0}$. If rank $(\boldsymbol{A}) = r < n$, then the homogeneous system $\boldsymbol{Ax} = \boldsymbol{0}$ has $(n - r)$ linearly independent solutions.

The system of equations can be written in the form

$$x_1\boldsymbol{a}_{.1} + x_2\boldsymbol{a}_{.2} + \ldots + x_n\boldsymbol{a}_{.n} = \boldsymbol{b},$$

which expresses the right-hand side $\boldsymbol{b}$ as a linear combination of the column vectors of $\boldsymbol{A}$. From this representation, one can see that the condition that the equations have a solution (solvability criterion) is that $\boldsymbol{b} \in \boldsymbol{R}(\boldsymbol{A})$, where $\boldsymbol{R}(\boldsymbol{A})$ is the subspace spanned by the columns of $\boldsymbol{A}$. This can also be expressed as rank $(\boldsymbol{A}, \boldsymbol{b}) =$ rank $(\boldsymbol{A})$. If $m = n = r$, then $\boldsymbol{R}(\boldsymbol{A}) = \boldsymbol{R}_n$ and the solvability criterion is satisfied for every vector $\boldsymbol{b}$. In this case the solution is uniquely determined, and given by $\boldsymbol{x} = \boldsymbol{A}^{-1}\boldsymbol{b}$. If the solvability criterion is satisfied but with rank $(\boldsymbol{A}) = r < n$, then the nonhomogeneous equation $\boldsymbol{Ax} = \boldsymbol{b}$ has an $(n - r)$-dimensional manifold of solutions. All of the above results are also valid in the complex case for $\boldsymbol{C}_n$ instead of $\boldsymbol{R}_n$.

5.2.4. Eigenvalues and Similarity Transformations

If for a given number λ and a vector $\boldsymbol{x} \neq \boldsymbol{0}$ we have that $\boldsymbol{Ax} = \lambda\boldsymbol{x}$, then λ is said to be an *eigenvalue* of $\boldsymbol{A}$ and $\boldsymbol{x}$ an *eigenvector* belonging to λ. The equation $\boldsymbol{Ax} = \lambda\boldsymbol{x}$ can be written as $(\boldsymbol{A} - \lambda\boldsymbol{I})\boldsymbol{x} = \boldsymbol{0}$ and thus is a homogeneous system of equations for $\boldsymbol{x}$. This system has a solution $\boldsymbol{x} \neq \boldsymbol{0}$ only if det $(\boldsymbol{A} - \lambda\boldsymbol{I}) = 0$; this equation is called the *characteristic equation* for $\boldsymbol{A}$. Since det $(\boldsymbol{A} - \lambda\boldsymbol{I})$ is an nth-degree polynomial in λ, there are exactly n real or complex roots $\lambda_1, \lambda_2, \ldots, \lambda_n$ if every root is counted a number of times equal to its multiplicity.

Let $\boldsymbol{C}$ be a nonsingular matrix. Then the matrix $\boldsymbol{B} = \boldsymbol{C}^{-1}\boldsymbol{AC}$ is said to be *similar* to $\boldsymbol{A}$ and the transformation is called a *similarity transformation* of $\boldsymbol{A}$. Suppose now that λ and $\boldsymbol{x}$ are an eigenvalue and corresponding eigenvector of $\boldsymbol{A}$. Then we have

$$\boldsymbol{Ax} = \lambda\boldsymbol{x}, \quad \text{from which} \quad (\boldsymbol{C}^{-1}\boldsymbol{AC})\boldsymbol{C}^{-1}\boldsymbol{x} = \lambda\boldsymbol{C}^{-1}\boldsymbol{x},$$

i.e., λ is also an eigenvalue of $\boldsymbol{B}$ and the corresponding eigenvector is $\boldsymbol{y} = \boldsymbol{C}^{-1}\boldsymbol{x}$.

An arbitrary square matrix $\boldsymbol{A}$ has n eigenvalues and eigenvectors, $\boldsymbol{Ax}_i = \lambda_i\boldsymbol{x}_i$, $i = 1, 2, \ldots, n$. These n equations can be summarized as

$$\boldsymbol{AX} = \boldsymbol{X\Lambda}, \quad \text{where} \quad \boldsymbol{X} = (\boldsymbol{x}_1, \boldsymbol{x}_2, \ldots, \boldsymbol{x}_n), \qquad \boldsymbol{\Lambda} = \text{diag}\,(\lambda_1, \lambda_2, \ldots, \lambda_n).$$

If the eigenvectors $x_1, x_2, \ldots, x_n$ are linearly independent, then the matrix X is nonsingular, and we get $\Lambda = X^{-1}AX$. Thus a similarity transformation by X transforms A to *diagonal form;* in this case A is said to be diagonalizable. If all the eigenvalues are distinct—i.e., $\lambda_i \neq \lambda_j$ for $i \neq j$—then the eigenvectors are always linearly independent. However, there exist matrices with eigenvalues of multiplicity > 1 which cannot be diagonalized. The simplest example of such a matrix is

$$A = \begin{pmatrix} 0 & 1 \\ 0 & 0 \end{pmatrix}.$$

All of the eigenvalues of a *real, symmetric* matrix are real. Furthermore, the eigenvectors x_i and x_j corresponding to two noncoinciding eigenvalues $\lambda_i \neq \lambda_j$ are orthogonal—i.e., $x_i^T x_j = 0$. Also, one can always choose eigenvectors corresponding to coincident eigenvalues such that the eigenvectors are orthogonal. (*Example:* the unit matrix I has every vector as eigenvector corresponding to the eigenvalue 1.) Thus *for every symmetric matrix, there is an orthogonal matrix* X, $X^T X = I$, *such that* $\Lambda = X^T AX$. A corresponding result holds also in the complex case, for Hermitian matrices. Hermitian matrices always have real eigenvalues, and there exists a *unitary* matrix U, $U^H U = I$, which diagonalizes A by a similarity transformation, $\Lambda = U^H AU$.

From $Ax = \lambda x$, it follows that $(A + cI)x = (\lambda + c)x$, $A^2x = \lambda^2 x$, and if $\lambda \neq 0$, $A^{-1}x = (1/\lambda)x$. Thus the matrices $A^{\pm n}$ have eigenvalues $\lambda^{\pm n}$ and the matrix $(A + cI)$ has eigenvalues $\lambda + c$. If $P(z) = a_0 z^n + a_1 z^{n-1} + \ldots + a_n$, then the matrix $P(A)$ has the eigenvalues $P(\lambda)$.

5.2.5. Singular-Value Decomposition and Pseudo-Inverse

Finally, we shall mention two modern concepts of growing importance for numerical analysis—namely, **singular-value decomposition** and **pseudo-inverse**.

Let A be an $m \times n$ matrix of rank r. Then there exists an $m \times m$ unitary matrix U, an $n \times n$ unitary matrix V, and an $r \times r$ diagonal matrix D with strictly positive elements, called the **singular values** of A, such that

$$A = U\Sigma V^H, \qquad \Sigma = \begin{bmatrix} D & 0 \\ 0 & 0 \end{bmatrix}. \tag{5.2.1}$$

(Σ is an $m \times n$ matrix.)† If $r = m$ or $r = n$, then there will be no zero matrices in Σ at the bottom or at the right, respectively. If $r = m = n$, then $\Sigma = D$. In many discussions this decomposition can be used instead of the diagonal form previously mentioned, one advantage being that the singular values are fairly insensitive to perturbations in the matrix elements, while the eigen-

†Sometimes the zeros on the prolonged diagonal D are also included in the singular values.

values of certain unsymmetric matrices are very sensitive; see Problem 4 in this section. Note that

$$AA^H = U\Sigma^2 U^H, \qquad A^H A = V\Sigma^2 V^H.$$

Hence, in principle, the singular-value decomposition can be reduced to the eigenvalue problem for Hermitian matrices, but there exist better algorithms; see the article by Golub and Reinsch in [86, pp. 134–51], where applications are also mentioned.

Let the singular-value decomposition of A be given by Eq. (5.2.1). Then the **pseudo-inverse** A^I is the $n \times m$ matrix defined by the formula

$$A^I = V\Sigma^I U^H, \quad \Sigma^I = \begin{bmatrix} D^{-1} & 0 \\ 0 & 0 \end{bmatrix} \tag{5.2.2}$$

It is easy to show (see Problem 5 in this section) that the following properties are valid for $G = A^I$:

$$AGA = A, \quad GAG = G, \quad GA \text{ and } AG \text{ are Hermitian.} \tag{5.2.3}$$

Penrose has shown that $G = A^I$ *is the only matrix that has all these properties.* This is very useful. For example, in order to establish the important result that *if* $r = n$, *then*

$$A^I = (A^H A)^{-1} \cdot A^H, \tag{5.2.4}$$

one has only to verify (5.2.3); see Problem 5(b). Some properties of the usual inverse can be generalized—e.g.,

$$(A^I)^I = A, \qquad (A^H)^I = (A^I)^H. \tag{5.2.5}$$

However *there are matrices for which* $(AB)^I \neq B^I A^I$. But equality holds in several interesting cases; see Problem 9.

More about generalized inverses and applications can be found in Bjerhammar[62] and Rao and Mitra [77]. Both work with a more general concept, which can be specialized depending on the purpose. Actually, in their most general formulation only the first of the relations in (5.2.3) is required, though it does not determine G uniquely. It is shown in [77, p. 20] that this is equivalent to defining a generalized inverse as an $n \times m$ matrix G such that $x = Gy$ is a solution of $Ax = y$ for any y, which makes that equation consistent. In the terminology of [77], the pseudo-inverse defined above is called the **Moore-Penrose inverse**.

PROBLEMS

1. (a) Show that the determinant of a matrix is equal to the product of its eigenvalues.

(b) Show that the matrix

$$\begin{bmatrix} 0 & 1 \\ 0 & 0 \end{bmatrix}$$

cannot be transformed into diagonal form by a similarity transformation.

(c) Are there any diagonal or diagonalizable matrices with multiple eigenvalues?

2. (a) Suppose that $A = X\Lambda X^{-1}$, where Λ is a diagonal matrix. Find a similarity transformation Y that diagonalizes the matrix

$$\begin{bmatrix} 0 & A \\ A & 0 \end{bmatrix}.$$

(b) Show that the same transformation diagonalizes any matrix of the form

$$B = \begin{bmatrix} P(A) & Q(A) \\ Q(A) & P(A) \end{bmatrix},$$

where P, Q are arbitrary polynomials and express the eigenvalues of B in terms of the eigenvalues of A.

3. Let A be a given **square** matrix. Show that if $Ax = y$ has **at least one** solution for any y, then it has **exactly one** solution for any y. (This is a useful formulation for showing uniqueness of approximation formulas; see Problem 9 of Sec. 7.7.)

4. (a) Show that the so-called **companion matrix**

$$A = \begin{bmatrix} 0 & 1 & 0 & \cdots & 0 \\ 0 & 0 & 1 & \cdots & 0 \\ \cdots & & & \cdots & \\ 0 & 0 & 0 & \cdots & 1 \\ -a_k & -a_{k-1} & -a_{k-2} & \cdots & -a_1 \end{bmatrix}.$$

has eigenvectors of the form

$$x = (1, \lambda, \lambda^2, \ldots, \lambda^{k-1})^T,$$

where λ is an eigenvalue. Show also that the characteristic equation is

$$\lambda^k + a_1\lambda^{k-1} + \ldots + a_k = 0.$$

Are there any other eigenvectors, when λ is a multiple root?

(b) Suppose that $k = 20$, $a_1 = a_2 = \ldots = a_{19} = 0$. Compute the eigenvalues when $a_{20} = 0$ and when $a_{20} = -2^{-40} \approx -10^{-12}$.

(c) Construct an example where a perturbation of order of magnitude 2^{-40} in a matrix element that is equal to 1 has as large an effect in an eigenvalue as in (b).

5. Show that the relations in (5.2.3) are satisfied by $G = A^I$:

(a) when A^I is defined by Eq. (5.2.2);

(b) when A^I is given by Eq. (5.2.4);

(c) Derive Eq. (5.2.4) from Eq. (5.2.2), when $r = n$.

6. Calculate the pseudo-inverse of

(a) a column vector x;

(b) an $m \times n$ matrix B, where $r = m$.

7. (a) Prove the formulas in (5.2.5).

(b) Show that

$$A^I A = I \Longleftrightarrow r = n, \qquad AA^I = I \Longleftrightarrow r = m.$$

8. (a) Show that $A^I = A^{-1}$, when A is a nonsingular square matrix.
(b) Construct an example where $G \neq A^I$ despite the fact that $GA = I$.

9. (a) Construct an example where $(AB)^I \neq B^I A^I$.
(b) Show that if A is an $m \times r$ matrix, B an $r \times n$ matrix, rank $(A) =$ rank $(B) = r$, then

$$(AB)^I = B^I A^I.$$

10. Let A be a real matrix, and denote by $R(A)$ the subspace of R_m spanned by the columns of A. Put $P = AA^I$. Show the following relations:
(a) $P = P^H$;
(b) $P \cdot (I - P) = 0$;
(c) $x \in R(A) \Longrightarrow Px = x$;
(d) $x \in R(A) \wedge y \in R_m \Longrightarrow x^H(y - Py) = 0$.
Because of the last relations, P is called the **orthogonal projector** onto $R(A)$.

5.3. DIRECT METHODS FOR SOLVING SYSTEMS OF LINEAR EQUATIONS

By a **direct method** for solving a system of linear equations we mean a method which after a certain finite number of steps gives the exact solution, disregarding rounding errors. For systems $Ax = b$ where the matrix A is **full** (that is, most of the elements of A are nonzero), direct elimination methods are almost always the most efficient. However, when A is **sparse** (that is, a large proportion of the elements of A are zero), **iterative methods** offer certain advantages, and for some very large sparse systems they are indispensable. Iterative methods give a sequence of approximate solutions, converging when the number of steps tends to infinity. They may give useful results with fewer arithmetic operations than direct methods, but this is true only for systems with special properties. Thus the choice between direct and iterative methods depends on the proportion and distribution as well as sign and size of the nonzero elements of A.

In this chapter we shall mainly consider direct elimination methods. Iterative methods are treated briefly in Sec. 5.6.1.

5.3.1. Triangular Systems

Linear systems of equations where the matrix is triangular are particularly simple to solve. A system $Ux = b$, where the matrix U is upper-triangular, has the form

$$\left.\begin{aligned} u_{11}x_1 + \ldots + u_{1,n-1}x_{n-1} + u_{1n}x_n &= b_1 \\ \ldots \qquad\qquad & \\ u_{n-1,n-1}x_{n-1} + u_{n-1,n}x_n &= b_{n-1} \\ u_{nn}x_n &= b_n \end{aligned}\right\}$$

If we assume that $u_{ii} \neq 0$, $i = 1, 2, \ldots, n$, then the unknowns can be computed in the order $x_n, x_{n-1} \ldots, x_1$ from

$$\begin{cases} x_n = \dfrac{b_n}{u_{nn}} \\ x_{n-1} = \dfrac{b_{n-1} - u_{n-1,n}x_n}{u_{n-1,n-1}} \\ \cdots \\ x_1 = \dfrac{b_1 - u_{1n}x_n - u_{1,n-1}x_{n-1} - \ldots - u_{12}x_2}{u_{11}} \end{cases}$$

This can be written in more compact form as

$$x_i = \frac{b_i - \sum_{k=i+1}^{n} u_{ik}x_k}{u_{ii}}, \quad i = n, n-1, \ldots, 1 \tag{5.3.1}$$

Since the unknowns are solved for in backward order, this algorithm is called **back substitution**.

A linear system which has lower-triangular form $\boldsymbol{Lx} = \boldsymbol{b}$ can be solved in a similar way. Assuming that $l_{ii} \neq 0$, $i = 1, 2, \ldots, n$, the unknowns can be solved for by **forward substitution**,

$$x_i = \frac{b_i - \sum_{k=1}^{i-1} l_{ik}x_k}{l_{ii}}, \quad i = 1, 2, \ldots, n. \tag{5.3.2}$$

From the formulas above it follows that the solution of a triangular system of equations takes n divisions and

$$\sum_{i=1}^{n} (i-1) = \tfrac{1}{2}n(n-1) \approx \tfrac{1}{2}n^2$$

additions and multiplications. This is almost exactly the same amount of work as for multiplying a vector by a triangular matrix !

5.3.2. Gaussian Elimination

Most important among the direct methods for solving a general linear system is Gaussian elimination. The idea behind this method is to eliminate the unknowns in a systematic way, so that we end up with a triangular system, which we know how to solve. Consider the system

$$\left.\begin{aligned} a_{11}x_1 + a_{12}x_2 + \ldots + a_{1n}x_n &= b_1 \\ a_{21}x_1 + a_{22}x_2 + \ldots + a_{2n}x_n &= b_2 \\ \cdots \\ a_{n1}x_1 + a_{n2}x_2 + \ldots + a_{nn}x_n &= b_n \end{aligned}\right\} \tag{5.3.3}$$

We assume in the following that the matrix $\boldsymbol{A} = (a_{ij})$ is nonsingular. Then the system $\boldsymbol{Ax} = \boldsymbol{b}$ has a unique solution.

Now assume that $a_{11} \neq 0$. Then we can eliminate x_1 from the last $(n-1)$ equations by subtracting from the ith equation the multiple

$$m_{i1} = \frac{a_{i1}}{a_{11}}, \quad i = 2, 3, \ldots, n$$

of the first equation. Then the last $(n-1)$ equations become

$$\left.\begin{aligned} a_{22}^{(2)}x_2 + \ldots + a_{2n}^{(2)}x_n &= b_2^{(2)} \\ \ldots \\ a_{n2}^{(2)}x_2 + \ldots + a_{nn}^{(2)}x_n &= b_n^{(2)} \end{aligned}\right\}$$

where the new coefficients are given by

$$a_{ij}^{(2)} = a_{ij} - m_{i1}a_{1j}, \qquad b_i^{(2)} = b_i - m_{i1}b_1, \qquad i = 2, 3, \ldots, n.$$

This is a system of $(n-1)$ equations in the $(n-1)$ unknowns $x_2, x_3, \ldots, x_n$. If $a_{22}^{(2)} \neq 0$, we can in a similar way eliminate x_2 from the last $(n-2)$ of these equations. We then get a system of $(n-2)$ equations in the unknowns $x_3, \ldots, x_n$. If we put

$$m_{i2} = \frac{a_{i2}^{(2)}}{a_{22}^{(2)}}, \quad i = 3, \ldots, n,$$

the coefficients of this system are given by

$$a_{ij}^{(3)} = a_{ij}^{(2)} - m_{i2}a_{2j}^{(2)}, \qquad b_i^{(3)} = b_i^{(2)} - m_{i2}b_2^{(2)}, \qquad i = 3, \ldots, n.$$

The elements $a_{11}, a_{22}^{(2)}, a_{33}^{(3)}, \ldots$ which appear during the elimination are called **pivotal elements.** If all these are nonzero, we can continue the elimination, until after $(n-1)$ steps we get the single equation

$$a_{nn}^{(n)}x_n = b_n^{(n)}.$$

We now collect the first equation from each step and get

$$\left.\begin{aligned} a_{11}^{(1)}x_1 + a_{12}^{(1)}x_2 + \cdots + a_{1n}^{(1)}x_n &= b_1^{(1)} \\ a_{22}^{(2)}x_2 + \ldots + a_{nn}^{(2)}x_n &= b_2^{(2)} \\ \ldots \\ a_{nn}^{(n)}x_n &= b_n^{(n)} \end{aligned}\right\}, \tag{5.3.4}$$

where we have introduced the notations $a_{ij}^{(1)} = a_{ij}$, $b_i^{(1)} = b_i$ for the coefficients in the original system. Thus we have reduced Eq. (5.3.3) to the triangular system of Eq. (5.3.4), which as we have seen can easily be solved.

We note that the right-hand side $\boldsymbol{b}$ is transformed in exactly the same way as the columns in $\boldsymbol{A}$. Therefore, the description of the elimination is simplified if we consider $\boldsymbol{b}$ as the last column of $\boldsymbol{A}$ and put

$$a_{i,n+1}^{(k)} = b_i^{(k)}, \quad i, k = 1, 2, \ldots, n.$$

Then the formulas can be summarized as follows. The elimination is performed in $(n-1)$ steps, $k = 1, 2, \ldots, n-1$. In step k the elements $a_{ij}^{(k)}$ with

$i, j > k$ are transformed according to

$$m_{ik} = \frac{a_{ik}^{(k)}}{a_{kk}^{(k)}}, \qquad a_{ij}^{(k+1)} \triangleq a_{ij}^{(k)} - m_{ik}a_{kj}^{(k)},$$
$$i = k+1, k+2, \dots, n, \qquad j = k+1, \dots, n, n+1. \tag{5.3.5}$$

We note here that if we have several systems of equations with the same matrix A,

$$A\boldsymbol{x}_1 = \boldsymbol{b}_1, \qquad A\boldsymbol{x}_2 = \boldsymbol{b}_2, \qquad \dots, \qquad A\boldsymbol{x}_p = \boldsymbol{b}_p,$$

these can be treated simultaneously, by adjoining $\boldsymbol{b}_j$ as $(n+j)$th column to A. The only difference in the algorithm of Eq. (5.3.5) will be that the index j will take the values $k+1, \dots, n, \dots, n+p$. After the elimination we will have p triangular systems to solve. A special case of this is the computation of the inverse matrix A^{-1}, which is treated in Sec. 5.3.6.

We shall now estimate the number of arithmetic operations required by Gaussian elimination to reduce a system with p right-hand sides to triangular form. From Eq. (5.3.5) it follows that step k takes $(n-k)$ divisions and $(n-k)(n-k+p)$ multiplications and additions. If we are only interested in large values of n, then obviously the divisions can be neglected. Thus the total number of operations is approximately

$$\sum_{k=1}^{n-1} (n-k)(n-k+p)$$
$$= \tfrac{1}{3}n(n^2-1) + \tfrac{1}{2}n(n-1)(p-1) \approx \tfrac{1}{3}n^3 + \tfrac{1}{2}n^2(p-1), \tag{5.3.6}$$

where one operation equals one addition and one multiplication. Since the number of operations needed to solve the final triangular system is $\frac{1}{2}n^2$, we note that when $p = 1$ and n is large the main work lies in the reduction to triangular form.

Suppose that the time for one operation is 15 μs for a certain computer. Then the time in seconds for the elimination and back substitution is given as a function of n in the table below, for a full matrix.

n	Elimination	Back Substitution
10	0.0050 s	0.0008 s
100	5.000	0.075
1,000	5,000.0	7.5

From this table it is seen that solving a linear system of equations with 100 unknowns is a simple task using a modern computer. However, a full 1,000 × 1,000 matrix is already near the limit of what can be solved at a reasonable cost. Here another difficulty arises, since the 10^6 elements in the matrix A cannot be stored in the primary memory. (This may not be true in 1980!) We must use drums, discs, or tapes, and the administration of the transfer to and from the primary memory must be done carefully (see Sec. 5.4.3).

5.3.3. Pivoting Strategies

We have seen that if in Gaussian elimination the pivotal element $a_{kk}^{(k)} = 0$ for some k, then the method breaks down. A simple example of such a system is

$$\left.\begin{aligned} x_1 + x_2 + x_3 &= 1 \\ x_1 + x_2 + 2x_3 &= 2 \\ x_1 + 2x_2 + 2x_3 &= 1. \end{aligned}\right\} \tag{5.3.7}$$

This system is nonsingular and has the unique solution $x_1 = -x_2 = x_3 = 1$. However, after the first step in the elimination we get

$$\left.\begin{aligned} x_3 &= 1 \\ x_2 + x_3 &= 0, \end{aligned}\right\}$$

so that $a_{22}^{(2)} = 0$, and we cannot proceed as usual. The remedy is obviously to interchange equations 2 and 3 before the next step, which in this case actually gives us the triangular system directly. We could also have interchanged columns 2 and 3, but then the order of the unknowns is also changed.

Now consider the general case that in step k we have $a_{kk}^{(k)} = 0$. (The equations may have been reordered in previous steps, but we assume that the notations have been changed accordingly.) Then some other element $a_{ik}^{(k)}$, $i = k, k+1, \ldots, n$ in column k must be nonzero because otherwise the first k columns are linearly dependent and A is singular. Assume that $a_{rk}^{(k)} \neq 0$. Then we can interchange rows k and r and proceed with the elimination. It follows that any nonsingular system of equations can be reduced to triangular form by Gaussian elimination combined with row interchanges.

To ensure numerical stability it will often be necessary to perform row interchanges not only when a pivotal element is **exactly** zero, but also when it is nearly zero. Suppose that in the system of Eq. (5.3.7) the element a_{22} is changed to 1.0001 and Gaussian elimination carried through without interchanges. The resulting triangular system becomes

$$\left.\begin{aligned} x_1 + x_2 + x_3 &= 1 \\ 0.0001x_2 + x_3 &= 1 \\ 9{,}999x_3 &= 10{,}000. \end{aligned}\right\}$$

If we perform the back substitution using three-figure floating-point arithmetic, the computed solution becomes

$$\bar{x}_1 = 0, \qquad \bar{x}_2 = 0, \qquad \bar{x}_3 = 1.000,$$

whereas the true solution rounded to four decimals is

$$x_1 = 1.0000, \qquad -x_2 = x_3 = 1.0001.$$

It is left to the reader to verify that if we interchange rows 2 and 3, then using the same precision the computed solution becomes $x_1 = -x_2 = x_3 = 1.000$, which is correct to three decimals.

We shall return to the question of rounding errors in Sec. 5.5.4, but note here that in order to prevent possible catastrophic errors as illustrated by the example above, it is usually necessary to choose the pivotal element in step k by one of the following strategies:

Partial Pivoting. Choose r as the smallest integer for which

$$|a_{rk}^{(k)}| = \max |a_{ik}^{(k)}|, \quad k \leq i \leq n.$$

and interchange rows k and r.

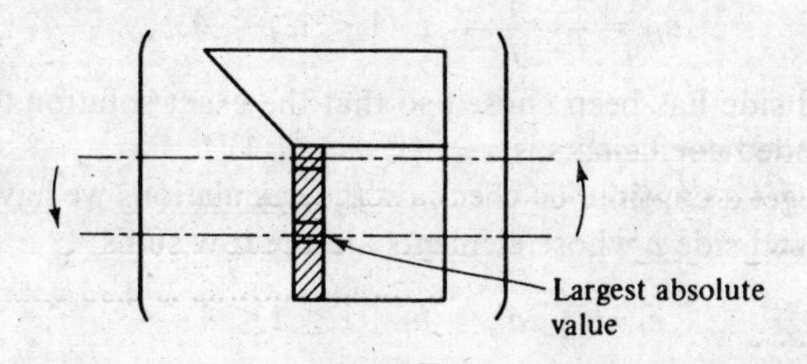

Complete Pivoting. Choose r and s as the smallest integers for which

$$|a_{rs}^{(k)}| = \max |a_{ij}^{(k)}|, \quad k \leq i, j \leq n.$$

and interchange rows k and r and columns k and s.

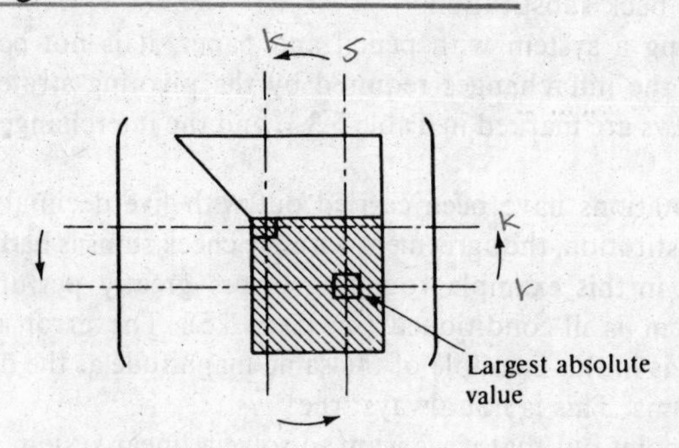

Thus in complete pivoting we select as pivot, at each stage, the element of largest absolute value in the whole relevant part of the matrix. In practice, partial pivoting is generally satsifactory and, because of the much larger amount of search required, complete pivoting is not used much.

There are two important cases when Gaussian elimination can be carried out without row or column interchanges. These are systems where the matrix A is of one of the following types:

Diagonally dominant—i.e.,

$$|a_{ii}| \geq \sum_{\substack{j=1\\ j\neq i}}^{n} |a_{ij}|, \quad i = 1, 2, \ldots, n.$$

Symmetric and positive-definite—i.e.,

$$A^T = A \quad \text{and} \quad x^T A x > 0 \quad \text{for all} \quad x \neq 0.$$

(See also Sec. 5.4.1.)

Example 5.3.1

In Table 5.3.1 we have carried through Gaussian elimination with partial pivoting for a system $Ax = b$, where

$$a_{ij} = \frac{1}{i+j-1}, \quad 1 \leq i, j \leq 4. \tag{5.3.8}$$

The right-hand side has been chosen so that the exact solution to the equations with rounded coefficients is $x = (1, -1, 1, 1)^T$.

In order to get a continuous check on the calculations we have formed a second right-hand side c whose elements are the row sums

$$c_i = \sum_{j=1}^{4} a_{ij} + b_i, \quad 1 \leq i \leq 4.$$

Then in step k of the elimination the relation,

$$c_i^{(k)} = \sum_{j=k}^{4} a_{ij}^{(k)} + b_i^{(k)}, \quad 1 \leq i \leq 4$$

should hold. Also, the solution to $Ay = c$ satisfies $y_j = x_j + 1$, which gives a check on the back substitution.

When solving a system with pencil and paper, it is not convenient to actually make the interchanges required by the pivoting strategy. Instead, the pivoting rows are marked in Table 5.3.1, and the interchanges made only implicitly.

The computations have been carried out with five decimals. Note that in the back substitution, the agreement with the check sums is bad. The reason for this is that in this example rounding errors greatly perturb the solution—the system is ill-conditioned (see Sec. 5.5). The error in the computed solution is in this example of the same magnitude as the disagreement in the check sums. This is not always true!

We finally point out that if we want to solve a linear system using a desk calculator, then one of the compact schemes for Gaussian elimination in Sec. 5.3.5 should be used.

5.3.4. *LU*-Decomposition

We have seen that several right-hand sides can be treated simultaneously by Gaussian elimination. In many situations however, all the right-hand sides are not available from the beginning. We may, for example, want to solve

Table 5.3.1†

a_{11}	a_{12}	a_{13}	a_{14}	b_1	c_1	**1.00000**	**0.50000**	**0.33333**	**0.25000**	**0.58333**	**2.66666**
a_{21}	a_{22}	a_{23}	a_{24}	b_2	c_2	0.50000	0.33333	0.25000	0.20000	0.21667	1.50000
a_{31}	a_{32}	a_{33}	a_{34}	b_3	c_3	0.33333	0.25000	0.20000	0.16667	0.11666	1.06666
a_{41}	a_{42}	a_{43}	a_{44}	b_4	c_4	0.25000	0.20000	0.16667	0.14286	0.07381	0.83334
m_{21}	$a_{22}^{(2)}$	$a_{23}^{(2)}$	$a_{24}^{(2)}$	$b_2^{(2)}$	$c_2^{(2)}$	0.50000	**0.08333**	**0.08333**	**0.07500**	**−0.07500**	**0.16667**
m_{31}	$a_{32}^{(2)}$	$a_{33}^{(2)}$	$a_{34}^{(2)}$	$b_3^{(2)}$	$c_3^{(2)}$	0.33333	0.08333	0.08889	0.08333	−0.07778	0.17778
m_{41}	$a_{42}^{(2)}$	$a_{43}^{(2)}$	$a_{44}^{(2)}$	$b_4^{(2)}$	$c_4^{(2)}$	0.25000	0.07500	0.08333	0.08036	−0.07202	0.16667
	m_{32}	$a_{33}^{(3)}$	$a_{34}^{(3)}$	$b_3^{(3)}$	$c_3^{(3)}$		1.00000	0.00556	0.00833	−0.00278	0.01111
	m_{42}	$a_{43}^{(3)}$	$a_{44}^{(3)}$	$b_4^{(3)}$	$c_4^{(3)}$		0.90000	**0.00833**	**0.01286**	**−0.00452**	**0.01667**
		m_{33}	$a_{34}^{(4)}$	$b_3^{(4)}$	$c_3^{(4)}$			**0.66747**	**−0.00025**	**0.000237**	**0.00002**
x_1	x_2	x_3	x_4			0.99877	−0.97717	0.93945	−0.96000		
y_1	y_2	y_3	y_4			1.99720	0.04713	1.87770	0.08000		

†The pivotal rows have been printed in boldface.

$Ax_1 = b_1$ and $Ax_2 = b_2$, where b_2 is some function of x_1. Then it may seem necessary to repeat the elimination from the beginning, at a considerable cost in number of operations. We shall now show how to avoid this.

Suppose we can find a decomposition of A into a lower- and an upper-triangular matrix,

$$A = LU.$$

Then the system $Ax = b$ is equivalent to $LUx = b$, which decomposes into two triangular systems

$$Ly = b \quad \text{and} \quad Ux = y. \tag{5.3.9}$$

Thus if we knew L and U we could solve $Ax = b$ with $2 \cdot \frac{1}{2}n^2 = n^2$ operations compared with $\frac{1}{3}n^3$ operations required to perform Gaussian elimination on A. We now state a fundamental theorem.

THEOREM 5.3.1. *LU-theorem*

Let A be a given $n \times n$ matrix, and denote by A_k the $k \times k$ matrix formed by the intersection of the first k rows and columns in A. If $\det(A_k) \neq 0$, $k = 1, 2, \ldots, n-1$, then there exist a unique lower-triangular matrix $L = (m_{ij})$ with $m_{ii} = 1$, $i = 1, 2, \ldots, n$ and a unique upper-triangular matrix $U = (u_{ij})$ so that $LU = A$.

We prove this by induction in n. For $n = 1$, the decomposition $a_{11} = 1 \cdot u_{11}$ is unique. Suppose the theorem is true for $n = k - 1$. For $n = k$ we partition A_k, L_k, and U_k according to

$$A_k = \begin{pmatrix} A_{k-1} & b \\ c^{\mathrm{T}} & a_{kk} \end{pmatrix}, \quad L_k = \begin{pmatrix} L_{k-1} & 0 \\ m^{\mathrm{T}} & 1 \end{pmatrix}, \quad U_k = \begin{pmatrix} U_{k-1} & u \\ 0 & u_{kk} \end{pmatrix},$$

where b, c, m, and u are column vectors with $k - 1$ components. If we form the product $L_k U_k$ and identify with A_k, we get

$$L_{k-1}U_{k-1} = A_{k-1}, \qquad L_{k-1}u = b,$$
$$m^{\mathrm{T}}U_{k-1} = c^{\mathrm{T}}, \qquad m^{\mathrm{T}}u + u_{kk} = a_{kk}.$$

By the induction hypothesis, L_{k-1} and U_{k-1} are uniquely determined, and since $\det(L_{k-1}) \cdot \det(U_{k-1}) = \det(A_{k-1}) \neq 0$, they are nonsingular. It follows that u and m are uniquely determined by the triangular systems $L_{k-1}u = b$ and $U_{k-1}^{\mathrm{T}}m = c$. Finally $u_{kk} = a_{kk} - m^{\mathrm{T}}u$. Thus L_k and U_k are uniquely determined, which concludes the proof.

If, for some k, $\det(A_k) = 0$, there may not exist an LU-decomposition of A. A simple example of this is the matrix

$$A = \begin{pmatrix} 0 & 1 \\ 1 & 1 \end{pmatrix}.$$

Suppose that $A = LU$, where

$$LU = \begin{pmatrix} m_{11} & 0 \\ m_{21} & m_{22} \end{pmatrix}\begin{pmatrix} u_{11} & u_{12} \\ 0 & u_{21} \end{pmatrix} = \begin{pmatrix} m_{11}u_{11} & m_{11}u_{12} \\ m_{21}u_{11} & m_{21}u_{12} + m_{22}u_{22} \end{pmatrix}.$$

It follows that either m_{11} or u_{11} equals zero. But then either the first row or the first column in LU becomes zero, so $A \neq LU$.

We note that if in this example the two rows in A are interchanged, then the matrix becomes triangular, so an LU-decomposition trivially exists. In fact, for any nonsingular matrix A, the rows can be reordered so that an LU-decomposition exists. This will follow from the equivalence between Gaussian elimination and LU-decomposition, which we will now establish.

We first assume that the matrix A is such that Gaussian elimination can be carried out without row or column interchanges. We can think of the elimination as determining a sequence of matrices $A = A^{(1)}, A^{(2)}, \ldots, A^{(n)}$, where $A^{(k)} = (a_{ij}^{(k)})$ is equal to

$$A^{(k)} = \begin{pmatrix} a_{11}^{(1)} & a_{12}^{(1)} & \ldots & a_{1k}^{(1)} & \ldots & a_{1n}^{(1)} \\ & a_{22}^{(2)} & \ldots & a_{2k}^{(2)} & \ldots & a_{2n}^{(2)} \\ & & \ldots & & & \\ & 0 & & a_{kk}^{(k)} & \ldots & a_{kn}^{(k)} \\ & & & \ldots & & \\ & & & a_{nk}^{(k)} & \ldots & a_{nn}^{(k)} \end{pmatrix}.$$

We now consider a certain element a_{ij} during the elimination. If a_{ij} is on or above the principal diagonal—i.e., if $i \leq j$—then

$$a_{ij}^{(n)} = \ldots = a_{ij}^{(i+1)} = a_{ij}^{(i)}, \quad i \leq j.$$

If, on the other hand, a_{ij} is below the principal diagonal, then

$$a_{ij}^{(n)} = \ldots = a_{ij}^{(j+1)} = 0, \quad i > j.$$

Thus the elements a_{ij} are transformed for $k = 1, 2, \ldots, r = \min(i - 1, j)$ by Eq. (5.3.5):

$$a_{ij}^{(k+1)} = a_{ij}^{(k)} - m_{ik}a_{kj}^{(k)}. \tag{5.3.10}$$

If these equations are summed for $k = 1, 2, \ldots, r$, we obtain

$$\sum_{k=1}^{r} a_{ij}^{(k+1)} - \sum_{k=1}^{r} a_{ij}^{(k)} = a_{ij}^{(r+1)} - a_{ij} = -\sum_{k=1}^{r} m_{ik}a_{kj}^{(k)}.$$

This can be written

$$a_{ij} = \begin{cases} a_{ij}^{(i)} + \sum_{k=1}^{i-1} m_{ik}a_{kj}^{(k)}, & i \leq j \\ 0 + \sum_{k=1}^{j} m_{ik}a_{kj}^{(k)}, & i > j, \end{cases}$$

or, if we define $m_{ii} = 1$, $i = 1, 2, \ldots, n$,

$$a_{ij} = \sum_{k=1}^{p} m_{ik}a_{kj}^{(k)}, \quad p = \min(i, j). \tag{5.3.11}$$

However, these equations are equivalent to the matrix equation $A = LU$, where the nonzero elements in L and U are given by

$$(L)_{ik} = m_{ik}, \quad i \geq k, \qquad (U)_{kj} = a_{kj}^{(k)}, \quad k \leq j.$$

We conclude that the elements in $\boldsymbol{L}$ are the multipliers and the matrix $\boldsymbol{U}$ the final triangular matrix obtained by Gaussian elimination. Thus to get the matrix $\boldsymbol{L}$ we must save the multipliers m_{ik}. This can be done very conveniently on a computer. Since

$$m_{ik} = \frac{a_{ik}^{(k)}}{a_{kk}^{(k)}}$$

is determined in such a way that $a_{ik}^{(k+1)}$ becomes zero, we can let m_{ik} overwrite $a_{ik}^{(k)}$. Also, the trivial diagonal elements in $\boldsymbol{L}$ need not be stored. Thus no extra memory space is actually needed, and the effect of the elimination can be pictured as

$$\begin{pmatrix} a_{11} & a_{12} & \cdots & a_{1,n-1} & a_{1n} \\ a_{21} & a_{22} & \cdots & a_{2,n-1} & a_{2n} \\ \cdots & & & & \\ a_{n1} & a_{n2} & \cdots & a_{n,n-1} & a_{nn} \end{pmatrix} \Longrightarrow \begin{pmatrix} u_{11} & u_{12} & \cdots & u_{1,n-1} & u_{1n} \\ m_{21} & u_{22} & \cdots & u_{2,n-1} & u_{2n} \\ \cdots & & & & \\ m_{n1} & m_{n2} & & m_{n,n-1} & u_{nn} \end{pmatrix}.$$

Further, we note that since $\det(L_k) = 1$, we have

$$\det(A_k) = a_{11}^{(1)} a_{22}^{(2)} \ldots a_{kk}^{(k)}, \quad k = 1, 2, \ldots, n, \tag{5.3.12}$$

i.e., the product of the first k pivots. Therefore, Gaussian elimination can be carried out without row or column interchanges if and only if $\det(A_k) \neq 0$, $k = 1, 2, \ldots, n-1$, which is the condition in the LU-theorem. Now we know that Gaussian elimination may be unstable without partial pivoting, so usually row interchanges are performed during the elimination. It is important to note that these row interchanges just mean that we get triangular factors $\boldsymbol{L}$ and $\boldsymbol{U}$ such that $\boldsymbol{LU} = A'$, where A' is the matrix which results if the row interchanges are performed in the same order on the initial matrix A.

Example 5.3.2

Consider the same matrix as in Example 5.3.1. If Gaussian elimination is performed exactly and with partial pivoting we get the decomposition

$$\begin{pmatrix} 1 & & & \\ \frac{1}{2} & 1 & & \\ \frac{1}{4} & \frac{1}{9} & 1 & \\ \frac{1}{3} & 1 & \frac{2}{3} & 1 \end{pmatrix} \begin{pmatrix} 1 & \frac{1}{2} & \frac{1}{3} & \frac{1}{4} \\ & \frac{1}{12} & \frac{1}{12} & \frac{3}{40} \\ & & \frac{1}{120} & \frac{9}{700} \\ & & & -\frac{1}{4200} \end{pmatrix} = \begin{pmatrix} 1 & \frac{1}{2} & \frac{1}{3} & \frac{1}{4} \\ \frac{1}{2} & \frac{1}{3} & \frac{1}{4} & \frac{1}{5} \\ \frac{1}{4} & \frac{1}{5} & \frac{1}{6} & \frac{1}{7} \\ \frac{1}{3} & \frac{1}{4} & \frac{1}{5} & \frac{1}{6} \end{pmatrix}.$$

(The pivotal rows are $p_1 = 1, p_2 = 2, p_3 = 4$.)

We now sum up how LU-decomposition is used in practice to solve a linear system $A\boldsymbol{x} = \boldsymbol{b}$. We perform Gaussian elimination with partial pivoting on A, and save the indices for the pivotal rows in a vector

$$(p_1, p_2, \ldots, p_{n-1}).$$

The multipliers m_{ik} are stored as outlined above, and we note that these shall also take part in the row interchanges. After the elimination we apply the row interchanges $k \leftrightarrow p_k$, $k = 1, 2, \ldots, n-1$ to $\boldsymbol{b}$, which gives us a vector $\boldsymbol{b}'$. Finally, we solve the two triangular systems $\boldsymbol{L}\boldsymbol{y} = \boldsymbol{b}'$, $\boldsymbol{U}\boldsymbol{x} = \boldsymbol{y}$.

We point out that the best way of computing the determinant of a matrix A is generally to perform Gaussian elimination on A and use Eq. (5.3.12) with $k = n$. Note, though, that if we have performed row or column interchanges, we must multiply the result by $(-1)^s$, where s is the total number of interchanges.

5.3.5. Compact Schemes for Gaussian Elimination

When solving a linear system by Gaussian elimination we have to write down approximately $n^3/3$ intermediate results—one for each multiplication. Even for rather small values of n this is very tedious and easily gives rise to many errors. However, it is possible to arrange the calculations so that the elements in $\boldsymbol{L}$ and $\boldsymbol{U}$ are determined directly.

As in Gaussian elimination, in step k the kth column in $\boldsymbol{L}$ and the kth row in $\boldsymbol{U}$ are determined, but in the compact method the elements a_{ij}, $i, j > k$, are still unchanged. The matrix equation $\boldsymbol{A} = \boldsymbol{L}\boldsymbol{U}$ is equivalent to the equations

$$a_{ij} = \sum_{p=1}^{r} m_{ip} u_{pj}, \quad r = \min(i, j).$$

This can be thought of as n^2 equations for the $n(n+1)$ unknowns in $\boldsymbol{L}$ and $\boldsymbol{U}$. For the kth step we use the following equations:

$$a_{kj} = \sum_{p=1}^{k} m_{kp} u_{pj}, \quad j \geq k, \qquad a_{ik} = \sum_{p=1}^{k} m_{ip} u_{pk}, \quad i > k.$$

If we put $m_{kk} = 1$, then the elements

$$u_{kk}, u_{k,k+1}, \ldots, u_{kn} \quad \text{and} \quad m_{k+1,k}, \ldots, m_{nk}$$

can be solved for in this order from

$$u_{kj} = a_{kj} - \sum_{p=1}^{k-1} m_{kp} u_{pj}, \quad j = k, k+1, \ldots, n,$$

$$m_{ik} = \frac{a_{ik} - \sum_{p=1}^{k-1} m_{ip} u_{pk}}{u_{kk}}, \quad i = k+1, \ldots, n. \tag{5.3.12}$$

This method is called **Doolittle's method**. It gives the same factors $\boldsymbol{L}$ and $\boldsymbol{U}$ as Gaussian elimination. In fact, the successive partial sums in Eq. (5.3.12) are the unwanted elements $a_{ij}^{(k)}$, $i, j > k$. It follows that if each term in Eq. (5.3.12) is rounded separately, the method is even numerically equivalent to Gaussian elimination. On most desk calculators, however, these product sums can be accumulated without writing down intermediate results. Also, for computers where product sums can be accumulated in double precision, Doolittle's method has advantages.

If we instead normalize so that $u_{kk} = 1$, $k = 1, 2, \ldots, n$, we get a slightly different method known as **Crout's method**. The equations for step k then become

$$m_{ik} = a_{ik} - \sum_{p=1}^{k-1} m_{ip}u_{pk}, \quad i = k, k+1, \ldots, n,$$
$$u_{kj} = \frac{a_{kj} - \sum_{p=1}^{k-1} m_{kp}u_{pj}}{m_{kk}}, \quad j = k+1, \ldots, n. \tag{5.3.13}$$

Until now we have assumed that no row interchanges are made. It is essential for the usefulness of Doolittle's and Crout's methods that they can be combined with partial pivoting in a straightforward way. However, it is *not* easy to do complete pivoting with these methods.

For symmetric, positive-definite matrices, the compact schemes become particularly attractive, since no pivoting is needed. Usually one chooses the diagonal elements in L to be real and so that $U = L^T$, which is always possible (see Theorem 5.4.3). Then $u_{kk} = m_{kk}$ and $u_{pk} = m_{kp}$ and the formulas (5.3.12) are simplified to

$$m_{kk} = \Big(a_{kk} - \sum_{p=1}^{k-1} m_{kp}^2\Big)^{1/2},$$
$$m_{ik} = \frac{a_{ik} - \sum_{p=1}^{k-1} m_{ip}m_{kp}}{m_{kk}}, \quad i = k+1, \ldots, n.$$

This very popular method is known as **Choleski's method** or the square-root method. It is very closely related to a symmetric version [Eq. (5.4.2)] of Gaussian elimination. If the matrix A is stored row by row, it is often better to use the following variant of Choleski's method which computes L row by row

$$m_{kj} = \frac{a_{kj} - \sum_{p=1}^{j-1} m_{kp}m_{jp}}{m_{jj}}, \quad j = 1, 2, \ldots, k-1,$$
$$m_{kk} = \Big(a_{kk} - \sum_{p=1}^{k-1} m_{kp}^2\Big)^{1/2}. \tag{5.3.14}$$

Example 5.3.3

The linear system in Example (5.3.1) is symmetric and positive-definite, and therefore Choleski's method can be used to solve it. If the product sums in Eq. (5.3.14) are computed exactly and then rounded to five decimals, we obtain the matrix $\bar{L}$:

$$\bar{L} = \begin{pmatrix} 1.00000 & & & \\ 0.50000 & 0.28867 & & \\ 0.33333 & 0.28869 & 0.07449 & \\ 0.25000 & 0.25981 & 0.11187 & 0.01854 \end{pmatrix}$$

To obtain the solution we solve the two triangular systems $\bar{L}y = b$ and $\bar{L}^T x = y$, again accumulating product sums. We then get

$$y = (0.58333, \ -0.25979, \ -0.03736, \ -0.01872)^T,$$
$$x = (1.00053, \ -1.00611, \quad 1.01485, \ -1.00971)^T.$$

The errors are seen to be about four times less than for the approximate solution computed in Example 5.3.1.

5.3.6. Inverse Matrices

If the inverse matrix A^{-1} is known, then the solution of $Ax = b$ can be directly computed as $A^{-1}b$. It may therefore be tempting to compute A^{-1}, especially when a system with several right-hand sides has to be solved. However, the solution can be computed with fewer operations and normally with greater accuracy using the LU-decomposition. The solution of the two systems of Eq. (5.3.9) requires only n^2 operations, which is exactly the same number as required to compute $A^{-1}b$. However, the amount of work to compute A^{-1} is approximately three times greater than that for computing L and U. Compare also the remark on inverses of band matrices in Sec. 5.4.2.

Sometimes the inverse is needed in its own right—for example, in connection with regression analysis in statistics. Also, as we shall see, the inverse is needed to get a strictly reliable error estimate for a computed solution to a linear system. In these cases the inverse can be computed as follows.

If we put $X = A^{-1}$, we have $AX = I$ or

$$Ax_j = e_j, \quad j = 1, 2, \ldots, n,$$

where x_j and e_j are the jth column in X and I, respectively. Thus the columns in A^{-1} are solutions to linear systems with right-hand sides equal to the columns in the unit matrix I. We know from Sec. 5.3.2 that with Gaussian elimination several right-hand sides can be treated simultaneously. According to the operation count of Eq. (5.3.6), the number of operations to compute A^{-1} with this method is $n^3/3 + n^3 = 4n^3/3$. However, by taking into account the zeros in the right-hand sides, it is possible to reduce this number to n^3 operations.

Inverses of triangular matrices are easy to compute. Let L be a left triangular matrix and put $Y = L^{-1}$. Then the column y_j in Y satisfies

$$Ly_j = e_j, \quad j = 1, 2, \ldots, n.$$

The components in y_j can be computed by the forward substitution of Eq. (5.3.2). From this it follows immediately that the first $(j-1)$ components in y_j are zero, and thus L^{-1} is also a lower triangular matrix. The elements in L^{-1} can therefore be computed recursively from

$$y_{ij} = \frac{\delta_{ij} - \sum_{k=j}^{i-1} l_{ik} y_{kj}}{l_{ii}}, \quad i = j, j+1, \ldots, n, \qquad (5.3.15)$$

where δ_{ij} are the elements in the unit matrix. The number of operations needed to compute L^{-1} is approximately $n^3/6$.

In a similar way one can show that if U is upper-triangular, then $Z = U^{-1}$ is an upper-triangular matrix, whose elements can be computed from

$$z_{ij} = \frac{\delta_{ij} - \sum_{k=i+1}^{j} u_{ik} z_{kj}}{u_{ii}}, \quad i = j, j-1, \ldots, 1. \tag{5.3.16}$$

If the LU-decomposition of a matrix A is known, then Eqs. (5.3.15) and (5.3.16) and the relation

$$A^{-1} = (LU)^{-1} = U^{-1}L^{-1}$$

can be used to compute A^{-1}. If L and U are known, this takes

$$\frac{n^3}{6} + \frac{n^3}{6} + \frac{n^3}{3} = \frac{2}{3}n^3$$

operations. Since the LU-decomposition requires $n^3/3$ operations, the total number of operations for the inversion is n^3. If we note that $y_{ii} = 1/m_{ii} = 1$, and carefully choose the order in which the elements in L^{-1}, U^{-1}, and A^{-1} are computed, it is possible to carry out the computation without using extra storage. Finally, we must take into account that, in general, row interchanges have been made when determining L and U. This means that $U^{-1}L^{-1}$ is the inverse of a matrix A' obtained by making row interchanges $k \leftrightarrow p_k$, $k = 1, 2, \ldots, n-1$ on A. To get A^{-1} we must therefore perform *column* interchanges $k \leftrightarrow p_k$, $k = n-1, \ldots, 2, 1$ on $(A')^{-1}$.

PROBLEMS

1. (a) Consider the triangular system $U_n(a)x = e_n$, where

$$U_n(a) = \begin{pmatrix} 1 & a & a & \cdots & a \\ & 1 & a & & a \\ & & 1 & \cdots & a \\ & & & \ddots & \vdots \\ & & & & 1 \end{pmatrix},$$

and e_n is the nth unit vector. Determine the solution x.

(b) Give a bound for x_i, where $x = (x_i)$ is the solution to the upper triangular system $Ux = e_n$, where $|u_{ij}| \leq 1$, $u_{ii} = 1$.

2. (a) Solve by Gaussian elimination without row or column interchanges the linear system $Ax = b$, where

$$A = \begin{pmatrix} 1 & 2 & 3 & 4 \\ 1 & 4 & 9 & 16 \\ 1 & 8 & 27 & 64 \\ 1 & 16 & 81 & 256 \end{pmatrix}, \quad b = \begin{pmatrix} 2 \\ 10 \\ 44 \\ 190 \end{pmatrix}.$$

(b) Give the LU-decomposition of A and compute $\det(A)$.

3. (a) Describe in Algol or Fortran the solution of a linear system $Ax = b$. Assume that the LU-decomposition of A can be determined without row or column interchanges. Compute L and U, and then solve the resulting two triangular systems.
(b) Make the necessary changes in (a) to include partial pivoting.

4. Let A be the 4×4 matrix with elements $a_{ij} = 1/(i + j - 1)$, and denote by $\bar{A}$ the corresponding matrix with elements rounded to five decimal places. In Example 5.3.3 we computed the lower-triangular matrix $\bar{L}$ in the LL^T-decomposition of $\bar{A}$. Compute the difference $(\bar{L}\bar{L}^T - \bar{A})$ and compare it with $(A - \bar{A})$.

5. Compute the inverse A^{-1}, where

$$A = \begin{pmatrix} 2 & 1 & 2 \\ 1 & 2 & 3 \\ 4 & 1 & 2 \end{pmatrix},$$

(a) by solving the system $AX = I$, using partial pivoting;
(b) by LU-decomposition and using $A^{-1} = U^{-1}L^{-1}$.

6. Let A be a given nonsingular $n \times n$ matrix, u and v given $n \times 1$ vectors.
(a) Show that $(A - uv^T)^{-1} = A^{-1} + \alpha A^{-1}uv^TA^{-1}$, where $\alpha = 1/(1 - v^TA^{-1}u)$. Give conditions for the existence of the inverse on the left-hand side (**Sherman-Morrison formula**).
(b) Suppose that A^{-1} is known and that B is a matrix coinciding with A except in one row. Show that B^{-1} (if it exists) can be computed by roughly $2n^2$ multiplications ($n \gg 1$).
(c) Use the formula in (a) to compute B^{-1} if

$$A = \begin{bmatrix} 1 & 0 & -2 & 0 \\ -5 & 1 & 11 & -1 \\ 287 & -67 & -630 & 65 \\ -416 & 97 & 913 & -94 \end{bmatrix}, \qquad A^{-1} = \begin{bmatrix} 13 & 14 & 6 & 4 \\ 8 & -1 & 13 & 9 \\ 6 & 7 & 3 & 2 \\ 9 & 5 & 16 & 11 \end{bmatrix},$$

and B equals A except that the element 913 is changed to 913.01.
(d) Generalize to a formula for $(A - UV^T)^{-1}$, where U, V are $n \times r$ matrices (**Woodbury formula**).

7. (a) Derive from the identities

$$\begin{bmatrix} A & B \\ C & D \end{bmatrix}\begin{bmatrix} X & Y \\ Z & V \end{bmatrix} = \begin{bmatrix} X & Y \\ Z & V \end{bmatrix}\begin{bmatrix} A & B \\ C & D \end{bmatrix} = \begin{bmatrix} I & 0 \\ 0 & I \end{bmatrix}$$

the following formulas for computing the inverse of a partitioned matrix. A and B are square matrices and it is assumed that A and $D - CA^{-1}B$ are nonsingular.

$$V = (D - CA^{-1}B)^{-1}, \qquad Y = -A^{-1}BV, \qquad Z = -VCA^{-1}$$
$$X = A^{-1} - A^{-1}BZ = A^{-1} - YCA^{-1}.$$

(b) How many multiplications are needed to compute the inverse by the formulas in (a) if A is $p \times p$ and D is $q \times q$?

5.4. SPECIAL MATRICES

In practice one often encounters systems $Ax = b$, where the matrix has special properties such that operations and memory space can be saved by modifying the algorithms. In the following we discuss some especially important cases.

5.4.1. Symmetric Positive-Definite Matrices

Let $A = (a_{ij})$ be a given symmetric matrix—i.e., $a_{ij} = a_{ji}$, $1 \leqslant i, j \leqslant n$. We shall prove that if Gaussian elimination is carried out without row or column interchanges, then

$$a_{ij}^{(k)} = a_{ji}^{(k)}, \quad k \leqslant i, j \leqslant n, \tag{5.4.1}$$

i.e., the transformed elements form symmetric matrices of order $(n + 1 - k)$ for $k = 2, \ldots, n$. The statement is true for $k = 1$. According to Eq. (5.3.5), the elements are transformed in step k by

$$a_{ij}^{(k+1)} = a_{ij}^{(k)} - m_{ik}a_{kj}^{(k)} = a_{ij}^{(k)} - \frac{a_{ik}^{(k)}}{a_{kk}^{(k)}} a_{kj}^{(k)}.$$

Thus if Eq. (5.4.1) holds for some k, then

$$a_{ij}^{(k+1)} = a_{ji}^{(k)} - \frac{a_{jk}^{(k)}}{a_{kk}^{(k)}} a_{ki}^{(k)} = a_{ji}^{(k)} - m_{jk}a_{ki}^{(k)} = a_{ji}^{(k+1)},$$

and the statement follows by induction.

Thus we see that if Gaussian elimination can be carried out without row or column interchanges for a symmetric matrix, then we only have to compute the elements in $A^{(k)}$ on or above the main diagonal. In **symmetric Gaussian elimination** we transform these elements (see Eq. (5.3.5)) for $k = 1, 2, \ldots, n - 1$ according to

$$m_{ik} = \frac{a_{ki}^{(k)}}{a_{kk}^{(k)}}, \qquad a_{ij}^{(k+1)} = a_{ij}^{(k)} - m_{ik}a_{kj}^{(k)}, \tag{5.4.2}$$

$$i = k + 1, k + 2, \ldots, n, \qquad j = i, i + 1, \ldots, n.$$

This means that the number of operations is approximately halved to $n^3/6$ and also nearly half the memory space is saved.

It is not always possible to perform Gaussian elimination on symmetric matrices without pivoting. Symmetry is preserved if we choose any pivot along the principal diagonal, and do complete pivoting, but, as shown by the matrix

$$\begin{bmatrix} 0 & 1 \\ 1 & \epsilon \end{bmatrix}, \quad |\epsilon| \ll 1,$$

even this is not always stable. On the other hand, a row interchange will generally destroy the symmetry, as is shown by the same example. We conlude that the symmetric Gaussian elimination of Eq. (5.4.2) cannot be used or all symmetric matrices.

For symmetric positive-definite matrices, Gaussian elimination without ow or column interchanges is always stable, and Eq. (5.4.2) can be used. One ossible way to determine whether a matrix is positive-definite is to use the ollowing theorem, which we state without proof.

THEOREM 5.4.1. *Sylvester's Criterion*

A symmetric $n \times n$ matrix A is positive-definite if and only if

$$\det(A_k) > 0, \quad k = 1, 2, \ldots, n,$$

where A_k is the $k \times k$ matrix formed by the intersection of the first k rows and columns of A.

From Eq. (5.3.12) it follows that this criterion is equivalent to $a_{kk}^{(k)} > 0, k = 1, 2, \ldots, n$. Thus if we perform Gaussian elimination without row or column interchanges on A, then all pivots are positive if and only if A is positive-definite. We note that from the last criterion it also follows that the reduced matrices $(a_{ij}^{(k)})$, $k \leqslant i, j \leqslant n$ are positive-definite.

Example 5.4.1

The matrix A, $a_{ij} = 1/(i + j - 1)$ in Example 5.3.1 is symmetric. If we perform Gaussian elimination without row interchanges we get the reduced matrices

$$\begin{bmatrix} \frac{1}{12} & \frac{1}{12} & \frac{3}{40} \\ \frac{1}{12} & \frac{4}{45} & \frac{1}{12} \\ \frac{3}{40} & \frac{1}{12} & \frac{9}{112} \end{bmatrix}, \quad \begin{bmatrix} \frac{1}{180} & \frac{1}{120} \\ \frac{1}{120} & \frac{9}{700} \end{bmatrix}, \quad \left[\frac{1}{2800}\right].$$

The pivots, 1, $\frac{1}{12}$, $\frac{1}{180}$, and $\frac{1}{2800}$ are all positive. Thus A is positive-definite. Note that the reduced matrices all are symmetric and positive-definite!

From Theorem 5.4.1 with $k = 1$ it immediately follows that the diagonal elements of a positive-definite matrix are positive. We also have

THEOREM 5.4.2

For a symmetric positive-definite matrix A.

$$|a_{ij}|^2 \leq a_{ii}\, a_{jj},$$

and thus the maximum element of A lies on the diagonal.

Proof: If the same permutation is applied to rows and columns of A, the resulting matrix is still symmetric and positive-definite. Now for $k = 2$ and a suitable permutation, Sylvester's criterion gives

$$0 < \det\begin{pmatrix} a_{ii} & a_{ij} \\ a_{ji} & a_{jj} \end{pmatrix} = a_{ii}a_{jj} - a_{ij}^2,$$

from which the theorem follows.

As can be seen from the positive-definite matrix

$$\begin{bmatrix} 0.0001 & 1 \\ 1 & 10001 \end{bmatrix},$$

we can get large multipliers also when Eq. (5.4.2) is applied to a positive-definite matrix. However, in Sec. 5.5.4 we will show that large multipliers will not induce instability unless they cause a growth of the transformed elements. Now, since all reduced matrices $(a_{ij}^{(k)})$ are positive-definite, their maximum elements lie on the diagonals. But for the diagonal elements we get from Eq. (5.4.2) and from Theorem 5.4.2 the inequality

$$|a_{ii}^{(k+1)}| \leq |a_{ii}^{(k)}| + \left| \frac{a_{ki}^{(k)}}{a_{kk}^{(k)}} a_{ki}^{(k)} \right| \leq 2 |a_{ii}^{(k)}|.$$

Thus we have a bound on the growth of the elements, which is independent of the size of the multipliers.

For positive-definite matrices we have the following variant of the LU theorem.

THEOREM 5.4.3

Let A be a given symmetric positive-definite matrix. Then there is a unique upper-triangular matrix R with positive diagonal elements such that $A = R^T R$.

From the LU-theorem we have $A = LU$, where $u_{11} = a_{11} > 0$ and

$$u_{kk} = \frac{\det(A_k)}{\det(A_{k-1})} > 0, \quad k = 2, 3, \ldots, n.$$

If we introduce the diagonal matrix

$$D = \operatorname{diag}(u_{11}, u_{22}, \ldots, u_{nn}),$$

we can write the decomposition

$$A = LDD^{-1}U = LDU', \quad U' = D^{-1}U,$$

where L and U' are unit-triangular and uniquely determined. Since A is symmetric, it follows that

$$A = A^T = (U')^T D L^T \quad \text{or} \quad L^T = U' = D^{-1}U.$$

Now, if we put $R = D^{-1/2}\,U$, where $D^{-1/2}$ has positive diagonal elements $(u_{kk})^{-1/2}$, we get

$$R^T R = U^T D^{-1} U = LU = A. \tag{5.4.3}$$

We point out that it is not necessary to save the multipliers in the symmetric Gaussian elimination of Eq. (5.4.2). The system $Ax = b$ in this case decom poses into the two triangular systems

$$U^T y = b, \qquad Ux = Dy. \tag{5.4.4}$$

In this way we avoid the square roots needed in Choleski's method.

5.4.2. Band Matrices

Many problems—e.g., boundary-value problems for ordinary differential equations—result in sparse linear systems, where the nonzero elements are located in a band centered along the principal diagonal. In general, a matrix A for which

$$a_{ij} = 0 \quad \text{if} \quad j > i + p \quad \text{or} \quad i > j + q,$$

is called a band matrix with **band width** $w = p + q + 1$. Obviously, the number of nonzero elements in any row or column of A does not exceed w, and if A is $n \times n$, the total number of nonzero elements is less than $w \cdot n$. For symmetric matrices, a different definition is sometimes used. A symmetric matrix A is said to have band width m if and only if $a_{ij} = 0 \Rightarrow |i - j| \leq m$.

Equations associated with a band matrix are particularly well-suited for Gaussian elimination, since the band structure is essentially preserved. If no row or column interchanges are made, then the triangular factors $L = (m_{ij})$ and $U = (u_{ij})$ will be band matrices with

$$m_{ij} = 0 \quad \text{if} \quad j > i \quad \text{or} \quad i > j + q,$$
$$u_{ij} = 0 \quad \text{if} \quad j > i + p \quad \text{or} \quad i > j.$$

This is illustrated in Fig. 5.4.1.

Unless A is diagonally dominant or positive-definite, partial pivoting should be used. This will not change the band width for L, but that of U will now be the same as for A—i.e.,

$$u_{ij} = 0 \quad \text{if} \quad j > i + p + q \quad \text{or} \quad i > j.$$

The reader is advised to study the example below, which for the case $p = 1$ and $q = 2$ schematically illustrates the statements above. Only nonzero elements are indicated; m represents final elements in L, u final elements in U,

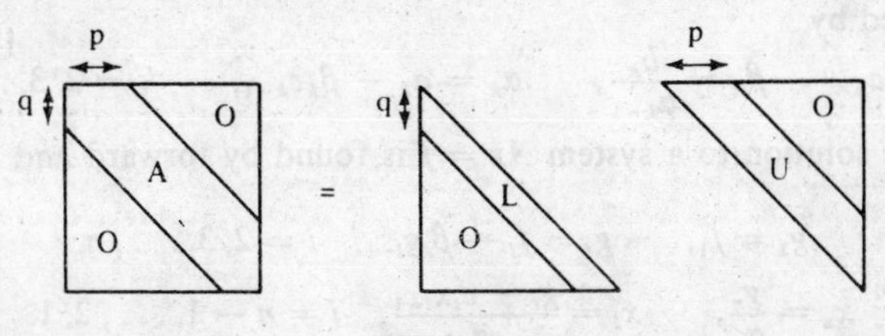

Fig. 5.4.1

and a' other transformed elements. The situation after the first elimination step is shown:

$$
\text{Without pivoting}\qquad
\begin{array}{ccccccc}
a & a & & & & \\
a & a & a & & & \\
a & a & a & a & & \\
 & a & a & a & a & \\
 & & a & a & a & a
\end{array}
\qquad
\begin{array}{cccccc}
u & u & & & & \\
m & a' & a & & & \\
m & a' & a & a & & \\
 & a & a & a & a & \\
 & & a & a & a & a
\end{array}
$$

$$
\begin{array}{l}\text{With pivoting}\\ \text{row } 1 \leftrightarrow \text{row } 3\end{array}\qquad
\begin{array}{cccccc}
a & a & a & a & & \\
a & a & a & & & \\
a & a & & & & \\
 & a & a & a & a & \\
 & & a & a & a & a
\end{array}
\qquad
\begin{array}{cccccc}
u & u & u & u & & \\
m & a' & a' & a' & & \\
m & a' & a' & a' & & \\
 & a & a & a & a & \\
 & & a & a & a & a
\end{array}
$$

The reduction in time and storage for band systems with $p \ll n$ and $q \ll n$ is considerable. In the following table we give the number of operations for elimination and back substitution in the case $p = q \ll n$, and for a full matrix—i.e., a matrix with no zero elements.

	Full Matrix	Band Matrix Without Pivoting	Band Matrix With Pivoting
Elimination	$\frac{n^3}{3}$	$np(p + 1)$	$np(2p + 1)$
Back substitution	n^2	$n(2p + 1)$	$n(3p + 1)$

Example 5.4.3

The special case when $p = q = 1$—i.e., when A is **tridiagonal**—arises frequently. If the LU-decomposition exists, then it can be written

$$
\begin{pmatrix}
a_1 & c_1 & & & \\
b_2 & a_2 & c_2 & & \\
 & \cdot & \cdot & \cdot & \\
 & \cdot & \cdot & \cdot & \\
 & \cdot & \cdot & \cdot & \\
 & & b_{n-1} & a_{n-1} & c_{n-1} \\
 & & & b_n & a_n
\end{pmatrix}
=
\begin{pmatrix}
1 & & & & \\
\beta_2 & 1 & & & \\
 & \beta_3 & \cdot & & \\
 & & \cdot & \cdot & \\
 & & \cdot & \cdot & \\
 & & & \beta_n & 1
\end{pmatrix}
\begin{pmatrix}
\alpha_1 & c_1 & & & \\
 & \alpha_2 & c_2 & & \\
 & & \cdot & \cdot & \\
 & & \cdot & \cdot & \\
 & & \cdot & \cdot & \\
 & & & & c_{n-1} \\
 & & & & \alpha_n
\end{pmatrix}.
$$

By equating elements we find that $\alpha_1, \alpha_2, \ldots, \alpha_n$ and $\beta_2, \beta_3, \ldots, \beta_n$ can be determined by

$$\alpha_1 = a_1, \qquad \beta_k = \frac{b_k}{\alpha_{k-1}}, \qquad \alpha_k = a_k - \beta_k c_{k-1}, \qquad k = 2, 3, \ldots, n.$$

Then, the solution to a system $Ax = f$ is found by forward and back substitution:

$$g_1 = f_1, \quad g_i = f_i - \beta_i g_{i-1}, \quad i = 2, 3, \ldots, n,$$

$$x_n = \frac{g_n}{\alpha_n}, \quad x_i = \frac{g_i - c_i x_{i+1}}{\alpha_i}, \quad i = n - 1, \ldots, 2, 1.$$

The total number of arithmetic operations needed is thus only $3(n-1)$ additions and multiplications and $(2n-1)$ divisions.

It is important to note that the inverse A^{-1} of a band matrix generally is full. Therefore, one should never try to compute the inverse of a band matrix explicitly, except for very small n. In a case when A is tridiagonal and, e.g., $n = 10{,}000$, A^{-1} has 10^8 nonzero elements, whereas L and U together only have some $3 \cdot 10^4$.

Example 5.4.4

Using the equations in Example 5.4.3, we can easily compute the LU-decomposition:

$$A = \begin{pmatrix} 1 & -1 & & & \\ -1 & 2 & -1 & & \\ & -1 & 2 & -1 & \\ & & -1 & 2 & -1 \\ & & & -1 & 2 \end{pmatrix},$$

$$U = L^T = \begin{pmatrix} 1 & -1 & & & \\ & 1 & -1 & & \\ & & 1 & -1 & \\ & & & 1 & -1 \\ & & & & 1 \end{pmatrix}.$$

Note that A is both diagonally dominant and positive-definite. The triangular factors have band structure, but the inverse is

$$A^{-1} = \begin{pmatrix} 5 & 4 & 3 & 2 & 1 \\ 4 & 4 & 3 & 2 & 1 \\ 3 & 3 & 3 & 2 & 1 \\ 2 & 2 & 2 & 2 & 1 \\ 1 & 1 & 1 & 1 & 1 \end{pmatrix},$$

which is easily verified.

Despite the usefulness of Gaussian elimination for many systems with band structure, serious limitations are sometimes imposed by the fact that the method does not preserve sparseness **within** the band. For example, when solving Poisson's equation (see Sec. 8.5.3) on a two-dimensional $N \times N$ lattice, we get a matrix with $n = N^2$ and band width $2N$. A itself has only $5N^2$ nonzero elements, but the elimination will fill up the band and L and U will have $2N^3$ nonzero elements. For such problems a different approach is needed which preserves the zero elements. This is a case when iterative methods are favorable.

5.4.3. Large-Scale Linear Systems

We have remarked that it is quite possible on a modern computer to solve a linear system $Ax = b$, where A is a full matrix of order $n = 1{,}000$. But on most computers only a fraction of the 10^6 nonzero elements can be stored simultaneously in the main store. If a secondary store of serial type (e.g., magnetic tape) is used, then approximately $n^3/3$ elements must be transferred back and forth during the elimination. We shall now show how Gaussian elimination can be modified so that the number of transfers is reduced by more than a factor n/N, where N is the available space in the main store.

We assume in the following that the matrix (A, b) is stored by rows, and first assume that n is so small that

$$\tfrac{1}{2}n(n+1) + n \leq N. \tag{5.4.5}$$

The elimination then proceeds in n steps, $k = 1, 2, \ldots, n$. Immediately before the kth step, only the first $(k-1)$ rows have been modified, and these have been reduced to trapezoidal form illustrated by

$$\begin{matrix} \times & \times & \cdots & \times & \cdots & \times \\ & \times & \cdots & \times & \cdots & \times \\ & & \cdot & & & \\ & & & \cdots & & \\ & & & \cdot & & \\ & & & \times & \cdots & \times. \end{matrix}$$

In step k equation number k is read into the main store, and the unknowns $x_1, \ldots, x_{k-1}$ are eliminated from this equation in $k-1$ minor steps, $i = 1, 2, \ldots, k-1$ as follows:

i. If $|a_{ki}| > |a_{ii}|$, then interchange row i and row k.
ii. If $a_{ii} \neq 0$, then compute $m_{ki} = a_{ki}/a_{ii}$ and subtract m_{ki} times row i from row k; otherwise, put $m_{ki} = 0$.

Step i insures that $|m_{ki}| \leqslant 1$, and therefore this elimination algorithm enjoys the same stability as Gaussian elimination with partial pivoting. Since in each minor step either an interchange is made or not, we can store information about these interchanges in $\frac{1}{2}n(n+1)$ Boolean variables. If this information and the multipliers are saved, then new right-hand sides can be treated without repeating the elimination. Note, however, that this algorithm, as opposed to ordinary Gaussian elimination, does not give rise to an *LU*-decomposition of some row permutation of A.

We mention in passing another advantage of the described elimination algorithm: it produces the values of the leading principal minors of A. This is of importance in methods for computing eigenvalues of matrices, based on values of this sequence of minors.

We now proceed to the general case when

$$\tfrac{1}{2}n(n+1)+n > N \geqslant 2(n+1).$$

We start the elimination as above. Since (5.4.5) is no longer satisfied, we see that when (say) $n_1 < n$ equations have been treated, there is no room for the next equation in the main store. Then, for $s = n_1, n_1+1, \ldots, n-1$, we proceed to read out equation s, read in equation $s+1$, and eliminate x_i, $i = 1, 2, \ldots, n_1 - 1$ from this.

After this sweep is finished, the first $n_1 - 1$ equations are no longer needed, and a system with $(n - n_1 + 1)$ equations and unknowns remains: A new sweep on this system is now performed in a similar way, and in this the equations and unknowns x_i, $i = n_1, n_1 + 1, \ldots, n_2 - 1$ will be eliminated. We continue in this way until all the unknowns have been eliminated. If in each sweep the main store is used fully, then approximately

$$s = \frac{1}{2}\frac{n^2}{N}$$

sweeps are needed, as can be seen from the schematic diagram below.

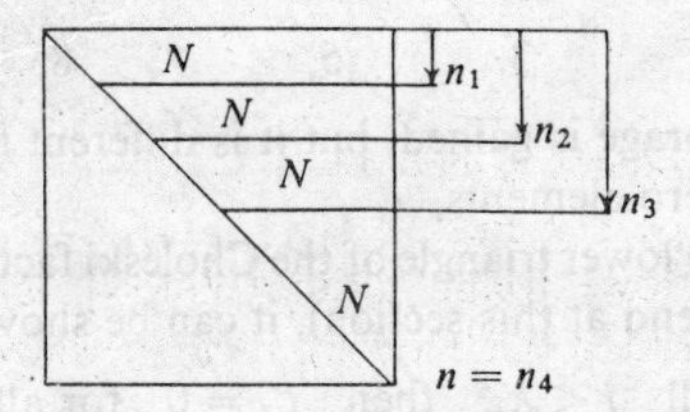

The number of transfers needed with this algorithm is approximately

$$n^2 + (n^2 - 2N) + (n^2 - 4N) + \ldots \approx n^2 s - 2N \cdot \frac{1}{2}s^2 = \frac{n^4}{4N},$$

since there are about $s = \tfrac{1}{2}n^2N$ terms in the sum. The reduction compared to ordinary Gaussian elimination is a factor $\tfrac{4}{3}(N/n)$, which in a typical case with $n = 1{,}000$ and $N = 50{,}000$ equals 67.

We finally mention that if instead the matrix (A, b) is stored by columns, the elimination can also be arranged in such a way that the same saving of memory transfers is made. This variant will more closely resemble ordinary Gaussian elimination.

5.4.4. Other Sparse Matrices

Among sparse matrices, the band matrices are easiest to handle. In some applications—e.g., finite-element calculations (see Sec. 8.6.5)—one encounters *positive-definite symmetric matrices*, where the location of the nonzero elements differs greatly from row to row. It can then be more efficient to use the following **profile-storage scheme**; see George [99].

The lower triangle of the symmetric $n \times n$ matrix A is stored row by row. In each row only the elements from the first nonzero entry to the diagonal are stored. (Note that the zeros contained between these two elements are stored.) A is thus represented by a vector, $s = (s_1, s_2, \ldots, s_{\mu(n)})$. Along with s, a *pointer vector*, $\boldsymbol{\mu} = (\mu(1), \mu(2), \ldots, \mu(n))$, is stored, where $\mu(i)$ points to the location in s of the diagonal element of the ith row of A. Hence

$$i - j < \mu(i) - \mu(i-1) \Longrightarrow a_{ij} = s_p, \quad \text{where} \quad p = \mu(i) - i + j. \tag{5.4.6}$$

For example:

$$A = \begin{bmatrix} 25 & 3 & 0 & 0 & 0 \\ 3 & 21 & 2 & 4 & 0 \\ 0 & 2 & 23 & 0 & 0 \\ 0 & 4 & 0 & 22 & 1 \\ 0 & 0 & 0 & 1 & 20 \end{bmatrix}.$$

$$s = (25,\ 3,\ 21,\ 2,\ 23,\ 4,\ 0,\ 22,\ 1,\ 20)$$

$$\boldsymbol{\mu} = (1,\quad 3,\quad 5,\quad 8,\quad 10).$$

In this example no storage is gained, but it is different for, say, a 100×100 matrix with 400 nonzero elements,

Let $L = (l_{ij})$ be the lower triangle of the Choleski factorization of A. Then (see Problem 6 at the end at this section), it can be shown that since

$$a_{ij} = 0 \quad \text{for all} \quad j < k_i, \quad \text{then} \quad l_{ij} = 0 \quad \text{for all} \quad j < k_i, \tag{5.4.7}$$
$$k_i = i + 1 + \mu(i-1) - \mu(i)$$

It is an interesting exercise to program operations for matrices stored in this fashion (see Problem 7).

By an appropriate **permutation** of the variables and the equations, the band width can often be reduced, which can mean a substantial saving of storage and computation. If symmetry is to be preserved, the same permutation must be applied to both. One can represent the sparsity structure of a symmetric matrix by a **graph**, where the *nodes* i and j are joined by a *line*, if and only if $a_{ij} \neq 0$. We then say that the nodes i and j are *adjacent*. The graph belonging to the matrix A of our example is shown in Fig 5.4.2. The number of lines incident to a node is called the *degree* of the node. A *permutation means a relabeling* of the nodes. (For a nonsymmetric matrix, a_{ij} and a_{ji} may not be simultaneously zero. In that case one must work with a directed graph—i.e., in which the lines have arrows.) Several algorithms for relabeling have been suggested for the symmetric case; see the article by Elizabeth Cuthill in [79]. In finite-element calculations, the **reverse Cuthill-McKee algorithm** has proven to be very successful: once a starting node has been labeled with an n, the algorithm consists in labeling successively, in

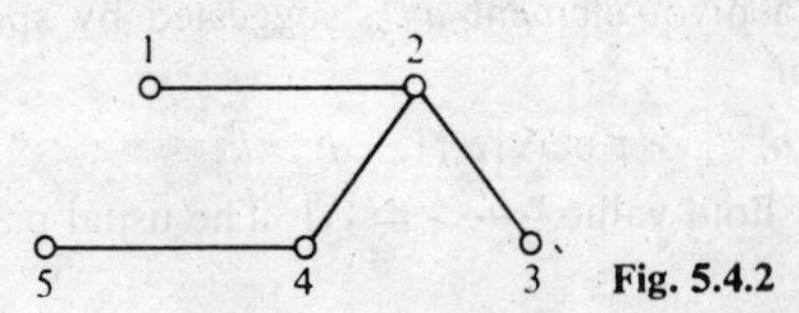

Fig. 5.4.2

order of increasing degree, those nodes not already labeled which are adjacent to the node with label i, for $i = n, n-1, \ldots, 2, 1$.

Very large *nonsymmetric matrices with highly irregular sparseness* are encountered in some applications—e.g., in the computational design of electrical networks. Then the profile-storage scheme may not be the most appropriate one. F. Gustavson [79, pp. 41–52] describes the data structure of a sparse matrix A by means of three vectors AN, JA, IA. AN contains the nonzero elements row by row; the zeros between nonzero elements need not be stored. The column number in A of the element AN(k) is given in JA(k), while IA(i) gives the position in arrays JA and AN of the first element of the ith row of A. For example:

$$A = \begin{bmatrix} 7 & 0 & -3 & 0 & -1 & 0 \\ 2 & 8 & 0 & 0 & 0 & 0 \\ 0 & 0 & 1 & 0 & 0 & 0 \\ -3 & 0 & 0 & 5 & 0 & 0 \\ 0 & -1 & 0 & 0 & 4 & 0 \\ 0 & 0 & 0 & -2 & 0 & 6 \end{bmatrix}$$

$$\begin{aligned} \text{AN} &= (7, \ -3, \ -1, \ 2, \ 8, \ 1, \ -3, \ 5, \ -1, \ 4, \ -2, \ 6) \\ \text{JA} &= (1, \ 3, \ 5, \ 1, \ 2, \ 3, \ 1, \ 4, \ 2, \ 5, \ 4, \ 6) \\ \text{IA} &= (1, \ 4, \ 6, \ 7, \ 9, \ 11, \ 13). \end{aligned}$$

The last element in IA is equal to the total number of elements in AN plus one.

Gustavson (loc. cit.) also shows how to save storage and computation when one has a large number of matrices with the same sparseness structure. An alternative solution to this important problem is given by Curtis and Reid, *Journal of the Institute of Mathematics and its Applications* 8 (1971): 344–53.

Having chosen an appropriate compact form for the matrix, it is clearly desirable to use an algorithm which is both numerically stable, and *sparseness-preserving*. By this we mean that the successive stages of the algorithm should not introduce too many "new" non-zero elements.

One difficulty is that the partial-pivoting strategy, recommended for stability reasons in Sec. 5.3.3, is usually not advantageous from the sparseness preserving point of view. One has, however, obtained good results by

threshold pivoting—i.e., a pivot element $a_{k'k}^{(k)}$, suggested by sparseness considerations, is accepted if

$$|a_{k'k}^{(k)}| \geq \tau \max |a_{ik}^{(k)}|, \quad i \geq k, \tag{5.4.8}$$

where τ is a preset "threshold value," $0 < \tau \leq 1$. The usual partial pivoting is obtained for $\tau = 1$.

Another difficulty is that some of those zeros of A which are contained between the first nonzero element in a column and the diagonal will be replaced by nonzero elements during the elimination. In the profile-storage scheme this is taken into account, though sometimes in an overly pessimistic way. **Iterative methods**, where the computer works with the original matrix A all the time, are therefore attractive alternatives to elimination, and will be considered in Sec. 5.6.

We shall briefly mention a direct method, the method of **conjugate gradients**, devised by Hestenes and Stiefel—see Reid [78, pp. 231–54]—which has the same attractive property. Given a system $\boldsymbol{Ax} = \boldsymbol{b}$, with a positive-definite $n \times n$ matrix $\boldsymbol{A}$ and a starting vector $\boldsymbol{x}_0$, compute $\boldsymbol{r}_0 = \boldsymbol{b} - \boldsymbol{Ax}_0$, $\boldsymbol{p}_0 = \boldsymbol{r}_0$. For $i = 1, 2, 3, \ldots, n$, compute

$$\begin{aligned} a_i &= \frac{\|\boldsymbol{r}_i\|_2^2}{\boldsymbol{p}_i^{\mathrm{T}}\boldsymbol{A}\boldsymbol{p}_i} \\ \boldsymbol{x}_{i+1} &= \boldsymbol{x}_i + a_i\boldsymbol{p}_i \\ \boldsymbol{r}_{i+1} &= \boldsymbol{r}_i - a_i\boldsymbol{A}\boldsymbol{p}_i \\ b_i &= \frac{\|\boldsymbol{r}_{i+1}\|^2}{\|\boldsymbol{r}_i\|^2} \\ \boldsymbol{p}_{i+1} &= \boldsymbol{r}_{i+1} + b_i\boldsymbol{p}_i. \end{aligned} \tag{5.4.9}$$

$\boldsymbol{x}_n$ will be the solution of the linear system, if rounding errors are neglected. Often a good approximation is obtained by $\boldsymbol{x}_i$ for some $i < n$. The method can be considered as a minimization method for the function $\frac{1}{2}\boldsymbol{x}^{\mathrm{T}}\boldsymbol{A}\boldsymbol{x} - \boldsymbol{b}^{\mathrm{T}}\boldsymbol{x}$. In the terminology of Sec. 10.5, the $\boldsymbol{p}_i$ are search directions; see Problems 1 and 5(a) of that section. One can show [78, p. 232] that $\boldsymbol{p}_i^{\mathrm{T}}\boldsymbol{r}_i = \|\boldsymbol{r}_i\|^2$ and that $\boldsymbol{p}_i^{\mathrm{T}}\boldsymbol{A}\boldsymbol{p}_j = 0$, whenever $i \neq j$. The latter condition is an n-dimensional generalization of the condition for conjugacy in the theory of conic sections and quadric surfaces.

The state of the art in 1972 of sparse matrices is well presented in [78] and [79].

PROBLEMS

1. (a) Show that the symmetric matrix

$$A = \begin{pmatrix} 10 & 7 & 8 & 7 \\ 7 & 5 & 6 & 5 \\ 8 & 6 & 10 & 9 \\ 7 & 5 & 9 & 10 \end{pmatrix}$$

is positive-definite.

(b) Determine an upper-triangular matrix R such that $A = R^T R$.

. Change the program in Problem 3 in Sec. 5.3 for solving a linear system $Ax = b$ so that the savings in number of operations and memory space for a positive-definite matrix A is realized.

. (a) Let A be an $n \times n$ tridiagonal matrix such that $\det(A_k) \neq 0$. Then the LU-decomposition of A exists and can be computed by the formulas given in Example 5.4.3. Use this to derive a recursion formula for computing $\det(A_k)$, $k = 1, 2, \ldots, n$.

(b) Determine the largest n for which the $n \times n$ matrix

$$A = \begin{pmatrix} 2 & 1.01 & & & \\ 1.01 & 2 & 1.01 & & \\ & 1.01 & \cdot & \cdot & \\ & & \cdot & \cdot & 1.01 \\ & & & 1.01 & 2 \end{pmatrix}$$

is positive-definite.

4. Use the formulas in Example 5.4.3 to compute the LU-decomposition of the matrix

$$A = \begin{pmatrix} 1 & 1 & & & \\ 1 & 2 & 1 & & \\ & 1 & 3 & 1 & \\ & & 1 & 4 & 1 \\ & & & 1 & 5 \end{pmatrix}.$$

5. (a) A matrix is called block-tridiagonal if it has the form

$$\begin{pmatrix} A_1 & C_1 & & & \\ B_2 & A_2 & C_2 & & \\ & \cdot & \cdot & \cdot & \\ & & \cdot & \cdot & C_{n-1} \\ & & & B_n & A_n \end{pmatrix},$$

Show that the formulas in Example 5.4.3 for solving $Ax = f$ in this case generalize to

$$\boldsymbol{\alpha}_1 = A_1, \quad \boldsymbol{\beta}_k = B_k \boldsymbol{\alpha}_{k-1}^{-1}, \quad \boldsymbol{\alpha}_k = A_k - \boldsymbol{\beta}_k C_{k-1}, \quad k = 2, 3, \ldots, n$$
$$g_1 = f_1, \quad g_i = f_i - \boldsymbol{\beta}_i g_{i-1}, \quad i = 2, 3, \ldots, n$$
$$x_n = \boldsymbol{\alpha}_n^{-1} g_n, \quad x_i = \boldsymbol{\alpha}_i^{-1}(g_i - C_i x_{i+1}), \quad i = n-1, \ldots, 2, 1,$$

where $\boldsymbol{\alpha}_k$ and $\boldsymbol{\beta}_k$ now are matrices and g_i, f_i, and x_i are vectors.

(b) Determine the number of multiplications needed to solve $Ax = f$ when A is block-tridiagonal and the matrices A_k, $k = 1, 2, \ldots, n$ all are $m \times m$.

6. Derive Eqs. (5.4.6), and (5.4.7).

7. Write a program for the operation $y = Ax$, where A is a profile-stored, positive-definite, symmetric matrix. x and y are vectors, stored in the usual way. (No secondary storage is to be used.)

5.5. ERROR ANALYSIS FOR LINEAR SYSTEMS

In the practical solution of a linear system, rounding errors will be introduced, whether we solve it by hand or use a computer. Also, in many cases the elements in A and b are not known, exactly, and we want to know the corresponding uncertainty in x. If we use a computer, most standard procedures will just give us a computed solution $\bar{x}$, without any error estimate at all. How do we proceed to estimate the error $\bar{x} - A^{-1}b$? Remember that for a large system several million operations have been carried out, each one involving a rounding error.

5.5.1. An Ill-Conditioned Example

If $\bar{x}$ is the computed solution to the system $Ax = b$, we define the **residual vector** r as

$$r = b - A\bar{x}. \tag{5.5.1}$$

Since $r = 0$ implies that $\bar{x} = A^{-1}b$, it is natural to expect that if r is small, then x is an accurate solution. However, as the following example constructed by W. Kahan shows, this may be far from the truth. Let

$$A = \begin{pmatrix} 1.2969 & 0.8648 \\ 0.2161 & 0.1441 \end{pmatrix}, \qquad b = \begin{pmatrix} 0.8642 \\ 0.1440 \end{pmatrix}, \tag{5.5.2}$$

and suppose we are given the approximate solution

$$\bar{x} = (0.9911, -0.4870)^T.$$

The residual vector corresponding to this $\bar{x}$ is exactly equal to

$$r = (-10^{-8}, 10^{-8})^T,$$

and therefore it seems reasonable to assume that the error in $\bar{x}$ is very small. However, not a single figure in $\bar{x}$ is meaningful! The exact solution is in fact

$$x = (2, -2)^T,$$

as can easily be verified by substitution.

In this case it is easy to recognize the extreme ill-conditioning of the system in Eq. (5.5.2). After eliminating x_1 we get $a_{22}^{(2)}x_2 = b_2^{(2)}$, where

$$a_{22}^{(2)} = 0.1441 - \frac{0.2161}{1.2969}0.8648 = 0.1441 - 0.1440999923 \approx 10^{-8}.$$

It is obvious that a very small change in the element 0.1441 will result in a large change in $a_{22}^{(2)}$, and therefore in x_2. Thus unless the coefficients in A and b

re given to a precision better than 10^{-8}, it is meaningless to talk about a olution to Eq. (5.5.2).

Even if the size of the residual vector gives no direct indication of the rror in $\bar{x}$, it is possible to use accurately computed residuals to estimate the rror, or even to correct the approximate solution. We return to this later in ec. 5.5.6.

.5.2. Vector and Matrix Norms

For the purpose of quantitatively discussing errors, it is convenient to be ble to associate with any vector or matrix a nonnegative scalar that in some ense measures its magnitude. Such measures which also satisfy some reasonble axioms are called **norms**. The most common vector norms are special ases of the family of L_p-norms (see Sec. 4.1.3)

$$\|x\|_p = (|x_1|^p + |x_2|^p + \ldots + |x_n|^p)^{1/p}, \quad 1 \leq p < \infty. \tag{5.5.3}$$

wo particular cases will be used often:

$p = 2$: **Euclidean norm:**

$$\|x\|_2 = (|x_1|^2 + |x_2|^2 + \ldots + |x_n|^2)^{1/2},$$

$p \to \infty$: **Maximum norm:**

$$\|x\|_\infty = \max_{1 \leq i \leq n} |x_i|. \tag{5.5.4}$$

ector norms must have the following properties, which are analogous to roperties of the usual concept of length:

1(a). $\|x\| > 0$ if $x \neq 0$, $\|x\| = 0$ if $x = 0$.
2(a). $\|\alpha x\| = |\alpha| \|x\|$, α a scalar.
3(a). $\|x + y\| \leq \|x\| + \|y\|$.

or the maximum norm these are all easy to verify.

Matrix norms shall have similar properties:

1(b). $\|A\| > 0$ if $A \neq 0$, $\|A\| = 0$ if $A = 0$.
2(b). $\|\alpha A\| = |\alpha| \|A\|$, α a scalar
3(b). $\|A + B\| \leq \|A\| + \|B\|$.
4(b). $\|AB\| \leq \|A\| \; \|B\|$.

f a matrix norm and a vector norm are related in such a way that

5(b). $\|Ax\| \leq \|A\| \; \|x\|$

satisfied for any A and x, then the two norms are said to be **consistent**. For ny vector norm, there exists a consistent matrix norm. In fact such a norm given by the **matrix-bound norm** subordinate to the vector norm

$$\|A\| \overset{\Delta}{=} \max_{x \neq 0} \frac{\|Ax\|}{\|x\|}. \tag{5.5.5}$$

In the following we shall mainly use the maximum vector norm of Eq. (5.5.4), and the subordinate matrix-bound norm, which can be shown to be

$$\|A\|_\infty = \max_{1 \le i \le n} \sum_{j=1}^{n} |a_{ij}|. \tag{5.5.6}$$

These have the advantage of being very simple to compute. Also, Eq. (5.5.6) has the attractive property that

$$\| \, |A| \, \| = \|A\|, \qquad |A| = (|a_{ij}|).$$

This is not true for the matrix-bound norm subordinate to the Euclidean norm, for which the best we can say is

$$\| \, |A| \, \| \le n^{1/2} \|A\|.$$

Example 5.5.1

The maximum norms for A and b in Eq. (5.5.2) are

$$\|A\|_\infty = \max (2.1617, 0.3602) = 2.1617, \qquad \|b\|_\infty = 0.8642.$$

5.5.3. Perturbation Analysis

We shall now investigate the condition of the problem of solving a nonsingular linear system $Ax = b$. We assume that the elements in A and b are the given data, and estimate the effect of perturbations in b and A. If we let

$$A(x + \delta x) = b + \delta b,$$

then $\delta x = A^{-1}\delta b$ and from 5(b) it follows that

$$\|\delta x\| \le \|A^{-1}\| \, \|\delta b\|. \tag{5.5.7}$$

This inequality is sharp in the sense that for any matrix-bound norm and for any A and b there exists a perturbation δb such that the equality sign holds.

Similarly, if we let

$$(A + \delta A)(x + \delta x) = b,$$

then

$$A\delta x + \delta A(x + \delta x) = 0.$$

It follows that $\delta x = -A^{-1}\delta A(x + \delta x)$, and thus we have the estimate

$$\|\delta x\| \le \|A^{-1}\| \, \|\delta A\| \, \|x + \delta x\|. \tag{5.5.8}$$

This is usually rewritten as

$$\frac{\|\delta x\|}{\|x + \delta x\|} \le \kappa(A) \frac{\|\delta A\|}{\|A\|}, \tag{5.5.9}$$

where

$$\kappa(A) = \|A\| \, \|A^{-1}\| \tag{5.5.10}$$

is the **condition number** of the matrix A with respect to the given norm. Again, for any A and b, there exists a perturbation δA such that the equality sign holds in (5.5.8) and (5.5.9).

From (5.5.7) and the inequality

$$\|\boldsymbol{b}\| = \|A\boldsymbol{x}\| \leq \|A\| \, \|\boldsymbol{x}\|$$

it follows that

$$\frac{\|\delta\boldsymbol{x}\|}{\|\boldsymbol{x}\|} \leq \kappa(A)\frac{\|\delta\boldsymbol{b}\|}{\|\boldsymbol{b}\|}, \tag{5.5.11}$$

but here *the equality sign will hold only for rather special right-hand sides* $\boldsymbol{b}$.

If $\kappa(A)$ is large, then small relative perturbations in A and $\boldsymbol{b}$ will produce large relative perturbations in $\boldsymbol{x}$, and the problem of solving $A\boldsymbol{x} = \boldsymbol{b}$ is ill-conditioned. However, to get $\kappa(A)$ we have to know the inverse A^{-1}, and this is generally not available.

Example 5.5.2

The matrix A in Eq. (5.5.2) can be shown to have the inverse (verify this!)

$$A^{-1} = 10^8\begin{pmatrix} 0.1441 & -0.8648 \\ -0.2161 & 1.2969 \end{pmatrix}$$

Thus

$$\|A^{-1}\|_\infty = 1.5130\cdot 10^8, \qquad \kappa(A) = 2.1617\cdot 1.5130\cdot 10^8 \approx 3.3\cdot 10^8,$$

which shows that the system is extremely ill-conditioned.

Even if the elements in A and $\boldsymbol{b}$ are known exactly, they usually will not be exactly represented as floating-point numbers in the computer. If we denote the corresponding rounded matrix by $\bar{A} = (\bar{a}_{ij})$, we have, from Theorem 2.3.1,

$$|\bar{a}_{ij} - a_{ij}| \leq u\,|a_{ij}|,$$

where u is the rounding unit of the floating-point representation. It follows that

$$\max_{1\leq i\leq n}\sum_{j=1}^{n}|\bar{a}_{ij} - a_{ij}| \leq u \max_{1\leq i\leq n}\sum_{j=1}^{n}|a_{ij}|.$$

A similar inequality holds for the error in the elements of $\bar{b}$, and thus

$$\|\bar{A} - A\|_\infty \leq u\|A\|_\infty, \qquad \|\bar{\boldsymbol{b}} - \boldsymbol{b}\|_\infty \leq u\|\boldsymbol{b}\|_\infty.$$

From (5.5.9) and (5.5.11) it follows that these rounding errors will produce a perturbation $\delta\boldsymbol{x}$ in $\boldsymbol{x}$, whose norm is bounded approximately by

$$\|\delta\boldsymbol{x}\|_\infty \leq 2u\kappa(A)\|\boldsymbol{x}\|_\infty.$$

5.5.4. Rounding Errors in Gaussian Elimination

A complete rounding-error analysis of elimination methods was first given by J. H. Wilkinson in 1961. Since this analysis gives the reasons for using the pivotal strategies we have recommended, it will here be briefly sketched in a somewhat modified form developed by J. K. Reid.

We have seen that in the kth step of the elimination, the elements ar transformed according to Eq. (5.3.5):

$$m_{ik} = \frac{a_{ik}^{(k)}}{a_{kk}^{(k)}}, \qquad a_{ij}^{(k+1)} = a_{ij}^{(k)} - m_{ik}a_{kj}^{(k)},$$

$$i, j = k+1, k+2, \ldots, n.$$

Because of rounding errors, the computed quantities, here denoted b an overbar, will instead satisfy, by Eq. (2.5.2),

$$\bar{m}_{ik} = \frac{\bar{a}_{ik}^{(k)}}{\bar{a}_{kk}^{(k)}}(1 + \delta_1) \tag{5.5.12a}$$

$$\bar{a}_{ij}^{(k+1)} = (\bar{a}_{ij}^{(k)} - \bar{m}_{ik}\bar{a}_{kj}^{(k)}(1 + \delta_2))(1 + \delta_3), \tag{5.5.12b}$$

where $|\delta_i| \leq u$, $i = 1, 2, 3$, and u is the rounding unit. Now it is possible t find perturbations $\epsilon_{ij}^{(k)}$ to $\bar{a}_{ij}^{(k)}$ such that if the **exact** operations are performe on the perturbed elements, then $\bar{m}_{ik}$ and $\bar{a}_{ij}^{(k+1)}$ are obtained—i.e.,

$$\bar{m}_{ik} = \frac{\bar{a}_{ik}^{(k)} + \epsilon_{ik}^{(k)}}{\bar{a}_{kk}^{(k)}}, \tag{5.5.13a}$$

$$\bar{a}_{ij}^{(k+1)} = \bar{a}_{ij}^{(k)} + \epsilon_{ij}^{(k)} - \bar{m}_{ik}\bar{a}_{kj}^{(k)}. \tag{5.5.13b}$$

From Eqs. (5.5.12a) and (5.5.13a) it follows immediately that

$$\epsilon_{ik}^{(k)} = \bar{a}_{ik}^{(k)} \cdot \delta_1.$$

If we write Eq. (5.5.12b) as

$$\bar{m}_{ik}\bar{a}_{kj}^{(k)} = \frac{\bar{a}_{ij}^{(k)} - \bar{a}_{ij}^{(k+1)}/(1 + \delta_3)}{1 + \delta_2}$$

and substitute in Eq. (5.5.13b), we obtain

$$\epsilon_{ij}^{(k)} = \bar{a}_{ij}^{(k+1)}(1 - (1 + \delta_3)^{-1}(1 + \delta_2)^{-1}) - \bar{a}_{ij}^{(k)}(1 - (1 + \delta_2)^{-1}).$$

Thus if higher powers of u are neglected, we get the estimates

$$|\epsilon_{ik}^{(k)}| \leq u\,|\bar{a}_{ik}^{(k)}|, \qquad |\epsilon_{ij}^{(k)}| \leq 3 \cdot u \max(|\bar{a}_{ij}^{(k)}|, |\bar{a}_{ij}^{(k+1)}|), \quad j > k.$$

If we take $\bar{m}_{ii} = 1$, and sum the equations in Eq. (5.5.13b), then in the sam way as Eq. (5.3.11) was derived we get

$$a_{ij} = \sum_{k=1}^{p} \bar{m}_{ik}\bar{a}_{kj}^{(k)} + e_{ij}, \quad e_{ij} = \sum_{k=1}^{r} \epsilon_{ij}^{(k)}, \tag{5.5.1}$$

$$p = \min(i, j), \qquad r = \min(i - 1, j).$$

This shows that the **computed matrices $\bar{L} = (\bar{m}_{ik})$ and $\bar{U} = (\bar{a}_{kj}^{(k)})$ are the exa triangular factors of the matrix $A + E$,**

$$\bar{L}\bar{U} = A + E, \quad E = (e_{ij}),$$

where, since e_{ij} is the sum of $\min(i - 1, j)$ quantities,

$$|e_{ij}| \leq 3 \cdot u \min(i - 1, j) \max_k |\bar{a}_{ij}^{(k)}|. \tag{5.5.1}$$

Since the right-hand side is transformed in exactly the same way as the columns in A, we also have

$$\bar{L}\bar{y} = b + c,$$

where

$$|c_i| \leq 3 \cdot u(i-1) \max_k |\bar{b}_i^{(k)}|. \tag{5.5.16}$$

These results hold without any assumption about the multipliers. **This shows that the purpose of any pivotal strategy is to avoid growth in the size of the elements $\bar{a}_{ij}^{(k)}$ and $\bar{b}_i^{(k)}$, and that the size of the multipliers is of no consequence** (see the remark on large multipliers for positive-definite matrices, Sec. 5.4.1).

Finally, we must take into account the rounding errors performed when solving $\bar{U}x = \bar{y}$. However, for most matrices occurring in practice it has been observed that it is the errors E and c in the elimination step that limit the accuracy of the computed solution of $Ax = b$. Although this can be proved analytically for only special classes of matrices, large errors almost never occur in the solution of the triangular set of equations $\bar{U}x = \bar{y}$.

If the quantities $\max_k |\bar{a}_{ij}^{(k)}|$ and $\max_k |\bar{b}_i^{(k)}|$ are observed during the elimination, then (5.5.15) and (5.5.16) give an a posteriori bound for the perturbations in A and b. In order to get a priori bounds we now assume that $|\bar{m}_{ik}| \leq 1$, which is true if partial or complete pivoting is used. It can be shown that with this assumption an estimate similar to Eq. (5.5.15) holds,

$$|e_{ij}| \leq 2 \cdot u \min(i-1, j) \max_{i,j,k} |\bar{a}_{ij}^{(k)}|.$$

Thus if we put

$$g_n = \frac{\max_{i,j,k} |\bar{a}_{ij}^{(k)}|}{\max_{i,j} |a_{ij}|}$$

then the error matrix can be bounded by

$$|E| \leq 2 \cdot g_n u \max_{i,j} |a_{ij}| \begin{pmatrix} 0 & 0 & 0 & \cdots & 0 & 0 \\ 1 & 1 & 1 & \cdots & 1 & 1 \\ 1 & 2 & 2 & \cdots & 2 & 2 \\ \vdots & & & & & \vdots \\ 1 & 2 & 3 & \cdots & n-2 & n-2 \\ 1 & 2 & 3 & \cdots & n-1 & n-1 \end{pmatrix}.$$

The maximum norm of the matrix appearing on the right-hand side is

$$(1 + 2 + \cdots + n - 1 + n) - 1 = \tfrac{1}{2}n(n+1) - 1,$$

and by slightly refining the estimate of E it can be shown that

$$\|E\|_\infty \leq n^2 g_n u \max_{i,j} |a_{ij}|.$$

Using the crude estimate $\max_{i,j} |a_{ij}| \leq \|A\|_\infty$, we get the theorem:

THEOREM 5.5.1

Let $\bar{L}$ and $\bar{U}$ be the computed triangular factors of A, obtained by using Gaussian elimination with partial or complete pivoting. Then if floating-point arithmetic with rounding unit u has been used, there is a matrix E satisfying

$$\|E\|_\infty \leq k_1(n)\cdot u\|A\|_\infty, \quad k_1(n) = n^2 g_n,$$

such that

$$\bar{L}\bar{U} = A + E.$$

By analyzing the rounding errors involved in solving the triangular systems $\bar{L}y = b$ and $\bar{U}x = \bar{y}$, it is possible to derive a similar result for the computed solution $\bar{x}$.

THEOREM 5.5.2

Let $\bar{x}$ denote the computed solution of the system $Ax = b$, obtained by forward and back substitution in $\bar{L}y = b$ and $\bar{U}x = \bar{y}$, where $\bar{L}$ and $\bar{U}$ have been obtained as in Theorem 5.5.1. Then there is a matrix δA, depending on both A and b, satisfying

$$\|\delta A\|_\infty \leq k_2(n)u\|A\|_\infty, \quad k_2(n) = (n^3 + 3n^2)g_n,$$

such that

$$(A + \delta A)\bar{x} = b.$$

The bounds obtained for E and δA in the theorems above are satisfactory, unless the ratio g_n is large. A priori the following estimates for g_n hold (Wilkinson [64]):

Partial pivoting: $g_n \leq 2^{n-1}$,
Complete pivoting: $g_n \leq 1.8n^{0.25 \ln n}$.

The upper limit for partial pivoting can be attained, and thus already for $n = 41$ we can have $g_n = 2^{40} \approx 10^{12}$. The limit for g_n with complete pivoting is much smaller for large values of n, and could justify the additional work involved. The gain is, however, mostly theoretical, since practical experience shows that it is extremely rare for g_n to exceed 8 even with partial pivoting!

Theorem 5.5.2 gives no direct estimate of the error in $\bar{x}$. To get such an estimate we must compute $\kappa(A)$ and use (5.5.9). However, the backward analysis above is often sufficient. For example, if all the elements in A already have relative errors larger than $u\cdot k_2(n)$, then it is usually not meaningful to try to compute a solution more accurate than $\bar{x}$.

Also, in many practical applications the error in $\bar{x}$ is not of primary importance. Rather, any solution $\bar{x}$ giving a sufficiently small residual $(b - A\bar{x})$ will do. Now, from Theorem 5.5.2,

$$b - A\bar{x} = (A + \delta A)\bar{x} - A\bar{x} = \delta A\bar{x},$$

and it follows that

$$\|b - A\bar{x}\|_\infty \leq k_2(n)u\|A\|_\infty\|\bar{x}\|_\infty.$$

Thus if $\|\bar{x}\|_\infty$ is small, then Gaussian elimination guarantees a small residual, even when A is ill-conditioned!

5.5.5. Scaling of Linear Systems

In a linear system $Ax = b$ the unknowns x_j are often physical quantities. If we change the units in which these are measured, this is equivalent to a scaling of the unknowns—say, $x_j = \alpha_j x'_j$. If we at the same time multiply the ith equation by β_i, then the original system $Ax = b$ will be transformed into a system $A'x' = b'$, where

$$A' = D_2 A D_1, \qquad b' = D_2 b, \qquad x = D_1 x'$$

and

$$D_1 = \text{diag}(\alpha_1, \alpha_2, \ldots, \alpha_n), \qquad D_2 = \text{diag}(\beta_1, \beta_2, \ldots, \beta_n).$$

Now it seems natural to expect that such a scaling should have no effect on the relative accuracy of the computed solution. This is in fact to a certain extent true, as the following theorem, designed by F. L. Bauer, shows.

THEOREM 5.5.3

Denote by x *and* x' *the computed solutions to the two systems* $Ax = b$ *and* $(D_2AD_1)x' = D_2b$. *Assume that* D_2 *and* D_1 *are diagonal matrices, whose elements are even powers of the base in the number system used, so that no rounding errors are introduced by the scaling. Then, if Gaussian elimination is performed in floating-point arithmetic on the two systems and* ***if the same choice of pivots is used****, all the results will differ only in the exponents, and we have exactly* $\bar{x} = D_1\bar{x}'$.

It follows that essentially the only effect a scaling can have is to influence the choice of pivots. Now assume that we use the partial-pivoting strategy. Obviously, for any given sequence of pivots (which does not give a pivot exactly equal to zero) there exists a scaling of the equations such that partial pivoting will select these pivots. It is clear, then, that an unsuitable scaling of the equations may lead to a very poor choice of pivots.

Example 5.5.3

Consider the linear system

$$\begin{pmatrix} 1 & 10{,}000 \\ 1 & 0.0001 \end{pmatrix} \begin{pmatrix} x_1 \\ x_2 \end{pmatrix} = \begin{pmatrix} 10{,}000 \\ 1 \end{pmatrix},$$

which has the solution $x_1 = x_2 = 0.9999$, correctly rounded to four decimals. Partial pivoting will here select a_{11} as pivot, and if we use three-figure floating-point arithmetic, the computed solution will be

$$\bar{x}_2 = 1.00, \qquad \bar{x}_1 = 0. \qquad \text{(Bad!)}$$

If the first equation is multiplied by 10^{-4}, we get the equivalent system

$$\begin{pmatrix} 0.0001 & 1 \\ 1 & 0.0001 \end{pmatrix}\begin{pmatrix} x_1 \\ x_2 \end{pmatrix} = \begin{pmatrix} 1 \\ 1 \end{pmatrix}.$$

Here a_{21} will become the pivot, and we shall get the computed solution

$$\bar{x}_2 = 1.00, \quad \bar{x}_1 = 1.00 \quad \text{(Good!)}$$

It is often recommended that if partial pivoting is used, then the equations should be **equilibrated** before the elimination. By this we mean that the matrix $A = (a_{ij})$ of the scaled system shall satisfy

$$\max_{1 \le j \le n} |a_{ij}| = 1, \quad i = 1, 2, \ldots, n.$$

We note that with equilibration we avoid the bad selection of pivots in Example 5.5.3. From Theorem 5.5.3 it follows that it is not necessary to perform the equilibration explicitly. Instead, we can modify the partial-pivoting strategy and in step k look for

$$\max_{k \le i \le n} \left| \frac{a_{ik}^{(k)}}{s_i} \right|, \quad \text{where} \quad s_i = \max_{1 \le j \le n} |a_{ij}|.$$

The scaling of the unknowns will obviously influence the equilibration, and therefore indirectly the choice of pivots. Thus an unfortunate choice of scaling of the unknowns can also have disastrous effects.

Example 5.5.4

Consider the equilibrated system $Ax = b$, where

$$A = \begin{pmatrix} \epsilon & -1 & 1 \\ -1 & 1 & 1 \\ 1 & 1 & 1 \end{pmatrix}, \qquad A^{-1} = \begin{pmatrix} 0 & -2 & 2 \\ -2 & 1-\epsilon & 1+\epsilon \\ 2 & 1+\epsilon & 1-\epsilon \end{pmatrix}, \qquad |\epsilon| \ll 1.$$

This is a well-conditioned matrix, $\kappa(A) = 3$ in maximum norm, and therefore Gaussian elimination with partial pivoting will give an accurate solution. However, the choice of $a_{11} = \epsilon$ as pivot will have a disastrous effect on the accuracy in the computed solution.

Now consider the scaling $x_2' = x_2/\epsilon$, $x_3' = x_3/\epsilon$. If the resulting system is again equilibrated, we get $A'x' = b'$, where

$$A' = \begin{pmatrix} 1 & -1 & 1 \\ -1 & \epsilon & \epsilon \\ 1 & \epsilon & \epsilon \end{pmatrix}.$$

Here partial, and even complete, pivoting will select $a_{11}' = 1$ as the first pivot. However, from Theorem 5.5.3, we know that this will have the same unfortunate effects on accuracy as did the same choice of pivots in the system $Ax = b$.

It is difficult to give a general rule as to how a linear system should be scaled before Gaussian elimination. It can be verified that the condition number in maximum norm for the matrices in the example above satisfies

$$\kappa(A') \approx \frac{3}{\epsilon} \gg \kappa(A).$$

It seems, then that a possible approach to the scaling is to determine D_1 and D_2 so that $\kappa(D_2 A D_1)$ is minimized. However, it turns out that these optimal D_1 and D_2 essentially depend on A^{-1}, which in practice is unknown. Another objection to this approach is that the scaling of the unknowns will change the norm in which the error is measured. Thus a sensible approach in most cases is to choose D_1 in a way which reflects the importance of the unknowns and to use D_2 to equilibrate the system.

5.5.6. Iterative Improvement of a Solution

We have seen that when A is ill-conditioned, the computed solution $\bar{x}$ may be inaccurate without any indication in the form of a large solution vector. This will happen for particular right-hand sides b, such that $A^{-1}b$ is small even though the norm of A^{-1} is large. If we compute the inverse matrix, we cannot be deceived in this way and ill-conditioning will show up in the form of large elements in the computed inverse $\bar{A}^{-1}$. If the computed condition number $\|A\|_\infty \|\bar{A}^{-1}\|_\infty$ is small, then $\bar{A}^{-1}$ certainly is close to the true inverse.

Unfortunately, the extra work involved in computing the inverse is often prohibitively large. We shall here describe an alternative approach, which requires very little extra work when n is large, and which also gives a correction to $\bar{x}$ and not just an estimate of the error. If $r = b - A\bar{x}$ is the residual vector to a computed solution $\bar{x}$, then

$$A(x - \bar{x}) = r.$$

Now assume that Gaussian elimination has given the approximate triangular factors $\bar{L}$ and $\bar{U}$. From Theorem 5.5.1 we know that $\bar{L}\bar{U} = A + E$, where E is small. We can therefore approximate the correction $x - \bar{x}$ with the solution to

$$\bar{L}(\bar{U}\delta x) = r,$$

which splits into the two triangular systems $\bar{L}y = r$ and $\bar{U}\delta x = y$. The computation of r and δx, therefore, takes only $n^2 + 2 \cdot \frac{1}{2}n^2 = 2n^2$ operations, which is an order of magnitude less than the $n^3/3$ operations required for computing $\bar{x}$.

New rounding errors are introduced in the computation of δx, and $\bar{x} + \delta x$ may not be a more accurate solution than $\bar{x}$. A more detailed analysis shows that, because of the cancellation which will take place in computing $r = b - A\bar{x}$, it is essential that it be computed with sufficient accuracy. It is often advisable to proceed as follows. The components in $\boldsymbol{r}$ are

$$r_i = b_i - \sum_{k=1}^{n} a_{ik}\bar{x}_k, \quad i = 1, 2, \ldots, n.$$

If a_{ik} and $\bar{x}_k$ are given with t digits, then the products $a_{ik}\bar{x}_k$ contain at most $2t$ digits. We compute these products exactly and accumulate the sum using $2t$ digits. Finally, $b_i - (A\bar{x})_i$ is computed and rounded to t digits. This can be done very conveniently on most computers, and will insure that the error from this part of the calculation is small.

The improved solution $\bar{\boldsymbol{x}} + \boldsymbol{\delta x}$ can, of course, be corrected in the same way, etc., and we can carry out the following iterative process:

Iterative improvement. Put $\boldsymbol{x}^{(1)} = \bar{\boldsymbol{x}}$ and compute $\boldsymbol{x}^{(s)}$, $s = 2, 3, \ldots$, from

$$\boldsymbol{r}^{(s)} = \boldsymbol{b} - A\boldsymbol{x}^{(s)}, \qquad \bar{L}(\bar{U}\boldsymbol{\delta x}^{(s)}) = \boldsymbol{r}^{(s)}, \quad \boldsymbol{x}^{(s+1)} = \boldsymbol{x}^{(s)} + \boldsymbol{\delta x}^{(s)}, \tag{5.5.17}$$

where only the computation of $\boldsymbol{r}^{(s)}$ requires double precision. If A is not too ill-conditioned—say,

$$nu\kappa(A) \leq 0.1,$$

—then $\boldsymbol{x}^{(s)}$ will converge rapidly to the correct solution rounded to single precision. We can also get a good estimate of $\kappa(A)$ from

$$\kappa(A) \leq \frac{1}{nu}\frac{\|\boldsymbol{\delta x}^{(1)}\|_\infty}{\|\boldsymbol{x}^{(2)}\|_\infty}. \tag{5.5.18}$$

If convergence is not obtained in practice, then we must assume that A is so ill-conditioned that higher-precision arithmetic throughout is unavoidable.

Example 5.5.5

We illustrate the method on the equations

$$\begin{pmatrix} 0.20000 & 0.16667 & 0.14286 \\ 0.16667 & 0.14286 & 0.12500 \\ 0.14286 & 0.12500 & 0.11111 \end{pmatrix}\begin{pmatrix} x_1 \\ x_2 \\ x_3 \end{pmatrix} = \begin{pmatrix} 0.50953 \\ 0.43453 \\ 0.37897 \end{pmatrix},$$

which have the exact solution $x_1 = x_2 = x_3 = 1$. If floating-point arithmetic with $t = 5$ digits is used, Gaussian elimination will give the computed triangular factors

$$\bar{L} = \begin{pmatrix} 1 & 0 & 0 \\ 0.83335 & 1 & 0 \\ 0.71430 & 1.49874 & 1 \end{pmatrix}, \quad \bar{U} = \begin{pmatrix} 0.20000 & 0.16667 & 0.14286 \\ 0 & 0.00397 & 0.00595 \\ 0 & 0 & 0.00015 \end{pmatrix},$$

and the computed solution

$$\bar{\boldsymbol{x}} = (1.03845, 0.89673, 1.06667)^T.$$

We compute first $A\bar{x}$ using $2t$ digits, then $r^{(1)} = b - A\bar{x}$,

$$A\bar{x} = \begin{pmatrix} 0.5095324653 \\ 0.4345190593 \\ 0.3789619207 \end{pmatrix}, \qquad r^{(1)} = 10^{-5}\begin{pmatrix} -0.24653 \\ 1.09407 \\ 0.80793 \end{pmatrix},$$

and then solve for $\delta x^{(1)}$. We get

$$\delta x^{(1)} = \begin{pmatrix} -0.03709 \\ 0.09955 \\ -0.06424 \end{pmatrix}, \quad x^{(2)} = \bar{x} + \delta x^{(1)} = \begin{pmatrix} 1.00136 \\ 0.99628 \\ 1.00243 \end{pmatrix}.$$

The errors in the corrected solution are about 30 times smaller than those in $\bar{x}$. The rapid convergence obtained if we continue the iterations is clearly illustrated in the tables below.

s	$x^{(s)}$		
1	1.03845	0.89673	1.06667
2	1.00136	0.99628	1.00243
3	1.00005	0.99986	1.00009
4	1.00000	1.00000	1.00000

s	$10^5 \cdot r^{(s)}$		
1	−0.24653	1.09407	0.80793
2	0.08626	0.10180	0.07131
3	0.04764	0.04169	0.03571

It is interesting to note that the residuals do not decrease at the same rate as the errors in successive $x^{(s)}$. Using (5.5.18), we can derive the estimate

$$\kappa(A) \approx \frac{1}{3 \cdot \frac{1}{2} \cdot 10^{-5}} \frac{0.1}{1} = 0.7 \cdot 10^4,$$

which agrees well with the known value.

PROBLEMS

1. To the linear system $Ax = b$,

$$A = \begin{pmatrix} 0.780 & 0.563 \\ 0.913 & 0.659 \end{pmatrix}, \qquad b = \begin{pmatrix} 0.217 \\ 0.254 \end{pmatrix},$$

two different approximate solutions are proposed:

$$\bar{x}_1 = \begin{pmatrix} 0.999 \\ -1.001 \end{pmatrix}, \qquad \bar{x}_2 = \begin{pmatrix} 0.341 \\ -0.087 \end{pmatrix}.$$

(a) Compute the residual vector corresponding to these two approximate solutions.

(b) Determine the errors in the two proposed solutions.

2. (a) Derive the formula for $\|A\|_\infty$ given in the text.
(b) Show that the inequalities

$$\|A+B\| \le \|A\| + \|B\|, \qquad \|A\cdot B\| \le \|A\|\cdot\|B\|,$$

are satisfied by any matrix-bound norm.

3. (a) Show that $\|A\|_2 = \|UAV\|_2$ if U and V are unitary matrices.
(b) Show that $\|A\|_2$ is equal to the largest singular value of A; see Sec. 5.2.5.

4. (a) Let T be a nonsingular matrix, and let $\|\cdot\|$ be a given vector norm and its subordinate matrix-bound norm. Show that the function $N(x) = \|Tx\|$ is a norm of the vector x.
(b) What is the matrix-bound norm, subordinate to $N(x)$?
(c) If $N(x) = \max_i |k_i x_i|$, what is the subordinate matrix-bound norm?
(d) Let G be a positive-definite Hermitian matrix. Show that the square root of $x^H Gx$ is a norm of x. (*Hint:* Think of the Choleski decomposition.)
(e) Show that $x^T Gy$ satisfies the axioms for a scalar product of two real vectors x, y if G is a real, positive-definite matrix. The axioms are given in Sec. 4.2.2.
(f) The spectral radius $\rho(A)$ of a matrix A is defined as the largest modulus of one of its eigenvalues. Show that $\|A\| \ge \rho(A)$ for any matrix-bound norm. Show also that, for any given diagonalizable matrix A, there exists a matrix-bound norm, such that $\|A\| = \rho(A)$. (This norm is not the same for all matrices.)

5. (a) Compute the inverse A^{-1} to the matrix in Problem 1, Sec. 5.4 and determine the solution to $Ax = b$ when

$$b = (4 \quad 3 \quad 3 \quad 1)^T.$$

(b) Assume that the right-hand side b is perturbed by a vector δb such that $\|\delta b\|_\infty \le 0.01$. Give an upper bound for $\|\delta x\|_\infty$, where δx is the corresponding perturbation in the solution.
(c) Compute the condition number $\kappa(A)$, and compare it with the bound for the quotient between $\|\delta x\|/\|x\|$ and $\|\delta b\|/\|b\|$ which can be derived from (b).

6. Define the condition number of a matrix A in the L_p-norm by

$$\kappa_p(A) = \|A\|_p \|A^{-1}\|_p.$$

(a) Prove that if the perturbed matrix $(A + \delta A)$ is singular, then we must have

$$\kappa_p(A) \ge \frac{\|A\|_p}{\|\delta A\|_p}.$$

(b) Show that with L_2-norm, equality in (a) holds for

$$\delta A = -\frac{yx^T}{x^T x}, \quad x = A^{-1}y,$$

where y is a vector for which $\|A^{-1}\|_2 \|y\|_2 = \|A^{-1}y\|_2$.

(c) Use the inequality in (a) to get a lower bound for $\kappa_\infty(A)$ for the matrix

$$A = \begin{pmatrix} 1 & -1 & 1 \\ -1 & \epsilon & \epsilon \\ 1 & \epsilon & \epsilon \end{pmatrix}, \quad 0 < |\epsilon| < 1.$$

7. (a) Prove the identity

$$A^{-1} - B^{-1} = A^{-1}(B - A)B^{-1},$$

and hence deduce that

$$\|A^{-1} - B^{-1}\| \le \|A^{-1}\|\|B - A\|\|B^{-1}\|.$$

(b) Prove that if

$$B = A + \delta A, \qquad \|\delta A\|\|B^{-1}\| = \delta < 1,$$

then it follows that

$$\|A^{-1}\| \le \frac{1}{1-\delta}\|B^{-1}\|, \qquad \|A^{-1} - B^{-1}\| \le \frac{\delta}{1-\delta}\|B^{-1}\|.$$

(c) Prove that if

$$x = A^{-1}b, \qquad x + \delta x = (A + \delta A)^{-1}b,$$

then

$$\|\delta x\| \le \frac{\delta}{1-\delta}\|x + \delta x\|, \quad \delta = \|\delta A\|\|B^{-1}\| < 1,$$

$$\|\delta x\| \le \frac{\epsilon}{1-\epsilon}\|x\|, \qquad \epsilon = \|\delta A\|\|A^{-1}\| < 1.$$

8. Show that for the matrix

$$A_5 = \begin{pmatrix} 1 & 0 & 0 & 0 & 1 \\ -1 & 1 & 0 & 0 & 1 \\ -1 & -1 & 1 & 0 & 1 \\ -1 & -1 & -1 & 1 & 1 \\ -1 & -1 & -1 & -1 & 1 \end{pmatrix}$$

we have $g_5 = 2^4$ using partial pivoting, where g_n is defined by

$$g_n = \frac{\max_{i,j,k} |a_{i,j}^{(k)}|}{\max_{i,j} |a_{ij}|}.$$

9. The linear system

$$\begin{pmatrix} 10^{-5} & 10^{-5} & 1 \\ 10^{-5} & -10^{-5} & 1 \\ 1 & 1 & 2 \end{pmatrix} x = \begin{pmatrix} 2\cdot 10^{-5} \\ -2\cdot 10^{-5} \\ 1 \end{pmatrix}$$

has the exact solution

$$x_3 = \frac{10^{-5}}{1 - 2\cdot 10^{-5}}, \quad x_2 = 2, \quad x_1 = \frac{-1}{1 - 2\cdot 10^{-5}}.$$

Solve the system using three-digit floating-point arithmetic
(a) without pivoting;
(b) with complete pivoting;
(c) with complete pivoting after the scaling $x'_3 = 10^5 x_3$ and equilibration.

10. Using a computer with six-digit floating-point arithmetic, the system $\boldsymbol{Ax} = \boldsymbol{b}$,

$$\boldsymbol{A} = \begin{pmatrix} 5 & 7 & 3 \\ 7 & 11 & 2 \\ 3 & 2 & 6 \end{pmatrix}, \quad \boldsymbol{b} = \begin{pmatrix} 0 \\ -1 \\ 0 \end{pmatrix},$$

was solved using Gaussian elimination and iterative improvement. The following results were obtained:

$$\boldsymbol{x}^{(1)} = \begin{pmatrix} -35.9671 \\ 20.9809 \\ 10.9899 \end{pmatrix}, \quad \boldsymbol{x}^{(2)} = \begin{pmatrix} -36.0000 \\ 21.0000 \\ 11.0000 \end{pmatrix}.$$

(a) Estimate from $\boldsymbol{x}^{(1)}$ and $\boldsymbol{x}^{(2)}$ the condition number of A.
(b) Compute A^{-1} and from this $\kappa_\infty(A)$.

5.6. ITERATIVE METHODS

The methods discussed so far for solving linear equations have been direct methods, involving a fixed finite number of operations. Unlike these, iterative methods start from a first approximation which is successively improved until a sufficiently accurate solution is obtained. (An iterative method for improving a solution has already been described in Sec. 5.5.6. There, however, the first approximation was computed by a direct method, and usually already had many correct digits.)

Iterative methods are used most often for large sparse systems of equations. Such systems appear, for example, in studies of electrical networks, economic-system models, and physical processes like diffusion, radiation, and elasticity. The latter give rise to boundary-value problems for partial-differential equations, which are often solved by finite-difference methods. Then, for example, in two dimensions the five-point operator for the Poisson equation (see Sec. 8.6.3) on a mesh with $N \times N$ points will give a matrix A, whose nonzero elements all lie on five diagonals according to Fig. 5.6.1. If Gaussian elimination is used, then nearly all zero elements within the band will be destroyed, and we have to store nearly $2N^3$ elements instead of the $5N^2$ in the original matrix A. The iterative methods we describe below will have the property that no extra nonzero elements are introduced.

Consider the linear system $\boldsymbol{Ax} = \boldsymbol{b}$, where we assume that $a_{ii} \neq 0$, $i =$

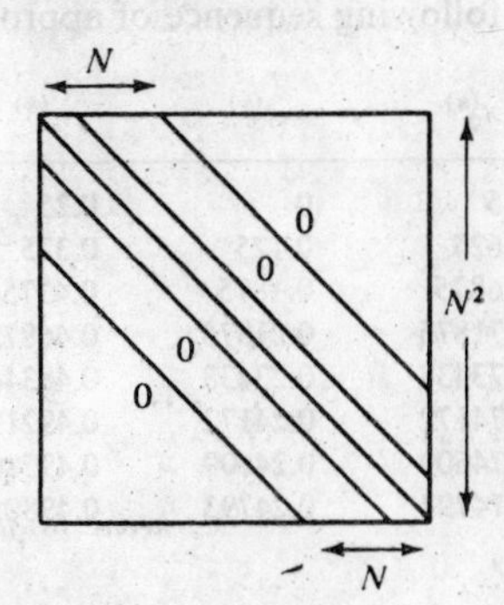

Fig. 5.6.1

1, 2, . . . , n. This can be written

$$x_i = \frac{-\sum_{\substack{j=1 \\ j \neq i}}^{n} a_{ij} x_j + b_i}{a_{ii}}, \quad i = 1, 2, \ldots, n.$$

In **Jacobi's method** we compute a sequence of approximations $\boldsymbol{x}^{(1)}, \boldsymbol{x}^{(2)}, \ldots$ by

$$x_i^{(k+1)} \triangleq \frac{-\sum_{\substack{j=1 \\ j \neq i}}^{n} a_{ij} x_j^{(k)} + b_i}{a_{ii}}, \quad i = 1, 2, \ldots, n. \tag{5.6.1}$$

The initial approximation is often taken to be $\boldsymbol{x}^{(0)} = \boldsymbol{0}$. By taking the limit of both sides in Eq. (5.6.1) it is seen that if $\lim_{k \to \infty} \boldsymbol{x}^{(k)} = \boldsymbol{x}$, then $\boldsymbol{x}$ is a solution to the original equation.

In Jacobi's method, one does not use the improved values until after a complete iteration. In the closely related **Gauss-Seidel's method** they are used as soon as they are computed. We have, then,

$$x_i^{(k+1)} = \frac{-\sum_{j=1}^{i-1} a_{ij} x_j^{(k+1)} - \sum_{j=i+1}^{n} a_{ij} x_j^{(k)} + b_i}{a_{ii}}, \tag{5.6.2}$$

$$i = 1, 2, \ldots, n.$$

Note that here only one approximation for each x_i needs to be stored at a time.

Example 5.6.1

We want to solve the system $A\boldsymbol{x} = \boldsymbol{b}$, where

$$A = \begin{pmatrix} 4 & -1 & -1 & 0 \\ -1 & 4 & 0 & -1 \\ -1 & 0 & 4 & -1 \\ 0 & -1 & -1 & 4 \end{pmatrix}, \quad \boldsymbol{b} = \begin{pmatrix} 1 \\ 2 \\ 0 \\ 1 \end{pmatrix}.$$

With Jacobi's method we get the following sequence of approximations:

k	$x_1^{(k)}$	$x_2^{(k)}$	$x_3^{(k)}$	$x_4^{(k)}$
1	0.25	0.5	0	0.25
2	0.375	0.625	0.125	0.375
3	0.4375	0.6875	0.1875	0.4375
4	0.46875	0.71875	0.21875	0.46875
5	0.48344	0.73438	0.23438	0.48344
6	0.49219	0.74172	0.24172	0.49219
7	0.49586	0.74609	0.24609	0.49586
8	0.49805	0.74793	0.24793	0.49805
.	. . .	. . .	. . .	. . .
∞	0.5	0.75	0.25	0.5

The iteration converges, but rather slowly.

If we instead use Gauss-Seidel's method, we obtain

k	$x_1^{(k)}$	$x_2^{(k)}$	$x_3^{(k)}$	$x_4^{(k)}$
1	0.25	0.5625	0.0625	0.40625
2	0.40625	0.70312	0.20312	0.47656
3	0.47656	0.73828	0.23828	0.49414
4	0.49414	0.74707	0.24707	0.49854
5	0.49854	0.74927	0.24927	0.49963
.	. . .	. . .	. . .	. . .

The convergence with Gauss-Seidel's method is, in this example, about twice as fast as with Jacobi's method. This is often, but not always, true. Indeed, there are examples for which Gauss-Seidel's method diverges and Jacobi's method converges.

We shall now discuss the convergence of iterative methods. We first show that Jacobi's and Gauss-Seidel's methods can be written in the form

$$x^{(k+1)} = Bx^{(k)} + c, \quad k = 0, 1, 2, \dots, \tag{5.6.3}$$

which is the general form for a **stationary iterative method**. It is called stationary because no variation occurs from iteration to iteration. We split the matrix A into a lower triangle, a diagonal, and an upper triangle,

$$A = D(L + I + U), \quad D = \text{diag}\,(a_{ii}).$$

Then Jacobi's method, Eq. (5.6.1), can be written

$$x^{(k+1)} = -(L + U)x^{(k)} + D^{-1}b,$$

and Gauss-Seidel's method, Eq. (5.6.2), becomes

$$x^{(k+1)} = -Lx^{(k+1)} - Ux^{(k)} + D^{-1}b.$$

Thus they both are of the form of Eq. (5.6.3), with

$$B_J = -(L + U), \qquad B_{GS} = -(I + L)^{-1}U. \tag{5.6.4}$$

A relation between the errors in successive approximations can be derived by subtracting from Eq. (5.6.3) the equation $x = Bx + c$:

$$x^{(k+1)} - x = B(x^{(k)} - x) = \ldots = B^{k+1}(x^{(0)} - x). \tag{5.6.5}$$

Now let B have eigenvalues $\lambda_1, \lambda_2, \ldots, \lambda_n$, and assume that the corresponding eigenvectors $u_1, u_2, \ldots, u_n$ are linearly independent. Then we can expand the initial error $x^{(0)} - x = \alpha_1 u_1 + \alpha_2 u_2 + \ldots + \alpha_n u_n$, and thus

$$x^{(k)} - x = \alpha_1 \lambda_1^k u_1 + \alpha_2 \lambda_2^k u_2 + \ldots + \alpha_n \lambda_n^k u_n. \tag{5.6.6}$$

From this it follows that the iterations converge from an arbitrary starting approximation if and only if we have $|\lambda_i| < 1$, $i = 1, 2, \ldots, n$. It can be shown (see Varga [81]) that this conclusion holds without the assumption on the eigenvectors and we have the following theorem.

THEOREM 5.6.1

A necessary and sufficient condition for a stationary iterative method $x^{(k+1)} = Bx^{(k)} + c$ to converge for an arbitrary initial approximation $x^{(0)}$ is that

$$\rho(B) = \max_{1 \le i \le n} |\lambda_i(B)| < 1,$$

where $\rho(B)$ is called the spectral radius of B.

From Eq. (5.6.6) we also see that to reduce the amplitude of the error component $\alpha_j u_j$ in $x^{(0)} - x$ by a factor of 10^{-m}, we have to make k iterations, where k is the smallest number such that

$$|\lambda_j|^k \le 10^{-m}, \quad \text{or} \quad k \ge \frac{m}{-\log_{10} |\lambda_j|}.$$

Asymptotically, the eigenvalue of largest modulus dominates, and this leads us to define the **asymptotic rate of convergence** for the iterative method $x^{(k+1)} = Bx^{(k)} + c$ as

$$R = -\log_{10} (\rho(B)).$$

In practice, the eigenvalues of B will not be known, and therefore Theorem 5.6.1 is difficult to use. However, from Eq. (5.6.5) we also get

$$\|x^{(k)} - x\| \le \|B^k\| \|x^{(0)} - x\| \le \|B\|^k \|x^{(0)} - x\|.$$

It follows that a sufficient condition for convergence is that $\|B\| < 1$, for some consistent matrix norm.

To get an estimate for the error in $x^{(k)}$ we use the relation

$$x^{(k)} - x = -B(x^{(k)} - x^{(k-1)}) + B(x^{(k)} - x).$$

If $\|B\| = \beta < 1$, we get

$$\|x^{(k)} - x\| \leq \frac{\beta}{1-\beta}\|x^{(k)} - x^{(k-1)}\|. \tag{5.6.7}$$

We note that only when $\beta \leq 0.5$ is it correct to break off the iterations when the last correction is smaller than the error we can accept. Also note that Eq. (5.6.7) does not estimate rounding errors, and these are not avoided even though the elements in A are used in unmodified form.

In Jacobi's method, the iteration matrix B_J has the elements

$$b_{ij} = \frac{a_{ij}}{a_{ii}}, \qquad i \neq j, \quad b_{ii} = 0,$$

and thus

$$\|B_J\|_\infty = \max_{1 \leq i \leq n} \sum_{\substack{j=1 \\ j \neq i}}^{n} \frac{|a_{ij}|}{|a_{ii}|}.$$

It follows that for strictly diagonally dominant matrices $\|B_J\|_\infty < 1$, and Jacobi's method converges.

By the definition of Eq. (5.5.5) we have for Gauss-Seidel's method

$$\|B_{GS}\|_\infty = \max_{x \neq 0} \frac{\|y\|_\infty}{\|x\|_\infty}, \quad y = B_{GS}x.$$

Now choose k so that $\|y\|_\infty = |y_k|$. Then, from the kth equation in $y = -Ly - Ux$, we get

$$|y_k| = \|y\|_\infty \leq s_k\|y\|_\infty + r_k\|x\|_\infty,$$

where

$$r_i = \sum_{j=i+1}^{n} \frac{|a_{ij}|}{|a_{ii}|}, \qquad s_i = \sum_{j=1}^{i-1} \frac{|a_{ij}|}{|a_{ii}|}.$$

Thus we have

$$\|B_{GS}\|_\infty \leq \max_{1 \leq i \leq n} \frac{r_i}{1 - s_i},$$

and it follows that Gauss-Seidel's method is also convergent when A is strictly diagonally dominant.

By a simple modification of Gauss-Seidel's method it is often possible to make a substantial improvement in the rate of convergence. We note that Eq. (5.6.2) can be written $x_i^{(k+1)} = x_i^{(k)} + r_i^{(k)}$, where $r_i^{(k)}$ is the current residual of the ith equation

$$r_i^{(k)} = \frac{-\sum_{j=1}^{i-1} a_{ij}x_j^{(k+1)} - \sum_{j=i}^{n} a_{ij}x_j^{(k)} + b_i}{a_{ii}}. \tag{5.6.8}$$

The iterative method,

$$x_i^{(k+1)} = x_i^{(k)} + \omega r_i^{(k)},$$

is the **successive overrelaxation (SOR) method**. Here ω, the relaxation parameter, should be chosen so that the rate of convergence is maximized. For $\omega = 1$, the method obviously reduces to Gauss-Seidel's method. The iteration matrix for the SOR method is

$$B_\omega = (I + \omega L)^{-1}[(1 - \omega)I - \omega U].$$

We now show that only values of ω, $0 < \omega < 2$, are of interest.

THEOREM 5.6.2

For the iteration matrix in the SOR method we have

$$\rho(B_\omega) \geq |\omega - 1|,$$

so the method can only converge for $0 < \omega < 2$.

Proof. Since the determinant of a triangular matrix is the product of its diagonal elements we get

$$\det(B_\omega) = \det(I + \omega L)^{-1} \det((1 - \omega)I - \omega U) = (1 - \omega)^n.$$

Now, from $\det(B_\omega) = \lambda_1 \lambda_2 \ldots \lambda_n$ it follows that

$$\max_i |\lambda_i| \geq |1 - \omega|.$$

It can also be shown that for real matrices A, which are symmetric and positive-definite, the SOR method converges for all ω, $0 < \omega < 2$. For some classes of matrices of practical importance the optimal value of ω is known. We state without proof (see Young [116]) the important theorem:

THEOREM 5.6.3

Let the real matrix A be symmetric, positive-definite, and of the block-tridiagonal form

$$A = \begin{pmatrix} D_1 & U_1 & & & \\ L_2 & D_2 & U_2 & & \\ & L_3 & \cdot & \cdot & \\ & & \cdot & \cdot & \cdot \\ & & & \cdot & \cdot & U_{n-1} \\ & & & & L_n & D_n \end{pmatrix},$$

where D_i are ***diagonal*** *submatrices. Then $\rho(B_{GS}) = \rho^2(B_G)$ and the optimal relaxation factor $\bar{\omega}$ in SOR is given by*

$$\bar{\omega} = \frac{2}{1 + (1 - \rho(B_{GS}))^{1/2}}, \quad \rho(B_{GS}) < 1, \tag{5.6.9}$$

where $\rho(B_J)$ is the spectral radius of the Jacobi iteration matrix corresponding to A. The optimal value of $\rho(B_\omega)$ is

$$\rho(B_{\bar{\omega}}) = \bar{\omega} - 1.$$

Example 5.6.2

The matrix A in Example 5.6.1 arises from the Laplace equation on a square with a mesh of $(N+1) \times (N+1)$ points where $N = 3$. A is positive-definite, and if we partition it—

$$A = \left(\begin{array}{c|cc|c} 4 & -1 & -1 & 0 \\ \hline -1 & 4 & 0 & -1 \\ -1 & 0 & 4 & -1 \\ \hline 0 & -1 & -1 & 4 \end{array}\right)$$

—it is seen to be of the particular block-tridiagonal form in Theorem 5.6.3. It can be shown that for this model problem we have

$$\rho(B_J) = \cos\frac{\pi}{N}.$$

For $N = 3$ the formula (5.6.9) gives $\bar{\omega} = 1.0718$, and below we have solved $Ax = b$ using this optimal value in SOR.

k	$x_1^{(k)}$	$x_2^{(k)}$	$x_3^{(k)}$	$x_4^{(k)}$
1	0.26795	0.60770	0.07180	0.45002
2	0.43078	0.72828	0.23086	0.49264
3	0.49402	0.74798	0.24780	0.49940
4	0.49930	0.74980	0.24981	0.49994

The rate of convergence is evidently much better than with Gauss-Seidel's method, which corresponds to $\omega = 1$.

For large systems the improvement in the rate of convergence by using $\bar{\omega}$ is much more spectacular than in Example 5.6.2. For large values of N in the model problem we have

$$\rho(B_{\bar{\omega}}) = \frac{2}{1+\sin(\pi/N)} - 1 \approx 1 - 2\frac{\pi}{N},$$

and the rate of convergence becomes

$$R_{\bar{\omega}} \approx -\log_{10}\left(1 - 2\frac{\pi}{N}\right) \approx 0.4343 \cdot 2\frac{\pi}{N}.$$

We compare this with Gauss-Seidel's method ($\omega = 1$), for which

$$\rho(B_{GS}) = \rho^2(B_J) = \cos^2\frac{\pi}{N} \approx 1 - \left(\frac{\pi}{N}\right)^2$$

and

$$R_{GS} \approx -\log_{10}\left(1 - \left(\frac{\pi}{N}\right)^2\right) \approx 0.4343\left(\frac{\pi}{N}\right)^2.$$

Thus when $N \gg 1$, we need a factor of

$$\frac{R_{\bar{\omega}}}{R_{GS}} \approx \frac{2N}{\pi}$$

fewer iterations with optimal SOR than with Gauss-Seidel's method.

PROBLEMS

1. Use the error estimate of Eq. (5.6.7) to derive error bounds for the last computed approximations with Jacobi's and Gauss-Seidel's method in Example 5.6.1.

2. Compute an approximate solution, with an error less than $2 \cdot 10^{-3}$ in maximum norm, to the linear system $Ax = b$, where

$$A = \begin{pmatrix} 0.95 & 0.07 & 0 & 0 & 0.05 & 0.01 \\ 0.07 & 0.95 & 0.07 & 0 & 0 & 0.04 \\ 0 & 0.07 & 0.95 & 0.06 & 0 & 0 \\ 0 & 0 & 0.06 & 0.95 & 0.06 & 0 \\ 0.05 & 0 & 0 & 0.06 & 0.95 & 0.06 \\ 0.01 & 0.04 & 0 & 0 & 0.06 & 0.95 \end{pmatrix}, \quad b = \begin{pmatrix} 0.5 \\ 0.5 \\ 0.5 \\ 0.5 \\ 0.5 \\ 0.5 \end{pmatrix}.$$

The coefficients are exact. *Hint:* Don't divide by the original diagonal elements!

3. Let A be a given nonsingular $n \times n$ matrix, and X_0 an arbitrary $n \times n$ matrix. We define a sequence of matrices by

$$X_{k+1} = X_k + X_k(I - AX_k), \quad k = 0, 1, 2, \ldots.$$

(a) Prove that $\lim_{k \to \infty} X_k = A^{-1}$ if and only if $\rho(I - AX_0) < 1$.

(b) Use the iterations in (a) to compute the inverse A^{-1}, where

$$A = \begin{pmatrix} 1 & 1 \\ 1 & 2 \end{pmatrix}, \quad X_0 = \begin{pmatrix} 1.9 & -0.9 \\ -0.9 & 0.9 \end{pmatrix}.$$

4. The iterative method $x^{(k+1)} = Bx^{(k)} + c$ can be accelerated by computing the sequence $\{z^{(k)}\}$ by putting $z^{(0)} = x^{(0)}$, and computing

$$x^{(k+1)} = Bz^{(k)} + c, \quad z^{(k+1)} = \alpha z^{(k)} + (1 - \alpha)x^{(k+1)}, \quad k \geq 0.$$

Let $\mathfrak{M}$ denote the set of matrices with eigenvalues $\lambda_i \in [a, b]$, $a < b$. Determine the parameter α so that the minimum rate of convergence for $B \in \mathfrak{M}$ is maximized.

5. The linear system of equations

$$\begin{pmatrix} 1 & -a \\ -a & 1 \end{pmatrix} x = b,$$

where a is real, can under certain conditions be solved by the iterative method

$$\begin{pmatrix} 1 & 0 \\ -\omega a & 1 \end{pmatrix} x^{(k+1)} = \begin{pmatrix} 1 - \omega & \omega a \\ 0 & 1 - \omega \end{pmatrix} x^{(k)} + \omega b.$$

(a) For which values of a is the method convergent for $\omega = 1$?
(b) For $a = 0.5$, find the value of $\omega \in \{0.8, 0.9, 1.0, 1.1, 1.2, 1.3\}$ which minimizes the spectral radius of the matrix

$$\begin{pmatrix} 1 & 0 \\ -\omega a & 1 \end{pmatrix}^{-1} \begin{pmatrix} 1-\omega & \omega a \\ 0 & 1-\omega \end{pmatrix}.$$

6. Many iterative methods for solving $Ax = b$ can be described as follows. Let $x^{(k)}$ be an approximate solution and denote the error vector by $e^{(k)} = x^{(k)} - A^{-1}b$. We choose a direction $p^{(k)}$ and a distance γ_k such that the error vector associated with

$$x^{(k+1)} = x^{(k)} + \gamma_k p^{(k)}$$

is minimized in some suitable norm.
(a) Let A be symmetric and positive-definite. Show that if we define the norm by $\|e\|^2 = e^T A e$, then $\|e^{(k+1)}\|$ is minimized by taking

$$\gamma_k = \frac{-(p^{(k)})^T(Ax^{(k)} - b)}{(p^{(k)})^T A p^{(k)}}.$$

(b) Show that it is possible to choose in (a) a sequence of vectors $p^{(0)}, p^{(1)}, p^{(2)}, \ldots$ so that Gauss-Seidel's method results.

5.7. OVERDETERMINED LINEAR SYSTEMS

Assume that one wants to fit a linear mathematical model to given data. Then, in order to reduce the influence of errors resulting from measurements, one often makes a greater number of measurements than the number of unknowns. The resulting problem is to "solve" an overdetermined system, and it can be formulated in matrix notations as:

Given an $m \times n$ matrix A, where $m \geq n$, and an $m \times 1$ vector b, find a vector x such that Ax is the "best" approximation to b.

Since the system $Ax = b$ is overdetermined, it cannot generally be satisfied exactly, and there are many possible ways of defining the "best" solution. A choice which can often be justified for statistical reasons (see Sec. 4.5.5), and which also leads to particularly simple calculations, is the following.

A **least-squares solution** to the overdetermined system of equations $Ax = b$ is a vector x which minimizes the Euclidean length of the residual vector—i.e., minimizes

$$\|r\|_2 = (r^T r)^{1/2}, \quad r = b - Ax. \tag{5.7.1}$$

If the elements in b are subject to random errors, then the equations should be scaled so that the variance in b_i, $i = 1, 2, \ldots, n$ is constant. For the case when the errors in b_i are independent random variables, Gauss placed the method of least squares on a sound theoretical basis in 1821.

In some applications it may be more natural to use some other norm in Eq. (5.7.1)—e.g., $\|r\|_1$ or $\|r\|_\infty$. However, then the solution leads to a more difficult combinatorial problem. Also, the least-squares solution is often a good approximation to the solution in other norms. We shall, therefore, discuss here only methods for computing least-squares solutions.

5.7.1. The Normal Equations

The least-squares solution to a system of equations is characterized by the following theorem.

THEOREM 5.7.1

Let A be a given real $m \times n$ matrix and $\boldsymbol{b}$ a given $m \times 1$ vector. Then if $\boldsymbol{x}$ satisfies

$$A^{\mathrm{T}}(\boldsymbol{b} - A\boldsymbol{x}) = \boldsymbol{0}, \tag{5.7.2}$$

we have, for any vector $\boldsymbol{y}$,

$$\|\boldsymbol{b} - A\boldsymbol{x}\|_2 \leq \|\boldsymbol{b} - A\boldsymbol{y}\|_2.$$

Proof. If we put $\boldsymbol{r}_x = \boldsymbol{b} - A\boldsymbol{x}$ and $\boldsymbol{r}_y = \boldsymbol{b} - A\boldsymbol{y}$, then

$$\boldsymbol{r}_y = (\boldsymbol{b} - A\boldsymbol{x}) + (A\boldsymbol{x} - A\boldsymbol{y}) = \boldsymbol{r}_x + A(\boldsymbol{x} - \boldsymbol{y}).$$

Squaring this and using Eq. (5.7.2), which can be written $A^{\mathrm{T}}\boldsymbol{r}_x = \boldsymbol{0}$, we obtain

$$\boldsymbol{r}_y^{\mathrm{T}}\boldsymbol{r}_y = \boldsymbol{r}_x^{\mathrm{T}}\boldsymbol{r}_x + (\boldsymbol{x} - \boldsymbol{y})^{\mathrm{T}}A^{\mathrm{T}}A(\boldsymbol{x} - \boldsymbol{y}),$$

and thus

$$\|\boldsymbol{r}_y\|_2^2 = \|\boldsymbol{r}_x\|_2^2 + \|A(\boldsymbol{x} - \boldsymbol{y})\|_2^2 \geq \|\boldsymbol{r}_x\|_2^2,$$

which concludes the proof.

From Eq. (5.7.2) it follows that for any vector $\boldsymbol{z}$ we have

$$(A\boldsymbol{z})^{\mathrm{T}}(\boldsymbol{b} - A\boldsymbol{x}) = 0.$$

Hence the residual vector $(\boldsymbol{b} - A\boldsymbol{x})$ to a least-squares solution $\boldsymbol{x}$ is orthogonal to all vectors in the space $R(A)$ spanned by the columns of A. A least-squares solution splits the vector $\boldsymbol{b}$ into two components

$$\boldsymbol{b} = A\boldsymbol{x} + \boldsymbol{r},$$

where $A\boldsymbol{x}$ is the orthogonal projection of $\boldsymbol{b}$ onto $R(A)$ and $\boldsymbol{r}$ is orthogonal to $R(A)$. This geometric interpretation is illustrated in Fig. 5.7.1.

From Eq. (5.7.2) it also follows that the least-squares solution satisfies the **normal equations**

$$(A^{\mathrm{T}}A)\boldsymbol{x} = A^{\mathrm{T}}\boldsymbol{b}. \tag{5.7.3}$$

Here $C = A^{\mathrm{T}}A$ is a symmetric $n \times n$ matrix with elements

$$c_{ij} = \boldsymbol{a}_i^{\mathrm{T}}\boldsymbol{a}_j, \qquad A = (\boldsymbol{a}_1, \boldsymbol{a}_2, \ldots, \boldsymbol{a}_n),$$

and Eq. (5.7.3) gives n linear equations for the unknown elements in $\boldsymbol{x}$.

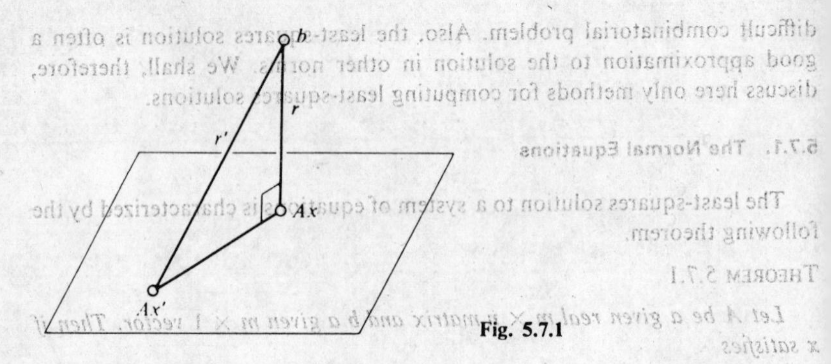

Fig. 5.7.1

Since A^TA is symmetric, we can compute A^TA and A^Tb in

$$\tfrac{1}{2}n(n + 1)m + nm = \tfrac{1}{2}mn(n + 3)$$

operations. When A^TA is nonsingular, we can then solve the normal equations by Choleski's method, or by symmetric Gaussian elimination in approximately $n^3/6$ operations. Thus most of the work lies in forming the normal equations.

THEOREM 5.7.2

The matrix A^TA is nonsingular if and only if the columns of A are linearly independent.

Proof. If the columns in A are linearly independent, then

$$\boldsymbol{x} \neq \boldsymbol{0} \Longrightarrow A\boldsymbol{x} \neq \boldsymbol{0},$$

and therefore

$$\boldsymbol{x} \neq \boldsymbol{0} \Longrightarrow \boldsymbol{x}^T(A^TA)\boldsymbol{x} = (A\boldsymbol{x})^TA\boldsymbol{x} = \|A\boldsymbol{x}\|_2^2 > 0.$$

Hence A^TA is positive-definite. But then by Theorem 5.4.1 det $(A^TA) > 0$, and thus A^TA is nonsingular.

If, on the other hand, the columns in A are linearly dependent, then for some $\boldsymbol{x}_0 \neq \boldsymbol{0}$ we have $A\boldsymbol{x}_0 = \boldsymbol{0}$. But then also $A^TA\boldsymbol{x}_0 = \boldsymbol{0}$, which shows that A^TA is singular.

From this theorem it follows that when the columns of A are linearly independent, then the least-squares solution is unique, and

$$\boldsymbol{x} = A^I\boldsymbol{b}, \quad A^I = (A^TA)^{-1}A^T. \tag{5.7.4}$$

Here the $n \times m$ matrix A^I is the **pseudo-inverse** of A, see Eq. (5.2.4). Also,

$$\boldsymbol{r} = (I - P_A)\boldsymbol{b}, \quad P_A = AA^I = A(A^TA)^{-1}A^T, \tag{5.7.5}$$

where P_A is the **orthogonal projector** onto the column space (range) of A; see Problem 10 of Sec. 5.2. When the columns of A are linearly dependent, then there are many least-squares solutions $\boldsymbol{x}$. Since Eq. (5.7.2) must be satisfied and the orthogonal projection of $\boldsymbol{b}$ onto $R(A)$ is unique, they all give the same residual vector $\boldsymbol{r}$. We return to the case of linearly dependent columns in Sec. 5.7.2.

Example 5.7.1

In order to estimate the height above sea level for three points, A, B, and C, the difference in altitude was measured according to Fig. 5.7.2. The points, D, E, and F lie at sea level.

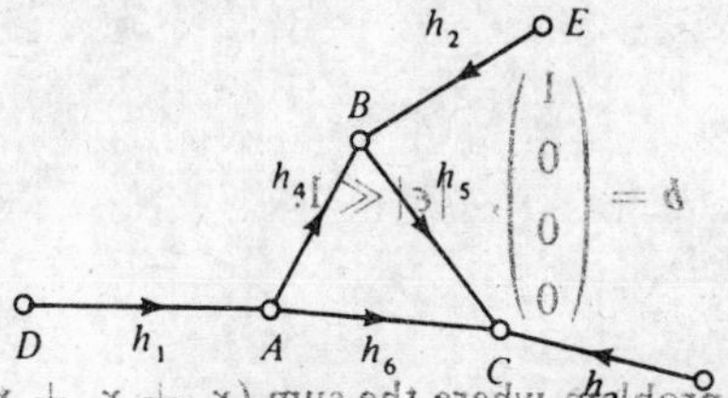

$h_1 = 1,\quad h_2 = 2,\quad h_3 = 3,$

$h_4 = 1,\quad h_5 = 2,\quad h_6 = 1.$

Fig. 5.7.2

Each measurement gives a linear relation between the heights x_A, x_B, and x_C of A, B, and C,

$$\begin{pmatrix} 1 & 0 & 0 \\ 0 & 1 & 0 \\ 0 & 0 & 1 \\ -1 & 1 & 0 \\ 0 & -1 & 1 \\ -1 & 0 & 1 \end{pmatrix} \begin{pmatrix} x_A \\ x_B \\ x_C \end{pmatrix} = \begin{pmatrix} 1 \\ 2 \\ 3 \\ 1 \\ 2 \\ 1 \end{pmatrix}$$

The normal equations are

$$\begin{pmatrix} 3 & -1 & -1 \\ -1 & 3 & -1 \\ -1 & -1 & 3 \end{pmatrix} \begin{pmatrix} x_A \\ x_B \\ x_C \end{pmatrix} = \begin{pmatrix} -1 \\ 1 \\ 6 \end{pmatrix}$$

Using symmetric Gaussian elimination we get the triangular system

$$\begin{pmatrix} 3 & -1 & -1 \\ & \frac{8}{3} & -\frac{4}{3} \\ & & 2 \end{pmatrix} \begin{pmatrix} x_A \\ x_B \\ x_C \end{pmatrix} = \begin{pmatrix} -1 \\ \frac{2}{3} \\ 6 \end{pmatrix},$$

and, by back substitution,

$$x_C = 3, \qquad x_B = \tfrac{7}{4}, \qquad x_A = \tfrac{5}{4}.$$

It is easily verified that the corresponding residual vector

$$\tfrac{1}{4}(-1 \quad 1 \quad 0 \quad 2 \quad 3 \quad -3)^{\mathsf{T}}$$

is orthogonal to all columns in A.

In practice, the columns in A are often **nearly** linearly dependent. The next example shows that in this case the normal equations have certain drawbacks.

Example 5.7.2

Consider the problem of computing the least-squares solution to the equations $Ax = b$ where

$$A = \begin{pmatrix} 1 & 1 & 1 \\ \epsilon & 0 & 0 \\ 0 & \epsilon & 0 \\ 0 & 0 & \epsilon \end{pmatrix}, \qquad b = \begin{pmatrix} 1 \\ 0 \\ 0 \\ 0 \end{pmatrix}, \quad |\epsilon| \ll 1.$$

This may correspond to a practical problem where the sum $(x_1 + x_2 + x_3)$ can be measured much more accurately than the individual components. We have, exactly,

$$A^{\mathsf{T}}A = \begin{pmatrix} 1+\epsilon^2 & 1 & 1 \\ 1 & 1+\epsilon^2 & 1 \\ 1 & 1 & 1+\epsilon^2 \end{pmatrix}, \qquad A^{\mathsf{T}}b = \begin{pmatrix} 1 \\ 1 \\ 1 \end{pmatrix},$$

and $x^{\mathsf{T}} = (3+\epsilon^2)^{-1}(1, 1, 1)$.

Now assume that $\epsilon = 10^{-4}$, and that we use eight-digit floating-point arithmetic. Then $1 + \epsilon^2 = 1.00000001$ rounds to 1, and the computed matrix $A^{\mathsf{T}}A$ will be singular. No matter what we do, we cannot retrieve the information which was lost in forming the normal equations.

The last example shows that when A is ill-conditioned, we should use double precision in forming and solving the normal equations in order to avoid loss of significant information. Sometimes ill-conditioning can be caused by an unsuitable formulation of the problem. For example, an ill-conditioned approximation problem may be made well-conditioned by use of orthogonal, or nearly orthogonal, base functions.

Example 5.7.3

Fit a linear function $x_0 + x_1(t - c)$ by the least-squares method to the following data

t	1	3	4	6	7
$f(t)$	−2.1	−0.9	−0.6	0.6	0.9

If we make the (unsuitable) choice $c = 1{,}000$, the normal equations

$$\begin{pmatrix} 5 & 4{,}979 \\ 4{,}979 & 4{,}958{,}111 \end{pmatrix}\begin{pmatrix} x_0 \\ x_1 \end{pmatrix} = \begin{pmatrix} -2.1 \\ -2{,}097.3 \end{pmatrix}$$

become very ill-conditioned. If we round the element 4,958,111 to $4{,}958 \cdot 10^3$, then the component x_1 is perturbed from the correct value $x_1 = 0.5053$ to $\bar{x}_1 = -0.1306$ (!).

A much better choice of base functions is obtained by taking c equal to the midpoint of the interval (1, 7)—i.e., $c = 4$. The normal equations are then

$$\begin{pmatrix} 5 & 1 \\ 1 & 23 \end{pmatrix}\begin{pmatrix} x_0 \\ x_1 \end{pmatrix} = \begin{pmatrix} -2.1 \\ 11.1 \end{pmatrix},$$

which is a very-well-conditioned system.

Note that when the elements in $\boldsymbol{A}$ and $\boldsymbol{b}$ are the original data, then we cannot avoid ill-conditioning in this way.

5.7.2. Orthogonalization Methods

We have pointed out the possible difficulties in computing the least-squares solution from the normal equations. In this section we shall develop other methods, which avoid forming the normal equations. These are based on the decomposition of a rectangular matrix $\boldsymbol{A}$ into the product of a matrix $\boldsymbol{Q}$ with orthogonal columns and an upper-triangular matrix $\boldsymbol{R}$.

THEOREM 5.7.3. *QR-theorem*

Let $\boldsymbol{A}$ be a given $m \times n$ matrix with $m \geq n$ and linearly independent columns. Then there exists a unique $m \times n$ matrix $\boldsymbol{Q}$, with

$$\boldsymbol{Q}^T\boldsymbol{Q} = \boldsymbol{D}, \qquad \boldsymbol{D} = \operatorname{diag}(d_1, \ldots, d_n), \quad d_k > 0, \quad k = 1, \ldots, n,$$

and a unique upper-triangular matrix $\boldsymbol{R}$, with $r_{kk} = 1$, $k = 1, \ldots, n$ such that $\boldsymbol{A} = \boldsymbol{Q}\boldsymbol{R}$.

We shall give a constructive proof of this theorem later. First we illustrate the use of this decomposition for the least-squares problem. The least-squares solution was characterized by Eq. (5.7.2), which can now be written

$$\boldsymbol{R}^T\boldsymbol{Q}^T(\boldsymbol{b} - \boldsymbol{A}\boldsymbol{x}) = 0.$$

But since $\boldsymbol{R}$ is nonsingular and $\boldsymbol{Q}^T\boldsymbol{A} = \boldsymbol{Q}^T\boldsymbol{Q}\boldsymbol{R} = \boldsymbol{D}\boldsymbol{R}$, this is equivalent to

$$\boldsymbol{R}\boldsymbol{x} = \boldsymbol{y}, \qquad \boldsymbol{y} = \boldsymbol{D}^{-1}\boldsymbol{Q}^T\boldsymbol{b}. \tag{5.7.6}$$

Thus if $\boldsymbol{Q}$ and $\boldsymbol{R}$ are known, we get the least-squares solution by solving a triangular system of equations. We now describe the **modified Gram-Schmidt method** for computing $\boldsymbol{Q}$, $\boldsymbol{R}$, and $\boldsymbol{y}$.

We compute a sequence of matrices, $A = A^{(1)}, A^{(2)}, \ldots, A^{(n+1)} = Q$ where $A^{(k)}$ has the form

$$A^{(k)} = (q_1, \ldots, q_{k-1}, a_k^{(k)}, \ldots, a_n^{(k)}).$$

Here the first $(k-1)$ columns equal the first $(k-1)$ columns in Q, an $a_k^{(k)}, \ldots, a_n^{(k)}$ have been made orthogonal to $q_1, \ldots, q_{k-1}$. In the kth ste we orthogonalize $a_{k+1}^{(k)}, \ldots, a_n^{(k)}$ against q_k:

$$q_k = a_k^{(k)}, \quad d_k = q_k^T q_k, \quad r_{kk} = 1,$$
$$a_j^{(k+1)} = a_j^{(k)} - r_{kj} q_k, \quad r_{kj} = \frac{q_k^T a_j^{(k)}}{d_k}, \tag{5.7.7}$$
$$j = k+1, \ldots, n.$$

Thus in this step the kth **row** of R is determined.

The vector b is transformed in the same way, $b = b^{(1)}, b^{(2)}, \ldots, b^{(n+1)}$ where

$$b^{(k+1)} = b^{(k)} - y_k q_k, \quad y_k = \frac{q_k^T b^{(k)}}{d_k}. \tag{5.7.8}$$

Here $b^{(n+1)}$ will be the part of b orthogonal to $R(A)$ and therefore equals th residual vector r. After n steps, $k = 1, 2, \ldots, n$, we obtain

$$Q = (q_1, \ldots, q_n), \qquad R = (r_{kj}), \qquad y = (y_1, \ldots, y_n)^T$$

such that

$$Q^T Q = \operatorname{diag}(d_k), \qquad A = QR, \qquad b = Qy + r.$$

We can now easily solve for x from $Rx = y$.

To compute R and y requires approximately

$$2m \sum_{k=1}^{n} (n - k + 1) = 2m \cdot \tfrac{1}{2} n(n+1) = mn(n+1)$$

operations, and to solve $Rx = y$ only $\frac{1}{2}n(n+1)$. The total amount of work i therefore, about twice as much as required to form the normal equation

So far we have assumed that the columns in A are linearly independen From the relation $Q = AR^{-1}$, where $S = R^{-1}$ is unit upper-triangular*, follows that q_k is a linear combination of $a_1, a_2, \ldots, a_k$,

$$q_k = s_{1k} a_1 + s_{2k} a_2 + \cdots + a_k.$$

Now assume that $a_1, \ldots, a_{k-1}$ are linearly independent, but that a_k is linear combination of $a_1, \ldots, a_{k-1}$ and therefore also of $q_1, \ldots, q_{k-1}$. The we must get $a_k^{(k)} = 0$, and the orthogonalization process breaks down. How ever, if rank $(A) > k - 1$, there must be a vector $a_j^{(k)} \neq 0$ for $k \leq j \leq n$. W can then interchange columns k and j and proceed until all remaining colum are linearly dependent on the computed q vectors. This suggests that w

*An upper-triangular matrix S is unit upper-triangular if $s_{ii} = 1$ for all i.

gment the modified Gram-Schmidt method by column pivoting so that step k we choose s as the smallest integer for which

$$\|a_s^{(k)}\|_2 = \max_{k \le j \le n} \|a_j^{(k)}\|_2,$$

d then interchange columns k and s.

With column pivoting the method works also when A has linearly dependent columns—i.e., when rank $(A) = r < n$. We then get the decomposition

$$Q = (q_1, \ldots, q_r), \qquad A = Q(R, S), \qquad b = Qy + r,$$

here the unit upper-triangular matrix R is $r \times r$ and S is $r \times (n - r)$. (For mplicity we have assumed that the column permutations have been carried ut on A in advance.) In this case the general least-squares solution satisfies $(R, S)x = y$, and can be written

$$x = (x_1, x_2)^T, \quad x_1 = R^{-1}y - R^{-1}Sx_2, \tag{5.7.9}$$

here x_2 is an arbitrary vector with $(n - r)$ components. Thus a particular olution is $x_1 = R^{-1}y$, $x_2 = 0$.

We note that the **least-squares solution $\hat{x}$ of minimum L_2-norm** will always e unique. Since the general solution x can be interpreted as the residual ector of the system

$$\begin{pmatrix} R^{-1}S \\ -I \end{pmatrix} x_2 = \begin{pmatrix} R^{-1}y \\ 0 \end{pmatrix}, \tag{5.7.10}$$

e get $\hat{x}$ by choosing x_2 as the least-squares solution to the overdetermined ystem of Eq. (5.7.10). Since the corresponding matrix has linearly independent columns, this x_2 is always unique and can be computed by the modified ram-Schmidt method.

xample 5.7.4

Consider the same overdetermined system as in Example 5.7.2, and asume that $|\epsilon|$ is so small that ϵ^2 can be neglected compared to 1. Using modied Gram-Schmidt, we get

$$Q = \begin{pmatrix} 1 & 0 & 0 \\ \epsilon & -\epsilon & -\frac{\epsilon}{2} \\ 0 & \epsilon & -\frac{\epsilon}{2} \\ 0 & 0 & \epsilon \end{pmatrix}, \qquad r = -\frac{\epsilon}{3}\begin{pmatrix} 0 \\ 1 \\ 1 \\ 1 \end{pmatrix}$$

nd

$$R = \begin{pmatrix} 1 & 1 & 1 \\ 0 & 1 & \frac{1}{2} \\ 0 & 0 & 1 \end{pmatrix}, \qquad y = \begin{pmatrix} 1 \\ \frac{1}{2} \\ \frac{1}{3} \end{pmatrix}.$$

Solving $Rx = y$ we get $x_1 = x_2 = x_3 = \frac{1}{3}$, which agrees well with the exac solution $1/(3 + \epsilon^2)$.

It is important to note that the computed matrix Q will not be orthogona to working accuracy when A is ill-conditioned. Therefore, it is essential that R and y be computed exactly by Eqs. (5.7.7) and (5.7.8). If in the example abov we use $y = D^{-1}Q^T b$ instead of Eq. (5.7.8), we get $y = (1, 0, 0)^T$ and $x =$ $(1, 0, 0)$, which is far from correct!

If in Eq. (5.7.7) we normalize the vectors q_k by taking

$$q_k = \frac{a_k^{(k)}}{r_{kk}}, \qquad r_{kk} = \frac{1}{\| a_k^{(k)} \|_2}, \qquad d_k = 1,$$

then $A^T A = R^T Q^T Q R = R^T R$. From the uniqueness of the matrix R in Theo rem 5.4.3 it follows that this QR-decomposition gives **mathematically** th same triangular matrix R as Choleski's method applied to the matrix $A^T A$ i the normal equations. However, this normalization requires the computatio of n square roots, and has no advantages if the only purpose is to compute th least-squares solution.

The modified Gram-Schmidt method differs from the classical only in th order in which the operations are performed. Thus in the classical method w compute in the kth step the kth **column** in R by

$$q_k = a_k - \sum_{i=1}^{k-1} r_{ik} q_i, \qquad r_{ik} = \frac{q_i^T a_k}{d_i}, \quad i = 1, \ldots, k-1.$$

Mathematically this is equivalent to the modified method, but numericall the classical method will not give accurate least-squares solutions. The poo numerical performance results essentially from using a_k instead of $a_k^{(k)}$ in th equation for r_{ik}. To appreciate the difference, the reader is advised to solv Problem 4 at the end of this section.

Unlike the case when $m = n$, in least-squares problems we cannot scal the equations arbitrarily, since the scaling by definition will change the solu tion. As remarked before, using a model where A is exactly known and b are random variables with standard deviations σ_i, the ith equation should b scaled by σ_i^{-1}. The unknowns may be scaled arbitrarily, but unless pivotin is used both the normal equations and the Gram-Schmidt method will giv numerically the same solution, independent of such a scaling. (We assum that the scaling does not introduce any rounding errors; see Sec. 5.5.5.) How ever, if pivoting is used in solving the normal equations or in the Gram Schmidt method, then the scaling influences the choice of pivot and thereb the computed solution. In this case it can be recommended that the unknown be scaled so that the columns in A get equal L_2-norms.

5.7.3. Improvement of Least-Squares Solutions

In Sec. 5.5.6 we described a simple and efficient method of improving a computed solution of a linear system of equations. This method of iterative improvement can also be generalized to linear least-squares solutions. We shall outline the algorithm here without proof.

The essential difference is that, since now the residual vector $(b - Ax)$ is not zero for the exact solution, we must simultaneously improve both x and r. It is easily seen from Eq. (5.7.2) that r and x satisfy the augmented system of $(m + n)$ equations

$$\begin{pmatrix} I & A \\ A^T & 0 \end{pmatrix}\begin{pmatrix} r \\ x \end{pmatrix} = \begin{pmatrix} b \\ 0 \end{pmatrix}, \tag{5.7.11}$$

and we shall use accurately computed residual vectors to this system for the improvement.

Assume that the modified Gram-Schmidt method has given the computed matrices

$$\bar{R}, \qquad \bar{Q} = (\bar{q}_1, \bar{q}_2, \ldots, \bar{q}_n), \qquad \bar{D} = \operatorname{diag}(\bar{d}_k)$$

and the computed solution $\bar{x}$ and $\bar{r}$. We put $x^{(1)} = \bar{x}$, $r^{(1)} = \bar{r}$, and compute for $s = 2, 3, \ldots$ the improved solutions $x^{(s)}$ and $r^{(s)}$ as follows:

i. Compute the residuals of the system of Eq. (5.7.11):

$$c^{(s)} = (b - r^{(s)} - Ax^{(s)}), \quad d^{(s)} = -A^T r^{(s)}.$$

ii. (a) Solve for $v^{(s)}$ from

$$\bar{R}^T v^{(s)} = d^{(s)}.$$

(b) Put $b_1^{(s)} = c^{(s)}$ and compute, for $k = 1, 2, \ldots, n$,

$$b_{k+1}^{(s)} = b_k^{(s)} - \delta y_k^{(s)} \bar{q}_k, \quad \delta y_k^{(s)} = \frac{\bar{q}_k^T b_k^{(s)} - v_k^{(s)}}{\bar{d}_k}.$$

(c) Solve for $\delta x^{(s)}$ from

$$\bar{R} \delta x^{(s)} = \delta y^{(s)}.$$

iii. Correct $x^{(s)}$ and $r^{(s)}$ by

$$x^{(s+1)} = x^{(s)} + \delta x^{(s)}, \qquad r^{(s+1)} = r^{(s)} + b_{n+1}^{(s)}.$$

It is easily verified that i–iii require only $4nm + n^2$ operations, which is an order of magnitude less than the nm^2 operations required to compute $\bar{x}$. Double precision only has to be used in step i.

Usually one or two steps will suffice to get the exact solution. Even in a physical problem, where the exact solution has little significance, it may be worthwhile carrying out one step, which will give a good indication of the number of figures in the solution worth quoting.

5.7.4. Least-Squares Problems with Linear Constraints

Often, the least-squares solution x has to satisfy some linear equations exactly—i.e., we have a least-squares problem with linear constraints. This can be formulated as follows:

Given a $p \times n$ matrix G, $p \leq n$, and an $m \times n$ matrix A, $m \geq n - p$, determine a vector x which minimizes

$$\|b - Ax\|_2 \quad \text{subject to} \quad Gx = h.$$

It can be shown that this problem has a unique solution for all vectors h and b if and only if

$$\text{rank}\,(G) = p, \qquad \text{rank}\begin{pmatrix} G \\ A \end{pmatrix} = n. \tag{5.7.12}$$

To solve this problem we first compute the general solution to the **underdetermined system** $Gx = h$. We note that Eq. (5.7.9), which we developed for computing the general least-squares solution, is also valid for an underdetermined system. We can thus apply the modified Gram-Schmidt method with column pivoting to $Gx = h$, to get the decomposition (we again suppress the column permutations from our notations)

$$G = Q_1(R_1, S_1), \qquad h = Q_1 y_1 + r_1, \tag{5.7.13}$$

where Q_1 is $p \times p$, and, since the system is compatible, $r_1 = 0$. The general solution to $Gx = h$ is, then,

$$x_1 = \hat{x}_1 - Tx_2, \qquad \hat{x}_1 = R_1^{-1}y_1, \quad T = R_1^{-1}S_1, \tag{5.7.14}$$

where x_2 is arbitrary. We assume that the column permutations have been carried out also on A, partition A, and get

$$b - Ax = b - (A_1, A_2)\begin{pmatrix} x_1 \\ x_2 \end{pmatrix} = b - A_1(\hat{x}_1 - Tx_2) - A_2x_2 = b' - A'x_2,$$

where

$$b' = b - A_1\hat{x}_1, \qquad A' = A_2 - A_1T \tag{5.7.15}$$

Now we have transformed the problem to the unconstrained least-squares problem of minimizing $\|b' - A'x_2\|_2$, where A' is an $m \times (n - p)$ matrix. From the assumptions it follows that A' has independent columns, and we can solve this problem by once again using the modified Gram-Schmidt method.

We remark that, by using Gaussian elimination with **complete** pivoting, we can compute a decomposition similar to Eq. (5.7.13),

$$G = L_1(R_1, S_1), \qquad h = L_1y_1,$$

with fewer operations, which can be used to eliminate the constraints.

PROBLEMS

1. Nitrogen and oxygen have the atomic weights N $\approx$ 14 and O $\approx$ 16. Use the molecular weights of the six nitrogen oxides given below to compute the atomic weights for nitrogen and oxygen to four decimal places.

NO	30.006	N_2O	44.013	NO_2	46.006
N_2O_3	76.012	N_2O_5	108.010	N_2O_4	92.011

2. The recently discovered comet 1968 Tentax is supposed to move within the solar system. The following observations of its position in a certain polar coordinate system have been made

r	2.70	2.00	1.61	1.20	1.02
φ	48°	67°	83°	108°	126°

By Kepler's first law the comet should move in a plane orbit of elliptic or hyperbolic form, if the perturbations from the planets are neglected. Then the coordinates satisfy

$$r = \frac{p}{1 - e\cos\varphi},$$

where p is a parameter and e the eccentricity. Estimate p and e by the method of least squares from the given observations. *Hint:* The parameters must appear linearly for the method to be applicable.

3. Describe in Algol or Fortran the modified Gram-Schmidt method for solving the overdetermined linear system $Ax = b$. It can be assumed that the columns of A are linearly independent.

4. Use the classical Gram-Schmidt method and Eq. (5.7.6) to solve the overdetermined system in Example 5.7.2, assuming that $|\epsilon|$ is so small that $1 + \epsilon^2$ is rounded to 1. *Hint:* Note that the first two steps are indentical with the modified method.

5. The Euclidean condition number $\kappa_2(A)$ for a rectangular $m \times n$ matrix A with linearly independent columns is defined by

$$\kappa_2(A) = \frac{\max\limits_{\|x\|_2=1} \|Ax\|_2}{\min\limits_{\|x\|_2=1} \|Ax\|_2}.$$

(a) Show that when $m = n$, this is equivalent to the ordinary definition—i.e.,

$$\kappa_2(A) = \|A\|_2 \|A^{-1}\|_2.$$

(b) Prove that

$$\kappa_2^2(A) = \kappa(A^T A) = \frac{\max\limits_i \lambda_i}{\min\limits_i \lambda_i},$$

where λ_i are the eigenvalues of $A^T A$.

(c) Compute $\kappa_2(A)$ for the $(n+1) \times n$ matrix

$$A = \begin{pmatrix} 1 & 1 & \cdots & 1 \\ \epsilon & 0 & \cdots & 0 \\ 0 & \epsilon & \cdots & 0 \\ \cdot & & & \\ 0 & 0 & \cdots & \epsilon \end{pmatrix}.$$

5.8. COMPUTATION OF EIGENVALUES AND EIGENVECTORS

We shall here give a brief survey of a few of the most important numerical methods for computing eigenvalues and eigenvectors of square matrices. It is interesting to observe that with few exceptions the methods used today were not known fifteen years ago. Algol programs for nearly all methods mentioned here can be found in the handbook edited by Reinsch and Wilkinson [86]. For simplicity we shall here restrict ourselves to real matrices. The methods described can, however, also be applied to complex matrices, or in the case of methods for symmetric matrices to complex Hermitian matrices.

The eigenvalues λ_i and eigenvectors x_i of an $n \times n$ matrix A satisfy

$$(A - \lambda_i I)x_i = 0, \quad x_i \neq 0. \tag{5.8.1}$$

It follows that the eigenvalues λ_i are the n roots of the characteristic equation

$$p_A(\lambda) = \det(A - \lambda I) = 0, \tag{5.8.2}$$

where $p_A(\lambda)$, the characteristic polynomial, is of nth degree in λ. The eigenvalues can thus be computed separately. If one eigenvalue λ_i is known, the corresponding eigenvector is a solution to the linear homogenous system of Eq. (5.8.1). On the other hand, if some eigenvector x_i is known, then we immediately get λ_i from

$$\lambda_i = \frac{x_i^T A x_i}{x_i^T x_i}. \tag{5.8.3}$$

In theory, one possible method for computing the eigenvalues is to determine the coefficients of the characteristic polynomial

$$p_A(\lambda) = (-1)^n \lambda^n + b_1 \lambda^{n-1} + \ldots + b_n,$$

and then solve $p_A(\lambda) = 0$ by some numerical method for algebraic equations. We emphasize the importance of *not* proceeding in this way unless the matrix A is of very low order and has well-separated eigenvalues. It is often the case that the eigenvalues are very sensitive to perturbations in the coefficients $b_1, b_2, \ldots, b_n$ even when they are very accurately determined by the matrix A. In fact, for a symmetric matrix A, small perturbations in the elements of A only cause perturbations of the same magnitude in the eigenvalues, even when A has multiple or close eigenvalues. If, for example, for a symmetric matrix with eigenvalues $1, 2, \ldots, 20$ we compute the coefficients of $p_A(\lambda)$, then we transform a well-conditioned problem into a very-ill-conditioned one (see Example 6.8.2).

A simple theorem which can be used to localize eigenvalues is the following.

'HEOREM 5.8.1. (Gerschgorin)

Let the $n \times n$ matrix A have eigenvalues λ_i, $i = 1, \ldots, n$. Then each λ_i es in the union of the circles

$$|z - a_{ii}| \leq r_i, \quad r_i = \sum_{\substack{j=1 \\ j \neq i}}^{n} |a_{ij}|.$$

Proof. If λ is an eigenvalue, then there is an eigenvector $x \neq 0$ such that $x = \lambda x$, or

$$(\lambda - a_{ii})x_i = \sum_{\substack{j=1 \\ j \neq i}}^{n} a_{ij}x_j, \quad i = 1, 2, \ldots, n.$$

Choose i so that $|x_i| = \|x\|_\infty$. Then

$$|\lambda - a_{ii}| \leq \sum_{\substack{j=1 \\ j \neq i}}^{n} \frac{|a_{ij}||x_j|}{|x_i|} \leq r_i.$$

We can use the Gerschgorin theorem to study the effect of perturbations the elements of A. Assume that the $n \times n$ matrix A has n linearly inde-endent eigenvectors, and consider the perturbed matrix $A(\epsilon) = A + \epsilon B$. Ve can write the equations $Ax_i = \lambda_i x_i$, $i = 1, 2, \ldots, n$ as

$$AX = X\Lambda, \qquad X = (x_1, \ldots, x_n), \quad \Lambda = \operatorname{diag}(\lambda_1, \ldots, \lambda_n),$$

'here X is nonsingular. It follows that

$$X^{-1}A(\epsilon)X = \Lambda + \epsilon C, \quad C = X^{-1}BX,$$

nd $X^{-1}A(\epsilon)X$ has the same eigenvalues $\lambda_i(\epsilon)$ as $A(\epsilon)$. Using Gerschgorin's ıeorem, we get the result.

$$|\lambda_i(\epsilon) - \lambda_i| = O(\epsilon).$$

A is symmetric, then X can be chosen orthogonal, and then the elements C are of the same magnitude as those in B. This proves that *the eigenvalue roblem for a* **symmetric** *matrix is always well-conditioned.*

.8.1. The Power Method

When only a few eigenvalues and eigenvectors are needed, then the **power ıethod** is often the simplest to use. Let z_0 be an arbitrarily chosen initial ector, and form the sequence of vectors

$$z_{k+1} = Az_k = A^{k+1}z_0, \quad k = 0, 1, 2, \ldots$$

Ve assume now that A has n linearly independent eigenvectors and a unique ıgenvalue of maximum magnitude—i.e.,

$$|\lambda_1| > |\lambda_2| \geq |\lambda_3| \geq \cdots \geq |\lambda_n|.$$

we expand the initial vector $z_0 = \alpha_1 x_1 + \alpha_2 x_2 + \ldots + \alpha_n x_n$, then it follows that

$$z_k = \sum_{j=1}^{n} \lambda_j^k \alpha_j x_j = \lambda_1^k \left(\alpha_1 x_1 + \sum_{j=2}^{n} \left(\frac{\lambda_j}{\lambda_1} \right)^k \alpha_j x_j \right), \quad k = 0, 1, 2, \ldots. \tag{5.8.4}$$

Now, since $|\lambda_j/\lambda_1| < 1, j \geq 2$, the direction of the vector z_k tends to that of x_1, provided only that $\alpha_1 \neq 0$. The **Rayleigh quotient** Eq. (5.8.3), of z_k

$$\sigma_k = \frac{z_k^T A z_k}{z_k^T z_k},$$

will then tend to λ_1.

In practice, since λ_1^k often tends to zero or becomes unbounded, it is necessary to scale the sequence $\{z_k\}$. We can do this simply by using instead a sequence $\{y_k\}$, with $\| y_k \|_2 = 1$, formed by

$$y_k = \frac{z_k}{\| z_k \|_2}, \qquad z_{k+1} = A y_k, \quad k = 0, 1, 2, \ldots, \tag{5.8.5}$$

whence

$$\sigma_k = \frac{y_k^T A y_k}{y_k^T y_k} = y_k^T z_{k+1}.$$

From Eq. (5.8.4) it is readily seen that the errors in y_k and σ_k tend to zero as $|\lambda_2/\lambda_1|^k$. (If A is symmetric, then the error in σ_k will instead be proportional to $|\lambda_2/\lambda_1|^{2k}$.) The convergence, which is slow when $|\lambda_1| \approx |\lambda_2|$, can, therefore, be accelerated by Aitken extrapolation (Sec. 3.2.3).

If the power method is used with the matrix A^{-1}, then in a similar way the eigenvalue of smallest magnitude is determined, provided that $|\lambda_n| < |\lambda_{n-1}|$. More generally, if we form the sequence

$$y_k = \frac{z_k}{\| z_k \|_2}, \qquad (A - \lambda^* I) z_{k+1} = y_k, \quad k = 0, 1, 2, \ldots, \tag{5.8.6}$$

this is equivalent to the power method with the matrix $(A - \lambda^* I)^{-1}$. Since this matrix has eigenvalues $1/(\lambda_i - \lambda^*)$, this method, called **inverse iteration**, will yield the eigenvalue closest to λ^* and the corresponding eigenvector. It is best performed by first computing the LU-decomposition $A - \lambda^* I = L^* U^*$. Then each step in (5.8.6) will involve only the solution of two triangular systems, and use only n^2 operations just as the simple power method. If λ^* is a good approximation of λ_i, then inverse iteration will give an accurate eigenvector x_i in just one or two iterations, and is often the best method to use for this. The fact that $(A - \lambda^* I)$ is very near a singular matrix when $\lambda^* \approx \lambda_i$ will **not** affect the accuracy of x_i, if only some care is taken in choosing the initial vector z_0 (see [85]).

The power method can be developed into a method for computing several eigenvalues if it is combined with some method of **deflation**. By this we mean that if the eigenvalue and vector λ_1 and x_1 of A are known, then we want to compute a matrix A' which has the same eigenvalues as A except for λ_1. This usually done as follows. Suppose we can find a nonsingular matrix P such that $Px_1 = e_1$, where e_1 is the first unit vector. Then from $Ax_1 = \lambda_1 x_1$ we get

$$PAP^{-1}Px_1 = \lambda_1 Px_1, \quad \text{or} \quad (PAP^{-1})e_1 = \lambda_1 e_1.$$

But then the matrix PAP^{-1}, which has the same eigenvalues as A, must be of the form

$$PAP^{-1} = \left(\begin{array}{c|c} \lambda_1 & b^T \\ \hline 0 & \\ \vdots & A' \\ 0 & \end{array}\right),$$

and it follows that the matrix in the lower-right corner of PAP^{-1} can be taken as the desired matrix A'. The transformation matrix P can be determined in several ways. For example, P can always be chosen to be an orthogonal matrix of the form $P = I - 2ww^T$, $w^T w = 1$ (see the next section).

5.8.2. Methods Based on Similarity Transformations

If P is any nonsingular matrix, then we know that the matrices A and PAP^{-1} have the same eigenvalues. Indeed, they have the same characteristic equation, since

$$\det(\lambda I - PAP^{-1}) = \det(P)\det(\lambda I - A)\det(P^{-1}) = \det(\lambda I - A).$$

Also, since $Ax = \lambda x$ implies that

$$(PAP^{-1})Px = \lambda(Px),$$

it follows that if x is an eigenvector of A, then Px is an eigenvector of PAP^{-1}. The matrices A and PAP^{-1} are said to be similar and the transformation PAP^{-1} of A is called a similarity transformation. If the matrix P is orthogonal, then the condition of the eigenproblem is not affected.

Many methods for solving the eigenvalue problem are based on a sequence of similarity transformations with orthogonal matrices. A sequence of matrices $A = A_0, A_1, A_2, \ldots$ is then formed by

$$A_k = Q_k^T A_{k-1} Q_k, \qquad Q_k^T Q_k = I, \quad k = 1, 2, \ldots. \tag{5.8.7}$$

The matrix A_k is similar to A and the corresponding eigenvectors are related by

$$x = Q_1 Q_2 \ldots Q_k x_k.$$

Note that the sequence of transformations in Eq. (5.8.7) preserves symmetry since

$$A_k^T = (Q_k^T A_{k-1} Q_k)^T = Q_k^T A_{k-1}^T Q_k, \quad k = 1, 2, \ldots.$$

The orthogonal transformations used in practice are mainly of two different types: **plane rotations** and **reflections**. A plane rotation in the (p, q)-plane is defined by the matrix $R_{pq}(\varphi)$, $|\varphi| \leq \pi$, which is equal to the unit matrix except for the elements

$$r_{pp} = r_{qq} = \cos\varphi, \quad r_{pq} = -r_{qp} = \sin\varphi.$$

We note that $R^{\mathrm{T}}_{pq}(\varphi) = R_{pq}(-\varphi)$ is a matrix of the same type. If for a given matrix A we form the product $A' = AR_{pq}(\varphi)$, then only the elements in **columns** p and q will change. We have $a'_{ij} = a_{ij}$ if $j \neq p$ and $j \neq q$, and

$$\left.\begin{aligned} a'_{ip} &= \cos\varphi a_{ip} - \sin\varphi a_{iq}, \\ a'_{iq} &= \sin\varphi a_{ip} + \cos\varphi a_{iq}. \end{aligned}\right\} \tag{5.8.8}$$

In the same way, when we complete the similarity transformation by forming $A'' = R_{pq}(-\varphi)A'$, then only ***rows p*** and q will change.

The second type of transformation, reflection, uses orthogonal matrices of the form

$$P(w) = I - 2ww^{\mathrm{T}}, \qquad w^{\mathrm{T}}w = 1.$$

This matrix is orthogonal, since

$$(I - 2ww^{\mathrm{T}})(I - 2ww^{\mathrm{T}}) = I - 4ww^{\mathrm{T}} + 4ww^{\mathrm{T}}ww^{\mathrm{T}} = I,$$

and it reflects the space in the hyperplane through the origin orthogonal to w. Obviously, $P(w)$ is also a symmetric matrix, and therefore

$$P^{-1}(w) = P^{\mathrm{T}}(w) = P(w).$$

If we premultiply a matrix $A = (a_1, a_2, \ldots, a_n)$ by $P(w)$, $A' = P(w)A$, then each **column** is transformed independently by

$$a'_k = (I - 2ww^{\mathrm{T}})a_k = a_k - 2(w^{\mathrm{T}}a_k)w. \tag{5.8.9}$$

In the same way in postmultiplication, $A'' = A'P(w)$, the **rows** are transformed independently.

In **Jacobi's method** (1846) for a symmetric matrix A, a sequence of similarity transformations as in (5.8.7) is used, where Q_k are plane rotations chosen so that A_k tends to diagonal form. In the kth step of the **classical** Jacobi method we search A_k for the largest nondiagonal element. If this is $a^{(k)}_{pq}$, then we make the similarity transformation

$$A_{k+1} = R_{pq}(-\varphi)A_k R_{pq}(\varphi),$$

where the angle φ is chosen so that the element $a^{(k+1)}_{pq}$ is reduced to zero. Renaming $A_{k+1} = A'$ and $A_k = A$, it is easily verified that

$$a'_{pq} = (a_{qq} - a_{pp})\cos\varphi\sin\varphi + a_{pq}(\cos^2\varphi - \sin^2\varphi),$$

and thus the angle φ satisfies

$$\cot 2\varphi = \frac{a_{pp} - a_{qq}}{2a_{pq}}.$$

Only elements in rows p and q and columns p and q will be changed, and since symmetry is preserved, only the upper triangle of each $A^{(k)}$ has to be computed.

We can always take φ in the range $|\varphi| \leq \pi/4$, by choosing $t = \tan \varphi$ as the smaller root (in modulus) of the equation

$$t^2 + 2t \cot 2\varphi = 1.$$

Then $c = \cos \varphi$ and $s = \sin \varphi$ are easily determined and, if we put $r = s/(1 + c)$, the new elements in A' are given by

$$a'_{pp} = a_{pq} + ta_{pq}, \qquad a'_{qq} = a_{qq} - ta_{pq},$$
$$a'_{jp} = a'_{pj} = a_{pj} + s(a_{qj} + ra_{pj}),$$
$$a'_{jq} = a'_{qj} = a_{qj} - s(a_{pj} - ra_{qj}).$$

We have used here formulas designed (by H. Rutishauser) to diminish the accumulation of rounding errors.

Instead of searching in each step for the largest nondiagonal element, one can let (p, q) repeatedly run through the values $(1, 2), (1, 3), \ldots, (1, n), (2, 3), \ldots, (2, n), \ldots, (n - 1, n)$. The method is then called the **cyclic** Jacobi method, and such a set of $N = \frac{1}{2}n(n - 1)$ rotations is called a sweep. It is important that the convergence of both the classical and the cyclic Jacobi method can be shown to be ultimately quadratic. Denote by $\tau^2(A)$ the sum of squares of the nondiagonal elements—i.e.,

$$\tau^2(A) = \sum_i \sum_{k \neq i} a_{ik}^2.$$

Then for k large enough, we have

$$\tau^2(A_{k+N}) \leq K(\tau^2(A_k))^2,$$

for some constant K. We can interrupt the process when all nondiagonal elements are less than a prescribed tolerance. Then the diagonal elements approximate the eigenvalues of A with an error which can be estimated by Gerschgorin's theorem (Theorem 5.8.1).

In practice, with the cyclic Jacobi method usually not more than five sweeps are needed to attain highly accurate eigenvalues. In each rotation approximately $4n$ elements are transformed, and for each of these two operations are required. Taking symmetry into account it can be seen that the total number of operations for computing all the eigenvalues of A are of the order

$$5 \cdot \tfrac{1}{2}n(n - 1) \cdot 4n \approx 10n^3.$$

If we want the eigenvectors, then we must also compute

$$X_k = Q_1 Q_2 \ldots Q_k = X_{k-1} Q_k, \quad k = 1, 2, \ldots.$$

Here X_k will tend to the eigenvector matrix $X = (x_1, x_2, \ldots, x_n)$. An advantage with this formula is that X will be nearly orthogonal even when A has multiple or close eigenvalues. Since only two columns in X_k will change in each step, only $4n$ operations per step are required. This will exactly double the operation count above for the Jacobi method.

It is in general not possible to reduce a given matrix to diagonal form by a

finite sequence of similarity transformations. Many methods for the eigenproblem therefore start with an initial transformation of the matrix A to some other compact form, which can be attained in a finite number of steps. For symmetric matrices a suitable form is **symmetric tridiagonal** form

$$\boldsymbol{C} = \begin{pmatrix} a_1 & b_2 & & & \\ b_2 & a_2 & b_3 & & \\ & b_3 & \cdot & \cdot & \\ & & \cdot & \cdot & b_n \\ & & & b_n & a_n \end{pmatrix}.$$

For nonsymmetric matrices, the corresponding form is almost triangular or **Hessenberg** form,

$$\boldsymbol{H} = \begin{pmatrix} h_{11} & h_{12} & \cdots & h_{1n} \\ h_{21} & h_{22} & \cdots & h_{2n} \\ & h_{32} & & \cdot \\ & & \cdot & \cdot \\ & & h_{n,n-1} & h_{nn} \end{pmatrix}.$$

The reduction to either of these forms can be performed by a finite sequence of either plane rotations, **Givens' method**, or reflections, **Householder's method**. Since Householder's method is the more effective, we shall discuss it alone.

In the nonsymmetric case we use a sequence of $(n - 2)$ reflections—i.e., we take $\boldsymbol{A} = \boldsymbol{A}_0$ and compute

$$\boldsymbol{A}_k = \boldsymbol{P}(\boldsymbol{w}_k)\boldsymbol{A}_{k-1}\boldsymbol{P}(\boldsymbol{w}_k), \quad k = 1, 2, \ldots, n-2.$$

Here $\boldsymbol{A}_{k-1}$ already has zeros in the appropriate places in columns $1, 2, \ldots, k-1$, and $\boldsymbol{P}(\boldsymbol{w}_k)$ is chosen so that in the product $\boldsymbol{P}(\boldsymbol{w}_k)\boldsymbol{A}_{k-1}$ the elements in positions

$$(k+2, k), (k+3, k), \ldots, (n, k)$$

become zero. It can be shown that this is realized with a vector $\boldsymbol{w}_k$ which has zeros in the **first** k elements. Thus premultiplication by $\boldsymbol{P}(\boldsymbol{w}_k)$ leaves the **first** $k - 1$ columns unchanged, and postmultiplication by $\boldsymbol{P}(\boldsymbol{w}_k)$ leaves the **first** k columns unchanged. It follows that the zero elements introduced in previous steps will not be influenced.

If the described transformations are applied to a symmetric matrix A, then all transformed matrices $\boldsymbol{A}_k$ will also be symmetric. In particular, the matrix $\boldsymbol{A}_{n-2}$ will be a symmetric Hessenberg matrix, and therefore tridiagonal. Thus the transformations $\boldsymbol{P}(\boldsymbol{w}_k)$ are determined in the same way. We should, however, now take full advantage of symmetry when computing the sequence $\boldsymbol{A}_k$. If this is done, the reduction to symmetric tridiagonal form requires $2n^3/3$ operations, whereas the transformation to Hessenberg form in the nonsymmetric case requires $5n^3/3$ operations.

The reduction to the compact forms described leads to a remarkable simplification of the eigenvalue problem. How to compute eigenvalues and eigenvectors of symmetric tridiagonal and Hessenberg matrices is discussed in the remaining sections of this chapter. We note that if the eigenvectors of A are wanted, then the transformations performed in the reduction to compact form should be saved. Then if y is an eigenvector of the tridiagonal (Hessenberg) matrix,

$$x = P(w_1)P(w_2)\dots P(w_{n-2})y$$

is an eigenvector of A.

5.8.3. Eigenvalues by Equation Solving

The eigenvalues satisfy the characteristic equation $p_A(\lambda) = \det(A - \lambda I) = 0$. Therefore, if a stable method is found for evaluating function values and possibly derivatives of $p_A(\lambda)$, for numerical values of λ, then any method for solving nonlinear equations can be used to compute the eigenvalues of A. The computation of $\det(A - \lambda I)$ can be done by Gaussian elimination, but then requires $n^3/3$ operations for a full matrix A. This will clearly be inefficient. However, for the compact forms introduced in the previous section, the work will be much less!

Let C be a symmetric tridiagonal matrix with diagonal elements $a_1, a_2, \dots, a_n$ and off-diagonal elements $b_2, \dots, b_n$. We can assume that $b_i \neq 0$, $i = 2, \dots, n$, since otherwise C can be replaced by the direct sum of smaller tridiagonal matrices which can be treated separately. For any number λ we define the sequence $p_0(\lambda), p_1(\lambda), \dots, p_n(\lambda)$ by the recursion

$$\begin{gathered} p_0(\lambda) = 1, \qquad p_1(\lambda) = a_1 - \lambda, \\ p_i(\lambda) = (a_i - \lambda)p_{i-1}(\lambda) - b_i^2 p_{i-2}(\lambda), \quad i = 2, \dots, n. \end{gathered} \tag{5.8.10}$$

It is easily shown that this sequence yields the leading principal minors of $\det(C - \lambda I)$, and thus in particular we have $\det(C - \lambda I) = p_n(\lambda)$. Using this recursion, we can compute a function value in only about $3n$ operations.

The computation of the eigenvalues of C can be based on the following property of the computed sequence. Let $s(\lambda)$ be the number of agreements in sign between consecutive members in $p_0(\lambda), p_1(\lambda), \dots, p_n(\lambda)$. Then it can be shown that the number of eigenvalues greater than λ equals $s(\lambda)$. This enables us to compute any particular eigenvalue λ_k by the method of bisection (see Sec. 6.2.2). Assume that we have found an interval $[a, b]$, such that $s(a) \geq k > s(b)$. Then by computing $s(\lambda)$ for $\lambda = \frac{1}{2}(a + b)$, we can locate λ_k in an interval of width $\frac{1}{2}(b - a)$, and continuing, after p bisections in an interval of width $2^{-p}(b - a)$. Even with this rather primitive method, the total work for determining all eigenvalues of C will be less than for reducing the initial matrix A to tridiagonal form.

Often it may be worthwhile to use a more efficient method to locate the eigenvalues. We could use, for example, the Illinois method described in Sec. 6.4.4. It is also possible to compute derivatives of $p_n(\lambda)$ by differentiating the relations of Eq. (5.8.10). This enables us to use, for instance, Newton's method.

In the nonsymmetric case we have a matrix $\boldsymbol{H}$ of upper Hessenberg form. There is a similar recursion formula for computing $p_H(\lambda)$. Again, it is no restriction to assume that the elements $h_{i+1,i}$, $i = 1, 2, \ldots, n-1$ are nonzero. Thus we can work with the equation $q_H(\lambda) = 0$ instead, where $q_H(\lambda)$ is defined by

$$(-1)^{n-1} h_{21} h_{32} \ldots h_{n,n-1} q_H(\lambda) = p_H(\lambda).$$

For numerical values of λ, $q_H(\lambda) = q_n(\lambda)$ can now be computed by the recursion

$$q_0 = 1, \qquad q_1(\lambda) = \frac{-(h_{11} - \lambda)}{h_{21}},$$

$$q_i(\lambda) = \frac{-(h_{1i} q_0 + \cdots + h_{i-1,i} q_{i-2}(\lambda) + (h_{ii} - \lambda) q_{i-1}(\lambda))}{h_{i+1,i}} \qquad (5.8.11)$$

$$i = 2, 3, \ldots, n,$$

where we have put $h_{n+1,n} = 1$. One function evaluation by this recursion formula requires $\frac{1}{2}n^2$ operations. Therefore, it is here essential, more so than in the symmetric case, to use an efficient method for equation solving. Derivatives of $q_H(\lambda)$ can again be computed by differentiating the recursion formula. In spite of the fact that it uses second derivatives, Laguerre's method (Sec. 6.8.1), seems to be one of the best choices.

The methods described here for evaluating function values and derivatives are very stable. Unfortunately, even for matrices of small order it is not uncommon that underflow or overflow occurs in the recursion. In the symmetric case this problem can be overcome; the nonsymmetric case is much more difficult. Generally, if ***all*** the eigenvalues are wanted, it is probably better to use the QR-method described below.

5.8.4. The QR-Algorithm

This recently developed algorithm (Francis, Kublanovskaya, 1961) is probably the most efficient method known today for solving the complete eigenproblem for symmetric or nonsymmetric matrices.

In the basic QR-algorithm, a sequence of matrices $A = A_0, A_1, A_2, \ldots$ is computed by

$$A_s = Q_s R_s, \qquad R_s Q_s = A_{s+1}, \quad s = 0, 1, 2, \ldots,$$

where Q_s is orthogonal and R_s is upper-triangular. Notice that from the uniqueness of the QR-decomposition (Theorem 5.7.3) it follows that the seqence A_s, $s = 0, 1, 2, \ldots$ is essentially uniquely defined. Since

$$A_{s+1} = R_s Q_s = Q_s^T A_s Q_s,$$

each step in the QR-algorithm is a similarity transformation. It can be shown that under rather general conditions A_s will tend to an upper-triangular matrix with the eigenvalues on the diagonal. Since the QR-algorithm preserves symmetry, it follows that for symmetric matrices the limiting form is a diagonal matrix. A proof of these statements can be found in Wilkinson [85].

If some ratios $|\lambda_{k+1}/\lambda_k|$ are near 1, then convergence will be slow just as in the power method. The QR-algorithm must, therefore, be modified to incorporate shifts as follows:

$$A_s - k_s I = Q_s R_s, \qquad R_s Q_s + k_s I = A_{s+1}, \quad s = 0, 1, 2, \ldots.$$

It is easily seen that A_{s+1} is still similar to A_s. Here k_s is a shift parameter, which is used to accelerate the convergence.

For a full matrix A, one QR-iteration will take $4n^3/3$ operations, which is too much to be of practical use. However, if the matrix is initially transformed to Hessenberg or symmetric tridiagonal form, this form will be preserved and the work per iteration will be much less. That Hessenberg form is preserved follows from the relation

$$A_{s+1} = R_s A_s R_s^{-1}$$

and from the fact that the product of a triangular and a Hessenberg matrix is again a Hessenberg matrix. Since symmetry also is preserved, the QR-iterations also preserve symmetric tridiagonal form. Then the work for one QR-iteration is reduced to $4n^2$ in the nonsymmetric case and about $12n$ in the symmetric case.

It is important to note that the QR-iterations cannot be carried out by the Gram-Schmidt method, since the matrices Q_s then will not be sufficiently orthogonal, and therefore the transformations will not be accurate similarity transformations. Instead, we use, as in the reduction to compact form, plane rotations or reflections to carry out the QR-decomposition. For a detailed description of how a QR-iteration is performed and of how the shift parameters should be chosen, the reader is referred to the handbook [86]. With suitably chosen shifts, experimental results show that in the symmetric case fewer than two QR-iterations per eigenvalue are needed.

6 NONLINEAR EQUATIONS

6.1. INTRODUCTION

The roots of a nonlinear equation $f(x) = 0$ cannot in general be expressed in closed form. (Even when this is possible, the expression is often so complicated that it is not practical to use it!) Thus in order to solve nonlinear equations, we are obliged to use approximate methods. These methods are usually based on the idea of successive approximation or on linearization (see Chap. 1). Such methods are *iterative*; that is, starting from one or more initial approximations to the root, they produce a sequence $x_0, x_1, x_2, \ldots$ which presumably converges to the desired root. With certain methods, it is sufficient (for convergence) to know an interval $[a, b]$ which contains the root. Other methods require an initial approximation which is close to the desired root; in return, these methods converge more quickly. Thus it is often suitable to begin with a rough method and then change to a more rapidly convergent method in the final stage.

For simplicity, we shall for the most part limit ourselves to the problem of determining a real *simple root* α to the equation $f(x) = 0$; i.e., we assume that $f'(\alpha) \neq 0$. Multiple roots are discussed in Sec. 6.7.

An important practical case is the determination of all the zeros (real and complex) of a polynomial $P(x)$ with real coefficients. Special observations on this problem are taken up in Sec. 6.8.

Solving a *system* of nonlinear equations is in general a far more difficult problem. Even though many of the methods for a single equation are easily generalized, they presume that good approximations to the roots are known. Relatively little is known about how one should attack the problem if one has no a priori information as to the location of the roots.

6.2. INITIAL APPROXIMATIONS; STARTING METHODS

6.2.1. Introduction

Initial approximations to the roots of an equation $f(x) = 0$ can often be obtained by graphing the function $f(x)$.

Example 6.2.1

Obtain a rough estimate of the positive root of the equation

$$\left(\frac{x}{2}\right)^2 - \sin x = 0.$$

From the figure, it is clear that the root α lies in the interval $(\pi/2, 2)$, probably close to $x_0 = 1.9$.

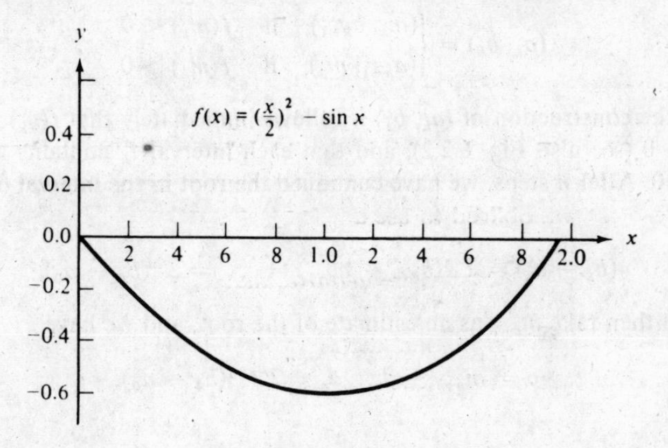

Fig. 6.2.1

Another possibility is to make a table of some values of the function $f(x)$. In the above example, we can easily set up the following table:

x	$\left(\frac{x}{2}\right)^2$	$\sin x$	$f(x)$
1.6	0.64	0.9996	<0
1.8	0.81	0.974	<0
2.0	1.00	0.909	>0

From this we can conclude that $\alpha \in (1.8, 2.0)$. In general, if a function $f(x)$ is continuous and $f(a) \cdot f(b) < 0$, then the equation $f(x) = 0$ has at least one real zero in the interval (a, b).

6.2.2. The Bisection Method

A more systematic use of the above tabulation method can be made in the following way. Suppose that $f(x)$ is continuous in the interval (a_0, b_0), and that $f(a_0)\cdot f(b_0) < 0$. We shall determine a sequence of intervals $(a_1, b_1) \supset (a_2, b_2) \supset (a_3, b_3) \ldots$, which all contain a root of the equation $f(x) = 0$. Suppose in particular that $f(a_0) < 0$ and thus $f(b_0) > 0$. (This is no limitation, since we could otherwise consider the equation $-f(x) = 0$.) The intervals $I_k = (a_k, b_k)$, $k = 1, 2, 3, \ldots$ are then determined recursively as follows. The midpoint of the interval I_{k-1} is

$$m_k = \tfrac{1}{2}(a_{k-1} + b_{k-1}).$$

We can assume that $f(m_k) \neq 0$, since otherwise we have found a root of the equation. *Compute $f(m_k)$ and take*

$$(a_k, b_k) = \begin{cases} (m_k, b_{k-1}), & \text{if} \quad f(m_k) < 0 \\ (a_{k-1}, m_k), & \text{if} \quad f(m_k) > 0. \end{cases} \tag{6.2.1}$$

From the construction of (a_k, b_k) it follows immediately that $f(a_k) < 0$ and $f(b_k) > 0$ (see also Fig. 6.2.2), and that each interval I_k contains a root of $f(x) = 0$. After n steps, we have contained the root in the interval (a_n, b_n) of length

$$(b_n - a_n) = 2^{-1}(b_{n-1} - a_{n-1}) = \ldots = 2^{-n}(b_0 - a_0).$$

We can then take m_{n+1} as an estimate of the root, and we have

$$\alpha = m_{n+1} \pm d_n, \quad d_n = 2^{-n-1}(b_0 - a_0).$$

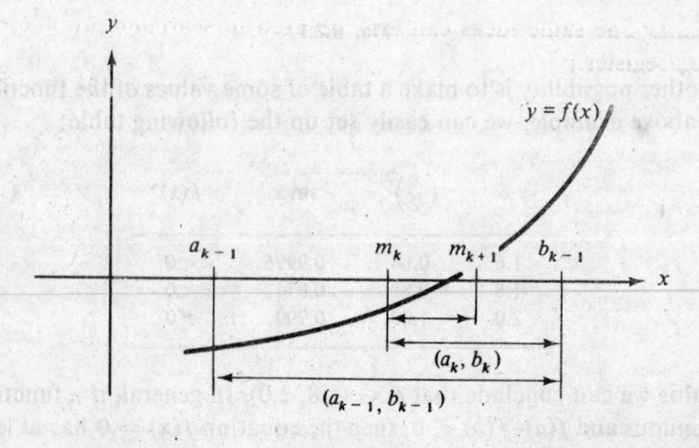

Fig. 6.2.2

Example 6.2.2

The bisection method applied to the equation $(x/2)^2 - \sin x = 0$, with $I_0 = (1.5, 2)$ gives the sequence of intervals:

k	a_{k-1}	b_{k-1}	m_k	$f(m_k)$
1	1.5	2	1.75	<0
2	1.75	2	1.875	<0
3	1.875	2	1.9375	>0
4	1.875	1.9375	1.90625	<0
5	1.90625	1.9375		

The convergence of the bisection method is slow. *At each step we gain one binary digit in accuracy.* Since $10^{-1} \approx 2^{-3.3}$, we gain on the average one decimal digit per 3.3 steps. Notice that the rate of convergence is completely independent of the function $f(x)$. This is because we only make use of the sign of the computed function values. By making more effective use of these values (and possibly also the values of the derivative of $f(x)$) we can attain significantly faster convergence. These more refined methods require in general, however, good initial approximations to guarantee convergence; suitable initial approximations can often be determined by the method of bisection.

[*Comment*: An important problem in administrative information processing is *searching in an ordered register*—e.g., a register of employees ordered according to increasing Social Security number. If the nth number in the register is denoted by $f(n)$ (where f here is an increasing, discontinuous function), then searching for a certain number a means that the equation $f(n) = a$ is to be solved. The application (with slight modifications) of the bisection method to this problem is known as *logarithmic searching*. Methods analogous to other classical methods of solving equations are also used in searching problems. The same ideas can also be used in searching an *alphabetically* ordered register.]

PROBLEMS

1. Use graphic representation to determine the zeros of the following functions to one correct decimal:
(a) $4 \sin x + 1 - x$; (b) $1 - x - e^{-2x}$; (c) $(x + 1)e^{x-1} - 1$;
(d) $x^4 - 4x^3 + 2x^2 - 8$; (e) $e^x + x^2 + x$; (f) $e^x - x^2 - 2x - 2$;
(g) $3x^2 + \tan x$.

2. The following equations all have a root in the interval $(0, 1.6)$. Determine these with an error less than 0.02 using the bisection method.
(a) $x \cdot \cos x = \ln x$; (b) $2x - e^{-x} = 0$; (c) $e^{-2x} = 1 - x$.

3. Describe the bisection method in Fortran or Algol.

6.3. NEWTON-RAPHSON'S METHOD

The idea behind Newton-Raphson's method for solving an equation $f(x) = 0$ has been described previously in Chap. 1. Starting with a given initial approximation x_0, a sequence $x_1, x_2, x_3, \ldots$ is computed, where x_{n+1} is determined in the following way. The function $f(x)$ is approximated by its tangent at the point $(x_n, f(x_n))$, and x_{n+1} is taken as the abscissa of the point of intersection of the tangent with the x-axis (see Fig. 1.2.7). Thus for determining x_{n+1}, we get the equation

$$f(x_n) + (x_{n+1} - x_n)f'(x_n) = 0.$$

Newton-Raphson's method (or Newton's method) is defined by the following iteration formula:

$$x_{n+1} \triangleq x_n + h_n, \quad h_n = \frac{-f(x_n)}{f'(x_n)}. \tag{6.3.1}$$

The iterations can be broken off when $|h_n|$ has become less than the largest error one is willing to permit in the root. (In this connection, one must also take account of rounding errors and other errors made in computing h_n.)

One can show that the above criterion for terminating the iterations gives the desired accuracy if $|Kh_n| \leq \frac{1}{2}$, where K is an upper bound for $|f''/f'|$ in the neighborhood of the root; see Sec. 6.6. (This restriction is seldom of practical importance.)

One can as a rule save one iteration, if one instead bases the termination criterion on the error estimate given in Sec. 6.6.

Example 6.3.1

Given $f(x) = \sin x - (x/2)^2$, $f'(x) = \cos x - x/2$. We want to compute the positive root to five correct decimals; see Example 6.2.1. Set $x_0 = 1.5$. (Note that the last column can't be filled in until the root has been determined.)

n	x_n	$f(x_n)$	$f'(x_n)$	h_n	$x_n - \alpha$
0	1.5	0.434995	−0.67926	0.64039	−0.43375
1	2.14039	−0.303197	−1.60948	−0.18838	+0.20664
2	1.95201	−0.024372	−1.34805	−0.01808	+0.01826
3	1.93393	−0.000233	−1.32217	−0.00018	+0.00018
4	1.93375	0.000005	−1.32191	$<\frac{1}{2}\cdot 10^{-5}$	$<\frac{1}{2}\cdot 10^{-5}$

The desired root is 1.93375. Only four iterations were needed, even though the initial approximation was fairly bad. We see also that $|h_n|$ decreases more and more quickly, until rounding errors begin to dominate.

Of course, when one is far away from the root, then one need not (in hand computation) work with so many decimals. In the above example it is only necessary to compute $f(x_n)$ to as many decimals as might be correct in x_{n+1} if h_n were computed exactly. More importantly, it is the derivative which has been computed to unnecessary accuracy. This is because $f'(x_n)$ is only a means for computing

$$h_n = \frac{-f(x_n)}{f'(x_n)}.$$

Thus it is unnecessary to compute $f'(x_n)$ to much greater *relative* precision than that of $f(x_n)$ (see Example 2.3.2). Since the relative precision of $f(x_n)$ is low as x_n approaches the root, one could (in the above example) just use $f'(x_2)$ even for $n = 3$ and $n = 4$. Such shortcuts can often be important even when working with a computer; see the comments at the end of this section.

Exercise. Treat the previous example, making use of the simplifications mentioned above. Use a slide rule and linear interpolation in tables, or else an electronic desk calculator (with a "sin" button), if you have one handy.

We shall now study the **convergence properties** of Newton-Raphson's method. We assume in what follows that $f(x)$ has two continuous derivatives, and that the root α which we are seeking is a simple root. Then we have $f'(\alpha) \neq 0$ and thus $f'(x) \neq 0$ for all x in a certain neighborhood of the root α. Let ϵ_n be the error in the estimate x_n; i.e.,

$$\epsilon_n \overset{\Delta}{=} x_n - \alpha.$$

Under the above assumptions, we shall derive a relation between ϵ_n and ϵ_{n+1}. Expanding f in a Taylor series about x_n we get

$$0 = f(\alpha) = f(x_n) + (\alpha - x_n)f'(x_n) + \tfrac{1}{2}(\alpha - x_n)^2 f''(\xi), \quad \xi \in \operatorname{int}(x_n, \alpha),$$

or, after dividing by $f'(x_n)$,

$$\frac{f(x_n)}{f'(x_n)} + \alpha - x_n = \alpha - x_{n+1} = \frac{-\frac{1}{2}(\alpha - x_n)^2 f''(\xi)}{f'(x_n)}.$$

Thus we have

$$\epsilon_{n+1} = \frac{1}{2}\epsilon_n^2 \frac{f''(\xi)}{f'(x_n)}, \tag{6.3.2}$$

and, as $x_n \longrightarrow \alpha$,

$$\frac{\epsilon_{n+1}}{\epsilon_n^2} \longrightarrow \frac{1}{2}\frac{f''(\alpha)}{f'(\alpha)}.$$

Exercise. Compute $\epsilon_{n+1}/\epsilon_n^2$ for $n = 0, 1, 2$, and the limit, as $n \longrightarrow \infty$, in Example 6.3.1.

Exercise. Show that if $x_n \longrightarrow \alpha$, then $\epsilon_n/h_n \longrightarrow -1$.

Since ϵ_{n+1} is approximately proportional to the square of ϵ_n, Newton-Raphson's method is said to be *quadratically convergent* or to be a *second-order method*. A more precise definition can be given as follows:

DEFINITION

Let $x_0, x_1, x_2, \ldots$ be a sequence which converges to α and set $\epsilon_n = x_n - \alpha$. If there exists a number p and a constant $C \neq 0$ such that

$$\lim_{n\to\infty} \frac{|\epsilon_{n+1}|}{|\epsilon_n|^p} = C, \tag{6.3.3}$$

then p is called the **order of convergence** of the sequence and C the **asymptotic error constant**. For $p = 1, 2, 3$, the convergence is said to be *linear*, *quadratic*, and *cubic*, respectively.

Now suppose that Newton-Raphson's method gives a sequence such that $\lim_{n\to\infty} x_n = \alpha$. Then, from the above we have that

$$\lim_{n\to\infty} \frac{|\epsilon_{n+1}|}{\epsilon_n^2} = C, \quad C = \frac{1}{2}\frac{|f''(\alpha)|}{|f'(\alpha)|}, \tag{6.3.4}$$

where $C \neq 0$ if $f''(\alpha) \neq 0$. This shows that Newton-Raphson's method in general gives rise to a second-order sequence.

Now suppose that I is a neighborhood of the root α such that

$$\frac{1}{2}\left|\frac{f''(y)}{f'(x)}\right| \leq m \quad \text{for all} \quad x \in I, \quad y \in I.$$

If $x_n \in I$, then from Eq. (6.3.2) it follows that $|\epsilon_{n+1}| \leq m\epsilon_n^2$. This can also be written as $|m\epsilon_{n+1}| \leq (m\epsilon_n)^2$. Suppose that $|m\epsilon_0| < 1$ and that the interval $[\alpha - |\epsilon_0|, \alpha + |\epsilon_0|]$ is a subset of I. By induction, one can show that $x_n \in I$ for all n, and that

$$|\epsilon_n| \leq \frac{1}{m}(m\epsilon_0)^{2^n}.$$

From this it follows that *Newton-Raphson's method always converges (to a simple root) assuming that x_0 has been chosen sufficiently close to the root α*—specifically, if

$$|m\epsilon_0| = m|x_0 - \alpha| < 1.$$

Example 6.3.2

Even if $m|\epsilon_0|$ is only somewhat less than 1, we get, after a few steps, very fast decrease in the error. If, for example, $m|\epsilon_0| = 0.9$ and $m = 1$, then for $n = 1, 2, \ldots, |\epsilon_n|$ is bounded by

0.81, 0.66, 0.44, 0.19, 0.036, 0.0013, 0.000016, ...

For $n \geq 6$ the number of significant decimals is approximately doubled at each iteration.

In practice, α is of course unknown, and the above condition is thus difficult to verify. Often one does not even bother to investigate convergence beforehand, since eventual divergence is very quickly detected in the iterations. The theorem below, given here without proof, gives a practical criterion for convergence. (For the proof, see Ostrowski [75], Chap. 7.)

THEOREM 6.3.1

Let x_0 be a given initial approximation and define x_n and h_n according to Eq. (6.3.1). Denote by I_0 the interval $\operatorname{int}(x_0, x_0 + 2h_0)$ *and suppose that*

$$2|h_0|M \leq |f'(x_0)|, \quad M = \max_{x \in I_0} |f''(x)|.$$

Then $x_n \in I_0$, $n = 1, 2, 3, \ldots$ and $\lim_{n \to \infty} x_n = \alpha$, where α is the only root of $f(x) = 0$ in I_0.

Another criterion, which is occasionally easier to apply, is given in the following theorem:

THEOREM 6.3.2

Suppose that $f'(x) \neq 0$, that $f''(x)$ does not change sign in the interval $[a, b]$, and that $f(a) \cdot f(b) < 0$. If

$$\left|\frac{f(a)}{f'(a)}\right| < b - a, \quad \left|\frac{f(b)}{f'(b)}\right| < b - a,$$

then Newton-Raphson's method converges from an arbitrary initial approximation $x_0 \in [a, b]$. That the theorem is true can be seen easily from Fig. 6.3.1.

Exercise. Devise an example where Newton-Raphson's method diverges, even though the equation has real roots.

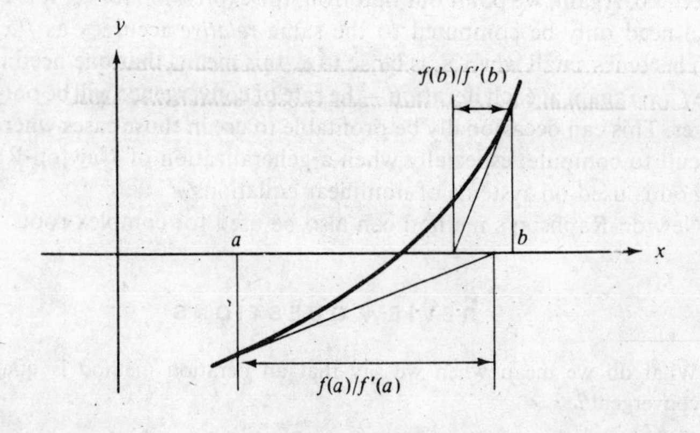

Fig. 6.3.1

Example 6.3.3

To compute $c^{1/p}$, $p = \pm 1, \pm 2, \ldots$, one can apply Newton-Raphson's method to the equation $f(x) = x^p - c = 0$. The sequence $x_1, x_2, x_3, \ldots$ is thus computed recursively from

$$x_{n+1} = x_n - \frac{x_n^p - c}{px_n^{p-1}},$$

which can be written as

$$x_{n+1} = \frac{1}{p}\left[(p-1)x_n + \frac{c}{x_n^{p-1}}\right] = \frac{x_n}{(-p)}[(1-p) - cx_n^{-p}]. \qquad (6.3.5)$$

It is convenient to use the first expression in Eq. (6.3.5) when $p > 0$ and the second when $p < 0$. This iteration formula is often used for calculating, e.g. $\sqrt{c}$, $\sqrt[3]{c}$, and $1/\sqrt{c}$ ($p = 2$, 3, and -2 respectively) on a computer. Even $1/c$ ($p = -1$) can be computed in this way if division is not built in on the machine.

Exercise. For $p = 2$, we get the same method for computing square roots which was given in Example 1.2.1. Show that in this case

$$x_{n+1} - \sqrt{c} = \frac{1}{2x_n}(x_n - \sqrt{c})^2.$$

Use this relation to show that for all x_0, $0 < x_0 < \infty$, we have $x_1 \geq x_2 \geq x_3 \geq \ldots \geq \sqrt{c}$ and that $\lim_{n\to\infty} x_n = \sqrt{c}$ (see Fig. 1.2.5).

Finally, we note that *the relation of Eq. (6.3.2) between the errors only holds as long as the round-off errors in the calculations can be ignored.* The limiting factor for the accuracy which can be achieved in calculating the root is the accuracy of the computation of $f(x_n)$. This will be discussed more fully in Sec. 6.6. Again, we point out that from the expression for h_n, it is clear that $f'(x_n)$ need only be computed to the same *relative* accuracy as $f(x_n)$. Since $f(x_n)$ becomes small when x_n is close to α, this means that one need not compute $f'(x_n)$ again at each iteration—the rate of convergence will be only slightly slower. This can occasionally be profitable to use in those cases where $f'(x)$ is difficult to compute, especially when a generalization of Newton-Raphson's method is used on systems of nonlinear equations.

Newton-Raphson's method can also be used for complex roots.

REVIEW QUESTIONS

1. What do we mean when we say that an iteration method is quadratically convergent?
2. Under what assumptions is Newton-Raphson's method quadratically convergent?

PROBLEMS

1. Study the convergence of Newton-Raphson's method by applying it to the equation $x^2 - 1 = 0$. Choose $x_0 = 2$ as a starting value, and compute the estimates of the root with successively increasing accuracy.

2. (a) The real root α to the equation $x^3 = x + 4$ can be written in the form

$$\alpha = \sqrt[3]{2 + \tfrac{1}{9}\sqrt{321}} + \sqrt[3]{2 - \tfrac{1}{9}\sqrt{321}}$$

Use this expression to compute α to four correct decimals.
(b) Compute α to the same accuracy by Newton's method, using $x_0 = 2$.

3. Use Newton-Raphson's method to determine the nonzero root of $x = 1 - e^{-2x}$ to four correct decimals.

4. Use Newton-Raphson's method to determine a root of $x \ln x - 1 = 0$ to five decimals.

5. Even complex roots of an equation can be determined using iterative methods. Apply Newton-Raphson's method to determine a root of the equation

$$z^2 + 1 = 0.$$

Start with $z_0 = 1 + i$.

6.4. THE SECANT METHOD

6.4.1. Description of the Method

The secant method can be derived from Newton-Raphson's method by approximating the derivative $f'(x_n)$ by the quotient $(f_n - f_{n-1})/(x_n - x_{n-1})$, where we denote $f(x_n)$ by f_n. This leads to the following method. Compute, from given initial approximations x_0 and x_1, the sequence $x_2, x_3, \ldots$ recursively from

$$x_{n+1} = x_n + h_n, \quad h_n = -f_n \frac{x_n - x_{n-1}}{f_n - f_{n-1}}, \quad f_n \neq f_{n-1}. \tag{6.4.1}$$

The geometrical interpretation of this method is that x_{n+1} is determined as the abscissa of the point of intersection between the secant through (x_{n-1}, f_{n-1}) and (x_n, f_n) and the x-axis. Notice that the secant method, unlike Newton-Raphson's method, requires two initial approximations, but that only *one* new function evaluation is made per step.

Example 6.4.1

$f(x) = \sin x - (x/2)^2$, $x_0 = 1$, $x_1 = 2$. We want to compute α to five decimals.

n	x_n	$f(x_n)$	$x_n - \alpha$
0	1	+0.59147	−0.93
1	2	−0.090703	+0.066
2	1.86704	+0.084980	−0.067
3	1.93135	+0.003177	−0.0024
4	1.93384	−0.000114	+0.00009
5	1.93375	+0.000005	$<\frac{1}{2}\cdot 10^{-5}$
6	1.93375		

In this example, the secant method gives the desired accuracy with the sam number of iterations as Newton-Raphson's method (see Example 6.3.1 This is because one of the initial approximations (x_1) was very close to th root.

When $|x_n - x_{n-1}|$ is small, the quotient $(x_n - x_{n-1})/(f_n - f_{n-1})$ will in gen eral be determined with poor relative accuracy. If, for example, we choos the initial approximations x_0 and x_1 very close to α, then because of round ing errors, $|x_2 - \alpha|$ can be quite large. However, it follows from the erro analysis below that the secant method gives, in general, a sequence such tha $|x_n - x_{n-1}| \gg |\alpha - x_n|$. Then a closer analysis shows that the dominar contribution to the relative error in h_n comes from the error of $f(x_n)$; *poo accuracy in the other factor is of less importance.*

Notice however, that if Eq. (6.4.1) is rewritten in the form

$$x_{n+1} = \frac{x_{n-1}f_n - x_n f_{n-1}}{f_n - f_{n-1}},$$

then we can get difficulty with *cancellation* when $x_n \approx x_{n-1}$ and $f_n f_{n-1} >$ (Hence one should *not* rewrite Eq. (6.4.1) in this form. One can run into a lo of trouble if one overlooks such details in a computer program.

The choice between the secant method and Newton-Raphson's metho depends on the amount of work required to compute $f'(x)$. Suppose th amount of work to compute $f'(x)$ is θ times the amount of work to compute value of $f(x)$. Then an asymptotic analysis can be used to motivate the rule *if $\theta > 0.44$, then use the secant method; otherwise, use Newton-Raphson method.*

6.4.2. Error Analysis for the Secant Method

We shall now derive an asymptotic formula for the connection betwee the errors in successive estimates produced by the secant method. Accordin to Newton's interpolation formula with error term (Sec. 7.3), we have

$$f(x) = f_n + (x - x_n)f[x_{n-1}, x_n] + (x - x_{n-1})(x - x_n)\tfrac{1}{2}f''(\xi), \qquad (6.4.2)$$

where

$$f[x_{n-1}, x_n] = \frac{f_n - f_{n-1}}{x_n - x_{n-1}} \quad \text{and} \quad \xi \in \text{int}\,(x, x_{n-1}, x_n).$$

f we ignore the remainder term, we get the equation of the secant. Thus x_{n+1} atisfies the equation

$$0 = f_n + (x_{n+1} - x_n) f[x_{n-1}, x_n].$$

Now put $x = \alpha$ in Eq. (6.4.2) and subtract the equation for x_{n+1}. Since $f(\alpha) = 0$, we get

$$(\alpha - x_{n+1}) f[x_{n-1}, x_n] + \tfrac{1}{2}(\alpha - x_{n-1})(\alpha - x_n) f''(\xi) = 0.$$

According to the mean-value theorem, we have

$$f[x_{n-1}, x_n] = f'(\xi'), \quad \xi' \in \text{int}\,(x_{n-1}, x_n),$$

and we get

$$\epsilon_{n+1} = \frac{f''(\xi)}{2 \cdot f'(\xi')} \epsilon_n \cdot \epsilon_{n-1}. \tag{6.4.3}$$

From this relation it follows that the secant method converges, for sufficiently good initial approximations x_0 and x_1, if $f'(\alpha) \neq 0$ and if $f(x)$ has a continuous second derivative. Notice that if we let $x_{n-1} \longrightarrow x_n$, then the formula for one step with the secant method becomes the formula for one step with Newton-Raphson's method, and we get Eq. (6.3.2) again.

Exercise. Compute $\epsilon_{n+1}/\epsilon_n \epsilon_{n-1}$ for $n = 1, 2, 3$ in Example 6.4.1.

Suppose now that the secant method converges. When n is large, we have $\xi \approx \alpha$, $\xi' \approx \alpha$, and

$$|\epsilon_{n+1}| \approx C|\epsilon_n| \cdot |\epsilon_{n-1}| \tag{6.4.4}$$

(with the same C as in Eq. (6.3.4)). We now attempt to determine the order of the secant method and try putting

$$|\epsilon_{n+1}| \approx K|\epsilon_n|^p, \qquad |\epsilon_n| \approx K|\epsilon_{n-1}|^p.$$

Substitution in Eq. (6.4.4) gives

$$K|\epsilon_n|^p \approx C|\epsilon_n| K^{-1/p} |\epsilon_n|^{1/p}.$$

This relation can hold only if $p = 1 + 1/p$—i.e., if $p = \frac{1}{2}(1 \pm \sqrt{5})$ and $C = K^{1+1/p} = K^p$. One can show that the root with least absolute value here can be ignored and that

$$|\epsilon_{n+1}| \approx C^{1/p} |\epsilon_n|^p, \quad p = \tfrac{1}{2}(1 + \sqrt{5}) = 1.618\ldots \quad (n \gg 1).$$

The above heuristic discussion can be developed into a strict proof; see Ostrowski [74], Chap. 12.

6.4.3. Regula Falsi

Régula falsi is a variant of the secant method where one instead choose the secant through (x_n, f_n) and $(x_{n'}, f_{n'})$ where n' is the largest index, $n' < n$ for which $f_n \cdot f_{n'} < 0$. The initial approximations x_0 and x_1 must of course b chosen so that $f_0 \cdot f_1 < 0$. The advantage of regula falsi is that, like the bisec tion method, it is always convergent for continuous functions $f(x)$. In contras to the secant method, however, the regula falsi is in general a first-orde method. That this is the case when $f(x)$ is convex on $[x_0, x_1]$ can be see from Fig. 6.4.1. We see that in this case the successive secants will all pas through the point (x_0, f_0)—i.e., the point (x_0, f_0) is retained—and Eq. (6.4.4 becomes

$$\lim_{n \to \infty} \frac{|\epsilon_{n+1}|}{|\epsilon_n|} = C|\epsilon_0| = C'.$$

Thus $p = 1$ and we have *linear convergence*. Regula falsi is a good "start" method, but it should not be used near a root. It may, however, be used a one part of a "hybrid" method which has good convergence properties nea a root; see Sec. 6.4.4.

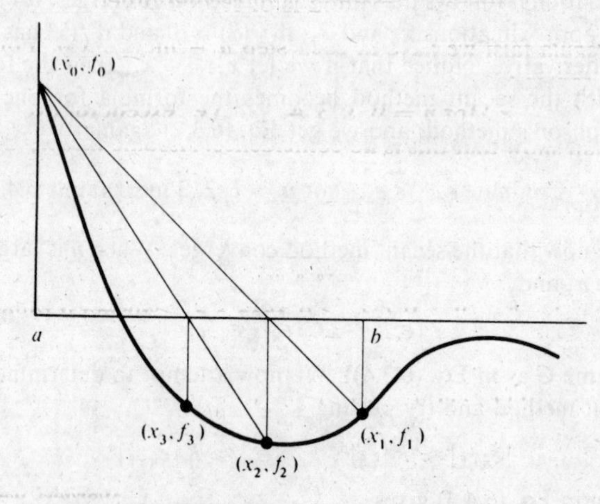

Fig. 6.4.1

6.4.4. Other Related Methods

The secant method does not achieve quadratic convergence. Indeed, it has been shown under very weak restrictions that no iterative method using only one function evaluation per step can have $p = 2$. However, **Steffensen's method**,

$$x_{n+1} = x_n - \frac{f(x_n)}{g(x_n)}, \quad g(x_n) = \frac{f(x_n + f(x_n)) - f(x_n)}{f(x_n)}, \qquad (6.4.5)$$

which requires two function evaluations but no derivatives, will be shown to be of second order. This method, which is closely related to the secant method, is of particular interest for solving systems of nonlinear equations in several variables (see Sec. 6.9).

If we put $\beta_n = f(x_n)$ and expand $g(x_n)$ in a Taylor series about x_n, we get

$$g(x_n) = \frac{f(x_n + \beta_n) - f(x_n)}{\beta_n} = f'(x_n)\Big(1 - \frac{1}{2}h_n f''(x_n) + O(\beta_n^2)\Big),$$

where $h_n = -f(x_n)/f'(x_n)$ is the Newton correction. Thus

$$x_{n+1} = x_n + h_n(1 + \tfrac{1}{2}h_n f''(x_n) + O(\beta_n^2)),$$

and using the error equation Eq. (6.3.2), for Newton's method, which can be written

$$h_n = -\epsilon_n + \frac{1}{2}\epsilon_n^2 \frac{f''(\xi)}{f'(x_n)}, \quad \epsilon_n = x_n - \alpha,$$

we get

$$\frac{\epsilon_{n+1}}{\epsilon_n^2} \longrightarrow \frac{1}{2}\frac{f''(\alpha)}{f'(\alpha)}(1 + f'(\alpha)).$$

This proves that Steffensen's method is of second order.

Using regula falsi we have in each step $\alpha \in \operatorname{int}(x_n, x_{n'})$. For the secant method we saw in Example 6.4.1 that the error in successive steps changed so that $\alpha \in \operatorname{int}(x_{n+1}, x_n)$ for $n = 0, 1, 3, 4, \ldots$, i.e., except for every third value of n. We shall show that this is no coincidence. Assume that $x_n \in (a, b)$, $n = 0, 1, 2, \ldots$, and that $f'(x)$ and $f''(x)$ have constant sign in (a, b). Then from Eq. (6.4.3) it follows that

$$\frac{\epsilon_{n+1}}{\epsilon_n \epsilon_{n-1}}$$

has constant sign for all n. If $f_0 f_1 < 0$, then $\epsilon_0 \epsilon_1 < 0$, and it follows that sign (ϵ_n) must vary according to one of the following two schemes

(i) $\ldots + - + + - + + - + \ldots;$

(ii) $\ldots + - - + - - + - - + \ldots.$

This shows that for the secant method (asymptotically) only every third step is not a regula falsi step. When the secant method converges, we shall get a strict error estimate easily (whereas Newton's method under the same assumptions converges monotonically). This observation can be used to improve the rate of convergence for regula falsi without violating the global convergence property. We choose again x_0 and x_1 so that $f_0 f_1 < 0$. Now assume that for a certain n we have $f_{n-1} f_n < 0$, and that therefore x_{n+1} is computed by Eq. (6.4.1). If now $f_{n+1} f_n < 0$, then we can for the next step use the secant through (x_{n+1}, f_{n+1}) and (x_n, f_n). If, on the other hand, we have $f_{n+1} f_n > 0$, then certainly $f_{n+1} f_{n-1} < 0$. We then make a modified step and take x_{n+2} as the root of the linear function through (x_{n+1}, f_{n+1}) and $(x_{n-1}, \alpha f_{n-1})$, where

$0 < \alpha < 1$ is a parameter. Cleary $\alpha = 1$ would correspond to a regula fal step and usually give $f_{n+2}f_{n+1} > 0$. On the other hand, $\alpha = 0$ gives $x_{n+2} =$ x_{n-1}, and thus $f_{n+2}f_{n+1} < 0$. A suitable choice of α will evidently give an x_{n} $\neq x_{n-1}$ such that $f_{n+2}f_{n+1} < 0$, and then the next step can be an unmodifie secant step. This device will prevent the retention of a point, which causes th linear convergence of regula falsi.

An asymptotic analysis of a modified step shows that

$$\epsilon_{n+1} \approx -\epsilon_n \quad \text{for} \quad \alpha = \tfrac{1}{2}. \tag{6.4.6}$$

From the sign variation in the secant method it follows that asymptotical we have to take a modified step every third time. Assume that x_n has bee computed by a modified step; then by Eqs. (6.4.6) and (6.4.3),

$$\epsilon_n = -\epsilon_{n-1}, \qquad \epsilon_{n+1} = -c\epsilon_{n-1}^2, \qquad \epsilon_{n+2} = c^2 \cdot \epsilon_{n-1}^3.$$

Thus we can say that this algorithm, sometimes called the **Illinois algorithm** gives cubic convergence with three function evaluations. It might still happe that after a modified step with $\alpha = \frac{1}{2}$ we get $f_{n+2}f_{n+1} > 0$. We can, then, fo the next step use the line through (x_{n+2}, f_{n+2}) and $(x_{n-1}, \frac{1}{4}f_{n-1})$. More general ly, we take successively $\alpha = (\frac{1}{2})^k$, $k = 1, 2, 3, \ldots$, until we get the change o sign in f.

PROBLEMS

1. Use the secant method to determine the roots of the following equations to five correct decimals:
(a) $2x = e^{-x}$; (b) $\tan x + \cosh x = 0$.

2. Describe the Illinois method in Algol or Fortran.

3. Carry out the Illinois algorithm graphically on the equation in Fig. 6.4.1, starting from $x_0 = a$, $x_1 = b$. Compare with the performance of the secant method and regula falsi on the same equation.

4. The secant method was derived by estimating $f'(x_n)$ by linear interpolation through (x_{n-1}, f_{n-1}) and (x_n, f_n). If we also use the information (x_{n-2}, f_{n-2}), then $f'(x_n)$ can be estimated by quadratic interpolation.
(a) Show, using Newton's interpolation formula (Sec. 7.3.3), that the resulting iterative method can be written

$$x_{n+1} = x_n - \frac{f_n}{\omega},$$

where

$$\omega = f[x_n, x_{n-1}] + (x_n - x_{n-1})f[x_n, x_{n-1}, x_{n-2}].$$

(b) Show that if, instead, x_{n+1} is taken as the zero of the same interpolating parabola, then

$$x_{n+1} = x_n - \frac{2f_n}{\omega \pm (\omega^2 - 4f_n f[x_n, x_{n-1}, x_{n-2}])^{1/2}}$$

This is the *Muller-Traub* method.

The Muller-Traub method uses three points to determine the coefficients of an interpolating parabola. The same points could also be used to determine the three coefficients in the rational function

$$g(x) = \frac{x - a}{bx + c}.$$

In an iterative method we could then take x_{n+1} as the zero of $g(x) = 0$—i.e., $x_{n+1} = a$. A theorem in projective geometry, according to which the cross ratio of any four values of x is equal to the cross ratio of the corresponding values of $g(x)$, implies (see Householder [72], p. 159) that

$$\frac{(0 - f_n)/(0 - f_{n-2})}{(f_{n-1} - f_n)/(f_{n-1} - f_{n-2})} = \frac{(x_{n+1} - x_n)/(x_{n+1} - x_{n-2})}{(x_{n-1} - x_n)/(x_{n-1} - x_{n-2})}.$$

(a) Show that this is equivalent to calculating x_{n+1} from the "modified secant method"

$$x_{n+1} = x_n - f_n \frac{x_n - x_{n-2}}{f_n - \bar{f}_{n-2}}, \quad \bar{f}_{n-2} = f_{n-2} \frac{f[x_n, x_{n-1}]}{f[x_{n-1}, x_{n-2}]}.$$

(b) Use the result in (a) to show that when $\text{sign}(f_n) = -\text{sign}(f_{n-2})$ and $\text{sign}(f[x_n, x_{n-1}]) = \text{sign}(f[x_{n-1}, x_{n-2}])$, then $x_{n+1} \in \text{int}(x_{n-2}, x_n)$.

(c) The results above suggest that in the Illinois method we take

$$\alpha = \begin{cases} \beta, & \beta > 0 \\ \frac{1}{2}, & \beta \leq 0 \end{cases}, \quad \beta = \frac{f[x_{n+1}, x_n]}{f[x_n, x_{n-1}]}.$$

Try this modification in Example 6.4.1.

.5. GENERAL THEORY OF ITERATION METHODS

Newton-Raphson's method and the secant method can both be considered s special cases of the following more general iteration method. Let x_{n+1} be etermined by function values and values of the derivative of $f(x)$ at m points $_n$, $x_{n-1}, \ldots, x_{n-m+1}$ and put

$$x_{n+1} = \varphi(x_n, x_{n-1}, \ldots, x_{n-m+1}).$$

Ve call φ an **iteration function.**

xample 6.5.1

Newton-Raphson's method:

$$\varphi(x) = x - \frac{f(x)}{f'(x)}, \quad m = 1.$$

The secant method:

$$\varphi(x, y) = x - f(x)\frac{x - y}{f(x) - f(y)}, \quad m = 2.$$

The general theory of iteration methods is simplest when $m = 1$. In thi case we have

$$x_{n+1} = \varphi(x_n),$$

which is called a **one-point iteration** method. Since we shall only study th case $m = 1$ in this section, we shall omit the prefix "one-point" in what fo lows. Suppose now that we have a sequence $\{x_n\}$, generated by a certain initia value x_0, and that $\lim_{n\to\infty} x_n = \alpha$. If $\varphi(x)$ is continuous, then we have that

$$\alpha = \lim_{n\to\infty} x_{n+1} = \lim_{n\to\infty} \varphi(x_n) = \varphi(\alpha),$$

i.e., the limiting value α is a root of the equation $x = \varphi(x)$. Reversing this, w see that to construct an iterative method for solving an equation $f(x) = ($ we can try rewriting it in the form $x = \varphi(x)$; this then defines an iteratio method, $x_{n+1} = \varphi(x_n)$.

Example 6.5.2

The equation $f(x) = x^3 - x - 5$ can, for example, be written a $x = \varphi_i(x)$, where $\varphi_1(x) = x^3 - 5$, $\varphi_2(x) = \sqrt[3]{x+5}$; $\varphi_3(x) = 5/(x^2 - 1)$.

Rewriting an equation in the form $x = \varphi(x)$ can clearly be done in man ways. As was shown in Chap. 1, we cannot always be sure that we get method which converges. The sequence $\{x_n\}$ can diverge even if x_0 is chose arbitrarily close to the root. A sufficient condition for convergence is given i the following theorem:

THEOREM 6.5.1

Suppose that the equation $x = \varphi(x)$ has a root α, and that in the interva

$$J = \{x : |x - \alpha| \leq \rho\}$$

$\varphi'(x)$ exists and satisfies the condition

$$|\varphi'(x)| \leq m < 1.$$

Then for all $x_0 \in J$:

(a) $x_n \in J$, $n = 0, 1, 2, \ldots$;
(b) $\lim_{n\to\infty} x_n = \alpha$;
(c) *α is the only root in J of $x = \varphi(x)$.*

Proof. We first prove assertion (*a*), by induction. Suppose that $x_{n-1} \in$ J. By the mean-value theorem we can show that

$$x_n - \alpha = \varphi(x_{n-1}) - \varphi(\alpha) = \varphi'(\xi_n)(x_{n-1} - \alpha), \quad \xi_n \in J.$$

whence

$$|x_n - \alpha| \leq m|x_{n-1} - \alpha| \leq m\rho.$$

Hence $x_n \in J$ and (*a*) is proved.

Repeated use of the inequality above gives

$$|x_n - \alpha| \leq m|x_{n-1} - \alpha| \leq \ldots \leq m^n |x_0 - \alpha|,$$

nd since $m < 1$, the result (*b*) follows. Suppose, finally, that $x = \varphi(x)$ has nother root β, $\beta \neq \alpha$, $\beta \in J$. Then

$$\alpha - \beta = \varphi(\alpha) - \varphi(\beta) = \varphi'(\xi)(\alpha - \beta) \qquad \xi \in J$$

nd thus

$$|\alpha - \beta| \leq m|\alpha - \beta| < |\alpha - \beta|,$$

contradiction; thus (*c*) follows.

Exercise. The equation in Example 6.5.2 has a root near $x_0 = 1.9$. Show ıat of the three proposed iteration methods $x_{n+1} = \varphi_i(x_n)$, $i = 1, 2, 3$, only = 2 gives a convergent method for this root.

In Theorem 6.5.1 we assumed the existence of a root α. The theorem can, owever, be modified so that it can be used to *prove* the existence of a root of ıe equation $x = \varphi(x)$. For a proof see Ostrowski [75, p. 46].

HEOREM 6.5.2

Let J be a closed interval on which $\varphi'(x)$ exists and satisfies the inequality

$$|\varphi'(x)| \leq m < 1.$$

et the sequence $x_1, x_2, x_3, \ldots$ be defined by $x_{n+1} = \varphi(x_n)$, $x_0 \in J$. If

$$x_1 \pm \frac{m}{1-m}|x_1 - x_0| \in J.$$

hen (a), (b), and (c) of Theorem 6.5.1 are true.

From the proof of Theorem 6.5.1 it follows that

$$\frac{x_n - \alpha}{x_{n-1} - \alpha} \approx \varphi'(\alpha),$$

nd if $\varphi'(\alpha) \neq 0$, then $x_n - \alpha$ approximately forms terms in a geometric series. hen the sequence $\{x_n\}$ can be transformed into a more rapidly convergent equence $\{x'_n\}$ by **Aitken extrapolation** (Sec. 3.2.3),

$$x'_n = x_n - \frac{(x_n - x_{n-1})^2}{x_n - 2x_{n-1} + x_{n-2}}.$$

Jsing the difference operator introduced in Chap. 1, this can also be writ- en as

$$x'_n = x_n - \frac{(\Delta x_{n-1})^2}{\Delta^2 x_{n-2}}.$$

This transformation is so simple that it should be regularly used if the con- ergence is known to be *linear*; otherwise $\{x'_n\}$ will usually converge slower han $\{x_n\}$.

Example 6.5.3

The equation $x = e^{-x}$ has one root $\alpha \approx 0.567$. We shall compute it to six decimals using the iterative method $x_{n+1} = e^{-x_n}$, and use Aitken extrapolation to accelerate the convergence. We get the following sequence of approximations:

n	x_n	$\Delta x_{n-1}\cdot 10^6$	$\Delta^2 x_{n-2}\cdot 10^6$	x'_n
0	0.567 000			
1	0.567 225	225		
2	0.567 097	−128	−353	0.567 143
3	0.567 170	73	201	0.567 143
4	0.567 129	−41	−114	0.567 144
5	0.567 152	23	64	0.567 144
6	0.567 139	−13	−36	0.567 144

It is seen that the extrapolated sequence $\{x'_n\}$ converges much more rapidly in this example.

The iteration method $x_{n+1} = \varphi(x_n)$ is in general a first-order method unless $\varphi(x)$ is chosen in some special way. Suppose now that $\varphi(x)$ is p times continuously differentiable in a neighborhood of α, where $\alpha = \varphi(\alpha)$ and

$$\varphi^{(j)}(\alpha) = 0, \quad j = 1, 2, \ldots, p-1 \qquad \varphi^{(p)}(\alpha) \neq 0. \tag{6.5.1}$$

According to Taylor's theorem, then, we have

$$x_{n+1} = \varphi(x_n) = \alpha + \frac{1}{p!}\varphi^{(p)}(\xi_n)(x_n - \alpha)^p, \quad \xi_n \in \text{int}\,(x_n, \alpha).$$

If $\lim_{n\to\infty} x_n = \alpha$, then we have

$$\lim_{n\to\infty} \frac{|\epsilon_{n+1}|}{|\epsilon_n|^p} = \frac{1}{p!}|\varphi^{(p)}(\alpha)| \neq 0, \quad \epsilon_n = x_n - \alpha.$$

The iteration method $x_{n+1} = \varphi(x_n)$ is of order p for the root α if Eq. (6.5.1) *holds.*

Example 6.5.4

The above gives an alternative proof for the fact that Newton-Raphson's method is of at least second order for simple roots. We have

$$\varphi(x) = x - \frac{f(x)}{f'(x)}, \quad \varphi'(x) = \frac{f(x)f''(x)}{(f'(x))^2}.$$

If α is a simple root, $f'(\alpha) \neq 0$, and thus $\varphi'(\alpha) = 0$.

Exercise. Show that if we also have $f''(\alpha) = 0$, then Newton-Raphson's method is at least of third order.

One might think that it is difficult to construct iteration methods of arbitrarily high order for solving the equation $f(x) = 0$. There are, however, many general methods for this. One possibility (due to E. Schröder, 1865) is to put $\varphi(x) = x + h(x)$, where

$$h(x) = -u(x) + \sum_{k=2}^{p-1} c_k(x)u^k(x), \quad u(x) = \frac{f(x)}{f'(x)}.$$

One can determine $c_k(x)$, independent of p, so that the iteration method $x_{n+1} = \varphi(x_n)$ is of order p. In particular,

$$c_2 = -a_2, \qquad c_3 = -(2a_2^2 - a_3), \quad \text{where} \quad a_i = \frac{f^{(i)}(x)}{i!\,f'(x)},$$

which give methods of order 3 and 4. One can show that a one-point iterative method of order p always requires computation of the p quantities $f(x_n)$, $f'(x_n), \ldots, f^{(p-1)}(x_n)$. For proofs of the above assertions, see Traub [80]. Methods with $p > 3$ are in general of practical interest only in cases where higher derivatives of $f(x)$ can be easily computed.

PROBLEMS

1. One wants to solve the equation $x + \ln x = 0$, whose root is $\alpha \approx 0.5$, by iteration, and one chooses among the following iteration formulas:

$$\text{I1:} \quad x_{n+1} = -\ln x_n; \qquad \text{I2:} \quad x_{n+1} = e^{-x_n}; \qquad \text{I3:} \quad x_{n+1} = \frac{x_n + e^{-x_n}}{2}.$$

(a) Which of the formulas *can* be used?
(b) Which formula *should* be used?
(c) Give an even better formula.

2. One wants to use the iteration formula

$$x_{n+1} = 2^{x_n - 1}$$

to solve the equation $2x = 2^x$. Investigate if and to what the iteration sequence converges for various choices of x_0.

3. Let the function $f(x)$ be four times continuously differentiable and have a *simple* zero ξ. Successive approximations x_n, $n = 1, 2, \ldots$ to ξ are computed from $x_{n+1} = (x'_{n+1} + x''_{n+1})/2$, where

$$x'_{n+1} = x_n - \frac{f(x_n)}{f'(x_n)},$$

$$x''_{n+1} = x_n - \frac{u(x_n)}{u'(x_n)}, \quad u(x) = \frac{f(x)}{f'(x)}.$$

Prove that if the sequence $\{x_n\}$ converges to ξ, then the convergence is cubic.

4. Determine p, q, and r so that the order of the iterative method

$$x_{n+1} = px_n + \frac{qa}{x_n^2} + \frac{ra^2}{x_n^5}$$

for $a^{1/3}$ becomes as high as possible. For this choice of p, q, and r, indicate how the error in x_{n-1} depends on the error in x_n.

5. In Example 6.5.3 Aitken extrapolation was used in a *passive* way to transform the sequence $\{x_n\}$ into $\{x'_n\}$. It is also possible to use Aitken extrapolation in an *active* way as follows. We start as before by computing

$$x_1 = \varphi(x_0), \qquad x_2 = \varphi(x_1),$$

and applying the Aitken formula to compute x'_2. Now we take x'_2 as a new starting value—i.e., compute

$$x_3 = \varphi(x'_2), \qquad x_4 = \varphi(x_3),$$

and extrapolate from x'_2, x_3, and x_4 to get x'_4, etc.

(a) Show that the sequence $z_n = x'_{2n}$ can be generated by $z_0 = x_0$,

$$z_{n+1} = \psi(z_n), \quad \psi(z) = z - \frac{(\varphi(z) - z)^2}{\varphi(\varphi(z)) - 2\varphi(z) + z}.$$

(b) Prove that the iteration $z_{n+1} = \psi(z_n)$ is equivalent to applying Steffensen's method to the equation $f(z) = 0$, where $f(z) = \varphi(z) - z$.

(c) From the result in (b) follows that active Aitken extrapolation may work even when the basic iteration $x_{n+1} = \varphi(x_n)$ diverges. Use this method to compute the smallest root of the equation $x = 5 \ln x$ with an error $< \frac{1}{2} \cdot 10^{-4}$ starting with $x_0 = 1.3$.

6.6. ERROR ESTIMATION AND ATTAINABLE ACCURACY IN ITERATION METHODS

6.6.1. Error Estimation

In the previous section we studied the asymptotic behavior of the error in x_n, as $n \longrightarrow \infty$, for an iteration method $x_{n+1} = \varphi(x_n)$. We shall now derive an estimate of the error after a finite number of iterations; this estimate will also take into account the fact that the calculated value of $\varphi(x)$ is usually afflicted with error—for example, round-off errors.

Denote by $\bar{x}_1, \bar{x}_2, \bar{x}_3$ the *computed* sequence, and let the error in the computation of $\varphi(\bar{x}_n)$ be δ_n. Then

$$\bar{x}_{n+1} = \varphi(\bar{x}_n) + \delta_n, \quad n = 0, 1, 2, \ldots.$$

If we subtract the equation $\alpha = \varphi(\alpha)$ from the above, we get, using the mean-value theorem,

$$\bar{x}_{n+1} - \alpha = \varphi'(\xi_n)(\bar{x}_n - \alpha) + \delta_n, \quad \xi_n \in \text{int}\,(\bar{x}_n, \alpha).$$

After a simple manipulation we get

$$(1 - \varphi'(\xi_n))(\bar{x}_{n+1} - \alpha) = \varphi'(\xi_n)(\bar{x}_n - \bar{x}_{n+1}) + \delta_n.$$

Suppose now that

$$|\varphi'(\xi_n)| \le m < 1 \quad \text{and} \quad |\delta_n| < \delta.$$

Then we get the inequality

$$|\bar{x}_{n+1} - \alpha| < \frac{m}{1-m}|\bar{x}_{n+1} - \bar{x}_n| + \frac{1}{1-m}\delta. \qquad (6.6.1)$$

which gives a strict estimate of the error in $\bar{x}_{n+1}$ using known quantities. The first term on the right-hand side estimates the truncation error and the second estimates the computational error.

Note that the final computational error, $\delta/(1-m)$, only depends on the error δ_n in the *last* iteration, and that $\bar{x}_n$ should be considered as an exact quantity when estimating δ_n. This means that it is not necessary to compute with full accuracy in the first few iterations, since roundings (and even minor mistakes in calculation) made in these iterations have no influence on the final accuracy. In this sense, iteration methods are self-correcting. For sufficiently large values of n, the computational error is the dominant contribution to $|\bar{x}_{n+1} - \alpha|$. Further iterations do not give any further improvement in accuracy; the error then behaves irregularly with constant magnitude of about $\delta/(1-m)$.

Example 6.6.1

The iteration method $x_{n+1} = 1 - e^{-2x_n}$ produces the sequence

$$\bar{x}_1 = 0.798, \quad \bar{x}_2 = 0.7973, \quad \bar{x}_3 = 0.79701, \quad \bar{x}_4 = 0.79689,$$

where correct rounding has been performed in each step. We have $\varphi'(x) = 2 \cdot e^{-2x}$ and thus

$$|\varphi'(x)| < |\varphi'(0.8)| < 0.41, \quad x \in \text{int}\,(\bar{x}_4, \alpha).$$

The error estimate Eq. (6.6.1), with $m = 0.41$ and $\delta = 0.5 \cdot 10^{-5}$ gives, then,

$$|\bar{x}_4 - \alpha| \leq \frac{0.41}{0.59} 12 \cdot 10^{-5} + \frac{1}{0.59} 0.5 \cdot 10^{-5} < 9.2 \cdot 10^{-5}.$$

Newton-Raphson's method can be rewritten in the form $x_{n+1} = \varphi(x_n)$, with

$$\varphi(x) = x + h(x), \quad \text{where} \quad h(x) = \frac{-f(x)}{f'(x)}, \quad \varphi'(x) = -h(x) \cdot \frac{f''(x)}{f'(x)}.$$

The error estimate given by (6.6.1) is also applicable for this method. The following somewhat modified formulation, easily proved, is more comfortable to use: *Put $h_n \triangleq h(\bar{x}_n)$ and suppose that*

$$|Kh_n| \leq m < 1, \quad \text{where} \quad \left|\frac{f''(x)}{f'(x)}\right| \leq K, \quad x \in \text{int}\,(\bar{x}_n, \alpha). \qquad (6.6.2)$$

Then for $\bar{x}_{n+1} = \bar{x}_n + h_n + \delta_n$, we have the error estimate of (6.6.1), where δ is an upper bound for the error in computation (δ_n) of h_n.

Example 6.6.2

In Example 6.3.1, Newton-Raphson's method gave, for $f(x) = \sin x - (x/2)^2$:

$$\bar{x}_3 = 1.93393, \quad h_3 = -0.00018 \pm \tfrac{1}{2} \cdot 10^{-5}.$$

Since $f'(x) = \cos x - x/2$, $f''(x) = -\sin x - \frac{1}{2}$, we get

$$|f'(x)| \geq 1, \quad x \in \text{int}\,(\bar{x}_4, \alpha), \qquad |f''(x)| \leq \tfrac{3}{2}.$$

Thus $|Kh_3| \leq \frac{3}{2}\cdot 1.85\cdot 10^{-4} < 3\cdot 10^{-4}$ and

$$|\bar{x}_4 - \alpha| < \frac{3\cdot 10^{-4}\cdot 18\cdot 10^{-5} + \frac{1}{2}\cdot 10^{-5}}{1 - 3\cdot 10^{-4}} < 0.506\cdot 10^{-5}$$

We also see that the truncation error is negligible in comparison to the round-off error in h_3.

6.6.2. Attainable Accuracy; Termination Criteria

Suppose that x_n is an approximation to a simple root α of the equation $f(x) = 0$. Then, from the mean-value theorem,

$$f(x_n) = (x_n - \alpha)f'(\xi), \quad \xi \in \text{int}\,(x_n, \alpha) = J.$$

From this we get the **method-independent error estimate**

$$|x_n - \alpha| \leq \frac{|f(x_n)|}{M_1}, \qquad |f'(x)| \geq M_1, \quad x \in J. \tag{6.6.3}$$

Denote by $\bar{f}(x_n)$ the computed value of the function, and let

$$\bar{f}(x_n) = f(x_n) + \delta(x_n),$$

where $|\delta(x)| \leq \delta$ independent of x. The accuracy to which α can be determined is limited by δ. The best we can hope for is to find an x_n such that $\bar{f}(x_n) = 0$. In this case, the exact value of the function satisfies $|f(x_n)| \leq \delta$. Supposing that $f'(x)$ does not vary greatly near $x = \alpha$, we get, from (6.6.3),

$$|x_n - \alpha| \leq \frac{\delta}{M_1} \approx \epsilon_\alpha, \quad \text{where} \quad \epsilon_\alpha = \frac{\delta}{|f'(\alpha)|}. \tag{6.6.4}$$

The best error bound for *any* method is thus ϵ_α, which we call the *attainable accuracy for the root* α. We note that if $|f'(\alpha)|$ is small, then ϵ_α is large. In this case, the problem of computing α is ill-conditioned (see Fig. 6.6.1).

A similar analysis for a root α of multiplicity p gives the limiting accuracy as

$$\epsilon_\alpha = \left(\frac{\delta\cdot p!}{|f^{(p)}(\alpha)|}\right)^{1/p} \tag{6.6.5}$$

Because of the exponent $1/p$, we see that multiple roots are in general ill-conditioned (so long as the function f is not given in such a form that its value can be computed with less absolute error as x approaches α).

Example 6.6.3

The equation $f(x) = (x - 2)x + 1 = 0$ has a double root $x = 1$. The value of the function for $x = 1 + \epsilon$ is

$$f(x) = (\epsilon - 1)(1 + \epsilon) + 1 = -(1 - \epsilon^2) + 1 = \epsilon^2.$$

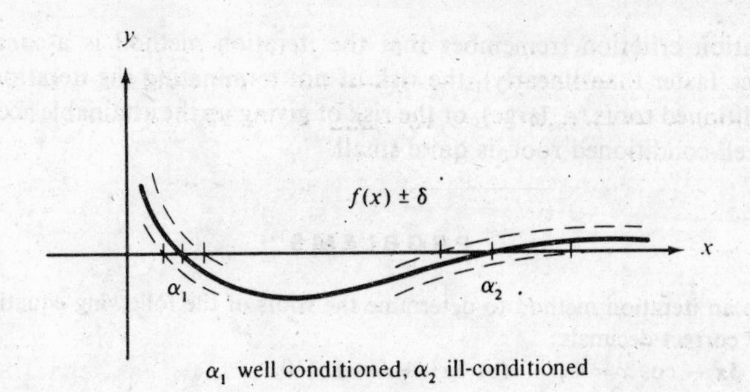

α_1 well conditioned, α_2 ill-conditioned

Fig. 6.6.1

Now suppose that we have floating-point arithmetic with eight decimals in the mantissa. Then

$$fl(1-\epsilon^2) = 1 \quad \text{if} \quad |\epsilon| < \tfrac{1}{2}\sqrt{2}\cdot 10^{-4},$$

hence the computed value of the function $f(x)$ will be zero for all x such that $0.99992929 \le x \le 1.0000707$.

Suppose that we want to determine a root α to a certain prescribed accuracy ϵ. One possible way of doing this is to interrupt the iterations on the basis of some error estimate such as (6.6.1) or (6.6.3). This can be especially advisable in hand computation, assuming that the necessary derivatives are easy to estimate. On a computer, however, it is often more advantageous to iterate a few extra times rather than to use a complicated formula for error estimation.

The limiting accuracy ϵ_α is usually not known a priori. If we happen to demand an accuracy $\epsilon < \epsilon_\alpha$, then the termination criterion will probably never be satisfied. Instead of specifying a tolerance ϵ, it is often more suitable to determine the root to its limiting accuracy ϵ_α, and to estimate ϵ_α. Usually, for iteration methods which converge faster than linearly only a few extra iterations are needed to attain the limiting accuracy even if $\epsilon \gg \epsilon_\alpha$. In such cases one can use a termination criterion based on the following simple observation: When the sequence $\{x_n\}$ converges to α, we know that from some n onward the differences $|x_n - x_{n-1}|$ decrease until $|x_n - \alpha| \approx \epsilon_\alpha$. Thereafter, rounding errors dominate and the differences vary irregularly. We can terminate the iterations and accept x_n as the root when the following two conditions are satisfied simultaneously:

$$|x_{n+1} - x_n| \ge |x_n - x_{n-1}| \quad \text{and} \quad |x_n - x_{n-1}| < \delta. \tag{6.6.6}$$

Here δ is some coarse tolerance, used only to prevent the iterations from being terminated before x_n has even come close to α. When (6.6.6) is satisfied, the quantity $|x_{n+1} - x_n|$ is usually a good estimate of $|x_n - \alpha|$. With the above termination criterion (remember that the iteration method is assumed to converge faster than linearly), the risk of not terminating the iterations for ill-conditioned roots (ϵ_α large), or the risk of giving up the attainable accuracy for a well-conditioned root, is quite small.

PROBLEMS

1. Use an iteration method to determine the roots of the following equations to five correct decimals:
   (a) $3x - \cos x = 0$; (b) $3x^2 + \tan x = 0$.

2. One wants to determine a root of the equation $x = \varphi(x)$ with an error less than $0.5 \cdot 10^{-4}$. One has computed
   $$x_4 = 0.43789, \qquad x_5 = 0.43814,$$
   and one knows that $|\varphi'(x)| \leq 0.4$. How many more iterations are needed to be sure to attain the desired accuracy?

3. Write an Algol procedure, **procedure** IT($x0$, F, $x1$, eps); **real** $x0$, F, $x1$, eps; or Fortran subroutine which solves the equation $x = F(x)$ with the iteration method $x_{n+1} = F(x_n)$. $x0$ is the initial approximation and $x1$ the accepted root. The iterations are to be terminated when
   $$\frac{m\,|x_{n+1} - x_n|}{(1 - m)} < eps,$$
   where m is estimated by
   $$\left|\frac{x_{n+1} - x_n}{x_n - x_{n-1}}\right|.$$

6.7. MULTIPLE ROOTS

In the discussion of Newton-Raphson's method and the secant method, we assumed that the desired root was a simple root. In general, a root α of the equation $f(x) = 0$ is said to have *multiplicity* q if

$$0 \neq |g(\alpha)| < \infty, \quad \text{where} \quad g(x) = (x - \alpha)^{-q} f(x). \tag{6.7.1}$$

If the root has multiplicity $q > 1$, then the previous results in order of convergence are no longer valid. For example, Newton-Raphson's method is only linearly convergent for multiple roots, and its asymptotic error constant is $C = (q - 1)/q$. Already for $q = 2$, the convergence is asymptotically the same as with the bisection method!

The modified iteration method

$$x_{n+1} = x_n + qh_n, \quad h_n = \frac{-f(x_n)}{f'(x_n)} \tag{6.7.2}$$

gets back the quadratic convergence; however, it assumes that the multiplicity q is known a priori. This is usually not the case, so it would be desirable to have iteration methods for which the order of convergence is independent of q. Such methods are in fact easy to derive. Suppose $f(x)$ is q times continuously differentiable in a neighborhood of a root α of multiplicity q. Then $f^{(j)}(\alpha) = 0, j < q$, and Taylor's formula gives

$$f(x) = \frac{1}{q!}(x - a)^q f^{(q)}(\xi), \qquad f'(x) = \frac{1}{(q-1)!}(x - \alpha)^{q-1} f^{(q)}(\xi'),$$

where $\xi, \xi' \in \text{int}(x, \alpha)$. Put $u(x) = f(x)/f'(x)$. Then we have

$$\lim_{x \to \alpha} \frac{u(x)}{x - \alpha} = \frac{1}{q} \cdot$$

Thus the equation $u(x) = 0$ *has a simple root at* $x = \alpha$ (see Eq. (6.7.1)), and all of our previous methods can be applied to this equation. In particular, Newton-Raphson's method gives

$$x_{n+1} = x_n - \frac{u(x_n)}{u'(x_n)}, \quad u'(x_n) = 1 - \frac{f''(x_n)}{f'(x_n)} u(x_n), \tag{6.7.3}$$

and the secant method

$$x_{n+1} = x_n - u(x_n) \frac{x_n - x_{n-1}}{u(x_n) - u(x_{n-1})} \cdot \tag{6.7.4}$$

Note, however, that these methods are less effective in the case of a simple root, since one more derivative of f must be computed.

6.8. ALGEBRAIC EQUATIONS

6.8.1. Introduction

In this section we shall make a more detailed study of the important problem of determining all the zeros of an algebraic equation given in the form

$$p(z) = a_0 z^n + a_1 z^{n-1} + \ldots + a_n = 0, \quad a_0 \neq 0. \tag{6.8.1}$$

According to the fundamental theorem of algebra, this equation has exactly n roots, $\alpha_1, \alpha_2, \ldots, \alpha_n$ and $p(z) = a_0(z - \alpha_1)(z - \alpha_2) \ldots (z - \alpha_n)$. If the coefficients $a_0, a_1, \ldots, a_n$ are real, then eventual complex roots occur in conjugate pairs.

Newton-Raphson's method and the secant method are also applicable to this problem. The necessary function values and derivatives can be conveniently computed by repeated synthetic division (see Sec. 1.3.2) using the formulas

$$p(z_k) = b_n, \qquad p'(z_k) = c_{n-1}, \qquad p''(z_k) = 2!\, d_{n-2}, \qquad \text{etc.,}$$

where

$$\begin{aligned} b_0 &= a_0, & b_i &= b_{i-1} z_k + a_i, & i &= 1, 2, \ldots, n, \\ c_0 &= b_0, & c_i &= c_{i-1} z_k + b_i, & i &= 1, 2, \ldots, n-1, \\ d_0 &= c_0, & d_i &= d_{i-1} z_k + c_i, & i &= 1, 2, \ldots, n-2, \text{ etc.} \end{aligned} \tag{6.8.2}$$

The number of arithmetic operations required to compute $p'(z_k)$ is about the same as for $p(z_k)$. Thus according to the rule given in Sec. 6.4, the secant method should be preferable to Newton-Raphson's method for this problem, assuming that sufficiently good intitial estimates of all the roots are available. However, this is often not the case, and then neither of these two methods is especially applicable. It is important, then, to choose some method which has good **global convergence** properties.

One method which is superior in this respect is **Laguerre's iteration method**,

$$z_{k+1} = z_k - \frac{np(z_k)}{p'(z_k) \pm \sqrt{H(z_k)}},$$

where

$$H(z) = (n-1)[(n-1)(p'(z))^2 - np(z)p''(z)], \tag{6.8.3}$$

where n is the degree of the polynomial. We shall not derive the method here; the interested reader can see, e.g., Householder [72]. The sign in the denominator should be chosen so that $|z_{k+1} - z_k|$ is as small as possible. Laguerre's method requires the computation of $p(z_k)$, $p'(z_k)$, and $p''(z_k)$ at each step, and can be shown to be *cubically convergent* for simple roots (real or complex). For algebraic equations with only real roots, Laguerre's method is convergent for *every* choice of real initial estimate. Suppose the roots are ordered such that $\alpha_1 \leq \alpha_2 \leq \ldots \leq \alpha_n$. If the initial estimate $z_0 \in (\alpha_{j-1}, \alpha_j)$, then Laguerre's method converges to one of the roots α_j, α_{j-1} $(j = 2, 3, \ldots, n)$. Furthermore, if $z_0 < \alpha_1$ or $z_0 > \alpha_n$, then convergence is to α_1 or α_n respectively.

All of the above methods (secant, Newton-Raphson, Laguerre) can be used for complex roots. In the important special case where all the coefficients $a_0, a_1, \ldots, a_n$ are *real*, the formulas of Eq. (6.8.2) are somewhat inefficient. One can save a considerable number of operations by synthetic division with the quadratic factor

$$(z - z_k)(z - \bar{z}_k) = z^2 - 2z \operatorname{Re}(z_k) + |z_k|^2,$$

which has real coefficients (see Problem 1 at the end of this section).

For algebraic equations with complex roots, it is no longer true that Laguerre's method converges for every choice of initial estimate. However, experience has shown that the global convergence properties are also good in this case.

Notice that in the case of real coefficients, $p(z_k)$, $p'(z_k)$, $\ldots$ are real for z_k real. This means that Newton-Raphson's method and the secant method cannot converge to a complex root from a real initial estimate. With Laguerre's method, however, z_{k+1} can be complex even if z_k is real, since we can have $H(z_k) < 0$. Thus this method can begin searching the complex plane "on its own."

Example 6.8.1

Let $p(z) = a_0z^n + \ldots + a_{n-2}z^2 + a_{n-1}z + a_n$, and suppose $a_n \neq 0$ (this is no limitation, since we can always factor out one or more roots $\alpha = 0$). Now suppose that at least one of a_{n-1} and a_{n-2} are $\neq 0$, and put $z_0 = 0$ in Laguerre's method. A simple calculation gives

$$z_1 = \frac{-na_n}{a_{n-1} \pm \sqrt{H(z_0)}}, \quad H(z_0) = (n-1)^2a_{n-1}^2 - 2n(n-1)a_na_{n-2}.$$

In particular, for $n = 2$, $H(z_0)$ is the discriminant of $p(z)$ and z_1 is the root with least modulus (if the sign in the denominator is chosen as mentioned previously). If the roots are complex, then we have two roots of least modulus and we can just choose one of them. For example, the equation

$$p(z) = z^3 - 2z^2 + z - 2$$

has roots $\pm i$ and 2. Using the above formula for z_1 ($n = 3$), we get

$$z_1 = \frac{3 \cdot 2}{1 \pm i \cdot 2\sqrt{11}} = \frac{2}{15} \pm i\frac{4\sqrt{11}}{15}.$$

Notice that in this case also we get convergence to one of the two roots of least modulus $\pm i$.

6.8.2. Deflation

The computation of values of $p(z)$ according to Eq. (6.8.2) is based on the relation

$$p(z) = (z - z_k)g_1(z) + b_n, \quad b_n = p(z_k),$$

where

$$g_1(z) = b_0z^{n-1} + \ldots + b_{n-2}z + b_{n-1}.$$

Now suppose we have found a root $z_k = \alpha$ to $p(z) = 0$. Then $b_n = 0$, and the remaining roots are also roots of $g_1(z) = 0$. To compute these roots, we can replace $p(z)$ by the quotient polynomial $g_1(z)$. This is called *deflation*. This process can be repeated; as soon as a root is found, we can factor it out. In this way, we can work with polynomials of lower and lower degree.

Clearly, deflation saves arithmetic operations. Another advantage is that the iterations cannot converge to the same simple root more than once. In those cases where information about the roots is poor, one can proceed in the following way: Choose more or less randomly an initial approximation. The probability is great that we get convergence to *some* root. This root can then be factored out and we can continue in the same way until all the roots are determined.

Thus far we have ignored the fact that z_k is not an exact root of $p(z) = 0$. We have also ignored rounding errors in the computation of $b_0, b_1, \ldots, b_{n-1}$. Clearly there is a risk that these errors can have the effect that the zeros of the successive quotient polynomials deviate more and more from the zeros of $p(z)$. A closer analysis (see Wilkinson [84]) shows that the errors resulting from deflation are negligible assuming that:

1. the roots are determined in order of *increasing* absolute value; and
2. every root is determined to its limiting accuracy.

The first requirement can, of course, be difficult to satisfy. With Laguerre's method, however, it is quite probable that we get convergence to the root with least absolute value if we use $z_0 = 0$ (see Example 6.8.1). To be safe, one can first determine all the roots (using deflation in the process) and then, for each root, make one last iteration using the *original* polynomial $p(z)$.

6.8.3. Ill-Conditioned Algebraic Equations

We have previously noted that multiple roots are in general ill-conditioned—i.e., sensitive to perturbations from, for example, round-off errors. Thus it is natural that if the equation $p(z) = 0$ has a few roots which are very close to each other (*nearly* multiple roots) then these roots are sensitive to perturbations in the coefficients of $p(z)$. That the roots can be very ill-conditioned even though they *seem* to be well-separated is perhaps more surprising. The following example shows that this can happen.

Example 6.8.2

Consider the polynomial

$$p(z) = (z - 1)(z - 2) \dots (z - 20) = z^{20} - 210z^{19} + \dots + 20!,$$

whose zeros are 1, 2, . . . , 20. Let $\bar{p}(z)$ be the polynomial which is obtained when the coefficient $a_1 = -210$ in $p(z)$ is replaced by

$$-(210 + 2^{-23}) = -210.000000119\dots,$$

while the rest of the coefficients remain unchanged. Even though the relative change in a_1 is of order 10^{-10}, many of the roots of $\bar{p}(z) = 0$ deviate greatly from the roots of $p(z) = 0$. For example, two of the roots of $\bar{p}(z) = 0$ are (correct to 9 decimals)

$$16.730737466 \pm i\, 2.812624894 \text{ (!)}.$$

One way of interpreting the above example is to say that our intuitive notion of what is "well-separated" may be insufficient. If the coefficients a are given with full machine accuracy (floating point), then the error in the computed value of $p(x)$ can be as large as $\epsilon = \sum |(2i + 1)a_{n-i}x^i| \cdot 1.06u$ (see Sec. 2.4). On the other hand, if the error in $p(x)$ is equal to ϵ, then the relative error in the root can be as large as $\epsilon/|\alpha p'(\alpha)|$. Hence the condition number of the determination of a root of an equation *which is given in the form*

$$\sum_{i=0}^{n} a_{n-i}x^i = 0$$

s (ignoring the factor 1.06)

$$C = \frac{\sum |(2i+1)a_{n-i}\alpha^i|}{|\alpha p'(\alpha)|} = \frac{\sum |(2i+1)a_{n-i}\alpha^i|}{|\sum i a_{n-i}\alpha^i|}.$$

Thus in the case of $p(z)$ given above,

$$C = 1.35 \cdot 10^{15} \quad \text{for} \quad \alpha = 14.$$

If the roots are $\alpha_i = 2^{i-21}$, $i = 1, 2, \ldots, 20$, then the condition number or each of the roots is less than $3.83 \cdot 10^3$. Hence, the latter equation causes ess trouble than the former, despite the fact that the roots seem to be better eparated in the former case. An "explanation" of this paradox is that the elevant interpretation of "separation" should be the *relative separation* of he roots. For a group of k real roots, $\alpha_1 < \alpha_2 < \ldots < \alpha_k$, the product

$$s = \prod_{i=1}^{k-1} \frac{\alpha_{i+1} - \alpha_i}{|\alpha_i|}, \quad (\alpha_i \neq 0)$$

ould be taken as a measure of separation; if s is small, then the group of oots is not well-separated. If we apply this to $\alpha_i = i$, $i = 1, 2, \ldots, 20$, then $= 8.2 \cdot 10^{-18}$. However, for $\alpha_i = 2^{i-21}$, $i = 1, \ldots, 20$, we have $s = 1$ (veriy!), indicating that in this case the roots are in fact better-separated.

Example 6.8.2 was taken from Wilkinson [84]. This book contains an xtensive discussion of numerical problems in determining roots of algebraic quations. Often in these problems one must use double precision in order to get desired accuracy in the roots. *If the coefficients of $p(z)$ are not given as riginal data, then it is often better to avoid computing them.* An important exmple of this is the determination of eigenvalues of matrices; the eigenvalues re of course the zeros of a certain characteristic polynomial, see also Example 2.4.5.

Example 6.8.3

The largest positive root of the equation

$$(x+2)(x^2-1)^6 = 3 \cdot 10^{-6} \cdot x^{11}$$

s to be computed. On a machine which uses floating arithmetic and seven lecimals, the equation, in standard form,

$$x^{13} + 2x^{12} - (12 - 3 \cdot 10^{-6})x^{11} + \ldots + 2 = 0,$$

s stored in the machine as

$$x^{11} + 2x^{12} - 12.00000x^{11} + \ldots + 2 = 0.$$

Thus the machine will treat the equation $(x+2)(x^2-1)^6 = 0$, whose exact positive root is 1. (With more complicated coefficients one would probably also have large errors even in solving the equation.)

This is a poor result. One can get the root to full machine accuracy by writing the equation in the form

$$x = 1 + \frac{1}{x+1}\frac{1}{10}\left(\frac{3x^{11}}{x+2}\right)^{1/6},$$

and solving this by iteration. If $x_0 = 1$, then $x_1 = 1.05$. The sequence $\{x_k\}$ is monotone increasing; hence $\alpha > 1.05$. Thus the relative error in the result obtained using the standard form is greater than 5 percent.

PROBLEMS

1. Given a polynomial

$$p(z) = a_0 z^n + a_1 z^{n-1} + \ldots + a_n, \quad a_0 \neq 0,$$

with real coefficients a_i, $i = 0, 1, \ldots, n$:

(a) Count the number of (real) additions and multiplications needed to compute a value $p(z_0)$ by synthetic division of $p(z)$ by $(z - z_0)$ when z_0 is (1) a real number, (2) a complex number.

(b) For complex numbers $z_0 = x_0 + iy_0$, $p(z_0)$ can also be computed by performing the synthetic division

$$p(z) = q(z)d(z) + r_0 z + r_1$$

with the divisor

$$d(z) = (z - z_0)(z - \bar{z}_0) = z^2 - 2x_0 z + x_0^2 + y_0^2$$

and the quotient

$$q(z) = b_0 z^{n-2} + b_1 z^{n-3} + \ldots + b_{n-2}.$$

Construct a scheme for the calculation and count the number of additions and multiplications needed.

Also, show how to compute $p'(z_0)$.

2. Determine all the roots of the equations

(a) $2z^3 + 21z^2 - 26z - 240 = 0$;

(b) $2z^3 + 21z^2 - 10z - 210 = 0$.

Compute using six decimals on a desk calculator.

6.9. SYSTEMS OF NONLINEAR EQUATIONS

Several of the methods which we have presented for a single nonlinear equation can be generalized to systems of nonlinear equations. Consider a general system of n nonlinear equations in n unknowns

$$f_i(x_1, x_2, \ldots, x_n) = 0, \quad i = 1, 2, \ldots, n. \tag{6.9.1}$$

6.9.1. Iteration

A one-point iteration method for solving (6.9.1) can be constructed by rewriting the system in the form

$$x_i = \varphi_i(x_1, x_2, \ldots, x_n), \quad i = 1, 2, \ldots, n,$$

which suggests the iterative method

$$x_i^{(k+1)} = \varphi_i(x_1^{(k)}, x_2^{(k)}, \ldots, x_n^{(k)}), \quad i = 1, 2, \ldots, n.$$

The formal similarity to the case $n = 1$ is more easily seen if we introduce the vector notations

$$\mathbf{x} = (x_1, x_2, \ldots, x_n)^T, \qquad \boldsymbol{\varphi}(\mathbf{x}) = (\varphi_1(\mathbf{x}), \varphi_2(\mathbf{x}), \ldots, \varphi_n(\mathbf{x}))^T.$$

The iteration method can then be written

$$\mathbf{x}^{(k+1)} = \boldsymbol{\varphi}(\mathbf{x}^{(k)}), \quad k = 0, 1, 2, \ldots. \tag{6.9.2}$$

A convergence criterion similar to that given in Theorem 6.5.1 can also be shown for Eq. (6.9.2). Suppose that $\boldsymbol{\alpha} = \boldsymbol{\varphi}(\boldsymbol{\alpha})$ and that the partial derivatives

$$d_{ij}(\mathbf{x}) = \frac{\partial \varphi_i}{\partial x_j}(\mathbf{x}), \quad 1 \leq i, j \leq n$$

exist for $\mathbf{x} \in R$ where $R = \{\mathbf{x}: \|\mathbf{x} - \boldsymbol{\alpha}\| < \rho\}$. Let $\mathbf{D}(\mathbf{x})$ be an $n \times n$ matrix with elements $d_{ij}(\mathbf{x})$. *Then a sufficient condition for the iteration method of* Eq. (6.9.2) *to converge* for every $\mathbf{x}_0 \in R$ is that for some choice of norm we have

$$\|\mathbf{D}(\mathbf{x})\| \leq m < 1, \quad \mathbf{x} \in R. \tag{6.9.3}$$

One says that $\boldsymbol{\varphi}$ is a **contraction mapping,** when (6.9.3) is satisfied, because the condition implies the inequality $\|\boldsymbol{\varphi}(\mathbf{x}) - \boldsymbol{\varphi}(\mathbf{y})\| \leq m\|\mathbf{x} - \mathbf{y}\|$, for all $\mathbf{x}, \mathbf{y} \in R$.

A necessary condition for Eq. (6.9.2) *to converge is that the spectral radius* (see Chap. 5) *of* $\mathbf{D}(\boldsymbol{\alpha})$ *be less than or equal to* 1. The rate of convergence depends linearly on m, and we have

$$\|\mathbf{x}^{(k+1)} - \boldsymbol{\alpha}\| = \|\boldsymbol{\varphi}(\mathbf{x}^{(k)}) - \boldsymbol{\varphi}(\boldsymbol{\alpha})\| \leq m\|\mathbf{x}^{(k)} - \boldsymbol{\alpha}\|, \quad k = 0, 1, 2, \ldots.$$

In many important applications one wants to solve a system of the form

$$\mathbf{x} = \mathbf{a} + h\boldsymbol{\varphi}(\mathbf{x}),$$

where $\mathbf{a}$ is a constant vector and h a parameter, $0 < h \ll 1$. If $\boldsymbol{\varphi}(\mathbf{x})$ has bounded partial derivatives, then the convergence criterion, (6.9.3), is always satisfied, for sufficiently small h, in the iteration method

$$\mathbf{x}^{(k+1)} = \mathbf{a} + h\boldsymbol{\varphi}(\mathbf{x}^{(k)}), \quad k = 0, 1, 2, \ldots. \tag{6.9.4}$$

6.9.2. Newton-Raphson's Method and Some Modifications

Newton-Raphson's method (Newton's method) can also be generalized to n dimensions. For $n = 1$, we derived this method from Taylor's formula,

$$f(x) = f(x_k) + (x - x_k)f'(x_k) + O(|x - x_k|^2).$$

By neglecting the quadratic term, we obtained (for $f(x) = 0$) the iteration method

$$f'(x_k)(x_{k+1} - x_k) + f(x_k) = 0, \quad k = 0, 1, 2, \ldots .$$

Analogously, Taylor's formula in n dimensions gives

$$\boldsymbol{f}(\boldsymbol{x}) = \boldsymbol{f}(\boldsymbol{x}^{(k)}) + \boldsymbol{f}'(\boldsymbol{x}^{(k)})(\boldsymbol{x} - \boldsymbol{x}^{(k)}) + O(\|\boldsymbol{x} - \boldsymbol{x}^{(k)}\|^2),$$

where $\boldsymbol{f}'(\boldsymbol{x})$ is an $n \times n$ matrix, the **Jacobian** (sometimes denoted by $\boldsymbol{J}$), with elements

$$f'_{ij}(\boldsymbol{x}) = \frac{\partial f_i}{\partial x_j}(\boldsymbol{x}), \quad 1 \leq i, j \leq n.$$

This leads to Newton-Raphson's method in n dimensions,

$$\boldsymbol{f}'(\boldsymbol{x}^{(k)})(\boldsymbol{x}^{(k+1)} - \boldsymbol{x}^{(k)}) + \boldsymbol{f}(\boldsymbol{x}^{(k)}) = \boldsymbol{0}. \tag{6.9.5}$$

This is a linear system of equations for $\boldsymbol{x}^{(k+1)}$, and if $\boldsymbol{f}'(\boldsymbol{x}^{(k)})$ is nonsingular, it can be solved by the methods described in Chap. 5. In case n is very large and $\boldsymbol{f}'(\boldsymbol{x})$ sparse, an iterative method may have to be used. Newton's method in n dimensions is of second order—i.e., there is a constant $C = C(\rho)$ such that for all $\boldsymbol{x}^{(k)}$, $\|\boldsymbol{x}^{(k)} - \boldsymbol{\alpha}\| \leq \rho$, we have

$$\|\boldsymbol{x}^{(k+1)} - \boldsymbol{\alpha}\| \leq C\|\boldsymbol{x}^{(k)} - \boldsymbol{\alpha}\|^2.$$

Hence it converges provided that $\boldsymbol{x}^{(0)}$ lies sufficiently close to $\boldsymbol{\alpha}$ (cf. p. 224).

Each step in Newton's method requires the solution of the linear system of Eq. (6.9.5), which, in problems where n is very large, may be a difficult task. Moreover, at each step the n^2 entries of $\boldsymbol{f}'(\boldsymbol{x}^{(k)})$ have to be computed. This might be impractical, unless these entries have a simple functional form. As has been remarked before, it might be preferable to reevaluate $\boldsymbol{f}'(\boldsymbol{x})$ only occasionally using

$$\boldsymbol{f}'(\boldsymbol{x}^{(p)})(\boldsymbol{x}^{(k+1)} - \boldsymbol{x}^{(k)}) + \boldsymbol{f}(\boldsymbol{x}^{(k)}) = \boldsymbol{0}, \quad k = p, \ldots, p + m.$$

Methods not using derivatives can be derived by estimating the partial derivatives in the n-dimensional Newton's method in the same way as the secant method was derived by estimating $f'(x)$ by a difference quotient. A frequently used difference approximation is

$$\frac{\partial f_i(\boldsymbol{x})}{\partial x_j} \approx \Delta_{ij}(\boldsymbol{x}, \boldsymbol{h}) = \frac{f(\boldsymbol{x} + h_j \boldsymbol{e}_j) - f(\boldsymbol{x})}{h_j},$$

where $\boldsymbol{e}_j$ is the jth coordinate vector and $\boldsymbol{h}$ an n-dimensional parameter vector with components $h_j \neq 0$, $j = 1, 2, \ldots, n$. Then if we let $\boldsymbol{J}(\boldsymbol{x}, \boldsymbol{h})$ be the $n \times n$ matrix with elements $\Delta_{ij}(\boldsymbol{x}, \boldsymbol{h})$, we get an n-dimensional discretized Newton method:

$$\boldsymbol{J}(\boldsymbol{x}^{(k)}, \boldsymbol{h})(\boldsymbol{x}^{(k+1)} - \boldsymbol{x}^{(k)}) + \boldsymbol{f}(\boldsymbol{x}^{(k)}) = \boldsymbol{0}. \tag{6.9.6}$$

We can obviously write

$$\boldsymbol{J}(\boldsymbol{x}, \boldsymbol{h}) = (\boldsymbol{f}(\boldsymbol{x} + h_1 \boldsymbol{e}_1) - \boldsymbol{f}(\boldsymbol{x}), \ldots, \boldsymbol{f}(\boldsymbol{x} + h_n \boldsymbol{e}_n) - \boldsymbol{f}(\boldsymbol{x}))\boldsymbol{H}^{-1},$$

where

$$H = \text{diag}\,(h_1, h_2, \ldots, h_n).$$

This shows that $J(x^{(k)}, h)$ is nonsingular and thus $x^{(k+1)}$ defined if and only if the vectors $f(x^{(k)} + h_j e_j) - f(x^{(k)}), j = 1, 2, \ldots, n$, are linearly independe

This method requires $f(x)$ to be evaluated at the $(n + 1)$ points $x^{(k)}$, $x^{(k)} + h_1 e_1, \ldots, x^{(k)} + h_n e_n$. Thus the amount of computation is comparable to that of Newton's method if the evaluation of $f_i(x)$ takes as much work as that of $\partial f_i / \partial x_j$.

In order to get the same rapid rate of convergence as the ordinary secant method for $n = 1$, it is necessary to choose h in some special way. The direct generalization is obviously to take

$$h_j = x_j^{(k-1)} - x_j^{(k)}, \quad j = 1, 2, \ldots, n.$$

This choice assumes that $x_j^{(k-1)} - x_j^{(k)} \neq 0,\ i = 1, \ldots, n,\ k = 1, 2, \ldots$. No useful results in this direction are known. However, if this condition is satisfied, then under suitable differentiability conditions on $f(x)$ the method converges when $\| x^{(0)} - \alpha \|$ is sufficiently small. The order of convergence can be shown to be $1.618\ldots$, as it is for $n = 1$. If, in Eq. (6.9.6), we take

$$h_j = f_j(x^{(k)}),$$

then we get a generalization of **Steffensen's method**. Here we have to evaluate $f(x)$ at the $(n + 1)$ points $x^{(k)}$, $x^{(k)} + f_j(x^{(k)}) e_j, j = 1, 2, \ldots, n$. Thus when $n > 1$, the number of function evaluations for Steffensen's method is the same as for the discretized Newton method, but the order of convergence equals 2.

The secant method can be modified in different ways so that the number of function evaluations per step is decreased. For example, one can use values of $f(x)$ at the points $x^{(k-n)}, \ldots, x^{(k-1)}, x^{(k)}$ to estimate the derivatives. Then $f(x)$ has only to be computed at the one point $x^{(k)}$ for the computation of $x^{(k+1)}$. This method is, however, more prone to unstable bevavior than other methods described here.

6.9.3. Other Methods

Another approach for solving $f(x) = 0$ which is sometimes appropriate is to use the ith equation to solve for $x_i^{(k+1)}$—i.e., for $i = 1, 2, \ldots, n$ the one-dimensional equation

$$f_i(x_1^{(k+1)}, \ldots, x_{i-1}^{(k+1)}, x_i, x_{i+1}^{(k)}, \ldots, x_n^{(k)}) = 0$$

is solved. This may be called a **nonlinear Gauss-Seidel method**, and similarly nonlinear Jacobi and SOR methods can be defined.

Recently a class of methods known as **matrix-updating procedures** has been suggested. The following method is due to Broyden. In each iteration a new approximation J_i to the Jacobian is obtained by adding a rank-one matrix to the previous approximation J_{i-1} in order to satisfy the following relations

$$J_i(x_i - x_{i-1}) = f(x_i) - f(x_{i-1}),$$
$$J_i p = J_{i-1} p, \quad \text{when} \quad (x_i - x_{i-1})^T p = 0. \tag{6.9.7}$$

Put $J_\nu^{-1} = H_\nu$, $x_i - x_{i-1} = q$, $f(x_i) - f(x_{i-1}) = y$. It is left as an exercise (see Problem 3 at the end of this section) to verify that these requirements lead to the following formulas:

$$J_i = J_{i-1} - \frac{(J_{i-1}q - y)q^T}{q^T q},$$
$$H_i = H_{i-1} - \frac{(H_{i-1}y - q)q^T H_{i-1}}{q^T H_{i-1} y}. \tag{6.9.8}$$

The latter is obtained by an application of the Sherman-Morrison formula (see Problem 6 of Sec. 5.3). Then x_{i+1} is obtained by the formula

$$x_{i+1} = x_i - \lambda_i H_i f(x_i). \tag{6.9.9}$$

The parameter λ_i is normally equal to 1, but if this choice makes $\|f(x_{i+1})\| \geq \|f(x_i)\|_2$, λ_i will be halved repeatedly, until $\|f(x_{i+1})\|_2 < \|f(x_i)\|_2$.

The equations in (6.9.8) look complicated, but in fact the number of multiplications needed for calculating H_i is only about $3n^2$. Similar ideas are used in nonlinear optimization; see Sec. 10.5. Note the close relation between the nonlinear system $f(x) = 0$ and the problem of minimizing $\|f(x)\|$. We do not, however, recommend the direct application of the minimization methods of Sec. 10.5 to the solution of nonlinear systems.

When it is hard to find an initial approximation, one can use the important idea of **imbedding** the problem in a family of problems,

$$f(x, k) = 0, \tag{6.9.10}$$

where $f(x, 1) \equiv f(x)$, and where the system $f(x, 0) = 0$ can be solved without difficulty. The solution $x = x(k)$ of Eq. (6.9.10) is determined for an increasing sequence of values of k, $k_0 = 0, k_1, k_2, \ldots, k_n = 1$. The first approximation $x_0(k_i)$ is obtained from the previous results—e.g.,

$$x_0(k_i) = x(k_{i-1}), \tag{6.9.11}$$

or, if $i \geq 2$, by linear extrapolation:

$$x_0(k_i) = x(k_{i-1}) + (x(k_{i-1}) - x(k_{i-2}))\cdot\frac{k_i - k_{i-1}}{k_{i-1} - k_{i-2}}. \tag{6.9.12}$$

The step size $k_i - k_{i-1}$ is to be adjusted automatically—it is the total number of iterations rather than the number of steps that counts.† This technique can be used in connection with any of the methods previously mentioned. Also

†There can be trouble with singular points for the function $x(k)$, in which case the imbedding family has to be changed, or some other special measure must be taken.

note the possibility of using the same Jacobian in several successive steps. Imbedding has important applications to the nonlinear systems encountered when finite-difference methods or finite-element methods are applied to nonlinear boundary-value problems; see Sec. 8.4 and 8.6. It can also be used in nonlinear optimization; see Sec. 10.5.4.

In principle, an imbedding can be achieved by putting

$$f(x, k) \equiv k \cdot f(x) + (1 - k) \cdot g(x),$$

where $g(x)$ is any function the zeros of which are known. Often a better choice can be made where the systems to be solved for $k_i \neq 1$ also contribute to the insight into the questions which originally led to the system. In elasticity, this technique is known as the method of **incremental loading**, because $k = 0$ may correspond to an unloaded construction for which the solution is known, while $k = 1$ corresponds to the actual loading. The technique is also called the **continuation** method.

Sometimes it is recommended to differentiate Eq. (6.9.10)—

$$\frac{\partial f}{\partial x} \cdot \frac{dx}{dk} + \frac{\partial f}{\partial k} = 0,$$

$$\frac{dx}{dk} = -\left(\frac{\partial f}{\partial x}\right)^{-1} \frac{\partial f}{\partial k},$$

—and to use one of the numerical methods of Chap. 8 to solve the differential equation. We are, however, not convinced that this would be more efficient than using Eq. (6.9.12).

A thorough theoretical treatment of nonlinear systems is given in the book by Ortega and Rheinboldt [74].

PROBLEMS

1. The system of equations

$$\begin{cases} x = 1 + h^2(e^{y\sqrt{x}} + 3x^2) \\ y = 0.5 + h^2 \tan(e^x + y^2) \end{cases}$$

can, for small values of h, be solved by iteration. Put $x_0 = 1$, $y_0 = 0.5$. These values are put into the right-hand side, whereupon a new approximation (x_1, y_1) is obtained. The process is repeated until the changes in x and y are sufficiently small. Write an Algol or Fortran program which uses this method to solve the system for $h = 0(0.01)0.10$. The iterations should be broken off when the changes in x and y are $< 0.1h^4$ and the root (x, y) printed out.

2. (a) Show that if

$$y = x - \lambda[f'(x)]^{-1}f(x), \quad \lambda > 0,$$

$f(x) \neq 0$, then $\|f(y)\|_2 < \|f(x)\|_2$ if λ is sufficiently small.

(b) If a parameter λ_i is introduced in Newton-Raphson's method, which is determined in the same way as the parameter of Eq. (6.9.9), what conclusion can be drawn about this method in the light of the result in (a) above?

3. Show that the requirements in Eq. (6.9.7) lead to Broyden's formulas in Eq. (6.9.8).

7 FINITE DIFFERENCES WITH APPLICATIONS TO NUMERICAL INTEGRATION, DIFFERENTIATION, AND INTERPOLATION

7.1. DIFFERENCE OPERATORS AND THEIR SIMPLEST PROPERTIES

Let y denote a sequence $\{y_n\}$. Then we define the **shifting operator** E and the forward **difference operator** Δ by the relations

$$Ey = \{y_{n+1}\}, \qquad \Delta y = \{y_{n+1} - y_n\}$$

(see Sec. 1.2). E and Δ are thus operators which map one sequence to another sequence. One can define powers of the operators E and Δ recursively by

$$E^k y = \{y_{n+k}\}, \qquad \Delta^k y = \Delta(\Delta^{k-1} y), \tag{7.1.1}$$

where $\Delta^k y$ is called the ***k*th difference** of the sequence y. From the definition, it follows, for example, that

$$\begin{aligned} \Delta^2 y &= \Delta\{(y_{n+1} - y_n)\} = \{(y_{n+2} - y_{n+1}) - (y_{n+1} - y_n)\} \\ &= \{y_{n+2} - 2y_{n+1} + y_n\}. \end{aligned} \tag{7.1.2}$$

We abbreviate the notation and write, for example,

$$\Delta y_n = y_{n+1} - y_n, \qquad \Delta y_{n-1} = y_n - y_{n-1}, \qquad \Delta^2 y_n = y_{n+2} - 2y_{n+1} + y_n,$$
$$\Delta y_3 = y_4 - y_3, \qquad \Delta^2 y_0 = y_2 - 2y_1 + y_0,$$

but it is important to remember that Δ *operates on sequences* and not on elements of sequences. Thus, strictly speaking, the above notation is incorrect.

THEOREM 7.1.1

$$\Delta^k y_n = y_{n+k} - \binom{k}{1} y_{n+k-1} + \binom{k}{2} y_{n+k-2} + \ldots + (-1)^k y_n.$$

Proof. (By induction.) From the definition, the formula holds for $k = 1$. Suppose that it holds for $k = p$. For $k = p + 1$ we have then

$$\Delta^{p+1}y_n = \Delta^p y_{n+1} - \Delta^p y_n = y_{n+1+p} - \binom{p}{1} y_{n+p} + \binom{p}{2} y_{n+p-1}$$
$$+ \ldots + (-1)^p y_{n+1} - \left[y_{n+p} - \binom{p}{1} y_{n+p-1} \right.$$
$$\left. + \binom{p}{2} y_{n+p-2} + \ldots + (-1)^p y_n \right].$$

The coefficients of $y_{n+p+1-\nu}$ $(1 \leq \nu \leq p)$ are here equal to

$$(-1)^\nu \left[\binom{p}{\nu} + \binom{p}{\nu - 1} \right] = (-1)^\nu \binom{p+1}{\nu}$$

according to a well-known theorem about the binomial coefficients (the rule for building Pascal's triangle). Since in addition $\binom{p+1}{0} = 1$ and $\binom{p+1}{p+1} = 1$, then it holds generally that the coefficient of $y_{n+p+1-\nu}$ in the expansion of $\Delta^{p+1}y_n$ is $(-1)^\nu \binom{p+1}{\nu}$.

The following theorem can be proved in an analogous way:

THEOREM 7.1.2.

$$y_{n+k} = y_n + \binom{k}{1} \Delta y_n + \binom{k}{2} \Delta^2 y_n + \ldots + \binom{k}{k} \Delta^k y_n.$$

Theorems 7.1.1 and 7.1.2 can be rewritten as "symbolical" binomial theorems:

$$\Delta^k = (E - 1)^k, \qquad E^k = (1 + \Delta)^k. \tag{7.1.3}$$

A systematic treatment of such operator formulas will be given in Sec. 7.6.

A **difference scheme** consists of a sequence and its difference sequences, arranged in the following way:

$$\begin{array}{ccccc}
y_0 & & & & \\
 & \Delta y_0 & & & \\
y_1 & & \Delta^2 y_0 & & \\
 & \Delta y_1 & & \Delta^3 y_0 & \\
y_2 & & \Delta^2 y_1 & & \Delta^4 y_0 \\
 & \Delta y_2 & & \Delta^3 y_1 & \\
y_3 & & \Delta^2 y_2 & & \\
 & \Delta y_3 & & & \\
y_4 & & & &
\end{array}$$

A difference scheme is best computed by successive subtractions (the formula of Theorem 7.1.1 is used mostly in theoretical contexts).

Example 7.1.1

Compute the difference scheme for the sequence

$$y = (0, 0, 0, 0, 1, 0, 0, 0, 0)$$

y								
0								
	0							
0		0						
	0		0					
0		0		1				
	0		1		−5			
0		1		−4		15		
	1		−3		10		−35	
1		−2		6		−20		70
	−1		3		−10		35	
0		1		−4		15		
	0		−1		5			
0		0		1				
	0		0					
0		0						
	0							
0								

This example shows the *effect of a disturbance in one element* of the sequence on the higher differences. Because the effect of a disturbance broadens out and grows quickly, the difference scheme is useful in the investigation (and correction) of computational errors, so-called *difference checks* (see Problem 2 at the end of this section). The values in the above scheme look like binomial coefficients. (Explain this!) Notice also that a kind of *superposition principle* holds here; since the first and higher differences are *linear* functions of the elements of a sequence, the effect of errors can be seen by studying simple sequences such as the one above.

Example 7.1.2

The difference scheme for a section of a table of the function tan x (the step $h = 0.01$) is as follows:

x	y	Δy	$\Delta^2 y$	$\Delta^3 y$	$\Delta^4 y$	$\Delta^5 y$	$\Delta^6 y$
1.30	3.602						
		145					
1.31	3.747		11				
		156		2			
1.32	3.903		13		0		
		169		2		−2	
1.33	4.072		15		−2		8
		184		0		6	
1.34	4.256		15		4		
		199		4			
1.35	4.455		19				
		218					
1.36	4.673						

We see that the differences decrease rapidly as the order of differences in creases until the round-off error takes over. Such behavior is typical of th difference schemes of "well-behaved" functions, when h is sufficiently smal (In fact, to some degree this can be taken as the definition of h being "suffi ciently" small.) For $k \geq 4$, $\Delta^k y$ in the above scheme is completely determine by the round-off errors in the table values of y.

Even though difference schemes do not have the same importance toda that they had in the days of hand calculation or calculation with desk calcu lators, they are still important conceptually. They are also useful in makin difference checks and in preliminary investigations of methods to be used t approximate a given function, as well as in making quick estimates of variou types—for example, in interpolation and numerical differentiation.

Uses of the type illustrated in Example 7.1.2 make it natural to also con sider *difference operations on functions* (not just on sequences). E and Δ ma the function f onto functions whose values at the point x are

$$Ef(x) = f(x + h), \qquad \Delta f(x) = f(x + h) - f(x), \tag{7.1.4}$$

where h is the step in a tabulation of f. With obvious modifications, the abov remarks also hold for operators on functions—for example:

$$\begin{aligned} \Delta^2 f(x) &= f(x + 2h) - 2f(x + h) + f(x), \\ \Delta^2 f(x - h) &= f(x + h) - 2f(x) + f(x - h). \end{aligned} \tag{7.1.5}$$

Of course, Δf depends on the step h; in some cases this should be indicated i the notation. One can, for example, write $\Delta_h f(x)$, or $\Delta f(x; h)$.

Difference operators are in many respects similar to differentiation oper ators; in fact, the main significance of difference operators is connected wit the fact that differentiation is a kind of limiting case of forming difference

THEOREM 7.1.3

If f is a polynomial of degree m, then $\Delta^k f$, for $1 \le k \le m$, is a polynomial of degree $m - k$, and $\Delta^{m+1} f = 0$.

Proof. We prove the theorem for $k = 1$. According to Taylor's theorem,

$$\Delta f(x) = f(x + h) - f(x) = hf'(x) + \frac{1}{2}h^2 f''(x) + \ldots + \frac{1}{m!}h^m f^{(m)}(x),$$

where the last expression is clearly a polynomial of degree $m - 1$.

The result for arbitrary k follows using induction.

One can develop formulas for differences which are analogous to but usually more complicated than formulas for derivatives:

Example 7.1.3

$$\Delta^k(a^x) = (a^h - 1)^k \cdot a^x.$$

Proof. Let c be a given constant. For $k = 1$ we have

$$\Delta(ca^x) = ca^{x+h} - ca^x = ca^x a^h - ca^x = c(a^h - 1)a^x.$$

The general result follows easily now using induction.

COROLLARY

For sequences ($h = 1$), it holds that:

$$\Delta^k\{a^n\} = (a - 1)^k \cdot \{a^n\}, \qquad \Delta^k\{2^n\} = \{2^n\}.$$

Example 7.1.4. *The Difference of a Product*

$$\Delta(u_n v_n) = u_{n+1}v_{n+1} - u_n v_n$$
$$= u_n(v_{n+1} - v_n) + (u_{n+1} - u_n)v_{n+1},$$
$$\Delta(u_n v_n) = u_n \Delta v_n + \Delta u_n \cdot v_{n+1}. \tag{7.1.6}$$

Compare the above result with the formula for differentials, $d(uv) = u\,dv + v\,du$—but note that we have v_{n+1} (not v_n) on the right-hand side.

Example 7.1.5. *Summation by Parts*

$$\sum_{n=0}^{N-1} u_n \Delta v_n = u_N v_N - u_0 v_0 - \sum_{n=0}^{N-1} \Delta u_n \cdot v_{n+1}. \tag{7.1.7}$$

Proof. (Compare the rule for integration by parts and its proof!) Notice that

$$\sum_{n=0}^{N-1} \Delta w_n = (w_1 - w_0) + (w_2 - w_1) + (w_3 - w_2) + \ldots + (w_N - w_{N-1})$$
$$= w_N - w_0.$$

Use this on $w_n = u_n v_n$. From Eq. (7.1.6) one gets, then, after summation

$$u_N v_N - u_0 v_0 = \sum_{n=0}^{N-1} u_n \Delta v_n + \sum_{n=0}^{N-1} \Delta u_n \cdot v_{n+1},$$

and the result, Eq. (7.1.7), follows.

From the kth differences of a function one can get approximations to its kth derivative:

THEOREM 7.1.4

$\Delta^k f(x) = h^k f^{(k)}(\xi)$, where $\xi \in [x, x + kh]$ assuming that all the derivatives of f up to kth order are continuous.

Proof. For $k = 1$, the theorem is equivalent to the mean-value theorem

$$f(x + h) - f(x) = hf'(\xi).$$

For $k = 2$ one has, using Taylor's formula (with $a = x + h$),

$$f(a + h) - f(a) = hf'(a) + \tfrac{1}{2}h^2 f''(\xi_1),$$
$$f(a - h) - f(a) = -hf'(a) + \tfrac{1}{2}h^2 f''(\xi_2),$$

where $x = a - h \leq \xi_i \leq a + h = x + 2h$, $i = 1, 2$. Adding the above formulas, we get

$$f(a + h) - 2f(a) + f(a - h) = h^2 \cdot \tfrac{1}{2}(f''(\xi_1) + f''(\xi_2)).$$

Use Eq. (7.1.5) on the left-hand side. Since f'' was assumed to be continuous, the mean of $f''(\xi_1)$ and $f''(\xi_2)$ is equal to the value of f'' at some point ξ between ξ_1 and ξ_2, which thus lies in the interval $[a - h, a + h] = [x, x + 2h]$. Hence

$$\Delta^2 f(x) = \Delta^2 f(a - h) = h^2 f''(\xi).$$

We refrain from developing this method of proof for $k > 2$, since we shall later get Theorem 7.1.4 as a consequence of a more general result; see the note to Theorem 7.3.5.

Thus we have that $h^{-k}\Delta^k f(x)$ is an approximation to $f^{(k)}(x)$; the error of this approximation approaches zero as $h \to 0$ (i.e., as $\xi \to x$). As a rule, the error is approximately proportional to h. In the computation of $\Delta^k f(x)$ one uses the values of f at the points $x, x + h, x + 2h, \ldots, x + kh$; they lie symmetrically about the point $a = x + \tfrac{1}{2}kh$. The next example and the theorem which follows it show that $h^{-k}\Delta^k f(x)$ is a much better approximation to $f^{(k)}(a)$ than to $f^{(k)}(x)$—with an error which is $O(h^2)$.

Example 7.1.6

For $f(x) = e^x$, $h = 0.1$,

$$h^{-1}\Delta f(0) = \frac{e^{0.1} - e^0}{0.1} = 1.05171.$$

Compare $f'(0) = 1$, $f'(0.05) = 1.05127$.

We introduce the following operators and notations:

$$\delta f(x) = f(x + \tfrac{1}{2}h) - f(x - \tfrac{1}{2}h) \qquad \text{Central difference operator}$$

$$\mu f(x) = \tfrac{1}{2}[f(x + \tfrac{1}{2}h) + f(x - \tfrac{1}{2}h)] \qquad \text{Averaging operator.}$$

Thus, in the above example, we could write

$$h^{-1}\delta f(0.05) = 1.05171 \approx f'(0.05).$$

The relation between the forward difference notation and the central difference notation is shown in the following table:

$$\begin{array}{lllllllll}
f_{-2} & & & & & f_{-2} & & & \\
 & \Delta f_{-2} & & & & & \delta f_{-1.5} & & \\
f_{-1} & & \Delta^2 f_{-2} & & & f_{-1} & & \delta^2 f_{-1} & \\
 & \Delta f_{-1} & & \Delta^3 f_{-2} & & & \delta f_{-0.5} & & \delta^3 f_{-0.5} \\
f_0 & & \Delta^2 f_{-1} & & \Delta^4 f_{-2} & f_0 & & \delta^2 f_0 & & \delta^4 f_0 \\
 & \Delta f_0 & & \Delta^3 f_{-1} & & & \delta f_{+0.5} & & \delta^3 f_{+0.5} \\
f_1 & & \Delta^2 f_0 & & & f_1 & & \delta^2 f_1 & \\
 & \Delta f_1 & & & & & \delta f_{+1.5} & & \\
f_2 & & & & & f_2 & & &
\end{array}$$

In the central difference notation subscripts are constant along a horizontal row, while in the forward difference notation subscripts are constant along a diagonal.

Example 7.1.7

The following formulas occur frequently:

$$\delta^2 f(a) = \Delta^2 f(a - h) = f(a + h) - 2f(a) + f(a - h),$$

$$\mu\delta f(a) = \tfrac{1}{2}[f(a + h) - f(a)] + \tfrac{1}{2}[f(a) - f(a - h)]$$

$$= \tfrac{1}{2}[f(a + h) - f(a - h)].$$

THEOREM 7.1.5

$$f^{(k)}(a) = h^{-k}\delta^k f(a) + c_1 h^2 f^{(k+2)}(a) + c_2 h^4 f^{(k+4)}(a) + \cdots.$$

In the series expansion on the right-hand side (beginning with $c_1 h^2 f^{(k+2)}(a)$), there occur only even powers of h; this is characteristic of symmetric difference approximations. The coefficients c_j depend on k but are independent of the function f.

Proof. (For the cases $k = 1$ and $k = 2$.) Using Taylor's formula, we have

$$f(a + t) - f(a) = tf'(a) + \frac{t^2}{2!}f''(a) + \frac{t^3}{3!}f'''(a) + \frac{t^4}{4!}f^{(4)}(a) + \cdots, \tag{7.1.8}$$

$$f(a - t) - f(a) = -tf'(a) + \frac{t^2}{2!}f''(a) - \frac{t^3}{3!}f'''(a) + \frac{t^4}{4!}f^{(4)}(a) + \cdots. \tag{7.1.9}$$

For $k = 1$, subtract Eq. (7.1.9) from Eq. (7.1.8):

$$f(a+t) - f(a-t) = 2tf'(a) + 2\frac{t^3}{3!}f'''(a) + 2\frac{t^5}{5!}f^{(5)}(a) + \dots \tag{7.1.10}$$

Divide by $2t$ and set $t = \frac{1}{2}h$:

$$\begin{aligned} h^{-1}\delta f(a) &= h^{-1}[f(a+\tfrac{1}{2}h) - f(a-\tfrac{1}{2}h)] \\ &= f'(a) + \tfrac{1}{24}h^2 f'''(a) + \tfrac{1}{1920}h^4 f^{(5)}(a) + \dots \end{aligned} \tag{7.1.11}$$

For $k = 2$, add Eq. (7.1.8) to Eq. (7.1.9):

$$f(a+t) - 2f(a) + f(a-t) = 2\left(\frac{t^2}{2!}f''(a) + \frac{t^4}{4!}f^{(4)}(a) + \frac{t^6}{6!}f^{(6)}(a) + \dots\right). \tag{7.1.12}$$

Divide by t^2 and set $t = h$:

$$h^{-2}\delta^2 f(a) = f''(a) + \tfrac{1}{12}h^2 f^{(4)}(a) + \tfrac{1}{360}h^4 f^{(6)}(a) + \dots \tag{7.1.13}$$

The proof for general k is more complicated, but proceeds analogously. One uses Theorems 7.1.2 and 7.1.1, together with the symmetry properties of the binomial coefficients, $\binom{k}{p} = \binom{k}{k-p}$.

In the practical application of Theorems 7.1.4 and 7.1.5, one must take account of the large effect which round-off errors in the values of the function can have on the higher differences.

Example 7.1.8

For all *second-degree polynomials* it holds that

$$f'(a) = \mu\delta f(a) = \frac{f(a+h) - f(a-h)}{2h},$$

$$f''(a) = \frac{\delta^2 f(a)}{h^2} = \frac{f(a+h) - 2f(a) + f(a-h)}{h^2},$$

since one need only put $t = h$ in Eqs. (7.1.10) and (7.1.12) and recall that $f^{(m)} = 0$ for $m \geq 3$. The formula for f'' holds, in fact, even for *third-degree polynomials*.

REVIEW QUESTIONS

1. Express $\Delta^k y_n$ in terms of the elements of the sequence $\{y_n\}$.
2. Express y_{n+k} in terms of y_n, Δy_n, $\Delta^2 y_n, \dots$
3. *Why* does the difference scheme for a sequence $y = (0, 0, 0, 0, 1, 0, 0, 0, 0)$ give an indication of the effect of an error in z_5 in the sequence $(z_1, z_2, z_3, z_4, z_5, z_6, z_7, z_8, z_9)$ on the differences of the sequence $\{z_i\}_{i=1}^9$?
4. Derive the formula for summation by parts.

PROBLEMS

1. Give the difference schemes for the following sequences:
(a) 1, 2, 4, 8, 16; (b) 1, −1, 1, −1, 1.

2. Locate and correct the misprint in the following table of the values of a "well-behaved" function:

0852 2251 3651 5045 6458 7864 9272

3. The table below gives the position x of a particle at various times t:

t	74		75		76		77		78		79		80		81		82
x	75032		79426		83807		88177		93057		99433		105839		112276		118746
Δ		4394		4381		4370		4880		6376		6406		6437		6470	

From the differences one can see that the particle at a certain time received an impulse which made dx/dt discontinuous. Determine the time at which the impulse occurred and the size of the discontinuity. (*Hint:* How would such a disturbance spread in the difference scheme?)

4. Show that if $u_1 = u_N = v_1 = v_N = 0$, then

$$\sum_{n=1}^{N-1} u_n \Delta^2 v_{n-1} = -\sum_{n=1}^{N-1} \Delta u_n \Delta v_n = \sum_{n=1}^{N-1} v_n \cdot \Delta^2 u_{n-1}.$$

5. What is $\Delta^4 y_n$ and $\delta^4 y_n$ if
(a) $y_n = n^3 - n^2 + 17n - 1,\ h = 1$? (b) $y_n = 2^{-n},\ h = 1$?

7.2. SIMPLE METHODS FOR DERIVING APPROXIMATION FORMULAS AND ERROR ESTIMATES

7.2.1. Statement of the Problems and Some Typical Examples

Throughout the remainder of this chapter the following notations will be used, where f is a function defined on a net $(x_0, x_1, \ldots, x_n)$. (The points need not be equidistant or even monotonely ordered.)

int $(x_0, x_1, \ldots, x_n)$ = Smallest interval containing the points $x_0, x_1, \ldots, x_n$,

$$f(x_i) = f_i, \quad f'(x_i) = f'_i, \quad f''(x_i) = f''_i, \quad \ldots \tag{7.2.1}$$

R_T denotes truncation error, R_X round-off error due to uncertainty in the input data, R_C round-off errors made during the computations. R_C and R_X are occasionally combined in the summarizing term R_A.

We shall treat a number of **problems** where one knows certain data about a function—for example, the value of the function and possibly some of its derivatives on a net. One wishes to compute other data about the function—for example:

(a) the value of the function at some point x which does not belong to the net—*interpolation* if $x \in \text{int}(x_0, x_1, \ldots, x_n)$, *extrapolation* if $x \notin \text{int}(x_0, x_1, \ldots, x_n)$;
(b) the value of some derivative of the function at a point (*numerical differentiation*);
(c) the value of the integral of the function over an interval (*numerical integration*, occasionally called *numerical quadrature*).

For functions which with reasonable accuracy can be locally approximated by a polynomial, it is natural to use formulas which are exact for functions which (piecewise) are polynomials of some given degree. Such formulas often have a simple structure.

There are many useful **ways of deriving such formulas**:

(a) the method of *undetermined coefficients*—see Example 7.2.1;
(b) replacing the function (piecewise) by an *approximating polynomial* which can be expressed using the given data, and then computing the desired data for that polynomial—see Examples 7.2.3 and 7.2.4;
(c) *repeated Richardson extrapolation* (the deferred approach to the limit) —Sec. 7.2.2;
(d) *operator series* (or *generating functions*)—Sec. 7.6.2;
(e) making use of known properties of polynomials in a freer fashion—Example 7.2.2.

Concerning the *estimation of the discretization error* (or *truncation error*), there are two main levels of ambition:

I. **strict** error estimates;
II. approximate error estimates, usually **asymptotically correct** error estimates.

The relation between the actual error and such an error estimate usually becomes closer as the step length approaches 0.

Strict error estimates can occasionally be derived by ad hoc uses of the mean-value theorems of differential and integral calculus. (More systematically, one can start from Theorems 4.5.2 and 4.5.1, or from Taylor's theorem with remainder in integral form, or from the remainder term in the interpolation polynomial, Theorem 4.3.3. However, in this book we shall not go deeper into this aspect of the subject.)

Approximate error estimates often begin with approximating the first neglected term in a series expansion in powers of the step length (Example

7.2.1) or by an expansion in differences of increasing order. It is especially common that in such a case a strict remainder term of the form $ch^n f^{(n)}(\xi)$ is estimated by $c\Delta^n f$.

We shall now *illustrate various ways of deriving formulas* and remainder terms for a series of examples, where *the results are also of interest.*

Example 7.2.1

Derive a formula,

$$f_{1/2} \approx af_0 + bf_1 + cf'_0 + df'_1, \qquad x_i = x_0 + ih,$$

which is exact for polynomials of the highest possible degree. Derive an approximate error estimate. Use the formula and the error estimate for the case $f(x) = e^x$, $x_0 = 0$, $x_1 = 0.2$.

Set $\frac{1}{2}h = \alpha$; thus $x_0 - x_{1/2} = -\alpha$, $x_1 - x_{1/2} = \alpha$. There are four coefficients to determine. One can determine them with the help of the functions $f(x) = (x - x_{1/2})^n$, $n = 0, 1, 2, 3$. In choosing "test functions" of this type, one should make use of the symmetry aspects of the problem as much as possible:

$$\begin{aligned}
f(x) &= 1: & 1 &= a + b. \\
f(x) &= x - x_{1/2}: & 0 &= (-\alpha)a + \alpha b + c + d. \\
f(x) &= (x - x_{1/2})^2: & 0 &= \alpha^2 a + \alpha^2 b + (-2\alpha)c + (2\alpha)d. \\
f(x) &= (x - x_{1/2})^3: & 0 &= (-\alpha)^3 a + \alpha^3 b + 3\alpha^2 c + 3\alpha^2 d.
\end{aligned}$$

After some manipulations,

$$\begin{aligned}
\alpha(-a + b) + (c + d) &= 0, \\
\alpha(-a + b) + 3(c + d) &= 0, \\
a + b &= 1, \\
\alpha(a + b) + 2(-c + d) &= 0.
\end{aligned}$$

The solution of the system of equations is:

$$a = b = \tfrac{1}{2}, \qquad c = -d = \tfrac{1}{4}\alpha = \tfrac{1}{8}h.$$

The desired formula is thus

$$f_{1/2} \approx \tfrac{1}{2}(f_0 + f_1) + \tfrac{1}{8}h(f'_0 - f'_1), \tag{7.2.2}$$

which is exact for all third-degree polynomials. (Why?)

For $f(x) = (x - x_{1/2})^4$, the truncation error becomes

$$\begin{aligned}
R_T &= af_0 + bf_1 + cf'_0 + df'_1 - f_{1/2} \\
&= \frac{1}{2}\alpha^4 + \frac{1}{2}\alpha^4 - \frac{1}{4}\alpha \cdot 4\alpha^3 - \frac{1}{4}\alpha \cdot 4\alpha^3 - 0 = -\alpha^4 = \frac{-h^4}{16}.
\end{aligned}$$

For every function having five continuous derivatives we have, using Taylor's formula:

$$f(x) = \sum_{i=0}^{3} \frac{f^{(i)}(x_{1/2})}{i!}(x - x_{1/2})^i + \frac{f^{(4)}(x_{1/2})}{4!}(x - x_{1/2})^4 + O(h^5)$$

$$= \text{third-degree polynomial} + \frac{f^{(4)}(x_{1/2})}{24} \cdot (x - x_{1/2})^4 + O(h^5).$$

The desired error estimate is thus

$$R_T = -\frac{h^4}{16} \cdot \frac{f^{(4)}(x_{1/2})}{24} + O(h^5). \tag{7.2.3}$$

For $f(x) = e^x$, $x_0 = 0$, $x_1 = 0.2$, we get (using six decimals)

$$f_{1/2} \approx \frac{1}{2}(1 + 1.221403) + \frac{0.2}{8}(1 - 1.221403) = 1.105166,$$

$$R_T \approx -5 \cdot 10^{-6},$$

$$|R_C| \leq (|a| + |b| + |c| + |d|) \cdot \tfrac{1}{2} \cdot 10^{-6} + \tfrac{1}{2} \cdot 10^{-6} \lesssim 10^{-6}.$$

Using a table, $e^{0.1} = 1.105171$. Hence both the result and the error estimate agree perfectly with the predictions of the theory.

Note: the formula

$$f_{n+1/2} \approx af_n + bf_{n+1} + cf'_n + df'_{n+1}$$

holds for arbitrary n, with the same coefficients. This formula is useful for interpolating a function whose derivative is easily computed.

Example 7.2.2

Derive a formula

$$\int_{x_{n-1}}^{x_{n+1}} f(x)\,dx \approx h(af_{n-1} + bf_n + cf_{n+1}), \quad (x_n = x_0 + nh),$$

which is exact for polynomials of highest possible degree. Give an approximate error estimate.

For simplicity in the derivation, we can set $n = 0$, $x_n = 0$, with no loss of generality. (Why?) According to Taylor's formula we have, for $|x| \leq h$,

$$f(x) = f_0 + xf'_0 + \frac{x^2}{2}f''_0 + \frac{x^3}{3!}f'''_0 + O(h^4),$$

where the remainder is zero for all polynomials of degree 3 (or less). Thus

$$\int_{-h}^{h} f(x)\,dx = f_0 \int_{-h}^{h} dx + f'_0 \int_{-h}^{h} x\,dx + \tfrac{1}{2}f''_0 \int_{-h}^{h} x^2\,dx$$

$$+ \tfrac{1}{6}f'''_0 \int_{-h}^{h} x^3\,dx + O(h^5)$$

$$= 2hf_0 + 0 + \tfrac{1}{3}h^3f''_0 + 0 + O(h^5).$$

But, using Theorem 7.1.6, we have that

$$f''_0 = h^{-2}(f_{-1} - 2f_0 + f_1) + O(h^2),$$

where the remainder term is zero for all third-degree polynomials (see Example 7.1.8). Hence

$$\int_{-h}^{h} f(x)\,dx = 2hf_0 + \tfrac{1}{3}h(f_{-1} - 2f_0 + f_1) + O(h^5)$$
$$= \tfrac{1}{3}h(f_{-1} + 4f_0 + f_1) + O(h^5).$$

The desired formula is thus

$$\int_{x_{n-1}}^{x_{n+1}} f(x)\,dx = \tfrac{1}{3}h(f_{n-1} + 4f_n + f_{n+1}) + O(h^5), \tag{7.2.4}$$

where the remainder term is zero for all third-degree polynomials. The above formula is one of the classical formulas for numerical integration, **Simpson's formula.**

In the same way as in Example 7.2.1, we first determine R_T for $f(x) = x^4$ ($n = 0$, $x_n = 0$):

$$\frac{1}{3}h(h^4 + 0 + h^4) - \int_{-h}^{h} x^4\,dx = \frac{2h^5}{3} - \frac{2h^5}{5} = \frac{4h^5}{15},$$

from which it follows generally that

$$R_T = \frac{4h^5}{15} \cdot \frac{f^{(4)}(x_n)}{24} + O(h^6) = \frac{h^5}{90} f^{(4)}(x_n) + O($$

Here R_T is an **asymptotic** error estimate. One can prove (see Fröberg [7, p. 200]) the following **strict** error estimate:

$$R_T = \frac{h^5}{90} f^{(4)}(\xi), \quad \xi \in [x_{n-1}, x_{n+1}]. \tag{7.2.5}$$

In the practical use of Simpson's formula for computing

$$\int_a^b f(x)\,dx,$$

one divides the interval $[a, b]$ into $2m$ steps of length h, $x_i = x_0 + ih$, $a = x_0$, $b = x_{2m}$. Simpson's formula is then used on each of the m double steps—i.e., with $n = 2i + 1$, $i = 0, 1, 2, \ldots, m - 1$. (An explicit formula for this is given in Problem 1 at the end of this section.) The remainder is then (using $2m = (b - a)/h$):

$$R_T = \sum_{i=0}^{m-1} \frac{h^5}{90} f^{(4)}(\xi_i) = \frac{h^5}{90} m f^{(4)}(\xi),$$
$$= \frac{(b-a)h^4}{180} f^{(4)}(\xi), \quad (\xi \in [a, b]). \tag{7.2.6}$$

Hence one says that the **global** error (i.e., the error over the entire interval) for Simpson's formula is $O(h^4)$, while the **local** error (Eq. 7.2.5) is $O(h^5)$.

The method of undetermined coefficients could also have been used in the derivation. Then the three functions $f(x) = 1, x, x^2$ would suffice to determine the three coefficients a, b, c. One finds, however, that the formula holds also for $f(x) = x^3$; hence one must go to $f(x) = x^4$ to get the error estimate. Generally speaking, one can expect similar experiences in problems which have a certain symmetry.

Example 7.2.3

We know the values of a function at the equidistant points $x_n = x_0 + nh$. Suppose that for a certain n we have that $f_n \le f_{n-1}$ and $f_n \le f_{n+1}$. Determine an approximate value for both the minimum point $\hat{x}$ and the minimum value of the function f in $[x_{n-1}, x_{n+1}]$.

Approximate f using a second-degree polynomial $\bar{f}$,

$$\bar{f}(x) = a + b(x - x_n) + \tfrac{1}{2}c(x - x_n)^2,$$

where we have (from Taylor's formula and Example 7.1.7),

$$a = \bar{f}_n,$$

$$b = \bar{f}'_n = \frac{\bar{f}_{n+1} - \bar{f}_{n-1}}{2h},$$

$$c = \bar{f}''_n = \frac{\bar{f}_{n+1} - 2\bar{f}_n + \bar{f}_{n-1}}{h^2}.$$

$$\bar{f}'(x) = b + c(x - x_n) = 0 \quad \text{for} \quad x = \hat{x} = x_n - \frac{b}{c},$$

$$\bar{f}(\hat{x}) = a - \frac{b^2}{c} + \frac{1}{2}c\left(\frac{b}{c}\right)^2 = a - \frac{1}{2}\frac{b^2}{c}.$$

Thus, having chosen $\bar{f}$ to be the interpolating polynomial which agrees with f at $x = x_{n-1}, x_n, x_{n+1}$, we get the *result*

$$\min f(x) \approx a - \frac{1}{2}\frac{b^2}{c} \quad \text{for} \quad x \approx \hat{x} = x_n - \frac{b}{c}, \tag{7.2.7}$$

where

$$a = f_n, \qquad b = \frac{f_{n+1} - f_{n-1}}{2h}, \qquad c = \frac{f_{n+1} - 2f_n + f_{n-1}}{h^2}.$$

As an error estimate for the minimum value of f we get:

$$|\min f(x) - \min \bar{f}(x)| \le \max |f(x) - \bar{f}(x)|, \quad x \in [x_{n-1}, x_{n+1}].$$

Using the remainder term for interpolation (Theorem 4.3.3), we have, if $|f'''(x)| < M$ (see Problem 3 at the end of this section),

$$\max |f(x) - \bar{f}(x)| \le \frac{1}{6} M \max |(x - x_n - h)(x - x_n)(x - x_n + h)|$$

$$= \frac{Mh^3}{(243)^{1/2}}.$$

M can be estimated by using third differences; see Theorem 7.1.4.

xample 7.2.4

Estimate f'_0, when $f_{-2}, f_{-1}, f_0, f_1, f_2$ are known $(x_n = x_0 + nh)$. The unction f is judged to be so regular that a local fit with a second-degree olynomial (in $[x_{-2}, x_2]$) is reasonable, but the values of the function are contaminated by random errors; hence the method of least squares is motivated. Approximate $f(x)$ by $\bar{f}(x) = \sum_{j=0}^{2} c_j\varphi_j(x)$, where $\varphi_j(x) = (x - x_0)^j$. Then $f'_0 \approx \bar{f}'_0 = c_1$. Use the scalar product

$$(f, g) = \sum_{i=-2}^{2} f_i g_i.$$

From the normal equations of Eq. (4.2.6), we get:

$$(\varphi_0, \varphi_1)c_0 + (\varphi_1, \varphi_1)c_1 + (\varphi_2, \varphi_1)c_2 = (f, \varphi_1).$$

Since $(\varphi_0, \varphi_1) = (\varphi_2, \varphi_1) = 0$, we have

$$c_1 = \frac{(f, \varphi_1)}{(\varphi_1, \varphi_1)} = \frac{\sum_{i=-2}^{2} hif_i}{\sum_{i=-2}^{2} h^2 i^2}.$$

Hence

$$f'_0 \approx c_1 = \frac{2(f_2 - f_{-2}) + (f_1 - f_{-1})}{10h}. \tag{7.2.8}$$

Later in this chapter we shall see other approximate formulas for numerical differentiation.

7.2.2. Repeated Richardson Extrapolation

In many calculations what one would really like to know is the limiting value of a certain quantity as the step length approaches zero. Let $F(h)$ denote the value of the quantity obtained with step length h. The work to compute $F(h)$ often increases sharply as $h \to 0$. In addition, the effects of round-off errors often set a practical bound for how small h can be chosen.

Often, one has some knowledge of how the truncation error $F(h) - F(0)$ behaves when $h \to 0$. If

$$F(h) = a_0 + a_1 h^p + O(h^r), \quad (h \to 0, r > p),$$

where $a_0 = F(0)$ is the quantity we are trying to compute and a_1 is unknown, then a_0 and a_1 can be estimated if we compute F for two step lengths, h and qh, $(q > 1)$:

$$F(h) = a_0 + a_1 h^p + O(h^r),$$

$$F(qh) = a_0 + a_1 (qh)^p + O(h^r),$$

from which we get

$$F(0) = a_0 = F(h) + \frac{F(h) - F(qh)}{q^p - 1} + O(h^r). \tag{7.2.9}$$

This process is called **Richardson extrapolation**, or *the deferred approach to the limit*. An example of this was mentioned in Chap. 1—the application of the above process to the trapezoidal rule for numerical integration (where $p = 2, q = 2$).

Suppose that, as in Theorem 7.1.5, one knows the form of a more complete expansion of $F(h)$ in powers of h. Then one can, even if the values of the coefficients in the expansion are unknown, **repeat the use of Richardson extrapolation** in a way to be described below. *This process is, in many numerical problems—especially in the numerical treatment of integrals and differential equations—the simplest way to get results which have negligible truncation error.* The application of this process becomes especially simple when the step lengths form a geometric series,

$$h_0, q^{-1}h_0, q^{-2}h_0, \ldots.$$

THEOREM 7.2.1

Suppose that

$$F(h) = a_0 + a_1 h^{p_1} + a_2 h^{p_2} + a_3 h^{p_3} + \ldots, \tag{7.2.10}$$

where $p_1 < p_2 < p_3 < \ldots$, *and set*

$$F_1(h) = F(h), \qquad F_{k+1}(h) = F_k(h) + \frac{F_k(h) - F_k(qh)}{q^{p_k} - 1}. \tag{7.2.11}$$

Then $F_n(h)$ *has an expansion of the form*

$$F_n(h) = a_0 + a_n^{(n)} h^{p_n} + a_{n+1}^{(n)} h^{p_{n+1}} + \ldots.$$

Proof. (By induction.) From Eq. (7.2.10), the theorem holds for $n = 1$. Suppose the above expansion holds for $n = k$. Then, using Eq. (7.2.11), we see that $F_{k+1}(h)$ has an expansion containing the same powers of h as the expansion for $F_k(h)$. In the expansion for $F_{k+1}(h)$, the coefficient of h^{p_k} is

$$a_k^{(k)} + \frac{a_k^{(k)} - a_k^{(k)} q^{p_k}}{q^{p_k} - 1} = a_k^{(k)} - a_k^{(k)} = 0.$$

Thus the theorem holds for $n = k + 1$, and the proof is complete.

Hence if an expansion of the form of Eq. (7.2.10) is known, the above theorem gives a way to compute increasingly better estimates of a_0. The first term in the expression for the truncation error in $F_n(h)$ is $a_n^{(n)} h^{p_n}$, so we get truncation errors which begin with increasingly higher powers of h, as n increases (recall $p_1 < p_2 < p_3 < \ldots$).

A moment's reflection on Eq. (7.2.11) will convince the reader that (using the notation of the theorem) $F_{k+1}(h)$ is determined by the $k + 1$ values $F_1(h)$, $F_1(qh), \ldots, F_1(q^k h)$. One gets (with some slight changes in notation) the following algorithm:

Algorithm. For $m = 0, 1, 2, \dots$, set $A_{m,0} = F(q^{-m}h_0)$, and compute, or $k = 1, 2, \dots, m$,

$$A_{m,k} = A_{m,k-1} + \frac{A_{m,k-1} - A_{m-1,k-1}}{q^{p_k} - 1}. \tag{7.2.12}$$

'he value $A_{m,k+1}$ is accepted as an estimate of a_0 when $|A_{m,k} - A_{m-1,k}|$ is less han the permissible error. The computations can be conveniently set up in he scheme:

$$\begin{array}{cccc}
 & \dfrac{\Delta}{q^{p_1} - 1} & \dfrac{\Delta}{q^{p_2} - 1} & \dfrac{\Delta}{q^{p_3} - 1} \\
A_{00} & & & \\
 & \square & & \\
A_{10} & A_{11} & & \\
A_{20} & A_{21} & A_{22} & \\
 & & \square & \\
A_{30} & A_{31} & A_{32} & A_{33}
\end{array} \tag{7.2.13}$$

Thus one extrapolates until two values *in the same column* agree to the desired accuracy. In most situations, the magnitude of the difference between two values in the same column gives (if h is sufficiently small), with a large margin, a bound for the truncation error in the lower of the two values. One cannot, however, get a guaranteed error bound in all situations.

The most common special case is to take $q = 2$ when one has an expansion of the form

$$F(h) = a_0 + a_1 h^2 + a_2 h^4 + a_3 h^6 + \dots, \tag{7.2.14}$$

where clearly $p_k = 2k$. Then in (7.2.13), the headings of the columns become

$$\frac{\Delta}{3}, \frac{\Delta}{15}, \frac{\Delta}{63}, \dots, \quad \text{when} \quad p_k = 2k, \quad q = 2. \tag{7.2.15}$$

We now illustrate the application of the above in *numerical differentiation*. Using Theorem 7.1.5, one has an expansion of the form

$$\frac{f(a+h) - f(a-h)}{2h} = f'(a) + a_1 h^2 + a_2 h^4 + \dots.$$

Example 7.2.5

Compute $f'(3)$ for $f(x) = \ln(x)$ using values for $\ln(x)$ taken from a six-place table. Choose $h_0 = 0.8$. Then,

$$A_{m0} = \frac{\ln(3+h) - \ln(3-h)}{2h} \quad \text{with} \quad h = 2^{-m}h_0.$$

h		$\frac{\Delta}{3}$		$\frac{\Delta}{15}$		$\frac{\Delta}{63}$
0.8	$A_{00} = 0.341590$					
		$-2,087$				
0.4	$A_{10} = 0.335330$		$A_{11} = 0.333243$			
		-500		$+6$		
0.2	$A_{20} = 0.333830$		$A_{21} = 0.333330$		$A_{22} = 0.333336$	
		-125		0		0
0.1	$A_{30} = 0.333455$		$A_{31} = 0.333330$		$A_{32} = 0.333330$	

Using the stopping criterion of the algorithm, one accepts $A_{32} = 0.33333($ where $|R_T| \lesssim \frac{1}{2}\cdot 10^{-6}$. Since $f'(x) = 1/x$, the correct answer is $f'(3) =$ 0.333333. The actual error is thus $-3\cdot 10^{-6}$. Here, round-off error is th dominant source of error, something more typical of numerical differentia tion than of Richardson extrapolation.

One can show (for $p_k = 2k, q = 2$) that *if the values in the first column—i.e $A_{00}, A_{10}, A_{20}, \ldots$—are afflicted with errors whose magnitudes are less than then the errors caused later in the extrapolation scheme have magnitudes whic nowhere exceed 2ϵ.* The reader is recommended to verify this, at least for $k =$ 1, 2, 3. In the example above, the error in A_{m0} is at most $10^{-6}/2h \leq 5\cdot 10^{-}$ which gives $|R_A| \leq 10^{-5}$. When *choosing the precision to be retained* in th values in the first column, one should consider *what precision one hopes t attain in the final result of the extrapolations.* In the example, the truncatio error in A_{10} is—if the value is understood as an approximation to the deriv tive—about 0.002, but it would be wasteful to round A_{10} to three decimal The extrapolation process also uses the information contained in the digi which are afflicted with truncation error.

The *idea* of a deferred approach to the limit (Richardson extrapolatio is much more general than the theorem and the algorithm given above. is, for instance, not necessary that the step sizes form a geometric progressio Note that if $p_j = j\cdot p$ in Eq. (7.2.10), then the partial sums of the expansio are **polynomial** functions of h^p. If $k+1$ values $F(q_0h), F(q_1h), \ldots, F(q_k$ are known, then by Theorem 4.3.2 a kth-degree interpolation polynomial determined uniquely by the conditions:

$$Q((q_ih)^p) = F(q_ih), \quad i = 0, 1, 2, \ldots, k.$$

One can prove that $Q(0) - F(0) = O(h^{(k+1)p}), h \to 0$.

There are many other variations. It is essentially a problem of estimatin the coefficients in some theoretically motivated expression for $F(h)$, whe some values of F are numerically known. The program for the extrapolatio in these generalizations is more complicated, though quite practicable. O generalization that has proved advantageous in the numerical solution of differential equations (see the Bulirsch-Stoer method, Sec. 8.3.1) is to fit a **rational** function with almost the same degree in the numerator and the denominator to the given values of F.

Moreover, it is not at all necessary that the parameter be a step size. For instance, the same idea can be used when one has some theoretical knowledge of how the remainder in an infinite series depends asymptotically on the number of terms. Aitken extrapolation (see Sec. 3.2.3) is in this sense an application of the same basic idea.

Finally, the idea of a deferred approach to the limit is sometimes used in the experimental sciences—for example, when some quantity is to be measured in complete vacuum (difficult or expensive to produce). It can then be more practical to measure the quantity for several different values of the pressure. Expansions analogous to Eq. (7.2.10) can sometimes be motivated by the kinetic theory of gases, and the deferred approach to the limit can be used.

REVIEW QUESTION

Give the theory behind repeated Richardson extrapolation and explain its use in numerical differentiation.

PROBLEMS

1. Simpson's rule is occasionally written in the form:

$$\int_a^b f(x)\,dx \approx \frac{h}{3}(f_0 + 4U + 2E + f_n),$$

where $U = f_1 + f_3 + \ldots + f_{n-1}$, $E = f_2 + f_4 + \ldots + f_{n-2}$, for n even. Show that this agrees with the formula given in Example 7.2.2.

2. (a) Derive a formula

$$(2h)^{-1/2}\int_0^{2h} x^{-1/2} f(x)\,dx \approx (A_0 f(0) + A_1 f(h) + A_2 f(2h))$$

which is exact when $f(x)$ is any second-degree polynomial.

(b) Give an asymptotically correct error term.

3. In Example 7.2.3 it is asserted that

$$\max \frac{1}{6}|(x - x_n - h)(x - x_n)(x - x_n + h)| = \frac{h^3}{243^{1/2}}, \quad |x - x_n| < h.$$

Prove this.

4. $\{f_n\}$ is a sequence of function values at equidistant points. Set

$$g_n = af_{n+1} + bf_n + cf_{n-1}.$$

(a) For what values of a, b, c does it hold that $g_n = f_n$ for all n when f is a first-degree polynomial?
(b) For which of the formulas satisfying the condition in (a) is $a^2 + b^2 + c^2$ smallest?
(c) Determine a formula which fulfills the conditions in (a) and for which $g_n = 0$ when $f_n = k \cdot (-1)^n$, k an arbitrary constant.
Note: Formulas of the above type, **moving averages**, are occasionally used when one wishes to eliminate errors in measurement or other "noise."

5. The following table of the function e^x is given:

x	0.00	0.25	0.50	0.75	1.00	1.25	1.50	1.75	2.00
e^x	1.0000	1.2840	1.6487	2.1170	2.7183	3.4903	4.4817	5.7546	7.3891

Compute $(d/dx)(e^x)_{x=1}$ by performing repeated Richardson extrapolation.

6. The following table of values of a function $f(x)$ is given:

x	0.6	0.8	0.9	1.0	1.1	1.2	1.4
$f(x)$	1.820365	1.501258	1.327313	1.143957	0.951849	0.752084	0.335920

Compute, using repeated Richardson extrapolation,
(a) $f'(1.0)$; (b) $f''(1.0)$.

7. A physical quantity X is assumed to depend on the pressure of a gas according to

$$X = c_0 + c_1 P^2 + c_2 P^3 + c_3 P^6,$$

where c_0, c_1, c_2, c_3 are nonzero constants. Determine the value for X in a vacuum ($P = 0$), using the following series of measurements:

P (mm Hg)	0.8	0.4	0.2	0.1	0.05
X (units)	740	487	475	485	489

8. Even the ancient Greeks computed approximate values of the circumference of the unit circle, 2π, by inscribing a regular polygon and computing its perimeter. Archimedes considered the inscribed 96-sided regular polygon, whose perimeter is 6.2821.

A regular n-sided polygon inscribed in a circle with radius 1 has circumference

$$c_n = 2n \sin \frac{\pi}{n}.$$

Put $h = 1/n$.
(a) Show that $c(h) = c_{1/h}$ satisfies the assumptions for repeated Richardson extrapolation.
(b) Compute c_2 (= twice the diameter), c_3, and c_6, and perform repeated Richardson extrapolation. Use four decimals.

(c) Look at the program below which computes 2π to nine correct decimals by performing repeated Richardson extrapolation on $c_6, c_{12}, c_{24}, \ldots$. Nine correct decimals are considered to be attained when the difference between two successive elements in a column is less than $0.5 \cdot 10^{-9}$. Derive the recursion formula for c_i used in the program. (Note that no trigonometric functions are needed.)

Program:

```
begin real x, R; array C[0: 10]; integer i, k, n, p;
  k: = 0; n: = 3; C[0]: = 6;
  OUTPUT(61, '/4Z, B − D.9D', 2*n, C[0]);
ANCORA: k: = k + 1; n: = 2*n; p: = 1;
  x: = sqrt(8*n↑2 − 4*n*sqrt(4*n↑2 − C[0]↑2));
  OUTPUT(61, '/4Z, B − D.9D', 2*n, x);
  for i: = 0 step 1 until k − 1 do
  begin R: = x − C[i]; p: = 4*p;
    C[i]: = x;
    x: = x + R/(p − 1);
    OUTPUT(61, 'B − D.9D', x);
    if abs(R/(p − 1)) < 0.5₁₀ − 9 then goto FINITO
  end;
  C[k]: = x; if k < 10 then goto ANCORA;
FINITO:
end
```

(d) The program produced the following output:

6	6.000000000				
12	6.211657082	6.282209443			
24	6.265257228	6.283123943	6.283184909		
48	6.278700411	6.283181472	6.283185307	6.283185313	
96	6.282063901	6.283185065	6.283185304	6.283185304	6.283185304
192	6.282905019	6.283185391	6.283185413	6.283185415	6.283185415
384	6.283115395	6.283185520	6.283185528	6.283185530	6.283185531
768	6.283168822	6.283186632	6.283186706	6.283186724	6.283186729
1536	6.283183393	6.283188250	6.283188358	6.283188384	6.283188391
3072	6.283154251	6.283144537	6.283141623	6.283140881	6.283140695
6144	6.283193107	6.283206059	6.283210161	6.283211249	6.283211525

The convergence is very bad from $N = 192$ onward. What do you think is the trouble? What statement in the program should be changed, and how?

7.3. INTERPOLATION

7.3.1. Introduction

From Theorem 4.3.2 and 4.3.3 (pp. 99–100), we know that the **interpolation problem** of determining a polynomial Q of degree m such that

$$Q(x_i) = f_i \quad i = 0, 1, 2, \ldots, m, \tag{7.3.1}$$

has a unique solution, and that

$$f(x) - Q(x) = \frac{f^{(m+1)}(\xi)}{(m+1)!}(x - x_0)(x - x_1)\dots(x - x_m), \qquad (7.3.2)$$

$$\xi \in \text{int}(x, x_0, x_1, \dots, x_m).$$

In Chap. 4 we also brought up the fact that equidistant interpolation can be an ill-conditioned operation (see Example 4.3.2). However, this applies only to polynomials of high degree and then only in the peripheral parts of int $(x_0, x_1, \dots, x_m)$. *Equidistant interpolation in the central part of the interval* is, on the other hand, a fairly well-conditioned and very useful operation.

7.3.2. When is Linear Interpolation Sufficient?

THEOREM 7.3.1

In a section of a table of equidistant, correctly rounded function values, where $|\Delta^2 f| \leq 4U$, *(U = one unit in the last digit of the function values), the total error in linear interpolation can only slightly exceed U in magnitude.*

Here we are referring to the sum of the truncation error R_T and the rounding error R_X which arises from rounding the table values. The rounding errors during the computation, R_C, are ignored. (One should use an extra decimal.) The error in the argument is assumed to be zero.

Proof. Denote the step size in the table by h. We want to compute the value $f(x)$, where $x = x_0 + ph$, $0 \leq p \leq 1$. p is called the *normalized interpolation argument*. From Eq. (7.3.2) it follows that

$$|R_T| = \left|\frac{f''(\xi)(x - x_0)(x - x_1)}{2}\right| \approx \left|\frac{\Delta^2 f_0}{h^2} \cdot \frac{ph \cdot (p-1)h}{2}\right|$$

$$\approx \frac{|\Delta^2 f_0| \cdot p(1-p)}{2} \leq \frac{|\Delta^2 f_0|}{8}, \qquad (7.3.3)$$

since $p(1-p)$ takes on its maximum value $\frac{1}{4}$ for $p = \frac{1}{2}$. In linear interpolation,

$$Q(x) = f_0 + p\Delta f_0 = (1-p)f_0 + pf_1.$$

If the values f_0 and f_1 have errors $\epsilon_0, \epsilon_1, |\epsilon_i| \leq \frac{1}{2}U$ (by the assumption of correct rounding), then, for $0 \leq p \leq 1$,

$$|R_X| = |(1-p)\epsilon_0 + p\epsilon_1| \leq (1-p)\cdot\tfrac{1}{2}U + p\cdot\tfrac{1}{2}U = \tfrac{1}{2}U \qquad (7.3.4)$$

Thus if $|\Delta^2 f_0| \leq 4U$, then

$$|R_T| + |R_X| \lesssim \frac{4U}{8} + \frac{1}{2}U = U,$$

which was to be shown. (Note that the formulas (7.3.3) and (7.3.4) are themselves of interest.)

The inequality in (7.3.4) is a good illustration of the distinction between actual errors (which are positive or negative) and error bounds (never negative). The following line of reasoning is, unfortunately, quite common. If f_0 and f_1 have "error" $\frac{1}{2}U$, then $\Delta f_0 = f_1 - f_0$ has "error" U and $Q(x) = f_0 + p\Delta f_0$ then has "error" $\frac{1}{2}U + |p|\,U$, which for $p \approx 1$ is close to $1.5U$. The reader should investigate why this way of reasoning gives a result much coarser than that of (7.3.4).

7.3.3. Newton's General Interpolation Formula

By Theorem 4.3.2, we can write the solution to the interpolation problem (see also Sec. 7.3.1) in the form

$$f(x) = c_0 + c_1(x - x_0) + \ldots + c_m(x - x_0)(x - x_1)\ldots(x - x_{m-1}) + A(x)(x - x_0)(x - x_1)\ldots(x - x_m), \tag{7.3.5}$$

where, according to Eq. (7.3.2),

$$A(x) = \frac{f^{(m+1)}(\xi)}{(m+1)!}.$$

Note that $A(x)$ is bounded when $f^{(m+1)}$ is continuous in $\text{int}\,(x_0, x_1, \ldots, x_m, x)$. For $x = x_0$ we get $f_0 = c_0$. Set

$$f[x_0, x] = \frac{f(x) - f(x_0)}{x - x_0}. \tag{7.3.6}$$

Thus

$$f[x_0, x] = c_1 + c_2(x - x_1) + \ldots + c_m(x - x_1)\ldots(x - x_{m-1}) + A(x)(x - x_1)\ldots(x - x_m),$$

$$f[x_0, x_1] = c_1.$$

We now define, recursively, **divided differences**

$$f[x_0, x_1, \ldots, x_{k-1}, x_k, x] = \frac{f[x_0, x_1, \ldots, x_{k-1}, x] - f[x_0, x_1, \ldots, x_{k-1}, x_k]}{x - x_k}. \tag{7.3.7}$$

One gets, for $k = 1$,

$$f[x_0, x_1, x] = c_2 + c_3(x - x_2) + \ldots + c_m(x - x_2)\ldots(x - x_{m-1}) + A(x)(x - x_2)\ldots(x - x_m),$$

$$c_2 = f[x_0, x_1, x_2].$$

By induction one can show that

$$c_k = f[x_0, x_1, \ldots, x_{k-1}, x_k], \quad k \leq m. \tag{7.3.8}$$

For $k = m$ we get the following important result:

THEOREM 7.3.2

$$f[x_0, x_1, \ldots, x_m, x] = A(x) = \frac{f^{(m+1)}(\xi)}{(m+1)!},$$

$$\xi \in \text{int}\,(x_0, x_1, \ldots, x_m, x).$$

We summarize the above results in a theorem:

THEOREM 7.3.3 *Newton's General Interpolation Formula*

The interpolation problem of Eq. (7.3.1) has the solution

$$Q(x) = f_0 + \sum_{j=1}^{m} f[x_0, x_1, \ldots, x_j](x - x_0)(x - x_1)\ldots(x - x_{j-1})$$

with remainder

$$f(x) - Q(x) = f[x_0, x_1, \ldots, x_m, x](x - x_0)(x - x_1)\ldots(x - x_m)$$

$$= \frac{f^{(m+1)}(\xi)(x - x_0)(x - x_1)\ldots(x - x_m)}{(m+1)!},$$

$$\xi \in \operatorname{int}(x_0, x_1, \ldots, x_m, x).$$

THEOREM 7.3.4

The divided difference $f[x_0, x_1, \ldots, x_m]$ is a symmetric function of its $m + 1$ arguments. (Here m is an arbitrary natural number.)

Proof. By Eq. (7.3.8), $f[x_0, x_1, \ldots, x_m]$ is equal to c_m. But c_m is the leading coefficient of that polynomial which is the unique solution to the interpolation problem of Eq. (7.3.1). There we made no assumptions as to the order of enumeration of the points—e.g., they need not be in increasing (or any other) order. The interpolation problem is independent of how the points are numbered; hence c_m is a symmetric function of the points.

Divided differences are most easily computed recursively using the formula,

$$f[x_i, x_{i+1}, \ldots, x_{k-1}, x_k] = \frac{f[x_{i+1}, \ldots, x_{k-1}, x_k] - f[x_i, x_{i+1}, \ldots, x_{k-1}]}{x_k - x_i}. \tag{7.3.9}$$

Derivation of Eq. (7.3.9). Use Eq. (7.3.7), but call the first point x_{i+1} instead of x_0:

$$f[x_{i+1}, \ldots, x_{k-1}, x_k, x] = \frac{f[x_{i+1}, \ldots, x_{k-1}, x] - f[x_{i+1}, \ldots, x_{k-1}, x_k]}{x - x_k}.$$

Then set $x = x_i$ and use the symmetry property (Theorem 7.3.4).

The computation of differences according to (7.3.9) can be arranged in a scheme—e.g.,

$$\begin{array}{cc|ccc}
x_0 & f_0 & & & \\
 & & f[x_0, x_1] & & \\
x_1 & f_1 & & f[x_0, x_1, x_2] & \\
 & & f[x_1, x_2] & & f[x_0, x_1, x_2, x_3] \\
x_2 & f_2 & & f[x_1, x_2, x_3] & \\
 & & f[x_2, x_3] & & \\
x_3 & f_3 & & &
\end{array}$$

Example 7.3.1

Compute the interpolation polynomial for the following table. Carry out the calculations using Eq. (7.3.9).

	f	Div.diff.		
$x_0 = 1$	0			
		2		
$x_1 = 2$	2		1	
		5		0
$x_2 = 4$	12		1	
		8		
$x_3 = 5$	20			

Thus

$$Q(x) = 2(x-1) + (x-1)(x-2) = (x-1)\cdot x.$$

There are other recursive algorithms for interpolation—for example, Aitken's algorithm (see Problem 8 in this section) and Neville's algorithm. Both of these are described in Isaacson-Keller [12]. See also Eq. (4.3.5) in Chap. 4.

7.3.4. Formulas for Equidistant Interpolation

THEOREM 7.3.5 *The Connection Between Ordinary and Divided Differences*

If $x_\nu = x_0 + \nu h$, then

$$f[x_i, x_{i+1}, \ldots, x_{i+j}] = \frac{\Delta^j f_i}{h^j j!}$$

(Notice the factorial in the denominator!)

Proof. (By induction, with the use of Eq. (7.3.9).) Set $k = i + j + 1$. The details are left to the reader.

Note: Combining this result with Theorem 7.3.2 (set $m = j - 1$, $x = x_j$) we get a general proof for the formula $\Delta^j f_0 = h^j f^{(j)}(\xi)$, (Theorem 7.1.4).

THEOREM 7.3.6 *Newton's Interpolation Formula, Equidistant Case*

$$f(x_0 + ph) = f_0 + \sum_{j=1}^{m} \Delta^j f_0 \frac{p(p-1)\ldots(p-j+1)}{j!} + h^{m+1} f^{(m+1)}(\xi)\cdot\frac{p(p-1)\ldots(p-m)}{(m+1)!}.$$

Proof. For $x = x_0 + ph$ we have the relation

$$(x-x_0)(x-x_1)\ldots(x-x_{j-1}) = ph(p-1)h\ldots(p-j+1)h = h^j p(p-1)\ldots(p-j+1).$$

The formula for $f(x_0 + ph)$ given above follows, then, using (successively) Theorems 7.3.5 and 7.3.3.

Note: Theorem 7.1.2 is a special case of the above theorem—i.e., the case $p = k =$ an integer. The above theorem is easy to remember if one sees it as a *symbolic binomial theorem* (see Sec. 7.6):

$$E^p = (1 + \Delta)^p$$

The Newton interpolation formula uses differences along a diagonal in the difference scheme. It is often more practical, when interpolating in the interval $[x_0, x_1]$, to use differences which lie on about the same horizontal lines as f_0, f_1. The following formula is one of the many possible formulas with this desirable property.

Bessel's formula.

$$f(x_0 + ph) \approx f_0 + p\,\Delta f_0 + B''_p(\Delta^2 f_{-1} + \Delta^2 f_0) + B'''_p \Delta^3 f_{-1} + B^{iv}_p(\Delta^4 f_{-2} + \Delta^4 f_{-1}) + B^v_p \Delta^5 f_{-2} + \dots \tag{7.3.10}$$

$$B''_p = \frac{-p(1-p)}{4},$$

$$B'''_p = \frac{p(p-1)(2p-1)}{12},$$

$$B^{iv}_p = \frac{(p+1)p(p-1)(p-2)}{48},$$

$$B^v_p = \frac{(p+1)p(p-1)(p-2)(2p-1)}{240}.$$

In central difference notation,

$$f(x_0 + ph) \approx f_0 + p\,\delta f_{1/2} + 2B''_p \mu\, \delta^2 f_{1/2} + B'''_p\, \delta^3 f_{1/2} + 2B^{iv}_p \mu\, \delta^4 f_{1/2} + \dots$$

If the series is broken off after a difference of odd order, $2n - 1$, then the ordinary remainder formula for polynomial interpolation holds; in the above notation,

$$R_{2n} = B^{(2n)}_p \cdot 2h^{2n} f^{(2n)}(\xi), \quad \xi \in [x_{-m}, x_m]. \tag{7.3.11}$$

Thus, from Theorem 7.1.4,

$$R_{2n} \approx \textit{First neglected term in the expansion in Eq. (7.3.10).}$$

If the expansion is broken off after a difference of even order, then the error term is most easily estimated by the first *two* neglected terms. Below we give the values of $\max_{0 \le p \le 1} |B^{(n)}_p|$ and bounds for how large a given difference can be in order that the corresponding term in Eq. (7.3.10) has magnitude $\le \frac{1}{2}U$.

n	2	3	4	5	6
$\max \lvert B^{(n)} \rvert$	0.0625	0.0080	0.0117	0.0009	0.0024
$\max \lvert \Delta^n \rvert$	4	60	20	500	100

(7.3.12)

The coefficients B'', B^{vi}, B^{iv} are tabulated, for example, in Chambers [31, p. 384].

Example 7.3.2

Compute tan 1.322 using the table in Example 7.1.2.

$$p = 0.2; \quad B''_p = -0.04;$$
$$\Delta^2 f_{-1} = 13U; \quad \Delta^2 f_0 = 15U; \quad U = 0.001.$$

$\Delta^3 \approx 2U$ and $\Delta^4 \approx -2U$ can be neglected. Thus

$$\tan 1.322 = 3.903 + 0.2 \cdot 0.169 - 0.04 \cdot (13 + 15) \cdot 0.001 = 3.936.$$

(See [31, p. 198].)

Derivation of Bessel's formula. Newton's general interpolation formula (Theorem 7.3.3) for interpolation in the points $x'_0, x'_1, x'_2, \ldots, x'_m$ is

$$Q(x) = f(x'_0) + \sum_{j=1}^{m} f[x'_0, x'_1, \ldots, x'_j](x - x'_0) \ldots (x - x'_{j-1}).$$

Set, as we have done previously, $x_\nu = x_0 + \nu \cdot h$, but use points x'_0, x'_1, x'_2 according to the ordering:

$$x_0, x_1, x_{-1}, x_2, x_{-2}, x_3, x_{-3}, \ldots$$

Then we have

$$f[x'_0, x'_1, x'_2] = f[x_{-1}, x_0, x_1] = \frac{\Delta^2 f_{-1}}{2!\, h^2},$$
$$f[x'_0, x'_1, x'_2\, x'_3] = f[x_{-1}, x_0, x_1, x_2] = \frac{\Delta^3 f_{-1}}{3!\, h^3},$$
$$(x - x'_0)(x - x'_1)(x - x'_2) = h^3 p(p-1)(p+1),$$
$$(x - x'_0)(x - x'_1)(x - x'_2)(x - x'_3) = h^4 p(p-1)(p+1)(p-2).$$

Recall the symmetry property (Theorem 7.3.4).

From these and similar formulas, it follows that

$$f(x + ph) \approx f_0 + \frac{p}{1!}\Delta f_0 + \frac{p(p-1)}{2!}\Delta^2 f_{-1} + \frac{(p+1)p(p-1)}{3!}\Delta^3 f_{-1}$$
$$+ \frac{(p+1)p(p-1)(p-2)}{4!}\Delta^4 f_{-2} + \ldots \tag{7.3.13}$$

If, instead, we order the points as follows,

$$x_0, x_1, x_2; x_{-1}, x_3, x_{-2}, x_4, \ldots$$

then we get

$$f(x_0 + ph) \approx f_0 + \frac{p}{1!}\Delta f_0 + \frac{p(p-1)}{2!}\Delta^2 f_0 + \frac{p(p-1)(p-2)}{3!}\Delta^3 f_{-1}$$
$$+ \frac{(p+1)p(p-1)(p-2)}{4!}\Delta^4 f_{-1} + \cdots. \qquad (7.3.14$$

If the expansions are truncated after an odd order difference, then one get the *same* interpolation polynomial in Eq. (7.3.13) as in Eq. (7.3.14); the ex- pansions look different, but the same function values have been used. Th same does *not* hold, however, if one truncates the expansions after a differenc of even order. For example, if one truncates after Δ^2, then x_{-1}, x_0, and x are used in Eq. (7.3.13), while x_0, x_1, x_2, are used in Eq. (7.3.14).

The above formulas are sometimes called Newton-Gauss formulas *Bessel's formula is the mean of Eqs.* (*7.3.13*) *and* (*7.3.14*).

7.3.5. Complementary Remarks on Interpolation

Interpolation is of interest as a means of deriving formulas for variou purposes. By differentiating or integrating the interpolation polynomial one gets useful formulas for numerical computation of derivatives and in- tegrals, respectively.

In addition, higher-degree interpolation is of direct practical interest since a table which is constructed so that higher-degree interpolation is pos- sible can be made much smaller than a table for which linear interpolation i intended to be sufficient.

Example 7.3.3

(Taken from [28], exercise 10.5.7.) A table of the function ln x, $1 \leq x \leq 5$ given to five decimals must contain 450 function values if one wishes to in- terpolate linearly, but only 100 table values are needed if one intends to us quadratic interpolation. (Here we have assumed certain reasonable practica restrictions on the choice of step length and how often it is changed, and re- quire that the total error in interpolation shall not exceed 10^{-5}.)

For tables which are used daily, however, the requirement that they b easy to use overweighs the desire for compactness. But for tables of less com- mon functions and ten- or fifteen-place tables, compactness is often th dominant requirement. (Tables of the latter type are occasionally useful i testing computer programs.)

Also, in a digital computer, sparse tables and the use of higher-degree in- terpolation (piecewise) can be a useful alternative when one must deal with a function for which there is no known, easily used formula. In such cases, th need to keep memory requirements low is often more important than th desire to keep the number of arithmetic operations (required to evaluat the given function) at a minimum.

The interpolation problem gives a good illustration of the distinctions made between various types of error in numerical calculation. For the **truncation error** R_T we have strict error estimates—e.g., Theorem 7.3.3 and Eq. (7.3.11). These error estimates are difficult to apply strictly, since they involve higher-order derivatives; these derivatives may be unknown or very tedious to calculate. In practice, such derivatives are often estimated by higher differences according to Theorem 7.3.2, where one sets $x = x_{m+1}$, a point where $f(x_{m+1})$ is known but not used in the interpolation. If the interpolation formula is considered as the first $m + 1$ terms in a longer expansion (where the number of interpolation points changes), then this means that *the remainder term is estimated by the first neglected term.* If the differences in the same column vary greatly, then one should use, in the error estimate, the difference which has the largest absolute value in the given column and which lies within a few rows of the difference in question. Even this precaution, however, does not always give a guaranteed error bound.

If the table is sparse, then it is possible that the remainder term never becomes small, no matter how many terms one takes. In this case, the table gives insufficient information about the function. This is often indicated by a strong variation in the higher differences. However, even this indication can sometimes be absent. An extreme example is given by a table of the function $\sin \pi x$, for $x = 0, 1, 2, \ldots$; all the table values and all the differences are zero. A less trivial example is given by the functions

$$f(x) = \sum_{n=1}^{20} a_n \sin(2\pi nx), \qquad g(x) = \sum_{n=1}^{10} (a_n + a_{10+n}) \sin(2\pi nx).$$

They have identical tables for $x = i/10$ ($i = 0, 1, 2, \ldots$). Hence one gets the same results using interpolation in the two tables, in spite of the fact that $f(x) \neq g(x)$ in general, when $x \neq i/10$.

Many different types of **rounding errors** occur in interpolation:

R_{XA} = Errors in output data arising from error in the **argument**,

R_{XF} = Errors in output data arising from errors in the **table values**,

R_C = Effect of rounding errors during the **computations**.

R_C should be made negligible by taking along an extra decimal in the calculations, if this is not too inconvenient.

Suppose that x has error δx. Then the error in the normalized interpolation argument p is $\delta p = \delta x/h$. Hence

$$R_{XA} = f(x + \delta x) - f(x) \approx f'(x)\,\delta x \approx \Delta f_0 \cdot \delta p. \qquad (7.3.15)$$

THEOREM 7.3.7

If $0 < p < 1$, *and if no difference of order higher than* 5 *is used, then for interpolation with Bessel's formula, we have* $|R_{XF}| \leq 0.7U$. (U = one unit in the last significant decimal of the tabulated values.)

We shall be satisfied to derive a bound which is only a bit coarser than that of the theorem, $0.742U$ instead of $0.7U$. For *linear interpolation* we have, from Eq. (7.3.4),

$$|R_{XF}| \leq 0.5U.$$

Thus, for interpolation, we have generally that

$$\begin{aligned}&(|R_{XF}| \text{ for the result of interpolation})\\ &\qquad \leq 0.5U + (|R_{XF}| \text{ for the terms of higher order}). \qquad (7.3.16)\end{aligned}$$

To estimate $|R_{XF}|$ for the higher-order terms, suppose that f_i has error ϵ_i, $|\epsilon_i| \leq \frac{1}{2}U$. Then, using Theorem 7.1.1,

$$\begin{aligned}(|R_{XF}| \quad \text{for} \quad \Delta^2 f_{-1} + \Delta^2 f_0) &= |(\epsilon_0 - 2\epsilon_1 + \epsilon_2) + (\epsilon_1 - 2\epsilon_2 + \epsilon_3)|\\ &= |\epsilon_0 - \epsilon_1 - \epsilon_2 + \epsilon_3| \leq 4 \cdot \tfrac{1}{2}U = 2U,\\ (|R_{XF}| \quad \text{for} \quad \Delta^3 f_{-1}) &= |\epsilon_0 - 3\epsilon_1 + 3\epsilon_2 - \epsilon_3|\\ &\leq (1 + 3 + 3 + 1) \cdot \tfrac{1}{2}U = 4U,\\ (|R_{XF}| \quad \text{for} \quad \Delta^4 f_{-2} + \Delta^4 f_{-1}) &\leq 6U.\\ (|R_{XF}| \quad \text{for} \quad \Delta^5 f_{-2}) &\leq 16U.\end{aligned}$$

Using the maximal magnitudes of B'', B''', . . . , given in (7.3.12), we get

$$\begin{aligned}&(|R_{XF}| \quad \text{for the result of interpolation with Bessel's formula})\\ &\qquad \leq 0.5U + 0.0625 \cdot 2U + 0.0080 \cdot 4U + 0.0117 \cdot 6U + 0.0009 \cdot 16U\\ &\qquad = 0.742U.\end{aligned}$$

For *Newton's general interpolation formula* the bound $0.7U$ will *not* hold in general, but the formula of (7.3.16) can still be applied. Bounds for the magnitudes of the perturbations in the higher-order terms can be computed recursively, by considering perturbations ϵ_i in the table values, $|\epsilon_i| \leq \frac{1}{2}U$, and computing bounds for the magnitudes of the various differences in the difference scheme for the ϵ_i.

7.3.6. Lagrange's Interpolation Formula

The solution to the interpolation problem defined in Sec. 7.3.1 can be expressed in the following simple formula, *Lagrange's interpolation formula:*

$$Q(x) = \sum_{i=0}^{m} f_i \delta_i(x) \qquad (7.3.17)$$

This formula is more suitable for deriving theoretical results than for practical computation. In Eq. (7.3.17), δ_i is the polynomial of degree m which satisfies the relations

$$\delta_i(x_j) = \begin{cases} 0, & j \neq i \\ 1, & j = i \end{cases} \qquad (7.3.18)$$

for $j = 0, 1, 2, \ldots, m$. Hence

$$\delta_i(x) = \frac{\prod_{\substack{j=0 \\ j \neq i}}^{m} (x - x_j)}{\prod_{\substack{j=0 \\ j \neq i}}^{m} (x_i - x_j)}. \tag{7.3.19}$$

It is left to the reader to show that

$$Q(x_j) = f_j, \quad j = 0, 1, 2, \ldots, m.$$

One interesting thing about Lagrange's formula is that it can easily be generalized for interpolation with classes of functions other than polynomials, for instance, spline functions (see Sec. 4.6). As an exercise, the reader should verify that if one sets

$$\Phi(x) = \prod_{j=0}^{m} (x - x_j), \tag{7.3.20}$$

then

$$\delta_i(x) = \frac{\Phi(x)}{(x - x_i)\Phi'(x_i)}. \tag{7.3.21}$$

7.3.7. Hermite Interpolation

We define divided differences with coincident values of the argument by means of a passage to the limit—e.g.,

$$f[x_0, x_0] = \lim_{x_1 \to x_0} \frac{f(x_1) - f(x_0)}{x_1 - x_0} = f'(x_0),$$

$$f[x_0, x_0, x_1] = \frac{f[x_0, x_0] - f[x_0, x_1]}{x_0 - x_1} = \frac{f'(x_0) - f[x_0, x_1]}{x_0 - x_1}.$$

From Theorem 7.3.2, it follows that

$$\underbrace{f[x_0, x_0, \ldots, x_0]}_{n+1 \text{ arguments}} = \frac{f^{(n)}(x_0)}{n!}. \tag{7.3.22}$$

Newton's interpolation formula can also be used in generalized interpolation problems, where one or more derivatives are known at the interpolation points; this is referred to as *Hermite interpolation*.

Example 7.3.4

Compute a third-degree polynomial for which $Q(0) = 0$, $Q'(0) = 1$, $Q(1) = 3$, $Q'(1) = 6$. In the difference scheme below, the given data are underlined.

	x			
x_0	0	$\underline{0}$		
			$\underline{1}$	
x_1	0	$\underline{0}$		2
			3		1
x_2	1	$\underline{3}$		3
			$\underline{6}$	
x_3	1	$\underline{3}$		

Thus

$$Q(x) = 0 + 1 \cdot x + 2x^2 + 1 \cdot x^2(x - 1).$$

Check, as an exercise, that the value of $Q(\frac{1}{2})$ agrees with the value obtained using Eq. (7.2.2).

7.3.8. Inverse Interpolation

Problem. $f(x)$ is given in a table. Solve the equation $f(x) = c$, assuming that $f(x_0) - c$ and $f(x_1) - c$ have different signs.

If the table is equidistant, then $f(x)$ can be approximated using Bessel's interpolation formula Eq. (7.3.10). Set $x = x_0 + ph$. The equation

$$f_0 + p\,\Delta f_0 + B_p''(\Delta^2 f_{-1} + \Delta^2 f_0) + B_p'''\,\Delta^3 f_{-1} + \dots = c$$

can be written in the form

$$p = \varphi(p),$$

where

$$\varphi(p) = \frac{1}{\Delta f_0}(c - f_0 - B_p''(\Delta^2 f_{-1} + \Delta^2 f_0) - B_p'''\,\Delta^3 f_{-1} - \dots).$$

This equation is then solved by *iteration:*

$$p_0 = \frac{c - f_0}{\Delta f_0}, \qquad p_{i+1} = \varphi(p_i).$$

Notice that the initial approximation (p_0) was obtained by *linear inverse interpolation.*

An error estimate can be obtained using Eq. (6.6.1), with

$$m = \max|\varphi'(p)| \approx \max\left|\frac{dB_p''}{dp} \cdot \frac{\Delta^2 f_{-1} + \Delta^2 f_0}{\Delta f_0}\right| = \frac{\Delta^2 f_{-1} + \Delta^2 f_0}{4\,\Delta f_0}, \qquad (0 \le p \le 1),$$

$$\delta = \frac{|R_A + R_T|}{|\Delta f_0|} + |R_C|,$$

where R_A and R_T are the rounding and truncation errors in the computation of the numerator of $\varphi(p)$, and R_C arises from the division.

Convergence is generally fast, since

$$m \approx \left|\frac{2h^2 f''}{4hf'}\right| = h\left|\frac{f''}{2f'}\right|.$$

If $|f'|$ is small, then the problem is ill-conditioned.

With nonequidistant tables one can proceed analogously to the above, using Newton's general interpolation formula.

The original problem can also be posed in another way. The given table can be seen as a collection of pairs of numbers, where $y_i = f(x_i)$ (see Fig. 7.3.1):

$$\dots, (x_{-1}, y_{-1}), (x_0, y_0), (x_1, y_1), \dots.$$

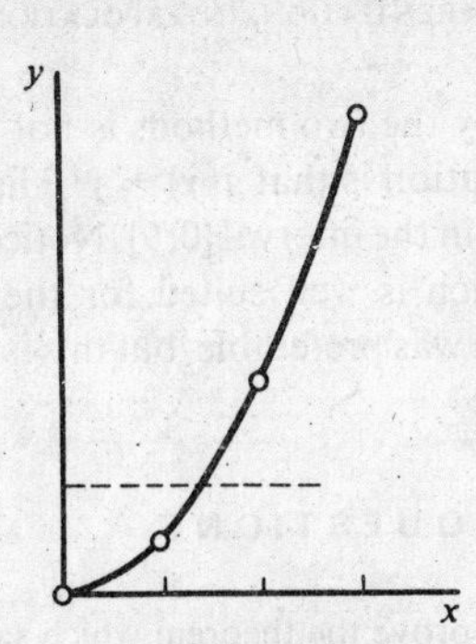

Fig. 7.3.1

This can also be interpreted as a table of $g(y)$, the inverse function of f. If the x-values are equidistant, then the y-values are as a rule not equidistant. One can then solve the problem by applying Newton's interpolation formula to the function $g(y)$. *This procedure is fundamentally different from the one given previously, and gives, as a rule, different results.* The first method gives a polynomial in x, the second a polynomial in y. The two methods are identical only in the case of linear interpolation.

Example 7.3.5

Solve the equation $f(x) = 2$, where $f(x)$ is defined by the following table.

x	$f(x)$	Δ^1	Δ^2
0	0		
		1	
$x_0 = 1$	1		2
		3	
2	4		2
		5	
3	9		

The first process (the iterative process based on Bessel's formula) gives $x = 1.414$. (Verify this!)

The second process gives the following divided differences:

y	$g(y)$			
0	0			
		1		
1	1		$-\frac{1}{6}$	
		$\frac{1}{3}$		$\frac{1}{60}$
4	2		$-\frac{1}{60}$	
		$\frac{1}{5}$		
9	3			

$$g(y) \approx 0 + 1 \cdot y - \tfrac{1}{6}y(y-1) + \tfrac{1}{60}y(y-1)(y-4),$$

$$\therefore \quad x = 2 - \tfrac{2}{6} - \tfrac{4}{60} = 1.600.$$

The difference in the results given by the two methods is not usually so great as in the above example. The explanation is that $g(y) = y^{1/2}$ here, which is not well-approximated by a polynomial in the interval [0, 9]. Notice that $g'(0) = \infty$. On the other hand $f(x) = x^2$, which is well suited for the use of Bessel's formula. Therefore, the first method was preferable, but this is not always true.

REVIEW QUESTIONS

1. Give a precise formulation of and prove the theorem which says that the interpolation problem for polynomials has a unique solution (see Sec. 4.3).
2. When is linear interpolation sufficient?
3. Derive Newton's general interpolation formula and Theorem 7.3.2. (The formula for the remainder term in polynomial interpolation is assumed known.)
4. Derive Newton's interpolation formula for the equidistant case, starting from Newton's general interpolation formula. How is this formula easily remembered?
5. Why is there no contradiction between the warning for equidistant interpolation which was given in Chap. 4 and the "propaganda" for its use given in this chapter?
6. Investigate the effect of round-off errors in the tabulated values on Bessel's formula (degree at most 5), $0 \leq p \leq 1$.
7. Derive the Lagrange interpolation formula.
8. How does one determine a polynomial $Q(x)$, of degree $2n - 1$, such that

$$Q(x_j) = f(x_j), \qquad Q'(x_j) = f'(x_j), \quad j = 1, 2, \ldots, n$$

   Give the formula for the remainder.
9. Describe two methods for inverse interpolation.

PROBLEMS

1. (a) Compute $f(3)$ by quadratic interpolation in the following table.

x	1	2	4	5
$f(x)$	0	2	12	21

   Use the points 1, 2, and 4, and the points 2, 4, and 5, and compare the results.

   (b) Compute $f(3)$ by cubic interpolation.
2. (a) Compute $f(0)$ using Newton's interpolation formula on the following table:

x	0.1	0.2	0.4	0.8
$f(x)$	64,987	62,055	56,074	43,609

The interpolation formula is used here for *extrapolation.*

(b) If the tabulated values have errors not exceeding $\frac{1}{2}$, show that the resulting error in $f(0)$ does not exceed $\frac{7}{2}$.

(c) Compute $f(0)$ by repeated Richardson extrapolation, assuming that

$$f(x) = c_0 + c_1x + c_2x^2 + c_3x^3 + \dots.$$

3. Compute

(a) $\sum_{i=0}^{m} \delta_i(x)$, (b) $\sum_{i=0}^{m} x_i\delta_i(x)$, $(m \geq 1)$,

where $\delta_i(x)$ is the polynomial which is defined in Eq. (7.3.18).

4. Show that

$$f[x_0, x_1, \dots, x_m] = \sum_{i=0}^{m} \frac{f(x_i)}{\Phi'(x_i)},$$

where $\Phi(x)$ is defined by Eq. (7.3.20). *Hint:* Compare the coefficients of x^m in Newton's and Lagrange's expressions for the interpolation polynomial.

5. Show that

$$\begin{bmatrix} 1 & x_0 & x_0^2 & \dots & x_0^m \\ 1 & x_1 & x_1^2 & \dots & x_1^m \\ \dots \\ 1 & x_m & x_m^2 & \dots & x_m^m \end{bmatrix}^{-1} = \begin{bmatrix} c_{00} & c_{01} & \dots & c_{0m} \\ c_{10} & c_{11} & \dots & c_{1m} \\ \dots \\ c_{m0} & c_{m1} & \dots & c_{mm} \end{bmatrix},$$

where $\sum_{i=0}^{m} c_{ij}x^i \equiv \delta_j(x)$. ($\delta_j(x)$ is defined in Eq..(7.3.18).)

6. Determine the fourth-degree polynomial $Q(x)$ for which

$$Q(0) = Q'(0) = 0, \qquad Q(1) = Q'(1) = 1, \qquad Q(2) = 1.$$

7. Verify the result ($x = 1.414$) of the first part of Example 7.3.5 (inverse interpolation based on Bessel's formula).

8. (Aitken's interpolation algorithm) Let

$$P_{i,0} = f(x_i), \quad i = 0, 1, \dots, n \quad (x_i \neq x_j \quad \text{for} \quad i \neq j),$$

and set

$$P_{i,k+1} = \frac{(x - x_k)P_{i,k} - (x - x_i)P_{k,k}}{x_i - x_k},$$

$$i = 0, 1, \dots, n,$$
$$k = 0, 1, \dots, i - 1.$$

Notice that the $P_{i,j}$ ($i = 0, 1, \dots, n$; $j = 0, 1, \dots, i$) form a triangular array and that $P_{i,j}$ is a polynomial of degree $\leq j$.

Show that $P_{n,n}$ is the interpolation polynomial of degree $\leq n$ through $(x_0, f(x_0)), \dots, (x_n, f(x_n))$. *Hint:* Proceed column by column and show that if P_{ij} (for $i \geq j$, with j fixed) interpolates f at $A = \{x_i, x_0, x_1, \dots, x_{j-1}\}$ ($j \geq 1$), $A = \{x_i\}$ for $j = 0$, then $P_{i,j+1}$, $i \geq j + 1$, (as defined above) interpolates f at

$x_i, x_0, x_1, \ldots, x_j$ (and thus $P_{j+1,j+1}$ interpolates f at $x_{j+1}, x_0, x_1, \ldots, x_j$). Th result follows, then, from the case $j = 0$.

9. Several formulas have been suggested for the calculation of $f(x_0 + ph)$ b quadratic interpolation—for example:

   A. Newton, equidistant case with forward differences:

$$f(x_0 + ph) \approx f_0 + p\,\Delta f_0 + \frac{p(p-1)}{2}\Delta^2 f_0.$$

   B. Bessel:

$$f(x_0 + ph) \approx f_0 + p\,\delta f_{1/2} + \frac{p(p-1)}{4}(\delta^2 f_0 + \delta^2 f_1).$$

   C. Newton, general, with f_{-1}, f_0, f_1 as tabular values:

$$f(x_0 + ph) \approx f_0 + (x - x_0)f[x_0, x_1] + (x - x_0)(x - x_1)f[x_0, x_1, x_{-1}$$

   D. Taylor's formula with central differences instead of derivatives (se Example 7.1.8):

$$f(x_0 + ph) \approx f_0 + \frac{f_1 - f_{-1}}{2}p + \frac{f_1 - 2f_0 + f_{-1}}{2}p^2.$$

Test the formulas numerically on the case, $f(x) = \exp(x)$, $x_0 = 0$, $h = 0.$ $p = 0.5$. Which tabular values are requested by the formulas? Which of thes formulas are mathematically equivalent?

7.4. NUMERICAL INTEGRATION

Problem. Compute an approximation to

$$\int_a^b f(x)\,dx$$

by using values of f at equidistant points. (Later in this section we shall touc upon variants of the above problem)

Set $x_0 = a$, $x_i = x_0 + ih$, $x_n = b$; thus $n = (b - a)/h$.

7.4.1. The Rectangle Rule, the Trapezoidal Rule, and Romberg's Method

One obvious way to solve the above problem is to approximate $f(x)$ by simple, easily integrated function. We now study two methods which a based on approximating the function $f(x)$ piecewise by straight lines.

The rectangle rule (Fig. 7.4.1):

$$\int_a^b f(x)\,dx \approx \hat{R}(h) = h\sum_{i=1}^{n} f_{i-1/2}, \tag{7.4.}$$

(Compare the above approximation with the *definition* of the definite int gral.) Notice that Eq. (7.4.1) uses the values of f at the *midpoints* of each su interval (x_{i-1}, x_i). This is essential for good accuracy, as can be seen fro Fig. 7.4.1.

The trapezoidal rule (Fig. 7.4.2). This rule was mentioned already i

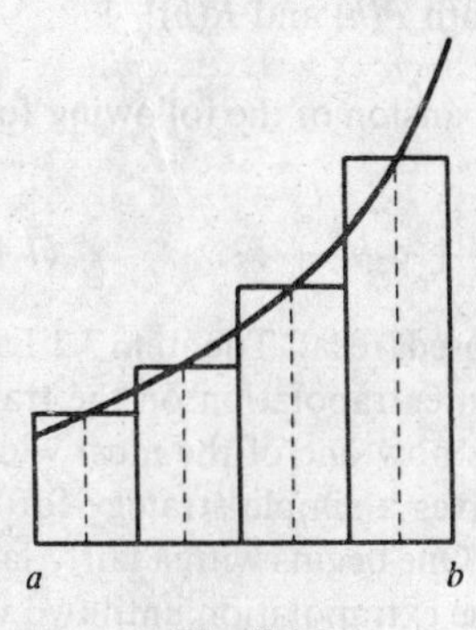

Fig. 7.4.1

Fig. 7.4.2

Sec. 1.2:

$$\int_a^b f(x)\,dx \approx \hat{T}(h) = h(\tfrac{1}{2}f_0 + f_1 + f_2 + \ldots + f_{n-1} + \tfrac{1}{2}f_n). \qquad (7.4.2)$$

This formula is based on piecewise linear interpolation in the intervals (x_{i-1}, x_i), $i = 1, 2, \ldots, n$.

If U = one unit in the last decimal place of the values of the function, then for both formulas we have

$$|R_A| \le h\cdot n\cdot\tfrac{1}{2}U = (b-a)\cdot\tfrac{1}{2}U. \qquad (7.4.3)$$

Verify that

$$\hat{T}(h) = \tfrac{1}{2}(\hat{T}(2h) + \hat{R}(2h)). \qquad (7.4.4)$$

Example 7.4.1

Compute approximately

$$\int_0^{0.8} \frac{\sin x}{x}\,dx.$$

As an exercise, the reader should check some of the rectangle and trapezoidal sums given below. Use Eq. (7.4.4).

x	$f(x)$		
0	1		$\hat{T}(0.8) = 0.758680$
0.1	0.99833	$\hat{R}(0.8) = 0.778840$	
0.2	0.99334		$\hat{T}(0.4) = 0.768760$
0.3	0.98507	$\hat{R}(0.4) = 0.773764$	
0.4	0.97355		$\hat{T}(0.2) = 0.771262$
0.5	0.95885	$\hat{R}(0.2) = 0.772512$	
0.6	0.94107		$\hat{T}(0.1) = 0.771887$
0.7	0.92031		
0.8	0.89670	$\lvert R_A\rvert \le 0.8\cdot\frac{1}{2}\cdot 10^{-5} = 4\cdot 10^{-6}$	

The correct value, to four decimals, is 0.7721. Verify that in this example th error is approximately proportional to h^2 for both $\hat{T}(h)$ and $\hat{R}(h)$!

We shall later (Theorem 7.4.2) derive an expansion of the following for for $\hat{T}(h)$:

$$\hat{T}(h) = \int_a^b f(x)\,dx + a_1 h^2 + a_2 h^4 + a_3 h^6 + \dots, \qquad (7.4..$$

Thus repeated Richardson extrapolation can be used (recall Theorem 7.2.1 an Example 7.2.5). This application of Richardson extrapolation on the trap zoidal rule is known as **Romberg's method.** It is now one of the most wide used methods, since, among other things, it gives a simple strategy for th automatic determination of a suitable step size. One begins with a fairly larg step, and then one can halve the step size and use extrapolation until two va ues in the same column agree to the desired accuracy. With this strategy o must also check to see that the agreement of two values in the same colum is not a chance agreement—i.e., one must make sure that h is so small th the expansion of Eq. (7.4.5) is applicable.

Example 7.4.2. *The Use of Romberg's Method on the Values $\hat{T}(h)$ of Example 7.4*

Concerning the choice of computational precision, see the commen after Eq. (7.2.12).

$$\begin{array}{lcccc}
 & \frac{\Delta}{3} & & \frac{\Delta}{15} & \\
A_{00} = 0.758680 & & & & \\
 & 3{,}360 & & & \\
A_{10} = 0.768760 & & A_{11} = 772{,}120 & & \\
 & 834 & & -1 & \\
A_{20} = 0.771262 & & A_{21} = \underline{772{,}096} & & (772{,}095) \\
 & 208 & & 0 & \\
A_{30} = 0.771887 & & A_{31} = \underline{772{,}095} & & (772{,}095)
\end{array}$$

$A_{21} - A_{31} = 10^{-6}$. We accept A_{32}, and estimate that $|R_T| \le 10^{-6}$. Accordi to Theorem 7.4.1, $|R_A| \le 0.8 \cdot \frac{1}{2} \cdot 10^{-5} = 4 \cdot 10^{-6}$. Thus

$$\int_0^{0.8} \frac{\sin x}{x}\,dx = 0.772095 \pm 0.000005.$$

The correct result is 0.772095.

We shall see that *the second column of the Romberg scheme gives the sar result as Simpson's rule.* For the integral from $x_{2(i-1)}$ to x_{2i} we get, with t trapezoidal rule,

$$\text{with step } 2h: \qquad h(f_{2i-2} + f_{2i}),$$

$$\text{with step } h: \qquad h(\tfrac{1}{2}f_{2i-2} + f_{2i-1} + \tfrac{1}{2}f_{2i}),$$

Thus the second column gives:

$$h\left(\frac{1}{2}f_{2i-2} + f_{2i-1} + \frac{1}{2}f_{2i} + \frac{1}{3}\left(-\frac{1}{2}f_{2i-2} + f_{2i-1} - \frac{1}{2}f_{2i}\right)\right)$$
$$= h\left(\frac{f_{2i-2}}{3} + \frac{4f_{2i-1}}{3} + \frac{f_{2i}}{3}\right),$$

which agrees with Simpson's formula in Example 7.2.2. The second column is thus exact for all third-degree polynomials.

One can show generally, using the Euler-Maclaurin formula (Theorem 7.4.2), that the kth column gives a formula of the type,

$$I_k = ph\sum_{i=1}^{n/p}(\alpha_0 f_{pi} + \alpha_1 f_{pi-1} + \ldots + \alpha_p f_{pi-p}),\ nh = b - a,\ p = 2^{k-1}, \tag{7.4.6}$$

which is exact for all polynomials of degree $2k - 1$. One can also show that

$$\alpha_i > 0, \qquad \alpha_0 + \alpha_1 + \ldots + \alpha_p = 1. \tag{7.4.7}$$

The first of these relations is easy to prove for low values of k. The general proof is, however, more complicated. The second relation follows from the requirement that the formula be exact for $f(x) \equiv 1$; this requires that

$$ph \cdot \frac{n}{p}(\alpha_0 + \alpha_1 + \ldots + \alpha_p) = \int_a^b 1 \cdot dx = (b - a) = nh.$$

THEOREM 7.4.1.

If the magnitude of the error in the function values does not exceed $\frac{1}{2}U$, then for every column of the Romberg scheme we have that $|R_{XF}| \leq (b - a)\cdot\frac{1}{2}U$, independent of h.

Proof. From Eq. (7.4.6) it follows that

$$|R_X| \leq ph\sum_{i=1}^{n/p}\left(|\alpha_0| \cdot \frac{1}{2}U + |\alpha_1|\frac{1}{2}U + \ldots + |\alpha_p|\frac{1}{2}U\right)$$
$$= ph \cdot \frac{n}{p}\frac{1}{2}U\sum_{i=1}^{n/p}|\alpha_i|.$$

The theorem then follows by applying (7.4.7).

7.4.2. The Truncation Error of the Trapezoidal Rule

It is easy to get strict estimates for the truncation error (discretization error) of both the rectangle rule and the trapezoidal rule. Suppose f'' is continuous in $[a, b]$. *For one step* of the trapezoidal rule, we have (using Eq. (7.3.2)—the formula for the truncation error in linear interpolation):

$$\epsilon_i = \hat{T}(h) - \int_{x_{i-1}}^{x_i} f(x)\,dx = -\int_{x_{i-1}}^{x_i} \frac{f''(\xi)}{2}(x - x_{i-1})(x - x_i)\,dx,$$
$$\xi \in [x_{i-1}, x_i] \quad \text{depends on } x.$$

But $(x - x_{i-1})(x - x_i) < 0$ for $x \in [x_{i-1}, x_i]$. Using (a generalized form of the mean-value theorem of integral calculus, we then have

$$\epsilon_i = -\tfrac{1}{2}f''(\xi_i)\int_{x_{i-1}}^{x_i}(x - x_{i-1})(x - x_i)\,dx, \quad \xi_i \in [x_{i-1}, x_i].$$

Setting $x = x_{i-1} + ht$, we get

$$\epsilon_i = -\frac{1}{2}f''(\xi_i)\int_0^1 ht\cdot h(t-1)h\cdot dt = \frac{h^3 f''(\xi_i)}{12}.$$

The **global** truncation error is just the sum of these **local** truncation errors *For the trapezoidal rule:*

$$R_T = \sum_{i=1}^{n}\epsilon_i = \frac{h^3}{12}\sum_{i=1}^{n}f''(\xi_i) = \frac{nh^3 f''(\xi)}{12} = \frac{(b-a)h^2}{12}f''(\xi), \quad \xi \in [a, b]. \tag{7.4.8}$$

For the rectangle rule, we can get analogously—using the remainder term in Taylor's formula—

$$R_T = -\frac{(b-a)h^2}{24}f''(\xi). \tag{7.4.9}$$

Thus the rectangle rule is more accurate than the trapezoidal rule, but the latter is more economical when repeated Richardson extrapolation is used. (Recall that about half of the values of the function needed for $\hat{T}(h)$ were already computed and used in $\hat{T}(2h)$; see Eq. (7.4.4).)

7.4.3. Some Difficulties and Possibilities in Numerical Integration

The extent to which repeated Richardson extrapolation, in connection with numerical integration, is successful depends on how well the function can locally be approximated by a polynomial. It is often profitable to investigate whether or not one can transform or modify the given problem in some way so that the resulting integrand is more suitable for numerical integration.

If the integrand becomes infinite at a point, such a modification is *necessary*. Even if some low-order derivative of the function is infinite at some point in or near the interval of integration, one *should* make such a modification. It is not uncommon that a single step taken close to a point where, for example, the derivative of the integrand is infinite gives a larger error than all the other steps combined, when using the methods previously described.

Some common situations and remedies will now be given. We first consider a few cases where **the integrand has a singularity** or is "almost singular."

Example 7.4.3. *Substitution*

$I = \int_0^1 x^{-1/2}e^x\,dx$. The function is infinite at the origin. Set $x = t^2$.

$$I = \int_0^1 \exp(t^2)\cdot 2\,dt.$$

This integral can be treated without difficulty by, for example, Romberg's method.

Example 7.4.4. *Integration by Parts*

$$I = \int_0^1 x^{-1/2}e^x\,dx = [2x^{1/2}e^x]_0^1 - 2\int_0^1 x^{1/2}e^x\,dx$$
$$= 2e - 2[\tfrac{2}{3}x^{3/2}e^x]_0^1 + \tfrac{4}{3}\int_0^1 x^{3/2}e^x\,dx = \tfrac{2}{3}e + \tfrac{4}{3}\int_0^1 x^{3/2}e^x\,dx.$$

The last integral has a mild singularity at the origin. If one wants high accuracy, then it is advisable to integrate by parts a few more times previous to the numerical treatment.

With the rectangle rule and Romberg's method, a sufficient condition that the method converges as $h \longrightarrow 0$ is that the integrand be continuous, but to get rapid convergence more is required.

Example 7.4.5. *Simple Comparison Problem*

$I = \int_{0.1}^1 x^{-3}e^x\,dx$. The integrand is infinite near the left end point.

$$I = \int_{0.1}^1 x^{-3}\left(1 + x + \frac{x^2}{2}\right)dx + \int_{0.1}^1 x^{-3}\left(e^x - 1 - x - \frac{x^2}{2}\right)dx.$$

The first integral can be computed analytically ("simple comparison problem"). The second integral can be treated numerically. The integrand and all its derivatives are of moderate size. Note, however, the cancellation in the evaluation of the integrand.

It would also work if we computed the integral entirely *with a series expansion* or used *series expansion in a subinterval*—e.g., in [0.1, 0.3]—and then used Romberg's method in [0.3, 1].

Example 7.4.6. *Special Integration Formula*

$\int_0^1 (1 + x)(e^x - 1)^{-1/2}\,dx$. The integrand is of the form $x^{-1/2}f(x)$, where $f(x)$ is well-suited for local approximation by a polynomial. One can then use the method of undetermined coefficients (see Example 7.2.1) to derive a formula for such integrals—e.g.,

$$\int_0^{3h} x^{-1/2}f(x)\,dx = A_0f(0) + A_1f(h) + A_2f(2h) + A_3f(3h),$$

which is exact when $f(x)$ is a third-degree polynomial—and then use Romberg's method on the interval $[3h, 1]$. (The substitution $x = t^2$ also works on the above problem.)

Infinite intervals of integration occur often in practical problems. For integrals of the form

$$\int_{-\infty}^{\infty} f(x)\,dx,$$

the trapezoidal rule or the rectangle rule often give surprisingly good accuracy if one integrates over the interval $[-R_1, R_2]$, assuming that $f(x)$ and its lower derivatives are quite small for $x \leq -R_1$ and $x \geq R_2$.

Example 7.4.7

Compute $\int_{-\infty}^{\infty} \exp(-x^2)\,dx$. For $x = \pm 4$, the integrand is less than $\frac{1}{2}\cdot 10^{-6}$. Using the trapezoidal rule for the integral $\int_{-4}^{4}$ we get the estimate 1.772636 with $h = 1$ and 1.772453 with $h = 0.5$. (The values of the function have been taken from six-place tables.) The correct value is $\pi^{1/2} = 1.772454$. The reason that the trapezoidal rule gives such a good result is explained partially in Sec. 7.4.5(c). The truncation error in the value of the integral is here less than 1/10,000 of the truncation error in the largest term of the trapezoidal sum—a superb example of "cancellation of truncation error"!

The error which is committed when we replace ∞ by 4 can be estimated in the following way:

$$|R| = 2\int_4^{\infty} \exp(-x^2)\,dx = 2\int_{16}^{\infty} \exp(-t)\cdot\tfrac{1}{2}t^{-1/2}\,dt$$

$$< 2\cdot\tfrac{1}{2}16^{-1/2}\int_{16}^{\infty} \exp(-t)\,dt = \tfrac{1}{4}e^{-16} < 10^{-7}.$$

Example 7.4.8

$I = \int_0^{\infty} (1 + x^2)^{-4/3}\,dx$. If one wants five decimals in the result, then $\int_R^{\infty}$ is not negligible until $R \approx 1{,}000$. But one can instead expand the integrand in powers of x^{-1} and integrate termwise,

$$\int_R^{\infty} (1 + x^2)^{-4/3}\,dx = \int x^{-8/3}(1 + x^{-2})^{-4/3}\,dx$$

$$= \int_R^{\infty} (x^{-8/3} - \tfrac{4}{3}x^{-14/3} + \tfrac{14}{9}x^{-20/3} - \dots)\,dx$$

$$= \tfrac{3}{5}R^{-5/3} - \tfrac{4}{11}R^{-11/3} + \tfrac{14}{51}R^{-17/3} - \dots$$

$$= R^{-5/3}(\tfrac{3}{5} - \tfrac{4}{11}R^{-2} + \tfrac{14}{51}R^{-4} - \dots).$$

If this expansion is used, then one need only apply numerical integration on the interval (0, 8).

One can also try some substitution which maps the interval $(0, \infty)$ to (0, 1)—e.g., $t = \exp(-x)$ or $t = 1/(1 + x)$. However, in such cases one must be careful not to introduce an unpleasant singularity into the integrand instead.

Example 7.4.9

The same integral as in Example 7.4.8. Make the substitution $t = 1/(1 + x)$, $I = \int_0^1 [t^2 + (1 - t)^2]^{-4/3} t^{2/3}\,dt$. The integrand now has an infinite derivative at the origin. This singularity is eliminated by making the substitution $t = u^3$. One gets

$$I = \int_0^1 [u^6 + (1 - u^3)^2]^{-4/3} 3u^4 \, du,$$

which can be computed with, for example, Romberg's method.

Integration by parts, the method of a "simple comparison problem," and special integration formulas (derived by the method of undetermined coefficients) are also useful for problems with infinite intervals of integration.

Integrands (or their derivatives) often have strongly varying orders of magnitude in various parts of the interval of integration. One should then choose *different step sizes in different parts of the interval*. Since

$$\int_a^b = \int_a^{c_1} + \int_{c_1}^{c_2} + \ldots + \int_{c_{k-1}}^{c_k} + \int_{c_k}^b,$$

we can treat the integrals on the right-hand side as independent problems, with different choices of step length. It would be a useful exercise for the reader to formulate a strategy for automatization of the choice of subdivision into subintervals, and, within each subinterval, a strategy for the choice of step length.

If the integrand is *oscillating*, then with ordinary integration methods one must choose a step size which is small with respect to the wave length; this is considered to be an irritating limitation in many applications. The techniques previously mentioned (simple comparison problem, special integration formula, etc.) are sometimes effective in such situations. In addition, the following method can be used on integrals of the form

$$I = \int_0^\infty f(x) \sin (g(x)) \, dx,$$

where $g(x)$ is an increasing function, and both $f(x)$ and $g(x)$ can be locally approximated by a polynomial. Set

$$I = \sum_{n=0}^{\infty} (-1)^n u_n, \quad u_n = \int_{x_n}^{x_{n+1}} f(x) |\sin g(x)| \, dx,$$

where $x_0, x_1, x_2, \ldots$ are the successive zeros of $\sin (g(x))$. The convergence of this alternating series can then be improved with the help of repeated averaging; see Sec. 3.2.1.

7.4.4. The Euler-Maclaurin Summation Formula

THEOREM 7.4.2 *Euler-Maclaurin Summation Formula*

Let $\hat{T}(h)$ be the trapezoidal sum which was defined in Eq. (7.4.2). Then

$$\hat{T}(h) = \int_a^b f(x) \, dx + \frac{h^2}{12}[f'(b) - f'(a)] - \frac{h^4}{720}[f'''(b) - f'''(a)] + \frac{h^6}{30{,}240}[f^{v}(b) - f^{v}(a)] + \ldots + c_{2r} h^{2r}[f^{(2r-1)}(b) - f^{(2r-1)}(a)] + \underset{(h \to 0)}{O(h^{2r+2})}. \tag{7.4.10}$$

The coefficients $\{c_{2r}\}$ have the generating function

$$1 + c_2h^2 + c_4h^4 + c_6h^6 + \ldots = \frac{h}{2} \cdot \frac{e^h + 1}{e^h - 1}. \tag{7.4.11}$$

If the expansion on the right of Eq. (7.4.10) is broken off after the term containing derivatives of order $2r - 1$, then the remainder term has the same sign as the first neglected term, assuming that $f^{(2r+2)}(x)$ does not change sign in the interval (x_0, x_n).

If, in addition, the sign of the successive terms is alternating, then we can get an upper and lower bound for the sum; see Theorem 3.1.3. An error estimate that holds without these assumptions is given in formula (7.4.14).

Derivation. We first note that *if* an expansion of the form

$$\hat{T}(h) = \int_a^b f(x)\,dx + \sum_{j=0}^{p-2} c_{j+1}h^{j+1}[f^{(j)}(b) - f^{(j)}(a)] + O(h^p) \tag{7.4.12}$$

exists, then the *generating function* is easily obtained by considering the special case $f(x) = e^x$, $a = 0$, $b = 1$ $(nh = 1)$. Then

$$\hat{T}(h) = h\left(-\frac{1}{2} + \sum_{i=0}^{n-1} e^{ih} + \frac{1}{2}e\right) = h\left(\frac{e^{nh} - 1}{e^h - 1} + \frac{e - 1}{2}\right)$$

$$= \frac{h(e - 1)(e^h + 1)}{2(e^h - 1)}.$$

Thus, Eq. (7.4.12) gives

$$\frac{h(e - 1)(e^h + 1)}{2(e^h - 1)} = e - 1 + \sum_{j=0}^{p-2} c_{j+1}h^{j+1}(e - 1) + O(h^p),$$

or

$$\frac{h}{2}\frac{e^h + 1}{e^h - 1} = 1 + \sum_{j=0}^{p-2} c_{j+1}h^{j+1} + O(h^p).$$

In Example 3.1.7 we showed how the coefficients in this expansion can be computed. By dividing the numerator and denominator of the left-hand side by $e^{h/2}$, we see that the function is an even function of h. *Thus the right-hand side contains only even powers of* h. The coefficients can be expressed with the help of the *Bernoulli numbers* B_r (see [31, p. 387]):

$$c_{2r} = \frac{(-1)^{r+1}B_r}{(2r)!}$$

In order to *prove that an expansion of the form of Eq. (7.4.12) exists*, we shall use a formula for *repeated integration by parts*. This formula is also useful in many other situations. (Another derivation, based on operator techniques, is given in Sec. 7.6).

Suppose that F has p continuous derivatives in $[0, 1]$. Let $G_0, G_1, G_2, \ldots$ be a sequence of continuous functions which satisfy the relations

$$G'_{j+1}(t) = G_j(t), \quad (j = 0, 1, 2, 3, \ldots).$$

Then, using induction, one can easily prove that

$$\int_0^1 F(t)G_0(t)\,dt = \left[\sum_{j=0}^{p-1} (-1)^j F^{(j)}(t)G_{j+1}(t)\right]_{t=0}^{t=1} + (-1)^p \int_0^1 F^{(p)}(t)G_p(t)\,dt.$$

Choose $G_0(t) = 1$. The constants of integration in the integrations which give $G_1, G_2, G_3, \ldots$ are then chosen so that

$$\int_0^1 G_j(t)\,dt = 0 \quad \text{for} \quad j \geq 1.$$

Thus

$$G_1(t) = t - \tfrac{1}{2},$$

$$G_{j+1}(1) = G_{j+1}(0) \quad \text{for} \quad j \geq 1.$$

Now set

$$G_{j+1}(0) = (-1)^{j+1} c_{j+1}.$$

Then we get an expansion (reminiscent of Eq. (7.4.12)),

$$\int_0^1 F(t)\,dt = \tfrac{1}{2}(F(1) + F(0)) - \sum_{j=1}^{p-1} c_{j+1}(F^{(j)}(1) - F^{(j)}(0)) + (-1)^p \int_0^1 F^{(p)}(t)G_p(t)\,dt.$$

The following form is more convenient for a discussion of the remainder (assume p is even):

$$\int_0^1 F(t)\,dt = \tfrac{1}{2}(F(1) + F(0)) - \sum_{j=1}^{p-2} c_{j+1}(F^{(j)}(1) - F^{(j)}(0)) - c_p(F^{(p-1)}(1) - F^{(p-1)}(0)) + \int_0^1 F^{(p)}(t)G_p(t)\,dt.$$

Now set $x = x_{i-1} + ht$, $F(t) = f(x)$. Then $F^{(j)}(t) = h^j f^{(j)}(x)$.

$$\int_{x_{i-1}}^{x_i} f(x)\,dx = h\int_0^1 F(t)\,dt = \tfrac{1}{2}h(f_{i-1} + f_i) - \sum_{j=1}^{p-2} c_{j+1}(f_i^{(j)} - f_{i-1}^{(j)})h^{j+1} - R_i,$$

$$R_i = +hc_p[F^{(p-1)}(1) - F^{(p-1)}(0)] - h\int_0^1 F^{(p)}(t)G_p(t)\,dt = O(h^{p+1}). \tag{7.4.13}$$

Now add these formulas, for $i = 1$ to n. We get

$$\int_a^b f(x)\,dx = \hat{T}(h) - \sum_{j=1}^{p-2} c_{j+1}h^{j+1}(f^{(j)}(b) - f^{(j)}(a)) - \sum_{i=1}^{n} R_i,$$

where $\sum_{i=1}^n R_i = n \cdot O(h^{p+1}) = (b-a)O(h^p)$. This is equivalent to Eq. (7.4.12).

Discussion of the remainder term. From Eq. (7.4.13) it follows that

$$R_i = h\int_0^1 F^{(p)}(t)(c_p - G_p(t))\,dt.$$

If p is even, then $c_p - G_p(t)$ *has the same sign as* c_p. (The proof of this is not easy. One elegant method, which gives many interesting by-products, is to expand $G_1(t)$ in a Fourier series on the interval (0, 1) and then integrate this series $(p-1)$ times; see Problem 3 of Sec. 9.2.) From this it follows that *if* $F^{(p)}(t)$ *does not change sign on* $(0, 1)$, then

$$\text{sign } R_i = \text{sign} \int_0^1 hc_p F^{(p)}(t)\, dt = \text{sign}\,[hc_p(F^{(p-1)}(1) - F^{(p-1)}(0))]$$
$$= \text{sign}\,[c_p h^p (f_i^{(p-1)} - f_{i-1}^{(p-1)})].$$

The assertion (of Theorem 7.4.2) regarding the sign of the remainder term now follows, if we set $p = 2r + 2$ and add, for $i = 1$ to n.

In the same way, one can also get the *error estimate*,

$$\left|\sum_{i=1}^{n} R_i\right| \le |2c_{2r+2}h^{2r+2}| \cdot \int_a^b |f^{(2r+2)}(t)|\, dt. \tag{7.4.14}$$

which holds *without* the assumption that $f^{(2r+2)}$ has constant sign.

7.4.5. Uses of the Euler-Maclaurin Formula

The Euler-Maclaurin formula has many uses:

(a) It is the theoretical basis of Romberg's method. Note that the validity depends on the differentiability properties of f.

(b) It can be used for highly accurate numerical integration when the values of the derivatives of f are known at $x = a$ and $x = b$.

(c) It shows that the trapezoidal rule gives high accuracy when

$$f'(a) = f'(b), \qquad f'''(a) = f'''(b), \qquad f^v(a) = f^v(b), \ldots \tag{7.4.15}$$

The *trapezoidal rule* (and also the rectangle rule) thus *gives high accuracy* when used for integration of *periodic functions over an entire period.* One could be led to believe that the trapezoidal rule gives the exact value of the integral in this situation—all the terms in the remainder given by the Euler-Maclaurin expansion are zero! However, the expansion on the right-hand side of Eq. (7.4.10) does not always converge to the expression on the left-hand side as $r \to \infty$; see Sec. 3.2.5. The bound for the remainder in (7.4.14) shows that the error of the trapezoidal rule need not approach zero *as* $r \to \infty$ for a given h. But in the special case of a periodic function with period $b - a$, it approaches zero *as* $h \to 0$ faster than any power of h.

Similar observations are true for the use of the trapezoidal or rectangle rules for computing integrals of the form

$$\int_{-\infty}^{\infty} f(x)\, dx$$

when the magnitudes of $f(x)$ and its derivatives decrease rapidly as $|x| \to \infty$; see Example 7.4.7. In these two situations, one gains nothing by using *Romberg's method;* there is an enormous *cancellation of errors* in the respective terms.

(d) The most important use of the Euler-Maclaurin summation formula is, however, not numerical integration, but the inverse problem: to *estimate sums*—numerically or asymptotically—when the corresponding integral can be computed in other ways; see Sec. 3.2.2.

Example 7.4.10

Stirling's asymptotic formula for $\ln(M!)$ is:

$$\ln(M!) = K + (M + \tfrac{1}{2}) \ln M - M + \tfrac{1}{12}M^{-1} - \tfrac{1}{360}M^{-3} + \tfrac{1}{1260}M^{-5} - \tfrac{1}{1680}M^{-7} - \dots, \tag{7.4.16}$$

where $K = \lim_{N \to \infty} \ln(N!) - (N + \tfrac{1}{2}) \ln N + N \approx 0.918938$.

It can be shown, using methods of elementary analysis (see Courant [2, p. 363]) that $K = \tfrac{1}{2} \ln 2\pi$. (We shall determine it numerically below.) The successive partial sums are alternately larger and smaller than the correct value. The infinite expansion is in fact divergent for any M, but by stopping at the smallest term one can obtain remarkable accuracy.

To derive the expansion, we use the Euler-Maclaurin formula for the case

$$f(x) = \ln x, \qquad h = 1.$$

Then

$$f^{(2r+1)}(x) = (2r)!\, x^{-2r-1} > 0.$$

Hence the error can be strictly estimated with the help of the first neglected term. Now consider

$$\int_M^N \ln x \, dx = N \ln N - N - M \ln M + M.$$

The Euler-Maclaurin formula gives:

$$\sum_{M+1}^{N} \ln n + \frac{1}{2} \ln M - \frac{1}{2} \ln N = N \ln N - N - M \ln M + M$$

$$- \frac{1}{12}(M^{-1} - N^{-1}) + \frac{2!}{720}(M^{-3} - N^{-3})$$

$$- \frac{4!}{30{,}240}(M^{-5} - N^{-5}) + \frac{6!}{1{,}209{,}600}(M^{-7} - N^{-7}) - \dots.$$

But since

$$\ln(N!) - \ln(M!) = \sum_{n=M+1}^{N} \ln n,$$

we have

$$\ln(N!) - \frac{1}{2} \ln N - N \ln N + N + O(N^{-1})$$

$$= \ln(M!) - \frac{1}{2} \ln M - M \ln M + M - \frac{M^{-1}}{12} + \frac{M^{-3}}{360} - \frac{M^{-5}}{1{,}260} + \dots.$$

The result, Eq. (7.4.16), now follows by letting $N \to \infty$.

Determination of K.

(Choose $M = 10$ in Eq. (7.4.16).)

$$
\begin{array}{rl}
15.104412 & \ln(10!) \\
-23.025851 & 10 \ln 10 \\
1.151293 & \tfrac{1}{2}\ln 10 \\
10.000000 & \\
-\ 0.008333 & -10^{-1}/12 \\
+\ 0.000003 & +10^{-3}/360 \\
\hline
K = 0.918938 &
\end{array}
$$

7.4.6. Other Methods for Numerical Integration

Let $w(x)$ be a given nonnegative integrable function. By integrating the Lagrange interpolation formula (7.3.17) with remainder given by Eq. (7.3.2), one finds that the formula

$$\int_a^b f(x)w(x)\,dx \approx A_0 f_0 + A_1 f_1 + \ldots + A_m f_m, \quad A_i = \int_a^b \delta_i(x)w(x)\,dx \tag{7.4.17}$$

is *exact when $f(x)$ is a polynomial of degree m*, and that the truncation error is

$$|R_T| = \frac{1}{(m+1)!}\left|\int_a^b \Phi(x) f^{(m+1)}(\xi_x) w(x)\,dx\right| \le \frac{1}{(m+1)!}\|f^{(m+1)}\|_\infty \cdot \int_a^b |\Phi(x)|w(x)\,dx$$

where $\Phi(x) = (x - x_0)(x - x_1)\ldots(x - x_m)$.

The coefficients A_i depend only on the weight function w and on the distribution of the points $\{x_i\}_0^m$. In practice, the coefficients are often more easily computed using the method of undetermined coefficients rather than by integrating $\delta_i(x)$. When the points $\{x_i\}$ are equidistant, the integration formulas are called **Cotes formulas.**

In certain cases, Eq. (7.4.17) is exact for all polynomials of degree m' for some $m' > m$. This is true, for example, of Simpson's rule, where $m = 2$, $m' = 3$.

In the following theorem we consider the problem of how to make a good choice of the distribution of the points $x_0, x_1, \ldots, x_m$.

THEOREM 7.4.3 *Gauss Quadrature*

Let $w(x)$ be a positive weight function on the interval $[a, b]$. If $x_0, x_1, \ldots, x_m$ are chosen as the zeros of the polynomial φ_{m+1} of degree $m + 1$ in the family of orthogonal polynomials associated with $w(x)$, then the formula

$$\int_a^b f(x)w(x)\,dx \approx A_0 f_0 + A_1 f_1 + \ldots + A_m f_m,$$

is exact for all polynomials of degree $2m + 1$ or less, if the coefficients are computed as in (7.4.17).

Proof. Let f be a polynomial of degree $2m + 1$ (or lower), and denote by q and r the quotient and remainder, respectively, after the division f/φ_{m+1}, i.e.,

$$f = q\varphi_{m+1} + r,$$

where both q and r are polynomials of (at most) degree m. Then we have

$$\int_a^b f(x)w(x)\,dx = \int_a^b q(x)\varphi_{m+1}(x)w(x)\,dx + \int_a^b r(x)w(x)\,dx$$
$$= \int_a^b r(x)w(x)\,dx$$

because of the orthogonality property. We also have

$$\sum_{i=0}^{m} A_i f_i = \sum_{i=0}^{m} A_i q(x_i)\varphi_{m+1}(x_i) + \sum_{i=0}^{m} A_i r(x_i) = \sum_{i=0}^{m} A_i r(x_i),$$

since x_i is a zero of φ_{m+1}, $i = 0, 1, \ldots, m$. But

$$\int_a^b r(x)w(x)\,dx = \sum_{i=0}^{m} A_i r(x_i),$$

since the coefficients were chosen such that the formula was exact for all polynomials of degree m. This completes the proof.

THEOREM 7.4.4

The coefficients A_i in the Gauss quadrature formulas are positive.

Proof. The formula is exact for $f(x) = [\delta_i(x)]^2$, since this is a polynomial of degree $2m$. But $f_j = 0$ for $j \neq i$. Thus

$$\int_a^b (\delta_i(x))^2 w(x)\,dx = A_i(\delta_i(x_i))^2, \qquad A_i = \frac{\int_a^b \delta_i(x)^2 w(x)\,dx}{[\delta_i(x_i)]^2} > 0.$$

THEOREM 7.4.5

The remainder term in Gauss quadrature is given by the formula

$$\frac{f^{(2m+2)}(\xi)}{(2m+2)!}\int_a^b [\varphi_{m+1}(x)]^2 w(x)\,dx = c_m f^{(2m+2)}(\xi),$$

where

$$\varphi_{m+1} = \prod_{i=0}^{m} (x - x_i).$$

The constant c_m can be determined by applying the formula to some polynomial of degree $2m + 2$.

Proof. Denote by $Q(x)$ the polynomial of degree $2m + 1$ which solves the Hermite interpolation problem (see Sec. 7.3.7), $Q(x_i) = f_i$, $Q'(x_i) = f'_i$, $i = 0, 1, 2, \ldots, m$. The Gauss quadrature formula is exact for $Q(x)$. Hence

$$\int_a^b Q(x)w(x)\,dx = \sum A_i Q(x_i) = \sum A_i f_i.$$

Thus

$$\sum A_i f_i - \int_a^b f(x)w(x)\,dx = \int_a^b (Q(x) - f(x))w(x)\,dx.$$

But since the Hermite interpolation problem is a boundary case (where each point is counted twice) of ordinary interpolation, we have, using Eq. (7.3.2),

$$f(x) - Q(x) = \frac{f^{(2m+2)}(\xi)}{(2m+2)!}[\varphi_{m+1}(x)]^2.$$

From this the theorem follows.

Abscissae and coefficients for Gauss quadrature formulas can be found in, e.g., [27].

Example 7.4.11

Derive a two-point Gauss formula for

$$\int_{-1}^{1} f(x)\,dx.$$

Here $w \equiv 1$, and by Sec. 4.4.4 the relevant orthogonal polynomial ($m = 1$) is the Legendre polynomial $P_2(x) = \frac{1}{2}(3x^2 - 1)$. Hence $x_0 = -3^{-1/2}$, $x_1 = 3^{-1/2}$. The coefficients are determined from the application of the formula to $f(x) = 1$ and $f(x) = x$, respectively, i.e.,

$$A_0 + A_1 = 2$$
$$-3^{-1/2}A_0 + 3^{-1/2}A_1 = 0,$$

with the solution $A_0 = A_1 = 1$. Hence the formula

$$\int_{-1}^{1} f(x)\,dx \approx f(3^{-1/2}) + f(-3^{-1/2}) = f(0.577350) + f(-0.577350)$$

is exact for polynomials of degree less than or equal to $2m + 1 = 3$.

The corresponding formula for the interval $[a, b]$ reads

$$\int_a^b f(x)\,dx \approx \frac{b-a}{2}\left[f\left(\frac{b+a}{2} + \frac{b-a}{2}\cdot 3^{-1/2}\right) + f\left(\frac{b+a}{2} - \frac{b-a}{2}\cdot 3^{-1/2}\right)\right].$$

For instance,

$$\int_{-0.25}^{0.25} \exp(x)\,dx \approx \tfrac{1}{4}[\exp(0.144338) + \exp(-0.144338)] = 0.505215.$$

The exact result is $2\sinh(0.25) = 0.505224$. For comparison, Simpson's rule (with *three* function evaluations) gives 0.505235.

Occasionally one needs formulas where some of the points x_i are fixed while others are to be chosen in an optimal way. In simple cases, such formulas can be derived using the method of undetermined coefficients, where one should first seek the coefficients of the polynomial which is to have the "free" points x_i as its zeros.

REVIEW QUESTIONS

1. Give an account of Romberg's method (its theoretical background and its use).
2. Describe Simpson's rule, its remainder term, and its connection with Romberg's method.
3. Derive a strict remainder term for the trapezoidal rule.
4. Describe at least three different ways to deal with
   (a) singularities,
   (b) infinite interval of integration,
   in numerical integration. Give examples.
5. (a) What pieces of information appear in the Euler-Maclaurin formula? Give the generating function for the coefficients. What do you know about the remainder term?
   (b) Give at least three important uses of the Euler-Maclaurin formula.
6. Give an account of Gauss quadrature formulas (accuracy, how the points are determined, some important property of the coefficients).

PROBLEMS

1. Use Romberg's method to compute the integral

$$\int_0^4 f(x)\,dx,$$

where $f(x)$ is defined by the following table. Need all the values be used? (The values are correctly rounded.)

x	0.0	0.5	1.0	1.5	2.0	2.5	3.0	3.5	4.0
$f(x)$	−4271	−2522	−0499	1795	4358	7187	10279	13633	17247

2. Derive the remainder term (in Eq. (7.4.9)) for the rectangle rule.
3. Compute the integral

$$\frac{1}{2\pi}\int_0^{2\pi} \exp\,(2^{-1/2}\sin x)\,dx$$

by the trapezoidal rule, $h = \pi/2$, $h = \pi/4$.

4. Suppose we have two estimates, $F_1(h)$ and $F_2(h)$, of a quantity $F(0)$. Suppose also that the asymptotic truncation errors are c_1h^p and c_2h^p, respectively, where c_2/c_1 is known.
   (a) Determine a linear combination of $F_1(h)$ and $F_2(h)$ which has a truncation error which is $o(h^p)$. (The symbol "o" was defined in Sec. 2.1.1.)
   (b) Apply the above result to the rectangle rule and the trapezoidal rule. Is the resultant linear combination identical to some known method?

5. Propose a suitable plan (using a computer) for computing the following integrals, for $s = 0.5, 0.6, 0.7, \ldots, 3.0$:

(a) $\int_0^\infty (x^3 + sx)^{-1/2}\, dx$;

(b) $\int_0^\infty (x^2 + 1)^{-1/2} e^{-sx}\, dx$, error $< 10^{-6}$;

(c) $\int_\pi^\infty (s + x)^{-1/3} \sin x\, dx$.

6. (a) Show that the **trapezoidal rule,** with $h = 2\pi/(n + 1)$, is exact for all **trigonometric polynomials** of period 2π—i.e., for functions of the form

$$\sum_{k=-n}^{n} c_k e^{ikt}, \quad (i = \text{Imaginary unit})$$

—when it is used for integration over a whole period.

(b) Show that if $f(t)$ can be approximated by a trigonometric polynomial of degree n so that the magnitude of the error is less than ϵ, ($t \in (0, 2\pi)$), then the error with the use of the trapezoidal rule (with $h = 2\pi/(n + 1)$) on the integral

$$\frac{1}{2\pi}\int_0^{2\pi} f(t)\, dt$$

is less than 2ϵ.

(c) Use the above to explain the sensationally good result in Problem 3 above, when $h = \pi/4$. *Hint:* First use Theorem 4.5.1 to determine how well the function $g(x) = \exp(2^{-1/2}x)$ can be approximated by a seventh-degree algebraic polynomial for $x \in [-1, 1]$.

7. Prove that the formula

$$\int_{-1}^{1} f(x)\, dx \approx \tfrac{1}{9}[5f(\sqrt{0.6}) + 8f(0) + 5f(-\sqrt{0.6})]$$

is exact for polynomials of degree 5, and apply it to the computation of

$$\int_0^1 \frac{\sin x}{1 + x}\, dx.$$

8. (a) Derive a two-point integration formula for integrals of the form

$$\int_{-1}^{1} f(x)(1 + x^2)\, dx,$$

which is exact when $f(x)$ is a polynomial of degree 3. *Hint:* Use the result of Problem 2 in Sec. 4.4 or the method of undetermined coefficients.

(b) Apply the formula to $f(x) = x^4$. Use the result to derive a remainder term.

9. Show that the formula

$$\int_{-1}^{1} f(x)(1 - x^2)^{-1/2}\, dx = \frac{\pi}{n}\sum_{k=1}^{n} f\left(\cos\frac{2k - 1}{2n}\pi\right)$$

is exact for all polynomials of degree $2n - 1$.

10. Euler's constant is defined by

$$\gamma = \lim_{N\to\infty} F(N),$$

where

$$F(N) = 1 + \frac{1}{2} + \frac{1}{3} + \ldots + \frac{1}{N-1} + \frac{1}{2N} - \log N.$$

(a) Show that, for any integer M,

$$\gamma = F(M) + \tfrac{1}{12}M^{-2} - \tfrac{6}{720}M^{-4} + \tfrac{120}{30240}M^{-6} + \ldots,$$

where every other partial sum is larger than γ, and every other is smaller.

(b) Compute γ to seven decimal places, using $M = 10$, $\sum_1^{10} n^{-1} = 2.92896825$, $\ln 10 = 2.30258509$.

(c) Show how Richardson extrapolation can be used to compute γ from the following values:

M	1	2	4	8
$F(M)$	0.50000	0.55685	0.57204	0.57592

7.5. NUMERICAL DIFFERENTIATION

We have already met the problem of computing approximate values for the derivative of a function which is defined by a table, and which is *assumed to be well-suited for local approximation by a polynomial.*

We saw that derivatives can be approximated by differences—the equidistant case in Theorem 7.1.5, the general case (less accurate) in Theorem 7.3.2. In the equidistant case, we saw that repeated Richardson extrapolation could be used to reduce the truncation error; see Example 7.2.5.

The general considerations regarding the derivation of formulas, given in Sec. 7.2.1, are also applicable to numerical differentiation. If $Q(x)$ is a good polynomial approximation to $f(x)$, then $Q'(x)$ can be used as a polynomial approximation to $f'(x)$; however, the accuracy of the values of the derivative is, as a rule, worse than the accuracy of the values of the function obtained in this way.

Example 7.5.1

Set $f(x) = \exp(x)$, $Q(x) = 1 + x + \frac{1}{2}x^2$. Then $f = f' = f'' = f'''$, $Q'(x) = 1 + x$, $Q''(x) = 1$, $Q'''(x) = 0$. For $x \in [-0.1, 0.1]$ we have

$$\max|f(x) - Q(x)| \approx 2\cdot 10^{-4}, \qquad \max|f'(x) - Q'(x)| \approx 5\cdot 10^{-3},$$

$$\max|f''(x) - Q''(x)| \approx 10^{-1}, \qquad \max|f'''(x) - Q'''(x)| \approx 1.$$

In the following two examples, some formulas will be derived by differentiation of the interpolation polynomial. A more elegant derivation (by operator techniques) will be given in Sec. 7.6.

Example 7.5.2

If we set $p = r + \frac{1}{2}$, hence $x = x_0 + (r + \frac{1}{2})h$, then we can write Bessel's interpolation formula Eq. (7.3.10a),

$$f(x) = f_0 + \left(r + \frac{1}{2}\right)\Delta f_0 + \frac{r^2 - \frac{1}{4}}{4}(\Delta^2 f_{-1} + \Delta^2 f_0)$$
$$+ \frac{r(r^2 - \frac{1}{4})}{6}\Delta^3 f_{-1} + \frac{r^4 - 10r^2/4 + 9/16}{48}(\Delta^4 f_{-2} + \Delta^4 f_{-1})$$
$$+ \frac{r^5 - 10r^3/4 + 9r/16}{120}\Delta^5 f_{-2} + \dots$$

Differentiate, $dx/dr = h$,

$$hf'(x) = \Delta f_0 + \frac{r}{2}(\Delta^2 f_{-1} + \Delta^2 f_0) + \frac{3r^2 - \frac{1}{4}}{6}\Delta^3 f_{-1}$$
$$+ \frac{4r^3 - 5r}{48}(\Delta^4 f_{-2} + \Delta^4 f_{-1}) + \frac{5r^4 - 15r^2/2 + 9/16}{120}\Delta^5 f_{-2} + \dots$$

For $r = 0$ we get

$$\begin{aligned} hf'(x_{0.5}) &= \Delta f_0 - \tfrac{1}{24}\Delta^3 f_{-1} + \tfrac{3}{640}\Delta^5 f_{-2} - \dots \\ &= \delta f_{+0.5} - \tfrac{1}{24}\delta^3 f_{+0.5} + \tfrac{3}{640}\delta^5 f_{+0.5} - \dots \end{aligned} \tag{7.5.1}$$

In these expansions, the truncation error can be estimated by the first neglected term. Since round-off errors in the tabulated values cause *large relative errors* in the differences when h is small, one should often use a rather large h in numerical differentiation; see Example 7.5.4.

Example 7.5.3

Another formula for numerical differentiation is given (with $x = x_0 + ph$) by

$$\begin{aligned} hf'(x) = {} & \mu\delta f_0 - \tfrac{1}{6}(1 - 3p^2)\,\mu\delta^3 f_0 + \tfrac{1}{120}(4 - 15p^2 + 5p^4)\,\mu\delta^5 f_0 \\ & + \dots + (p\,\delta^2 f_0 - \tfrac{1}{12}(p - 2p^3)\,\delta^4 f_0 \\ & + \tfrac{1}{360}(4p - 10p^3 + 3p^5)\,\delta^6 f_0 + \dots). \end{aligned} \tag{7.5.2}$$

(This formula can be derived by differentiating Stirling's interpolation formula the derivation of which is similar to that of Bessel's formula in Sec. 7.3.4. One takes the average of Eq. (7.3.13) and Newton's formula with interpolation points $x'_0, x'_1, x'_2, \dots$ according to the ordering (where $x_i = x_0 + ih$):

$$x_0, x_{-1}, x_1, x_{-2}, x_2, x_{-3}, x_3, \dots,$$

The details are left as an exercise to the reader.)

From the previous formula and its derivative we get, after setting $p = 0$

$$hf'_0 = \mu\delta f_0 - \tfrac{1}{6}\mu\delta^3 f_0 + \tfrac{1}{30}\mu\delta^5 f_0 - \tfrac{1}{140}\mu\delta^7 f_0 + \dots, \tag{7.5.3}$$
$$h^2 f''_0 = \delta^2 f_0 - \tfrac{1}{12}\delta^4 f_0 + \tfrac{1}{90}\delta^6 f_0 - \tfrac{1}{560}\delta^8 f_0 + \dots \tag{7.5.4}$$

Here also, we can estimate the truncation error by the first neglected term.

Errors in the values of the function are of much greater importance in numerical differentiation than in interpolation and integration. The higher the order of the derivative, the more the calculations are influenced by cancellation. Suppose that the values of the function have errors whose magnitude does not exceed $\frac{1}{2}U$. Then the error bound for $\mu\delta f_0 = \frac{1}{2}(f_1 - f_{-1})$ is also equal to $\frac{1}{2}U$. By considerations similar to those used in the proof of Theorem 7.3.7, one gets the following error bounds:

	$\mu\delta f_0$	$\mu\delta^3 f_0$	$\mu\delta^5 f_0$	$\mu\delta^7 f_0$
R_{XF}	$0.5U$	$1.5U$	$5U$	$17.5U$

Thus, one gets the bounds $U/(2h)$, $3U/(4h)$, and $11U/(12h)$ for R_{XF} in f_0', if one, two, and three terms, respectively, of Eq. (7.5.3) are used.

Example 7.5.4

Study the truncation error and round-off error when Eq. (7.5.3) is used to compute $f'(3)$, where $f(x) = \ln(x)$, $U = 10^{-6}$. Notice that for the truncation error we have $\mu\delta^{2k+1}f_0 \approx h^{2k+1}f^{(2k+1)}(3) = (2k)!(h/3)^{2k+1}$. (See Example 7.2.5). k = number of terms:

k	1	2	3
R_T	$0.012h^2$	$0.0033h^4$	$0.0024h^6$
R_{XF}	$0.5\cdot 10^{-6}h^{-1}$	$0.75\cdot 10^{-6}h^{-1}$	$0.92\cdot 10^{-6}h^{-1}$

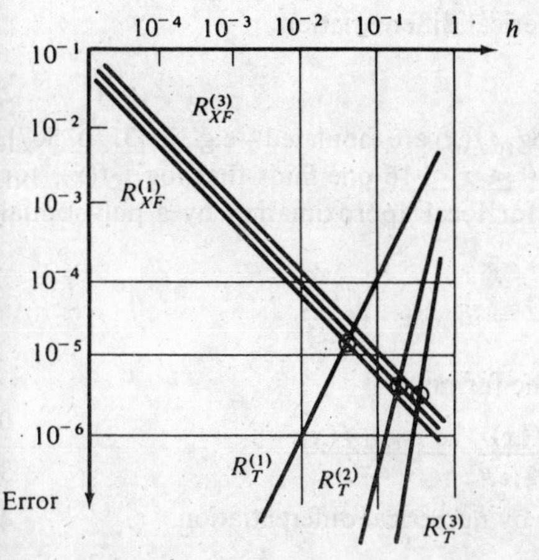

Fig. 7.5.1

Notice that R_T grows with h, whereas R_{XF} is a decreasing function of h. Figure 7.5.1 is a logarithmic diagram of the relation between R_T, R_{XF}, and h, for $k = 1, 2, 3$. The optimal choice of h is that for which R_T and R_{XF} are about equal. Since the R_T-lines are somewhat steeper than the R_{XF}-lines, h should be chosen somewhat to the left of the intersection between the relevant R_{XF}-line and the relevant R_T-line. For $k = 2$, $h = 0.2$ one gets an error bound of about $9 \cdot 10^{-6}$, and not much better for $k = 3$. With $k = 1$, $h = 0.03$ is a good choice—the error bound is about $3 \cdot 10^{-5}$.

Example 7.5.5

To use Richardson extrapolation once, in numerical differentiation (see Example 7.2.5), means that the formula

$$f'_0 \approx \frac{f_1 - f_{-1}}{2h} + \frac{1}{3}\left(\frac{f_1 - f_{-1}}{2h} - \frac{f_2 - f_{-2}}{4h}\right) = \frac{1}{h}\left(\mu\delta f_0 - \frac{1}{6}\mu\delta^3 f_0\right)$$

is used. This is equivalent to taking two terms in the expansion of Eq. (7.5.3). Equation (7.2.8), which was derived by the method of least squares, can be written as

$$f'_0 \approx \frac{1}{h}\left(\mu\delta f_0 + \frac{2}{5}\mu\delta^3 f_0\right).$$

In the two formulas one is using the same data for different purposes: f_2 and f_{-2} are used to reduce truncation error in the first case, while in the second case these values are used to reduce the effect of *random* errors in the values of the function. It is interesting to note that the third-difference term (which can be considered as a correction to the simplest difference approximation for a derivative) has different signs in the two formulas!

By making some simple transformation, one can often get a function which is more suitable than f for local approximation by a polynomial. This principle is easy to apply to numerical differentiation.

Example 7.5.6

The functions $f(x) = x!$ and $\log_{10} f(x)$ are tabulated—e.g., in [31, p. 307]. From the difference scheme for $10 \le x \le 16$ one finds that $\log_{10} f(x)$, for $x \in [10, 16]$, is far better suited for local approximation by a polynomial than is $f(x)$. (Verify this!) Since

$$\frac{d \log_{10} f(x)}{dx} = \log_{10} e \cdot \frac{f'(x)}{f(x)},$$

one should compute $f'(x)$ using the formula

$$f'(x) = \frac{f(x)}{\log_{10} e} \cdot \frac{d \log_{10} f(x)}{dx},$$

where the last factor is computed by numerical differentiation.

REVIEW QUESTION

Discuss how various sources of error influence the choice of step length in numerical differentiation (see Fig. 7.5.1).

PROBLEMS

1. Compute $f'(1)$ using two terms in Eq. (7.5.3), for $f(x) = \cosh(x)$, $h = 0.2$. Compare with the exact value.

2. Carry out the details in the derivation of Eq. (7.5.3).

3. Derive an approximate formula for $f'(x_0)$ when the values $f(x_{-1}), f(x_0), f(x_1)$ are given at three nonequidistant points. Give an approximate remainder term.

4. Let $x_i = x_0 + ih$; let $P(x)$ be the least-squares approximation to f by a second degree polynomial on the grid $\{x_i\}_{i=-2}^{2}$. Show that

$$h^2 f''(x_0) \approx h^2 P''(x_0) = \delta^2 f_0 + \tfrac{2}{7}\delta^4 f_0,$$
$$hP'(x_0) = \tfrac{1}{2}(f_1 - f_{-1}) + \tfrac{1}{5}[f_2 - 2f_1 + 2f_{-1} - f_{-2}],$$
$$P(x_0) = \tfrac{1}{5}\sum f_i - h^2 P''(x_0).$$

7.6. THE CALCULUS OF OPERATORS

7.6.1. Operator Algebra

We have previously indicated the possibility of computing with difference operators. Theorems 7.1.1 and 7.1.2 were written in the form

$$\Delta^k = (E - 1)^k, \qquad E^k = (1 + \Delta)^k, \tag{7.6.1}$$

where the right-hand side is expanded by the binomial theorem. Newton's interpolation formula (equidistant case—Theorem 7.3.6) gave a generalization of the second of these formulas to noninteger values of k.

Formal calculations with operators, using the rules of algebra and analysis, are often an elegant means of assistance in finding approximation formulas. A strict treatment of operator calculus is outside the scope of this book. Formulas which have been *discovered* with these operator methods should afterward be *proved* in some other way. For simplicity, we shall assume that the space of functions on which the operators are defined is $C^\infty(-\infty, \infty)$—i.e., the functions are infinitely differentiable on $(-\infty, +\infty)$. We define the following operators:

$Ef(x) = f(x + h)$	Displacement operator
$\Delta f(x) = f(x + h) - f(x)$	Forward difference operator
$Df(x) = f'(x)$	Differentiation operator
$\delta f(x) = f(x + \frac{1}{2}h) - f(x - \frac{1}{2}h)$	Central difference operator
$\mu f(x) = \frac{1}{2}[f(x + \frac{1}{2}h) + f(x - \frac{1}{2}h)]$	Averaging operator
$\nabla f(x) = f(x) - f(x - h)$	Backward difference operator.

An operator P is said to be a *linear* operator if

$$P(\alpha f + \beta g) = \alpha Pf + \beta Pg$$

holds for arbitrary constants α, β, and arbitrary functions f, g. The above six operators are all linear. The operation of multiplying by a constant α, is also a linear operator.

If P and Q are two operators, then their sum, product, etc. can be defined in the following way:

$$(P + Q)f = Pf + Qf$$
$$(P - Q)f = Pf - Qf$$
$$(PQ)f = P(Qf)$$
$$(\alpha P)f = \alpha(Pf)$$
$$P^n f = P \cdot P \cdot \ldots \cdot Pf, \qquad (n \text{ factors}).$$

Notice that

$$\Delta = E - 1.$$

(Two operators are equal, $P = Q$, if $Pf = Qf$, for all f). One can show that the following rules hold for all linear operators:

$$P + Q = Q + P, \qquad P + (Q + R) = (P + Q) + R,$$
$$P(Q + R) = PQ + PR, \qquad P(QR) = (PQ)R.$$

All of the operators which we shall consider are *commutative*, $PQ = QP$. Also, the rest of the axioms for a *ring* can be shown to hold. From algebra, it is known that the binomial theorem can be strictly derived from the basic axioms for such a ring. Thus when k is a positive integer, the results of Eq. (7.6.1) are special cases of more general theorems in algebra.

7.6.2. Operator Series with Applications

We shall now, however, go outside of algebra and also consider *infinite series of operators*. The Taylor series

$$f(x + h) = f(x) + hf'(x) + \frac{h^2}{2!}f''(x) + \frac{h^3}{3!}f'''(x) + \ldots$$

can be written symbolically as

$$Ef(x) = \left(1 + hD + \frac{(hD)^2}{2!} + \frac{(hD)^3}{3!} + \ldots\right)f(x).$$

Thus, the following theorem.

THEOREM 7.6.1

$$E = e^{hD}. \tag{7.6.2}$$

This beautiful relation is fundamental for the results which follow.

Example 7.6.1

From Eqs. (7.6.1) and (7.6.2) it follows that

$$e^{hD} = 1 + \Delta.$$

Now take the logarithm of both sides! We get

$$\begin{aligned} hD &= \ln(1+\Delta) = \Delta - \tfrac{1}{2}\Delta^2 + \tfrac{1}{3}\Delta^3 - \ldots, \\ hf'(x) &= (\Delta - \tfrac{1}{2}\Delta^2 + \tfrac{1}{3}\Delta^3 - \ldots)f(x). \end{aligned} \tag{7.6.3}$$

This is a typical example of the astounding calculations which can be made with operator series. In order to check the correctness of the result, we first consider the special case $f(x) = e^{\alpha x}$, where α is an arbitrary (complex) constant. Then

$$\begin{aligned} Df(x) &= \alpha e^{\alpha x} = \alpha f(x), \\ \Delta f(x) &= e^{\alpha(x+h)} - e^{\alpha x} = (e^{\alpha h} - 1)f(x), \\ \Delta^p f(x) &= (e^{\alpha h} - 1)^p f(x), \\ \sum_{p=1}^{n} (-1)^p \frac{\Delta^p f(x)}{p} &= \sum_{p=1}^{n} (-1)^p \frac{(e^{\alpha h} - 1)^p}{p} f(x). \end{aligned}$$

If $|e^{\alpha h} - 1| < 1$, then as $n \to \infty$ the right-hand side converges to

$$\ln(1 + e^{\alpha h} - 1)f(x) = \alpha h e^{\alpha x} = hDf(x).$$

Thus, Eq. (7.6.3) holds for $f(x) = e^{\alpha x}$, assuming that $|e^{\alpha h} - 1| < 1$, i.e.,

$$hDe^{\alpha x} = (\Delta - \tfrac{1}{2}\Delta^2 + \tfrac{1}{3}\Delta^3 \ldots)e^{\alpha x}. \tag{7.6.4}$$

We shall also show that *the relation of Eq. (7.6.3) holds for all polynomials.* In this case, the expansion on the right-hand side has only a finite number of terms. *The first n terms give, then, a formula for numerical differentiation which is exact for all polynomials of degree n. This is characteristic of formulas which are derived with the operator method.*

Now expand the two sides of Eq. (7.6.4) into powers of α, and compare coefficients of $\alpha^p/p!$. We get

$$hDx^p = (\Delta - \tfrac{1}{2}\Delta^2 + \tfrac{1}{3}\Delta^3 - \ldots)x^p, \qquad p = 0, 1, 2, \ldots. \tag{7.6.5}$$

(One can show that the inversion of the order of summation presents no principal difficulties.) By linear combination of these equations for various values of p, then, it follows that Eq. (7.6.3) is valid for all polynomials, which was to be shown.

In fact, we also have from Eq. (7.6.4) the result that Eq. (7.6.3) *holds for all functions of the form*

$$f(x) = \sum_{j=0}^{n} c_j \exp(\alpha_j x), \quad \text{if} \quad |\exp(\alpha_j h) - 1| < 1.$$

The methods in Example 7.6.1 can often be used to get at the real content of relations between operators. Many examples of useful relations between operators can be found in [7, Chap. 7]. See also the problems at the end of this section.

Example 7.6.2. *Central Difference Formulas for Numerical Differentiation*

From the definitions in Sec. 7.6.1 and Eq. (7.6.2) we have

$$\delta f(x) = f(x + \tfrac{1}{2}h) - f(x - \tfrac{1}{2}h) = (e^{hD/2} - e^{-hD/2})f(x),$$

i.e.,

$$\delta = 2 \sinh (\tfrac{1}{2}hD). \tag{7.6.6}$$

From, e.g., Chambers [31, p. 204],

$$x = \sinh x - \frac{1}{2}\frac{\sinh^3 x}{3} + \frac{1 \cdot 3}{2 \cdot 4}\frac{\sinh^5 x}{5} - \frac{1 \cdot 3 \cdot 5}{2 \cdot 4 \cdot 6}\frac{\sinh^7 x}{7} + \cdots.$$

Hence

$$\frac{1}{2}hD = \frac{\delta}{2} - \frac{1}{2}\frac{(\delta/2)^3}{3} + \frac{1 \cdot 3}{2 \cdot 4}\frac{(\delta/2)^5}{5} - \frac{1 \cdot 3 \cdot 5}{2 \cdot 4 \cdot 6}\frac{(\delta/2)^7}{7} + \cdots$$

or

$$hD = \delta - \frac{\delta^3}{24} + \frac{3\delta^5}{640} - \frac{5\delta^7}{7{,}168} + \cdots. \tag{7.6.7}$$

See Eq. (7.5.1). By squaring the above relation, one gets

$$(hD)^2 = \delta^2 - \frac{\delta^4}{12} + \frac{\delta^6}{90} - \frac{\delta^8}{560} + \cdots.$$

See Eq. (7.5.4). Using the same reasoning as in Example 7.6.1, one can convince oneself that the above relations hold for all polynomials.

Equation (7.6.7) cannot, however, be used to compute f'_0, since $\delta f_0 = f_{0.5} - f_{-0.5}$, $\delta^3 f_0$ etc. are not directly defined by the table of values of the function. This difficulty is overcome by using the next theorem, and multiplying the right-hand side by

$$\mu\left(1 + \frac{1}{4}\delta^2\right)^{-1/2} = \mu\left(1 - \frac{\delta^2}{8} + \frac{3\delta^4}{128} - \frac{5\delta^6}{1{,}024} + \cdots\right). \tag{7.6.8}$$

THEOREM 7.6.2

$$\mu(1 + \tfrac{1}{4}\delta^2)^{-1/2} = 1.$$

Proof. From the definitions in Sec. 7.6.1 we have

$$\mu f(x) = \frac{1}{2}f\left(x + \frac{1}{2}h\right) + \frac{1}{2}f\left(x - \frac{1}{2}h\right)$$

$$= \left[\frac{1}{2}e^{hD/2} + \frac{1}{2}e^{-hD/2}\right]f(x) = \cosh\left(\frac{hD}{2}\right)f(x).$$

Hence

$$\mu = \cosh\frac{hD}{2}, \qquad \frac{\delta}{2} = \sinh\frac{hD}{2}$$

from Eq. (7.6.6). Since

$$\cosh^2 \alpha - \sinh^2 \alpha = 1,$$

then

$$\mu^2 - \tfrac{1}{4}\delta^2 = 1$$

or

$$\mu^2 = 1 + \tfrac{1}{4}\delta^2,$$

from which the theorem follows.

Example 7.6.3

From Eqs. (7.6.7) and (7.6.8) we have

$$hD = \mu\left(1 - \frac{\delta^2}{8} + \frac{3\delta^4}{128} + \dots\right)\left(\delta - \frac{\delta^3}{24} + \frac{3\delta^5}{640} + \dots\right)$$

$$= \mu\delta - \frac{\mu\delta^3}{6} + \frac{\mu\delta^5}{30} + \dots .$$

See Eq. (7.5.3).

Example 7.6.4. *An Operator Derivation of the Euler-Maclaurin Formula (see Sec. 7.4.4)*

The trapezoidal sum, Eq. (7.4.2),

$$\hat{T}(h) = h(\tfrac{1}{2}f_0 + f_1 + f_2 + \dots + f_{n-1} + \tfrac{1}{2}f_n),$$

can be written in operator form as

$$\hat{T}(h) = h(\tfrac{1}{2} + e^{hD} + e^{2hD} + \dots + e^{(n-1)hD} + \tfrac{1}{2}e^{nhD})f(x_0)$$

$$= h(-\tfrac{1}{2} + \sum_{j=0}^{n-1} e^{jhD} + \tfrac{1}{2}e^{nhD})f(x_0).$$

Using the formula for the sum of a geometric series,

$$\hat{T}(h) = \left(-\frac{1}{2} + \frac{e^{nhD} - 1}{e^{hD} - 1} + \frac{1}{2}e^{nhD}\right)hf(x_0)$$

$$= \left(\frac{1}{e^{hD} - 1} + \frac{1}{2}\right)(e^{nhD} - 1)hf(x_0) = \frac{e^{hD} + 1}{2(e^{hD} - 1)}(e^{nhD} - 1)hf(x_0).$$

Now set (see Example 3.1.7)

$$\frac{t}{2}\frac{e^t + 1}{e^t - 1} = 1 + \sum_{j=1}^{\infty} c_j t^j,$$

$$\hat{T}(h) = (1 + \sum_{j=1}^{\infty} c_j (hD)^j)\frac{e^{nhD} - 1}{hD}hf(x_0)$$

$$= \frac{e^{nhD} - 1}{D}f(x_0) + \sum_{j=1}^{\infty} c_j (hD)^{j-1}(e^{nhD} - 1)hf(x_0).$$

As an exercise, the reader should verify that this means that

$$\hat{T}(h) = \int_{x_0}^{x_n} f(x)\,dx + \sum_{j=1}^{\infty} c_j h^j (f^{(j-1)}(x_n) - f^{(j-1)}(x_0)),$$

which motivates the approach taken in the proof of Theorem 7.4.2.

The following table (adapted from an article written by W. G. Bickley in 1948) summarizes some useful relations between operators:

	E	Δ	δ	∇	D
E	E	$1+\Delta$	$1+\frac{1}{2}\delta^2+\delta\sqrt{1+\frac{1}{4}\delta^2}$	$\frac{1}{1-\nabla}$	e^{hD}
Δ	$E-1$	Δ	$\delta\sqrt{1+\frac{1}{4}\delta^2}+\frac{1}{2}\delta^2$	$\frac{\nabla}{1-\nabla}$	$e^{hD}-1$
δ	$E^{1/2}-E^{-1/2}$	$\Delta(1+\Delta)^{-1/2}$	δ	$\nabla(1-\nabla)^{-1/2}$	$2\sinh\frac{1}{2}hD$
∇	$1-E^{-1}$	$\frac{\Delta}{1+\Delta}$	$\delta\sqrt{1+\frac{1}{4}\delta^2}-\frac{1}{2}\delta^2$	∇	$1-e^{-hD}$
D	$\frac{1}{h}\ln E$	$\frac{1}{h}\ln(1+\Delta)$	$\frac{2}{h}\sinh^{-1}\frac{1}{2}\delta$	$\frac{1}{h}\ln\left(\frac{1}{1-\nabla}\right)$	D
μ	$\frac{1}{2}(E^{1/2}+E^{-1/2})$	$\left(1+\frac{1}{2}\Delta\right)(1+\Delta)^{-1/2}$	$\sqrt{1+\frac{1}{4}\delta^2}$	$\left(1-\frac{1}{2}\nabla\right)(1-\nabla)^{-1/2}$	$\cosh\frac{1}{2}hD$

REVIEW QUESTIONS

1. How does Taylor's series look in operator form?
2. Expand $(hD)^2$ in powers of δ^2.
3. Derive the Euler-Maclaurin summation formula (formally, without remainder term) with the help of operators.

PROBLEMS

1. Determine a such that the formula

$$h^{-2}(y_{n+1} - 2y_n + y_{n-1}) = y_n'' + a(y_{n+1}'' - 2y_n'' + y_{n-1}'')$$

is exact for polynomials of as high degree as possible. Determine an asymptotic expression for the error.

2. Define the operators Δ and Δ_1 by the equations

$$\Delta f(x) = f(x+h) - f(x),$$

$$\Delta_1 f(x) = f\left(x + \frac{h}{5}\right) - f(x), \qquad \text{(temporary notation).}$$

(a) Determine coefficients c_{rs} for $r + s \leq 3$ in the expansions

$$\Delta_1^r = c_{r0}\Delta^r + c_{r1}\Delta^{r+1} + c_{r2}\Delta^{r+2} + \ldots.$$

(b) Investigate how formulas of this type can be used to tabulate a function with a step length which is $\frac{1}{5}$ the step length in a given table (so-called **subtabulation**).

(c) Use the formulas to get values of e^x for $x = 0.01, 0.02, 0.03, 0.04$, given a table of e^x (to six decimals) at $x = 0, 0.05, 0.10, 0.15$. The magnitude of the error should not exceed 10^{-6}. (Assume that $\Delta^4 f_{-1}$ is zero in the interval.)
Note: Analogous formulas with central differences are even more practical.

3. (a) Derive **Euler's transformation**,

$$\sum_{j=n}^{\infty} u_j z^j = \frac{z^n}{1-z} \sum_{s=0}^{\infty} \left(\frac{z}{1-z}\right)^s \Delta^s u_n,$$

formally, using operators. A discussion of convergence is not required; z may be complex.

(b) If $u_j = a^j$, $z = e^{i\varphi}$, $\varphi \in [0, \pi]$, for which real values of $a \in [0, 1]$ does the series on the right converge faster than the series on the left?

(c) Set

$$M_{n,k} = \sum_{j=0}^{n-1} u_j z^j + \frac{z^n}{1-z} \sum_{s=0}^{k-1} \left(\frac{z}{1-z}\right)^s \Delta^s u_n.$$

When $n = 0$ or $k = 0$, the corresponding sums have the value zero. Show that

$$M_{n,k+1} = \frac{M_{n+1,k} - zM_{n,k}}{1-z}$$

$$= M_{n,k} + \frac{M_{n+1,k} - M_{n,k}}{1-z}, \quad M_{n,0} = \sum_{j=0}^{n-1} u_j z^j.$$

(d) Does the algorithm which one gets in the special case $z = -1$ in (c) seem familiar?

(e) Show that if $|z| \le 1$ and $\sum u_j z^j$ converges, and if $\Delta^{k-1} u_j$ tends monotonically to zero as $j \to \infty$, then

$$\left|M_{n,k} - \sum_{0}^{\infty} u_j z^j\right| < |M_{n,k} - M_{n,k-1}| \cdot |z|.$$

Hint: Show first, by induction in p, that

$$M_{n+p,k} - M_{n,k} = \left(\frac{z}{1-z}\right)^k \cdot \sum_{j=n}^{n+p-1} z^j \Delta^k u_j.$$

4. The following relations are of interest in the numerical integration of ordinary differential equations. Derive the generating function and compute the first three coefficients of each expansion.

(a) Adams-Bashforth formula:

$$y_{n+1} - y_n = h(a_0 y'_n + a_1 \nabla y'_n + a_2 \nabla^2 y'_n + \ldots).$$

(b) Adams-Moulton formula:

$$y_{n+1} - y_n = h(b_0 y'_{n+1} + b_1 \nabla y'_{n+1} + b_2 \nabla^2 y'_{n+1} + \ldots).$$

5. (a) A sequence $x_0, x_1, \ldots, x_N$ is given. Show that if

$$y_0 = x_0, \quad y_n = (-\Delta)^n x_0, \quad (n = 1, 2, \ldots, N),$$

then

$$x_0 = y_0, \quad x_n = (-\Delta)^n y_0, \quad (n = 1, 2, \ldots, N).$$

(b) A matrix $A = [a_{ij}]$, $0 \leq i, j \leq N$, is given, where

$$a_{ij} = \begin{cases} (-1)^j \binom{i}{j} & \text{for } 0 \leq j \leq i \leq N, \ \binom{i}{0} = 1 \text{ for } i \geq 0, \\ 0 & \text{for } i < j. \end{cases}$$

Show that $A^{-1} = A$.

7.7 FUNCTIONS OF SEVERAL VARIABLES

The ideas of numerical integration can be generalized to multiple integrals, but the amount of work will increase rapidly with the number of dimensions. It is therefore advisable to try to reduce the number of dimensions by applying analytic techniques to parts of the task.

Example 7.7.1

The following triple integral can be reduced to a single integral:

$$\int_0^\infty \int_0^\infty \int_0^\infty e^{-(x+y+z)} \sin(xz) \sin(yx)\, dx\, dy\, dz$$

$$= \int_0^\infty e^{-x}\, dx \int_0^\infty e^{-y} \sin(yx)\, dy \int_0^\infty e^{-z} \sin(zx)\, dz$$

$$= \int_0^\infty \left(\frac{x}{1+x^2}\right)^2 e^{-x}\, dx.$$

because

$$\int_0^\infty e^{-z} \sin(zx)\, dz = \int_0^\infty e^{-y} \sin(yx)\, dy = \frac{x}{1+x^2}.$$

The remaining single integral is simply evaluated by the techniques previously studied. (See Examples 7.4.8 and 3.2.6.)

Often a transformation of variables is needed for such a reduction (see Problem 1 at the end of this section), but sometimes that does not help either. For simplicity we restrict ourselves to the two-dimensional case, although the ideas are more general. Several approaches are then possible, such as:

(a) numerical integration in one direction at a time—see Sec. 7.7.1;
(b) the use of a rectangular grid, mainly if the boundary of the region of integration is composed of straight lines—see Sec. 7.7.2;
(c) the use of an irregular triangular grid—possible for more general boundaries—see Sec. 7.7.3;
(d) Monte Carlo method, mainly for problems with complicated boundaries and a large number of dimensions—see Chap. 11.

We shall give examples of the first three approaches to numerical integration, as well as to interpolation and differentiation.

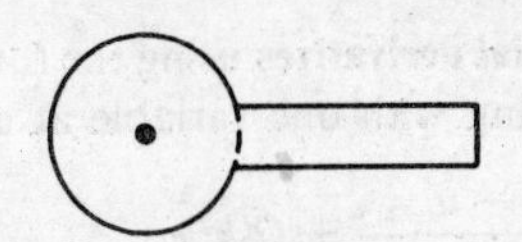

Fig. 7.7.1

7.7.1. Working with One Variable at a Time

Example 7.7.2

Compute

$$I = \iint_D \sin^2 y \cdot \sin^2 x \cdot (1 + x^2 + y^2)^{-1/2}\, dx\, dy,$$

where $D = \{(x, y): x^2 + y^2 \leq 1\} \cup \{(x, y): 1 \leq x \leq 3, |y| \leq 0.5\}$. Put

$$f(x) = \begin{cases} (1 - x^2)^{1/2}, & x \leq \tfrac{1}{2}\sqrt{3} \\ \tfrac{1}{2}, & x \geq \tfrac{1}{2}\sqrt{3}. \end{cases}$$

$$\varphi(x) = \int_{-f(x)}^{f(x)} \sin^2 y \cdot (1 + x^2 + y^2)^{-1/2}\, dy. \tag{7.7.1}$$

Then

$$J = \int_{-1}^{3} \varphi(x) \sin^2 x\, dx. \tag{7.7.2}$$

The values of $\varphi(x)$ are obtained by the application of Romberg's method to Eq. (7.7.1), and numerical integration applied to Eq. (7.7.2) yields the value $J = .13202 \pm 10^{-5}$. Ninety-six values of x are needed, and for each value of x, twenty function evaluations are needed, on the average. The grid is chosen so that $x = \frac{1}{2}\sqrt{3}$, where $\varphi'(x)$ is discontinuous, is a grid point.

A similar idea can also be used for other numerical problems. For example, the following formula for **bilinear interpolation,**

$$u(x_0 + ph, y_0 + qk) \approx \varphi(y_0) + q[\varphi(y_0 + k) - \varphi(y_0)], \tag{7.7.3}$$

where

$$\varphi(y) = u(x_0, y) + p[u(x_0 + h, y) - u(x_0, y)],$$

is exact for all functions of the form $u(x, y) = a + bx + cy + dxy$. When $0 \leq p \leq 1$, $0 \leq q \leq 1$, we have the following error bound (see Sec. 7.3.2):

$$\max_{(x,y)\in R} \tfrac{1}{2}[p(1 - p)h^2 |u''_{xx}| + q(1 - q)k^2 |u''_{yy}|],$$

where R is the rectangle $\{(x, y): x_0 \leq x \leq x_0 + h,\ y_0 \leq y \leq y_0 + k\}$.

7.7.2. Rectangular Grids

A **rectangular grid** in the (x, y)-plane with **grid spacings** h, k in the x and y directions, respectively consists of points

$$x_i = x_0 + ih, \qquad y_j = y_0 + jk.$$

Difference approximations for **partial derivatives** using the function values on the grid can be obtained by working with one variable at a time—e.g.,

$$\frac{\partial u}{\partial x} = \frac{u_{i+1,j} - u_{i-1,j}}{2h} + O(h^2),$$
$$\frac{\partial u}{\partial y} = \frac{u_{i,j+1} - u_{i,j-1}}{2k} + O(k^2). \tag{7.7.4}$$

Unsymmetric difference approximations are also used occasionally—e.g.,

$$\frac{\partial u}{\partial x} = \frac{u_{i+1,j} - u_{i,j}}{h} + O(h). \tag{7.7.5}$$

A differential operator of great importance in physics is the *Laplacean* ∇^2,

$$\nabla^2 u = \frac{\partial^2 u}{\partial x^2} + \frac{\partial^2 u}{\partial y^2}. \tag{7.7.6}$$

(Note that ∇ does *not* mean backward difference operator here, but it means the gradient operator $(\partial/\partial x, \partial/\partial y)$.)

∇^2 can be approximated by difference operators in many useful ways. *Suppose* $h = k$. The simplest approximation of the Laplacean is then the **five-point operator** ∇_5^2,

$$\nabla_5^2 u_{ij} \equiv \frac{u_{i+1,j} - 2u_{ij} + u_{i-1,j}}{h^2} + \frac{u_{i,j+1} - 2u_{ij} + u_{i,j-1}}{h^2}. \tag{7.7.7}$$

Figure 7.7.2 gives a so-called **computational molecule** for $h^2\nabla_5^2$. From Eq. (7.1.12) it follows that

$$\nabla_5^2 u_{ij} = \nabla^2 u + \frac{1}{12}h^2\left(\frac{\partial^4 u}{\partial x^4} + \frac{\partial^4 u}{\partial y^4}\right) + O(h^4).$$

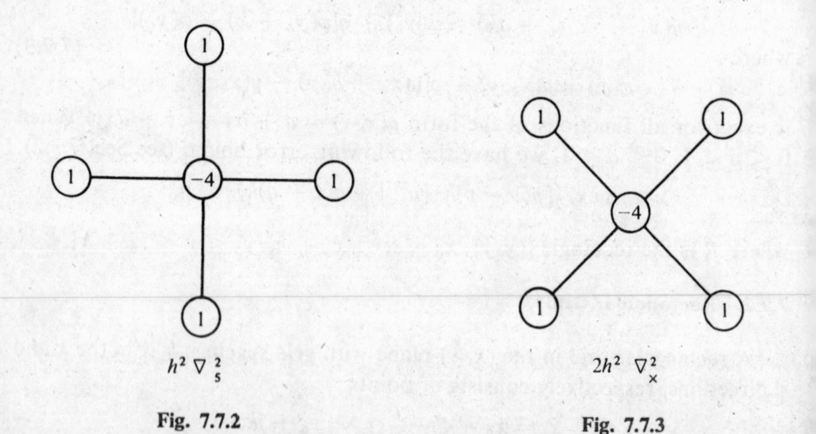

Fig. 7.7.2

Fig. 7.7.3

Another difference operator, denoted $\nabla_\times^2$ (see Fig. 7.7.3), is

$$\nabla_\times^2 u_{ij} \equiv \frac{u_{i+1,j+1} + u_{i-1,j-1} + u_{i+1,j-1} + u_{i-1,j+1} - 4u_{ij}}{2h^2}.$$

One can show that

$$\nabla_\times^2 u_{ij} = \nabla^2 u + h^2\left[\frac{1}{2}\frac{\partial^4 u}{\partial x^2\,\partial y^2} + \frac{1}{12}\left(\frac{\partial^4 u}{\partial x^4} + \frac{\partial^4 u}{\partial y^4}\right)\right] + O(h^4).$$

By a linear combination we get the following approximation, the *nine-point operator* ∇_9^2:

$$\nabla_9^2 = \tfrac{2}{3}\nabla_5^2 + \tfrac{1}{3}\nabla_\times^2.$$

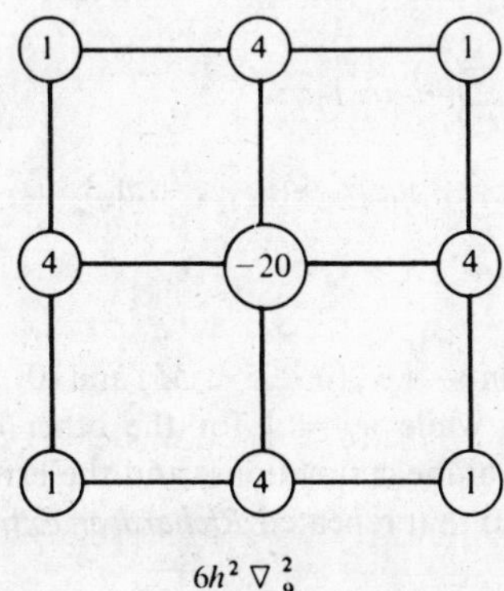

Fig. 7.7.4

Verify that the computational molecule is as shown in Fig. 7.7.4. This formula is advantageous in many situations (see the Problems following Sec. 8.5) because of the special form of the remainder term

$$\nabla_9^2 u_{ij} = \nabla^2 u + \tfrac{1}{12}h^2\nabla^4 u + O(h^4).$$

$$\nabla^4 = \left(\frac{\partial^2}{\partial x^2} + \frac{\partial^2}{\partial y^2}\right)^2 = \frac{\partial^4}{\partial x^4} + \frac{2\partial^4}{\partial x^2\,\partial y^2} + \frac{\partial^4}{\partial y^4}. \tag{7.7.8}$$

For more complicated differential expressions it is often easy to construct $O(h^2)$-approximations. For example (see Eqs. (8.4.9) and (8.4.10)),

$$f(x, y)\frac{\partial}{\partial y}\left(g(x, y)\frac{\partial u}{\partial y}\right) = \pi(u) + O(h^2), \tag{7.7.9}$$

where the difference operator π is defined by

$$k^2\pi(u)_{ij} \equiv f_{ij}[g_{i,j+1/2}(u_{i,j+1} - u_{ij}) - g_{i,j-1/2}(u_{ij} - u_{i,j-1})],$$

(and analogously for derivatives with respect to x).

These formulas have important applications to the numerical solution of partial differential equations.

Consider a **double integral** over a rectangular region (Fig. 7.7.5). Let $x_0 = a$, $h = (b - a)/M$, $y_0 = c$, $k = (d - c)/N$, $u_{ij} = u(x_i, y_j)$. Then the following

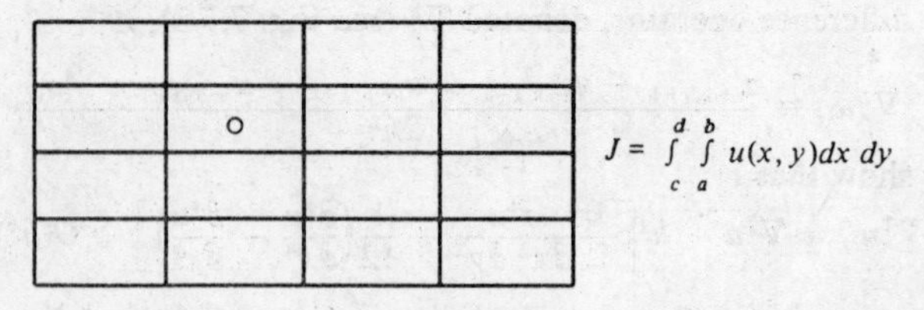

Fig. 7.7.5

formulas can be used, generalizing the rectangle rule and the trapezoidal rule, respectively:

$$J \approx hk \sum_{i=1}^{M} \sum_{j=1}^{N} u_{i-1/2,\,j-1/2}, \tag{7.7.10}$$

$$\begin{aligned} J &\approx hk \sum_{i=1}^{M} \sum_{j=1}^{N} \tfrac{1}{4}[u_{i-1,j-1} + u_{i-1,j} + u_{i,j-1} + u_{ij}] \\ &= hk \sum_{i=0}^{M} \sum_{j=0}^{N} w_{ij} u_{ij}. \end{aligned} \tag{7.7.11}$$

Here $w_{ij} = 1$ for the interior grid points—i.e., $(0 < i < M)$ and $(0 < j < N)$, $w_{ij} = \frac{1}{4}$ for the four corner points, while $w_{ij} = \frac{1}{2}$ for the other boundary points. Both formulas are exact for bilinear functions, and the error can be expanded into even powers of h, k so that repeated *Richardson extrapolation* can be used.

Formulas of higher accuracy can also be obtained by *operator techniques* based on an operator formulation of Taylor's expansion (see Theorem 7.6.1):

$$u(x_0 + h, y_0 + k) = \exp\left(h\frac{\partial}{\partial x} + k\frac{\partial}{\partial y}\right)u(x_0, y_0). \tag{7.7.12}$$

See also Problem 6 in this section.

For nonrectangular regions, the rectangular lattice may be bordered by triangles or "triangles" with one curved side, which may be treated with the techniques outlined in the next section.

7.7.3. Irregular Triangular Grids

A grid of triangles of arbitrary form is a convenient means for approximating a complicated plane region. It is fairly easy to program a computer to refine a coarse triangular grid automatically; see Fig. 7.7.6. It is also easy to adapt the density of points to the behaviour of the function.

Triangular grids are thus more flexible than rectangular ones. On the other hand, the administration of a rectangular grid requires less storage and a simpler program than a triangular, and sometimes the approximation formulas are also a little simpler. In recent years triangular grids have obtained an important application in the **finite-element method** for problems in

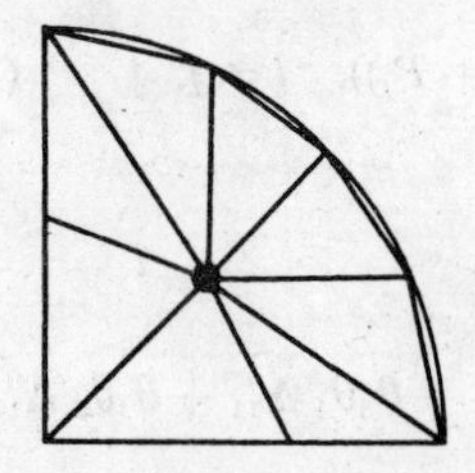
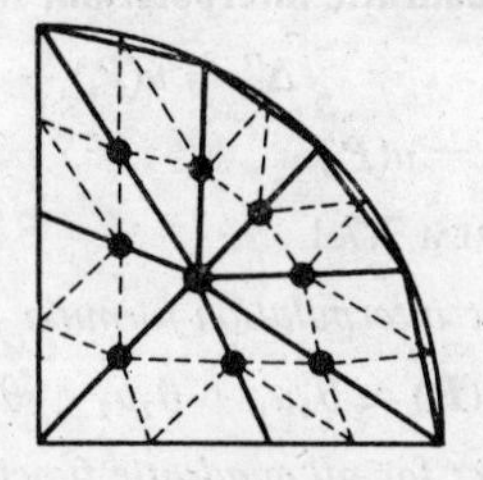

Fig. 7.7.6

continuum mechanics and other applications of partial differential equations; see Sec. 8.6.5. Let $P_i = (x_i, y_i)$, $i = 1, 2, 3$, be the vertices of a triangle T. Then any point $P = (x, y)$ in the plane can be uniquely expressed by the vector equation

$$P = \theta_1 P_1 + \theta_2 P_2 + \theta_3 P_3 \quad \text{where} \quad \theta_1 + \theta_2 + \theta_3 = 1. \tag{7.7.13}$$

In fact, the θ_i, which are called the **barycentric coordinates** of P, are determined from the following nonsingular set of equations:

$$\begin{aligned} \theta_1 x_1 + \theta_2 x_2 + \theta_3 x_3 &= x, \\ \theta_1 y_1 + \theta_2 y_2 + \theta_3 y_3 &= y, \\ \theta_1 \quad + \theta_2 \quad + \theta_3 \quad &= 1. \end{aligned} \tag{7.7.14}$$

The interior of the triangle is characterized by the inequalities

$$\theta_i > 0, \quad i = 1, 2, 3.$$

In this case P is the center of mass (centroid) of the three masses $\theta_1, \theta_2, \theta_3$ located at the vertices of the triangle. (This explains the term "barycentric coordinates.") $\theta_1 = 0$ is the equation for the side P_2P_3, and similarly for the other sides.

If u is a **nonhomogeneous linear function** of P—i.e., if $u(P) = a^T P + b$—then the reader can verify that

$$u(P) = \theta_1 u(P_1) + \theta_2 u(P_2) + \theta_3 u(P_3). \tag{7.7.15}$$

This is a formula for **linear interpolation** in triangular grids. In order to ob-

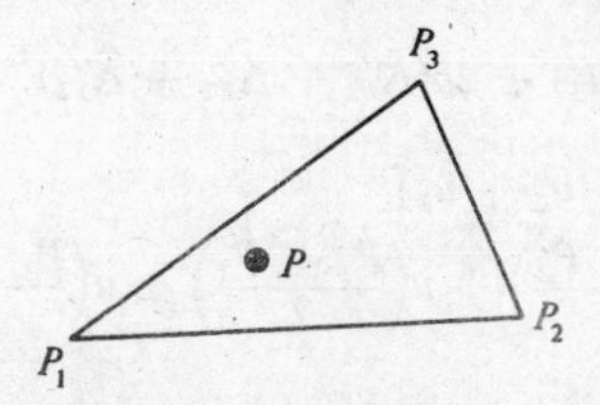

Fig. 7.7.7

tain **quadratic interpolation,** we define

$$\Delta''_{ij} = u(\boldsymbol{P}_i) + u(\boldsymbol{P}_j) - 2u(\tfrac{1}{2}(\boldsymbol{P}_i + \boldsymbol{P}_j)), \quad i \neq j. \tag{7.7.16}$$

Let $u_i = u(\boldsymbol{P}_i)$.

THEOREM 7.7.1

The interpolation formula

$$u(\boldsymbol{P}) \approx \theta_1 u_1 + \theta_2 u_2 + \theta_3 u_3 - 2(\theta_2\theta_3\, \Delta''_{23} + \theta_3\theta_1\, \Delta''_{31} + \theta_1\theta_2\, \Delta''_{12})$$

is exact for all quadratic functions.

Proof. The right-hand side is a quadratic function of $\boldsymbol{P}$, since it follows from Eq. (7.7.14) that the θ_i are (nonhomogeneous) linear functions of x, y. (See also Problem 8.) It remains to show that the right-hand side (RHS) is equal to $u(\boldsymbol{P})$ for $\boldsymbol{P} = \boldsymbol{P}_i$ and $\boldsymbol{P} = (\boldsymbol{P}_i + \boldsymbol{P}_j)/2$, $i, j = 1, 2, 3$.

For $\boldsymbol{P} = \boldsymbol{P}_i$,

$$\theta_i = 1, \theta_j = 0, \qquad j \neq i,$$

hence RHS $= u_i$. For $\boldsymbol{P} = (\boldsymbol{P}_i + \boldsymbol{P}_j)/2$,

$$\theta_i = \theta_j = \tfrac{1}{2}, \theta_k = 0 \qquad (k \neq i, k \neq j),$$

and hence

$$\text{RHS} = \tfrac{1}{2}u_i + \tfrac{1}{2}u_j - 2\cdot\tfrac{1}{2}\cdot\tfrac{1}{2}[u_i + u_j - 2u(\tfrac{1}{2}(\boldsymbol{P}_i + \boldsymbol{P}_j))] = u(\tfrac{1}{2}(\boldsymbol{P}_i + \boldsymbol{P}_j)).$$

The following theorem is equivalent to a rule which has been used in mechanics for the computation of moments of inertia since the nineteenth century (usually formulated in terms of equimomental systems of mass points).

THEOREM 7.7.2

The integration formula

$$\iint_T u(x, y)\, dx\, dy = \frac{1}{3}A \cdot \left(u\left(\frac{\boldsymbol{P}_1 + \boldsymbol{P}_2}{2}\right) + u\left(\frac{\boldsymbol{P}_2 + \boldsymbol{P}_3}{2}\right) + u\left(\frac{\boldsymbol{P}_3 + \boldsymbol{P}_1}{2}\right)\right) \tag{7.7.17}$$

(where A = area of T) *is exact for all quadratic functions.*

Proof. By symmetry, $\iint_T \theta_i\, dx\, dy$ is the same for $i = 1, 2, 3$, and similarly $\iint_T \theta_i\theta_j\, dx\, dy$ is the same for all the three (i, j)-combinations. Hence for a quadratic function,

$$\begin{aligned}\iint_T u(x, y)\, dx\, dy &= a(u_1 + u_2 + u_3) - 2b(\Delta''_{23} + \Delta''_{31} + \Delta''_{12}) \\ &= (a - 4b)(u_1 + u_2 + u_3) \\ &\quad + 4b\left[u\left(\frac{\boldsymbol{P}_2 + \boldsymbol{P}_3}{2}\right) + u\left(\frac{\boldsymbol{P}_3 + \boldsymbol{P}_1}{2}\right) + u\left(\frac{\boldsymbol{P}_1 + \boldsymbol{P}_2}{2}\right)\right],\end{aligned} \tag{7.7.18}$$

here

$$a = \iint_T \theta_1 \, dx \, dy, \qquad b = \iint_T \theta_1\theta_2 \, dx \, dy.$$

'e shall introduce θ_1, θ_2 as new variables of integration. By Eq. (7.7.14) and ıe relation $\theta_3 = 1 - \theta_1 - \theta_2$,

$$\begin{aligned} x &= \theta_1(x_1 - x_3) + \theta_2(x_2 - x_3) + x_3, \\ y &= \theta_1(y_1 - y_3) + \theta_2(y_2 - y_3) + y_3. \end{aligned} \tag{7.7.19}$$

[ence the functional determinant is equal to

$$\begin{vmatrix} x_1 - x_3 & x_2 - x_3 \\ y_1 - y_3 & y_2 - y_3 \end{vmatrix} = 2A.$$

[ence (check the limits of integration!)

$$a = \int_{\theta_1=0}^{1} \int_{\theta_2=0}^{1-\theta_1} \theta_1 \cdot 2A \, d\theta_1 \, d\theta_2 = 2A \int_0^1 \theta_1(1 - \theta_1) \, d\theta_1 = \frac{A}{3},$$

$$b = \int_{\theta_1=0}^{1} \int_{\theta_2=0}^{1-\theta_1} \theta_1\theta_2 \cdot 2A \, d\theta_1 \, d\theta_2 = 2A \int_0^1 \theta_1 \frac{(1 - \theta_1)^2}{2} \, d\theta_1 = \frac{A}{12}.$$

'he result now follows by the insertion of this into Eq. (7.7.18).

Numerical method. Cover the region by triangles, apply Eq. (7.7.17) to ach of them and add the results. For each curved boundary segment (Fig. .7.8), the correction

$$\tfrac{4}{3}u(S) \cdot (\text{Area of triangle } PRQ) \tag{7.7.20}$$

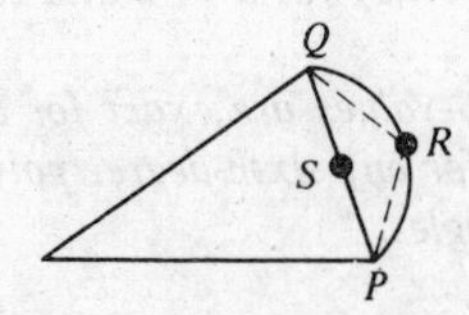

Fig. 7.7.8

; to be added. The error of this correction can be shown to be $O(\| Q - P \|^5)$ or each segment, if R is close to the midpoint of the arc PQ. If the boundary ; given in parametric form,

$$x = x(t), \qquad y = y(t),$$

'here x and y are differentiable twice on the arc PQ, then one should choose $_R = \frac{1}{2}(t_P + t_Q)$. Concerning the use of **Richardson extrapolation,** see the ex- mples.

Example 7.7.3

Consider the integral

$$I = \iint (x^2 + y^2)^k \, dx \, dy$$

over the region shown in Fig. 7.7.9. Let I_n be the result obtained with n trian-gles. The grid for I_4 and I_{16} are shown in Fig. 7.7.9. Put

$$\begin{aligned} R'_n &= I_{4n} + \tfrac{1}{15}(I_{4n} - I_n), \\ R''_n &= R'_{4n} + \tfrac{1}{63}(R'_{4n} - R'_n). \end{aligned} \tag{7.7.21}$$

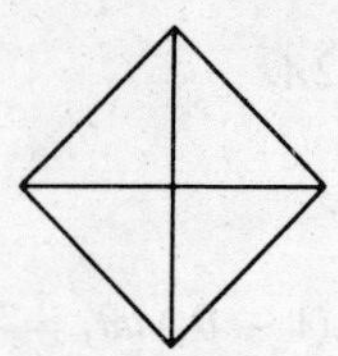

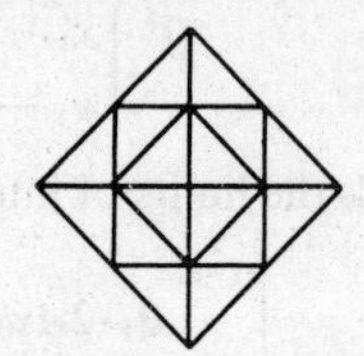

Fig. 7.7.9

The following results were obtained. In this case the work could be reduced by a factor of 4, because of symmetry.

k	I_4	I_{16}	I_{64}	R'_4	R'_{16}	R''_4	Exact
2	22.5/90	27.6562/90	27.9785/90	28.0000/90	28.0000/90	28.0000/90	28/90
3	0.104167	0.161784	0.170741	0.165625	0.171338	0.171429	0.171429
4	0.046875	0.0906779	0.104094	0.093598	0.104988	0.105169	0.105397

It is seen that R'-values have full accuracy for $k = 2$ and that the R''-values have high accuracy even for $k = 4$.

In fact, it can be shown that *R'-values are exact for any fourth-degree polynomial and R''-values are exact for any sixth-degree polynomial, when the region is covered exactly by the triangles.*

Example 7.7.4

The integral

$$a \cdot \iint (a^2 - y^2)^{-1/2} y \, dx \, dy$$

over a quarter of the unit circle is computed with the grids shown in Fig. 7.7.6, and with boundary corrections according to Eq. (7.7.20). The following results (using the notation of the previous example) were obtained and compared with the exact values:

a	I_8	I_{32}	R'_8	Exact
2	0.351995	0.352077	0.352082	0.352082
4	0.337492	0.337608	0.337615	0.337616
6	0.335084	0.335200	0.335207	0.335208
8	0.334259	0.334374	0.334382	0.334382

Note, however, that Richardson extrapolation may not always give improvement. There are cases analogous to the one-dimensional ones mentioned in Sec. 7.4.5(c), where the rate of convergence of the I_n is more rapid than usual.

For a more thorough treatment of multiple integrals the reader is referred to a recent book by Stroud [114].

REVIEW QUESTIONS

1. How is bilinear interpolation performed? What is the order of accuracy?
2. Give the five-point formula for the Laplacian operator. What is the order of accuracy?
3. Define barycentric coordinates, and give the formula for linear interpolation in a triangular grid.
4. Describe the methods for numerical integration with rectangular or triangular grids.

PROBLEMS

1. Let D be the unit circle. Introduce polar coordinates in the integral

$$I = \iint_D \frac{y \sin(ky)\, dx\, dy}{x^2 + y^2}$$

and reduce it analytically to a single integral.

2. Let E be the ellipse $\{(x, y) : x^2/a^2 + y^2/b^2 \leq 1\}$. Transform

$$I = \iint_E u(x, y)\, dx\, dy$$

into an integral over a rectangle in the (r, t)-plane with the transformation

$$x = ar \cos t, \qquad y = br \sin t.$$

3. Look up the algorithm D1 198 of the list given in Sec. 13.9. How is the call to be done for the solution of Example 7.7.2? The singularity at $x = \frac{1}{2}\sqrt{3}$ is to be taken into account.

4. Compute by bilinear interpolation $u(0.5, 0.25)$ when

$$u(0, 0) = 1, \quad u(1, 0) = 2,$$
$$u(0, 1) = 3, \quad u(1, 1) = 5.$$

5. (a) Derive a formula for $u''_{xy}(0, 0)$ using u_{ij}, $|i| \leq 1$, $|j| \leq 1$, which is vali for all quadratic functions.

(b) Show that the interpolation formula

$$\begin{aligned} u(x_0 + ph, y_0 + qk) \approx u_{0,0} &+ \tfrac{1}{2}p(u_{1,0} - u_{-1,0}) + \tfrac{1}{2}q(u_{0,1} - u_{0,-1}) \\ &+ \tfrac{1}{2}p^2(u_{1,0} - 2u_{0,0} + u_{-1,0}) \\ &+ \tfrac{1}{4}pq(u_{1,1} - u_{1,-1} - u_{-1,1} + u_{-1,-1}) \\ &+ \tfrac{1}{2}q^2(u_{0,1} - 2u_{0,0} + u_{0,-1}) \end{aligned}$$

is valid for all quadratic functions.

6. (a) Verify (7.7.8) and the computation molecule in Fig. 7.7.4.

(b) Show–for example, by operator techniques—that

$$\sum_{i=1}^{6} u_i - 6u_0 = \frac{3h^2}{2}\nabla^2 u_0 + \frac{3h^4}{32}\nabla^4 u_0 + O(h^6),$$

where ∇^2 = Laplacian, and the u_i are the values in the neighbors P_i, $i =$ 1, 2, . . . , 6 of P_0 in a regular hexagonal lattice with side h.

$\circ P_3$ $\quad$ $\circ P_2$

$\circ P_4$ $\quad$ $\circ P_0$ $\quad$ $\circ P_1$

$\circ P_5$ $\quad$ $\circ P_6$

$\longleftarrow h \longrightarrow$

7. Show that, using the notation for rectangular grids, the formula

$$\int_{x_0-h}^{x_0+h}\int_{y_0-k}^{y_0+k} u(x, y)\,dx\,dy = \frac{4hk}{6}[u_{1,0} + u_{0,1} + u_{-1,0} + u_{0,-1} + 2u_{0,0}]$$

is exact for all cubic polynomials.

8. If θ_i and θ'_i $(i = 1, 2, 3)$ are the barycentric coordinates of P and P', respec tively, what are the barycentric coordinates of $\alpha P + (1 - \alpha)P'$?

9. Is a quadratic polynomial uniquely determined, given six function values a the vertices and midpoints of the sides of a triangle?

10. Show that the boundary correction of Eq. (7.7.20) is exact, if $u \equiv 1$ and if the arc is a parabola where the tangent at R is parallel to PQ.

11. Formulate generalizations to several dimensions of the theorems of Sec. 7.7 and convince yourself of their validity. *Hint:* The integral formula is mos simply expressed in terms of the values in the vertices and in the centroid of a simplex.

2. Show that the Laplacian of the interpolation polynomial of Theorem 7.7.1 is equal to

$$\nabla^2 u = \frac{c_1 \Delta''_{23} + c_2 \Delta''_{31} + c_3 \Delta''_{12}}{a^2},$$

where a = the area of the triangle $P_1 P_2 P_3$, and

$$c_1 = (x_2 - x_1)(x_3 - x_1) + (y_2 - y_1)(y_3 - y_1).$$

The other c_i are obtained by cyclic permutation of subscripts. *Hint:* Express the θ_i in terms of x, y.

8 DIFFERENTIAL EQUATIONS

8.1. THEORETICAL BACKGROUND

8.1.1. Initial-Value Problems for Ordinary Differential Equations

To start with, we shall study the following problem: given a function $f(x, y)$, find a function $y(x)$ which for $a \leq x \leq b$ is an approximate solution to the **ordinary differential equation**

$$\frac{dy}{dx} = f(x, y) \quad \text{with the } \textbf{initial condition} \quad y(a) = c. \tag{8.1.1}$$

In mathematics courses, one learns how to determine the exact solution to this problem for certain special choices of the function f. An important special case is when f is a linear function of y. However, for most differential equations, one must be content with approximate solutions.

Many problems in science and technology lead to differential equations. Often, the variable x means time, and the differential equation expresses the rule or law of nature which governs the changes in the system being studied. However, x can equally often be a spatial coordinate. It is quite seldom, though, that one meets just one first-order differential equation. As a rule, one has a *system of many first-order differential equations* for many unknown functions (often more than ten), say $\eta_1, \eta_2, \ldots, \eta_s$, where

$$\frac{d\eta_i}{dx} = \varphi_i(x, \eta_1, \eta_2, \ldots, \eta_s), \quad (i = 1, 2, \ldots, s)$$

with initial conditions

$$\eta_i(a) = \gamma_i \quad (i = 1, 2, \ldots, s).$$

t is convenient to write such a **system** in **vector form**. Set

$$y = (\eta_1, \eta_2, \ldots, \eta_s)^T, \qquad c = (\gamma_1, \gamma_2, \ldots, \gamma_s)^T,$$
$$f = (\varphi_1, \varphi_2, \ldots, \varphi_s)^T.$$

Then the problem can be written as:

$$\frac{dy}{dx} = f(x, y), \quad y(a) = c. \tag{8.1.2}$$

When the vector form is used, it is just as easy to describe numerical methods for systems as it is for just one single equation.

Even a *higher-order differential equation can be rewritten as a system of first-order equations.*

Example 8.1.1

The differential equation

$$\frac{d^3y}{dx^3} = g\left(x, y, \frac{dy}{dx}, \frac{d^2y}{dx^2}\right)$$

with initial condition

$$y(0) = \gamma_1, \qquad y'(0) = \gamma_2, \qquad y''(0) = \gamma_3,$$

s, by the substitution

$$\eta_1 = y, \qquad \eta_2 = \frac{dy}{dx}, \qquad \eta_3 = \frac{d^2y}{dx^2},$$

transformed into the system

$$\frac{d\eta_1}{dx} = \eta_2, \qquad \eta_1(0) = \gamma_1,$$

$$\frac{d\eta_2}{dx} = \eta_3, \qquad \eta_2(0) = \gamma_2,$$

$$\frac{d\eta_3}{dx} = g(x, \eta_1, \eta_2, \eta_3), \quad \eta_3(0) = \gamma_3.$$

Thus systems of first-order equations are a suitable normal form even for initial-value problems for higher-order differential equations.

If x denotes *time*, then the differential equation, Eq. (8.1.2), determines the velocity vector of a particle as a function of the position vector **y**. Thus the differential equation defines a *velocity field.* A two-dimensional example is shown in Fig. 8.1.1. The solution of the equation describes the motion of a particle in that field. This interpretation is directly generalizable to three dimensions, and these geometric ideas are also suggestive for systems of more than three equations.

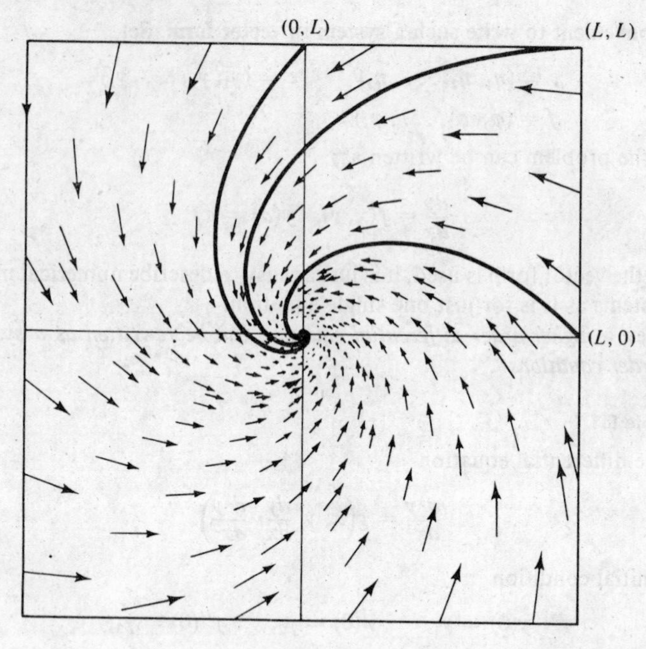

Fig. 8.1.1

Example 8.1.2

Consider the system

$$\frac{d\eta_1}{dt} = -(\eta_1 + \eta_2), \quad \eta_1(0) = A_1,$$

$$\frac{d\eta_2}{dt} = -(\eta_2 - \eta_1), \quad \eta_2(0) = A_2.$$

For various initial conditions, we get a whole *family of solution curves*. Severa such curves are shown in Fig. 8.1.1 (for $(A_1, A_2) = (L, 0), (L, L/2), (L, L)$ $(L/2, L), (0, L)$).

We shall *assume* that the functions φ_i are differentiable with respect to x and η_j $(i, j = 1, 2, \ldots, s)$ for all $x \in [a, b]$ and for all points y in a certain domain D that contains c as an interior point. In textbooks on differential equations—e.g., [90]—it is shown that under these assumptions the *initial-value problem defined by Eq. (8.1.2) has exactly one solution* (as long as $y(x) \in D$). In other words, the motion of the particle in a given velocity field is uniquely determined by its initial position. As long as the particle remains in the region where the regularity assumptions hold (see Example 8.1.2), the truth of this statement become very plausible when one considers Fig. 8.1.1;

he picture also suggests that if one chooses sufficiently small time steps, then he rule "Displacement = Time step × Mean velocity" can be used for a tep-by-step construction of an approximate solution. In fact, this is the basic dea of most methods for the numerical integration of differential equations.

More or less sophisticated constructions of the "mean velocity" over a ime step yield the different methods which are discussed in the following ections. These methods are sometimes presented under the name of *dynamic r continuous system simulation.*

Example 8.1.3

The initial-value problem

$$\frac{dy}{dx} = y^2, \quad y(0) = 1$$

has the solution $y(x) = 1/(1 - x)$. Note that $y(x) \longrightarrow \infty$ as $x \longrightarrow 1$. Hence the solution of the problem exists only when $x < 1$.

Many program packages have been written for systems of ordinary differential equations; see, e.g., Gear [98]. These often assume that the system is given in **autonomous form**—

$$\frac{dy}{dx} = f(y)$$

—i.e., f does not depend explicitly on x. A non-autonomous system is easily transformed into autonomous form by the addition of a trivial extra equation,

$$\frac{d\eta_{s+1}}{dx} = 1, \quad \eta_{s+1}(a) = 0,$$

which has the solution $\eta_{s+1}(x) = x$.

8.1.2. Error Propagation

The solution to the initial-value problem, Eq. (8.1.2), can be considered as a function $y(x, c)$, where c is the vector of initial conditions. Here again, one can visualize a **family of solution curves**, this time in the (x, y)-space, which relate the various solutions to the initial values $y(a) = c$. For the case $s = 1$, the family of solutions can, for example, look like Fig. 8.1.2 or Fig. 8.1.3. The dependence of the solution $y(x, c)$ on c is often of great interest for both the technical or scientific context it appears in and the numerical treatment of the differential equation.

A disturbance in the initial condition—e.g., a round-off error in the value of c—means that $y(x)$ is forced to follow "another track" in the family of solutions.

The methods we shall treat are such that one proceeds step by step to

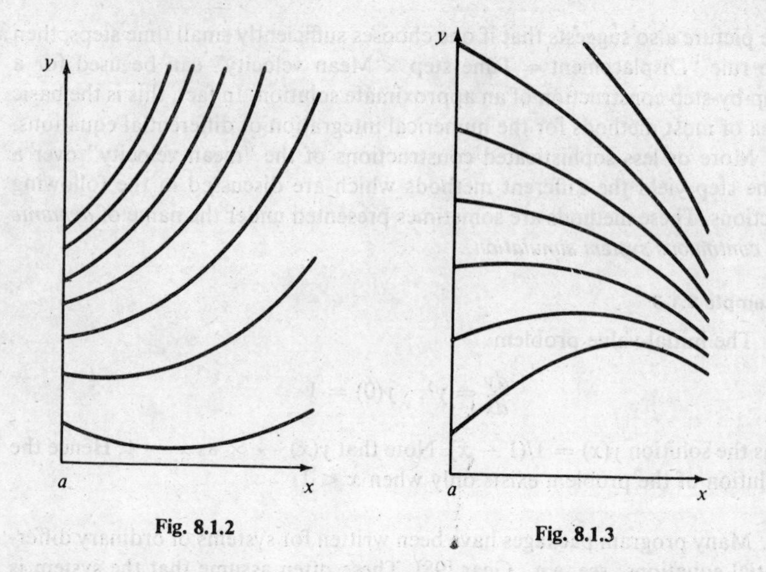

Fig. 8.1.2

Fig. 8.1.3

approximate values $y_1, y_2, \ldots$ for $y(x_1), y(x_2), \ldots$ in much the same way as in Euler's method (see Chap. 1). By a **one-step method** we mean a method whereby y_n is the only input data to the step in which y_{n+1} is to be computed, $n = 0, 1, 2, \ldots$. At each step there is a small disturbance—truncation error and/or rounding error—which produces a similar transition to "another track" in the family of solution curves. In Fig. 8.1.4, we give a greatly exaggerated view of what normally happens. One can compare the above process

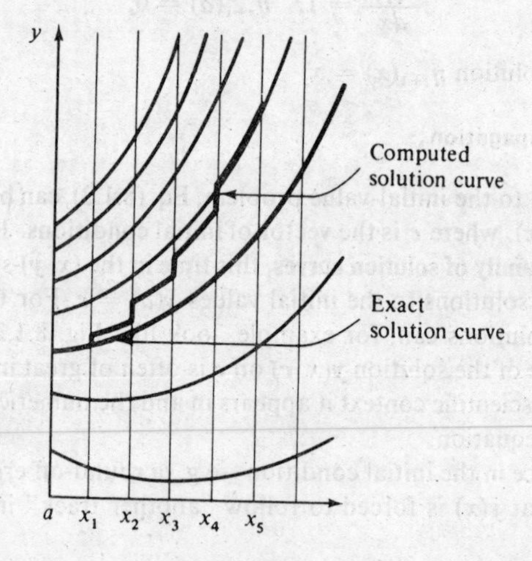

Fig. 8.1.4

o an *interest process*; in each stage there is "interest" on previously committed rrors. At the same time, a new "error capital" (local error) is put in. The interest rate" can, however, be negative (see Fig. 8.1.3), an advantage in his context!

If the curves in the family of solutions depart from each other quickly, hen the initial-value problem is ill-conditioned; otherwise, it is well-condi-ioned. (It is possible that a computational method which is not a one-step nethod can introduce other characteristics in the error-propagation mech-nism which are *not* inherent in the differential equation itself. However, we eave this aspect for a moment, and shall return to it in Example 8.3.1 and ec. 8.5.3.)

It is important to distinguish between **global and local truncation error**. he **global** error at the point x_{n+1} is $y_{n+1} - y(x_{n+1})$, where $y(x)$ is the exact olution to the initial-value problem. The **local** error at x_{n+1} is the difference etween the computed value y_{n+1} and the value at x_{n+1} of that solution of the ifferential equation which passes through the point (x_n, y_n). In other words, he local errors are the jumps in the staircase curve in Fig. 8.1.4.

The distinction between global and local truncation error is made already n numerical integration—which is, of course, a special case of the differential quation problem, where f is independent of y. There the global error is imply the sum of the local errors. We saw—e.g., in Sec. 7.4.2—that the lobal error of the trapezoidal rule is $O(h^2)$, while its local error is $O(h^3)$, lthough this terminology was not used there. The reason for the difference n order of magnitude of the two errors is that the number of local errors vhich are added is inversely proportional to h.

In differential equations, the error-propagation mechanism is, as we have een, more complicated; but even here it holds that *the difference between the xponents in the asymptotic dependence on h of the local and global errors is qual to* 1. (This is true for systems of first-order differential equations.)

We can look at the "interest process" more quantitatively in the scalar ase. Consider the function $u = \partial y(x, c)/\partial c$. It satisfies a linear differential quation, the **variational equation**

$$\frac{\partial u}{\partial x} = J(x) \cdot u, \quad \text{where} \quad J(x) = \left(\frac{\partial f}{\partial y}\right)_{y=y(x,c)}, \tag{8.1.3}$$

ince

$$\frac{\partial u}{\partial x} = \frac{\partial}{\partial x}\left(\frac{\partial y}{\partial c}\right) = \frac{\partial}{\partial c}\left(\frac{\partial y}{\partial x}\right) = \frac{\partial}{\partial c}(f(x, y(x, c))) = \frac{\partial f}{\partial y} \cdot \frac{\partial y}{\partial c}.$$

Ve can derive many other results from the above, since

$$y(x, c + \Delta c) - y(x, c) \approx \frac{\partial y}{\partial c} \cdot \Delta c.$$

he proofs of the theorems below are left as a series of exercises for the eader.

THEOREM 8.1.1

Closely lying curves in the family of solutions approach each other, as x increases, if $\partial f/\partial y < 0$ *and depart from each other if* $\partial f/\partial y > 0$. $\partial f/\partial y$ corresponds to the "rate of interest" in the analogy mentioned previously.

In the following we *assume that* $\partial f/\partial y \leq L$ *for all* (x, y) *in some plane region* D, $(a < x < b)$.

THEOREM 8.1.2

$$|u(x)| \leq |u(a)| \exp(L(x-a)), \quad a \leq x \leq b.$$

This holds even if L is negative.

THEOREM 8.1.3

Let $y(x_n)$ *be disturbed by a quantity* ϵ_n. *The effect of this disturbance on* $y(x)$, $x > x_n$, *will not exceed* $|\epsilon_n| \exp(L(x - x_n))$, *at least not as long as this bound guarantees that the perturbed solution curve remains in* D.

THEOREM 8.1.4

Suppose that, in a step-by-step solution, the magnitudes of the local errors at the points x_n, $n = 1, 2, \ldots$, *are less than* ϵ. *Then the global errors will satisfy the inequalities,*

$$|y_n - y(x_n)| \leq \epsilon \frac{e^{Lnh} - 1}{e^{Lh} - 1}$$

for $n = 1, 2, \ldots$ *as long as these bounds guarantee that the region between the staircase curve of Fig. 8.1.4 and the unperturbed solution curve is contained in* D.

If, in addition $L > 0$, *or if* $|Lh|$ *is small, then*

$$|y_n - y(x_n)| \lesssim \frac{\epsilon}{h} \frac{e^{Lnh} - 1}{L}.$$

For $L = 0$ *we have,* $|y_n - y(x_n)| \leq \epsilon n$.

The concept of variational equation can also be generalized to *systems* of differential equations. Introduce, as in Sec. 6.9.2, the Jacobian matrix, $f'(y) = [\partial\varphi_i/\partial\eta_j]$. Then, replace in Eq. (8.1.3) $J(x)$ and u by $J(x) = f'(y(x, c))$ and $u = \partial y/\partial \gamma_i$, respectively, where γ_i is any one of the components of c.

In order to generalize the last three theorems, we replace the condition that $\partial f/\partial y \leq L$ for all $(x, y) \in D$, by the condition

$$\mu(f'(y)) \leq L, \quad \text{for all } (x, y) \in D,$$

where D is a $(s + 1)$-dimensional region, and where $\mu(\cdot)$ is the logarithmic matrix norm, defined as follows:

DEFINITION

Let $\|\cdot\|$ denote some vector norm and its subordinate matrix norm. Then e subordinate **logarithmic norm** of the matrix A is given by,

$$\mu(A) = \lim_{\epsilon \downarrow 0} \frac{\|I + \epsilon A\| - 1}{\epsilon}.$$

See also Sec. 8.3.6 and Prob. 6 of Sec. 8.5 for practically interesting ternatives to Theorem 8.1.4. It must also be borne in mind that the validity Theorem 8.1.4 is related to the definition of local error given above. There e, in the literature, other definitions for which Theorem 8.1.4 is not valid.

1.3. Other Differential Equation Problems

Besides *initial-value problems*, there are, among others, *boundary-value oblems* and *eigenvalue problems* for differential equations, where the solu-on is to satisfy prescribed conditions at two (or more) points. An orienta-on on such problems is given in Sec. 8.4.

Many scientific and technical problems lead to *partial differential equa-ons*, where one seeks functions of more than one independent variable (e.g., ace coordinates x, y, z and the time t), where connections between certain the partial derivatives of the functions are given, together with additional nditions (initial values, boundary values). A short orientation on some ımerical methods in this important class of problems is given in Sec. 8.6.

REVIEW QUESTIONS

Explain, with an example, how a differential equation of higher order can be written as a system of first-order differential equations.

Define local and global truncation error, and explain (with a figure) the error propagation for a one-step method.

Define and derive the variational equation, and state a few results which can be derived from it.

Define logarithmic norm, and give one application.

PROBLEMS

Rewrite the system

$$y'' = x^2 - y' - z^2,$$
$$z'' = x + z' + y^3,$$

with initial conditions $y(0) = 0, y'(0) = 1, z(0) = 1, z'(0) = 0$, as an initial-value problem for a system of first-order equations.

Calculate $y(1) - z(1)$ where

$$y' = -y, \qquad y(0) = 1,$$
$$z' = -1.01z, \qquad z(0) = 0.99,$$

and compare with the estimate which can be obtained by Theorem 8.1.3.

3. (a) Let $A = [a_{ij}]$ be a real matrix and $\mu_\infty(A)$ be the logarithmic norm which is subordinate to the maximum norm. Show that

$$\mu_\infty(A) = \max_i \left(a_{ii} + \sum_{j,\, j \neq i} |a_{ij}| \right).$$

(b) Compute $\exp(\|A\|_\infty)$ and $\exp(\mu_\infty[A])$ for

$$A = \begin{bmatrix} -10 & 1 \\ 1 & -10 \end{bmatrix}.$$

8.2. EULER'S METHOD, WITH REPEATED RICHARDSON EXTRAPOLATION

The simplest approximation method for the initial-value problem in Eq. (8.1.2) is **Euler's method**, or the *tangent method*—already presented in Chap. 1. One divides the interval $[a, b]$ into subintervals of length h, $x_i = a + ih$, sets $y_0 = c$, and seeks approximate values $y_1, y_2, \ldots$ to the exact values $y(x_1), y(x_2), \ldots$ by approximating the derivative at the point (x_n, y_n) with the difference quotient $(y_{n+1} - y_n)/h$. This gives the difference equation

$$\frac{y_{n+1} - y_n}{h} = f(x_n, y_n),$$

i.e., the recursion formula

$$y_{n+1} = y_n + hf(x_n, y_n), \quad y_0 = c. \tag{8.2.1}$$

In this method it is of no importance whether the function f is defined by a complicated nonlinear expression. However, we content ourselves with presenting a simple example.

Example 8.2.1

$$\frac{dy}{dx} = y, \quad y(0) = 1. \tag{8.2.2}$$

Euler's method gives $y_{n+1} = y_n + hy_n$, $y_0 = 1$.

The solution is computed first with $h = 0.2$ and then with $h = 0.1$, and is compared to the exact solution $y(x) = e^x$:

			$h = 0.2$			$h = 0.1$	
x_n	$y(x_n)$	y_n	hf_n	error	y_n	hf_n	error
0	1.000	1.000	0.200	0.000	1.000	0.100	0.000
0.1	1.105				1.100	0.110	−0.005
0.2	1.221	1.200	0.240	−0.021	1.210	0.121	−0.011
0.3	1.350				1.331	0.133	−0.019
0.4	1.492	1.440	0.288	−0.052	1.464	0.146	−0.028
0.5	1.649				1.610	0.161	−0.039
0.6	1.822	1.728		−0.094	1.771		−0.051

The error seems to be approximately proportional to h. We see also that the error grows as x increases.

The weakness of Euler's method is that the step length must be chosen quite small in order to attain acceptable accuracy. The following theorem is the theoretical basis for the use of repeated Richardson extrapolation in connection with Euler's method. This theorem and some similar results mentioned later are particular cases of a theorem of H. J. Stetter [113, p. 25].

THEOREM 8.2.1

Denote by $y(x, h)$ the result of the use of Euler's method with step length h on the differential equation problem (8.1.2). Then

$$y(x, h) = y(x) + c_1(x)h + c_2(x)h^2 + c_3(x)h^3 + \ldots + c_p(x)h^p + O(h^{p+1}).$$

Repeated Richardson extrapolation can thus be performed according to (7.2.12) with $p_k = k$.

Notice that the expansion also contains the odd powers of h. Thus the headings in the extrapolation scheme of Eq. (7.2.13) become

$$\frac{\Delta}{1}, \frac{\Delta}{3}, \frac{\Delta}{7}, \frac{\Delta}{15}, \frac{\Delta}{31}, \ldots \quad \text{when} \quad q = 2.$$

If the round-off error in $y(x, h)$ has magnitude less than ϵ, then the resultant error in an extrapolated value is less than 8.26ϵ.

The theorem will not be proved here. It is related to the fact that there is a similar expansion for the error in the local approximation of the derivative upon which the method is based. From Taylor's formula we know that

$$\frac{y(x_{n+1}) - y(x_n)}{h} - y'(x_n) = \frac{h}{2}y''(x_n) + \frac{h^2}{3!}y'''(x_n) + \frac{h^3}{4!}y^{iv}(x_n) + \ldots.$$

Example 8.2.2

The value $A_{i,0}$ in the table means the result for $x = 1$ of integrating the differential equation $y' = -y$ with initial condition $y(0) = 1$ and step size $h = 0.25 \cdot 2^{-i}$.

	$\frac{\Delta}{1}$		$\frac{\Delta}{3}$		$\frac{\Delta}{7}$		$\frac{\Delta}{15}$	
$A_{00} = 0.316406$								
	27203							
$A_{10} = 0.343609$		.370812						
	12465		−758					
$A_{20} = 0.356074$		.368539		.367781				
	5981		−168		12			
$A_{30} = 0.362055$		.368036		.367868		.367880		
	2932		−39		2		0	
$A_{40} = 0.364987$		.367919		.367880		.367882		.367882

We accept $A_{44} = 0.367882$ and estimate R_T as $|A_{43} - A_{33}| = 0.000002$, $R_A \leq 5 \cdot 10^{-6}$.

The correct value is $y(1) = e^{-1} = 0.367879$.

There are many ways to apply Richardson extrapolation to initial-value problems. One way is called *active extrapolation*, which means that the result of the extrapolation is used as input data for the remaining calculations. Another way is called *passive extrapolation;* this means that the results of extrapolation are accepted as output data, but that they are not used in the remaining calculations. Thus a passive extrapolation can be performed after the problem has been solved from start to finish with a number of step sizes. In Example 8.2.2 passive extrapolation has been performed several times at the point $x = 1$.

Example 8.2.3

See the table in Example 8.2.1. Denote by $\bar{y}(x)$ the result of *one* passive Richardson extrapolation.

x	$y(x, 0.1) - y(x, 0.2)$	$\bar{y}(x)$	$\bar{y}(x) - y(x)$
0	0.000	1.000	0.000
0.2	0.010	1.220	−0.001
0.4	0.024	1.488	−0.004
0.6	0.043	1.814	−0.008

The accuracy in $\bar{y}(x)$ is thus much better than in $y(x, 0.1)$. If one wants an improved result in an intermediate point—e.g., $x = 0.3$—then one gets a suitable correction by interpolating linearly in the second column, i.e.,

$$y(0.3, 0.1) = 1.331; \qquad \bar{y}(0.3) = 1.331 + \tfrac{1}{2}(0.010 + 0.024) = 1.348.$$

The error in $\bar{y}(0.3)$ is -0.002, while the error in $y(0.3, 0.1)$ is -0.019.

One might think that active extrapolation should always be preferable, but with certain types of systems, passive extrapolation is better because it is numerically more stable.

In practice, it is usually unadvisable to use the same step length in the entire computation. In many computer programs, the step length is automatically adjusted—often by comparing the results obtained with two different step sizes. Extrapolation can be applied afterward, only under the condition that the *changes in step size* were done in the same way and at the same time (without exception) for the different initial step sizes. A situation where pa sive extrapolation is permissible for two different initial step sizes, h_0 and $h_0/2$, is illustrated in Fig. 8.2.1.

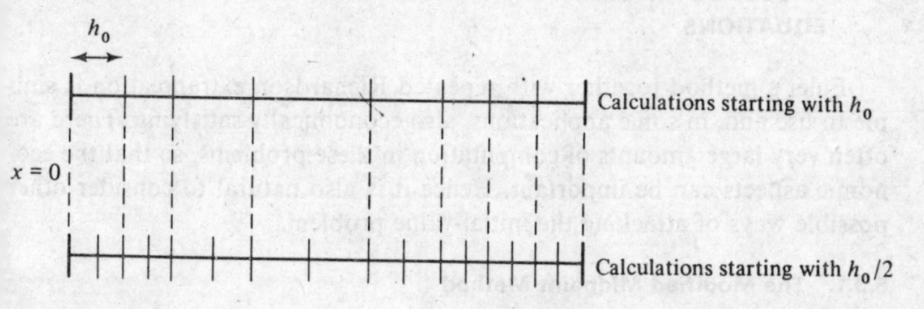

Fig. 8.2.1

PROBLEMS

1. Given: $y' = 1 + x^2y^2$, $y(0) = 0$. Compute $y(0.5)$ by Euler's method with repeated Richardson extrapolation. Use four decimals.

2. In a computation with Euler's method, the following results were obtained with various step sizes:

$h = 0.05$	$h = 0.1$	$h = 0.2$
1.22726	1.22595	1.22345

Compute a better value by extrapolation.

3. Write a segment of an Algol program for Euler's method with repeated Richardson extrapolation. Assume that $f(x, y)$ is given as a function procedure and that a sufficiently large array for the extrapolations is available.

4. Try Euler's method with step length h on the test problem

$$y' = -y, \quad y(0) = 1.$$

(a) Determine an explicit expression for y_n.
(b) For which values of h is the sequence $\{y_n\}_0^\infty$ bounded?
(c) Compute $\lim_{h\to 0} (y(x, h) - e^{-x})/h$.

5. Apply Euler's method to the system

$$y' = Ay, \quad y(0) = c.$$

A is a constant matrix which can be transformed to diagonal form. Its eigenvalues are assumed to lie in the interval $[-k, 0]$. What is the largest step size for which the sequence of vectors $\{y_n\}_0^\infty$ is bounded?

8.3. OTHER METHODS FOR INITIAL-VALUE PROBLEMS IN ORDINARY DIFFERENTIAL EQUATIONS

Euler's method together with repeated Richardson extrapolation is simple to use and, in some applications, also economically satisfying. There are often very large amounts of computation in these problems, so that the economic aspects can be important. Hence it is also natural to consider other possible ways of attacking the initial-value problem.

8.3.1. The Modified Midpoint Method

One way is to base the method on a *better formula for numerical differentiation*—e.g., the symmetrical formula

$$y'(x_n) = \frac{y(x_{n+1}) - y(x_{n-1})}{2h} + O(h^2).$$

This yields the recursion formula, the *explicit midpoint method* or *leap-frog method*:

$$y_{n+1} = y_{n-1} + 2hf(x_n, y_n). \tag{8.3.1}$$

It is a **two-step method**, since it requires two values from the past, namely y_{n-1} and y_n, in the step where y_{n+1} is computed. (See the notion of one-step method defined in Sec. 8.1.2.) It therefore requires a special formula for the calculation of y_1. The local truncation error in Eq. (8.3.1) is $O(h^3)$; hence the global error is $O(h^2)$. The following example shows that this method is not generally acceptable.

Example 8.3.1

Apply Eq. (8.3.1) to the equation $y' = -y$, $y(0) = 1$. Choose $h = 0.1$, $y_0 = 1$. The solution corresponding to $y_1 = 0.9$ is indicated by black circles in Fig. 8.3.1, while the solution corresponding to $y_1 = 0.85$ is shown with white circles. Note that the perturbation of the initial value gives rise to growing oscillations with a growth of approximately 10 percent per step. This phenomenon is sometimes called weak stability, sometimes weak instability; see Example 8.5.8.

In more realistic examples the oscillations become visible much later. For instance, if $y_1 = e^{-0.1}$ correct to ten decimals, we have (see Henrici [100, p. 241]) the table below:

(The exact solution is $y(x) = \exp(-x)$.)

x	0	0.1	0.2	...	5.0	5.1	5.2	5.3
y_n	1	0.90484	0.81903	...	0.01803	−0.00775	0.01958	−0.01166
$y_n - y(x_n)$	0.00000	0.00000	0.00030	...	0.01129	−0.01385	0.01406	−0.01665
$y(x_n)$	1.00000	0.90484	0.81873	...	0.00674	0.00610	0.00552	0.00499

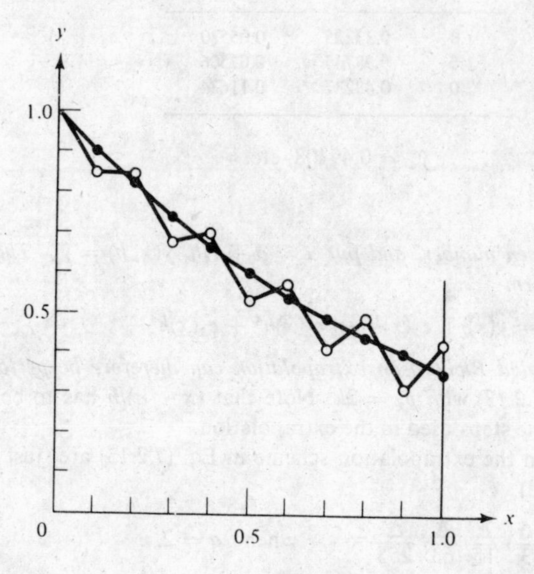

Fig. 8.3.1

Algorithm for the **modified midpoint method** (W. B. Gragg). *Put* $y_0 = y(a)$, $y_1 = y_0 + hf(x_0, y_0)$.

$$y_{n+1} = y_{n-1} + 2hf(x_n, y_n), \quad n = 1, 2, \ldots. \tag{8.3.2}$$

For certain values of x, one damps the oscillating error component by means of the following formula, where N is even:

$$\hat{y}_N = \tfrac{1}{2}(y_N + y_{N-1} + hf(x_N, y_N)), \quad n = 1, 2, \ldots. \tag{8.3.3}$$

The stepwise integration then proceeds by Eq. (8.3.2) from the point x_N *with*

$\hat{y}_N$, *and* $\hat{y}_{N+1} := \hat{y}_N + hf(x_N, \hat{y}_N)$ *as new starting values.*

For another way of writing Eq. (8.3.3) which illustrates the damping effect see Problem 5 on page 357.

Example 8.3.2

For the initial-value problem $dy/dx = y^2$, $y(0) = 0.25$ we get, with $h = 0.5$,

x	y	hf
0	0.25000	0.03125
0.5	0.28125	0.03955
1.0	0.32910	0.05415

$$\hat{y}_2 = \tfrac{1}{2}(0.32910 + 0.28125 + 0.05415) = 0.33225.$$

x	y	hf
1.0	0.33225	0.05520
1.5	0.38745	0.07506
2.0	0.48237	0.11634

$$\hat{y}_4 = 0.49308, \text{ etc.}$$

THEOREM 8.3.1

Let N be an even number, and put $x = a + Nh$, $y(x, h) = \hat{y}_N$. Then an expansion of the form

$$y(x, h) = y(x) + c_1(x)h^2 + c_2(x)h^4 + c_3(x)h^6 + \ldots$$

can be used. Repeated Richardson extrapolation can therefore be performed according to Eq. (7.2.12) with $p_k = 2k$. Note that $(x - a)/h$ has to be *even* for the largest of the steps used in the extrapolation.

The headings in the extrapolation scheme in Eq. (7.2.13) are (just as in Romberg's method)

$$\frac{\Delta}{3}, \frac{\Delta}{15}, \frac{\Delta}{63}, \frac{\Delta}{255}, \ldots, \quad \text{when} \quad q = 2.$$

We shall not prove the theorem here. Notice that the (asymptotically correct) expansion has the same form (but entirely different coefficients) as the symmetric approximation to the derivative upon which the method is based.

One can also take the extrapolated value as a new starting value (active extrapolation).

Example 8.3.3

For the same problem as in Example 8.3.2, the following results were obtained on a computer for various step sizes, at $x = 1$:

h	$y(1, h)$	$\frac{\Delta}{3}$		$\frac{\Delta}{15}$	
0.5	0.3322 52741				
		265,468			
0.25	0.3330 49146		0.3333 14614		
		70,726		1,162	
0.125	0.3332 61325		0.3333 32051		0.3333 33213

Since the exact solution of the problem is $y(x) = 1/(4 - x)$, we have $y(1) = 0.3333\ 33333$. The error in the extrapolated value is thus $\frac{1}{600}$ of the error in $y(1, 0.125)$.

Bulirsch and **Stoer** [89] have developed the application of an extrapolation based on the approximation of $y(x, h)$ by rational functions of h, which seems to be slightly more advantageous than the standard Richardson extrapolation. The algorithm is more complicated; see Gear [98, pp. 93–101], who also gives a program.

8.3.2. The Power-Series Method

In Example 3.1.2 we solved an initial-value problem by substituting a power series with undetermined coefficients. From the differential equation a recursion formula for the computation of the coefficients was derived.

This method can be extended to be a stepwise process. When an acceptable estimate for $y(x_n)$ is given, the coefficients in a series in powers of $(x - x_n)$ can be determined. This series is then used in approximately computing $y(x_{n+1})$, where the size of $(x_{n+1} - x_n)$ is chosen so that acceptable accuracy is obtained with a reasonable number of terms in the series.

The power-series method was previously used a great deal in hand computation, but was less popular during the first years of the computer age. Since then, programming techniques and languages have been improved, and the popularity of this method has risen again. (For the practical application of the power-series method on a computer, see Moore [56].)

The coefficients in the power series can also be computed recursively by differentiating the differential equation. According to Taylor's formula we have, of course, that the coefficient of $(x - x_n)^k$ is equal to $y^{(k)}(x_n)/(k!)$.

Example 8.3.4

Determine the first four Taylor coefficients for the function which is the solution to the initial-value problem

$$y' = 1 + xy + y^2, \quad y(0) = 0.$$

From the equation, we have $y'(0) = 1$.

Differentiating the equation, we get

$$y'' = y + xy' + 2yy' \Longrightarrow y''(0) = 0,$$

$$y''' = 2y' + 2(y')^2 + xy'' + 2yy'' \Longrightarrow y'''(0) = 4.$$

Thus

$$y(x) = x + \frac{4x^3}{6} + \ldots.$$

Notice that the computation of these recursion formulas for the Taylor coefficients need only be done once. The same recursion formulas can be used at each step (the numerical values of the Taylor coefficients are of course different at each step). It is difficult to make a general economic comparison between power-series methods and the methods previously mentioned.

Other types of series expansions have also been used with success in differential equation problems—for example, Tchebycheff series, see [45].

8.3.3. Runge-Kutta Methods

Suppose that we know $y(x_n)$ and we want to determine an approximation y_{n+1} to $y(x_n + h)$. The idea behind the Runge-Kutta methods is to compute the value of $f(x, y)$ at several strategically chosen points near the solution curve in the interval $(x_n, x_n + h)$, and to combine these values in such a way as to get good accuracy in the computed increment $y_{n+1} - y_n$.

The simplest example is called **Heun's method**. Here one computes

$$\begin{aligned} k_1 &= hf(x_n, y_n), \\ k_2 &= hf(x_n + h, y_n + k_1), \\ y_{n+1} &= y_n + \tfrac{1}{2}(k_1 + k_2). \end{aligned} \tag{8.3.4}$$

For $y(x, h)$, the result obtained at the point x with step h, we have

$$y(x, h) = y(x) + c_2(x)h^2 + c_3(x)h^3 + c_4(x)h^4 + c_5(x)h^5 + \dots. \tag{8.3.5}$$

Richardson extrapolation can be used. Notice the occurrence of odd powers of h. Because of this, the headings in the extrapolation scheme in (7.2.13) become $\Delta/3$, $\Delta/7$, $\Delta/15$, ...

The best-known Runge-Kutta method is defined by the formulas in Eq. (8.3.6), below. They seem quite complicated at first sight, but they are in fact very easy to program. The function f is computed four times in each step.

$$\begin{aligned} k_1 &= hf(x_n, y_n), \\ k_2 &= hf(x_n + \tfrac{1}{2}h, y_n + \tfrac{1}{2}k_1), \\ k_3 &= hf(x_n + \tfrac{1}{2}h, y_n + \tfrac{1}{2}k_2), \\ k_4 &= hf(x_n + h, y_n + k_3), \\ y_{n+1} &= y_n + \tfrac{1}{6}(k_1 + 2k_2 + 2k_3 + k_4). \end{aligned} \tag{8.3.6}$$

One can show that an expansion of the form

$$y(x, h) = y(x) + c_4(x)h^4 + c_5(x)h^5 + \dots$$

holds. If Richardson extrapolation is used, then the headings in the extrapolation scheme are $\Delta/15$, $\Delta/31$, ...

A complete derivation of the above formulas is outside the scope of this book. The following *heuristic discussion*, however, gives some background. For the exact solution to the differential equation, we have

$$y(x_{n+1}) - y(x_n) = \int_{x_n}^{x_n+h} \frac{dy}{dx}\,dx = \int_{x_n}^{x_n+h} f(x, y(x))\,dx.$$

The methods are based on approximating the integral using available data. In the special case where f is independent of y, Heun's method is identical to the trapezoidal rule for numerical integration—whose global truncation error is $O(h^2)$. The odd powers in Eq. (8.3.5) come from the fact that in the trapezoidal approximation for k_2, one has $y_n + k_1$ as argument instead of the unknown y_{n+1}. This approximation has local error

$$y(x_n) + hy'(x_n) - y(x_n + h) = -\sum_{k=2}^{\infty} \frac{h^k y^{(k)}(x_n)}{k!},$$

which contains both odd and even powers of h.

The Runge-Kutta method, Eq. (8.3.6), is, in the special case where f is independent of y, identical to Simpson's formula, Eq. (7.2.4), with h replaced by $h/2$,

$$\int_{x_n}^{x_n+h} f(x)\,dx = \frac{h}{6}\left(f(x_n) + 4f\left(x_n + \frac{1}{2}h\right) + f(x_n + h)\right),$$

since

$$k_1 = hf(x_n), \qquad k_2 = k_3 = hf(x_n + \tfrac{1}{2}h), \qquad k_4 = hf(x_n + h).$$

In this case the global truncation error is $O(h^4)$, which also holds for the general case.

There exist a large number of Runge-Kutta methods with different virtues. It is, for instance, possible to obtain fifth-order accuracy by six evaluations per step. It is beyond the scope of this book to treat these systematically; see [113].

8.3.4. Implicit Methods

The following formula, the **trapezoidal method,**

$$y_{n+1} - y_n = \tfrac{1}{2}h[f(x_n, y_n) + f(x_{n+1}, y_{n+1})], \quad n = 0, 1, 2, \ldots, \tag{8.3.7}$$

is an example of an **implicit** method, since y_{n+1}, which is to be computed (in the nth step), appears implicitly on the right-hand side. If f is a nonlinear function, one has to solve a nonlinear system at each step. This must be done by some iterative method—e.g., by the procedure

$$y_{n+1}^{(k+1)} = \tfrac{1}{2}hf(x_{n+1}, y_{n+1}^{(k)}) + u_n, \quad k = 0, 1, 2, \ldots, \tag{8.3.8}$$

where $u_n = y_n + \frac{1}{2}hf(x_n, y_n)$ is known from the previous time step. A sufficient criterion for convergence is, by Eq. (6.9.3),

$$\tfrac{1}{2}h\left\|\frac{\partial f}{\partial y_n}\right\| < 1, \tag{8.3.9}$$

and the smaller the quantity on the left-hand side of this inequality, the faster the convergence of the iterations. A fairly good initial guess, $y_{n+1}^{(0)}$, can be obtained from the values of the past—e.g., by the formula

$$y_{n+1}^{(0)} = y_n + hf(x_n, y_n).$$

Such a formula is usually called a **predictor**, while the implicit formula in Eq. (8.3.7) is called a **corrector**. The whole procedure is called a **predictor-corrector** method. The stopping of the iterations may be controlled *either* by comparing the difference $y_{n+1}^{(k+1)} - y_{n+1}^{(k)}$ to some preset tolerance ϵ *or* by predetermining the number of iterations. The latter is more common. For example, if we decide always to stop after the calculation of $y_{n+1}^{(1)}$, the method is identical to Heun's method, Eq. (8.3.4). It is advisable to choose a predictor such that one iteration will give almost the full accuracy of the corrector.

A classical predictor-corrector scheme is given by the following formulas, where $f_n = f(x_n, y_n)$ and ∇ is the backward difference operator.

Adams-Bashforth:

$$y_{n+1} - y_n = h\left(f_n + \frac{1}{2}\nabla f_n + \frac{5}{12}\nabla^2 f_n + \frac{3}{8}\nabla^3 f_n + \frac{251}{720}\nabla^4 f_n + \cdots\right). \tag{8.3.10}$$

Adams-Moulton:

$$y_{n+1} - y_n = h\left(f_{n+1} - \frac{1}{2}\nabla f_{n+1} - \frac{1}{12}\nabla^2 f_{n+1} - \frac{1}{24}\nabla^3 f_{n+1} - \frac{19}{720}\nabla^4 f_{n+1} - \cdots\right). \tag{8.3.11}$$

The basis is the formula:

$$y(x_{n+1}) - y(x_n) = \int_{x_n}^{x_n+h} \frac{dy}{dx}\,dx = \int_{x_n}^{x_n+h} f(x, y(x))\,dx,$$

where $f(x, y(x))$ is approximated by an interpolation polynomial, determined by the values $f_n, f_{n-1}, \ldots, f_{n-k}$ in the Adams-Bashforth case and by the values $f_{n+1}, f_n, \ldots, f_{n-k+1}$ in the Adams-Moulton case. The local errors are approximately equal to the first neglected terms. For an operator derivation, see Problem 4 of Sec. 7.6.

This method has recently been modified in order to facilitate a free variation of the step size. Gear [98, pp. 155ff.] gives a program for a modified Adams method with automatic control of both step size and the degree of the interpolation polynomial, very important features in a program.

This method and the Bulirsch-Stoer method are among the most successful ones in comparative tests carried out at various research centers. In choosing between algorithms, it is also important to realize that many problems can be solved with very little computing effort using any of these methods (possibly with the exception of the Euler method without extrapolation), so that considerations other than the choice of the fastest method may be more important—e.g., how convenient the program is for the user, how exceptional situations (singularities) are handled, etc. If a problem is to be solved over a long time period or for a very large number of parameter combinations, etc., then more careful preparations are called for, and one should consult a monograph—e.g., Gear [98].

8.3.5. Stiff Problems

There is, however, one class of problems where all of the previously mentioned methods require step sizes which are much shorter than what one would expect if one looks at the solution curve.

Example 8.3.6

$$\frac{dy}{dx} = 100(\sin x - y), \quad y(0) = 0.$$

Exact solution:

$$y(x) = \frac{\sin x - 0.01 \cos x + 0.01e^{-100x}}{1.0001}$$

Notice that the exponential term is less than 10^{-6} already for $x = 0.1$. One might expect that the step size $h = 0.1$ would be possible for $x > 0.1$. However, calculations with the Runge-Kutta method with various step sizes gave the following values for $x = 3$:

Step Size	0.015	0.020	0.025	0.030
Number of Steps	200	150	120	100
$y(3, h)$	0.151004	0.150996	0.150943	$6.7 \cdot 10^{11}$

We see that $h = 0.025$ gives a good result, whereas for $h = 0.030$ one has a frightful numerical instability.

Example 8.3.7

$$\frac{dy_1}{dx} = y_2, \qquad y_1(0) = 1,$$

$$\frac{dy_2}{dx} = -1{,}001y_2 - 1{,}000y_1, \quad y_2(0) = -1.$$

Exact solution: $y(x) = \exp(-x)$. (General solution: $A \exp(-x) + B \exp(-1{,}000x)$.) Here the Runge-Kutta method explodes for $h = 0.0027$ (approximately); this is a very unsatisfactory step size for describing the function $\exp(-x)$. An explanation for the explosion can be found in Problem 7 of this section.

The two examples above have in common that there are processes in the system (described by the differential equation) with significantly different time scales. The methods described require that one take account of the fastest process, even after the corresponding component has died out in the exact solution. Some explanation of this will be given in Sec. 8.5.3. More difficult variants—large nonlinear systems—occur in many very important applications—e.g., in simulation of chemical reactions, electronic networks, control systems, etc. These problems are sometimes called **stiff problems**. When this text was written, intensive research was still being carried out to find methods suitable for such problems. Here we can only give an orientation on this complex of problems; we refer the reader to Gear [98, Chap. 11] for a more complete discussion. A concept called $A(\alpha)$-stability which is appropriate for stiff problems will be introduced in Sec. 8.5.4.

The methods which have been most successful are *implicit*. For example, the *trapezoidal method* in Eq. (8.3.7), combined with smoothing and Richardson extrapolation, has been used with success; and so has the **implicit midpoint method,**

$$y_{n+1} - y_n = hf(\tfrac{1}{2}(y_n + y_{n+1})).$$

The iteration process in Eq. (8.3.8), however, is no good on stiff problems, since the convergence condition, Eq. (8.3.9), means exactly that the fastest components limit the rate of convergence and thus also the step size which can be used. This is unacceptable, and instead *one uses Newton-Raphson's method* (see Sec. 6.9) *or some variant of it in which the Jacobian need not be computed so often.*

Other implicit methods which have been used with success on stiff problems are based on the **backward differentiation** formula,

$$hD = -\ln(1 - \nabla) = \nabla + \tfrac{1}{2}\nabla^2 + \tfrac{1}{3}\nabla^3 + \cdots,$$

see the table at the end of Sec. 7.6. One example is the backward Euler method (good stability, low accuracy)

$$y_{n+1} - y_n = hf(y_{n+1}),$$

or more generally,

$$\nabla y_{n+1} + \frac{1}{2}\nabla^2 y_{n+1} + \frac{1}{3}\nabla^3 y_{n+1} + \cdots + \frac{1}{k}\nabla^k y_{n+1} = hf(y_{n+1}), \quad k \leq 6. \tag{8.3.12}$$

Other variants exist which facilitate variation of the step size. As k increases, the local truncation error decreases, but the stability properties become worse. The remark concerning *the necessity of Newton-like methods*, made in connection with the trapezoidal method, *applies also to these methods.*

8.3.6. Control of Step Size

It is usually uneconomical to keep the step size constant. The automatic control of it is therefore an important and interesting part of a program for initial-value problems. In addition to computing y and f there are in each step three things to be done:

I. Estimate the local error.
II. Decide whether the computed value of y can be accepted, or whether one has to take a shorter step from the previous point.
III. Determine the step size to be used next.

If the **order of accuracy** of the method is p—i.e., if the global error is $O(h^p)$—then the local error at x_{n+1} is approximately $c_n h^{p+1}$. For the Runge-Kutta methods, c_n depends on certain partial derivatives of f, which are not to be calculated. We assume that the value of c_n changes fairly little from one step to the next.

We shall describe some ways of **estimating the local error.** Starting at (x_n, y_n) we first take one step of length h, the result of which is denoted by (x_{n+1}, y^*_{n+1}). Then again, starting from (x_n, y_n) we take two steps, each of length $h/2$. The result is denoted by (x_{n+1}, y^{**}_{n+1}). The local error of y^{**}_{n+1} is, by Eq. (7.2.9), approximately bounded by

$$l_n \approx \left\| c_n \left(\frac{h}{2}\right)^{p+1} \right\| \approx \frac{\| y^*_{n+1} - y^{**}_{n+1} \|}{2^p - 1}. \tag{8.3.13}$$

The value y^R_{n+1} obtained by Richardson extrapolation (active or passive—see Sec. 8.2),

$$y^R_{n+1} = y^{**}_{n+1} + \frac{y^{**}_{n+1} - y^*_{n+1}}{2^p - 1},$$

is likely to be a better estimate of $y(x_{n+1})$ than y^{**}_{n+1}, but it is not easy to compute a sharper error bound than Eq. (8.3.13) on the basis of the available data.

Now, suppose l_n has been estimated. A common policy is to keep **the local error per unit step** below a given tolerance—say,

$$l_n \leq \epsilon \cdot (x_{n+1} - x_n). \tag{8.3.14}$$

By Theorem 8.1.3, and the generalization indicated at the end of Sec. 8.1.2,

$$\| y_N - y(x_N) \| \leq \sum_{n=0}^{N-1} \epsilon \cdot (x_{n+1} - x_n) \exp\left[L(x_N - x_n)\right]$$

$$\approx \epsilon \cdot \int_{x_0}^{x_N} \exp\left[L(x_N - t)\right] dt = \frac{\epsilon}{L} \cdot \{\exp\left[L(x_N - x_0)\right] - 1\}.$$

Therefore, if the primary requirement of accuracy is formulated in terms of the **global** error, the parameter ϵ needed in this policy can be set, provided that one has a rough guess about a bound for L. A minor disadvantage with this policy is that when the solution is comparatively rough in a short interval, the step size will become smaller than neccessary, when taking into account the importance of that short interval for the subsequent computation. This is the case, for example, with "stiff problems" and also when there are singularities. An alternative policy would be to require that

$$l_n \leq \max\{\epsilon \cdot (x_{n+1} - x_n), \epsilon'\}, \tag{8.3.15}$$

where some skill is needed for the choice of ϵ'. For $\epsilon' = 0$ we have, of course, the same policy as above. Gear [98, pp. 76ff.] chooses $\epsilon = 0$, on the basis of an optimization study, but he needs a rough guess on the number of steps in order to set ϵ'.

Therefore, y_{n+1}^{**} or y_{n+1}^{R} can be *accepted*, when

$$\left\| \frac{y_{n+1}^{*} - y_{n+1}^{**}}{2^p - 1} \right\| \leq \max \{\epsilon \cdot (x_{n+1} - x_n), \epsilon'\}.$$

Finally, we have the question of *choosing a step size h′ for the next step*—or else of recomputing from (x_n, y_n), if the inequality just given is not satisfied. This step size then has to satisfy the following condition, where θ, $\theta \leq 1$, is a preset safety factor to account for the circumstance that the error estimates are approximate and based on experience from the preceding interval, etc.:

$$c_n\left(\frac{h'}{2}\right)^{p+1} \leq \theta\epsilon h' \quad \text{or} \quad c_n\left(\frac{h'}{2}\right)^{p+1} \leq \theta\epsilon',$$

while, by Eq. (8.3.13),

$$c_n\left(\frac{h}{2}\right)^{p+1} \approx l_n.$$

Eliminating c_n, we obtain

$$h' \lesssim h \max \left\{ \left(\frac{\theta\epsilon h}{l_n}\right)^{1/p}, \left(\frac{\theta\epsilon'}{l_n}\right)^{1/(p+1)} \right\}.$$

(We see here that when p is large—say, $p \geq 4$—the choice of θ, ϵ, ϵ' is not very critical—take, for example, $\theta = 0.8$.)

In several programs the user is asked to set both upper and lower bounds for the permissible step size. There may also be other practical constraints on the step size—e.g., in order to obtain printing of results at regular intervals without interpolation.

For some methods the local error is $l_n \approx cy^{(p+1)}(x_n)h^{p+1}$, where c is a known constant. An alternative way of estimating the local error is, then, to approximate $y^{(p+1)}(x_n)$ by differences.

For some predictor-corrector methods the (local) error of the predicted value can be expressed similarly, with a constant c' instead of c. The difference between the predicted and the corrected value, multiplied by $c/(c' - c)$, is then a useful estimate of the local error of the corrected value. See Problem 4 of Sec. 7.4.

8.3.7. A Finite-Difference Method for a Second-Order Equation

Equations of the form

$$y'' = f(x, y) \tag{8.3.16}$$

with initial conditions

$$y(a) = \alpha, \qquad y'(a) = \gamma, \tag{8.3.17}$$

are often encountered in mathematical physics. The equation can be reduced to a set of two simultaneous first-order differential equations, as in Example 8.1.1, but there are also special methods. One can simply *replace the deriva-*

tives in the differential equation and initial conditions by symmetric difference approximations,

$$y'(x_n) \approx \frac{y_{n+1} - y_{n-1}}{2h}, \qquad y''(x_n) \approx \frac{y_{n+1} - 2y_n + y_{n-1}}{h^2}. \tag{8.3.18}$$

If we put $f_n = f(x_n, y_n)$, a method† is defined by the following relations:

$$y_{n+1} - 2y_n + y_{n-1} = h^2 f_n, \tag{8.3.19}$$

$$y_0 = \alpha, \qquad y_1 - y_{-1} = 2h\gamma.$$

y_{-1} can be eliminated by means of Eq. (8.3.19) with $n = 0$. The starting procedure is, therefore,

$$y_0 = \alpha, \qquad y_1 - y_0 = h\gamma + \tfrac{1}{2}h^2 f_0. \tag{8.3.20}$$

Then $y_2, y_3, y_4, \ldots$ can be computed successively by means of Eq. (8.3.19). Note that at each step there is an addition of the form $O(1) + O(h^2)$; this gives unfavorable rounding errors when h is small. One should instead use the following summed form of the method. Put $z_{i+1/2} = (y_{i+1} - y_i)/h$, and note that $z_{n+1/2} - z_{n-1/2} = (y_{n+1} - 2y_n + y_{n+1})/h = hf_n$. Then **the summed form** is defined by the formulas

$$\left.\begin{aligned} y_0 &= \alpha, \\ z_{1/2} &= \gamma + \tfrac{1}{2}hf_0, \\ y_n &= y_{n-1} + hz_{n-1/2}, \quad (n \geq 1), \\ z_{n+1/2} &= z_{n-1/2} + hf_n. \end{aligned}\right\} \tag{8.3.21}$$

The summed form is mathematically equivalent, but not numerically equivalent to the difference method in single-precision floating-point computations.

Example 8.3.8

Take the differential equation

$$y'' = k, \quad y(0) = 0.1, \qquad y'(0) = 0,$$

where $k = 0.654321$, $h = 0.01$, $n \leq 100$. Then one can easily show that $0.1 \leq y_n < 1$. Suppose that six-digit floating-decimal arithmetic is used. For the difference method each operation will be of the form:

$$\begin{array}{rl} 2y_n - y_{n-1} &= 0.\text{xxxxxx} \\ + \quad .h^2 f_n &= 0.000065\,4321 \\ \hline y_{n+1} &= 0.\text{yyyyyy} \end{array}$$

The last four digits have no influence. In fact, the equation $y'' = 0.65$ is treated instead of $y'' = 0.654321$.

†We shall call this the **explicit central difference method.** It is the simplest member of the **Störmer family** of methods.

For the summed form, the inequality $0.1 < z_n < 1$ holds for $n > 15$. Hence most operations will be of the form:

$$\begin{array}{llll} z_{n-1/2} & = 0.\text{uuuuuu} & y_n & = 0.\text{xxxxxx} \\ +hf_n & = 0.006543\,21 & +hz_{n+1/2} & = 0.00\text{vvvv vv} \\ \hline z_{n+1/2} & = 0.\text{vvvvvv} & y_{n+1} & = 0.\text{yyyyyy} \end{array}$$

In the calculating of z the last two digits of f are lost. This means roughly that for $x > 0.15$, the equation $y'' = 0.6543$ is treated. The resulting effect is roughly that the last two digits of z become unreliable; hence the shift of z in the addition for y is of minor importance. A clue to an explanation of the difference between the two algorithms is the role of z (which is computed by an addition of the form $O(1) + O(h)$) as a carrier of information from one step to the next. There is still a loss of accuracy in the summed form, but one has the same kind of loss, for instance, in numerical quadrature or with the methods for differential equations previously discussed. One could store z and y in double precision, while single precision is used in the computation of f, which is usually the most time-consuming part of the work. If such **partial double precision** is used, then the advantage of the summed form is reduced. See also Example 2.3.4 for a solution when double precision is not available.

The following expansion holds for the difference method just described (with the notation of the the previous section):

$$y(x, h) = y(x) + c_1(x)h^2 + c_2(x)h^4 + c_3(x)h^6 + \dots.$$

(As usual, the rounding errors are ignored in this expansion, and certain conditions about the differentiability of f have to be satisfied). The expansion shows that the global error is $O(h^2)$, and that higher accuracy can be obtained with **Richardson extrapolations** according to the scheme in (7.2.13) with headings $\Delta/3$, $\Delta/15$, $\Delta/63$, ...

There are, however, *other ways than Richardson extrapolation to improve the accuracy* by modifications of this method—e.g., by the following methods:

(a) **Implicit difference correction (Cowell's method)**—see Problem 7 in Sec. 7.2:

$$y_{n+1} - 2y_n + y_{n-1} = h^2\left[f_n + \frac{f_{n+1} - 2f_n + f_{n-1}}{12}\right]. \tag{8.3.22}$$

This method is known under several other names, e.g. Numerov's method. See Problem 8 of this section for a sufficiently accurate starting procedure.

(b) **Deferred difference correction** (suggested by L. Fox). One first computes a sequence y_n by solving the difference equation Eq

(8.3.19) (or the summed form), and computing a correction $C_n = h^2(f_{n+1} - 2f_n + f_{n-1})/12$. An improved sequence y_n is then obtained by solving the difference equation,

$$\hat{y}_{n+1} - 2\hat{y}_n + \hat{y}_{n-1} = h^2 f(x_n, \hat{y}_n) + C_n.$$

Note that the first sequence $\{y_n\}$ was used in the computation of C_n. The procedure can be iterated, and more sophisticated formulas for the correction C_n can be used.

(c) **Backward difference correction**—e.g.,

$$y_{n+1} - 2y_n + y_{n-1} = h^2\left[f_n + \frac{f_n - 2f_{n-1} + f_{n-2}}{12}\right].$$

The common basis of (a) and (b) is that $h^2(f_{n+1} - 2f_n + f_{n-1})/12$ is an accurate estimate of the local truncation error of the difference approximation in Eq. (8.3.19). Actually, the global error of the solution produced by these methods is $O(h^4)$. The estimate used in (c) is less accurate, and therefore the global error of method (c) is only $O(h^3)$. The price paid for the higher accuracy is more complicated computation. In (b), two difference equations are to be solved.

In Cowell's method one can proceed in the following way. Put $y_i - h^2 f_i/12 = u_i$. The difference equation then takes the form, similar to Eq. (8.3.20),

$$u_{n+1} - 2u_n + u_{n-1} = h^2 f_n, \tag{8.3.23}$$

although in order to compute f_n one has to solve for y_n from the equation

$$y_n - \frac{h^2 f(x_n, y_n)}{12} = u_n,$$

if the differential equation is nonlinear, this generally requires some iterative method; see Chap. 6. For the linear case, see Problem 10 in this section. The summed form of Cowell's method is similar to Eq. (8.3.21) with y replaced by u, and a more accurate starting procedure. Richardson extrapolation can be applied to Cowell's method. The headings are $\Delta/15$, $\Delta/63$, ...

The first method (Eq. (8.3.19) and its summed form) can be extended to equations of the form

$$y'' = f(x, y, y')$$

if one puts

$$y'_n \approx \frac{y_{n+1} - y_{n-1}}{2h} = \frac{z_{n+1/2} + z_{n-1/2}}{2}.$$

In this case the method becomes implicit.

The same choice between the four alternative ways to obtain higher accuracy occurs in many other problems—for example, boundary-value and eigenvalue problems and various problems with partial differential equations.

REVIEW QUESTIONS

1. What is a one-step method? Give an example. Give an example of a method which is not a one-step method.
2. How is repeated Richardson extrapolation used in (a) Euler's method? (b) the modified midpoint method?
3. Describe the modified midpoint method. What is the aim of the "modification"?
4. Explain in detail four different methods for numerical integration of ordinary differential equations. Comment on their advantages and disadvantages (amount of work, accuracy, stability properties).
5. What characterizes "stiff" differential equation problems?
6. What is meant by an implicit method and a predictor-corrector scheme?
7. Describe some ideas for the automatic control of step size.
8. What is meant by the summed form of, for example, the explicit central difference method, and what is its purpose?

PROBLEMS

1. Determine a Taylor expansion for the solution of the equation $y' = y^2$, $y(0) = 1$, about $x = 0$. Use this approximation to compute y for $x = 0.2$ and $x = 1.2$ to four decimals. Compare with the exact solution, and explain why the second case ($x = 1.2$) was unsuccessful.
2. Use the Runge-Kutta method to compute an approximation to $y(0.2)$, where $y(x)$ is the solution to the differential equation $y' = x + y$ with $y(0) = 1$. Compute with six decimals, for two different step sizes, $h = 0.2$ and $h = 0.1$. Extrapolate. Compare with the exact result.
3. The function $y(x)$ is defined by the problem
$$y' = x^2 - y^2, \quad y(0) = 1.$$
Compute $y(0.2)$ using the following methods:
(a) Euler-Richardson: $h = 0.1$ and $h = 0.2$;
(b) midpoint method: $h = 0.1$, with and without the modification;
(c) Runge-Kutta method: $h = 0.1$;
(d) Taylor series expansion up to and including the x^4-term;
(e) trapezoidal method: $h = 0.2$.
4. The following results were obtained for the previous differential equation problem using the modified midpoint method with different step sizes:

$h = 0.05$	$h = 0.1$
0.83602	0.83672.

Compute a better value using extrapolation.

5. Show that the damping of oscillations in the modified midpoint method can be expressed by the formula

$$\hat{y}_N = \tfrac{1}{4}(y_{N-1} + 2y_N + y_{N+1}),$$

where $x = a + Nh$ and y_{N+1} is computed according to Eq. (8.3.2). (See also Problem 4 of Sec. 7.2.)

6. We want to make a table of the function

$$y(x) = \int_0^\infty \frac{e^{-u^2}}{u + x}\, du$$

for various values of x. We can proceed in the following way: $y(x)$ is computed for $x = 1$ using some method for numerical integration; one finds $y(1) = 0.6051$. Show that y satisfies the differential equation

$$\frac{dy}{dx} + 2xy = -\frac{1}{x} + \sqrt{\pi}\,.$$

By solving the differential equation numerically with the initial value $y(1) = 0.6051$, more table values can be computed. Determine $y(x)$ for $x = 1.2$ and $x = 1.4$ by the Runge-Kutta method and estimate the error in $y(1.4)$.

7. Answer the questions corresponding to Problems 4 and 5 in Sec. 8.2, for the Runge-Kutta method.

8. (a) Put $hy'_0 = A\mu\delta y_0 + Bh^2\mu\delta y''_0 + R_T$, and determine A and B so that the remainder term R_T is of as high order as possible. Give an asymptotically correct expression for R_T.

   (b) Let $y(x)$ be the solution to the differential equation problem

$$y'' = f(x)y, \quad y(0) = \alpha, \quad y'(0) = \beta.$$

   Put $y(Kh) = y_K$, $K = -1, 0, 1, \ldots$. With Cowell's (Numeróv's) formula and the formula in (a), one gets a system of equations for determining y_{-1} and y_1. Give an asymptotically correct expression for the error in the determination of y_1 obtained from this system of equations.

   (c) Apply the formula in (b) to compute $y(0.1)$ in the case

$$y'' = e^x y, \quad y(0) = 0, \quad y'(0) = 1, \quad h = 0.1.$$

9. The function $y(x)$ is defined by the problem

$$y''' = yx, \quad y(0) = 1, \quad y'(0) = 0, \quad y''(0) = 1.$$

To determine $y(1)$, one puts $(x_n = n \cdot h)$

$$y'''(x_n) \approx \frac{\mu\delta^3 y_n}{h^3} = \frac{y_{n+2} - 2y_{n+1} + 2y_{n-1} - y_{n-2}}{2h^3}$$

and derives the recursion formula

$$y_{n+2} = 2y_{n+1} + 2h^3 x_n y_n - 2y_{n-1} + y_{n-2}, \quad y_0 = 1.$$

The initial values y_1, y_2, y_3 needed in the recursion formula can be obtained, for example, by using the Runge-Kutta method with step h.

(a) What is the order of accuracy of the method?

(b) The table below shows the results of calculations made according to th above method:

h	$h_0 = 0.2$	$\frac{h_0}{2}$	$\frac{h_0}{4}$	$\frac{h_0}{8}$	$\frac{h_0}{16}$	$\frac{h_0}{32}$	$\frac{h_0}{64}$
$y(1, h)$	1.54523	1.54568	1.54591	1.54593	1.54592	1.52803	1.5004

$\frac{h_0}{128}$	$\frac{h_0}{256}$
1.51262	1.48828

What is puzzling about these results? Explain what the trouble is.

10. Consider the application of Cowell's method to the linear equation

$$y'' = p(x)y + q(x).$$

Show that with the notation of Eq. (8.3.23),

$$f_n = \frac{p(x_n)u_n + q(x_n)}{1 - h^2 p(x_n)/12}.$$

11. Write programs for the solution of $y'' = f(x, y)$ with the difference method o Eq. (8.3.19) and the summed form of Eq. (8.3.21). Apply them to the equatio $y'' = -y$, compare with the exact solution, and print out the errors. Perforr a series of numerical experiments in order to get acquainted with the accurac obtained with Richardson extrapolations and to see the effect of roundin errors.

12. Consider the initial-value problem

$$\frac{d^2y}{dx^2} = (1 - x^2)y, \quad y(0) = 1, \quad \frac{dy}{dx} = 0 \quad \text{for} \quad x = 0.$$

(a) Show that the solution is symmetric about $x = 0$.
(b) Determine $y(0.4)$ using Cowell's method (without any special start for mula) with step lengths $h = 0.2$ and $h = 0.4$. Perform Richardson extra polation.

13. (a) Design a third order method for the solution of an ordinary differen tial equation $y' = f(x, y)$ based on the Adams-Bashforth coefficients o Eq. (8.3.10). Apply it to $y' = y^2$, $h = 0.1$, and compute y_3 and y_4 when

$$y_0 = 1.0000, \qquad y_1 = 1.1111, \qquad y_2 = 1.2500$$

are given. Use four decimals. Compare with the exact result.
(b) Improve the value of y_3 using the Adams-Moulton coefficients of Eq (8.3.11) truncated after the second difference term. Go on from this value compute y_4 using the Adams-Bashforth coefficients, compute f_4, an improve y_4 using the Adams-Moulton coefficients (predictor-correcto technique, Sec. 8.3.4). Improve y_4 by another iteration.
(c) What is the approximate convergence rate for the iterations in (b)?

8.4. ORIENTATION ON BOUNDARY AND EIGENVALUE PROBLEMS FOR ORDINARY DIFFERENTIAL EQUATIONS

8.4.1. Introduction

In this section we shall mainly consider boundary-value problems for a second-order scalar differential equation:

$$y'' = f(x, y, y'), \tag{8.4.1}$$

with **boundary conditions**

$$y(a) = \alpha, \qquad y(b) = \beta. \tag{8.4.2}$$

We assume that f has continuous partial derivatives of arbitrary order in the *closed* interval $[a, b]$. The ideas can be used on other types of equations and other types of boundary conditions—e.g.,

$$p_0 y(b) + p_1 y'(b) = p_2.$$

First, two methods will be described, the **shooting** method and the **band matrix** method. An important variant of the shooting method, **parallel shooting**, is mentioned briefly. In Secs. 8.6.4. and 8.6.5, some other methods for partial differential equations—e.g., the **collocation method** and the **Galerkin method**—are treated. These methods have also been applied to ordinary differential equations.

In contrast to the initial-value problem, it can happen that the boundary-value problem has several solutions or even no solution; compare the eigenvalue problem studied in Sec. 8.4.3.

8.4.2. The Shooting Method

There is also an initial-value problem for Eq. (8.4.1):

$$\begin{aligned} y'' &= f(x, y, y'), \\ y(a) &= \alpha, \quad y'(a) = \gamma. \end{aligned} \tag{8.4.3}$$

If γ is fixed, then the value of y at $x = b$ can be considered as a function of γ—say, $g(\gamma)$. The boundary-value problem of Eqs. (8.4.1) and (8.4.2) can then be written

$$g(\gamma) = \beta. \tag{8.4.4}$$

In the shooting method, one solves this equation with, for example, the secant method; see Sec. 6.4. One guesses a value γ_0, and computes (approximately) $g(\gamma_0)$ by solving the initial-value problem of Eq. (8.4.3) using one of the methods described in Sec. 8.2 or Sec. 8.3. One then chooses another value, γ_1, computes $g(\gamma_1)$ in the same way, and then, by Eq. (6.4.1):

$$\gamma_2 = \gamma_1 - (g(\gamma_1) - \beta)\cdot\frac{\gamma_1 - \gamma_0}{g(\gamma_1) - g(\gamma_0)},$$

$$\gamma_3 = \gamma_2 - (g(\gamma_2) - \beta)\cdot\frac{\gamma_2 - \gamma_1}{g(\gamma_2) - g(\gamma_1)}, \quad \text{etc.}$$

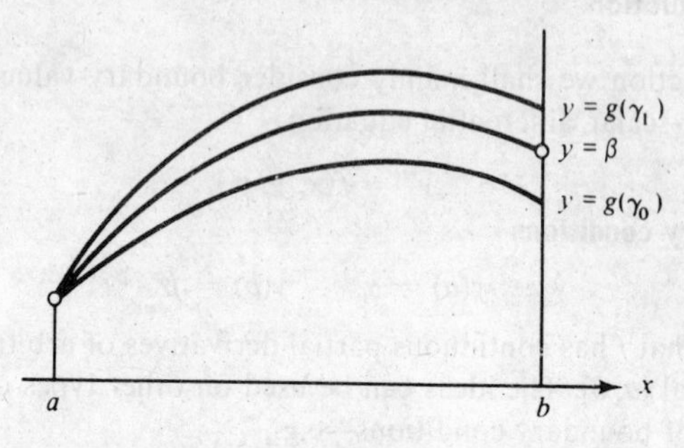

Fig. 8.4.1

One can show that Eq. (8.4.4) is a first-degree equation when Eq. (8.4.1 is a linear differential equation, even if the coefficients depend on x. In thi case, γ_2 is the solution to Eq. (8.4.4)—aside from the discretization errors and rounding errors in the computation of $g(\gamma_0)$ and $g(\gamma_1)$.

It sometimes happens that the initial-value problem, Eq. (8.4.3), is ill conditioned, even if the boundary-value problem, Eqs. (8.4.1) and (8.4.2), is well-conditioned. In such cases, the method described in Sec. 8.4.3 can be more advantageous. Another possibility is the parallel shooting method, also called multiple shooting; see Keller [102].

In **parallel shooting** the interval $[a, b]$ is divided into p subintervals $[x_{i-1}, x_i]$, $x_0 = a$, $x_p = b$. A new independent variable t is introduced, $t \in [0, 1]$. Then $y(x)$ is replaced by a $2p$-dimensional vector

$$\mathbf{y}(t) = [y_1(t), y_2(t), \ldots, y_p(t), y'_1(t), y'_2(t), \ldots, y'_p(t)]$$

such that $y_i(t) = y(x_{i-1} + t(x_i - x_{i-1}))$—i.e., y_i takes care of the variation in the subinterval $[x_{i-1}, x_i]$. The differential equation Eq. (8.4.3) is then replaced by the first-order system $\mathbf{y}' = \mathbf{f}(\mathbf{y}, t)$, where the continuity conditions, for $i = 1, 2, \ldots, p - 1$,

$$y_{i+1}(0) = y_i(1), \qquad \frac{y'_{i+1}(0)}{x_{i+1} - x_i} = \frac{y'_i(1)}{x_i - x_{i-1}}$$

appear as boundary conditions, in addition to the two conditions given in Eq. (8.4.3). (This is a particular case of the problem defined by Eqs. (8.4.11) and (8.4.12) in the next section.) When shooting is applied to this system, one obtains system of $2p$ equations for the vector $\mathbf{y}(0)$. The system is nonlinear if the differential equation is nonlinear. The techniques of Sec. 6.9 which do not require derivatives can then be applied. (It is also possible to use techiques where derivatives are needed, but then one has to solve the variational quation of the system $y' = f(y, t)$; see Sec. 8.1.2.)

4.3. The Band Matrix Method

Divide the interval $[a, b]$ into m equal parts and put $(b - a)/m = h$, $x_i = a + ih$. Let y_i denote the desired estimate of $y(x_i)$. Replace derivatives by difference quotients in all differential equations and boundary onditions involving derivatives as in Sec. 8.3.7, i.e.

$$y'(x_n) \approx \frac{y_{n+1} - y_{n-1}}{2h}, \qquad y''(x_n) \approx \frac{y_{n+1} - 2y_n + y_{n-1}}{h^2}.$$

n this way, the differential equation has been transformed into a **difference quation**, which in general yields a nonlinear system of equations,

$$y_{n+1} - 2y_n + y_{n-1} = h^2 f_n, \quad n = 1, 2, \dots, m - 1,$$
$$y_0 = \alpha, \qquad y_m = \beta,$$

where

$$f_n = f\left(x_n, y_n, \frac{y_{n+1} - y_{n-1}}{2h}\right),$$

or, in matrix form:

$$Ay = h^2 f(y) - r, \tag{8.4.5}$$

where

$$A = \begin{bmatrix} -2 & 1 & 0 & \cdots & 0 & 0 \\ 1 & -2 & 1 & \cdots & 0 & 0 \\ 0 & 1 & -2 & \cdots & 0 & 0 \\ \cdot & & & \cdot & & \\ \cdot & & & \cdot & & \\ \cdot & & & \cdot & & \\ 0 & 0 & 0 & \cdots & 1 & -2 \end{bmatrix}, \qquad r = \begin{bmatrix} \alpha \\ 0 \\ \cdot \\ \cdot \\ 0 \\ \beta \end{bmatrix}. \tag{8.4.6}$$

Thus A is a **band matrix** (and in fact a tridiagonal matrix in this example). If the differential equation is linear, then the system of equations is linear and tridiagonal—thus it can be solved with very little work. Note that the matrix of the system in Eq. (8.4.5) is not A, since f depends on y or y'.

For the error, even in nonlinear cases, we have

$$y(x, h) = y(x) + c_1(x)h^2 + c_2(x)h^4 + c_3(x)h^6 + \dots. \tag{8.4.7}$$

Richardson extrapolation can be used with the headings $\Delta/3$, $\Delta/15$, $\Delta/63, \dots$.

Example 8.4.1

The boundary-value problem

$$y'' + y = x, \quad y(0) = 1, \qquad y(\tfrac{1}{2}\pi) = \tfrac{1}{2}\pi - 1$$

has the exact solution $y(x) = \cos x - \sin x + x$. The difference equation becomes

$$(y_{n+1} - 2y_n + y_{n-1}) + h^2 y_n = h^2 x_n, \quad 1 \leq n \leq m-1,$$
$$y_0 = 1, \qquad y_m = \tfrac{1}{2}\pi - 1.$$

The solution of the system of equations for $m = 5$ and $m = 10$ is given by the following table:

$\frac{2x}{\pi}$	$y(x)$	$m = 5$ $y(x, 0.1\pi)$	$m = 10$ $y(x, 0.05\pi)$	$\frac{10^6\Delta}{3}$	Richard-son	$10^6 \times$ (Error in Richard-son)
0	1.000000	1.000000	1.000000	0	1.000000	0
0.1	0.988334		0.988402			
0.2	0.956199	0.956572	0.956291	−94	0.956197	−2
0.3	0.908255		0.908337			
0.4	0.849550	0.849741	0.849597	−48	0.849549	−1
0.5	0.785398		0.785398			
0.6	0.721246	0.721056	0.721199	+48	0.721247	1
0.7	0.662541		0.662460			
0.8	0.614598	0.614224	0.614505	+94	0.614599	1
0.9	0.582463		0.582395			
1.0	0.570796	0.570796	0.570796	0	0.570796	0

The methods for improving accuracy described in Sec. 8.3.7, can also be used for boundary problems. In particular, for equations of the form $y'' = f(x, y)$, **Cowell's method** gives $O(h^4)$-accuracy with about the same amount of computation as the $O(h^2)$-method just described.

If the equation is nonlinear, then one can use the techniques described in Sec. 6.9. As a first approximation one can—in the absence of a better proposal—choose

$$y_i^{(0)} = \alpha + \frac{(\beta - \alpha)i}{m}.$$

Then

$$y_0 = \alpha, \qquad y_m = \beta.$$

With **boundary conditions of the form**

$$p_0 y(b) + p_1 y'(b) = p_2,$$

one can introduce an *extra point*, $x_{m+1} = b + h$, and write the condition in the form

$$p_0 y_m + \frac{p_1(y_{m+1} - y_{m-1})}{2h} = p_2.$$

One can also instead put b *between* two net points. The form of the boundary conditions only affects the first and last rows in Eq. (8.4.5).

Finally, in boundary-value problems, eigenvalue problems, and partial differential equations one often encounters **differential expressions of the form**

$$\frac{d}{dx}\left(p(x)\frac{dy}{dx}\right). \tag{8.4.9}$$

These are approximated by

$$\frac{1}{h}\left(p_{n+1/2}\left(\frac{y_{n+1}-y_n}{h}\right)-p_{n-1/2}\left(\frac{y_n-y_{n-1}}{h}\right)\right). \tag{8.4.10}$$

The global error of this approximation is of the form $c_1(x)h^2 + c_2(x)h^4 + \ldots$.

For systems of first-order equations

$$\mathbf{y}' = \mathbf{f}(\mathbf{y}, x), \tag{8.4.11}$$

boundary conditions may be given in the form

$$A\mathbf{y}(a) + B\mathbf{y}(b) = \mathbf{c}, \tag{8.4.12}$$

where A, B are $s \times s$ matrices. If the trapezoidal method is used,

$$\mathbf{y}_{n+1} - \mathbf{y}_n = \tfrac{1}{2}h(\mathbf{f}(\mathbf{y}_n, x_n) + \mathbf{f}(\mathbf{y}_{n+1}, x_{n+1})),$$

one obtains a nonlinear system of simple structure. If no natural initial approximation is available, an **imbedding** technique (see Sec. 6.9.3) is recommended. One can introduce a parameter in the differential equation, or sometimes it may be sufficient to use the step size h of the difference methods as a parameter—i.e., one starts with a very crude net, and refines it successively in using, for instance, the previously obtained solution and interpolation as initial approximation for the solution on the next set of net points. *Shooting techniques* can also be applied to Eqs. (8.4.11) and (8.4.12); see Sec. 8.4.2.

8.4.4. Numerical Example of an Eigenvalue Problem

Example 8.4.2

For which values of λ does the boundary-value problem

$$\begin{gathered} y'' + \lambda y = 0, \\ y(0) = 0, \qquad y(1) = 0 \end{gathered} \tag{8.4.13}$$

have solutions other than $y = 0$?

The general solution to Eq. (8.4.13) is

$$y(x) = a\cos(x\sqrt{\lambda}) + b\sin(x\sqrt{\lambda}),$$

$$y(0) = 0 \Longrightarrow a = 0,$$

$$y(1) = 0 \Longrightarrow \sqrt{\lambda} = n\pi, \quad n = 0, \pm 1, \pm 2, \ldots.$$

Thus

$$\lambda = n^2\pi^2, \quad n = 1, 2, 3, \ldots.$$

($n = 0$ yields the trivial solution $y = 0$; $n = -k$ gives the same solution
$n = k$.) These values are called **eigenvalues** of the given problem (problem
differential equation + boundary conditions). The solutions of the equati
when λ is an eigenvalue are called **eigenfunctions**. In this example the eige
functions $b \sin(n\pi x)$ belong to the eigenvalue $\lambda = n^2\pi^2$.

Eigenvalue problems occur in most areas of classical and modern physi
(for eigenvibrations, etc.). Example 8.4.2 comes up, for example, in the co
putation of wave numbers for a vibrating string. Some other important pro
lems in partial differential equations from physics can be reduced, b
separation of variables, to eigenvalue problems for ordinary differenti
equations.

The **difference method** according to Sec. 8.4.2 gives an approximation $\lambda($
to the eigenvalues which satisfies

$$\lambda(h) = \lambda + c_2h^2 + c_3h^3 + c_4h^4 + \ldots,$$

where c_3 is zero in some cases (among others, in Example 8.4.3). (Note t
regularity assumptions made in Sec. 8.4.1.)

Example 8.4.3

For the problem in Example 8.4.2, the difference method (with $h =$
gives the system of equations

$$\frac{-2y_1 + y_2}{h^2} + \lambda y_1 = 0,$$

$$\frac{y_1 - 2y_2}{h^2} + \lambda y_2 = 0.$$

0 y_1 y_2 0
0 x_1 x_2 1

This is a homogenous system of equations, two equations, two unknown
which has a nontrivial solution if and only if

$$\begin{vmatrix} -2 + \lambda h^2 & 1 \\ 1 & -2 + \lambda h^2 \end{vmatrix} = 0.$$

Thus

$$\lambda h^2 - 2 = \pm 1, \quad (h = \tfrac{1}{3}),$$

with solutions

$$\lambda_1 = 9 \qquad \text{(Exact value } \pi^2 = 9.8696\text{)},$$

$$\lambda_2 = 27 \qquad \text{(Exact value } 4\pi^2 = 39.48\text{)}.$$

The higher eigenvalues cannot even be estimated when we have such a coars
net.

By similar calculations with various h we get the following results for th
smallest eigenvalue:

h			Richardson	Error
$\frac{1}{2}$	8			
$\frac{1}{3}$	9	$\frac{\Delta}{\frac{5}{4}} = 0.8$	9.8	−0.0696
$\frac{1}{4}$	9.3726	$\frac{\Delta}{\frac{7}{9}} = 0.4791$	9.8517	−0.0179

There are computational methods for solving eigenvalue problems for much larger matrices; see Sec. 5.8. By using Richardson extrapolation one can, however, obtain good accuracy with a reasonable number of points.

The same general procedure can be used with differential expressions of the form of Eq. (8.4.9) with the difference approximation of Eq. (8.4.10).

The **shooting method** can also be used on eigenvalue problems, see Problems 3 and 4 below.

REVIEW QUESTION

Describe in general terms the different ways to attack boundary-value problems presented in this section.

PROBLEMS

1. In using the shooting method on the problem

$$y'' = \frac{1}{2}y - \frac{2(y')^2}{y}, \quad y(0) = 1, \quad y(1) = 1.5,$$

the following results were obtained:

$$y'(0) = 0 \Longrightarrow y(1) = 1.543081,$$
$$y'(0) = -0.05 \Longrightarrow y(1) = 1.425560.$$

Which value of $y'(0)$ should be used on the next "shot"?

2. (From Collatz [91].) The displacement u of a loaded beam of length $2L$ satisfies (under certain assumptions) the differential equation

$$\frac{d^2}{ds^2}\left(EI(s)\frac{d^2u}{ds^2}\right) + Ku = q(s), \quad (-L \le s \le L),$$

with boundary conditions

$$u''(-L) = u'''(-L) = 0,$$
$$u''(\ L) = u'''(\ L) = 0.$$

For a certain beam we have:

$$I(s) = I_0\left(2 - \left(\frac{s}{L}\right)^2\right); \quad q(s) = q_0\left(2 - \left(\frac{s}{L}\right)^2\right); \quad K = \frac{40EI_0}{L^4}.$$

One wants to know the displacement at $s = 0$.

(a) Introduce more suitable variables in the problem, and write it as a system of two second-order equations with boundary conditions. Prove, and make use of the symmetry property $u(s) = u(-s)$. (Assume that the system has a unique solution.)

(b) Propose a difference approximation for this problem, where $h = L/N$, N an arbitrary positive integer. Count the number of equations and unknowns. Verify that for $N = 1$ one gets $u(0) = 13(c/280)$, where

$$c = \frac{q_0 L^4}{EI_0}.$$

(c) In a computation with $N = 2$, one obtained $u(0) = 0.045752c$, and with $N = 5$ one obtained $u(0) = 0.045332c$. Perform Richardson extrapolation first with $N = 1$ and $N = 2$, and then with $N = 2$ and $N = 5$.

(d) How should one number the variables to get an advantageous band width in the matrix?

3. (a) Compute, approximately, the least eigenvalue for which the problem

$$\frac{d}{dx}\left[(1 + x^2)\frac{dy}{dx}\right] + \lambda y = 0, \quad y(-1) = y(1) = 0$$

has a nontrivial solution. Use a difference method first with step size $h = \frac{2}{3}$ and then with $h = \frac{2}{5}$, and perform Richardson extrapolation. (*Hint:* Utilize the symmetry of the eigenfunction about $x = 0$.)

(b) Use the same difference method (with $h = \frac{1}{5}$) to treat the initial-value problem

$$y(0) = 1, \qquad y'(0) = 1, \qquad \text{(same differential equation as above)}$$

for $\lambda = 3.50$ and $\lambda = 3.75$. Use inverse interpolation to compute the intermediate value of λ for which $y(1) = 0$, and make a new computation with the value of λ so obtained. Then improve the estimate of the eigenvalue using Richardson extrapolation.

(c) Compute the next smallest eigenvalue in the same way. Note the connection between this approach and the method for computing matrix eigenvalues described in Sec. 5.8.3.

4. One seeks the solution of the eigenvalue problem

$$\frac{d}{dx}\left[\left(\frac{1}{1 + x}\right)\frac{dy}{dx}\right] + \lambda y = 0, \quad y(0) = y(1) = 0$$

by integrating, for a few values of λ, an equivalent system of two first-order differential equations with initial values $y(0) = 0$, $y'(0) = 1$, with the Runge-Kutta method. Computations using three different λ, each with three different step sizes, gave the following values of $y(1) \cdot 10^7$.

h \ λ	6.76	6.77	6.78
$\frac{1}{10}$	16,126	5,174	−5,752
$\frac{1}{15}$	15,396	4,441	−6,490
$\frac{1}{20}$	15,261	4,304	−6,627

Compute, for each value of λ, a better value for $y(1)$ using Richardson extrapolation. Then use inverse interpolation to determine the value of λ which gives $y(1) = 0$.

. The eigenvalue problem for a stretched circular membrane is

$$\frac{1}{r}\frac{d}{dr}\left(r\frac{du}{dr}\right) + \lambda u = 0,$$

with $u(1) = 0$, $u(0)$ and $u''(0)$ finite.

(a) Set up a difference equation with net points $r_i = (2i + 1)/(2N + 1)$, $i = -1, 0, 1, \ldots, N$. (Thus the origin lies between two net points and $h = 2/(2N + 1)$.

(b) Determine the smallest eigenvalue, first for $N = 1$ and then for $N = 2$. Perform Richardson extrapolation (under the assumption that the error is proportional to h^2).

(c) For large values of N, one would like to use a standard program for computing eigenvalues of symmetric matrices. How does one put the above problem in a suitable form?

Remark: The origin is a singular point for the differential equation, but not for its solution. The singularity causes no trouble in the above treatment.

5. Show that Eq. (8.4.4) is a first-degree equation when Eq. (8.4.1) is a linear differential equation, even if the coefficients depend on x. Show also that parallel shooting yields a linear system for the determination of $y(0)$ when the differential equation is linear.

8.5. DIFFERENCE EQUATIONS

Recursion formulas play a large part in numerical calculations. A recursion formula of the type

$$y_{n+k} = f(y_n, y_{n+1}, \ldots, y_{n+k-1}, n) \tag{8.5.1}$$

can, from Theorem 7.1.2, also be written in the form

$$\Delta^k y_n = g(y_n, \Delta y_n, \ldots, \Delta^{k-1} y_n, n). \tag{8.5.2}$$

An equation of the form of Eq. (8.5.2) or of Eq. (8.5.1) is called a ***k*th-order difference equation**. Its solution is uniquely determined when k initial values $y_0, y_1, \ldots, y_{k-1}$ are given, and these can be chosen arbitrarily—at least if the function f is defined in the whole space.

Example 8.5.1

From the difference equation $y_{n+2} - 5y_{n+1} + 6y_n = 0$ with initial conditions $y_0 = 0$, $y_1 = 1$, one gets recursively

$$y_2 = 5 \cdot 1 - 6 \cdot 0 = 5,$$

$$y_3 = 5 \cdot 5 - 6 \cdot 1 = 19,$$

$$y_4 = 5 \cdot 19 - 6 \cdot 5 = 65, \text{ etc.}$$

The term "recursion formula" brings to mind the step-by-step use of Eq. (8.5.1) when the initial values (as in Example 8.5.1) are known numerically. The term "difference equation" is also used to describe the situation where the solution is determined with the use of conditions other than initial conditions. *The general solution to a difference equation of* k*th order contains* k *arbitrary constants.*

This is reminiscent of the situation with a kth-order differential equation. There are many other parallels between the two types of equations. For some simple difference equations one can express y_n explicitly as a function of n. The resultant expressions and other results are greatly similar to results in related differential equation problems.

Such explicit expressions are perhaps not directly of great interest in numerical calculations, but they are very useful for studying how a numerical method functions (see Sec. 8.5.3). They are also of interest in many other areas of pure and applied mathematics.

8.5.1. Homogeneous Linear Difference Equations with Constant Coefficients

An equation of the form

$$y_{n+k} + a_1 y_{n+k-1} + \ldots + a_k y_n = 0 \tag{8.5.3}$$

is called a *homogeneous linear difference equation of* k*th order with constant coefficients.* It is satisfied by the sequence $\{y_j\}$, where $y_j = cu^j$ ($u \neq 0$, $c \neq 0$), if and only if

$$cu^{n+k} + a_1 cu^{n+k-1} + \ldots + a_k cu^n = 0,$$

that is, when

$$\varphi(u) \equiv u^k + a_1 u^{k-1} + \ldots + a_k = 0. \tag{8.5.4}$$

Equation (8.5.4) is called the *characteristic equation* of Eq. (8.5.3); $\varphi(u)$ is called the *characteristic polynomial.*

THEOREM 8.5.1

If the characteristic equation has k different roots, $u_1, u_2, \ldots, u_k$, then the general solution of Eq. (8.5.3) is given by the sequences $\{y_n\}$, where

$$y_n = c_1 u_1^n + c_2 u_2^n + \ldots + c_k u_k^n. \tag{8.5.5}$$

$c_1, c_2, \ldots, c_k$ are arbitrary constants.

Proof. That $\{y_n\}$ satisfies Eq. (8.5.3) follows from the previous comments and from the fact that the equation is linear. The parameters $c_1, c_2, \ldots, c_k$ can be adjusted to arbitrary initial conditions $y_0, y_1, \ldots, y_{k-1}$ by solving the system of equations

$$\begin{bmatrix} 1 & 1 & \cdots & 1 \\ u_1 & u_2 & \cdots & u_k \\ \cdot & \cdot & & \cdot \\ \cdot & \cdot & & \cdot \\ \cdot & \cdot & & \cdot \\ u_1^{k-1} & u_2^{k-1} & \cdots & u_k^{k-1} \end{bmatrix} \begin{bmatrix} c_1 \\ c_2 \\ \cdot \\ \cdot \\ \cdot \\ c_k \end{bmatrix} = \begin{bmatrix} y_0 \\ y_1 \\ \cdot \\ \cdot \\ \cdot \\ y_{k-1} \end{bmatrix}$$

That the matrix is nonsingular follows, for example, from Problem 5 of Sec. 7.3. (The determinant is of Vandermonde type and is thus equal to the product of all differences $(u_i - u_j)$, $i > j$, $1 < i \leq k$, which is nonzero.) Thus the general solution of the difference equation is given by Eq. (8.5.5).

Example 8.5.2

Difference equation:

$$y_{n+2} - 5y_{n+1} + 6y_n = 0.$$

Initial conditions: $y_0 = 0, \quad y_1 = 1$.

Characteristic equation: $u^2 - 5u + 6 = 0$, with roots $u_1 = 3$, $u_2 = 2$.

General solution: $u_n = c_1 \cdot 3^n + c_2 \cdot 2^n$.

The initial conditions give the system of equations

$$n = 0: \qquad c_1 + c_2 = 0,$$
$$n = 1: \qquad 3c_1 + 2c_2 = 1,$$

with solution $c_1 = 1$, $c_2 = -1$. Thus

$$y_n = 3^n - 2^n.$$

For example, one gets $y_4 = 81 - 16 = 65$, in agreement with Example 8.5.1.

Example 8.5.3

Difference equation:

$$T_{n+1}(x) - 2xT_n(x) + T_{n-1}(x) = 0, \quad n \geq 1, -1 < x < 1.$$

Initial conditions: $T_0(x) = 1, \quad T_1(x) = x$.

Characteristic equation:

$$u^2 - 2xu + 1 = 0, \text{ with roots } u = x \pm i\sqrt{1 - x^2}.$$

Set $x = \cos\varphi$. Then $u = \cos\varphi \pm i \sin\varphi$; thus

$$u_1 = \exp(i\varphi), \qquad u_2 = \exp(-i\varphi), \qquad u_1 \neq u_2.$$

General solution: $T_n(x) = c_1 \exp(in\varphi) + c_2 \exp(-in\varphi)$.

The initial conditions give:

$$n = 0: \qquad c_1 + c_2 = 1,$$
$$n = 1: \qquad c_1 \exp(i\varphi) + c_2 \exp(-i\varphi) = \cos\varphi,$$

with solution $c_1 = c_2 = \frac{1}{2}$. Thus

$$T_n(x) = \cos(n\varphi), \quad \text{where} \quad x = \cos\varphi.$$

Compare with the Tchebycheff polynomials, Sec. 4.4.1.

THEOREM 8.5.2

When u_i is an m-fold root of the characteristic equation, then the difference equation, Eq. (8.5.3), is satisfied by the sequence $\{y_j\}$, where

$$y_j = P(j)u_i^j.$$

P is an arbitrary polynomial of degree $(m-1)$.

The general solution of the difference equation is a linear combination of solutions of this form using all of the distinct roots of the characteristic equation.

Proof. We can write a polynomial P of degree $m-1$ in the form

$$P(j) = b_0 + b_1 j + b_2 j(j-1) + \ldots + b_{m-1} j(j-1)\ldots(j-m+2).$$

Thus it is sufficient to show that Eq. (8.5.3) is satisfied when

$$y_j = j(j-1)\ldots(j-p+1)u_i^j, \quad (p = 1, 2, \ldots, m-1), \qquad (8.5.6)$$

i.e., where y_j is equal to the value of $u^p \partial^p(u^j)/\partial u^p$ for $u = u_i$. Substitute this in the left-hand side of Eq. (8.5.3):

$$u^p \frac{\partial^p}{\partial u^p}[u^{n+k} + a_1 u^{n+k-1} + \ldots + a_k u^n] = u^p \frac{\partial^p}{\partial u^p}[\varphi(u)u^n]$$

$$= u^p\left[\varphi^{(p)}(u)u^n + \binom{p}{1}\varphi^{(p-1)}(u)nu^{n-1} + \ldots + \binom{p}{p}\varphi(u)\frac{\partial^p}{\partial u^p}(u^n)\right].$$

The last manipulation was made using Leibniz's rule.

Now φ and all the derivatives of φ which occur in the above expression are 0 for $u = u_i$, since u_i is an m-fold root. Thus $\{y_j\}$ in Eq. (8.5.6) satisfy the difference equation.

We obtain a solution with k parameters by adding solutions of this form derived from the different roots of the characteristic equation. It can be shown (see Henrici [100, p. 214]) that these solutions are linearly independent. Thus the theorem is proved.

Example 8.5.4

Difference equation: $y_{n+3} - 3y_{n+2} + 4y_n = 0.$

Characteristic equation: $u^3 - 3u^2 + 4 = 0$, with roots

$$u_1 = -1, \qquad u_2 = u_3 = 2.$$

General solution: $y_n = c_1(-1)^n + (c_2 + c_3 n)\cdot 2^n.$

8.5.2. General Linear Difference Equations

A general nonhomogeneous linear system of first-order difference equations can be written in the vector-matrix form

$$\boldsymbol{y}_{n+1} = \boldsymbol{B}_n \boldsymbol{y}_n + \boldsymbol{x}_n,$$

where $\{\boldsymbol{x}_n\}$ is a known sequence of vectors. If the initial value $\boldsymbol{y}_0$ is given, then by induction one gets

$$\boldsymbol{y}_n = \boldsymbol{P}_{n0}\boldsymbol{y}_0 + \sum_{j=1}^{n} \boldsymbol{P}_{nj}\boldsymbol{x}_{j-1}, \qquad \boldsymbol{P}_{nj} = \boldsymbol{B}_{n-1}\boldsymbol{B}_{n-2}\ldots\boldsymbol{B}_j, \quad \boldsymbol{P}_{nn} = \boldsymbol{I}. \tag{8.5.7}$$

In particular, if all $\boldsymbol{B}_n$ are equal to $\boldsymbol{B}$, then

$$\boldsymbol{y}_n = \boldsymbol{B}^n\boldsymbol{y}_0 + \sum_{j=1}^{n} \boldsymbol{B}^{n-j}\boldsymbol{x}_{j-1}. \tag{8.5.8}$$

Example 8.5.5

Difference equation: $y_{n+1} - 2y_n = a^n$.

Initial condition: $y_0 = 1$.

From Eq. (8.5.8) it follows that

$$y_n = 2^n + \sum_{j=1}^{n} 2^{n-j}a^{j-1}.$$

Difference equations of order k can be written as a system of first-order difference equations; see the corresponding result for differential equations in Sec. 8.1.1. One introduces, for example, the vector

$$\boldsymbol{y}_n = (y_n, y_{n+1}, \ldots, y_{n+k-1})^T.$$

For details, see Problem 5 at the end of this section.

Occasionally, however, it can be simpler not to go over to a system. For a nonhomogeneous system of order k, one can often determine a particular solution using the method of undetermined coefficients; thereafter, one can get the general solution by adding the general solution to the corresponding homogeneous difference equation.

Example 8.5.6

Difference equation: $y_{n+1} - 2y_n = a^n$.

Initial condition: $y_0 = 1$.

Try putting $y_j = ca^j$. One gets

$$ca^{n+1} - 2ca^n = a^n \qquad c = \frac{1}{a-2}, \quad a \neq 2.$$

Thus the general solution is:

$$y_n = \frac{a^n}{a-2} + c_1 \cdot 2^n.$$

For $n = 0$:

$$1 = \frac{1}{a-2} + c_1.$$

Thus

$$y_n = 2^n + \frac{a^n - 2^n}{a - 2} \quad (\text{for} \quad a \neq 2).$$

When $a \to 2$, l'Hospital's rule gives

$$y_n = 2^n + n2^{n-1} \quad (\text{for} \quad a = 2).$$

The reader is recommended to verify, as an exercise, that the results—despite appearances—agree with the result in Example 8.5.5. Also, notice how the substitution ca^j must be modified when $a = 2$, or, more generally, when a is a root of the characteristic equation.

Example 8.5.7

Difference equation: $y_{n+2} - 5y_{n+1} + 6y_n = 2n + 3(-1)^n$.
The characteristic equation has roots $u_1 = 3$, $u_2 = 2$; thus the general solution to the homogeneous equation is $c_1 3^n + c_2 2^n$. Try a particular solution of the form:

$$y_j = aj + b + c(-1)^j.$$

$$a[n + 2 - 5(n + 1) + 6n] + b(1 - 5 + 6) + c(-1)^n(1 + 5 + 6)$$
$$\equiv 2n + 3(-1)^n,$$
$$2an - 3a + 2b + 12c(-1)^n \equiv 2n + 3(-1)^n.$$

Match coefficients:

$$2a = 2,$$
$$-3a + 2b = 0,$$
$$12c = 3.$$

Thus $a = 1$, $b = \frac{3}{2}$, $c = \frac{1}{4}$—i.e., the general solution is:

$$y_n = n + \frac{3}{2} + \frac{(-1)^n}{4} + c_1 3^n + c_2 2^n.$$

8.5.3. Analysis of a Numerical Method with the Help of a Test Problem

One can often get good insight into how a numerical method functions by considering **test problems** which are simple enough to be analyzed theoretically but still so general that they can present some difficulty for a prospective numerical method.

The simple differential equation problem

$$\frac{dy}{dx} = qy, \quad y(0) = 1, \tag{8.5.9}$$

where q is a complex constant, is generally used as a test problem for method for the numerical treatment of ordinary differential equations. The exact solution of the problem is

$$y(x) = e^{qx}.$$

One can draw conclusions about how a method functions on the system $dy/dx = Ay$, where A is a constant, diagonalizable matrix, by checking its behavior on test equations with $q =$ an eigenvalue of A. (See Problem 5 of Sec. 8.2.)

Example 8.5.8

The **explicit midpoint method** Eq. (8.3.1), applied to the test equation gives the difference equation

$$y_{n+1} = y_{n-1} + 2hqy_n.$$

Put $hq = \alpha$.

Characteristic equation: $u^2 = 1 + 2\alpha u$, with roots

$$u_1 = \alpha + (1 + \alpha^2)^{1/2}, \qquad u_2 = \frac{-1}{u_1}.$$

Thus

$$u_1 = 1 + \alpha + \tfrac{1}{2}\alpha^2 + O(\alpha^4).$$

Compare

$$\exp(\alpha) = 1 + \alpha + \tfrac{1}{2}\alpha^2 + \tfrac{1}{6}\alpha^3 + O(\alpha^4).$$

Hence

$$u_1 = \exp(\alpha) - \tfrac{1}{6}\alpha^3 + O(\alpha^4)$$
$$= \exp(\alpha - \tfrac{1}{6}\alpha^3 + O(\alpha^4)). \qquad \text{(Check this!)} \qquad (8.5.10)$$

$$u_1^n = \exp[\alpha n(1 - \tfrac{1}{6}\alpha^2 + O(\alpha^3))].$$

For $x = nh$, $\alpha n = qhn = qx$,

$$u_1^n = \exp[qx(1 - \tfrac{1}{6}\alpha^2 + O(\alpha^3))].$$

If $|\alpha| \ll 1$, then u_1^n is close to the correct solution, when $x = nh$. Show that

$$u_1^n - \exp(qx) \approx -\tfrac{1}{6}\alpha^2 qx \exp(qx).$$

But u_2^n behaves in an entirely different way. Since $u_2 = -1/u_1$, it follows that

$$u_2^n = (-1)^n u_1^{-n} = (-1)^n \exp[-qx(1 - O(\alpha^2))].$$

If $q < 0$, then u_2^n produces exponentially growing oscillations, even though the solution to the differential equation decreases exponentially! This explains the **weak (in)stability** seen previously in Example 8.3.1.

The general solution is

$$y_n = c_1 u_1^n + c_2 u_2^n.$$

We shall now estimate the size of c_2 when the initial conditions are

$$y_0 = 1, \quad y_1 = \exp(qh),$$

$$n = 0: \quad c_1 + c_2 = 1,$$

$$n = 1: \quad c_1 u_1 + c_2 u_2 = \exp(\alpha),$$

$$c_2 = \frac{\exp(\alpha) - u_1}{u_2 - u_1} \approx \frac{\alpha^3/6}{-2} = \frac{-\alpha^3}{12},$$

using Eq. (8.5.10).

We shall now compare our results with Example 8.3.1. Thus put $q = -1$, $h = 0.1$, $x = 5$—i.e., $\alpha = -0.1$, $n = 50$:

$$|c_2 u_2^n| \approx 10^{-3} \frac{e^5}{12} = 0.0124,$$

which is in fact larger than the exact solution of the differential equation, which for $x = 5$, is 0.0067. When the oscillating-error component is dominant, then we should have

$$|c_2 u_2^n| \approx \tfrac{1}{2} |y_{51} - y_{50}| = \tfrac{1}{2}(0.01803 + 0.00775) = 0.0129.$$

Thus theory and experiment agree very well.

For most numerical methods, the long-range behavior of the solutions of the test problem depends on the quantity qh.

DEFINITION

The region of (absolute) stability of a numerical method for an initial-value problem is that set of (complex) values of qh for which all solutions of the test problem, Eq. (8.5.9), will remain bounded as $n \longrightarrow \infty$.

Example 8.5.9

For the midpoint method investigated in the previous example, the general solution (if $qh \neq 1$) is of the form $y_n = c_1 u_1^n + c_2 u_2^n$, where u_1, u_2 are the roots of the characteristic equation $u^2 - 2hqu - 1 = 0$. Since the product of the roots is equal to -1, stability implies that $|u_1| = |u_2| = 1$. Hence we may put $u = \exp(i\varphi)$, φ real, into the characteristic equation—i.e.,

$$e^{i2\varphi} - 2hqe^{i\varphi} - 1 = 0,$$

$$2hq = 2i \sin \varphi, \quad \varphi \text{ real}.$$

It follows that the region of stability of the midpoint method is the straight line from $-i$ to i.

Example 8.5.10

The trapezoidal method, Eq. (8.3.7), gives, in the test problem, the difference equation

$$y_{n+1} - y_n = \tfrac{1}{2}hq(y_{n+1} + y_n), \quad y_0 = 1.$$

Hence

$$(1 - \tfrac{1}{2}hq)y_{n+1} = (1 + \tfrac{1}{2}hq)y_n,$$

$$y_n = \left[\frac{1 + \tfrac{1}{2}hq}{1 - \tfrac{1}{2}hq}\right]^n \cdot$$

y_n is bounded if and only if $|\tfrac{1}{2}hq + 1| \leq |\tfrac{1}{2}hq - 1|$, and, by Fig. 8.5.1, this is equivalent to $\text{Re}(\tfrac{1}{2}hq) \leq 0$. *The left half-plane is the stability region of the*

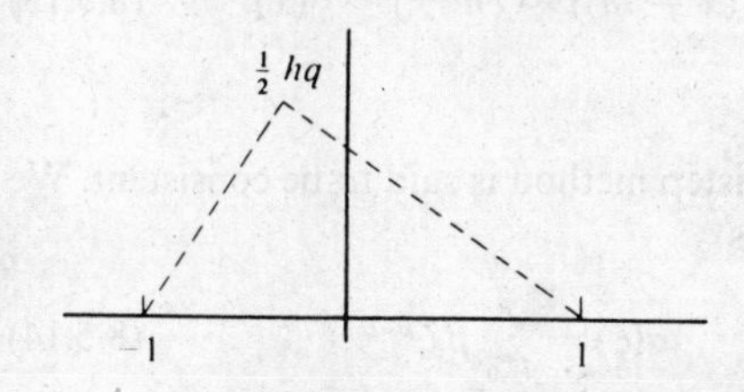

Fig. 8.5.1

trapezoidal method. Also note that if $\text{Re}(q) < 0$, then for any choice of step size the solution produced by the trapezoidal rule has the important property in common with the exact solution of the test problem that it tends to zero as nh becomes large.

If the trapezoidal method is applied to a linear system $dy/dx = Ay$, where some of the eigenvalues have large modulus and negative real part, it is relevant to consider $(1 + \tfrac{1}{2}qh)/(1 - \tfrac{1}{2}qh)$ when $|qh|$ is large. When $|qh| \to \infty$, this quantity tends to -1. This shows that there will appear slowly decreasing *oscillations*. The size of the oscillations can be reduced by the application of the smoothing formula given in Problem 4, Sec. 7.2. See also Sec. 8.3.5.

8.5.4. Linear Multistep Methods

A linear multistep method for the differential equation

$$\mathbf{y}' = \mathbf{f}(\mathbf{y}), \quad \mathbf{y}(a) = \mathbf{c} \tag{8.5.11}$$

is defined by a difference equation

$$\sum_{i=0}^{k} (\alpha_i \mathbf{y}_{n-i} - h\beta_i \mathbf{f}(\mathbf{y}_{n-i})) = \mathbf{0}. \tag{8.5.12}$$

This class includes several of the methods previously discussed, such as the midpoint method, Adams methods, and, with $k = 1$, Euler's method and the trapezoidal method. The $\mathbf{y}_n$ (which are intended to be estimates of $\mathbf{y}(a + nh)$) can be computed recursively from Eq. (8.5.12) if, in addition to the initial value $\mathbf{y}_0 = \mathbf{c}$, $k - 1$ more values $\mathbf{y}_1, \mathbf{y}_2, \ldots, \mathbf{y}_{k-1}$ are given. (If $\beta_0 \neq 0$, this may be true only for sufficiently small h.) These values can be obtained, for example, by Taylor expansion or by a one-step method. Suppose that for all polynomials y, of degree p,

$$\sum_{i=0}^{k} (\alpha_i y(x - ih) - h\beta_i y'(x - ih)) = 0.$$

The value of p can be determined by the techniques of Examples 7.2.1 and 7.2.2, which can also be used for the determination of the constant c in the following relation, valid for any function y with continuous derivatives of order $p + 1$:

$$\sum_{i=0}^{k} (\alpha_i y(x - ih) - h\beta_i y'(x - ih)) \sim ch^{p+1} y^{(p+1)}(x). \tag{8.5.13}$$

DEFINITIONS

Consistency. If $p \geq 1$, the multistep method is said to be consistent. We introduce the **generating polynomials**

$$\rho(\zeta) = \sum_{i=0}^{k} \alpha_i \zeta^{k-i}, \qquad \sigma(\zeta) = \sum_{i=0}^{k} \beta_i \zeta^{k-i}. \tag{8.5.14}$$

Consistency is equivalent to the equations (see Problem 17(b)),

$$\rho(1) = 0, \qquad \rho'(1) = \sigma(1).$$

Stability. *If the stability region* (defined in the previous section) *contains the origin, the method is said to be stable.* This is equivalent to the requirement that all solutions of the difference equation

$$\sum_{i=0}^{k} \alpha_i y_{n-i} = 0$$

should be bounded for all positive n. By Theorem 8.5.2 we have the following result.

THEOREM 8.5.3

The following **root condition** *is necessary and sufficient for stability of a linear multistep method: all roots of ρ should be located inside or on the unit circle, and the roots of unit modulus should be simple.*

In this terminology, even the midpoint method investigated in Examples 8.5.1 and 8.5.2 is stable, but the unpleasant properties displayed in the first example motivates the term *weakly stable* (or weakly unstable).

The relevance of the stability concept defined above is shown by the following theorem, which summarizes several theorems, proved in Henrici [100, Chap. 5]. (A hint to a proof is given by Problems 5 and 6(c) of this section.)

THEOREM 8.5.4

Suppose that $y(x)$ is a $p + 1$ times differentiable solution of the initial-value problem, Eq. (8.5.11), $\|y^{(p+1)}(x)\| \leq K_0$, $p \geq 1$, and suppose that f is differentiable for all y. Suppose further that $\{y_n\}$ is defined by the equations

$$y_j = y(x_j) + \epsilon_j \quad j = 0, 1, \ldots, k-1,$$

$$\sum_{i=0}^{k} (\alpha_i y_{n-i} - h\beta_i f(y_{n-i})) = \epsilon_n, \quad k \le n \le \frac{b-a}{h}.$$

If the multistep method is stable and satisfies Eq. (8.5.13), then there exist constants K_1, K_2, h_0 such that for all $x_n \in [a, b]$, $h \le h_0$,

$$\| y_n - y(x_n) \| \le \left(ch^p(x_n - a) \cdot K_0 + \sum_{i=0}^{n} \| \epsilon_i \| \right) K_1 e^{K_2(x_n - a)}. \tag{8.5.15}$$

K_1 depends only on the coefficients of the method, while K_2 also contains an upper bound for $\|f'(y)\|$.

In view of this result, the integer p is called the **order of accuracy** of the method.

It is sufficient to consider the trivial case $f(y) \equiv$ **const.** in order to show that stability and consistency are *necessary* for such a result, with $p > 0$. A corollary of the theorem and this remark (in a more precise formulation) is that

$$\text{Consistency} \wedge \text{Stability} \iff \text{Convergence}, \tag{8.5.16}$$

Convergence here includes uniform convergence in $[a, b]$, when $h \longrightarrow 0$, for all f which satisfy the assumptions made in Sec. 8.1.1, as well as a requirement that the effect of perturbations of the initial values should tend to zero when the perturbations do so themselves. The formulation given in Eq. (8.5.16) occurs in numerous other applications of finite-difference methods to ordinary and partial differential equations where the three concepts are defined appropriately for each problem area. "Consistency" usually means that the difference equation formally converges to the differential equation as $h \longrightarrow 0$, while "convergence" is related to the behavior of *the solutions* of the difference equations.

When $\beta_0 \neq 0$, the method is implicit; see Sec. 8.3.4. In general, some iterative solution of a nonlinear system is needed in each step. If the termination of the iteration is controlled by a tolerance on the residual, then an error bound can be obtained by the theorem. If a predictor-corrector technique with a fixed number of iterations is used, then this theorem does not guarantee that the $\|\epsilon_n\|$ do not grow. The stability properties of such a scheme can be different from the stability properites of the corrector method.

Predictor-corrector methods are considered in an extension of the multistep methods, named **multivalue methods**; see Gear [98, Chap. 9]. This extension is important in several other respects, but it is beyond the scope of this book.

It is important to distinguish between the stability question of the differential equation (see Figs. 8.1.2 and 8.1.3) and the stability questions for the numerical methods. The former is well-conditioned—e.g., if $\exp[L \cdot (b - a)]$ is of moderate size, where L is an upper bound for the logarithmic norm, defined at the end of Sec. 8.1.2. Compare the general distinction between an ill-conditioned problem and an ill-conditioned algorithm in Sec. 2.4.2.

The midpoint method exemplifies that K_2 may be positive even though L is negative in (8.5.15). Hence if $b - a$ is large, the error bound of Theorem 8.5.4 (as well as the actual error) can be large unless h and the perturbation level are very small. Therefore, *the stability concept just defined is insufficient*, in particular for the **stiff problems** of Sec. 8.3.5. For them the concept of **$A(\alpha)$-stability** is more adequate, which means that the stability region should include the sector $|\arg(qh) - \pi| < \alpha$. The special case of the left half-plane, $\alpha = \pi/2$, is called **A-stability**. The trapezoidal method is A-stable, according to Example 8.5.10. These definitions apply also to methods other than the linear multistep methods. There are several variations on the theme of stability; see Gear [98, Chap. 11].

The stability requirements impose restrictions on the attainable accuracy of linear multistep methods. Although for each k there exists a linear k-step method with $p = 2k$, one can show that *stability* implies $p \leq k + 2$ for even k and $p \leq k + 1$ for odd k. *A-stability* implies $p \leq 2$. Actually, the trapezoidal method has the smallest truncation error among all A-stable linear multistep methods. There are, however, several ways to get around the restriction $p \leq 2$ for A-stability, by a slight extension of the family of methods—e.g., an A-stable method of order 4 can be obtained by the use of passive Richardson extrapolation in connection with the trapezoidal method. Liniger and Odeh (*IBM Journal of Research and Development*, 1972) have shown that one can use several A-stable methods of order 2, and then obtain higher-order accuracy by taking a suitable linear combination of the results.

Finally, when $k > 1$, the $k - 1$ extra conditions necessary to define a solution of the difference equation Eq. (8.5.12), need not be initial values. One can also formulate a *boundary-value problem* for the difference equation, even when the given problem is an initial-value problem for the differential equation. The boundary-value problem can be stable for a difference equation, even though the root condition is not satisfied. This has been pointed out in an important article by J. C. P. Miller [25, pp. 63–98].

REVIEW QUESTIONS

1. Derive the expression for the general solution to a linear, homogeneous difference equation with constant coefficients. (Assume that the characteristic equation has only simple roots.)
2. Define the concepts stability region, $A(\alpha)$-stability, and A-stability. Why are the latter concepts appropriate for stiff problems?
3. Explain
   (a) the weak (in)stability of the midpoint method;

(b) the good stability properties of the trapezoidal rule.

4. Define consistency, order of accuracy, root condition, and stability for a linear multistep method. State in your own words the essentials of the theorems for linear multistep methods.

PROBLEMS

1. Solve the following difference equations;
   (a) $y_{n+2} - 2y_{n+1} - 3y_n = 0, \quad y_0 = 0, y_1 = 1;$
   (b) $y_{n+2} - 4y_{n+1} + 5y_n = 0; \quad y_0 = 0, y_1 = 2;$
   (c) $y_{n+2} - 2y_{n+1} - 3y_n = 0, \quad y_0 = 0, y_{10} = 1;$
   (d) $y_{n+2} + 2y_{n+1} + y_n = 0, \quad y_0 = 1, y_1 = 0.$

2. Solve the following difference equations:
   (a) $y_{n+1} - y_n = 2^n, \quad y_0 = 0;$
   (b) $y_{n+2} - 2y_{n+1} - 3y_n = 1, \quad y_0 = y_1 = 0;$
   (c) $y_{n+1} - y_n = n, \quad y_0 = 0;$
   (d) $y_{n+1} - 2y_n = n \cdot 2^n, \quad y_0 = 0.$

3. (a) Show that the characteristic equation for the difference equation

$$\sum_{i=0}^{k} b_i \Delta^i y_n = 0$$

is

$$\sum_{i=0}^{k} b_i (u-1)^i = 0.$$

   (b) Solve the difference equation $\Delta^2 y_n - 3\Delta y_n + 2y_n = 0$, $y_0 = 0$, $\Delta y_0 = 1$.
   (c) What is the characteristic equation for the difference equation

$$\sum_{i=0}^{k} b_i \nabla^i y_n = 0?$$

4. The Fibonacci numbers are defined by the recurrence relation

$$y_n = y_{n-1} + y_{n-2}, \quad y_0 = 0, y_1 = 1.$$

   (a) Calculate $\lim_{n \to \infty} y_{n+1}/y_n$.
   (b) The error of the secant method (see Sec. 6.4.2) satisfies approximately the difference equation $\epsilon_n = C\epsilon_{n-1}\epsilon_{n-2}$. Solve this difference equation. Determine p, such that $\epsilon_{n+1}/\epsilon_n^p$ tends to a finite nonzero limit as n tends to infinity. Calculate this limit.
   (c) Show that the sequence $\{y_{n+1}/y_n\}_0^\infty$ satisfies the condition for Aitken extrapolation. Compute a few terms, extrapolate, and check with the exact result.

5. (a) The difference equation

$$y_{n+k} + a_1 y_{n+k-1} + \ldots + a_k y_n = 0$$

can, after the substitution $\mathbf{y}_j = (y_j, y_{j+1}, \ldots, y_{j+k-1})^T$, be written in the form $\mathbf{y}_{n+1} = A\mathbf{y}_n$ What does the matrix A look like?

(b) The differential equation

$$y^{(k)} + a_1 y^{(k-1)} + \ldots + a_k y = 0$$

can, after the substitution $\mathbf{y} = (y, y', \ldots, y^{(k-1)})^T$, be written in the form $\mathbf{y}' = \mathbf{A}\mathbf{y}$. What does the matrix $\mathbf{A}$ look like?

6. (a) Suppose that a vector sequence $\{\mathbf{y}_n\}_0^\infty$ satisfies the inequality

$$\|\mathbf{y}_{n+1}\| \leq a\|\mathbf{y}_n\| + b, \quad a \geq 0, b \geq 0, n = 0, 1, 2, \ldots. \tag{8.5.17}$$

Show that

$$\|\mathbf{y}_n\| \leq a^n\|\mathbf{y}_0\| + \frac{b(1-a^n)}{1-a}, \quad \text{if} \quad a \neq 1,$$

$$\|\mathbf{y}_n\| \leq \|\mathbf{y}_0\| + nb, \qquad \text{if} \quad a = 1.$$

(b) Suppose that a vector sequence satisfies the recurrence relation

$$\mathbf{y}_{n+1} = \mathbf{A}_n\mathbf{y}_n + \mathbf{b}_n, \quad \|\mathbf{A}_n\| \leq m < 1, \|\mathbf{b}_n\| \leq b.$$

Show that, for all $n \geq 0$,

$$\|\mathbf{y}_n\| \leq \max\left\{\|\mathbf{y}_0\|, \frac{b}{1-m}\right\}.$$

(c) Suppose that $\mathbf{A}$ is a constant, diagonalizable matrix, the spectral radius of which less than or equal to 1. Show that if

$$\mathbf{y}_{n+1} = (\mathbf{A} + h\mathbf{B}_n)\mathbf{y}_n, \quad n = 0, 1, 2, 3, \ldots,$$

and if $\|\mathbf{B}_n\|_\infty \leq K$ for all n, then there exist constants K_1, K_2 such that $\|\mathbf{y}_n\|_\infty \leq K_1 \exp(K_2 nh) \cdot \|\mathbf{y}_0\|_\infty$ for all positive n.

(d) Modify the result in (a) for the case where b is replaced by ba^{n+1} in (8.5.17). *Note:* The vector version of Theorem 8.1.4 is obtained from (a) with $a = \exp(Lh)$, and (d) yields an interesting variant of this theorem.

7. (a) Give the general solution of the difference equation of Example 1.3.4,

$$y_n + 5y_{n-1} = \frac{1}{n}.$$

The particular solution

$$I_n = \int_0^1 \frac{x^n}{x+5}\,dx$$

was given in that example.

(b) Which solution satisfies the initial condition $y_0 = I_0 + \epsilon$?

(c) Which solution satisfies the condition $y_{10} = 0$?

(d) Verify the statement made in the example that the effect of all the rounding errors committed in the various steps of the computation is of less importance than the effect of the rounding of the initial value.

8. (a) Show that all solutions of the difference equation

$$y_{n+1} - 2\lambda y_n + y_{n-1} = 0$$

are bounded, as $n \longrightarrow \infty$, if $-1 < \lambda < 1$, while for any other λ in the complex plane there exists at least one solution which is unbounded.

(b) Let $\mathbf{A}$ be a diagonalizable matrix. Give, in terms of the eigenvalues of $\mathbf{A}$, a necessary and sufficient condition for the boundedness (as $n \longrightarrow \infty$) of all solutions of the difference equation

$$y_{n+1} - 2Ay_n + y_{n-1} = 0.$$

9. (a) For which values of the number λ does the boundary-value problem,

$$y_{n+1} - \lambda y_n + y_{n-1} = 0, \quad y_0 = y_{10} = 0,$$

have nontrivial solutions?

(b) Give the eigenvalues and eigenvectors of the tridiagonal 9×9 matrix with elements

$$a_{ij} = \begin{cases} 1, & \text{if } |i-j| = 1, \\ 0, & \text{otherwise.} \end{cases}$$

10. Consider an eigenvalue problem

$$\delta^2 y_n + \lambda N^{-2} y_n = 0, \quad y_0 = y_N = 0.$$

Let the eigenvalues be $\lambda_1(N) < \lambda_2(N) < \lambda_3(N) < \ldots$. Show that $\lambda_j(N)$ converges to the corresponding eigenvalue, $\lambda_j = (\pi j)^2$, of the differential equation treated in Example 8.4.2. Show also that

$$|\lambda_j(N) - \lambda_j| < \frac{\lambda_j(\pi j/N)^2}{12}$$

and that $\lambda_j(N)$ has an expansion in even powers of $1/N$.

11. Determine the stability region of

(a) the Euler method: $y_n - y_{n-1} = hf(y_{n-1})$;

(b) the backward Euler method: $y_n - y_{n-1} = hf(y_n)$.

(c) What is the order of accuracy of the latter method?

(d) Are the methods stable? Are they A-stable?

12. The stability region of the Runge-Kutta method, Eq. (8.3.6), is shown in Fig. 8.5.2. Calculate all the intersections of the region with the real axis and the

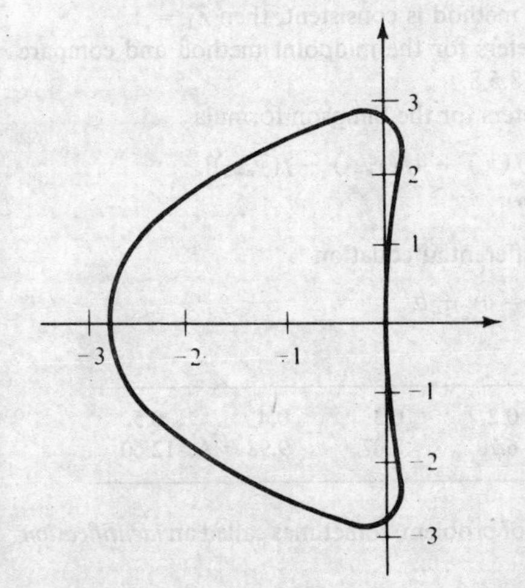

Fig. 8.5.2

imaginary axis. (The former can be found from the solution of Problem 7 of Sec. 8.3.)

13. For what positive values of the step size h does the Cowell method produce bounded solutions when applied to the differential equation

$$y'' + y = 0?$$

14. Show that the second-order backward differentiation method

$$\nabla y_n + \tfrac{1}{2}\nabla^2 y_n = hf(y_n)$$

is $A(0)$-stable. (Actually, it is A-stable.)

15. (a) Show that β_1, β_2 can be determined so that $p = 3$ for the multistep method,

$$y_n = -4y_{n-1} + 5y_{n-2} + h(\beta_1 f(y_{n-1}) + \beta_2 f(y_{n-2})).$$

Calculate the error constant c in Eq. (8.5.13).

(b) This method was applied with $h = 0.1$ to the problem $y' = y$, $y(0) = 1$. y_1 was equal to exp (0.1) with six correct decimals, and six decimals were used during the computation. The value y_{10} became negative! Explain why the error was so great.

16. Consider Example 8.5.8. Compute the effect of smoothing, according to Eq. (8.3.3), for $q = -1$, $h = 0.1$, $N = 10$.

17. Consider the application of the linear multistep method, defined by the polynomials ρ and σ, to the usual test problem $y' = qy$; $y(0) = 1$.

(a) Show that if ζ_j is a (simple) root of ρ, then the difference equation has a particular solution which is close to

$$y_n = \zeta_j^n \exp(\lambda_j q x_n), \quad (x_n = nh),$$

where the so-called *growth parameter* λ_j is given by

$$\lambda_j = \frac{\sigma(\zeta_j)}{\zeta_j \rho'(\zeta_j)}.$$

(b) Show that if $\zeta_1 = 1$ and the method is consistent, then $\lambda_1 = 1$.

(c) Compute the growth parameters for the midpoint method and compare with the results of Example 8.5.8.

(d) Compute the growth parameters for the Simpson formula

$$y_n - y_{n-2} = \tfrac{1}{3}h(f(y_n) + 4f(y_{n-1}) + f(y_{n-2})).$$

Will there be weak stability?

18. Fit the parameters a, b in the differential equation

$$y'' = ay + b$$

to the following data:

x	0.0	0.1	0.2	0.3	0.4	0.5
y	5.18	5.78	6.70	8.07	9.98	12.60

This is a simple example of a type of problem, sometimes called an *identification problem*, of great importance in control theory, chemistry, etc. A crude approach to this problem is to replace the differential equation by a difference equation—say, $h^{-2}\delta^2 y_n = \alpha y_n + \beta$—to fit the data to this for $n = 1$ and $n = 4$, and then to calculate a, b corresponding to α, β. (The parameter values can then be improved in many ways—e.g., by the techniques of Sec. 10.5).

(a) Give the formulas for the transformation from α, β to a, b.

(b) Carry out the data fitting according to the above, and check the difference equation for $n = 2, 3$.

8.6. PARTIAL DIFFERENTIAL EQUATIONS

8.6.1. Introduction

Computers have made it possible to treat many of the partial differential equations of physics, chemistry, and technology on a much larger scale than previously. The most common methods are **difference methods**—i.e., derivatives with respect to one or more of the variables are approximated by difference quotients. Some other methods, namely collocation, the Galerkin method, the Ritz method, and the finite-element method, will be presented in Secs. 8.6.4 and 8.6.5.

Numerical weather prediction is one of the most well-known applications of partial differential equations. Here, a model of flow in the atmosphere is described by a system of partial differential equations (with two or three space variables and a time variable). The most recently available observational data from a large area of the earth is the initial condition. On a large computer, it takes a few hours to compute the subsequent flow in the atmosphere for, say, three days ahead in time. The models of the atmosphere which are used are in many ways less complete than one would like them to be.

Partial differential equations with four independent variables x, y, z, t are one of the types of problems where one usually would be glad to have faster computers with more fast-access memory than is available today. Fortunately, one can often reduce the number of dimensions by making use of the *symmetry properties* of the problem or by so-called *separation of variables*. Problems with two independent variables are usually tractable.

For the most part, this chapter is only intended to give some insight into the possibilities that exist and some feeling for the difficulties with numerical stability which can occur. Some simple, general ideas are illustrated in the examples—these examples could be treated by more effective (but more difficult) methods.

A good introduction to numerical solution of *partial* differential equations is given in Smith [111]. More advanced treatments can be found in Richtmyer-Morton [110], Mitchell [107], Forsythe-Wasow [95], and Varga [81].

8.6.2. An Example of an Initial-Value Problem

In many applications, one wants to study some process in both time and space. In physics, a partial differential equation is often derived by first using laws of nature to derive approximate *difference* equations involving relevant physical quantities and small increments in time and space, and then passing to the limit to get partial differential equations. Construction of a numerical method often involves going "backward"—i.e., constructing difference equations from the differential equation. However, the art of the numerical analyst is to construct the difference scheme so that it not only is *consistent* with the differential equation, but is also *accurate* and *stable*. These concepts have already been discussed somewhat in Sec. 8.5.4.

Here we shall give a simple example which nevertheless is fairly typical: the problem is to compute the variation of temperature in a homogeneous wall, where initially the temperature is higher on one side than on the other. The curves which we shall compute—i.e., the temperatures in the wall at various times—will look approximately like those shown in Fig. 8.6.1. Here

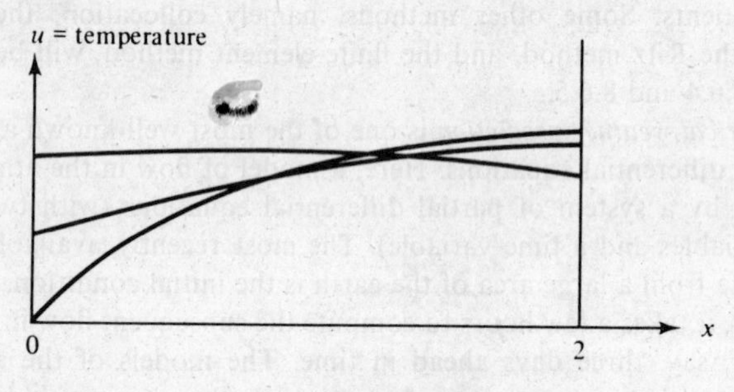

Fig. 8.6.1

$x = 0$ corresponds to one side of the wall, and $x = 2$ to the other. We now divide the wall into $N - 1$ layers of thickness $h = 2/N$, and denote the midpoints of the layers by $x_i = i \cdot h$ (see Fig. 8.6.2).

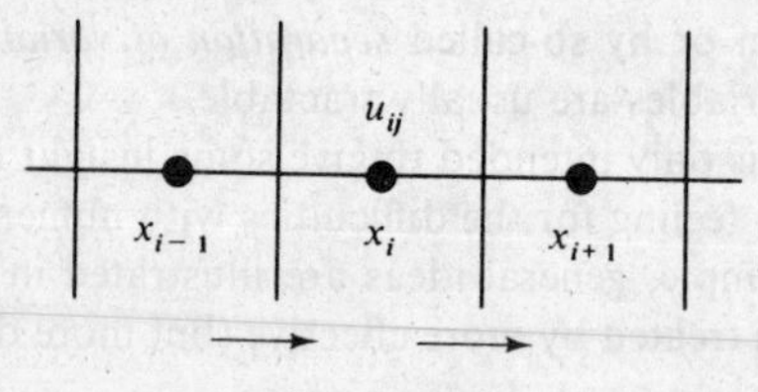

Fig. 8.6.2

We want to compute the temperature at equidistant times with time step k—i.e., at the times $t_j = j \cdot k$—and denote the temperature at the point x_i at time t_j by $u_{ij} = u(x_i, t_j)$. The first physical law which we shall use is that which says that the flow of heat past $x = \alpha$ is equal to a constant, κ, times $\partial u/\partial x$ at $x = \alpha$. The flow is measured, for example, in calories per second. The constant κ is called the heat conductivity. Applying this to the layer with center x_i, the amount of heat per second (see Fig. 8.6.2) flowing in at the left is $-\kappa \frac{\partial u}{\partial x}\Big|_{x=x_{i-1/2}}$, while the amount flowing out at the right is $-\kappa \frac{\partial u}{\partial x}\Big|_{x=x_{i+1/2}}$. Approximating these quantities by difference quotients, we get that the amount of heat flowing into the ith layer during $k = t_{j+1} - t_j$ seconds is

$$W = \left(-\kappa \frac{u_{ij} - u_{i-1,j}}{h} + \kappa \frac{u_{i+1,j} - u_{i,j}}{h}\right) \cdot k$$

$$= k\frac{\kappa}{h}(u_{i+1,j} - 2u_{ij} + u_{i-1,j}). \tag{8.6.1}$$

Now we shall use another physical law which says that the amount of heat which flows into a thin layer is approximately proportional to the change in temperature at the midpoint of the layer. In the layer with midpoint x_i the temperature has changed from u_{ij} at time t_j to $u_{i,j+1}$ at time t_{j+1}. The amount of heat W which comes in is thus $(u_{i,j+1} - u_{i,j})$ times a constant, which really consists of three constants—the specific heat c, the density ρ, and the thickness of the layer, h:

$$W = (u_{i,j+1} - u_{i,j}) \cdot c \cdot \rho \cdot h. \tag{8.6.2}$$

Equating Eqs. (8.6.1) and (8.6.2), we get

$$u_{i,j+1} - u_{i,j} = \frac{\kappa}{c \cdot \rho} \frac{k}{h^2}(u_{i+1,j} - 2u_{i,j} + u_{i-1,j}). \tag{8.6.3}$$

From this difference equation, approximate temperatures at time t_{j+1} can be computed if the temperature curve at time t_j is known, and if the temperature at the boundaries $x = 0$ and $x = 2$ is prescribed for all time t (boundary conditions). Also, dividing by k and letting h, $k \longrightarrow 0$, (and applying Eq. (1.2.5)) we get a partial differential equation, known as the **heat equation**,

$$\frac{\partial u}{\partial t} = \frac{\kappa}{c \cdot \rho} \frac{\partial^2 u}{\partial x^2}. \tag{8.6.4}$$

For simplicity in what follows, we assume that

$$\kappa = c \cdot \rho.$$

Boundary conditions can be prescribed in several ways, depending on the physical situation to be simulated. Here we shall assume that the temperature at $x = 0$ is given for all time t by $u(0, t) = 1{,}000 \sin(8\pi t/3)$, and also that at $x = 2$ we have perfect isolation—i.e., no heat flow at $x = 2$. Mathematically, we have then $(\partial u/\partial x)(2, t) = 0$. We also assume the **initial condition** $u(x, 0) = 1000 \sin(\pi x/4)$.

The first boundary condition is easy to put into Eq. (8.6.3)—i.e., we have

$u_{0,j} = 1{,}000 \sin (8\pi t_j/3)$. One way to put in the other boundary condition is to introduce a new point x_{N+1}; then

$$0 = \frac{\partial u}{\partial x}(2, t_j) \approx \frac{u_{N+1,j} - u_{N-1,j}}{2h} \quad \text{or} \quad u_{N+1,j} = u_{N-1,j} \quad \text{for all } j.$$

Since $u_{N-1,0}$ is known and $u_{N+1,0} = u_{N-1,0}$, it is clear that we can now compute $u(x_i, t_j)$ for all i, j using Eq. (8.6.3). If we choose $h = \frac{1}{4}$ and $k/h^2 = \frac{1}{2}$ (i.e., $k = \frac{1}{32}$), then we get the simple algorithm

$$u_{i,j+1} = \tfrac{1}{2}(u_{i-1,j} + u_{i+1,j}). \quad 1 \leq i \leq 8, \; j = 0, 1, 2, \ldots. \qquad (8.6.5)$$

Figure 8.6.3 shows the computational molecule in this case. With the above

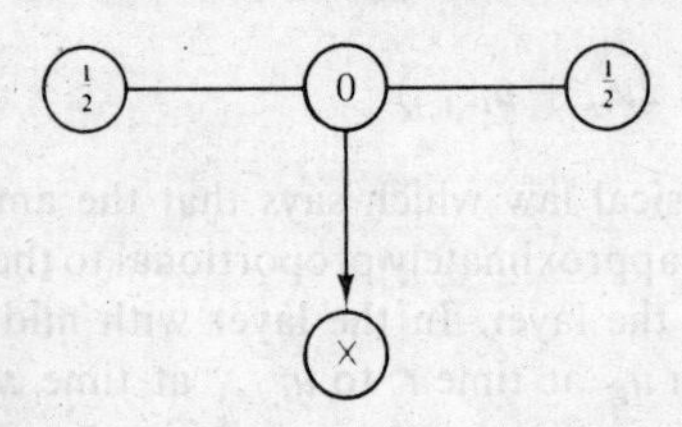

Fig. 8.6.3

values of k and h, the initial and boundary conditions become:

$$u_{i,0} = 1{,}000 \sin \left(\frac{\pi \cdot i}{16}\right),$$

$$u_{0,j} = 1{,}000 \sin \left(\frac{\pi j}{12}\right), \qquad u_{9,j} = u_{7,j}.$$

The computations can be arranged in a scheme as below; the reader is recommended to complete it.

	0	1	2	3	4	5	6	7	8	9
0	000	195	383	555	707	832	924	981	1000	981
1	258	192	375	545	694	816	906	962	981	962
2	500	316	368	534	680	800	889	944	962	944
3	707	434	425							
4	866									

One question which is especially important in the numerical solution of partial differential equations is: how is a disturbance propagated? To answer this, we can make an **experimental perturbational calculation**. Since the difference equation is linear, the difference between the perturbed and the unperturbed solution can also be computed by Eq. (8.6.5) with homogeneous boundary conditions. Thus a disturbance (perturbation) of magnitude 1 in ıe computation of $u_{3,1}$ is propagated according to the following scheme:

$\diagdown^{i}$	0	1	2	3	4	5	6	7	8	9
1	0	0	0	1	0	0	0	0	0	0
2	0	0	0.5	0	0.5	0	0	0	0	0
3	0	0.25	0	0.5	0	0.25	0	0	0	0
4	0	0	0.375	0	0.375	0	0.125	0	0	0
5	0	0.1875	0	0.375	0	0.25	0	0.0625	0	0.0625
6	0									

he effect of the disturbance is spread out and decreases. Notice the slight ymmetry due to the influence of the boundary values.

We can apply some of the ideas about stability and methods of ordinary fferential equations (see Sec. 8.3 and 8.5.3) by approximating Eq. (8.6.4) by system of ordinary differential equations. Again, put $h = 2/N$, $x_i = ih$, and t $u_i(t)$ denote an approximation to $u(x_i, t)$. Approximate $\partial^2 u/\partial x^2$ by a cen- al difference as previously, but retain the time derivative. (This procedure is metimes called the **method of lines**.) Then one gets a system of ordinary fferential equations:

$$\frac{du_i}{dt} = h^{-2}(u_{i-1} - 2u_i + u_{i+1}), \quad (i = 1, 2, \ldots, N). \tag{8.6.6}$$

he initial and boundary conditions give

$$u_0(t) = 1{,}000 \sin \frac{8\pi t}{3},$$

$$\frac{u_{N+1} - u_{N-1}}{2h} = 0, \quad \text{hence} \quad u_{N+1} = u_{N-1}.$$

ıese equations will now be written in vector form. Put $\boldsymbol{u} = (u_1, u_2, \ldots, u_N)^T$,

$$A = h^{-2}\begin{bmatrix} -2 & 1 & 0 & 0 & \cdots & 0 & 0 \\ 1 & -2 & 1 & 0 & \cdots & 0 & 0 \\ 0 & 1 & -2 & 1 & \cdots & 0 & 0 \\ \cdot & & & \cdot & & & \\ \cdot & & & \cdot & & & \\ \cdot & & & \cdot & & & \\ 0 & 0 & 0 & 0 & \cdots & -2 & 1 \\ 0 & 0 & 0 & 0 & \cdots & 2 & -2 \end{bmatrix}, \qquad \boldsymbol{f}(t) = h^{-2}\begin{bmatrix} u_0(t) \\ 0 \\ 0 \\ \cdot \\ \cdot \\ \cdot \\ 0 \\ 0 \end{bmatrix}.$$

ıen we get

$$\frac{d\boldsymbol{u}}{dt} = A\boldsymbol{u} + \boldsymbol{f}(t). \tag{8.6.7}$$

hen h is small, this is a "stiff" system. By Gerschgorin's theorem (Theorem 3.1) the eigenvalues of A are located in the circle:

$$\left|\lambda + \frac{2}{h^2}\right| \leq \frac{2}{h^2}. \tag{8.6.8}$$

We conclude that the eigenvalues have non-positive real parts and that the spectral radius is less than $4/h^2$. In fact, the eigenvalues are all real and negative, and one of them is close to $-4/h^2$.

Euler's method gives, on this system, exactly Eq. (8.6.3), (with $k/(c \cdot \rho) = 1$). It follows from Problem 5 of Sec. 8.2 that the computations are stable when $k \cdot 4/h^2 \leq 2$, i.e. *stable if* $k/h^2 \leq \frac{1}{2}$. A condition of this type was first obtained by **Courant, Friedrichs**, and **Lewy** in 1928 in a pioneering investigation of the *convergence* of the solutions of partial difference equations, as the grid spacings h, k tend to zero.

If one chooses $k/h^2 > \frac{1}{2}$, then with time the approximate solution becomes a more and more ragged function of x, which has very little in common with the solution of the differential equation.

The **trapezoidal rule** applied to the above system is called the **Crank-Nicolson method**. (Notice that the matrix A is **tridiagonal**, so elimination can be used conveniently when the equation is linear; for nonlinear equations an iterative process is not burdening, see Sec. 8.4.) This method is more accurate than Euler's method. If follows from Example 8.5.10 that it is *stable for all positive values of h and k*, whence it is possible to take longer time steps, but there may appear slowly decreasing oscillations in the computed values which require smoothing.

One might think that it would be better to use the explicit midpoint method, which here gives

$$u_{i,j+1} - u_{i,j-1} = \frac{2k}{h^2}(u_{i+1,j} - 2u_{ij} + u_{i-1,j}).$$

For $k/h^2 = \frac{1}{2}$ we get the computational molecule of Fig. 8.6.4. The "weak

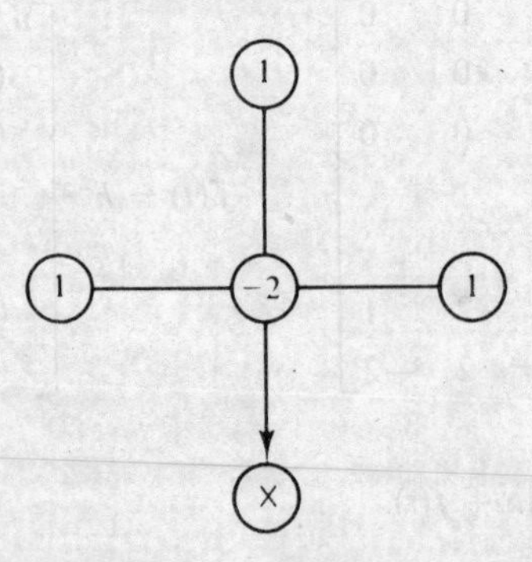

Fig. 8.6.4

instability" mentioned in Sec. 8.3.1 here leads to catastrophe. (Not even the oscillation-damping modification of the midpoint method would help here.)

$j \backslash i$	0	1	2	3	4	5	6	7	8	9
0	0	0	0	0	0	0	0	0	0	0
1	0	0	0	1	0	0	0	0	0	0
2	0	0	1	−2	1	0	0	0	0	0
3	0	1	−4	7	−4	1	0	0	0	0
4	0	−6	17	−24	17	−6	1	0	0	0
5	0	30	−68	89	−68	31	8	1	0	1
6	...			−338						

If there is a risk of numerical instability with a given method, it can usually e discovered by a perturbational calculation of this type.

The explicit midpoint method is useless in this problem because its tability region (see Example 8.5.9) is only a segment of the imaginary axis, vhile the matrix A has negative eigenvalues of large magnitude. The latter ; typical of **parabolic** problems, one example of which is the heat equation.

Nevertheless, the explicit midpoint method—or the **leap-frog method**, s it is usually called in this context—has found important applications to artial differential equations. There are many problems (**hyperbolic** systems) vhere the eigenvalues of the matrix which corresponds to the matrix A of his example, are all imaginary.

A systematic treatment of the design and analysis of stable and accurate lifference methods would require mathematical tools which are beyond the cope of this book. We refer the reader to the literature mentioned at the end of the previous section; see also Sec. 13.6. We hope, however, that the eader who encounters an initial value problem for a more difficult partial lifferential equation will derive some guidance from the points of view which ve have applied above. See also the formulas in Sec. 7.7.2 and the examples of nonlinear ordinary differential equations earlier in this chapter.

.6.3. An Example of a Boundary-Value Problem

One important equation of physics is **Poisson's equation**,

$$\nabla^2 u = f(x, y).$$

To get a solution to this equation we need **boundary conditions**: u or $\partial u/\partial n$, or some linear combination of u and $\partial u/\partial n$, are prescribed on the boundary of a closed region D. The solution is to be obtained for the points in the in- erior of D. ($\partial u/\partial n$ means the derivative of u taken in the direction of the normal to the boundary of D.) Poisson's equation and the special case when $\equiv 0$, **Laplace's equation**, are the best-known examples of **elliptic** differential equations. As a rule, boundary conditions are presented with such equations.

(Initial value problems are extremely ill-conditioned; see Problem 11 on page 402.)

Consider a square net with mesh width h—i.e., $k = h$. Suppose that the boundary consists of lines that are either parallel to the coordinate axes or form 45-degree angles with them—see Fig. 8.6.5. Even curved boundaries are

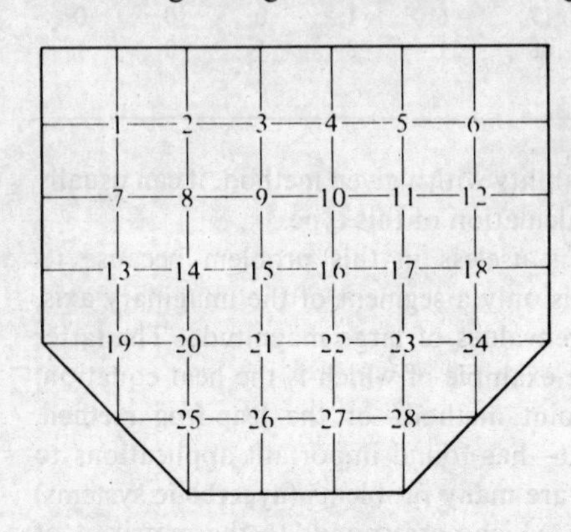

Fig. 8.6.5

treated in the special literature—see, e.g., Smith [111 p. 139]. If ∇^2 is approximated by ∇_5^2 (see Sec. 7.7.2), then one gets the difference equation

$$u_{i+1,j} + u_{i,j+1} + u_{i-1,j} + u_{i,j-1} - 4u_{i,j} = h^2 f_{ij}. \tag{8.6.9}$$

Since the difference equation is based on an approximation of the derivatives with a local truncation error which is $O(h^2)$ if $u \in C^4$, then one can expect that the truncation error in the solution is $O(h^2)$ and that Richardson extrapolation should be applicable. But this depends in general on the (given) boundary conditions.

Enumerate the points in the net as in Fig. 8.6.5, and form the vector

$$\boldsymbol{u} = (u_1, u_2, \ldots, u_n)^T$$

of the desired function values. The pair of integers (i, j) is replaced by the point's number. The difference equation can then be written in the form

$$A\boldsymbol{u} = h^2 \boldsymbol{f}.$$

For simplicity, we shall limit the discussion to the so-called **Dirichlet problem,** where only the values of the function u are prescribed at the boundary. The following hold:

1. f consists partly of values f_{ij} and partly of the values of u on the boundary.
2. All the diagonal elements in $A = \{a_{pq}\}$ are equal to -4.
3. $a_{pq} = 1$ if point p is a neighbor of point q. Otherwise, $a_{pq} = 0$. From this it follows that $a_{pq} = a_{qp}$—i.e., A is symmetric.
4. No point has more than four neighbors. The structure of A is evident

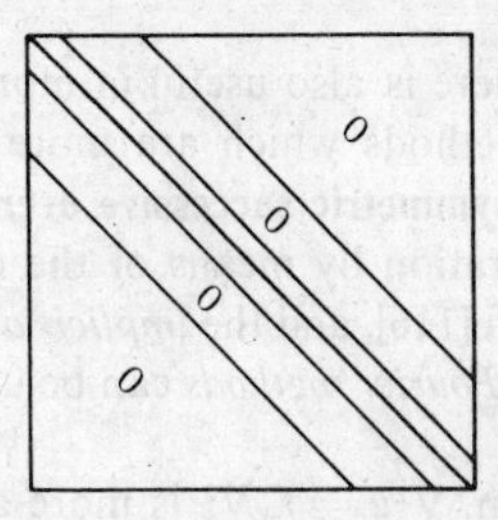

Fig. 8.6.6

from Fig. 8.6.6, where the lines parallel to the main diagonal indicate the location of the nonzero elements. Clearly, A is a **band matrix**.

In the particular example with 28 points which we discussed earlier (see Fig. 8.6.5), $a_{pq} = 0$ for $|q - p| > 6$. If h is halved, then the band width is approximately doubled, while the order of the matrix is (approximately) quadrupled. When the number of points is not too large, **Gaussian elimination** can be used; see Sec. 5.4.2. (Also, one can often reduce the order of the matrix by taking advantage of symmetry in the problem.) By using Richardson extrapolation one can often get sufficient accuracy with a reasonably small number of points.

It can happen, however, that h must be chosen so small that Gaussian elimination is no longer practicable.

In such cases, the **successive overrelaxation method** (SOR, see Sec. 5.6) is more attractive. In this case (see Eq. (8.6.9)), it means that the formula

$$u_{ij}^{(n+1)} = u_{ij}^{(n)} + \omega\cdot\tfrac{1}{4}(u_{i-1,j}^{(n+1)} + u_{i,j-1}^{(n+1)} + u_{i+1,j}^{(n)} + u_{i,j+1}^{(n)} - 4u_{ij}^{(n)} - h^2 f_{ij})$$

is used, where the points (x_i, y_j) are traversed in the same way as the numbering in Fig. 8.6.5. The process converges if $0 < \omega < 2$. For the case shown in the figure, $\omega \approx 1.4$ would give reasonably fast convergence. When there are more net points, then ω should be chosen closer to 2 (see Theorem 5.6.3). In the literature, certain rules are given for the ordering of the points (which is important if overrelaxation is used) and the choice of ω. In more difficult problems, one must experiment to find the best value, though this experimentation can be done in a systematic way. For more details, see the references given in Sec. 8.6.1. The attainable rate of convergence becomes lower the more points one has in the net.

A good strategy for boundary-value problems with linear differential equations is to *start with a coarse net*, and use *Gaussian elimination*. If the accuracy is insufficient, then one can use the results thereby obtained—possibly with extrapolation—to construct a good *initial approximation for an iterative method* with a finer net. In constructing this initial approximation one also needs values of u in the midpoints of those squares at whose corners u has been previously computed. To accomplish this, one can use the operator $\nabla_\times^2$; see Fig. 7.7.3 and Problem 5 at the end of this section.

The methodology described here is also useful in more complicated differential equations. There are methods which are more effective, but also more difficult—for example, the **symmetric successive overrelaxation (SSOR) methods** with convergence acceleration by means of the conjugate gradient method or Chebyshev acceleration [116], and the *implicit alternating-direction methods*. For Poisson's equation, *Fourier methods* can be used, and also some modern direct methods, see [79].

Also, for the Poisson equation, $\nabla^2 u = f$, ∇_9^2 is more advantageous than ∇_5^2, since from Eq. (7.7.8) we have

$$\nabla_9^2 u = \nabla^2 u + \frac{h^2 \nabla^4 u}{12} + O(h^4) = f + \frac{h^2 \nabla^2 f}{12} + O(h^4).$$

Thus if one uses the difference equation

$$\nabla_9^2 u_{ij} = f_{ij} + \frac{h^2 \nabla_9^2 f_{ij}}{12} \tag{8.6.10}$$

instead of Eq. (8.6.9), then one gets a truncation error which is $O(h^4)$—under certain smoothness conditions, on u and f. With this operator, however, the overrelaxation method must be slightly modified.

8.6.4. Methods of Undetermined Coefficients and Variational Methods

To begin with, we consider a *linear* differential equation, valid in a domain $D \subset R^k$, with linear boundary conditions. We write this symbolically in the form

$$Au = f \quad \text{in} \quad D \qquad \text{Differential equation} \tag{8.6.11}$$

$$Bu = g \quad \text{in} \quad \partial D \qquad \text{Boundary condition } (\partial D = \text{boundary of } D), \tag{8.6.12}$$

where A and B are linear operators. (Linearity means that $A(\alpha y + \beta v) = \alpha Ay + \beta Av$ for all constants α, β, and for all functions u, v for which Au, Av, Bu, Bv exist.) *We assume that this problem has a unique solution u^*.*

Example 8.6.1

$$-\frac{d^2 u}{dx^2} = f(x), \quad x \in [0, 1],$$

$$u(0) = u(1) = 0.$$

The boundary condition can be written: $u(x) = 0$, $x \in \{0\} \cup \{1\}$.

Example 8.6.2. *The Dirichlet problem for Poisson's Equation on a Region D*

$$-\nabla^2 u = f(x, y), \quad (x, y) \in D.$$

$$u(x, y) = g(x, y), \quad (x, y) \in \partial D.$$

or a linear problem, it is no restriction to *assume that either* $f \equiv 0$ or $g \equiv 0$. ▸ne only needs to replace u by $u - v$, where v satisfies the differential quation or the boundary condition; see Problem 13. (Actually, it is sufficient ıat either the differential equation or the boundary condition be linear.)

There are a great variety of methods where one seeks an approximate olution in the form

$$u \approx c_1\psi_1 + c_2\psi_2 + \dots + c_n\psi_n, \tag{8.6.13}$$

/here the ψ_i are n given linearly independent functions which span an n-imensional function space H_n. The methods differ in the choice of H_n as vell as in the method of fitting the coefficients to the problem. Put

$$\boldsymbol{u}_n = (c_1, c_2, \dots, c_n)^{\mathrm{T}}.$$

First, suppose that $g = 0$. It is convenient to choose H_n such that $Bu = 0$, $\forall u \in H_n$. In the classical applications the ψ_i were usually polynomials or rigonometric functions, multiplied by some appropriate function in order to atisfy the boundary conditions. Sometimes eigenfunctions of a simpler equa-ion with the same boundary conditions were chosen. A different choice of he ψ_i is used in the finite-element method; see Sec. 8.6.5.

It then remains to approximate f by a linear combination of the functions $\phi_i = A\psi_i$. In Chap. 4 the discussion of such **approximation problems** was ormulated for functions of one variable, but *the formalism and Theorem 4.2.5 ıave a more general validity* and can be used for **minimizing** $\|f - Au\|$, pro-vided that a scalar product (u, v) and the norm or seminorm $\|u\| = (u, u)^{1/2}$ ıave been defined which satisfy the conditions given in Sec. 4.1.3—e.g.,

$$(u, v) = \sum_{i=1}^{m} u(P_i)v(P_i) \quad P_i \in D, \tag{8.6.14}$$

or

$$(u, v) = \int_D\int u(x, y)v(x, y)\, dx\, dy. \tag{8.6.15}$$

If a discrete point set $\{P_i\}_1^m$ is used, the method is called **collocation**. We hen obtain a linear system with matrix elements

$$a_{ij} = A\psi_j(P_i),$$

overdetermined if $m > n$.

In the **Galerkin method**, the coefficients are determined by the requirement that *the residual function* $f - Au$ *should be orthogonal to all* $v \in H_n$, *in the sense of Eq. (8.6.15)*—i.e.,

$$(\psi_i, Au - f) = 0, \quad i = 1, 2, \dots, n.$$

This yields a linear system

$$A_n \boldsymbol{u}_n = f_n, \tag{8.6.16}$$

where the elements of A_n and f_n are $(\psi_i, A\psi_j)$ and (ψ_i, f), respectively. (The minimization of $\|f - Au\|$ mentioned earlier yields a similar system, but the matrix elements are $(A\psi_i, A\psi_j)$ and $(A\psi_i, f)$). (Verify these statements!)

If one wants to add one more term in Eq. (8.6.13), and if the LU-decomposition of A_n is given, the solution of the system $A_{n+1}u_{n+1} = f_{n+1}$ is obtained by $O(n^2)$ operations. (The same remark applies to collocation, if $m = n$.) Results for different values of n can be useful for estimating the order of magnitude of the error and, perhaps, for convergence acceleration.

An operator A in an inner-product space H is *symmetric* if $(u, Av) = (Au, v)$, $\forall u, v \in H$. A symmetric operator is *positive-definite* if $(u, Au) > 0$ when $u \neq 0$. In our case, all $u \in H$ satisfy the boundary conditions $Bu = 0$.

Example 8.6.3

Let

$$(u, v) = \int_0^1 u(x)v(x)\,dx, \quad u(0) = v(0) = u(1) = v(1) = 0.$$

Then $A = -d^2/dx^2$ is symmetric and positive-definite. We have, namely, $(Au, v) = -\int_0^1 u''(x)v(x)\,dx = [-u'(x)v(x)]_0^1 + \int_0^1 u'(x)v'(x)\,dx$. Because of the boundary conditions,

$$(Au, v) = \int_0^1 u'(x)v'(x)\,dx.$$

(u, Av) yields the same result. Evidently, $(Au, u) > 0$, unless $u'(x) = 0, \forall x$, but $u(0) = 0, u'(x) = 0, \forall x \Rightarrow u(x) = 0, \forall x$.

The same result is true for $A = -\nabla^2$, and we have the useful formula

$$(-\nabla^2 u, v) = (\nabla u, \nabla v). \tag{8.6.17}$$

More generally,

$$(-\text{div}\,(\rho\nabla u), v) = (\rho\nabla u, \nabla v). \tag{8.6.18}$$

If A is positive-definite, a new scalar product $\langle\cdot, \cdot\rangle$ and norm can be defined,

$$\langle u, v\rangle = (u, Av), \quad \|u\|^2 = \langle u, u\rangle. \tag{8.6.19}$$

(It is easy to verify the conditions of Sec. 4.2.2.)

THEOREM 8.6.1

Suppose that A is positive-definite. Let u^* be the solution of $Au^* = f$, $Bu^* = 0$. Then the Galerkin method gives the best approximation in H_n to u^*, measured in the $\langle\cdot, \cdot\rangle$ norm.

Proof. By Theorem 4.2.5, the solution of this minimization problem is given by the condition $\langle v, u - u^*\rangle = 0$, $\forall v \in H_n$. But $\langle v, u - u^*\rangle = (v, Au - Au^*) = (v, Au - f)$, and the Galerkin condition is $(v, Au - f) = 0$, $\forall v \in H_n$.

As an exercise, the reader can derive the relation

$$(u, Au) - 2(f, u) = \langle u - u^*, u - u^* \rangle - \langle u^*, u^* \rangle.$$

It follows from this that $u = u^*$ *minimizes the functional on the left-hand side among all functions u for which the functional exists.* It also follows that *the minimization of this functional in H_n is obtained by the Galerkin method.*

There exist **variational principles** in many areas of physics. This usually means that problems in that area can be formulated as minimization problems for some energy integral or some other functional. The approximate solution of such problems by a function containing a finite number of parameters is known as the **Ritz method**. By the previous remarks, the Ritz method is often equivalent to the Galerkin method, but this is not always the case; see [91].

Next, suppose that $f = 0$. Then, if possible, H_n should be built up by particular solutions of the differential equation—i.e., $Au = 0$, $\forall u \in H_n$. For the Laplace equation in two dimensions, the real and imaginary parts of any analytic function (of $x + iy$) are particular solutions. Then minimization of $\| Bu - g \|$, collocation, or the Galerkin method can be used to fit the boundary conditions. In the previous formulas, A, f should be replaced by B, g, and the scalar products will be integrals (or sums) on the boundary.

Finally, one can use the methods described with a space H_n where neither the differential equation nor the boundary conditions are satisfied, but then n must be large enough so that a good fit can be made to both the equations and the boundary values.

In this general form, the ideas presented above can be applied to problems where both the differential equations and the boundary conditions are *nonlinear*. If one of them is linear, H_n can be constructed according to the suggestions above. The nonlinear systems or minimization problems obtained can be treated by the methods of Sec. 6.9 or Sec. 10.5.

8.6.5. Finite-Element Methods

The finite-element method is a method for approximate treatment of physical problems, defined, for example, by differential equations with boundary conditions, or by variational principles. Compared to other methods it has been successful when the domain of the problem has a complicated shape or when the functions involved behave very differently in different parts of the domain. It is an important tool in structural mechanics, but there are numerous other applications; see Oden [108].

The domain is represented approximately by a collection of a finite number of connected subdomains of simple shape (e.g., triangles), called **finite elements**. Functions are approximated locally over each element by continuous functions uniquely defined in terms of the values of the function (and possibly some of its derivatives) at certain points, called **nodes**, inside or on the boundary of the element. Our treatment in Sec. 7.7.3 of interpolation and double integrals in triangulated domains is in the spirit of finite-element methods.

In most cases, finite-element methods can be looked upon as applications of *the Ritz or Galerkin methods with a particular type of "local" basis functions*. In order to explain this, we first consider the simplest particular case: piecewise linear interpolation in one variable, Fig. 8.6.7. The elements are here the

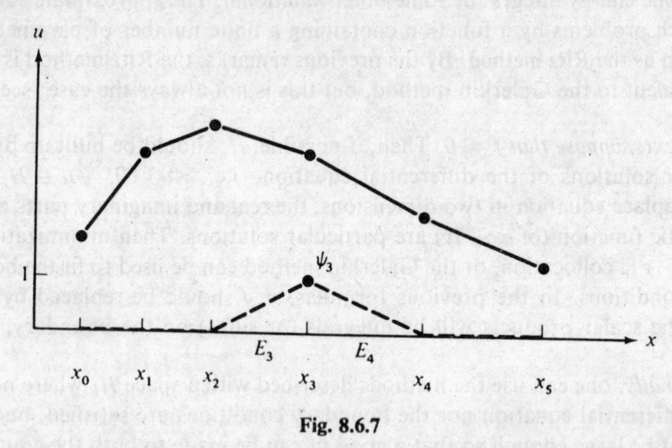

Fig. 8.6.7

intervals $E_i = [x_{i-1}, x_i]$, $i = 1, 2, \ldots, 5$. The nodes are the points x_i, $i = 0, 1, \ldots, 5$. In each element the function represented by the solid line is defined by linear interpolation of the values in the two nodes of the element. This function can be expressed in the form (see also spline functions, Sec. 4.6)

$$\sum_{i=0}^{5} u_i \psi_i(x), \tag{8.6.20}$$

where the $\psi_i(x)$ are piecewise linear functions, $\psi_i(x_j) = \delta_{ij}$ (the Kronecker delta). The "triangle-shaped" graph of ψ_3 is shown in Fig. 8.6.7.

In order to obtain higher-degree interpolation, we have to introduce internal nodes in the elements. A representation similar to Eq. (8.6.20) is obtained by the Lagrange interpolation formula. The ψ_i are piecewise polynomials, one for each node.

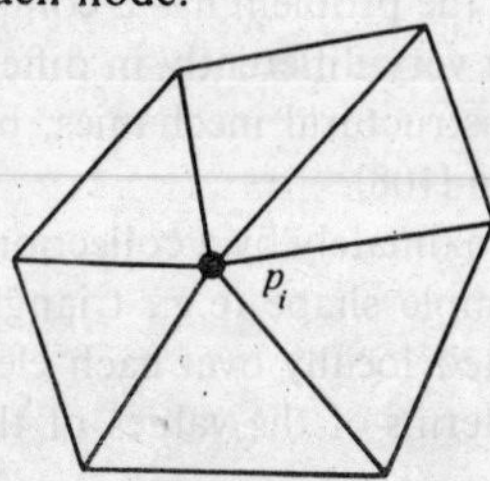

Fig. 8.6.8

Functions defined by piecewise linear interpolation in triangular grids (see Eq. (7.7.15)) can also be expressed in the form of Eq. (8.6.20), with one term ψ_i for each node P_i, $\psi_i(P_j) = \delta_{ij}$. $\psi_i(x) \neq 0$ only when x belongs to an element containing P_i (Fig. 8.6.8). The ψ_i are, for obvious reasons, called pyramid functions. In piecewise quadratic interpolation, Eq. (7.7.16) yields a similar representation, if the second differences are expressed in terms of the function values.

Now consider the finite-element solution to a linear boundary-value problem,

$$Au(P) = f, \quad P \in D,$$
$$u(P) = 0, \quad P \in \partial D.$$

The finite elements are run through in turn, and the contribution of each element to A_n and f_n in Eq. (8.6.16) is computed. In structural mechanics A_n is called the stiffness matrix. Note that A_n will be sparse when n is large, since $(\psi_i, A\psi_j) = 0$ unless there is an element containing both P_i and P_j. Equations similar to (8.6.18) are useful here—they are even necessary if the ψ_i have discontinuous first derivatives on the boundaries of the elements.

The mathematical and physical aspects of the method are treated in Strang and Fix [113a], and J. T. Oden [108]. The computer implementation is treated in a thesis of J. A. George, [99]. See also Sec. 5.4.4.

8.6.6. Integral Equations

Problems in partial differential equations can sometimes be transformed into integral equations. Many mathematical models—e.g., in statistical physics—also lead directly to integral equations or integro-differential equations.

The equations (where u is the unknown function)

$$\int_a^b K(x, y)u(y)\, dy = f(x), \tag{8.6.21}$$

$$u(x) - \int_a^b K(x, y)u(y)\, dy = f(x) \tag{8.6.22}$$

are called linear *Fredholm integral equations* of the *first kind* and *second kind*, respectively. The function K is called the *kernel*. For the mathematical theory, see Courant-Hilbert [3].

A natural approach is to use **numerical integration and collocation**. The trapezoidal method is accurate if the functions are periodic, with period $= b - a$. In the nonperiodic case the trapezoidal approximation with Richardson extrapolation has been successful for equations of the second kind. Gauss's quadrature has also been recommended. In any case, for Eq. (8.6.21) we obtain the linear system

$$\sum_{j=1}^{n} K(x_i, y_j)A_j u_j = f(x_i), \quad i = 1, 2, \ldots, m, \tag{8.6.23}$$

where the A_j are the coefficients in the quadrature formula ($m \geq n$). For Eq. (8.6.22) we obtain

$$u_i - \sum_{j=1}^{n} K(x_i, y_j)A_j u_j = f(x_i), \quad i = 1, 2, \ldots, n. \qquad (8.6.24)$$

For equations of the *first kind* with smooth kernels, the columns of Eq. (8.6.23) are likely to be approximately linearly dependent. It is therefore advisable to assume that the rank is equal to r and to use the concept of *least squares solution with minimum norm*, defined in Sec. 5.7.2. The rank can be estimated as the number of columns for which the size of the pivot element in the modified Gram-Schmidt process is well above the level of rounding errors. Alternatively, one can use the singular-value decomposition (Sec. 5.2.5), replace the very small singular values by zeros, and then compute the pseudo-inverse.

Equations of the *second kind* are usually less harmful. A simple example can illustrate this.

Example 8.6.4

Consider the "very smooth" kernel $K(x, y) = k$, $a = 0$, $b = 1$. The equation of the *first* kind reads

$$k \int_0^1 u(y)\, dy = f(x).$$

This equation has no solution unless f is a constant $f = c$, say, in which case any function for which $\int_0^1 u(y)\, dy = c/k$ is a solution. The equation of the *second* kind can be solved:

$$u(x) - k \int_0^1 u(y)\, dy = f(x),$$

$$u(x) = f(x) + k \cdot c,$$

$$c = \int_0^1 u(y)\, dy = \int_0^1 f(y)\, dy + k \cdot c,$$

$$u(x) = f(x) + \frac{k}{1-k} \int_0^1 f(y)\, dy.$$

Hence the solution is unique, unless $k = 1$. If $k = 1$, there is no solution unless $\int_0^1 f(y)\, dy = 0$, in which case the general solution is $u(x) = f(x) + p$, $p =$ arbitrary constant.

It is convenient to use operator notation and write the integral equations (8.6.21) and (8.6.22) in the form $\hat{K}u = f$ or $(I - \hat{K})u = f$, respectively. An *integral operator* $\hat{K}$ can be regarded as a continuous analog of a matrix.

If the homogeneous equation $\hat{K}u = \lambda u$ has a nontrivial solution, λ is called an *eigenvalue* of $\hat{K}$. If the kernel K is square-integrable, then, as a rule, the operator $\hat{K}$ has an enumerable infinity of eigenvalues with a limit point at the origin; see [1]. If the kernel is of the form

$$K(x, y) = \sum_{p=1}^{r} \chi_p(x)\psi_p(y), \tag{8.6.25}$$

where both sets of functions $\{\chi_p\}_1^r$ and $\{\psi_p\}_1^r$ are (internally) linearly independent, $\hat{K}$ spans an r-dimensional space; the rank of $\hat{K}$ is r. (In Example 8.6.4, $= 1$.) Zero is then an eigenvalue of infinite multiplicity, since $\hat{K}u = 0$ for any function u orthogonal to $\psi_1, \psi_2, \ldots, \psi_r$. The equation of the first kind $\hat{K}u = f$ is solvable, only if f is a linear combination of the χ_p.

Kernels can be approximated to the rounding-error level by sums of the form of Eq. (8.6.25); the smoother the kernel, the smaller r is. We may call that value of r the "numerical rank" of $\hat{K}$.

Therefore, one can expect *ill-conditioned* linear systems with any method for the numerical solution of equations of the *first kind* with smooth kernels. Equations of the *second kind*, however, cause no trouble, unless an eigenvalue of $\hat{K}$ happens to be close to unity.

The **Galerkin method** (see Sec. 8.6.4) can be applied to integral equations. For an equation of the *second kind*, put

$$u \approx f + c_1\varphi_1 + c_2\varphi_2 + \ldots + c_n\varphi_n. \tag{8.6.26}$$

Then the Galerkin equations $(\varphi_i, (I - \hat{K})u) = (\varphi_i, f)$, $i = 1, 2, \ldots, n$, can be expressed in matrix form,

$$\boldsymbol{A}_n\boldsymbol{u}_n = \boldsymbol{g}_n, \tag{8.6.27}$$

where (see Problem 18) the elements of $\boldsymbol{A}_n$, $\boldsymbol{u}_n$, $\boldsymbol{g}_n$ are $(\varphi_i, (I - \hat{K})\varphi_j)$, c_j, $(\varphi_i, \hat{K}f)$, respectively.

For equations of the *first kind* the f term is to be dropped from Eq. (8.6.26). We obtain the linear system

$$\boldsymbol{K}_n\boldsymbol{u}_n = \boldsymbol{f}_n, \tag{8.6.28}$$

where the elements of $\boldsymbol{K}_n$, $\boldsymbol{u}_n$, $\boldsymbol{f}_n$ are $(\varphi_i, \hat{K}\varphi_j)$, c_j, (φ_i, f), respectively. The system is ill-conditioned if n exceeds the "numerical rank" of $\hat{K}$, and the special techniques referred to above should be used.

The Galerkin method gives the *exact solution of an equation of the second kind, when the kernel is of the form of Eq. (8.6.25) with $n = r$, $\varphi_p = \chi_p$.*

The eigenvalues of $\hat{K}$ are approximated by the values of λ which make the matrix $\boldsymbol{K}_n - \lambda\{(\varphi_i, \varphi_j)\}$ singular.

The operator formalism makes the techniques of this section applicable to many linear operators other than integral operators.

Examples of integro-differential equations and nonlinear integral equations are given in the problems of this section. More information about the numerical treatment of integral equations and other operator equations can be found, e.g., in Collatz [91], Todd [24], and Rall [109]. Rall's book is an excellent guide on the bumpy road from a Banach space to a working computer program.

REVIEW QUESTIONS

1. Give a general description of a difference approximation for (a) the heat equation; (b) Poisson's equation.
2. Describe in your own words:
   (a) collocation; (b) the Galerkin method; (c) the Ritz method.
   The first two methods can be applied in two different ways to a linear boundary-value problem. How?
3. Describe an application of the finite-element method to a two-dimensional problem.
4. Describe two numerical methods for the solution of linear integral equations. Why are, as a rule, equations of the first kind (with smooth kernels) more troublesome than those of the second kind?

PROBLEMS

1. (a) Set up a difference equation for the same problem as in Sec. 8.6.2, but take $k/h^2 = 1$.
   (b) How is a disturbance of unit magnitude at the point $i = 3, j = 0$ propagated? How large is the effect for $i = j = 4$?
   (c) What conclusion can one draw concerning the use of this value of k/h^2?
2. Set up a difference approximation for the parabolic problem

$$\frac{\partial u}{\partial t} = \frac{\partial}{\partial x}\left((1 + x^2)\frac{\partial u}{\partial x}\right), \quad -1 \leq x \leq 1, \quad t \geq 0.$$

   Initial condition: $u(x, 0) = 1{,}000 - |1{,}000x|$.
   Boundary condition: $u(-1, t) = u(1, t) = 0, \quad t \geq 0$.
   Take $h = 0.4$, $k = 0.04$. Take advantage of the symmetry, and integrate to $t = 0.12$.
3. (a) Consider the linear system of $N - 1$ ordinary differential equations,

$$\frac{du}{dt} = Au. \tag{8.6.29}$$

   Suppose that A has $N - 1$ eigenvalues and linearly independent eigenvectors $\lambda_j, v_j, j = 1, 2, \ldots, N - 1$. Show that

$$u(0) = \sum \alpha_j v_j \Longrightarrow u(t) = \sum \alpha_j v_j \exp(\lambda_j t).$$

   (b) Consider the heat equation $\partial u/\partial t = \partial^2 u/\partial x^2$, with boundary condition $u(0, t) = u(1, t) = 0$. Use central difference approximation for $\partial^2 u/\partial x^2$,

$h = 1/N$. Show that this gives a system where the eigenvalues of A are $\lambda_j = -2N^2(1 - \cos(\pi j/N))$.

The following problems (c), (d) and (e) are concerned with the system defined in (b).

(c) Show that $d\|u\|_2^2/dt \leq 0$ and hence $\|u(t)\|_2 \leq \|u(0)\|_2$.

(d) Show that $\|u_{n+1}\|_2^2 \leq \|u_n\|_2^2$ if the trapezoidal method is used for solving the system. (This shows stability for any step size.)

(e) How many operations are needed for solving the system for u_{n+1} in each step by the trapezoidal method? Compare this to the explicit method.

(f) Generalize to arbitrary equations $\partial u/\partial t = \partial(p(x)\partial u/\partial x)/\partial x$, $p(x) > 0$, $u(0) = u(1) = 0$.

4. Consider the heat equation with reversed sign,

$$\frac{\partial u}{\partial t} = \frac{-\partial^2 u}{\partial x^2}, \quad u(0, t) = u(1, t) = 0.$$

(It corresponds to integrating the usual heat equation backward in time.)

(a) How is a disturbance of unit magnitude at (0.5, 0) propagated if a difference approximation, analogous to that in Sec. 8.6.2, is used, $k/h^2 = \frac{1}{2}$?

(b) Consider Eq. (8.6.29), where A is obtained by a central difference approximation of $-\partial^2 u/\partial x^2$. Show that the initial-value problem is very ill-conditioned when N is large, $h = 1/N$.

5. In a numerical treatment of the Laplace equation $\nabla^2 u = 0$ in a square net, the numerical values shown below were obtained in six of the net points. In order to obtain a first approximation in a net with half the grid size, one needs, among others, the values a, b, c indicated below. Compute this with an $O(h^4)$ error—disregarding the errors in the given values.

$$\begin{matrix} 101 & & 105 & & 117 \\ & a & b & c & \\ 109 & & 109 & & 117 \end{matrix}$$

6. Compute the optimum overrelaxation factor for the five-point difference approximation to the Laplace equation on a 20×20 square grid. (See Example 5.6.2.) Compute the asymptotic rate of convergence for Gauss-Seidel and optimal SOR.

7. Compute approximately the smallest value of λ for which the equation

$$\nabla^2 u + \lambda u = 0$$

has a nontrivial solution which vanishes on the boundary of the triangle with vertices (0, 0), (0, 1) and (1, 0) (the fundamental mode of a triangular membrane). Use a difference method with $h = \frac{1}{4}$ and $h = \frac{1}{5}$ and Richardson h^2-extrapolation. Compare with the exact result, $\lambda = 5\pi^2$.

8. Show that the difference equation

$$\nabla_9^2 u = \left(a + \frac{a^2 h^2}{12}\right)u, \quad a = \text{constant}$$

is an $O(h^4)$-approximation to the differential equation $\nabla^2 u = au$.

9. The displacement u of a loaded plate satisfies the differential equation

$$\nabla^4 u = f(x, y),$$

where f is the load density. Write the equation as a system (see Problem 2 of Sec. 8.4),

$$\nabla^2 u = v,$$
$$\nabla^2 v = f.$$

The boundary conditions are, in a certain symmetric loading case, for a rectangular plate whose four corners are $(-2, -1), (-2, 1), (2, 1), (2, -1)$.

$$u = \frac{\partial u}{\partial n} = 0 \qquad \text{on the sides } y = 1 \text{ and } x = 2,$$

$$\frac{\partial u}{\partial n} = \frac{\partial v}{\partial n} = 0 \qquad \text{on the symmetry lines } y = 0 \text{ and } x = 0.$$

Sketch a program for calculating the matrix of the system obtained when ∇^2 is approximated by ∇_5^2 on a square net with mesh width $1/N$. Consider in particular the boundary conditions. What is the order and the band width of the matrix?

10. By the Gauss integral theorem, the relation

$$\int_C \frac{\partial u}{\partial n}\, ds = 0$$

holds for any function which satisfies the Laplace equation everywhere inside a closed curve C. Formulate and prove a similar relation which holds for any solution of the difference equation $\nabla_5^2 u = 0$ in a rectangular domain.

11. (a) What happens if one tries to solve an initial-value problem for the Laplace equation by the usual difference method? Investigate how a disturbance of unit magnitude is propagated—e.g., solve recursively $u_{i,j+1}$ from the difference equation with initial and boundary conditions,

$$u_{i0} = 0, \quad u_{i1} = \begin{cases} 0, & i \neq 0, \\ 1, & i = 0, \end{cases}$$

$$u_{-4,j} = u_{4,j} = 0.$$

(b) Write the recurrence relation in the form

$$\mathbf{u}_{j+1} - 2A\mathbf{u}_j + \mathbf{u}_{j-1} = \mathbf{0}.$$

Compute the eigenvalues of A, and compare with the stability condition $-1 < \lambda < 1$ obtained in Problem 8 of Sec. 8.5.

12. Set up a difference approximation for the hyperbolic differential equation

$$\frac{\partial^2 u}{\partial t^2} = \frac{\partial^2 u}{\partial x^2}, \quad t \geq 0,$$

Initial conditions: u and $\partial u/\partial t$ given for $t = 0$. Boundary conditions: $u(0, t) = u(1, t) = 0$. Step sizes: h for x, k for t. Proceed as in Problem 11(b), and determine a stability condition.

13. Let D be the ellipse $x^2/a^2 + y^2/b^2 = 1$. Consider the boundary-value problem

$$\nabla^2 u = 1, \quad (x, y) \in D,$$
$$u = x^4 + y^4, \quad (x, y) \in \partial D.$$

(a) Reduce it to a problem with homogeneous boundary conditions.
(b) Reduce it to a Dirichlet problem for the Laplace equation.
(c) Which symmetry properties will the solution of this problem have? (You may assume that the problem has a unique solution.)

14. Let $p_n(x, y)$, $q_n(x, y)$ be the real and imaginary parts of $(x + iy)^n$, respectively. By analytic function theory, p_n, q_n satisfy the Laplace equation. Consider a truncated expansion,

$$u \approx \sum_{n=0}^{N} (a_n p_n + b_n q_n).$$

(a) Which coefficients will certainly be zero, if $u(x, y) = u(-x, y)$, $\forall\, x, y$?
(b) How can you reduce the number of coefficients in the application of this expansion to Problem 13(b) by utilizing the symmetry properties?
(c) Write a program for the calculation of the matrix and the right-hand side in an application of the collocation method to Problem 13(b) with M given boundary points, $M \geq \frac{1}{2}N + 1$.

15. In the notation of Sec. 8.6.4, is it true that $(v, Au) = v_n^T A_n u_n$, whenever $u \in H_n$, $v \in H_n$?

16. (a) In the notation of Theorem 8.6.1, show that

$$(u, Au) - 2(f, u) = \langle u - u*, u - u^* \rangle - \langle u^*, u^* \rangle.$$

(b) Show that the solution of the Poisson equation, $-\nabla^2 u = f$, with boundary condition $u = 0$ on ∂D, minimizes the integral

$$\iint_D [(\nabla u)^2 - 2f \cdot u]\, dx\, dy$$

among all functions, satisfying the boundary conditions.
(c) What differential equation corresponds to the minimization of the integral

$$\iint_D [p(\nabla u)^2 + g \cdot u^2 - 2f \cdot u]\, dx\, dy,$$

where p, f, g are given functions of x, y; p and g are positive. *Hint:* Use Eq. (8.6.18).

17. (a) Set up the equation for the finite-element solution of

$$-u''(x) = f(x), \quad x \in [0, 1],\ u(0) = u(1) = 0,$$

with the piecewise linear basis functions appearing in Eq. (8.6.20). Divide $[0, 1]$ into N subintervals of equal length. Treat f as a piecewise constant function. Compare with the finite-difference solution.
(b) The same question for the differential equation,

$$-\frac{d}{dx}\left(p\frac{du}{dx}\right) + gu - f = 0.$$

18. Check the formulas given in the text for the matrix elements of Eqs. (8.6.27) and (8.6.28).

19. (a) Determine the rank and the eigenvalues of the operator $\hat{K}$, defined by

$$(\hat{K}u)(x) = \int_0^{2\pi} \cos(x + y)u(y)\, dy.$$

(b) Solve the integral equation $(I - \hat{K})u = 1$.

(c) Give the general solution of the equation $(\hat{K}u)(x) = \cos x + ax$, when a solution exists. (a is constant.)

20. A kernel is symmetric when $K(x, y) = K(y, x)$ for all x, y. Are the matrices A_n and K_n symmetric in that case?

21. Approximate the integro-differential equation

$$\frac{\partial u}{\partial t} - \int_0^1 K(x, y)u(y, t)\, dy = f(x, t)$$

by a finite system of linear ordinary differential equations, using numerical integration and collocation.

22. Suggest a numerical method for the following nonlinear integral equation:

$$u(x) - \int_a^b K(x, y, u(x), u(y))\, dy = f(x).$$

23. Many integral equations can be solved in closed form; see Collatz [91, Chap. 6] or textbooks on integral equations. Suppose that equations of the form $\hat{A}v = f$ can be solved in closed form and that we want the solution of the equation

$$(\hat{A} - \hat{K})u = f. \tag{8.6.30}$$

(a) Set up the Galerkin equations for this problem, with the approach

$$u \approx v + \sum_1^n c_j \varphi_j,$$

where $\hat{A}v = f$.

(b) Show that the Galerkin method gives the exact solution of Eq. (8.6.30), if the operator $\hat{K}$ is of rank n and if the functions $\hat{A}\varphi_j$ are a basis for the n-dimensional space of functions which can be expressed in the form $\hat{K}g$ (g = arbitrary function).

9 FOURIER METHODS

9.1. INTRODUCTION

This chapter is to a great extent an application of the theory in the first two sections of Chap. 4. In fact, it might be useful to quickly review Sec. 4.1 and 4.2 at this point.

Many natural phenomena—for example, acoustical and optical phenomena—are of a periodic character. For instance, it is known that a musical sound is composed of regular oscillations, partly a fundamental tone with a certain frequency n, and partly overtones with frequencies $2n, 3n, 4n, \ldots$. The ratio of the strength of the fundamental tone to that of the overtones is decisive for our impression of the sound. Sounds which are free of overtones occur, for instance, in electronic music, where they are called pure **sine tones.**

In an electronic oscillator, a current is generated whose strength at time t depends on t according to the formula

$$r \sin (\omega t + v),$$

where r is called the *amplitude* of the oscillation; ω is called the *angular frequency*, and is equal to 2π times the frequency; v is a constant which defines the state at the time $t = 0$.

In a loudspeaker, variations of current are converted into variations in air pressure which—under ideal conditions—are described by the same function. In practice, however, there is always a certain distortion—overtones occur. Overtones of order $k - 1$ give a contribution of the form $r_k \sin (k\omega t + v_k)$. The variations in air pressure which reach the ear can be described from this viewpoint as a sum of the form

$$\sum_{k=0}^{\infty} r_k \sin (k\omega t + v_k). \tag{9.1.1}$$

The separation of a periodic phenomenon into a fundamental tone and overtones permeates not only acoustics, but also many other areas. It is related to an important, purely mathematical theorem, first given by Fourier (1758–1830). According to this theorem, every function $f(t)$ with period $2\pi/\omega$ can, under certain very general conditions, be expanded in a series of the form of Eq. (9.1.1). A more precise formulation of the theorem will be given later, in Theorem 9.2.2. By a function with period p, we mean that

$$f(t+p) = f(t) \quad \text{for all } t.$$

An expansion of the form of Eq. (9.1.1) can be expressed in many equivalent ways. If we set $a_k = r_k \sin v_k$, $b_k = r_k \cos v_k$, then using the addition theorem for the sine function we can write

$$f(t) = \sum_{k=0}^{\infty} (a_k \cos k\omega t + b_k \sin k\omega t), \tag{9.1.2}$$

where a_k, b_k are real constants. Another form, which is often the most convenient, can be found with the help of Euler's formulas,

$$\cos x = \frac{e^{ix} + e^{-ix}}{2}, \qquad \sin x = \frac{e^{ix} - e^{-ix}}{2i}, \quad (i = \sqrt{-1}).$$

If

$$c_0 = a_0, \qquad c_k = \tfrac{1}{2}(a_k - ib_k), \qquad c_{-k} = \tfrac{1}{2}(a_k + ib_k) \quad (k = 1, 2, \ldots), \tag{9.1.3}$$

then one gets

$$f(t) = \sum_{k=-\infty}^{\infty} c_k e^{ik\omega t}. \tag{9.1.4}$$

In the rest of this chapter we shall use the term **Fourier series** to denote an expansion of the form of Eq. (9.1.2) or Eq. (9.1.4). We shall call the partial sums of these series **trigonometric polynomials.** Fourier series are valuable aids in the study of phenomena which are periodic in time (vibrations, sound, light, alternating currents, etc.) or in space (waves, crystal structure, etc.). One recent and very important area of application is in statistical time series, which are used in communications theory, control theory, and the study of turbulence. Sometimes the term *spectral analysis* is used to describe the above methods. In certain applications, modifications of pure Fourier methods are used as a means of analyzing nonperiodic phenomena; see, e.g., Sec. 9.4 (periodic continuation of functions) and Sec. 9.5 (Fourier integrals).

9.2. BASIC FORMULAS AND THEOREMS IN FOURIER ANALYSIS

9.2.1. Functions of One Variable

We shall study functions with period 2π. If a function of t has period p, then the substitution $x = 2\pi t/p$ transforms the function to a function of x with period 2π. We assume that the function can have complex values, since the complex exponential function is convenient for manipulations. The inner product of two complex-valued functions f and g of period 2π is defined in the following way (the bar over g indicates complex conjugation).

Continuous case:

$$(f, g) = \int_{-\pi}^{\pi} f(x)\bar{g}(x)\,dx. \tag{9.2.1}$$

Discrete case:

$$(f, g) = \sum_{\alpha=0}^{M} f(x_\alpha)\bar{g}(x_\alpha), \quad \text{where} \quad x_\alpha = \frac{2\pi\alpha}{M+1}. \tag{9.2.2}$$

The **norm** of the function f is defined by

$$\| f \| = (f, f)^{1/2}.$$

One can make computations with these inner products in the same way as with the inner products defined in Sec. 4.2.1, with certain obvious modifications. Notice especially that $(g, f) = \overline{(f, g)}$. In the continuous case, it makes no difference what interval one uses, as long as it has length 2π—the value of the inner product is unchanged.

THEOREM 9.2.1

The following ***orthogonality relations*** *hold for the functions*

$$\varphi_j(x) = e^{ijx}, \quad j = 0, \pm 1, \pm 2, \ldots.$$

Continuous case:

$$(\varphi_j, \varphi_k) = \begin{cases} 2\pi, & \text{if } j = k, \\ 0, & \text{if } j \neq k. \end{cases}$$

Discrete case:

$$(\varphi_j, \varphi_k) = \begin{cases} M+1, & \text{if } \dfrac{j-k}{M+1} \text{ is an integer}, \\ 0, & \text{otherwise}. \end{cases}$$

Proof.

Continuous case:

For $j \neq k$:

$$(\varphi_j, \varphi_k) = \int_{-\pi}^{\pi} e^{ijx}e^{-ikx}\,dx = \left[\frac{e^{i(j-k)x}}{i(j-k)}\right]_{-\pi}^{\pi} = \frac{(-1)^{j-k} - (-1)^{j-k}}{i(j-k)} = 0.$$

For $j = k$:

$$(\varphi_j, \varphi_k) = \int_{-\pi}^{\pi} e^{ijx}e^{-ijx}\,dx = \int_{-\pi}^{\pi} 1\,dx = 2\pi.$$

Discrete case, $x_\alpha = 2\pi\alpha/(M+1)$:

$$(\varphi_j, \varphi_k) = \sum_{\alpha=0}^{M} \exp(ijx_\alpha)\cdot\exp(-ikx_\alpha) = \sum_{\alpha=0}^{M} \exp\left[i(j-k)\cdot\frac{2\pi\alpha}{M+1}\right].$$

This is a geometric series; the ratio between terms is

$$q = \exp\left[i(j-k)\frac{2\pi}{M+1}\right].$$

If $(j-k)/(M+1)$ is an integer, then $q = 1$ and the sum is $M+1$. Otherwise $q \neq 1$, but $q^{M+1} = \exp(i(j-k)2\pi) = 1$. But from the summation formula for a geometric series,

$$(\varphi_j, \varphi_k) = \frac{q^{M+1}-1}{q-1} = 0.$$

Thus the theorem is proved.

If one knows that f has an expansion of the form

$$f = \sum_{j=a}^{b} c_j \varphi_j,$$

where $a = -\infty$, $b = \infty$ in the continuous case and $a = 0$, $b = M$ in the discrete case, then it follows (formally) that

$$(f, \varphi_k) = \sum_{j=a}^{b} c_j(\varphi_j, \varphi_k) = c_k(\varphi_k, \varphi_k), \quad (\text{if } a \leq k \leq b),$$

since $(\varphi_j, \varphi_k) = 0$ for $j \neq k$. Thus (changing k to j) we have,

$$c_j = \frac{(f, \varphi_j)}{(\varphi_j, \varphi_j)} = \begin{cases} \dfrac{1}{2\pi}\displaystyle\int_{-\pi}^{\pi} f(x)e^{-ijx}\,dx & \text{in the continuous case,} \quad (9.2.3) \\ \dfrac{1}{M+1}\displaystyle\sum_{\alpha=0}^{M} f(x_\alpha)\exp(-ijx_\alpha) & \text{in the discrete case.} \quad (9.2.4) \end{cases}$$

These coefficients are called **Fourier coefficients** (see the more general case in Theorem 4.2.5). The purely formal treatment given above is, in the discrete case, justified by Theorem 4.2.6. In the continuous case, more advanced methods are required, but we shall not go into this further.

From a generalization of Theorem 4.2.5, we also know that

$$\|f - \sum_{j=-N}^{N} k_j \varphi_j\|, \quad (N < \infty)$$

becomes as small as possible if we choose $k_j = c_j$, $-N \leq j \leq N$.

Now consider the *continuous* case. In accordance with Eq. (9.1.3), set

$$a_j = c_j + c_{-j}, \qquad b_j = i(c_j - c_{-j}).$$

Then

$$c_0 = \tfrac{1}{2}a_0,$$

$$a_j = \pi^{-1}\int_{-\pi}^{\pi} f(x)\cos jx\,dx, \tag{9.2.5}$$

$$b_j = \pi^{-1}\int_{-\pi}^{\pi} f(x)\sin jx\,dx, \tag{9.2.6}$$

$$f(x) - \sum_{j=-N}^{N} c_j e^{ijx} = f(x) - \tfrac{1}{2}a_0 - \sum_{j=1}^{N} (c_j(\cos jx + i \sin jx) + c_{-j}(\cos jx - i \sin jx))$$
$$= f(x) - \tfrac{1}{2}a_0 - \sum_{j=1}^{N} (a_j \cos jx + b_j \sin jx).$$

(Notice that the factors preceding the integral are different in the expressions for c_j and for a_j, b_j, respectively.)

THEOREM 9.2.2 *Fourier Analysis, Continuous Case*

Every piecewise continuous function f with period 2π can be associated with a Fourier series in the following two ways:

$$\tfrac{1}{2}a_0 + \sum_{j=1}^{\infty} (a_j \cos jx + b_j \sin jx),$$

$$\sum_{j=-\infty}^{\infty} c_j e^{ijx}.$$

The coefficients a_j, b_j, and c_j can be computed using Eqs. (9.2.5) and (9.2.6) in the first case, and Eq. (9.2.3) in the second case. If f and its first derivative are everywhere continuous, then the series is everywhere convergent to $f(x)$. It is even permissible that f and f' have a finite number of jump discontinuities in each period. At such a discontinuity, the series gives the mean of the limiting values on the right and on the left of the relevant point. The partial sums of the above expansions give the best possible approximation to $f(x)$ by trigonometric polynomials, in the least-squares sense.

The proof of the convergence results is outside the scope of this book (see, however, the beginning of Sec. 9.4; see also [3]). The rest of the assertions follow from previously made calculations (in Theorem 9.2.1 and the comments following); see also the proof of Theorem 4.2.5.

THEOREM 9.2.3

If f is an even function—i.e., if $f(x) = f(-x)$ for all x—then $b_j = 0$ for all j; thus the Fourier series becomes a cosine series.

If f is an odd function—i.e., if $f(x) = -f(-x)$ for all x—then $a_j = 0$ for all j; thus the Fourier series becomes a sine series.

The proof is left as an exercise to the reader (use the formulas for the coefficients given in Eqs. (9.2.5) and (9.2.6)).

Example 9.2.1. *Fourier Expansion of a So-Called Rectangular Wave*

$$f(x) = \begin{cases} -1, & -\pi < x < 0, \\ 1, & 0 < x < \pi. \end{cases}$$

We make a periodic continuation of $f(x)$ outside the interval $(-\pi, \pi)$ (see Fig. 9.2.1). f is an odd function, so $a_j = 0$ for all j.

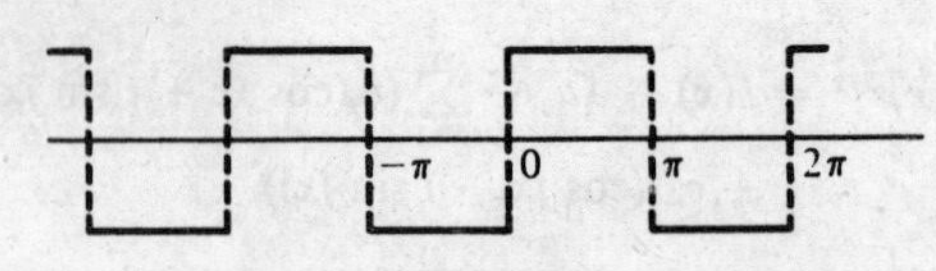

Fig. 9.2.1

$$b_j = -\pi^{-1}\int_{-\pi}^{0} \sin jx\,dx + \pi^{-1}\int_{0}^{\pi} \sin jx\,dx = 2\pi^{-1}\int_{0}^{\pi} \sin jx\,dx$$

$$= 2\pi^{-1}\frac{1-\cos j\pi}{j} = \begin{cases} 0 & \text{if } j \text{ is even,} \\ \dfrac{4}{j\pi} & \text{if } j \text{ is odd.} \end{cases}$$

Hence

$$f(x) = \frac{4}{\pi}\left(\sin x + \frac{\sin 3x}{3} + \frac{\sin 5x}{5} + \dots\right).$$

Notice that the sum of the series is zero at the points where f has a jump discontinuity; this agrees with the fact that the sum should be equal to the average of the limiting values to the left and to the right of the discontinuity. The Euler transformation can be used for accelerating the convergence of such series, except in the immediate vicinity of singular points (see Problem of Sec. 7.6).

Theorem 4.2.7 and its corollary, **Parseval's formula,**

$$2\pi \sum_{j=-\infty}^{\infty} |c_j|^2 = \| f \|^2 = \int_{-\pi}^{\pi} |f(x)|^2\,dx, \qquad (9.2.7)$$

are of great importance in many applications of Fourier analysis.

THEOREM 9.2.4 *Fourier Analysis, Discrete Case (Trigonometric Interpolation)*

Every function, defined on the net $\{x_0, x_1, \dots, x_M\}$, *where* $x_\alpha = 2\pi\alpha/(M+1)$ *can be interpolated by a trigonometric polynomial which can be written in the two forms*

$$f(x) = \sum_{j=0}^{M} c_j e^{ijx}, \qquad (9.2.8)$$

$$f(x) = \tfrac{1}{2}a_0 + \sum_{j=1}^{k} (a_j \cos jx + b_j \sin jx) + \tfrac{1}{2}\theta a_{k+1} \cos (k+1)x.$$

Here

$$\theta = 0,\ k = \tfrac{1}{2}M \quad \textit{if } M \textit{ is even,}$$

$$\theta = 1,\ k = \tfrac{1}{2}(M-1) \quad \textit{if } M \textit{ is odd,}$$

$$a_j = 2(M+1)^{-1} \sum_{\alpha=0}^{M} f(x_\alpha) \cos jx_\alpha, \qquad (9.2.$$

$$b_j = 2(M+1)^{-1} \sum_{\alpha=0}^{M} f(x_\alpha) \sin jx_\alpha,$$

$$c_j = (M+1)^{-1} \sum_{\alpha=0}^{M} f(x_\alpha) \exp(-ijx_\alpha).$$

If the sums in Eqs. (9.2.8) are terminated when $j = N$ (where $N < M$ or $N < k$ respectively), then one obtains the trigonometric polynomial which is the best least-squares approximation, among all trigonometric polynomials with the same number of terms, to f on the net.

Proof. The expression for c_j was formally derived previously (see Sec. 9.2.4), and it was mentioned that the derivation was justified by Theorem 4.2.6.

Since

$$(M + 1 - j)x_\alpha = 2\pi\alpha - jx_\alpha,$$

we have

$$\exp(-i(M + 1 - j)x_\alpha) = \exp(ijx_\alpha).$$

If a_j, b_j are defined as above, Eq. (9.2.9), then, for $1 \le j \le k$,

$$c_j = \tfrac{1}{2}(a_j - ib_j), \qquad c_{M+1-j} = \tfrac{1}{2}(a_j + ib_j).$$

Hence for all x on the net ($x = x_\alpha$),

$$\begin{aligned}\sum_{j=0}^{M} c_j e^{ijx} &= c_0 + \sum_{j=1}^{k} [c_j(\cos jx + i \sin jx) + c_{M+1-j}(\cos jx - i \sin jx)] \\ &\quad + \theta c_{k+1} e^{i(k+1)x} \\ &= \tfrac{1}{2}a_0 + \sum_{j=1}^{k} (a_j \cos jx + b_j \sin jx) + \tfrac{1}{2}\theta a_{k+1} \cos (k + 1)x,\end{aligned}$$

since for M odd, $\sin (k + 1)x_\alpha = \sin \pi\alpha = 0$—i.e., $e^{i(k+1)x} = \cos (k + 1)x$ on the net. (For even values of M, we have $\theta = 0$.)

Notice that the calculations required to compute the coefficients c_j according to Eq. (9.2.9) (**Fourier analysis**) are of essentially the same type as the calculations needed to tabulate $f(x)$ for $x = 2\pi\alpha/(M + 1)$, $\alpha = 0, 1, 2, \ldots, M$, when the expansion in Eq. (9.2.8) is known, so-called **Fourier synthesis.**

9.2.2. Functions of Several Variables

Functions of several variables are treated analogously. Quite simply, one takes one variable at a time. As an example, consider the discrete case, with two variables. Set

$$x_\alpha = \frac{2\pi\alpha}{M + 1}, \qquad y_\beta = \frac{2\pi\beta}{N + 1},$$

and assume that $f(x_\alpha, y_\beta)$ is known for $\alpha = 0, 1, 2, \ldots, M$; $\beta = 0, 1, 2, \ldots, N$. Set

$$c(j, y_\beta) = (M + 1)^{-1} \sum_{\alpha=0}^{M} f(x_\alpha, y_\beta) \exp(-ijx_\alpha),$$

$$c_{jk} = (N + 1)^{-1} \sum_{\beta=0}^{N} c(j, y_\beta) \exp(-iky_\beta).$$

From Theorem 9.2.4, then (with obvious changes in notation),

$$c(j, y_\beta) = \sum_{k=0}^{N} c_{jk} \exp(iky_\beta),$$

$$f(x_\alpha, y_\beta) = \sum_{j=0}^{M} c(j, y_\beta) \exp(ijx_\alpha) = \sum_{j=0}^{M} \sum_{k=0}^{N} c_{jk} \exp(ijx_\alpha + iky_\beta).$$

The above expansion is of considerable importance—e.g., in crystallography.

REVIEW QUESTIONS

1. Derive the orthogonality properties and coefficient formulas which are fundamental to Fourier analysis (for both the continuous and the discrete case).
2. Under what conditions does the Fourier series of the function f converge to f (in Theorem 9.2.2)?
3. How does one compute a Fourier expansion of a function of two variables?

PROBLEMS

1. (a) Prove Theorem 9.2.3 and the formula (9.4.2).
   (b) Give a simple characterization of the functions which have a sine expansion containing odd terms only.
2. Let f be an even function, with period 2π, such that

$$f(x) = \pi - x, \text{ when } 0 \leq x \leq \pi.$$

   (a) Plot the function $y = f(x)$ for $-3\pi \leq x \leq 3\pi$.
   (b) Expand f in a Fourier series.
   (c) Use this series to show that $1 + 3^{-2} + 5^{-2} + 7^{-2} + \ldots = \pi^2/8$.
   (d) Compute $S = 1 + 2^{-2} + 3^{-2} + 4^{-2} + 5^{-2} + \ldots$.
   (e) Compute (using Eq. (9.2.7)) the sum $1 + 3^{-4} + 5^{-4} + 7^{-4} + \ldots$.
   (f) Differentiate the Fourier series term by term, and compare with the result in Example 9.2.1.
3. Show that the function $G_1(t) = t - \frac{1}{2}, 0 < t < 1$, defined in Sec. 7.4.4, has the expansion,

$$G_1(t) = -\sum_{n=1}^{\infty} \frac{\sin 2n\pi t}{n\pi}.$$

   Derive, by termwise integration, the expansions for the functions $G_p(t)$, and show the statement (made in Sec. 7.4.4) that $c_p - G_p(t)$ has the same sign as c_p. Show also that $\sum_{n=1}^{\infty} n^{-p} = \frac{1}{2}|c_p|\cdot(2\pi)^p$, p even.

9.3. FAST FOURIER ANALYSIS

9.3.1. An Important Special Case

The problem is to compute the Fourier coefficients $\{c_j\}_0^{N-1}$ for a function

$$f(x) = \sum_{j=0}^{N-1} c_j \exp(ijx),$$

whose values are known at the points $2\pi\beta/N$, $\beta = 0, 1, 2, \ldots, N-1$. According to Theorem 9.2.4 (with obvious changes in notation) we have

$$c_j = N^{-1}\sum_{\beta=0}^{N-1} f\left(\frac{2\pi\beta}{N}\right) \exp\left(\frac{-ij2\pi\beta}{N}\right), \quad (j = 0, 1, 2, \ldots, N-1).$$

Set

$$w = \exp\left(\frac{-2\pi i}{N}\right), \qquad a_\beta = f\left(\frac{2\pi\beta}{N}\right)\cdot N^{-1}.$$

Then we can rewrite the problem as follows: for $j = 0, 1, 2, \ldots, N-1$, compute

$$c_j = \sum_{\beta=0}^{N-1} a_\beta w^{j\beta}, \quad \text{where} \quad w^N = 1. \tag{9.3.1}$$

If by an "operation" we mean one complex multiplication and one complex addition, then using Horner's scheme (see Example 1.3.3) one can solve this problem with N operations per coefficient—i.e., N^2 operations. *With fast Fourier analysis, however, one needs only* $N(r_1 + r_2 + \ldots + r_p)$ *operations if* $r_1 r_2 \ldots r_p = N$. Thus if $N = 2^7 = 128$, then the number of operations becomes $2^7 \cdot 14 = 1{,}792$ instead of $2^{14} = 16{,}384$. The algorithm was developed in 1965 by Cooley and Tukey ("Fast Fourier Transform"), but (as in many other cases) related ideas can be found in many previous works. In many areas of application (telecommunications, time-series analysis, etc.) this algorithm has caused a complete change of attitude toward what can be done using Fourier methods on a computer.

Consider now the special case $N = 2^k$. Set $\beta = 2\beta_1$, when β is even, and $\beta = 2\beta_1 + 1$ when β is odd, $0 \le \beta_1 \le \frac{1}{2}N - 1$. Then by Eq. (9.3.1),

$$c_j = \sum_{\beta_1=0}^{(1/2)N-1} a_{2\beta_1}(w^2)^{j\beta_1} + \sum_{\beta_1=0}^{(1/2)N-1} a_{2\beta_1+1}(w^2)^{j\beta_1}w^j.$$

Let α be the quotient and j_1 the remainder when j is divided by $\frac{1}{2}N$—i.e., $j = \alpha\frac{1}{2}N + j_1$. Then, since $w^N = 1$,

$$(w^2)^{j\beta_1} = (w^2)^{\alpha(1/2)N\beta_1}(w^2)^{j_1\beta_1} = (w^N)^{\alpha\beta_1}(w^2)^{j_1\beta_1} = (w^2)^{j_1\beta_1}.$$

Thus if we set

$$\begin{aligned} \varphi(j_1) &= \sum_{\beta_1=0}^{(1/2)N-1} a_{2\beta_1}(w^2)^{j_1\beta_1}, \quad (j_1 = 0, 1, 2, \ldots, \tfrac{1}{2}N - 1), \\ \psi(j_1) &= \sum_{\beta_1=0}^{(1/2)N-1} a_{2\beta_1+1}(w^2)^{j_1\beta_1}, \quad \text{where} \quad (w^2)^{(1/2)N} = 1, \end{aligned} \tag{9.3.2}$$

then

$$c_j = \varphi(j_1) + w^j\psi(j_1), \quad (j = 0, 1, \ldots, N-1).$$

Notice that the computation of $\varphi(j_1)$ and $\psi(j_1)$ means that one does two Fourier analyses with $\frac{1}{2}N = 2^{k-1}$ terms instead of one with $N = 2^k$ terms. Now apply the same idea to these two Fourier analyses. One then gets four Fourier analyses, each of which has 2^{k-2} terms; these can be further divided into eight Fourier analyses with 2^{k-3} terms, etc. The number of operations required to compute $\{c_j\}$, when $\{\varphi(j_1)\}$ and $\{\psi(j_1)\}$ have been computed, is at most $2 \cdot 2^k$. (Even more computation can be saved if the powers of w are stored beforehand in memory—i.e., if table look-up is used for these quantities.)

Denote by p_k the total number of operations needed to compute the coefficient when $N = 2^k$. Reasoning as above, we have

$$p_k \leq 2p_{k-1} + 2 \cdot 2^k, \quad (k = 1, 2, 3, \ldots).$$

Since $p_0 = 0$, it follows by induction that $p_k \leq 2k \cdot 2^k = 2N \cdot \log_2 N$.

Hence, *when N is a power of two the fast Fourier transform solves the problem with $2N \cdot \log_2 N$ operations.*

9.3.2. Fast Fourier Analysis, General Case

Suppose that

$$N = r_1 r_2 \ldots r_p,$$

and set

$$N_v = \prod_{i=v+1}^{p} r_i, \qquad N_p = 1.$$

Thus

$$N = r_1 r_2 \ldots r_v N_v, \quad N_0 = N.$$

The algorithm is based on two representations of integers which can be considered as generalizations of the position principle (Sec. 2.3.1):

I. Every integer j, $0 \leq j \leq N-1$ has a unique representation of the form

$$j = \alpha_1 N_1 + \alpha_2 N_2 + \ldots + \alpha_{p-1} N_{p-1} + \alpha_p, \quad 0 \leq \alpha_i \leq r_i - 1. \quad (9.3.3)$$

II. For every integer β, $0 \leq \beta \leq N-1$, β/N has a unique representation of the form

$$\frac{\beta}{N} = \frac{k_1}{N_0} + \frac{k_2}{N_1} + \frac{k_3}{N_2} + \ldots + \frac{k_p}{N_{p-1}}, \quad 0 \leq k_i \leq r_i - 1. \quad (9.3.4)$$

Set

$$j_v = \sum_{i=v+1}^{p} \alpha_i N_i, \qquad \frac{\beta_v}{N_v} = \sum_{i=v}^{p-1} \frac{k_{i+1}}{N_i}, \quad (j_v < N_v). \quad (9.3.5)$$

As an exercise, the reader can verify that the coefficients in the above representations can be recursively determined from the following algorithms:

1. $j_0 = j$, α_i is the quotient, and j_i the remainder in the division j_{i-1}/N_i;
2. $\beta_0 = \beta$, k_i is the remainder, and β_i the quotient in the division β_{i-1}/r_i.

From Eqs. (9.3.3), (9.3.4), and (9.3.5), it follows that, since N_i is divisible by N_v for $i \leq v$,

$$\frac{j\beta}{N} = \text{Integer} + \sum_{v=0}^{p-1} \frac{k_{v+1}}{N_v}\left(\sum_{i=v+1}^{p} \alpha_i N_i\right) = \sum_{v=0}^{p-1} \frac{k_{v+1} j_v}{N_v} + \text{Integer}.$$

From this, it follows that

$$w^{j\beta} = \exp\frac{2\pi i j\beta}{N} = \prod_{v=0}^{p-1} \exp\frac{k_{v+1} j_v 2\pi i}{N_v},$$

$$w^{j\beta} = \prod_{v=0}^{p-1} w_v^{j_v k_{v+1}}, \quad \text{where} \quad w_v = \exp\frac{2\pi i}{N_v}, \quad w_0 = w. \tag{9.3.6}$$

We now give an illustration of how the factorization in Eq. (9.3.6) can be utilized in fast Fourier analysis, for the case $p = 3$. Set, in accordance with Eq. (9.3.4),

$$a_\beta = c^{(0)}(k_1, k_2, k_3).$$

We have, then,

$$c_j = \sum_{\beta=0}^{N-1} a_\beta w^{j\beta} = \sum_{k_1=0}^{r_1-1} \sum_{k_2=0}^{r_2-1} \sum_{k_3=0}^{r_3-1} c^{(0)}(k_1, k_2, k_3) w_2^{j_2 k_3} w_1^{j_1 k_2} w^{j k_1}.$$

One can thus compute successively (see Eq. (9.3.5))

$$c^{(1)}(k_1, k_2, \alpha_3) = \sum_{k_3=0}^{r_3-1} c^{(0)}(k_1, k_2, k_3) w_2^{j_2 k_3} \quad (j_2 \text{ depends only on } \alpha_3),$$

$$c^{(2)}(k_1, \alpha_2, \alpha_3) = \sum_{k_2=0}^{r_2-1} c^{(1)}(k_1, k_2, \alpha_3) w_1^{j_1 k_2} \quad (j_1 \text{ depends only on } \alpha_2, \alpha_3),$$

$$c_j = c^{(3)}(\alpha_1, \alpha_2, \alpha_3) = \sum_{k_1=0}^{r_1-1} c^{(2)}(k_1, \alpha_2, \alpha_3) w^{j k_1}, \quad (j \text{ depends on } \alpha_1, \alpha_2, \alpha_3).$$

The quantities $c^{(i)}$ are computed for all $r_1 r_2 r_3 = N$ combinations of the values of the arguments. Thus the total number of operations for the entire Fourier analysis becomes (at most) $Nr_3 + Nr_2 + Nr_1$. The generalization to arbitrary p is obvious.

Note: The algorithms for integers described previously can, in the special case that $r_i = B$ for all i, be used for *converting integers or fractions* to the number system whose base is B.

REVIEW QUESTION

Give a brief summary of fast Fourier analysis and its computational advantages. (Take the special case where the number of terms is 2^k.)

PROBLEMS

1. Give a derivation of the two algorithms for α_i, k_i in Sec. 9.3.2.

2. The following Fortran program for the Fast Fourier Transform, valid when N is a power of 2, is quoted from Cooley, Lewis, and Welch, *IEEE Transactions* E-12 no. 1 (March 1965). Apply it to the function treated in Example 9.2.1. Choose, for instance, $N = 16, 32, 64$.

```
      SUBROUTINE FFT(A,M)
      COMPLEX A(1),U,W,T
      N=2**M
      NV2=N/2
      NM1=N-1
      J=1
      DO 7 I=1,NM1
      IF(I.GE.J) GO TO 5
      T=A(J)
      A(J)=A(I)
      A(I)=T
    5 K=NV2
    6 IF(K.GE.J) GO TO 7
      J=J-K
      K=K/2
      GO TO 6
    7 J=J+K
      DO 20 L=1,M
      LE=2**L
      LE1=LE/2
      U=(1.,0.)
      ANG=3.14159265358979/LE1
      W=CMPLX(COS(ANG),SIN(ANG))
      DO 20 J=1,LE1
      DO 10 I=J,N,LE
      IP=I+LE1
      T=A(IP)*U
      A(IP)=A(I)-T
   10 A(I)=A(I)+T
   20 U=U*W
      RETURN
      END
```

3. Find a program for the Fast Fourier Transform, where N is not a power of 2. See the list in Chap. 13.

9.4. PERIODIC CONTINUATION OF A NONPERIODIC FUNCTION

The more regular a function is, the faster its Fourier series converges. If f and its derivatives up to and including order k are periodic and *everywhere* continuous, and if $f^{(k+1)}$ is piecewise continuous, then

$$|c_j| \leq j^{-k-1} \| f^{(k+1)} \|_\infty. \tag{9.4.1}$$

(This useful result is relatively easy to prove using Eq. (9.2.3) and integrating by parts $k+1$ times.)

Sometimes Fourier series are used for functions which are defined only on the interval $(-\pi, \pi)$. The series then defines a *periodic continuation of the function* outside the interval (see Fig. 9.4.1). Thus the definition is extended so

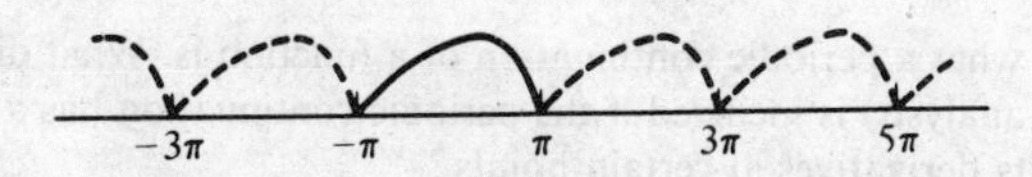

Fig. 9.4.1

that $f(x) = f(x + 2\pi)$ for all x. With this method, discontinuities in the values of the function or its derivatives can occur at $x = \pi$. This singularity can give rise to a slow rate of convergence, even for a function with good regularity properties in the open interval $(-\pi, \pi)$.

There are other ways to make a continuation of a function outside its interval of definition. If the function is defined in $[0, \pi]$, and if $f(0) = f(\pi) = 0$, then one can continue f to the interval $[-\pi, 0]$ by making the definition $f(x) = -f(-x)$; thereafter (outside $[-\pi, \pi]$) the function is periodically continued by $f(x) = f(x + 2\pi)$ (see Fig. 9.4.2). Since the resulting function

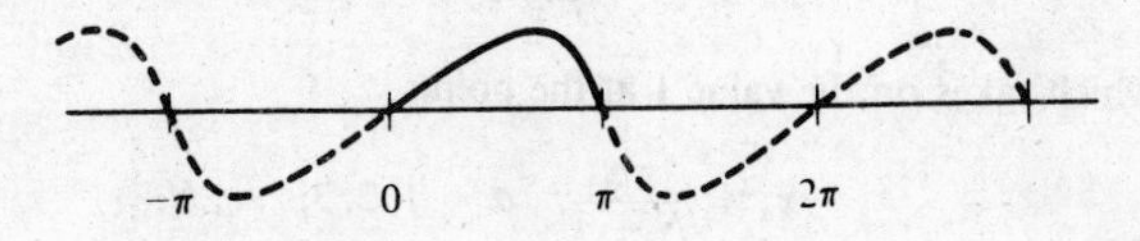

Fig. 9.4.2

is an odd function, its Fourier expansion becomes (using Theorems 9.2.3 and 9.2.4) a sine series:

Continuous case:

$$\sum_{j=1}^{\infty} b_j \sin jx, \quad b_j = \frac{2}{\pi} \int_0^{\pi} f(x) \sin jx \, dx. \tag{9.4.2}$$

Discrete case:

$$\sum_{j=1}^{N} b_j \sin jx, \quad b_j = \frac{2}{N+1} \sum_{\alpha=1}^{N} f(x_\alpha) \sin jx_\alpha, \tag{9.4.3}$$

where $x_\alpha = \pi\alpha/(N+1)$. Even if $f(0) \neq 0$ or $f(\pi) \neq 0$, one can still use suc a sine expansion, but it converges slowly.

One can, in an analogous way, make a continuation of a function in suc a way that one gets an even periodic function. Its Fourier series then become a pure cosine series (see Fig. 9.4.3).

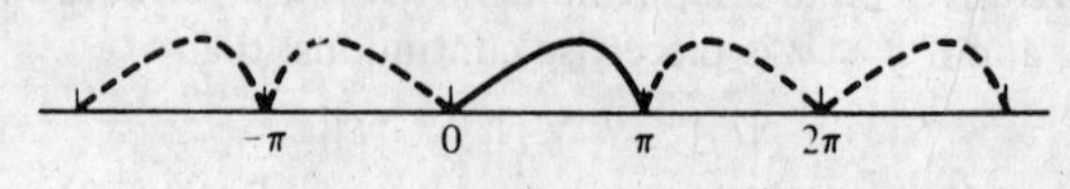

Fig. 9.4.3

REVIEW QUESTION

Explain what a periodic continuation of a function is. What disadvantage (fc Fourier analysis) is incurred if the periodic continuation has a discontinuity-e.g., in its derivatives at certain points?

PROBLEMS

1. (a) Prove that

$$\sum_{k=1}^{N} \sin \frac{\pi k}{N+1} = \cot \frac{\pi}{2(N+1)}.$$

Hint: $\sin x$ is the imaginary part of e^{ix}.

(b) Determine a sine polynomial

$$\sum_{j=1}^{N} b_j \sin jx,$$

which takes on the value 1 at the points

$$x_\alpha = \frac{\pi\alpha}{N+1}, \quad \alpha = 1, 2, 3, \ldots, N.$$

Hint: Use Eq. (9.4.2) or recall that the sine polynomial is an odd functio

(c) Compare the limiting value for b_j as $N \longrightarrow \infty$ with the result in Examp 9.2.1.

2. (a) Prove the inequality in Eq. (9.4.1)!

(b) Show, under the assumptions on f which hold in Eq. (9.4.1), that, for $k \geq$ f can be approximated by a trigonometric polynomial

$$\sum_{j=-n}^{n} c_j e^{ijx}$$

with an error whose maximum norm is less than $2\,\|f^{(k+1)}\|_\infty/(kn^k)$.

9.5. THE FOURIER INTEGRAL THEOREM

In Sec. 9.4 we showed how Fourier methods can be used on a nonperiodic function defined on a finite interval. Suppose now that the function $\varphi(\xi)$ is defined on the entire real axis, and that it satisfies the regularity properties which were required in Theorem 9.2.2. Set

$$x = \frac{2\pi\xi}{L}, \qquad \varphi(\xi) = f(x).$$

If

$$c_j = (2\pi)^{-1} \int_{-\pi}^{\pi} f(x)e^{-ijx}\, dx = L^{-1} \int_{-(1/2)L}^{(1/2)L} \varphi(\xi)e^{-2\pi i\xi j/L}\, d\xi,$$

then from Theorem 9.2.2 (for $x \in (-\pi, \pi)$—i.e., $\xi \in (-\frac{1}{2}L, \frac{1}{2}L)$),

$$f(x) = \sum_{j=-\infty}^{\infty} c_j e^{ijx}, \quad \text{and hence} \quad \varphi(\xi) = \sum_{j=-\infty}^{\infty} c_j e^{2\pi i\xi j/L}.$$

If we set

$$g_L(t) = \int_{-(1/2)L}^{(1/2)L} \varphi(\xi)e^{-2\pi i\xi t}\, d\xi, \tag{9.5.1}$$

then we get

$$\varphi(\xi) = L^{-1} \sum_{j=-\infty}^{\infty} g_L\left(\frac{j}{L}\right) e^{2\pi i\xi j/L}, \quad \xi \in (-\tfrac{1}{2}L, \tfrac{1}{2}L). \tag{9.5.2}$$

Now by passing to the limit $L \longrightarrow \infty$, one avoids making an artificial periodic continuation of the function outside a finite interval. The sum in Eq. (9.5.2) is a "sum of rectangles" similar to the sum which appears in the definition of a definite integral. However, here the argument varies from $-\infty$ to $+\infty$, and the function $g_L(t)$ depends on L. By a somewhat dubious passage to the limit, then, the pair of formulas of Eqs. (9.5.1) and (9.5.2) becomes the pair

$$g(t) = \int_{-\infty}^{\infty} \varphi(\xi)e^{-2\pi i\xi t}\, d\xi \Longleftrightarrow \varphi(\xi) = \int_{-\infty}^{\infty} g(t)e^{2\pi i\xi t}\, dt. \tag{9.5.3}$$

One can, in fact, after a rather complicated analysis, show that the above result is correct; see, e.g., Courant-Hilbert [3]. The proof requires, besides the previously mentioned "local" regularity conditions on f, the "global" assumption that

$$\int_{\infty}^{\infty} |\varphi(\xi)|\, d\xi$$

is convergent. The beautiful, (almost) symmetric relation of Eq. (9.5.3) is called the *Fourier integral theorem*. This theorem (and other versions of it, with varying assumptions under which they are valid) is one of the most important aids in both pure and applied mathematics. The function g is called the **Fourier transform** of φ. (The terminology in the literature varies somewhat as to the placement of the factor 2π—e.g., it can be taken out of the exponent by a simple change of variable.)

Example 9.5.1

$\varphi(\xi) = e^{-|\xi|}$ has the Fourier transform,

$$g(t) = \int_{-\infty}^{\infty} e^{-|\xi|} e^{-2\pi i \xi t}\, d\xi = \int_0^{\infty} (e^{-(1+2\pi i t)\xi} + e^{-(1-2\pi i t)\xi})\, d\xi$$

$$= \frac{1}{1 + 2it\pi} + \frac{1}{1 - 2it\pi} = \frac{2}{1 + 4\pi^2 t^2}.$$

From Eq. (9.5.3) it follows that

$$\int_{-\infty}^{\infty} 2(1 + 4\pi^2 t^2)^{-1} e^{2\pi i \xi t}\, dt = e^{-|\xi|},$$

or, after the substitution $2\pi t = x$,

$$\int_0^{\infty} (1 + x^2)^{-1} \cos \xi x\, dx = \frac{\pi}{2} e^{-|\xi|} \quad (\xi \text{ real}).$$

It is not so easy to prove this formula directly.

In many physical applications, the following relation, analogous to **Parseval's equality** (corollary to Theorem 4.2.7), is of great importance. If g the Fourier transform of φ, then

$$\int_{-\infty}^{\infty} |g(t)|^2\, dt = \int_{-\infty}^{\infty} |\varphi(\xi)|^2\, d\xi. \qquad (9.5.$$

Other useful properties are given in the problems which follow. For the numerical analyst, Fourier analysis is partly a very common computational task and partly an important aid in the analysis of the properties of numeric methods.

REVIEW QUESTION

Formulate the Fourier integral theorem.

PROBLEMS

In the following problems, we do not require any investigation of whether it is permissible to change the order of summations, integrations, differentiations, etc.; it is sufficient to treat the problems in a purely formal way.

1. The partial differential equation

$$\frac{\partial u}{\partial t} = \frac{\partial^2 u}{\partial x^2},$$

is called the heat equation. Show that the function

$$u(x, t) = \frac{4}{\pi} \sum_{k=0}^{\infty} \exp\,[-(2k + 1)^2 t] \frac{\sin (2k + 1)x}{2k + 1},$$

satisfies the differential equation for $t > 0$, $0 < x < \pi$, with boundary conditions $u(0, t) = u(\pi, t) = 0$ for $t > 0$, and initial condition $u(x, 0) = 1$ for $0 < x < \pi$. (See Example 9.2.1.)

2. Show that if $g(t)$ is the Fourier transform of $\varphi(\xi)$, then

(a) $e^{2\pi i\alpha t}g(t)$ is the Fourier transform of $\varphi(\xi + \alpha)$;

(b) $(2\pi it)^k g(t)$ is the Fourier transform of $\varphi^{(k)}(\xi)$, assuming that $\varphi(\xi)$ and its derivatives up to the kth order tend to zero, as $|\xi| \longrightarrow \infty$.

3. Show that if φ_1, φ_2 have Fourier transforms g_1 and g_2, respectively, then $\varphi(\xi) = \varphi_1(\xi)\varphi_2(\xi)$ has as Fourier transform the *convolution* of g_1 and g_2—i.e., the function

$$g(t) = \int_{-\infty}^{\infty} g_1(x)g_2(t - x)\, dx.$$

(This theorem is of great importance in the application of Fourier analysis—e.g., to differential equations and probability theory.)

4. Derive Parseval's equalities—i.e., Eqs. (9.2.7) and (9.5.4).

10 OPTIMIZATION

10.1. STATEMENT OF THE PROBLEM, DEFINITIONS, AND NORMAL FORM

Linear optimization (or **linear programming**) is a mathematical theory and method of calculation for *determining the maximum or minimum of a linear expression, where the domain of the variables is restricted by a system of linear inequalities*—or possibly also by a system of linear equations. Problems of this type come up mostly in economics, transportation problems, and production problems; however, there are also important applications in telecommunications, approximation theory, and other areas.

Example 10.1.1

In a given factory there are three machines M_1, M_2, M_3 used in making two products P_1, P_2. One unit of P_1 occupies M_1 5 minutes, M_2 3 minutes, and M_3 4 minutes. The corresponding figures for one unit of P_2 are: M_1 1 minute, M_2 4 minutes, M_3 3 minutes. The net profit per unit of P_1 produced is 30 dollars, and for P_2 20 dollars (independent of whether the machines are used to full capacity or not). What production plan gives the most profit?

Suppose that x_1 units of P_1 and x_2 units of P_2 are produced per hour. Then the problem is to

maximize

$$f = 30x_1 + 20x_2$$

subject to the constraints

$$
\begin{aligned}
5x_1 + x_2 &\leq 60 \quad (\text{for } M_1) \\
3x_1 + 4x_2 &\leq 60 \quad (\text{for } M_2) \\
4x_1 + 3x_2 &\leq 60 \quad (\text{for } M_3) \\
x_1 &\geq 0 \\
x_2 &\geq 0
\end{aligned}
\tag{10.1.1}
$$

The problem is illustrated geometrically in Fig. 10.1.1. The first of the inequalities above means that the point (x_1, x_2) must lie to the left of or on the line

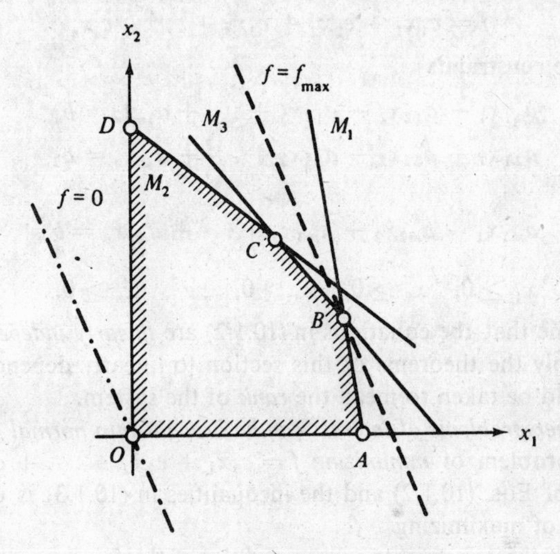

Fig. 10.1.1

AB whose equation is $5x_1 + x_2 = 60$. The other inequalities can be interpreted in a similar way. The point (x_1, x_2) must satisfy all of these inequalities. Thus it must lie within or on the boundary of the pentagon $OABCD$. The value of the function to be maximized is proportional to the distance between (x_1, x_2) and the dashed line $f(x_1, x_2) = 0$; it clearly takes on its largest value at the vertex B. Every vertex is the intersection of two sides of the pentagon. Thus at the vertices we must have equality in (at least) two of the inequalities. At the vertex B, equality holds in the inequality for M_1 and the inequality for M_3. In this way, one gets two equations for determining the two unknowns x_1 and x_2; the solution is $x_1 = 120/11$, $x_2 = 60/11$. The maximal profit is thus $4{,}800/11 = 436.36$ dollars/hour. M_1 and M_3 are to be used continuously, while M_2 is used only

$$
\frac{3 \cdot 120 + 4 \cdot 60}{11} = 54.55 \text{ min/hour.}
$$

A maximization problem of the above type cannot be solved by setting certain partial derivatives equal to zero. *The deciding factor is the domain in which the variables can vary.*

Practical problems as a rule involve many more variables. Hence we shall study a method of calculation which can be used when a graphical solution is impossible. We adopt the following definition of the **normal form** *for a linear optimization problem:*

Maximize

$$f = c_1x_1 + c_2x_2 + c_3x_3 + \dots + c_nx_n$$

subject to the constraints

$$\begin{aligned} a_{11}x_1 + a_{12}x_2 + a_{13}x_3 + \dots + a_{1n}x_n &= b_1 \\ a_{21}x_1 + a_{22}x_2 + a_{23}x_3 + \dots + a_{2n}x_n &= b_2 \\ \dots \\ a_{m1}x_1 + a_{m2}x_2 + a_{m3}x_3 + \dots + a_{mn}x_n &= b_m \end{aligned} \qquad (10.1.2)$$

$$x_1 \geq 0, \quad x_2 \geq 0, \quad x_3 \geq 0, \quad \dots, \quad x_n \geq 0. \qquad (10.1.3)$$

We assume that the equations in (10.1.2) are *linearly independent*. If one wants to apply the theorems of this section to linearly dependent systems, then m should be taken to mean the *rank* of the system.

Many other problems of interest can be brought into normal form; for example, the problem of *minimizing* $f = c_1x_1 + c_2x_2 + \dots + c_nx_n$ with the constraints of Eqs. (10.1.2) and the inequalities in (10.1.3) is equivalent to the problem of maximizing $-f$.

If some of the constraints are *inequalities* of the form

$$a_{i1}x_1 + a_{i2}x_2 + \dots + a_{in}x_n \leq b_i,$$

then one can introduce **"slack variables"** in such a way that one gets equalities

$$a_{i1}x_1 + a_{i2}x_2 + \dots + a_{in}x_n + x_{n+i} = b_i$$

together with inequalities of the form $x_{n+i} \geq 0$.

Example 10.1.2. *Continuation of Example 10.1.1*

The inequalities of Eq. (10.1.1) can be brought into normal form with the help of three slack variables x_3, x_4, x_5. We get

$$\begin{aligned} 5x_1 + x_2 + x_3 \qquad\qquad &= 60, \qquad (10.1.4) \\ 3x_1 + 4x_2 \qquad + x_4 \qquad &= 60, \\ 4x_1 + 3x_2 \qquad\qquad + x_5 &= 60. \end{aligned}$$

$$x_i \geq 0, \quad (i = 1, 2, 3, 4, 5). \qquad (10.1.5)$$

The three equations in Eq. (10.1.4) define a two-dimensional subspace (the lane in Fig. 10.1.1) in the five-dimensional space of points $(x_1, x_2, \ldots, x_5)$, ince $5 - 3 = 2$. Two of the coordinates—e.g., x_1, x_2—uniquely determine he remaining ones. Each side of the pentagon $OABCD$ has an equation of he form $x_i = 0$ $(i = 1, 2, 3, 4, 5)$. A vertex is the intersection between two ides; hence at a vertex two of the coordinates are zero, and the rest cannot e negative. (There are also five points of intersection between the extensions f the sides outside the pentagon. However, at these intersection points some f the quantities x_i are negative. Thus these points are not considered to be ertices.)

The geometrical ideas in the above example are also useful in the general ase: the m equations in Eq. (10.1.2) define an $(n - m)$-dimensional subspace n the n-dimensional space of points $(x_1, x_2, \ldots, x_n)$. It is possible to choose $n - m)$ coordinates and to give these coordinates arbitrary values; then the ystem of equations uniquely determines the rest of the coordinates. A point vhich satisfies the constraints of (10.1.2) and (10.1.3) is called a **feasible point,** or **feasible vector**. These points form a "polyhedron" in the $(n - m)$-limensional subspace. It is possible that there do not exist any feasible vectors. Also, it is possible that the polyhedron is unbounded in certain directions. *We shall, however, assume that there exist feasible vectors, and that f has a finite maximum.* (Then any eventual unboundedness of the polyhedron does not give rise to practical difficulties in the calculations.) These assumptions are as a rule satisfied in all practical problems which are properly formulated.

By a **feasible basic vector** (or point) we mean a feasible vector (point) for which at least $n - m$ of the n coordinates x_i are zero. It can happen that more than $n - m$ coordinates are zero at a feasible point—see, e.g., the top of the four-sided pyramid in Fig. 10.2.1 (p. 430). In such a case, one says that the point is a *degenerate feasible vector*.

An **optimal feasible vector** (or "optimal solution") is a feasible vector at which the function f takes on the largest possible value which it can take at a feasible vector. The following theorem, the validity of which the reader can easily convince himself of for $n - m \leq 3$, is fundamental. (For a general proof, see Gass [89, Chap. 3, theorems 2 and 4].) Notice, however, that the terminology in the above very excellent work and other works differs slightly from that which we use. For example, a point which we refer to as a feasible basic vector is sometimes called a "basic feasible solution". Similarly, the term "feasible solution" is sometimes encountered, whereas we use "feasible vector." Inasmuch as, strictly speaking, the term "feasible" refers to the fact that the constraints (including nonnegativity conditions) are satisfied for the point in question—and not, e.g., that the point is a "possible" solution to the problem as a whole—we prefer our usage. This is also the usage of Künzi et al. [136, trans. Rheinboldt]. In addition, we think that our usage is more suggestive of a geometrical point of view, which is useful in visualizing the relevant algorithms and theorems.

THEOREM 10.1.1

Some optimal feasible vector is also a feasible basic vector—i.e., at least $n - m$ *of its coordinates are zero. Put in another way, there is an optimal feasible solution for which at most* m *coordinates are strictly positive. The others are zero. The column vectors of Eq. (10.1.2) which correspond to the non-zero coordinates are linearly independent.*

There can exist many optimal solutions; not all of these need be feasible *basic* vectors. For example, in a problem like Example 10.1.1, one could have an f such that the line $f(x_1, x_2) = 0$ were parallel to one of the sides of the pentagon; see Example 10.2.3!

REVIEW QUESTIONS

1. Give the normal form for a linear programming problem. Define the terms "feasible vector" (feasible point), "feasible basic vector," and "slack variable."
2. State the basic theorem of linear optimization (Theorem 10.1.1). Can there be more than one optimal solution? Is every optimal solution a feasible basic vector?

10.2. THE SIMPLEX METHOD

The problem now is to find the right feasible basic vector—i.e., to find out *which* of the n coordinates are zero. In theory, one can consider trying all the $\binom{n}{n-m}$ possible ways of setting $n - m$ of the n given variables equal to zero, sorting out those combinations which do not give feasible vectors, and then choosing among the rest to find a point which gives the largest value of f. However, even for relatively small m and n, this is quite laborious. Instead, one can use the so-called **simplex** method. *This is an iterative method, whereby one goes from one feasible basic vector to another in such a way that f always increases.* It gives the exact result after a number of steps which is usually much less than $\binom{n}{n-m}$. Computational experience seems to indicate that the number of steps is a linear function of the number of constraints and essentially independent of the number of variables, e.g. $2m - 3m$.

We shall now illustrate the use of the simplex method on the previous example. Afterward, we shall treat some complications which can come up during the calculations.

I. First, look for a feasible basic vector at which the iterative process can begin—i.e., divide the variables into two classes: **right-hand variables** (the $n - m$ variables which are to have the value zero at the feasible basic vector) and **left-hand variables** (the m remaining variables). *Express the left-hand variables and f as functions of the right-hand variables by solving the system of equations in Eq. (10.1.2) (m equations, m unknowns).*

For Eq. (10.1.4) one gets, if x_1 and x_2 are taken as right-hand variables,

$$\begin{aligned} x_3 &= 60 - 5x_1 - x_2 \\ x_4 &= 60 - 3x_1 - 4x_2 \\ x_5 &= 60 - 4x_1 - 3x_2 \\ f &= 30x_1 + 20x_2. \end{aligned} \tag{10.2.1}$$

In favorable cases one gets the coordinates of a feasible basic vector by putting the right-hand variables equal to zero. But if this should give rise to negative values for some of the left-hand variables, then the point so obtained is not (by a previous definition) a feasible basic vector. In such a case one must try another set of right-hand variables.

In the present example, we get $x_3 = x_4 = x_5 = 60$ when $x_1 = x_2 = 0$. Thus we have found a feasible basic vector (the point O in Fig. 10.1.1).

One can often find a feasible basic vector directly (as in this example) or after only a few tries. A systematic method, which can be used in more difficult cases, will be given in Example 10.2.2. In principle, it is of no importance which feasible basic vector one starts at, but one can often decrease the number of iterations by guessing a point which lies close to an optimal feasible vector (e.g., by considering the content of the problem). In the examples in this chapter, however, we shall make no such "intelligent guesses."

II. See if the **maximum criterion** ("optimality criterion") is fulfilled. This means that one checks to see that *none of the coefficients in the expression of f as a function of the present right-hand variables is positive.*

If this criterion is fulfilled, then one has found an optimal solution: f cannot increase when one gives one (or more) of the right-hand variables positive values, and negative values are not permitted. One can show that this criterion is a necessary condition for an optimal solution.

III. Suppose that the maximum criterion is not fulfilled. Now check to see how much the value of f changes when one goes from the present feasible basic vector to one of its neighbors. To go to a neighboring feasible basic vector means to interchange one of the right-hand variables with one of the left-hand variables which is not zero at the point. We now give a method to see which of the right-hand variables is to be interchanged:

Consider a certain right-hand variable x_R whose coefficient in the expression for f is positive. Determine the largest positive increment one can give x_R without making any of the left-hand variables negative (the other right-hand variables are held equal to zero). Underline x_R in the equation which inhibited any further increase in x_R, and note the value of Δx_R and Δf—i.e., the corresponding increase of f. For example, in Eq. (10.2.1) the effect of increasing x_1 can be summarized as follows:

$$x_3 \geq 0 \text{ if } \Delta x_1 \leq \frac{60}{5} = 12$$

$$x_4 \geq 0 \text{ if } \Delta x_1 \leq \frac{60}{3} = 20$$

$$x_5 \geq 0 \text{ if } \Delta x_1 \leq \frac{60}{4} = 15.$$

Thus we underline x_1 in the equation for x_3 and note that $\Delta x_1 = 12$, $\Delta f = 30 \cdot \Delta x_1 = 360$. The reader should make a similar investigation for x_2. In general, this investigation is to be made for all the right-hand variables x_R which have positive coefficients in the expression for f. The other variables cannot of course produce any increase in f (when one only changes one variable at a time).

$$\begin{aligned}
x_3 &= 60 - \underline{5x_1} - x_2 \qquad \Delta x_1 = \frac{60}{5} = 12, \Delta f = 30 \cdot 12 = 360 \quad \text{(largest)} \\
x_4 &= 60 - 3x_1 - \underline{4x_2} \qquad \Delta x_2 = \frac{60}{4} = 15, \Delta f = 20 \cdot 15 = 300 \\
x_5 &= 60 - 4x_1 - 3x_2 \\
f &= \qquad 30x_1 + 20x_2.
\end{aligned}$$

IV. Exchange the right-hand variable x_R which gave the greatest value for Δf with the left-hand variable in the equation where x_R is underlined—i.e., x_R becomes a left-hand variable, and the relevant left-hand variable now becomes a right-hand variable. Thus in the present example we would write

$$x_1 = \frac{60 - x_3 - x_2}{5} \quad \text{(using 10.2.1)}$$

and then modify the rest of the equations (*and the expression for f*) accordingly. Geometrically, this means that one goes from 0 to A in Fig. 10.1.1. It is possible that the above criterion does not give a unique determination of the variables to the exchanged—see Example 10.2.1.

Now repeat II, III, *and* IV *until the maximum criterion is fulfilled.*

In the present example the calculations go as follows:

$$x_1 = \frac{60 - x_3 - x_2}{5}$$

$$x_4 = 60 - \frac{180 - 3x_3 - 3x_2}{5} - 4x_2$$

$$= \frac{120 + 3x_3 - 17x_2}{5}$$

$$x_5 = 60 - \frac{240 - 4x_3 - 4x_2}{5} - 3x_2$$

$$= \frac{60 + 4x_3 - 11x_2}{5}, \quad \Delta x_2 = \frac{60}{11}, \Delta f = \frac{840}{11}$$

$$f = (360 - 6x_3 - 6x_2) + 20x_2 = 360 + 14x_2 - 6x_3.$$

The coefficient of x_2 is positive. Thus the maximum criterion is not satisfied. (Why is it only necessary to investigate Δx_2?) Exchange x_2 and x_5—i.e., go from A to B in Fig. 10.1.1. We get, after calculations which the reader is recommended to carry out in detail,

$$x_2 = \frac{60 + 4x_3 - 5x_5}{11} \quad \text{(using the above equation for } x_5\text{)},$$

$$x_4 = \frac{60 - 7x_3 + 17x_5}{11}$$

$$x_1 = \frac{120 - 3x_3 + x_5}{11}$$

$$f = \frac{4{,}800 - 10x_3 - 70x_5}{11}.$$

No coefficient is positive. Thus the maximum criterion is fulfilled. The point with coordinates (120, 60, 0, 60, 0)/11—i.e., the point B—is thus an optimal solution, in agreement with the result obtained previously.

The simplex method (in the variant described here) gives, at each step, the largest possible increase in f, when just one right-hand coordinate is exchanged. It is possible, however, that even at a point which is not an optimal solution one cannot increase f by exchanging one single coordinate without coming into conflict with the constraints. This exceptional case occurs only when one of the left-hand variables is zero at the same time that the right-hand variables are zero—i.e., when one has found a feasible vector for which more than $n - m$ of its coordinates are zero. As we mentioned previously, such a point is called a **degenerate feasible vector.** In such a case one has to try exchanging a right-hand variable with one of the left-hand variables which is zero at the point. In more difficult cases, it may be necessary to make several such exchanges.

Example 10.2.1

Maximize

$$f = 2x_1 + 2x_2 + 3x_3$$

subject to the constraints

$$x_1 + x_3 \leq 1$$

$$x_2 + x_3 \leq 1$$

$$x_i \geq 0, \quad \text{for} \quad i = 1, 2, 3.$$

The feasible vectors form a four-sided pyramid in (x_1, x_2, x_3)-space; see Fig 10.2.1. Introduce the slack variables x_4, x_5 and take $\{x_1, x_2, x_3\}$ as right-hand variables. Notice that the equations $x_4 = 0$ and $x_5 = 0$ also mean sides of a

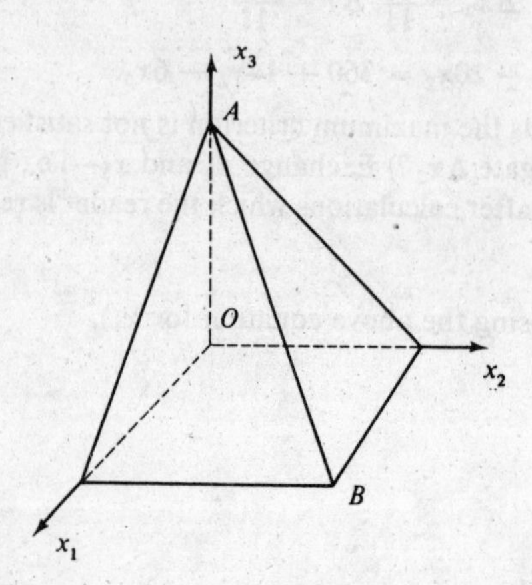

Fig. 10.2.1

pyramid. This gives a feasible vector, since $x_1 = x_2 = x_3 = 0$ (the point C in Fig. 10.2.1) satisfies the constraints. Then

$$\begin{aligned} x_4 &= 1 - \underline{x_1} - \underline{x_3}, && \Delta x_1 = 1, \Delta f = 2 \\ x_5 &= 1 - \underline{x_2} - \underline{x_3}, && \Delta x_2 = 1, \Delta f = 2 \\ f &= 2x_1 + 2x_2 + 3x_3 && \Delta x_3 = 1, \Delta f = 3 \quad \text{(largest)} \\ x_i &\geq 0 \quad \text{for} \quad i = 1, 2, 3, 4, 5. \end{aligned}$$

Notice also that x_3, which gave the largest increase in f, is underlined in two places; x_3 is to be exchanged with either x_4 or x_5. In such a situation one can choose either one—say, x_4. (Then one comes to the point A.)

$$\begin{aligned} x_3 &= 1 - x_1 - x_4, && \Delta x_2 = 0, \Delta f = 0 \\ x_5 &= x_1 - \underline{x_2} + x_4 \\ f &= 3 - x_1 + 2x_2 - 3x_4. \end{aligned}$$

The point A is a degenerate feasible vector; x_5 is also zero at this point. Four sides of a pyramid go through A. Since x_2 is the only right-hand coordinate which has a positive coefficient in f, we exchange x_2 and x_5. Thus, in this step, we remain at the point A. (Recall that in nondegenerate cases, one step in the iteration takes us to a neighboring basic point on the boundary of the admissible region.)

$$x_2 = x_1 + x_4 - x_5, \qquad \Delta x_1 = 1, \Delta f = 1$$
$$x_3 = 1 - \underline{x_1} - x_4$$
$$f = 3 + x_1 - x_4 - 2x_5.$$

Exchange x_1 and x_3. (We go from A to B.)

$$x_1 = 1 - x_3 - x_4$$
$$x_2 = 1 - x_3 - x_5$$
$$f = 4 - x_3 - 2x_4 - 2x_5.$$

The maximum criterion is fulfilled. $f_{max} = 4$. The optimal solution is the point (1, 1, 0, 0, 0).

The next example gives *a method of finding a* **starting point** *in difficult cases.*

Example 10.2.2

Maximize

$$f = x_1 - x_2$$

under the constraints

$$x_3 = -2 + 2x_1 - x_2$$
$$x_4 = \quad 2 - \ x_1 + 2x_2$$
$$x_5 = \quad 5 - \ x_1 - x_2$$
$$x_i \geq 0 \quad (i = 1, 2, 3, 4, 5).$$

x_1 and x_2 cannot be used as right-hand variables, since, when $x_1 = x_2 = 0$, x_3 is negative. It is not immediately obvious which pair of variables suffice as right-hand variables. In such a case, one can modify the problem by introducing a new **artificial variable** x_6, defined by the equation

$$x_3 = -2 + 2x_1 - x_2 + x_6,$$

with the constraint $x_6 \geq 0$. If we now take x_1, x_2, x_3 as right-hand variables, then x_4, x_5, x_6 are all nonnegative when the right-hand variables are set equal to zero. Thus we have found a feasible vector for an extended (six-dimensional) problem. In order for this problem to correspond, finally, to the original problem, we must make sure that the artificial variable (here x_6) is zero in the final solution. To accomplish this, we modify the function to be maximized by adding a term with a very large negative coefficient, say, $-M$, in front of the artificial variable (M is assumed to be much larger than all other numbers in the computations). Thus a positive value of an artificial variable will tend to make the function to be maximized quite small; in this way, the value of the artificial variable becomes zero in the final solution. Here, the modified problem becomes:

Maximize

$$\bar{f} = x_1 - x_2 - Mx_6 = (1 + 2M)x_1 - (1 + M)x_2 - Mx_3 - 2M$$

under the constraints

$$\begin{aligned} x_6 &= 2 - \underline{2x_1} + x_2 + x_3, \quad \Delta x_1 = 1, \Delta \bar{f} = 1 + 2M \\ x_4 &= 2 - x_1 + 2x_2 \\ x_5 &= 5 - x_1 - x_2 \\ x_i &\geq 0, \quad (i = 1, 2, 3, 4, 5, 6). \end{aligned}$$

Exchange x_1 and x_6:

$$x_1 = \frac{2 + x_2 + x_3 - x_6}{2}$$

$$x_4 = \frac{2 + 3x_2 - \underline{x_3} + x_6}{2}, \quad \Delta x_3 = 2, \Delta \bar{f} = 1$$

$$x_5 = \frac{8 - 3x_2 - x_3 + x_6}{2}$$

$$\bar{f} = \frac{2 - x_2 + x_3 + x_6}{2} - Mx_6.$$

Now the artificial variable has become a right-hand variable. This will happen sooner or later in any problem which satisfies the assumptions made in Sec 10.1. If one takes away x_6 and sets the rest of the right-hand variables equal to zero, then one gets a feasible vector to the original (five-dimensional) problem. *The artificial variable is no longer needed in the computations.* After two iterations (x_3 is exchanged with x_4, and thereafter x_2 with x_5), the maximum criterion is satisfied and one finds $f_{max} = 3$ for $(x_1, x_2, x_3, x_4, x_5) = (4, 1, 5, 0, 0)$. In more difficult situations, it may be necessary to introduce several artificial variables.

There can be **many optimal solutions to** a given problem, but of course f_{max} is uniquely determined. Such a situation is indicated in the calculations when one or more of the right-hand variables is absent in the expression for f when the maximum criterion is satisfied.

Example 10.2.3

Maximize

$$f = 2 - x_4$$

under the constraints

$$
\begin{aligned}
x_3 &= 6 - 2x_4 + x_2 \\
x_1 &= 2 - x_4 + x_2 \\
x_5 &= 3 + x_4 - 2x_2 \\
x_i &\geq 0.
\end{aligned}
$$

The maximum criterion is satisfied. Thus the point (2, 0, 6, 0, 3) is an optimal solution. But $f = f_{max} = 2$ for arbitrary values of x_2. In fact, x_2 can be increased to $\frac{3}{2}$ before any of the left-hand variables (in this case, x_5) become negative. If we exchange x_2 and x_5, and set x_4 and x_5 equal to zero, we get another optimal solution $(\frac{7}{2}, \frac{3}{2}, \frac{15}{2}, 0, 0)$. The general solution to the above problem is the line segment between the two solutions we have already found —i.e., the points $c(2, 0, 6, 0, 3) + (1 - c)(\frac{7}{2}, \frac{3}{2}, \frac{15}{2}, 0, 0)$, where $0 \leq c \leq 1$.

There are many variations of the simplex method. One can have different strategies for exchanging variables; for example, in treating a large problem on a computer it could be advantageous to stay at the first neighboring point which gave an increase in the value of f, instead of looking for the neighboring point which gives the *largest* increase in f.

The calculations can also be arranged in tables, in a way reminiscent of Gaussian elimination (see Example 5.3.1). The necessary formulas are not difficult to derive (see Problem 3 at the end of this section). These formulas can also be of use if one needs to write one's own program for linear optimization instead of using one of the standard programs in the literature or in a computer manufacturer's "subroutine package."

The simplex method is occasionally described in a different way; see, e.g., Gass [130]. One says that (in each iteration) m of the columns in the matrix $\{a_{ij}\}$ of Eq. (10.1.2) are chosen as basis vectors, and that the rest of the columns are expressed as linear combinations of the columns (vectors) in the basis. In each iteration, one exchanges one of the basis vectors with one of the columns not in the basis. The connection with our presentation is quite simply that *the basis vectors are those columns in the system of equations in Eq. (10.1.2) where the "present" left-hand variables appear*. The calculations themselves are completely analogous in both presentations.

REVIEW QUESTION

Describe the simplex method. What does one do in the case of a degenerate feasible vector?

PROBLEMS

1. (a) Find the maximum of $f = x_1 + 3x_2$ subject to the constraints

$$2x_1 + x_2 \leq 2 \quad (1)$$
$$x_1 + 2x_2 \leq 2 \quad (2)$$
$$x_1 \geq 0, x_2 \geq 0.$$

First solve the problem graphically, and then use the simplex method with $x_1 = x_2 = 0$ as initial point.

In the following variants, begin with the same right-hand variables as those you had at the optimal solution in (a).

(b) Find the maximum of $f = 2x_1 + 5x_2$ under the same constraints as in (a).

(c) Find the maximum of $f = x_1 + x_2$ with the same constraints as in (a).

(d) Find the maximum of $f = x_1 + 3x_2$ after changing the constraint (2) to $2x_1 + 2x_2 \leq 3$.

(e) What conclusion can be drawn from the problems (b)–(d)?

2. Suppose that for a certain computer there is a set of programs, which we call LP, that solves linear programming problems which are in normal form. Thus LP computes $\max f = \boldsymbol{c}^T\boldsymbol{x}$ under the constraints $A\boldsymbol{x} = \boldsymbol{b}$, $\boldsymbol{x} \geq \boldsymbol{0}$. Here, $\boldsymbol{c}$, $\boldsymbol{x}$, and $\boldsymbol{b}$ are column vectors, and A is a matrix. (The relation $\boldsymbol{x} \geq \boldsymbol{0}$ means that each component of $\boldsymbol{x}$ is nonnegative.) Only A, $\boldsymbol{b}$, and $\boldsymbol{c}$ need be read into the machine. One wants to treat the following problem:

Find $\min (x_1 + 2x_2 + 3x_3 + 4x_4 + 5x_5 + x_6 + x_7)$, where

$$|x_1 + x_2 + x_3 - 4| \leq 12,$$
$$3x_1 + x_2 + 5x_4 \leq 6,$$
$$x_1 + x_2 + 3x_3 \geq 3,$$
$$|x_1 - x_2 + 5x_7| \geq 1,$$
$$x_i \geq 0, \quad i = 1, 2, 3, 4, 5, 6, 7.$$

Give A, $\boldsymbol{b}$, and $\boldsymbol{c}$ in this case! (*Note:* One need not look for an $\boldsymbol{x}$ which satisfies the inequalities—this is done automatically by the set of programs.)

3. At a certain stage in the simplex method the left-hand variable x_i is exchanged with the right-hand variable x_j. (That is, after the exchange, x_i is a right-hand variable and x_j a left-hand variable.) Before the change, for each left-hand variable x_L there is a relation of the form

$$x_L = c_{Lj}x_j + \sum c_{LR}x_R \quad \text{(even for } x_L = x_i\text{)},$$

where the sum on the right is taken over all right-hand variables x_R except x_j. After the exchange, there is a relation of the form

$$x_L = c'_{Li}x_i + \sum c'_{LR}x_R \quad \text{(even for } x_L = x_j\text{)},$$

where the sum is now taken over all the right-hand variables x_R except x_i. Express the coefficients

$$c'_{Li} \quad c'_{LR}, \quad c'_{ji}, \quad c'_{jR}$$

in terms of

$$c_{Lj}, \quad c_{LR}, \quad c_{ij}, \quad c_{iR}.$$

(Note that there is no loss of generality by deleting the constant terms. These can be thought of as corresponding to a special right-hand variable $x_R = 1$.)

10.3. DUALITY

We formulate the linear programming problem of (10.1.2) and (10.1.3) in vector-matrix notation:

Maximize

$$f(\mathbf{x}) = \mathbf{c}^{\mathrm{T}}\mathbf{x} \tag{10.3.1}$$

subject to the constraints

$$A\mathbf{x} = \mathbf{b}, \tag{10.3.2}$$

$$\mathbf{x} \geq 0. \tag{10.3.3}$$

(The vector inequality here means that all components are nonnegative.) This problem will be called the **primal problem.** The columns of the $m \times n$ matrix A will be denoted $\mathbf{a}_1, \mathbf{a}_2, \ldots, \mathbf{a}_n$. We still *assume that there exist feasible vectors.* By Theorem 10.1.1, one of the optimal feasible vectors is a point $\hat{\mathbf{x}}$ where k ($k \leq m$) coordinates $\hat{x}_{i_1}, \ldots, \hat{x}_{i_k}$ are strictly positive and where the corresponding vectors $\mathbf{a}_{i_1}, \mathbf{a}_{i_2}, \ldots, \mathbf{a}_{i_k}$ are linearly independent. The reader can easily show that if $k < m$, the vector set can be extended to a set of m linearly independent vectors. In other words, *there exists a set S of m integers such that the columns $\mathbf{a}_i$ ($i \in S$) are linearly independent and such that $\hat{x}_j = 0$ when $j \notin S$.*

The **dual problem** to (10.3.1)–(10.3.3) is defined as follows:

Minimize

$$g(\mathbf{y}) = \mathbf{y}^{\mathrm{T}}\mathbf{b} \tag{10.3.4}$$

subject to the constraints

$$\mathbf{y}^{\mathrm{T}}A \geq \mathbf{c}^{\mathrm{T}}. \tag{10.3.5}$$

A vector $\mathbf{y}$ that satisfies (10.3.5) is called a feasible vector of the dual problem.

THEOREM 10.3.1

$$\min g(\mathbf{y}) = \max f(\mathbf{x}) \tag{10.3.6}$$

The minimum value is obtained at a point $\hat{\mathbf{y}}$ which is the solution of the m simultaneous equations

$$\hat{\mathbf{y}}^{\mathrm{T}}\mathbf{a}_i = c_i, \quad i \in S, \tag{10.3.7}$$

where S is the set of integers defined previously.

Proof. Let $\boldsymbol{x}$ and $\boldsymbol{y}$ be arbitrary feasible vectors for the primal and dual problems, respectively. Then

$$g(\boldsymbol{y}) = \boldsymbol{y}^{\mathsf{T}}\boldsymbol{b} = \boldsymbol{y}^{\mathsf{T}}A\boldsymbol{x} \leq \boldsymbol{c}^{\mathsf{T}}\boldsymbol{x} = f(\boldsymbol{x}).$$

Hence

$$\min g(\boldsymbol{y}) \geq \max f(\boldsymbol{x}). \tag{10.3.8}$$

Next we shall show that $\hat{\boldsymbol{y}}$, defined by Eq. (10.3.7), is a feasible vector. Since $A\boldsymbol{x} = \boldsymbol{b}$, we may write

$$f(\boldsymbol{x}) = \boldsymbol{c}^{\mathsf{T}}\boldsymbol{x} - \hat{\boldsymbol{y}}^{\mathsf{T}}(A\boldsymbol{x} - \boldsymbol{b}) = \hat{\boldsymbol{y}}^{\mathsf{T}}\boldsymbol{b} + [\boldsymbol{c}^{\mathsf{T}} - \hat{\boldsymbol{y}}^{\mathsf{T}}A]\boldsymbol{x}.$$

Hence, by Eq. (10.3.7),

$$f(\boldsymbol{x}) = \hat{\boldsymbol{y}}^{\mathsf{T}}\boldsymbol{b} + \sum_{j \notin S} (c_j - \hat{\boldsymbol{y}}^{\mathsf{T}}\boldsymbol{a}_j)x_j. \tag{10.3.9}$$

Now $f(\boldsymbol{x})$ is expressed in terms of the right-hand variables corresponding to the optimal solution of the primal problem. It then follows from the maximum criterion (see Sec. 10.2) that

$$c_j - \hat{\boldsymbol{y}}^{\mathsf{T}}\boldsymbol{a}_j \leq 0, \quad j \notin S.$$

This, together with Eq. (10.3.7), shows that $\hat{\boldsymbol{y}}$ is a feasible vector for the dual problem. Moreover, since $\hat{x}_j = 0, j \notin S$, then by Eq. (10.3.9)

$$f(\hat{\boldsymbol{x}}) = \hat{\boldsymbol{y}}^{\mathsf{T}}\boldsymbol{b} = g(\hat{\boldsymbol{y}}).$$

This is consistent with Eq. (10.3.8) only if $\min g(\boldsymbol{y}) = g(\hat{\boldsymbol{y}})$. Hence $\min g(\boldsymbol{y}) = \max f(\boldsymbol{x})$, and the theorem is proved.

PROBLEM

(a) Put the dual problem into the normal form, defined in Sec. 10.1 (note that there is no non-negativity condition on $\boldsymbol{y}$).
(b) Show that the dual problem of the dual problem is the primal problem.

10.4. THE TRANSPORTATION PROBLEM AND SOME OTHER OPTIMIZATION PROBLEMS

The **transportation problem** is one of the most well-known problems in optimization.

Suppose that a business concern has I factories which produce, during a given week, $a_1, a_2, \ldots, a_I$ units of a certain product. This product is sent to J consumers, who need (respectively) $b_1, b_2, \ldots, b_J$ units. We assume that the total number of units produced is equal to the total need—i.e.,

$$\sum_{i=1}^{I} a_i = \sum_{j=1}^{J} b_j. \tag{10.4.1}$$

It costs c_{ij} dollars to transport one unit from producer i to consumer j. The problem is to determine the quantities x_{ij} so that the total cost of transportation is minimized. This problem can be formulated as a linear optimization problem as follows (here we denote the variables with double indices):

Minimize

$$f = \sum_{i=1}^{I} \sum_{j=1}^{J} c_{ij} x_{ij} \tag{10.4.2}$$

under the constraints ($I + J$ equations, IJ inequalities)

$$\begin{aligned} &\sum_{j=1}^{J} x_{ij} = a_i, \quad (i = 1, 2, \ldots, I) \\ &\sum_{i=1}^{I} x_{ij} = b_j, \quad (j = 1, 2, \ldots, J) \end{aligned} \tag{10.4.3}$$

$$x_{ij} \geq 0. \tag{10.4.4}$$

There is a linear dependence between these equations, since

$$\sum_{i=1}^{I} \sum_{j=1}^{J} x_{ij} - \sum_{j=1}^{J} \sum_{i=1}^{I} x_{ij} = 0.$$

The number of linearly independent equations is thus (at most) $m = I + J - 1$. From Theorem 10.1.1, it follows that *there exists an optimal transportation scheme, where at most $I + J - 1$ of the IJ possible routes between producer and consumer are used.* In principle, the transportation problem can be solved using the simplex method; however, there are much more effective methods which make use of the special structure of the equations; see, e.g. Gass [130].

Many other problems can be formulated as transportation problems. One interesting example is the *personnel-assignment problem:* One wants to distribute I applicants to J jobs. The suitability c_{ij} of applicant i for job j is known. The problem is to distribute the applicants in such a way that the total "suitability" is as great as possible. This problem is clearly analogous to the transportation problem.

Linear optimization problems are the simplest of a wide variety of optimization problems which occur in operations analysis. In this field, one develops and uses mathematical models as a basis for strategic or organizational decisions. Many practical optimization problems are *nonlinear*—i.e., the function f and/or some of the constraints are nonlinear. In scheduling problems, for example, some of the variables are constrained to take on only *integer values*. The total number of variables can occasionally be as high as several thousand.

Techniques for the planning of multistage processes have been developed; these are often collected under the name dynamic optimization or *dynamic programming*. However, it would carry us too far afield to go into this important and theoretically interesting area; we refer the reader to Bellman and Kalaba [122] or Hadley [132b].

In the next section we shall discuss some nonlinear optimization problems which also occur quite frequently: the problem of "unconstrained optimization" or the minimization of a function of many variables (with no constraints), and the problems of constrained optimization and overdetermined nonlinear systems.

REVIEW QUESTION

Formulate the transportation problem. What conclusion can one make about the total number of routes used in an optimal scheme for transportation? (Use the main theorem of linear optimization.)

10.5. NONLINEAR OPTIMIZATION PROBLEMS

10.5.1. Basic Concepts and Introductory Examples

We shall discuss the problem of finding a local *minimum* of a real-valued function $\varphi(\mathbf{x})$, $\mathbf{x} = (x_1, x_2, \ldots, x_n)^T$. To begin with, we assume that there are *no constraints* on the variables. This problem is encountered in many applications such as operations research, control theory, chemical engineering, and all kinds of curve fitting or more general mathematical model fitting. Maximization problems can, of course, be transformed into minimization problems by reversing the sign of φ.

Example 10.5.1

Exponential fitting problems occur frequently—e.g., the parameter vector $\mathbf{x} = (x_1, x_2, x_3, x_4, x_5)^T$ in the expression

$$y(t, \mathbf{x}) = x_1 + x_2 e^{x_3 t} + x_4 e^{x_5 t}$$

is to be determined to give the best fit to m observed points (t_i, y_i), $i = 1, 2, \ldots, m$, $m > 5$. This problem cannot be handled directly by the methods of Chap. 4, because the expression is nonlinear in the parameters x_3, x_5. The problem can be considered as an *overdetermined nonlinear system*,

$$y(t_i, \mathbf{x}) \approx y_i, \quad i = 1, 2, \ldots, m.$$

If the observations have equal weight, the principle of least squares requires the minimization of the function

$$\varphi(\mathbf{x}) = \sum_{i=1}^{m} (y(t_i, \mathbf{x}) - y_i)^2. \tag{10.5.1}$$

Example 10.5.2

Let $t_1, t_2, \ldots, t_m$ be sample values from a population with a frequency function of the form $f(t, \boldsymbol{\alpha})$, where $\boldsymbol{\alpha}$ is an unknown parameter vector. According to the maximum-likelihood principle, $\boldsymbol{\alpha}$ is to be determined by a maximization of the function

$$\varphi(\boldsymbol{\alpha}) = \prod_{i=1}^{m} f(t_i, \boldsymbol{\alpha}).$$

This can be handled by analytic techniques only for a few special types of frequency functions.

We shall first consider general minimization problems, like the second example, and we assume that analytic expressions for the partial derivatives are available. (Special techniques for overdetermined systems will then be treated in Sec. 10.5.4.) The problem can then be reduced to a (usually nonlinear) system of equations,

$$g(x) = 0, \tag{10.5.2}$$

where $g(x) = (\partial\varphi/\partial x_1, \partial\varphi/\partial x_2, \ldots, \partial\varphi/\partial x_n)^T$ is the **gradient** of φ. The techniques of Sec. 6.9 are applicable—for example, Newton-Raphson's method

$$x_{\nu+1} = x_\nu - H(x_\nu)g(x_\nu), \tag{10.5.3}$$

where $H(x)$ is the inverse of the Jacobian $G(x)$ of g. $G(x)$ is also called the **Hessian** of φ,

$$H = G^{-1}, \qquad G(x) = \left[\frac{\partial^2 \varphi_i}{\partial x_i \partial x_j}\right]. \tag{10.5.4}$$

Note that G and H are *symmetric* matrices. A sufficient condition for a solution $\hat{x}$ of Eq. (10.5.2) to give a minimum is that $G(\hat{x})$ be positive-definite. It is necessary that $G(\hat{x})$ be at least nonnegative-definite. These facts together with the availability of values of the function φ create a more favorable situation than one encounters for general nonlinear systems. Note that the computation of g usually requires much more work than the computation of φ, and that the computation of G requires even more work, in particular when n is large.

We recall that Newton-Raphson's method has the advantage of *quadratic convergence*, which is to be matched against the following disadvantages (see also Sec. 10.5.3):

(a) A good initial approximation has to be found, unless g is close to a linear function (in other words, unless φ is close to a quadratic function).

(b) In orthodox Newton-Raphson, the Hessian must be computed (at least approximately by finite differences) and the linear system,

$$G(x)d(x) = g(x)$$

is to be solved in each iteration.

In order to avoid these difficulties, a great number of iterative methods have been developed. In general they are of the form

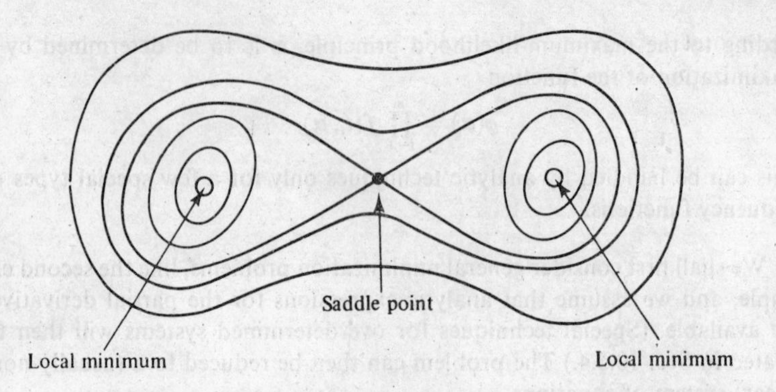

Fig. 10.5.1

$$x_{\nu+1} = x_\nu - \lambda_\nu d_\nu. \qquad (10.5.5)$$

Two things are to be done in each iteration:

1. Choose a **search direction** d_ν.
2. Carry out a **line search** in this direction.

A picturesque topographical terminology is often used, with words like "valleys," "descent," "troughs," and "search"—or (for maximization problems) "hill-climbing", "ridges," "peaks," etc.

10.5.2. Line Search

Put $\varphi(x_\nu - \lambda d_\nu) = \psi(\lambda)$. Assume that d_ν is a **descent direction**—i.e., that $\psi'(0) = -d_\nu^T g(x_\nu)$ is negative. *We want to find a* $\lambda_\nu \in [0, m]$ *such that*

$$\psi(\lambda_\nu) < \psi(0) \quad \text{and} \quad \psi(\lambda_\nu) \lesssim \min \psi(\lambda) + \epsilon, \qquad (10.5.6)$$

where ϵ is a prescribed tolerance, which may depend on ν.

Many algorithms are described in the literature; see Dixon [125, Chap. 1]. The following is a minor modification of a method devised by Powell. It is based on the use of the quadratic interpolation polynomial $Q(\lambda)$, defined by three points $(a, \psi(a))$, $(b, \psi(b))$, $(c, \psi(c))$.

Initially, we put $a = 0$, $b = k$, where k is a parameter, which gives the expected order of magnitude of λ_ν. (For the matrix-updating methods, which will be described in the next section, put $k = 1$.) If $\psi(k) > \psi(0)$, then put $c = -k$; otherwise, put $c = 2k$.

Let λ' be the turning point of the quadratic $Q(\lambda)$. In the divided difference notation of Sec. 7.3.3, it is given by the equations (see Problem 3 of this section)

$$\theta = \frac{\psi[a, b]}{\psi[a, b] - \psi[b, c]}, \qquad (10.5.7)$$
$$\lambda' = \tfrac{1}{2}[(a + b)(1 - \theta) + (b + c)\theta].$$

Q has no minimum at λ', if $\psi[a, b, c] \leq 0$. If this happens, or if $\lambda' \geq m$, then the point out of a, b, c which is furthest from m will be replaced by m, and the interpolation will be repeated. If a, b, or c is already equal to m, then put $\lambda_\nu = m$, and the line search is terminated.

Next, suppose that $\psi[a, b, c] > 0$ and $\lambda' < m$. Then Q has a minimum at λ'. $\psi(\lambda')$ is then calculated, and if

$$|\psi(\lambda') - \min\{\psi(a), \psi(b), \psi(c)\}| \leq \epsilon \quad \text{or} \quad |\psi(\lambda') - Q(\lambda')| \leq \epsilon,$$

then put $\lambda_\nu = \lambda'$. If none of these conditions is satisfied, the quadratic interpolation will be repeated, after a, b, or c has been replaced by λ'. The function value to be thrown away is normally the greatest, but it is not if rejecting a smaller one can yield a bracket on a minimum which would not be obtained otherwise. If the points are ordered such that $a < b < c$, then the condition for "a bracket on a minimum" is that $\psi(b) < \psi(a)$ and $\psi(b) < \psi(c)$.

In many applications, function evaluations are quite expensive. Then one should be careful not to choose ϵ too small. Some people recommend stopping the line search as soon as a bracket on a minimum has been found. A discussion of this problem is given by P. Wolfe, *SIAM Review*, v. 11, pp. 226–235 (1969).

If the gradient $g(x_\nu)$ is known, one can simplify starting the search. Put $a = b = 0$, $c = k$. Note that, by Sec. 7.3.7,

$$\psi[0, 0] = \psi'(0) = -d_\nu^T g(x_\nu), \qquad (10.5.8)$$

so that Eq. (10.5.7) can be applied.

10.5.3. Algorithms for Unconstrained Optimization

Steepest Descent Method (*Cauchy, 1847*). Here one chooses

$$d_\nu = g(x_\nu), \quad \lambda_\nu > 0.$$

Note that $g^T(x_\nu)d_\nu = \|g(x_\nu)\|^2 > 0$. The gradient direction is a descent direction. The method is recommended when one is far away from $\hat{x}$. An advantage is that the Hessian is not needed, and the method has a local optimum property, indicated in the name; see Problem 2 at the end of this section. In the long run, however, the convergence can be slow; see Fig. 10.5.2.

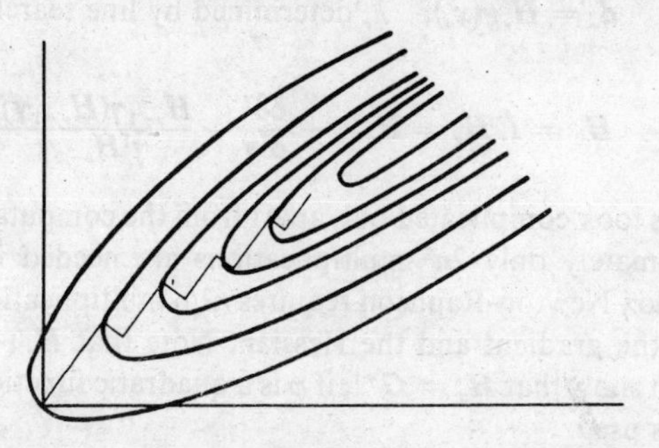

Fig. 10.5.2

Damped Newton-Raphson's Method (*Modification of an Algorithm of Biggs and Dixon, 1970*).

$$d_\nu = \begin{cases} G^{-1}(x_\nu)g(x_\nu) & \text{if } G^{-1} \text{ exists and } g^T G^{-1} g > \max\left[\dfrac{(g^T g)^2}{g^T G g}, 0\right], \\ g(x_\nu), & \text{otherwise.} \end{cases} \tag{10.5.9}$$

Levenberg (1944) suggested that one instead replace $G(x_\nu)$ by a positive definite matrix $G(x_\nu) + \mu I$. The main difficulty with this proposal is the choice of a suitable μ. In these algorithms line search is avoided. Instead, in the upper alternative one first tries $\lambda = 1$, but if $\varphi(x_\nu - \lambda d_\nu) \geq \varphi(x_\nu)$ then λ is halved. This is repeated until $\varphi(x_\nu - \lambda d_\nu) < \varphi(x_\nu)$, and the λ thus obtained is accepted as λ_ν. In the lower alternative, one first tries $\lambda = g^T g / g^T G g$, if $g^T G g > 0$; otherwise, put $\lambda = \max_{\mu < \nu} \lambda_\mu \| d_\mu \| / \| d_\nu \|$, if $\nu \neq 0$, or $\lambda = 1$, if $\nu = 0$. Again λ is repeatedly halved until $\varphi(x_\nu - \lambda d_\nu) < \varphi(x_\nu)$.

Note that the Newton-Raphson direction is not a descent direction if $g^T G^{-1} g \leq 0$. This situation is not likely to occur in the vicinity of $\hat{x}$, because of the positive (or at least nonnegative) definiteness of $G(\hat{x})$. Far away from $\hat{x}$, however, this can happen, and this is the reason for admitting the gradient as an alternative choice of direction—especially since there is a danger that the Newton-Raphson direction will lead to a saddle point.

The damped Newton-Raphson method is not so efficient when n is large—say, when $n > 5$—because the Hessian is needed.

There exists a class of methods, called **matrix-updating methods,** where the Hessian is never computed, but an approximate Hessian will be built up during the iterative process by successive additions of low-rank matrices. The **Davidon-Fletcher-Powell** method (1959, 1963) is one of the best-known of these methods. Put

$$\delta = x_\nu - x_{\nu-1}, \qquad \gamma = g(x_\nu) - g(x_{\nu-1}).$$

Then

$$d_\nu = H_\nu g(x_\nu), \quad \lambda_\nu \text{ determined by line search,}$$

where

$$H_0 = I, H_\nu = H_{\nu-1} + \frac{\delta\delta^T}{\delta^T\gamma} - \frac{H_{\nu-1}\gamma(H_{\nu-1}\gamma)^T}{\gamma^T H_{\nu-1}\gamma}. \tag{10.5.10}$$

The formulas look complicated but, apart from the computation of the gradient, approximately only $2n^2$ multiplications are needed in each iteration while orthodox Newton-Raphson requires $n^3/6$ multiplications plus the computation of the gradient and the Hessian. Note that $H_\nu - H_{\nu-1}$ is of rank two. One can show that $H_n = G^{-1}$, if φ is a quadratic function and if accurate line search is used.

The following method has performed better in some comparative tests:

Method of Broyden-Fletcher-Shanno (1969, 1970).

$$H_\nu = H_{\nu-1} + \frac{\delta\delta^T}{\delta^T\gamma}\cdot\left(1 + \frac{\gamma^T H_{\nu-1}\gamma}{\delta^T\gamma}\right) - \frac{\delta\gamma^T H_{\nu-1} + H_{\nu-1}\gamma\delta^T}{\delta^T\gamma}. \quad (10.5.11)$$

Broyden has also devised a promising method, without line search. It is based on the following matrix-updating formula (some safeguards, however, are necessary for satisfactory operation):

$$H_\nu = H_{\nu-1} + \frac{(\delta - H_{\nu-1}\gamma)(\delta - H_{\nu-1}\gamma)^T}{(\delta - H_{\nu-1}\gamma)^T\gamma}, \quad \lambda_\nu = 1. \quad (10.5.12)$$

The previous methods (except the steepest-descent method) assume that the functions can be locally approximated by a quadratic function, and all of them assume that analytic expressions for the gradient are available. The damped Newton-Raphson is recommended for small n, $n \leq 5$, say. The matrix-updating methods are better when n is larger. The methods can be modified to use a finite-difference approximation for the gradient, but for such problems the reader is advised to study the treatment of **direct search methods** in Dixon [125], and the same recommendation applies to problems where the gradient has discontinuities.

Minimization problems derived from the least-squares solution of over-determined systems can be better treated by special techniques; see the next section.

10.5.4. Overdetermined Nonlinear Systems

In Example 10.5.1 it was shown that the problem of exponential fitting could be formulated as an overdetermined nonlinear system with m equations, n unknowns, $m > n$,

$$f(x) \approx 0, \quad (10.5.13)$$

or as a minimization of

$$\varphi(x) = f(x)^T f(x). \quad (10.5.14)$$

The same applies to many other problems of *nonlinear approximation*—e.g., approximation by rational functions.

We linearize Eq. (10.5.13). By Taylor's formula,

$$f(x_\nu) + f'(x_\nu)\cdot(x - x_\nu) \approx 0,$$

where $f'(x)$ is the $m \times n$ Jacobian matrix of the vector-valued function f. A search direction is obtained by solving this linear overdetermined system for $x - x_\nu$, by the techniques of Sec. 5.8. The solution can be written

$$d = (f'(x)^T f'(x))^{-1} f'(x)^T f(x), \quad (10.5.15)$$

where we have dropped the subscript ν. More neatly—and more generally—

$$d_\nu = [f'(x_\nu)]^I f(x_\nu), \quad \lambda_\nu = 1 \tag{10.5.16}$$

where $[f'(x)]^I$ denotes the pseudo-inverse of the Jacobian; see Eqs. (5.2.2) and (5.2.4). This is called the **Gauss-Newton method.**

A **damped Gauss-Newton** method is obtained by putting

$$g(x) = f'(x)^T f(x), \qquad G(x) = f'(x)^T f'(x)$$

into Eq. (10.5.9). In this case, however, $G(x)$ is positive-definite or semidefinite everywhere, so the gradient direction will rarely be needed, but the control of λ described for the damped Newton-Raphson method may be useful.

If analytic expressions for $f'(x)$ are not available, finite-difference approximations can be used; see Sec. 6.9. If n is large, the solution of a linear system in each iteration is tedious; see Dixon [125, pp. 44ff.] concerning matrix-updating modifications.

The Gauss-Newton method is *not* identical to Newton-Raphson's method applied to Eq. (10.5.14). The latter would give the search direction (dropping subscripts)

$$d = [f''(x)^T f(x) + f'(x)^T f'(x)]^{-1} f'(x)^T f(x), \tag{10.5.17}$$

where $f''(x)^T f(x)$ denotes a matrix with elements

$$\sum_{k=1}^{m} \frac{\partial^2 f_k}{\partial x_i \partial x_j} f_k.$$

This is much more complicated than the Gauss-Newton direction, because of the second derivatives.

The difference can be of importance for the ultimate convergence, when $\| f''(\hat{x})^T f(\hat{x}) \|$ is not small, which roughly means that we have a strongly nonlinear model which cannot be fitted well to the given data.

10.5.5. Constrained Optimization

In many applications the vector x is subject to constraints, either equations or inequalities. In fact, two special cases have been encountered earlier:

	Minimize	Constraints
Sec. 10.1	linear function	linear inequalities
Sec. 5.8.3	quadratic function	linear equations

The simplex technique for linear programming has been extended to *quadratic programming* (see Dixon [125, p. 110]—i.e., quadratic function and linear inequalities.

In principle, these techniques can be extended to optimization problems with **general functions and general constraints,** by **piecewise approximation** by linear-quadratic problems. One has to add a few extra constraints, in order to make sure that the point does not leave the region where the approximation makes sense.

The possibility of reducing the number of variables and equations by **elimination** in the equality constraints should be investigated. Inequality constraints can sometimes be reduced by **substitution**—e.g., to satisfy the constraint $x_j \geq 0$ we can replace x_j by x_j^2.

There are other techniques available for the equality-constraint problem:

Minimize

$$\varphi(x), \quad \boldsymbol{x} \in \boldsymbol{R}_n \tag{10.5.18}$$

subject to the constraints

$$\boldsymbol{u}(x) = 0, \quad \boldsymbol{u}(x) \in \boldsymbol{R}_m.$$

The **Lagrange-multiplier** method reduces the problem to finding a stationary point of the function of $n + m$ variables

$$\varphi^*(x, \boldsymbol{\lambda}) = \varphi(x) + \boldsymbol{\lambda}^T \boldsymbol{u}(x), \quad \boldsymbol{\lambda} \in \boldsymbol{R}_m.$$

Unfortunately this stationary point is neither a minimum nor a maximum, so the typical minimization techniques will not work well. However, Newton-Raphson's method does not recognize the difference between different types of stationary points, so in principle it can be used if a satisfactory starting point can be found.

Another interesting idea is to replace the constraints by a **penalty function** —i.e., the problem is replaced by the unconstrained minimization of the function

$$\bar{\varphi}(x) = \varphi(x) + k^{-1} \cdot \boldsymbol{u}^T(x) \cdot \boldsymbol{u}(x), \tag{10.5.19}$$

where k is a very small number. However, the problem becomes ill-conditioned when k is very small, because in the gradient there will be a strong amplification of the errors in $\boldsymbol{u}(x)$.

Under certain conditions the solution can be expanded into powers of k,

$$\hat{x}(k) = \boldsymbol{c}_0 + k\boldsymbol{c}_1 + k^2\boldsymbol{c}_2 + \dots, \tag{10.5.20}$$

where $\boldsymbol{c}_0$ is the solution of Eq. (10.5.18). Therefore, there is a chance to solve the problem with several moderately small k, and then use **Richardson-extrapolation.** The use of a penalty function can be looked upon as an **imbedding** technique; see Sec. 6.9.3. The formulas of Eq. (6.9.11) or Eq. (6.9.12) can be used for initial approximations.

Example 10.5.1

Minimize $x_1^2 + x_2^2$ subject to the constraint

$$u(x) \equiv x_1^2 + 2x_1x_2 + 3x_2^2 - 1 = 0.$$

In other words, find the direction and length of the minor axis of the ellipse defined by the constraint. The exact solution is $\hat{x}_1 = 2^{-1/2} - 0.5 = 0.207107$,

$\hat{x}_2 = 0.5$. The following results were obtained by Newton-Raphson's method and Eq. (6.9.11) applied to the penalty function

$$\bar{\varphi}(x) = x_1^2 + x_2^2 + k^{-1}(x_1^2 + 2x_1x_2 + 3x_2^2 - 1)^2.$$

	Unextrapolated			After linear extrapolation to $k = 0$		
k	x_1	x_2	$u(x)$	x_1	x_2	$u(x)$
1	0.191342	0.461940	$-1.5\cdot10^{-1}$	—	—	—
$\frac{1}{2}$	0.199917	0.482642	$-6.8\cdot10^{-2}$	0.208492	0.503344	$-1.3\cdot10^{-2}$
$\frac{1}{4}$	0.203367	0.490971	$-3.6\cdot10^{-2}$	0.206817	0.499301	$-2.8\cdot10^{-3}$
$\frac{1}{8}$	0.205228	0.495463	$-1.8\cdot10^{-2}$	0.207088	0.499955	$-1.8\cdot10^{-4}$
$\frac{1}{16}$	0.206163	0.497722	$-9.1\cdot10^{-3}$	0.207098	0.499981	$-7.8\cdot10^{-5}$
$\frac{1}{32}$	0.206634	0.498858	$-4.6\cdot10^{-3}$	0.207105	0.499995	$-2.0\cdot10^{-5}$
$\frac{1}{64}$	0.206870	0.499429	$-2.3\cdot10^{-3}$	0.207106	0.499999	$-4.8\cdot10^{-6}$

For each k the iteration terminated when $\|x_{i+1} - x_i\|_\infty < 2\cdot10^{-7}$. When $k > 1$, only one iteration was needed. The computations were continued until the message "nearly singular matrix" was received for $k = 2^{-17}$. For $k < 2^{-15}$ the errors in the unextrapolated values were approximately 10^{-6}.

Penalty functions can also be used for inequality constraints, $h_j(x) \geq 0$, $j = 1, 2, \ldots, p$. Box, see [125, p. 100], has suggested the function

$$\bar{\varphi}(x) = \varphi(x) + k^{-1}\sum_{j=1}^{p} h_j^2 H(h_j(x)).$$

Here H is the Heaviside step function,

$$H(h_j) = \begin{cases} 0, & h_j \leq 0, \\ 1, & h_j > 0. \end{cases}$$

PROBLEMS

1. (a) Show that the general form for a quadratic function is

$$\varphi(x) = \tfrac{1}{2}x^T Gx - b^T x + c,$$

where G is a symmetric matrix and b is a column vector, and determine the gradient and Hessian. Also show that

$$\varphi(x) = \varphi(\hat{x}) + \tfrac{1}{2}(x - \hat{x})^T G(x - \hat{x}), \tag{10.5.21}$$

if $\varphi'(\hat{x}) = 0$. Does $\hat{x}$ always exist?

(b) Suppose that G is symmetric and nonsingular. Show that Newton-Raphson's method will find a stationary point of φ in one step, for all x_0. Under what condition is this a minimum point?

2. Put $\psi(\lambda) = \varphi(x_0 - \lambda d)$, and suppose that $\|d\|_2 = 1$. Show that $\psi'(0)$ is minimized when $d = g/\|g\|_2$, where g is the gradient at x_0. (Steepest descent!)

3. Verify that $Q'(\lambda') = 0$, if ψ is quadratic and λ' is defined by Eq. (10.5.7). *Hint:* See Sec. 7.5, Problem 3.

4. (a) Write a program for the line search described in Sec. 10.5.2.
   (b) Get acquainted with the programs of the list in Chap. 13 which are relevant for this section.

5. Suppose that φ is quadratic, and that $\boldsymbol{d}^T G \boldsymbol{d} > 0$, for a certain vector $\boldsymbol{d}$, but $\boldsymbol{G}$ need not be positive-definite.
   (a) Show that $\psi(\lambda) = \varphi(\boldsymbol{x}_0 - \lambda \boldsymbol{d})$ is minimized when
$$\lambda = \frac{\boldsymbol{g}^T \boldsymbol{d}}{\boldsymbol{d}^T \boldsymbol{G} \boldsymbol{d}},$$
   and that
$$\min \psi(\lambda) = \psi(0) - \frac{1}{2} \cdot \frac{(\boldsymbol{d}^T \boldsymbol{g})^2}{\boldsymbol{d}^T \boldsymbol{G} \boldsymbol{d}}.$$
   (b) Apply this to $\boldsymbol{d} = \boldsymbol{g}$ and to $\boldsymbol{d} = \boldsymbol{G}^{-1}\boldsymbol{g}$, and look at the damped Newton-Raphson method in the light of this.

6. (a) Show that $H_v \boldsymbol{\gamma} = \boldsymbol{\delta}$ for the three methods defined by Eqs. (10.5.10), (10.5 11), and (10.5.12).
   (b) Show that H_v is symmetric if H_0 is symmetric.

7. Show that the rank of $H_v - H_{v-1}$ is equal to one for the Broyden method defined by Eq. (10.5.12), and that it is the only symmetric, rank-one updating method which satisfies the requirement that $H_v \boldsymbol{\gamma} = \boldsymbol{\delta}$.

8. Assume that Eq. (10.5.20) is valid, that the rank of $\boldsymbol{u}'(\boldsymbol{c}_0)$ is equal to m, and that φ' and $\boldsymbol{u}'$ are continuous in the neighborhood of $\boldsymbol{c}_0$. Show that
$$\lim_{k \to 0} 2k^{-1} \boldsymbol{u}(\hat{\boldsymbol{x}}(k)) = \boldsymbol{\lambda},$$
   where $\boldsymbol{\lambda}$ is the Lagrange multiplier vector, mentioned in Sec. 10.5.5.

11 THE MONTE CARLO METHOD AND SIMULATION

11.1. INTRODUCTION

In most of the applications of probability theory one makes a mathematical formulation of a stochastic problem (i.e., a problem where chance plays some part), and then solves the problem by using analytical or numerical methods. In the **Monte Carlo method**, one does the opposite: a mathematical problem is given, and one constructs a **game of chance**, using numbers, which in some way leads to the given problem.

For example, the solution of the problem could be equivalent to the probability of some event, or to the mean or the distribution function for some variable in the game. In any case, after one has constructed the **game** or **experiment**, one plays it many times and then treats the results using traditional statistical methods. The various possibilities of the Monte Carlo method first began to be studied in the 1940s.

The picturesque name of the method, however, has become so popular that now the above definition is too narrow. For instance, in many of the problems where the "Monte Carlo method" is successful, there is already an element of chance in the system or process which one wants to study. Thus such games of chance can often be considered to be a **numerical simulation** of the most important aspects in the given system (process). We shall give some examples below. In the wider sense of the term, "Monte Carlo methods" were used as early as the 1800s (experimental sampling). The following are some areas where the method is applied.

(a) Problems in reactor and particle physics: for example, a neutron, because it collides with other particles, is forced to make a random

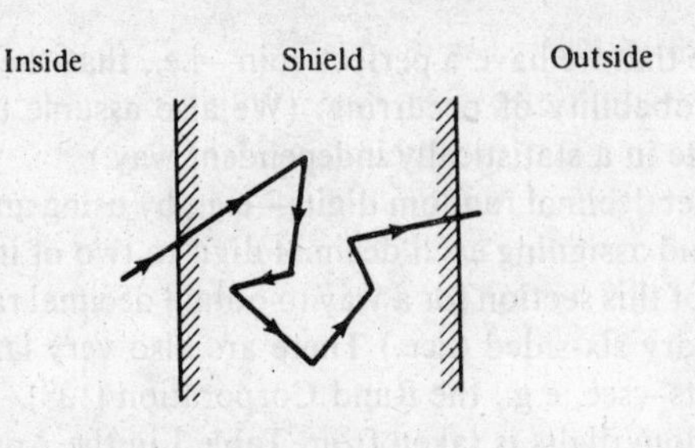

Fig. 11.1.1

journey; in infrequent but important cases the neutron can go through a layer of (say) shielding material (see Fig. 11.1.1).

(b) Technical problems concerning traffic (telecommunications, railway networks, regulation of traffic lights, etc.).
(c) Queueing problems.
(d) Models of conflicts.
(e) Study of different ways of operating large computer centers.
(f) Computation of multiple integrals.

In a simulation, one can study the result of various actions more cheaply, more quickly, and with less risk or organizational problems than if one were to take the corresponding actions on the actual system. In particular, for problems in operations analysis, it is quite common to take a shortcut from the actual system to a computer program for the game of chance without formulating any mathematical equations. The game is then a **model** of the system. In order for the term "Monte Carlo" to be correctly applied, however, we shall demand that:

(a) *the game be played with numbers*—i.e., physical experimental models are not included;
(b) *random choices* occur in the calculations.

The word "simulation" is used, however, even when an element of chance is absent. Simulation is now so important that special programming languages have been developed exclusively for its use. Three of the best-known are GPSS, Simscript, and Simula 67.

11.2. RANDOM DIGITS AND RANDOM NUMBERS

The series of twenty digits

11100 01001 10011 01100

is a record of twenty tosses of a coin where "heads" is denoted by 1 and "tails" by 0. Such digits are sometimes called (binary) **random digits**, assuming of course that we have a perfect coin—i.e., that "heads" and "tails" have the same probability of occurring. (We also assume that the tosses of the coin are made in a statistically independent way.)

We can get decimal random digits—e.g., by using an icosahedral (twenty sided) die, and assigning each decimal digit to two of its sides. (See Problem 1 at the end of this section for a way to obtain decimal random digits with the use of ordinary six-sided dice.) There are also very large tables of decimal random digits—see, e.g., the Rand Corporation [138]. The following row of decimal random digits is taken from Table I in the Appendix:

$$55693 \quad 02945 \quad 81723 \quad 43588 \quad 81350 \quad 76302 \quad \ldots \tag{11.2.1}$$

Random digits can be packed together to give series of equidistributed *integers*. The series in (11.2.1) can be considered as five-digit random numbers, where each element in the series has probability 10^{-5} of taking on the value 0, 1, 2, . . . , 99,999.

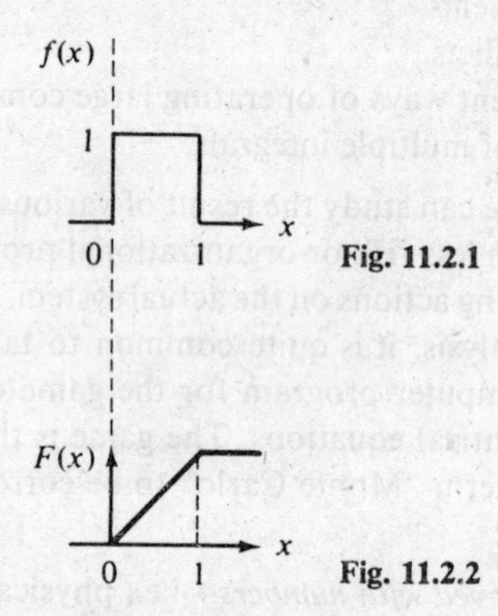

Fig. 11.2.1

Fig. 11.2.2

A variable whose frequency function looks like Fig. 11.2.1 is said to be *rectangularly distributed* (or "uniformly distributed") on the interval [0, 1]. Its distribution function $F(x)$ is shown in Fig. 11.2.2. From (11.2.1), one can also construct the following series

$$0.556935, \quad 0.029455, \quad 0.817235, \quad 0.435885, \quad \ldots,$$

which can be considered a good approximation to a series of independent observations of a rectangularly distributed variable on the interval [0, 1]. The 5s in the sixth decimal place have been added in order to get the correct mean (without them, the mean would be 0.499995 instead of 0.5). Often the accuracy requirements are so modest that one can forget about this slight correction. We call such a sequence (five-digit) **rectangularly distributed random numbers** (on the interval [0, 1]).

By arithmetic operations on random digits or rectangularly distributed random numbers one can form random numbers with other distributions. In computers, the basic source of random numbers is usually rectangularly distributed random numbers (or pseudorandom numbers; see Sec. 11.4).

Example 11.2.1

A variable whose frequency function is

$$f(x) = \begin{cases} \dfrac{1}{b-a} & a \le x \le b \\ 0 & x < a \quad \text{or} \quad x > b \end{cases}$$

is said to be rectangularly distributed on the interval $[a, b]$. If we have a variable R which is rectangularly distributed on $[0, 1]$, then $a + (b - a)R$ is rectangularly distributed on $[a, b]$.

The variable $R' = 1 - R$ is also rectangularly distributed in $[0, 1]$. From the series for R in (11.2.1) we get the following series for R':

$$0.443065, \quad 0.970545, \quad 0.182765, \quad 0.564115, \quad \ldots \qquad (11.2.2)$$

The series in (11.2.2) is said to be the **antithetic series** of the series derived from (11.2.1). Such series are an important means of reducing the variance in the estimates made using the Monte Carlo method; see Sec. 11.3.

Example 11.2.2. *Obtaining Random Numbers with Given Discrete Distribution*

Suppose that a random variable X takes on the values 0, 1, and 2 with probabilities 0.18, 0.44. and 0.38, respectively. Since the probabilities are given to two decimals, it is sufficient, in hand computation, to use two-digit rectangularly distributed random numbers R.

Rule:

R	X
$0 \le R < 0.18$	0
$0.18 \le R < 0.62$	1
$0.62 \le R < 1.00$	2

We use (11.2.1) to get values for $100R$, and get the following table:

	55	69	30	29	45	81	72	34	35	88	81	35	07	63	02
X	1	2	1	1	1	2	2	1	1	2	2	1	0	2	0

(Notice that we formulated the above rule in such a way that no "round-off correction" need be used.)

DEFINITION

By an expression of the form

$$P[X \le x]$$

we mean the probability that the random variable X takes on a value which is less than or equal to x.

Example 11.2.3. *A General Method for Generating Random Numbers with a Given Continuous (Cumulative) Distribution Function* F(x)

Rule: Take a rectangularly distributed random number R in the interval $[0, 1]$, and *solve the equation* $F(X) = R$. Then X has the desired distribution, and is called a random number with (cumulative) distribution $F(X)$; see Fig. 11.2.3.

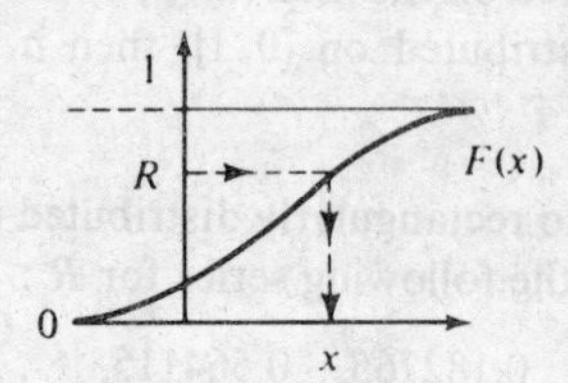

Fig. 11.2.3

Proof. Since $F(x)$ is a nondecreasing continuous function then

$$P[X \leq x] = P[F(X) \leq F(x)] = P[R \leq F(x)],$$

but the last expression is equal to $F(X)$, since by the definition of a rectangular distribution we have $P[R \leq r] = r$, where r is an arbitrary number between 0 and 1. Thus $P[X \leq x] = F(x)$, which (by definition) means that X has the distribution function $F(x)$. In practice, one often tabulates X for a sequence of values of R and then interpolates linearly between them.

Note: One can equally well solve the equation

$$F(X) = 1 - R.$$

Example 11.2.4. *Exponentially Distributed Random Numbers*

These are random numbers with distribution function

$$F(x) = 1 - e^{-\lambda x}, \quad \left(\text{mean} = \frac{1}{\lambda}\right). \tag{11.2.3}$$

Rule: Take a rectangularly distributed random number R in $[0, 1]$, and form

$$X = -\lambda^{-1} \ln R. \tag{11.2.4}$$

Proof. From the remark in Example 11.2.3 we see that X can be obtained by solving the equation

$$1 - e^{-\lambda X} = 1 - R.$$

Hence

$$e^{-\lambda X} = R, \quad -\lambda X = \ln R,$$

and we get the rule of Eq. (11.2.4).

A series of exponentially distributed random numbers *antithetic* to the above series can be obtained from

$$X' = -\lambda^{-1} \ln (1 - R). \tag{11.2.5}$$

One important use of exponentially distributed random numbers is in the *generation of so-called* **Poisson processes**. Such processes are often fundamental in models of telecommunications systems and other service systems. A Poisson process (with frequency parameter λ) is a series of events characterized by the property that the probability of occurrence of an event in a short time interval $(t, t + \Delta t)$ is equal to $\lambda \cdot \Delta t + o(\Delta t)$, independent of the sequence of events previous to time t.

The probability of more than one event in the above interval is $o(\Delta t)$. In applications, "event" can mean a call on a telephone line, or the arrival of a customer in a store, etc. Descriptions of Poisson processes are often formulated in such terms as: "Customers arrive in a random way, on the average of λ customers per minute".

For us the most important property of a Poisson process is that the *intervals of time between two successive events are independent random variables whose distribution function is given in Eq. (11.2.3)*. Thus Poisson processes can be generated using exponentially distributed random numbers.

Another interesting property of Poisson processes is that the probability that exactly n events occur during an interval of time L is given by

$$\frac{e^{-\lambda L}(\lambda L)^n}{n!} \quad \textbf{(Poisson distribution)}.$$

For the derivation and applications of Poisson processes and their properties we refer the reader to Feller [127, Chap. 17]. For use in the problems in this chapter, we have included, in the appendix, a series of one hundred exponentially distributed random numbers and a corresponding antithetic series.

The following example shows how one can obtain random numbers (in some important special cases) in an easier way than the general method given in Example 11.2.3. More material of this type can be found in the problems in this chapter, and in Hammersley and Handscomb [133].

Example 11.2.5. *Normally Distributed Random Numbers*

If one has two independent, rectangularly distributed random numbers R_1, R_2, in [0, 1], then one can obtain two independent, normally distributed random numbers N_1, N_2 (mean 0, standard deviation 1) using the following rule, **Box-Muller's transformation:**

$$N_1 = (-2 \ln R_1)^{1/2} \cos (2\pi R_2),$$

$$N_2 = (-2 \ln R_1)^{1/2} \sin (2\pi R_2).$$

We shall not derive the rule here. As a hint for the reader interested in deriving this result on his own we point out that N_1 and N_2 can be considered to be rectangular coordinates of a point whose polar coordinates (r, φ) are determined by the equations

$$r^2 = N_1^2 + N_2^2 = -2 \ln R_1, \qquad \varphi = 2\pi R_2.$$

Thus the problem is to show that the distribution function for a pair of independent, normally distributed random variables is rotationally symmetric (uniformly distributed angle) and that their sum of squares is exponentially distributed with mean 2 (see Example 11.2.4).

Notice also that we can get normally distributed random variables with mean m and standard deviation σ by forming $m + \sigma N_1$, $m + \sigma N_2$.

There are extensive tables of normally distributed random variables—see, e.g., [138]. Also, we give a series of one hundred such numbers in the Appendix. An *antithetic series* can be obtained simply by *reversing the sign* of all the numbers in the table.

REVIEW QUESTIONS

1. What is a rectangularly distributed random number?
2. Describe the general methods for obtaining random numbers with
   (a) given discrete distribution;
   (b) given continuous distribution.
   Give examples of their use.
3. What are the most important properties of a Poisson process? How can one generate a Poisson process with the help of random numbers?

PROBLEMS

1. Investigate whether the following two proposals for producing decimal random digits by casting two six-sided dice are correct. Denote the number of spots shown by the respective die by P and Q, and the decimal random digit by X.
   (a) If $P = 6$, then ignore the cast. Otherwise: if Q is even, set X equal to P; if Q is odd, set X equal to the last digit in $P + 5$.
   (b) If $P + Q < 3$, then ignore the cast. Otherwise, set X equal to $P + Q - 3$.
2. Random numbers with **triangular** frequency function. Show that if R_1, R_2 are two independent, rectangularly distributed random numbers on [0, 1], then min (R_1, R_2) has frequency function (see Fig. 11.2.4)

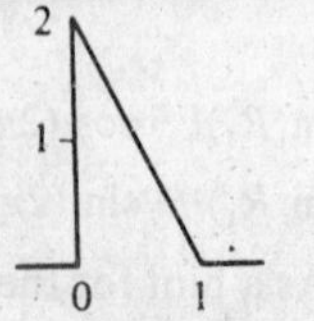

Fig. 11.2.4

$$f(x) = \begin{cases} 2 - 2x & \text{for } 0 \leq x \leq 1. \\ 0 & \text{for } x \notin [0, 1]. \end{cases}$$

3. Suppose that $f_1, f_2, \ldots, f_k$ are *k frequency functions* for which random numbers can easily be generated. Show that the random number X with frequency function

$$f(x) = \alpha_1 f_1(x) + \alpha_2 f_2(x) + \ldots + \alpha_k f_k(x), \quad (\alpha_j \geq 0, \Sigma\alpha_j = 1),$$

can be generated by the following algorithm. Take a rectangularly distributed random number R, and determine i such that

$$\alpha_1 + \alpha_2 + \ldots + \alpha_{i-1} < R \leq \alpha_1 + \alpha_2 + \ldots + \alpha_{i-1} + \alpha_i,$$
$$i = 1 \quad \text{if} \quad R \leq \alpha_1.$$

X is then obtained by taking a random number with frequency function $f_i(x)$.

4. Develop, using the results of Problems 2 and 3, an algorithm for generating random numbers with piecewise linear frequency function. (See Fig. 11.2.5.)

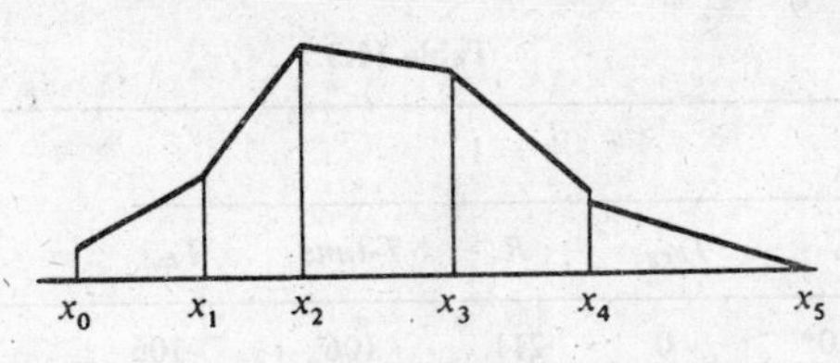

Fig. 11.2.5

11.3. APPLICATIONS; REDUCTION OF VARIANCE

One important area of application of Monte Carlo methods is queueing problems. Analytical methods are often insufficient for such problems. The example which follows is based on a study by Bailey [120] made in 1952. The question is how to give appointment times to patients at a polyclinic in such a way that there is a suitable balance between the mean waiting times of both patients and doctors. This problem was in fact solved analytically—much later—after Bailey had already gotten the results that he wanted; this situation is not uncommon when numerical methods (and especially Monte Carlo methods) have been used.

Example 11.3.1. *Queue at a Polyclinic*

Suppose that k patients have been booked at the time $t = 0$ (when the clinic opens), and that the rest of the patients (altogether 10) are booked at intervals of 50 time units thereafter. The time of treatment is assumed to be exponentially distributed with mean 50. (Bailey used a distribution function which was based on empirical data.) *Two alternatives, $k = 2$ and $k = 1$, are simulated using the same random numbers;* hence the same treatment times. In this way one gets a **reduced variance** *in the estimate of the change* in waiting times as k varies. Often, in simulation, one can save a lot of computational work by a similar use of random numbers. The computations are shown in Tables 11.3.1 and 11.3.2.

Abbreviations: $P =$ patient, $D =$ doctor, $T =$ treatment, $R =$ random number from a table. An *asterisk* indicates that P did not need to wait.

Computational procedure: P_{arr} follows from the rule for booking patients given previously. $T_{beg} =$ the larger of the numbers P_{arr} (on the same row) and T_{end} (in the row just above). R is an exponentially distributed random number with mean 100, taken from Table III(a), column 1 of the Appendix.

$$T_{time} = \frac{R \cdot 50}{100} \quad \text{(since the mean should be 50),}$$

$$T_{end} = T_{beg} + T_{time}.$$

Table 11.3.1

	$k = 1$					$k = 2$	
$P_{no.}$	P_{arr}	T_{beg}	R	T-time	T_{end}	P_{arr}	T_{end}
1	0*	0	211	106	106	0*	106
2	50	106	3	2	108	0	108
3	100	108	53	26	134	50	134
4	150*	150	159	80	230	100	214
5	200	230	24	12	242	150	226
6	250*	250	35	18	268	200	244
7	300*	300	54	27	327	250*	277
8	350*	350	39	20	370	300*	320
9	400*	400	44	22	422	350*	372
10	450*	450	13	6	456	400*	406
Σ	2,250			319	2,663	1,800	2,407

	D waited:	Total waiting time for all P:
From Table 11.3.1: $k = 1$	$456 - 319 = 137$	$2{,}663 - 2{,}250 - 319 = 94$
" " " : $k = 2$	$406 - 319 = 87$	$2{,}407 - 1{,}800 - 319 = 288$
After similar calculations: $k = 3$	28	$2{,}272 - 1{,}400 - 319 = 553$
" " : $k = 4$	0	$2{,}226 - 1{,}050 - 319 = 857$
If D never needs to wait (i.e., $k \geq 4$)		$-843 + 525k - 25k^2$ (Show this!)

Table 11.3.2

Antithetic experiment: R taken from the Appendix, Table III(b), column 1

	$k = 1$					$k = 2$	
$P_{no.}$	P_{arr}	T_{beg}	R	*T-time*	T_{end}	P_{arr}	T_{end}
1	0*	0	13	6	6	0*	6
2	50*	50	365	182	232	0	188
3	100	232	88	44	276	50	232
4	150	276	23	12	288	100	244
5	200	288	154	77	365	150	321
6	250	365	122	61	426	200	382
7	300	426	87	44	470	250	426
8	350	470	112	56	526	300	482
9	400	526	104	52	578	350	534
10	450	578	213	106	684	400	640
Σ	2,250			640	3,851	1,800	3,455

	D waited:	Total waiting time for all P:
From Table 11.3.2: $k = 1$	$684 - 640 = 44$	$3{,}851 - 2{,}250 - 640 = 961$
" " " $: k = 2$	0	$3{,}455 - 1{,}800 - 640 = 1{,}015$
If D never needs to wait ($k \geq 2$)		$65 + 525k - 25k^2$

The result of the pair of experiments is given as a set of **means:**

	D waited:	Total waiting time for all P:
$k = 1$	90	528
$k = 2$	44	652
$k = 3$	14	984
$k \geq 4$	0	$-389 + 525k - 25k^2$

Roughly speaking, since the influence of chance has opposing effects in the two (antithetic) experiments, one can presume that the effect of chance on the *means* is much less than the effect of chance in the original experiments. In Problem 3 following this section we give examples of how to make a quantitative estimate of the reduction of variance accomplished with the use of antithetic experiments.

One cannot, of course, draw any tenable conclusions from one experiment (pair of experiments). More experiments should be made in order to put the conclusions on statistically solid ground—e.g., if one wants to recommend a certain value for k. Bailey made 50 experiments with 25 patients in each, and summarized the relation between the doctors' and the patients' waiting times in a curve (Fig. 11.3.1).

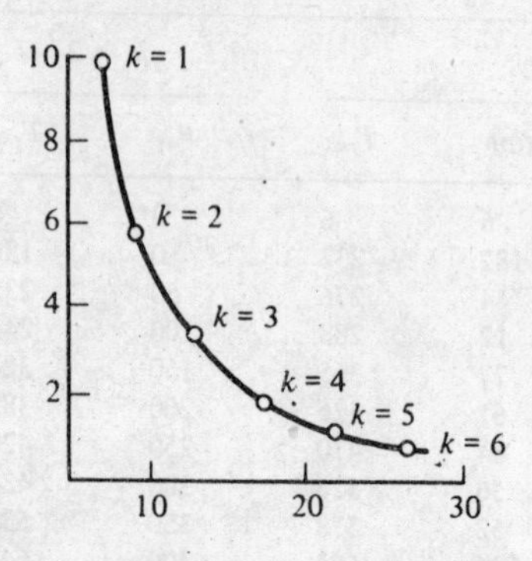

Fig. 11.3.1

Even isolated experiments, however, can give valuable suggestions for the planning of subsequent experiments, or perhaps suggestions of appropriate approximations to be made in an analytical treatment of the problem. The large-scale use of Monte Carlo methods with computers is a venture which requires careful planning. Otherwise, one can drown in enormous quantities of unintelligible results and expensive bills for machine time. One can hardly expect that Monte Carlo—or any other experimental activity for that matter—will give results without careful planning. One should specify what information one wants to obtain, which hypotheses are to be investigated, and which principles are to be used in determining when the series of trials is to stop. The above points are often disregarded; however, it is usually wise to consult a statistician about them.

Another thing to remember is that computations are based on a *model* of reality, not reality itself. This point is true in varying degrees about all applications of mathematics, but it is especially important when the computations form a basis for important decisions. A clear presentation of the assumptions made in the model is thus one of the most important parts of a report describing a mathematical investigation.

From statistics, we know that if one makes n independent observations of a quantity whose standard deviation is σ, then the standard error of the mean is $\sigma/\sqrt{n}$. In the Monte Carlo method, one can influence the value of σ by designing the experiment in various ways. One possible way was illustrated in the previous example—making the experiments in antithetic pairs and then computing the mean for each pair of experiments. If one has two ways (which require the same amount of work) of carrying out an experiment, and these experiments have standard deviations σ_1 and σ_2 associated with them, then if one repeats the experiments n_1 and n_2 times (respectively), the same precision will be obtained if

$$\frac{\sigma_1}{\sqrt{n_1}} = \frac{\sigma_2}{\sqrt{n_2}}$$

or

$$\frac{n_1}{n_2} = \left(\frac{\sigma_1}{\sigma_2}\right)^2 = \frac{\sigma_1^2}{\sigma_2^2} \tag{11.3.1}$$

Thus if a variance reduction by a factor k can be achieved, then the number of experiments needed is also reduced by the same factor k.

Example 11.3.2

In ten simulations of a service system the following values were obtained for the treatment time:

685 1,045 718 615 1,021 735 675 635 616 889.

From this material the mean for the treatment time is estimated as 763 ± 52. Using an antithetic series, one got the following values:

731 521 585 710 527 574 607 698 761 532.

The series of means is thus

708 783 652 662 774 654 641 666 688 710,

from which one gets the estimate 694 ± 16.

When one instead supplemented the first series with ten values (obtained with the use of completely independent random numbers), the estimate (using all twenty values) 704 ± 36 was obtained. These results indicate that, in this example, using antithetical series produces the desired accuracy with $(\frac{16}{36})^2 \approx \frac{1}{5}$ of the work required if completely independent random numbers are used. This quantitative estimate of the work saved is uncertain, since it is based on empirical estimates of the mean error from rather sparse data, but this rough estimate indicates that it would be profitable—in a much longer series of experiments—to use the technique of antithetic series.

Two methods for reduction of variance have now been introduced: **antithetic series of random numbers** and the technique of using (to as large an extent as possible) **the same random numbers in corresponding situations.** The latter technique is used when studying the changes in the behavior of a system when the value of a certain parameter is changed (e.g., the parameter k in Example 11.3.1). Many effective methods have been developed for reducing variance—e.g., "importance sampling" (see Example 11.3.3 below) and "splitting techniques" (see Hammersley and Handscomb [133]). One general rule is that, *if a part of a problem can be treated with analytic or traditional numerical methods, then one should use such methods*. There are many ways of combining analytical methods and Monte Carlo methods. More information on these topics can be found in Hammersley and Handscomb [133].

Example 11.3.3. *Numerical Integration*

One of the most important applications of the Monte Carlo method is in the numerical calculation of *multiple integrals.* We shall briefly describe some ideas which can be used in such problems. For simplicity, we consider integrals in *one* dimension, even though the Monte Carlo method cannot really compete with traditional numerical methods for this problem.

We shall compute

$$I = \int_0^1 f(x)\,dx.$$

Let $R_1, R_2, \ldots, R_n$ be a sequence of rectangularly distributed random numbers and set

$$I_1 = \sum_{i=1}^{n} \frac{f(R_i)}{n}.$$

One can show that the expectation of the variable I_1 is I, and that the standard deviation of this estimate decreases in proportion to $n^{-1/2}$. This is quite slow compared to the trapezoidal rule—where the error decreases as n^{-2}, or Simpson's rule, where the error decreases as n^{-4}.

The above estimate is a special case of a more general one. Suppose X_i, $i = 1, 2, \ldots, n$, has density function $g(x)$. Then

$$I_2 = \frac{1}{n} \sum_{i=1}^{n} \frac{f(X_i)}{g(X_i)}$$

has expected value I, since

$$E\left[\frac{f(X_i)}{g(X_i)}\right] = \int_0^1 \frac{f(x)}{g(x)} g(x)\,dx = \int_0^1 f(x)\,dx = I.$$

If one can find a frequency function $g(x)$ such that $f(x)/g(x)$ fluctuates less than $f(x)$, then I_2 will have smaller variance than I_1. This procedure is called **importance sampling;** it has proved very useful in particle-physics problems, where important phenomena (e.g., dangerous radiation which penetrates a shield) are associated with certain events with low probability. There are examples in which the variance has been reduced by several powers of 10 with the use of importance sampling.

In Chaps. 3 and 7 we mentioned the method of using a "simple comparison problem." The Monte Carlo variant of this method is called the **control variate method**. Suppose that $\varphi(x)$ is a function whose integral has a known value, k, and suppose that $f(x) - \varphi(x)$ fluctuates much less than $f(x)$. We know that

$$I = \int_0^1 \varphi(x)\,dx + \int_0^1 (f(x) - \varphi(x))\,dx = k + \int_0^1 [f(x) - \varphi(x)]\,dx.$$

The integral furthest to the right can be estimated by

$$I_3 = n^{-1} \sum_{i=1}^{n} (f(R_i) - \varphi(R_i)),$$

which has less variance than I_1. This idea can be used in many other applications of Monte Carlo.

REVIEW QUESTION

Give three methods for *reduction of variance* in estimates made with the Monte Carlo method, and explain what is meant by this term. What is the connection between reducing variance and decreasing the amount of computation needed in a given problem?

PROBLEMS

1. A target with depth $2b$ and very large width is to be shot at with a cannon. (The assumption that the target is very wide makes the problem one-dimensional.) The distance to the center of the target is unknown, but is estimated to be D. The difference between the actual distance and D is assumed to be a normally distributed random variable X with mean 0 and standard deviation σ_1.

   One shoots at the target with a salvo of three shots, which are expected to travel a distance $D - a$, D, and $D + a$, respectively. The difference between the actual distance traveled and the expected distance is assumed to be a random variable with mean 0 and standard deviation σ_2; the resulting error component in the three shots is denoted by Y_{-1}, Y_0, Y_1. We further assume that these three variables are stochastically independent of each other and of X.

   One wants to know how the probability of at least one "hit" in a given salvo depends on a and b.

   Use normally distributed random numbers (see the Appendix) to "shoot" ten salvos and determine, for each salvo, the least value of b for which there is at least one "hit" in the salvo. Show that this is equal to

$$\min_i |X - (Y_i + ia)|, \quad (i = -1, 0, 1).$$

   Fire an "antithetic salvo" for each salvo.

   Graph, using $\sigma_1 = 3$, $\sigma_2 = 1$, for both $a = 1$ and $a = 2$ (with the same random numbers) *curves which give the probability of a hit as a function of the depth of the target* (as in Fig. 11.3.2).

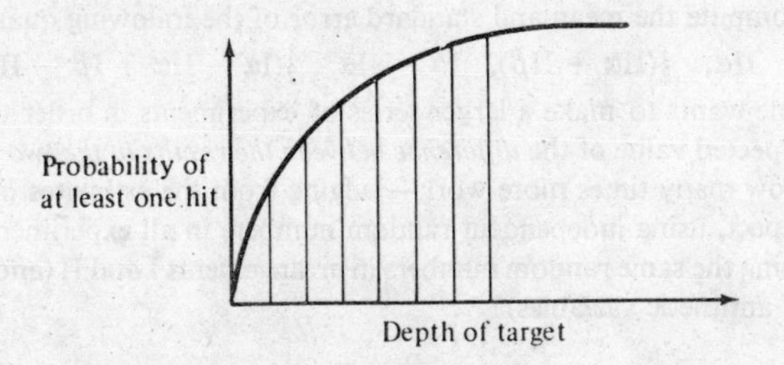

Fig. 11.3.2

2. Comparison of two service systems (see Fig. 11.3.3).

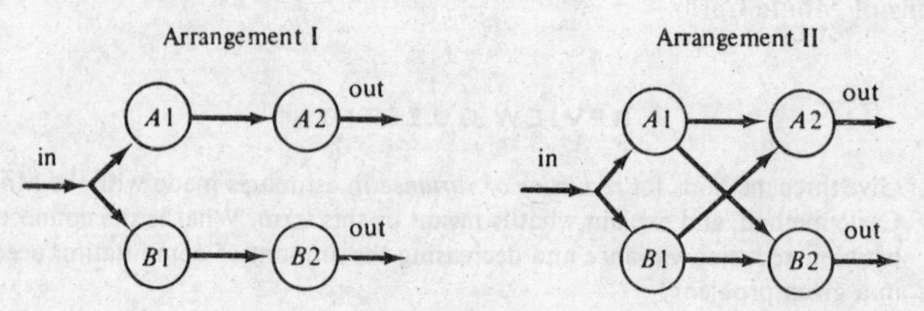

Fig. 11.3.3

Ten customers come into the system in a random way, on the average one per minute. Thus the arrival process is a Poisson process where the expected gap in time between two successive customers is 60 seconds. The service units $A1$ and $B1$ have a common queue. The customer whose turn it is chooses the unit which first becomes available. If there is no queue and if both units are available, then the unit which has been available longest is chosen.

In arrangement I, the customer proceeds from $A1$ to $A2$ or from $B1$ to $B2$. In arrangement II, a customer who comes out of $A1$ or $B1$ can choose between $A2$ and $B2$. The choice, then, is made using the same rules as for the choice between $A1$ and $B1$.

The treatment times at the units $A1$, $A2$, $B1$, and $B2$ are exponentially distributed, with mean 100 sec. Thirty exponentially distributed random numbers (and their antithetic series) are needed—use the table in the Appendix. The same random numbers are to be used in studying I and II. Both systems are to be studied in two antithetic experiments, α and β. Compute, for each of the four experiments, *the sum of the times that the ten customers are in the system.* Compare the setup of the polyclinic problem in Example 11.3.1 (make a similar table of the entire calculation).

3. Suppose that three experiments with the service system in Problem 2 gave the following results:

	Iα	Iβ	IIα	IIβ
Exp. 1	2,449	3,363	1,862	2,921
Exp. 2	2,232	4,350	2,157	3,651
Exp. 3	5,395	1,419	4,971	1,164

(a) Compute the mean and standard error of the following quantities:

$$\text{II}\alpha,\quad \tfrac{1}{2}(\text{II}\alpha + \text{II}\beta),\quad \text{I}\alpha - \text{II}\alpha,\quad \tfrac{1}{2}(\text{I}\alpha - \text{II}\alpha + \text{I}\beta - \text{II}\beta).$$

(b) One wants to make a larger series of experiments in order to estimate the expected value of the *difference between the results in the two configurations.* How many times more work—judging from the estimates in (a)—can one expect, using independent random numbers in all experiments rather than using the same random numbers in arrangements I and II (and the technique of antithetic variables)?

1.4. PSEUDORANDOM NUMBERS

On computers one almost always uses so-called **pseudorandom numbers** nstead of random numbers. Such numbers can be formed recursively, using n arithmetic process. The series of numbers obtained in this way is uniquely letermined by the starting value. The series is also **periodic**, but with a very ong period—of the order 2^t, where t is the number of binary digits in the nachine's arithmetic register. The numbers so obtained are thus not, strictly peaking, a series of random numbers. The most common methods of producing such numbers give numbers which satisfy some ingeniously constructed statistical tests for "randomness" in the same way that "genuine" eries of random numbers do.

The most commonly used method for producing pseudorandom numbers s called the **mixed congruential method**. Let P be a large integer, and let x_0, λ, μ be three positive integers which are less than P. Now generate recursively the integers $x_1, x_2, x_3, \ldots, 0 \leq x_n < P$, according to the formula

$$x_{n+1} \equiv \lambda x_n + \mu \bmod P. \tag{11.4.1}$$

Thus x_{n+1} is the remainder obtained when $\lambda x_n + \mu$ is divided by P. (As a rule, λ, μ, and P have simple form so that x_{n+1} in fact can be formed using faster arithmetic operations—e.g., shifts.) The sequence $\{R_n\}$ where $R_n = x_n/P$ is then called a sequence of rectangularly distributed pseudorandom numbers.

One can show that if $P = 2^t$ (which is natural on a binary machine), then *the period of the series is equal to* 2^t, assuming that μ is odd and that λ gives remainder 1 when divided by 4. Extensive theoretical and empirical investigations have, for $P = 2^{35}$, given good results—e.g., with

$$\lambda = 2^A + B, \quad 15 \leq A \leq 24, B = 1, 5, 9, \mu = 1.$$

More precise judgements can be found in a work by B. Jansson [134].

There is an **important practical difference** between "genuine" random numbers and pseudorandom numbers generated using Eq. (11.4.1). *The digits in pseudorandom numbers cannot be considered to be independent random digits.* To use (as in Example 11.2.2) digits in a pseudorandom number to form more numbers with fewer digits can be risky. There are other (more time-consuming) methods of producing pseudorandom numbers where the above operation is risk-free.

If one studies (e.g., in a sequence of pseudorandom numbers generated using Eq. (11.4.1) and $\mu = 0$, $P = 2^t$) the sequence of digits in a certain position (say the vth position counted from the left), then one finds that the larger v is, the shorter the period. The last digit is in fact always equal to 1 if x_0 and λ are odd!

REVIEW QUESTION

What is the mixed congruential method for generating pseudorandom numbers? What important difference is there between the numbers generated by this method and "genuine" random numbers?

PROBLEMS

1. The mixed congruential method with $\mu = 0$ is called the *multiplicative congruential method*. One can show that if x_0 is odd, and $P = 2^t$ $(t \geq 3)$, and $\lambda \equiv 3$ or $\lambda \equiv 5 \pmod 8$, then the period of the sequence generated is 2^{t-2}; this is the maximum period for such a method. It is often a good (somewhat faster) alternative to a mixed congruential generator in certain applications (see, however, the remarks at the end of Sec. 11.4). In particular, for most Fortran environments, the fact that no action is taken on integer overflow can be used to write an efficient subroutine as follows (we take, e.g., $P = 2^{31}$, $\lambda = 2^{16} + 3$ on a machine for which the largest integer is $2^{31} - 1$):

```
      FUNCTION RAND(I)
      I=I*65539
    1 IF(I.LT.0)I=I+2147483647+1
      RAND=I*.4656613E-9
      RETURN
      END
```

   (a) What does statement 1 in the above subroutine do? What further improvement in speed can be made when using the method of the above subroutine in a given program?
   (b) Write a Fortran program which uses the method of the above subroutine to compute $\int_0^1 e^x\,dx$ with the simple Monte Carlo method described in Sec. 11.3. Take 25, 100, 225, 400, and 625 points. (Use $I_0 = 123456789$ in each case.) How does the error depend (approximately) on the number of points?
   (c) Write a Fortran program for calculating the integral in (b) with the control variate method (see Sec. 11.3). Take $\varphi(x) = 1 + x + \frac{1}{2}x^2$. Use the same number of points and initial I_0 as in (b).

2. Write a Fortran program which uses rectangularly distributed random numbers to give a random "shuffle" of a deck of 52 cards.

3. (a) Let $x_1, \ldots, x_k$ be independent random variables with mean 0 and variance σ^2. Put $x = [x_1, x_2, \ldots, x_k]^T$, $B = [b_{ij}]$. Show that if

$$y = Bx, \quad y = [y_1, y_2, \ldots, y_k]^T,$$

   then the expectation of $y_i \cdot y_j$ is equal to the (i, j) element of the matrix $\sigma^2 BB^T$. This matrix is called the covariance matrix of y.
   (b) If you need to have random numbers with a given (positive-definite) covariance matrix V, how do you produce them, using ordinary independent random numbers?

12 SOLUTIONS TO PROBLEMS

For some problems, we give just the answers; for others we give hints for a second try, or, in some cases, complete solutions. Because of space limitations, we have excluded the solutions to certain problems.

CHAPTER 1

Section 1.2

1. $x_0 = 3$,
$x_1 = \frac{1}{2}(3 + 10/3) \approx 3.166667$,
$x_2 = \frac{1}{2}(3.166667 + 10/3.166667) \approx 3.16230$,
$x_3 = \frac{1}{2}(3.16230 + 3.16226) = 3.16228$ correct to $5D$.

2. (a) 0.648721; (b) 0.652096; (c) 0.648735; (d) about 1: 240.

3. Equation for tangent at x_n: $y(x) = f(x_n) + f'(x_n)(x - x_n)$. Set $y(x_{n+1}) = 0$; then $0 = f(x_n) + f'(x_n)(x_{n+1} - x_n)$ or $x_{n+1} = x_n - f(x_n)/f'(x_n)$. For $f(x) = x^2 - c$,

$$\begin{aligned} x_{n+1} &= x_n - (x_n^2 - c)/2x_n \\ &= x_n - x_n/2 + c/2x_n \\ &= \tfrac{1}{2}(x_n + c/x_n). \end{aligned}$$

4. (a) 1.44; (b) 1.4641; (c) about 8: 1; (d) about 30 steps.

5. Best with $h = 0.05$. Rounding error is approximately proportional to $1/h$, while truncation error is approximately proportional to h^2.

6. $F(x) = \frac{1}{2}(x + c/x)$, $F'(x) = \frac{1}{2} - c/(2x^2)$. At the root, $\alpha = F(\alpha) = \frac{1}{2}(\alpha + c/\alpha) \Rightarrow \alpha^2 = c$. Thus, $F'(\alpha) = \frac{1}{2} - c/2\alpha^2 = \frac{1}{2} - \frac{1}{2} = 0$.

Section 1.3

1. 22.

2. One way of using $x^i = x \cdot x^{i-1}$ is as follows (let $p(x) = \sum_{i=0}^{N} a_i x^{N-i}$ as in the text):

```
        s := 0;      t := 1;
    for i := n step -1 until 0 do
    begin
        s := s + a[i]*t;
        t := t*x;
    end
```

The final value of S is $p(x)$. This takes $2(n+1)$ multiplications and $(n+1)$ additions.

From Eq. (1.3.1), however, it is clear that evaluation of $p(x)$ by Horner's rule takes only n multiplications and n additions.

3. (a) Study the proof of Theorem 1.3.1 and notice that

$$p'(z) = \lim_{x \to z} \left[\frac{p(x) - p(z)}{x - z} \right].$$

(b) By Taylor's theorem,

$$p(x) = \sum_{i=0}^{n} c_i(z)(x - z)^i,$$

$$c_i(z) = \frac{p^{(i)}(z)}{i!}.$$

Now prove by induction that

$$g_k(x, z) = \sum_{i=0}^{n-k} c_{k+i}(z) \cdot (x - z)^i.$$

Then, putting $x = z$ in the above, $g_k(z, z) = c_k(z)$. Q.E.D.

4. Let $p(x)$ denote the polynomial $x^4 + 2x^3 - 3x^2 + 2$. Horner's rule with $z = 2$ gives the row 1 4 5 10 22, where $22 = p(2)$ and (see Theorem 1.3.1 and Problem 3)

$$1 \cdot x^3 + 4 \cdot x^2 + 5 \cdot x + 10 = [(x^4 + 2 \cdot x^3 - 3 \cdot x^2 + 0 \cdot x + 2) - 22]/(x - 2).$$

Using Horner's rule again (on this new polynomial with coefficients "b_i"), we get a new row, 1 6 17 44, where then $44 = p'(2)/1!$ (by Problem 3), and continuing in this way, $33 = p''(2)/2!$, $10 = p^{(3)}(2)/3!$; $p^{(4)}(2)/4! = 1$. Since

$$p(x) = p(2) + \frac{p'(2)}{1!}(x - 2) + \ldots + \frac{p^{(4)}(2)}{4!}(x - 2)^4,$$

we have $p(y + 2) = y^4 + 10y^3 + 33y^2 + 44y + 22$.

5. *Fortran*

```
      S = 0.
      DO 10  I = 1,N
   10 S = S + A(I) * B(I)
```

Algol

```
s := 0;
for i := 1 step 1 until n do
s := s + a[i]*b[i];
```

6. (a) The algorithm is clearly correct for $n = 1$. Suppose it is correct for all continued fractions of length N—i.e., for $n \leq N$. Now for a continued fraction with $N + 1$ terms, the algorithm gives

$$d_{N+1} = b_{N+1}$$
$$d_N = b_N + a_{N+1}/d_{N+1}.$$

Since the continued fraction with $N + 1$ terms can be written as a continued fraction with N terms by taking

$$b'_N = b_N + a_{N+1}/b_{N+1},$$

the result follows.

(b)

```
      DIMENSION A(10), B(10)
      READ (5, 901) N
      N1 = N + 1
      READ (5, 902) (B(I), I=1,N1),(A(I),I=2,N1)
      D = B(N1)
      DO 10 I=1,N
      K = N1 - I
   10 D = B(K) + A(K+1)/D
      WRITE (6, 903) D
      STOP
  901 FORMAT (I5)
  902 FORMAT (8E10.0)
  903 FORMAT (E14.6)
      END
```

With input data: 3, 3, 3, 5, 7, 2, 4, 6 the answer is 3.543046.

7. The result of the following (Algol) program segment is that $\Delta^k y_0$ gets put in the place where $y[k]$ was:

for $j := 0$ **step** 1 **until** $n - 1$ **do**

for $k := n$ **step** -1 **until** $j + 1$ **do** $y[k] := y[k] - y[k - 1]$;

In Fortran, the analogous program segment (except that we use $y(1)$, $y(2)$, $\ldots, y(n + 1)$) is

```
      NP1 = N+1
      DO 30 JP1 = 1, N
      DO 30 I = JP1, N
      K = NP1+JP1 - I
   30 Y(K) = Y(K) - Y(K-1)
```

8. Let $p(x) = f(x) \cdot g(x) = \sum_{k=1}^{m+n-1} c_k x^{k-1}$,

$$f(x) = \sum_{i=1}^{m} a_i x^{i-1},$$

$$g(x) = \sum_{j=1}^{n} b_j x^{j-1}.$$

The coefficient of x^{k-1} in $p(x)$ is

$$c_k = \sum_{i+j=k+1} a_i b_j, \quad 1 \leq k \leq m+n-1.$$

One way of programming this is, e.g.,

> **for** $i := 1$ **step** 1 **until** $m + n - 1$ **do** $c[i] = 0$;
> **for** $i := 1$ **step** 1 **until** m **do**
> **for** $j := 1$ **step** 1 **until** n **do**
> $c[i + j - 1] = c[i + j - 1] + a[i]*b[j]$;

9. We denote the greatest common divisor (by convention, positive) of two integers x and y, not both zero, by $\gcd(x, y)$. Notice that $\gcd(x, y) = \gcd(-x, y) = \gcd(-x, -y)$ so it is no restriction to consider $x > 0$, $y \geq 0$. The notation $a|b$ means a divides b. Put

$$r_{-1} = x, \qquad r_0 = y,$$

and determine r_1 and q_1 by

$$r_{-1} = q_1 r_0 + r_1, \qquad q_1 \geq 0, \quad r_1 < r_0.$$

If $r_1 = 0$ then $\gcd(r_{-1}, r_0) = \gcd(x, y) = r_0 = y$. Note

(i) any number which divides r_{-1} and r_0 must also divide r_1,

(ii) conversely, any number which divides r_1 and r_0 also divides r_{-1}.

This suggests the following algorithm (Euclid's algorithm). Compute the finite sequences $\{r_n\}$ and $\{q_n\}$ as follows:

$$r_{n-2} = q_n r_{n-1} + r_n, \qquad q_n \geq 0, \quad r_n < r_{n-1}, \quad n = 2, 3, \ldots, N.$$

The sequence is finite since $r_{n+1} < r_n$ (for $n \geq 0$) so it must terminate, with $r_n = 0$ for, say, $n = N$. But $r_N = 0$ implies that $r_{N-1}|r_{N-2}$ and $r_N|r_{N-1}$ (in fact $r_{N-1} = \gcd(r_{N-2}, r_{N-1})$). Also, $r_{N-1}|r_{N-3}$, by (ii) above. Continuing in this way we get $r_{N-1}|r_0$ and $r_{N-1}|r_{-1}$. That r_{N-1} is the *greatest* common divisor of r_0 and r_{-1} follows since (i) implies that any divisor of r_0 and r_{-1} must divide r_{N-1}.

The following is a Basic program for the second part of the question. It works for arbitrary fractions I1/I2.

```
10  PRINT "INPUT INTEGERS I1 AND I2, WITH I2#0"
20  INPUT I1,I2
30  S=SGN(I1)*SGN(I2)
40  R1=ABS(I1)
50  R2=ABS(I2)
60  Q=INT(R1/R2)
70  PRINT "INTEGER PART= "S*Q
80  R1=R1-Q*R2
90  REM- NOW USE EUCLID'S ALGORITHM (LINES 130-170) TO REDUCE
100 REM- THE FRACTIONAL PART (=S*(R1/R2)) TO LOWEST TERMS.
110 T1=R1
120 T2=R2
130 R3=R1-INT(R1/R2)*R2
140 IF R3=0 THEN 180
```

```
150 R1=R2
160 R2=R3
170 GOTO 130
180 PRINT "FRACTIONAL PART"S*T1/R2"/"T2/R2
190 END
```

(The function INT(X/Y) gives the largest integer $\leq (X/Y)$.) The reader can either run the BASIC program or step through the program to verify that it yields the following results for I1, I2 as given in the table:

I1	I2	Integer Part	Fractional Part
5	4	1	1/4
162	54	3	0/1
172	22	7	9/11
−44	154	0	−2/7
−7	7	−1	0/1
−65	−3	21	2/3

10. Recursion formula: $4y_{n+1} + y_n = 1/(n+1)$. The algorithm $y_0 = \frac{1}{4}\ln 5$, $y_{n+1} = \frac{1}{4}(1/(n+1) - y_n)$, $(n = 0, 1, 2, \ldots)$ works well, since previous errors are divided by -4 at each step.

Using the recursion formula "backward", however, does not work well here—e.g., for

$$y_{10} = 0, \qquad y_n = 1/(n+1) - 4y_{n+1}, \quad (n = 9, 8, 7, \ldots, 0)$$

previous errors are multiplied by -4 at each step.

CHAPTER 2

Section 2.1

1. 3.1415 ± 10^{-4} and $3.1416 \pm \frac{1}{2}\cdot 10^{-4}$, respectively.

Section 2.2

The notations R_X, R_C, and R_T are explained in Problem 8 on page 40.

1. (a) See Example 2.2.2. Take $\pi = 3.1416 \pm 1\cdot 10^{-5}$, $z = 4.1262$.

$$\left|\frac{\Delta z}{z}\right| \leq \left|\frac{\Delta x}{x}\right| + \left|\frac{\Delta y}{y}\right| \leq \frac{\frac{1}{2}\cdot 10^{-4}}{1} + \frac{10^{-5}}{3} < 6\cdot 10^{-5}, \quad \Delta z \leq 25\cdot 10^{-5}.$$

After rounding to $3D$, we have $z = 4.126$. (The error is less than $45\cdot 10^{-5}$.)

(b) By Example 2.2.2 four or five digits should be used in e. Try $e \approx 2.718 \pm 3\cdot 10^{-4}$. Then $z = 1.0222$.

$$\left|\frac{\Delta z}{z}\right| \leq \frac{\frac{1}{2}\cdot 10^{-4}}{0.376} + \frac{\frac{1}{2}\cdot 10^{-3}}{2.71} < 3.5\cdot 10^{-4}, \quad |\Delta z| < 4\cdot 10^{-4}.$$

The error of the rounded value $z = 1.022$ can hence be as large as $6\cdot 10^{-4}$. Answer, therefore, $z = 1.0222 \pm 0.0004$.

(c) $\pi \cdot e \approx 9$, relative error should not exceed $\frac{1}{2} \cdot 10^{-4}$. Take then, e.g., $\pi = 3.1416 \pm 10^{-5}$, $e = 2.7183 \pm 2 \cdot 10^{-5}$, $\pi \cdot e = 8.53981$.

$$\left|\frac{\Delta z}{z}\right| \leq \frac{10^{-5}}{3} + \frac{2 \cdot 10^{-5}}{2.7} \leq 1.2 \cdot 10^{-5}, \quad |\Delta z| \leq 1.3 \cdot 10^{-4}$$

Rounding the result to 8.540 is safe.

2. (a) $f = 3x + y - z$.

$\Delta f = 3\Delta x + \Delta y - \Delta z$ yields, using

$|\Delta x| \leq 0.5 \cdot 10^{-2}$, $|\Delta y| \leq 0.5 \cdot 10^{-2}$ and $|\Delta z| \leq 0.5 \cdot 10^{-2}$,

$|\Delta f| \leq 3|\Delta x| + |\Delta y| + |\Delta z|$,

$|R_X| \leq 5 \cdot 0.5 \cdot 10^{-2} = 2.5 \cdot 10^{-2}$.

$f = 3 \cdot 2.00 + 3.00 - 4.00 = 5.00, \quad R_B = 0$.

$$\left|\frac{R_X}{f}\right| \leq \frac{2.5 \cdot 10^{-2}}{5.00} = 0.5 \cdot 10^{-2}.$$

(b) $f = x \cdot y/z$ yields

$$\frac{\Delta f}{f} \approx \frac{\Delta x}{x} + \frac{\Delta y}{y} - \frac{\Delta z}{z}.$$

$$\left|\frac{\Delta f}{f}\right| \lesssim \left|\frac{\Delta x}{x}\right| + \left|\frac{\Delta y}{y}\right| + \left|\frac{\Delta z}{z}\right|.$$

Insert the bounds for the round-off errors:

$$\left|\frac{R_X}{f}\right| \lesssim (\tfrac{1}{2} + \tfrac{1}{3} + \tfrac{1}{4}) \cdot 0.5 \cdot 10^{-2} = \tfrac{13}{24} \cdot 10^{-2} < 0.55 \cdot 10^{-2},$$

$f = 1.5, \qquad R_C = 0.$

from which $|R_X| \lesssim 1.5 \cdot 0.55 \cdot 10^{-2} < 0.83 \cdot 10^{-2}$.

(c) $f = x \cdot \sin \frac{y}{40}, \qquad \ln f = \ln x + \ln \sin \frac{y}{40}.$

$$\frac{\Delta f}{f} \approx \frac{\Delta x}{x} + \frac{\Delta y}{40} \cot \frac{y}{40},$$

$$\left|\frac{\Delta f}{f}\right| \lesssim \left|\frac{\Delta x}{x}\right| + \left|\frac{\Delta y}{40} \cot \frac{y}{40}\right|,$$

$$\left|\frac{R_X}{f}\right| \lesssim (\tfrac{1}{2} + \tfrac{1}{40} \cdot \cot 0.075) \cdot 0.5 \cdot 10^{-2} = 0.42 \cdot 10^{-2},$$

$f = 2.00 \cdot \sin 0.075 = 2.00 \cdot 0.0749 = 0.1498, \qquad R_C = 0,$

$|R_X| \lesssim 0.15 \cdot 0.42 \cdot 10^{-2} = 0.63 \cdot 10^{-3},$

$|R_{\text{sin-tab}}| \leq 2.0 \cdot 0.5 \cdot 10^{-4} = 10^{-4},$

$|R| \lesssim 0.8 \cdot 10^{-3}, \qquad |R/f| \lesssim 0.6 \cdot 10^{-2}.$

Alternative solution: $f = x \cdot \sin \frac{y}{40}$ yields immediately

$$\Delta f \approx x \cdot \frac{\Delta y}{40} \cos \frac{y}{40} + \Delta x \cdot \sin \frac{y}{40}.$$

3. (a) $y = 18.0 \pm 4.2$, x_2 contributes most; (b) 2.72.

4.
$$\begin{cases} 3x + ay = 10, \\ 5x + by = 20. \end{cases}$$

Solve for $x + y$.

$$\begin{cases} 3(x + y) + (a - 3)y = 10, \\ 5(x + y) + (b - 5)y = 20. \end{cases}$$

$$x + y = 10 \cdot \frac{b - 2a + 1}{3b - 5a} = 2\left(2 + \frac{5 - b}{3b - 5a}\right) = \frac{10}{3}\left(1 + \frac{3 - a}{3b - 5a}\right),$$

$$\Delta(x + y) \approx \frac{1}{(3b - 5a)^2} \cdot \left(2 \cdot 5(5 - b)\Delta a - \frac{10}{3} \cdot 3(3 - a)\Delta b\right),$$

$$|\Delta(x + y)| \lesssim \frac{10}{0.6^2} \cdot (1.7|\Delta a| + 0.9|\Delta b|),$$

$$|\Delta(x + y)| \lesssim \frac{10}{0.6^2} \cdot (1.7 + 0.9) \cdot 0.5 \cdot 10^{-3} = 36 \cdot 10^{-3}.$$

5. Put the approximate value $= x$. Then we have

(1) $f = (x - 1)^6$,
$$|\Delta f| \lesssim 6(x - 1)^5 |\Delta x| \leq 6 \cdot 0.4^5 \cdot 0.015 < 9.3 \cdot 10^{-4}.$$

(2) $f = (x + 1)^{-6}$,
$$|\Delta f| \lesssim 6(x + 1)^{-7} |\Delta x| \leq 6 \cdot 2.4^{-7} \cdot 0.015 < 2.0 \cdot 10^{-4}.$$

(3) $f = (3 - 2x)^3$,
$$|\Delta f| \lesssim 6(3 - 2x)^2 |\Delta x| \leq 6 \cdot 0.2^2 \cdot 0.015 = 3.6 \cdot 10^{-3}.$$

(4) $f = (3 + 2x)^{-3}$,
$$|\Delta f| \lesssim 6(3 + 2x)^{-4} |\Delta x| \leq 6 \cdot 5.8^{-4} \cdot 0.015 < 8.1 \cdot 10^{-5}.$$

(5) $f = 99 - 70x$,
$$|\Delta f| = 70 |\Delta x| \leq 70 \cdot 0.015 = 1.05.$$

(6) $f = (99 + 70x)^{-1}$,
$$|\Delta f| \lesssim 70(99 + 70x)^{-2} |\Delta x| \leq 70 \cdot 197^{-2} \cdot 0.015 < 2.8 \cdot 10^{-5}.$$

Hence the last expression is preferable.

6. $f = \ln(x - \sqrt{x^2 - 1})$.

Put $\sqrt{x^2 - 1} = r$. $|\Delta r| \leq 0.5 \cdot 10^{-6} \cdot x$ is exact.

$$f = \ln(x - r) \Rightarrow |R_{\text{tab}}| \lesssim \frac{|\Delta r|}{x - r} \lesssim \frac{0.5 \cdot 10^{-6}}{30 - \sqrt{899}} = \frac{0.5 \cdot 10^{-6}}{0.017} < 3 \cdot 10^{-5}.$$

A more suitable expression for numerical computation is:

$$\ln(x - \sqrt{x^2 - 1}) = \ln \frac{(x - \sqrt{x^2 - 1})(x + \sqrt{x^2 - 1})}{x + \sqrt{x^2 - 1}} = -\ln(x + \sqrt{x^2 - 1}).$$

Then

$$f = -\ln(x + r) \Rightarrow |R_{\text{tab}}| \lesssim \frac{|\Delta r|}{x + r} \lesssim \frac{0.5 \cdot 10^{-6}}{60} < 9 \cdot 10^{-9}.$$

7. We denote the third side by d, and after using the cosine theorem we obtain

$$d^2 = r^2 + 2R(R + r)(1 - \cos \varphi), \quad \text{from which} \quad d \approx 250 \text{ km}.$$

Velocity: $v = d/5$ km/s.

$$\frac{\partial d^2}{\partial R} = 3.4, \qquad \frac{\partial d^2}{\partial r} = 252, \qquad \frac{\partial d^2}{\partial \varphi} = 1.36 \cdot 10^7, \quad \text{from which}$$

$|\Delta d^2| < 432.$

$$|\Delta v| = \left|\frac{\Delta d^2}{10d}\right| < 0.18 \text{ km/s}.$$

Expand the series

$$1 - \cos \varphi = \frac{\varphi^2}{2} - \frac{\varphi^4}{24} + \ldots.$$

If only the first term in the series is retained, then the resultant inaccuracy in $1 - \cos \varphi$ is at most $0.0075^4/24$ and the error in $d^2 < 0.25$, from which

$$|R_T| < 0.25/(10 \cdot 250) = 1 \cdot 10^{-4}.$$

Computation with two decimals gives $v = 51.56$ km/s, where

$$|R_C| < 0.5 \cdot 10^{-2},$$
$$|R_C + R_T| < 0.51 \cdot 10^{-2} < 0.2 \cdot 0.18 = 0.2|R_X|.$$

Thus we get full accuracy.

8. (a) $c = (a^2 + b^2 - 2ab \cos C)^{1/2}$.

c is determined approximately using the Pythagorean theorem.

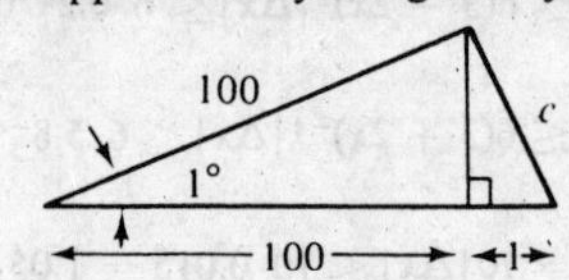

$$c^2 = 1^2 + (100 \cdot 0.01745)^2 \approx 4.045,$$
$$c \approx 2.01,$$
$$\frac{\partial c}{\partial a} = \frac{a - b \cos C}{c} = -0.49,$$
$$\frac{\partial c}{\partial b} = \frac{b - a \cos C}{c} = 0.51,$$
$$\frac{\partial c}{\partial C} = \frac{ab \sin C}{c} = 88,$$
$$|\Delta c| \lesssim \left|\frac{\partial c}{\partial a}\right||\Delta a| + \left|\frac{\partial c}{\partial b}\right||\Delta b| + \left|\frac{\partial c}{\partial C}\right||\Delta C| \leq 0.49 \cdot 0.1 + 0.51 \cdot 0.1$$
$$+ 88 \cdot 0.01 \cdot 0.01745 < 0.12.$$

(b) Put $\cos C = x$, where $x = 0.9998 \pm 0.5 \cdot 10^{-4}$.

$$c = (a^2 + b^2 - 2abx)^{1/2},$$

$$\left|\frac{\partial c}{\partial x}\right| = \frac{ab}{c} = \frac{100\cdot 101}{2.01} = 0.5\cdot 10^4,$$

$$|R_{\text{cos-tab}}| \lesssim \left|\frac{\partial c}{\partial x}\right| |\Delta x| \leq 0.5\cdot 10^4 \cdot 0.5\cdot 10^{-4} = 0.25,$$

whence $|R| \lesssim 0.37$.

(c) $\cos C = 1 - 2\sin^2 \frac{C}{2} \Rightarrow c = \left(a^2 + b^2 - 2ab + 4ab\sin^2\frac{C}{2}\right)^{1/2}$

$= \left((a-b)^2 + 4ab\sin^2\frac{C}{2}\right)^{1/2}$. Set $\sin\frac{C}{2} = y$, where $|\Delta y| \leq 0.5\cdot 10^{-4}$.

$$c = ((a-b)^2 + 4aby^2)^{1/2},$$

$$\left|\frac{\partial c}{\partial y}\right| = \frac{4aby}{c} = \frac{4\cdot 100\cdot 101\cdot 0.00873}{2.01} = 175,$$

$$|R_{\text{sin-tab}}| \lesssim \left|\frac{\partial c}{\partial y}\right|\cdot|\Delta y| \leq 175\cdot 0.5\cdot 10^{-4} < 0.009.$$

$$0.009 < 0.2\cdot 0.12$$

Full accuracy is obtained.

9. (c) $\bar{x} = 546.849$, $\quad s^2 = 22\cdot 10^{-4}$;

(d) Ten digits. No information is lost in the latter subtraction. Information is lost in the rounding of x_i^2.

(g) $n^{-1}\sum x_i' y_i' - (\bar{x} - \alpha)(\bar{y} - \beta)$.

10. $I_7 = 0.0212, 0.0214$;
$I_6 = 0.0242, 0.0244$.

11. To investigate the linearity in the dependence of I on a and b we form:

$$I(0.40, 0.34) - I(0.39, 0.34) = -0.0266,$$
$$I(0.41, 0.34) - I(0.40, 0.34) = -0.0255,$$
$$I(0.40, 0.34) - I(0.40, 0.32) = -0.0104,$$
$$I(0.40, 0.36) - I(0.40, 0.34) = -0.0103.$$

The values agree sufficiently well in order that the following approximations are reasonable for $a = 0.40$, $b = 0.34$.

$$\frac{\partial I}{\partial a} \approx \frac{I(0.41, 0.34) - I(0.39, 0.34)}{0.02} = -2.6,$$

$$\frac{\partial I}{\partial b} \approx \frac{I(0.40, 0.36) - I(0.40, 0.32)}{0.04} = -0.52,$$

$$|\Delta I| \lesssim \left|\frac{\partial I}{\partial a}\right||\Delta x| + \left|\frac{\partial I}{\partial b}\right||\Delta y|,$$

$$|\Delta I| \lesssim 2.6\cdot 0.003 + 0.52\cdot 0.005 < 0.011.$$

12. $\operatorname{var}(\bar{x}) = n^{-2}\operatorname{var}(\sum_{i=1}^{n} x_i) = n^{-2}\sum_{i=1}^{n}(\operatorname{var} x_i) = n^{-2}n\epsilon^2 = \epsilon^2/n.$

Section 2.3

1. (a) Numbers of the form $p \cdot 2^{-q}$ where p and q are integers;
(b) for example, one-tenth.

2. Put $a =$ ✗ , $b = \vee$, $c = \Delta$, and denote the base by R. Clearly we can assume $a \neq 0$, so we must have $a = 1$. The result of the addition means that

$$(aR + b) + (aR + b) = aR^2 + cR + a \quad \text{or}$$
$$2R + 2b = R^2 + cR + 1 \qquad (*)$$

Also we cannot have $b + b = a$ since $a = 1$, so $b + b = a + R$, or $2b = 1 + R$. Substituting in $(*)$ we get

$$3R + 1 = R^2 + cR + 1, \qquad 3 = R + c, \qquad R = 3 - c.$$

Now we also have that $0 \leq c < R$, so $R \leq 3$. But there are three distinct symbols, so $R \geq 3$. Thus $R = 3$ (and $c = 0$, $b = 2$, $a = 1$).

3. (a) 16; (b) 64; (c) 100; (d) 255; (e) 0.140625; (f) 0.1; (g) $2 \cdot 10^9$.

4.

(a)	+	2	7	1	8	2	8	1	8	2	8	5	+	0	0	1
(b)	−	1	0	7	3	7	4	1	8	2	4	0	+	0	1	0
(c)	+	5	7	7	2	1	6	0	0	0	0	0	+	0	0	0
(d)	−	1	2	3	0	0	0	0	0	0	0	0	−	0	4	2

5. $0.5 \cdot 10^{-10}$.

6. (a) $1.0 = \frac{1}{2} \cdot 2^1$, $p' = \frac{1}{2}$, $p'' = 1$. The exponent is represented by $p'' + 2^{10} = 1 + 2^{10} = (10000000001)_2$, since $p'' \geq 0$. The mantissa is represented by $p' = \frac{1}{2} = (0.100\ldots000)_2$. Thus the number is stored as

0	1 0 0 . . . 0 1	1 0 0 . . . 0
1	2 12	13 48

Note: CDC 3200 uses octal representation. The stored number thus becomes

2001 4000 0000 0000.

(b) $-0.0625 = -\frac{1}{2} \cdot 2^{-3}$, $p' = -\frac{1}{2}$, $p'' = -3$. Stored number: 6003 3777 7777 7777.

(c) $250.25 = \dfrac{1{,}001}{1{,}024} \cdot 2^8$, $p' = (0.1111101001)_2$, $p'' = 8$. Stored number: 2010 7644 0000 0000.

(d) $0.1 = 0.8 \cdot 2^{-3}$, $p' = 0.8 = (0.11001100\ldots)_2$, $p'' = -3$. stored number: 1774 6314 6314 6314.

(e) Largest number $= (1 - 2^{-36}) \cdot 2^{(210-1)} = 2^{1{,}023} \approx 8 \cdot 10^{307}$. Smallest positive number $= \frac{1}{2} \cdot 2^{(-210+1)} = 2^{-1{,}024} \approx 6 \cdot 10^{-309}$.

(f) $2^{-36} \approx 1.5 \cdot 10^{-11}$.

(a) $1.0 = \frac{1}{16} \cdot 16^1$ yields the representation

| 0100 0001 0001 0000 0000 0000 0000 0000 |,

or, in hexadecimal form, which is used on the IBM 360, 4110 0000;

(b) $0.5 = \frac{8}{16}\cdot 16^0$ gives 4080 0000;

(c) $-15.0 = -\frac{15}{16}\cdot 16^1$ gives C1F0 0000;

(d) the largest number which can be stored is $(1 - 2^{-24})\cdot 16^{63} \approx 7\cdot 10^{75}$ in single precision and $(1 - 2^{-56})\cdot 16^{63} \approx 7\cdot 10^{75}$ in double precision. The smallest positive number which can be stored is, in both single and double precision,

$$\tfrac{1}{16}\cdot 16^{-64} \approx 6\cdot 10^{-79};$$

(e) $2^{-21} \approx 5\cdot 10^{-7}$ in single precision, $2^{-53} \approx 1.1\cdot 10^{-16}$ in double precision. When chopping has been used, the error bounds are twice as large.

CHAPTER 3

Section 3.1

1. (a) $\sin x = x - x^3/3! + x^5/5! - \dots$. Alternating series with decreasing terms.

$$\therefore\quad |R_T| \le |x^3/3!|,$$

$$|R_T| \begin{cases} \le 0.5\cdot 10^{-4} & \text{if}\quad |x| < 0.067, \\ \le 0.5\cdot 10^{-6} & \text{if}\quad |x| < 0.014. \end{cases}$$

(b) $\cos x = 1 - x^2/2! + x^4/4! - \dots$

$$|R_T| \begin{cases} < 0.5\cdot 10^{-4} & \text{if}\quad |x| < 0.186, \\ < 0.5\cdot 10^{-6} & \text{if}\quad |x| < 0.059. \end{cases}$$

(c) $(1 - x^2)^{-1/2} = 1 + \frac{1}{2}x^2 + \frac{3}{8}x^4 + \frac{5}{16}x^6 + \dots$. Overestimate (majorize) the remainder using a geometric series with first term $3x^4/8$ and ratio x^2.

$$|R_T| < \frac{3x^4}{8(1 - x^2)} \begin{cases} < 0.5\cdot 10^{-4} & \text{if}\quad |x| < 0.107, \\ < 0.5\cdot 10^{-6} & \text{if}\quad |x| < 0.034. \end{cases}$$

2. (See Table 3.1.2 for the expansions needed.)

$$(\cos x)^{1/2} = (1 - x^2/2 + x^4/4! + O(x^6))^{1/2}$$
$$= 1 - x^2/4 - x^4/96 + O(x^6).$$

$$\exp((\cos x)^{1/2}) = \exp(1)\cdot\exp(-x^2/4 - x^4/96 + O(x^6))$$
$$= e\cdot(1 - x^2/4 + x^4/48) + O(x^6).$$

$$(1 + x^2)^{-1/2} = 1 - x^2/2 + 3x^4/8 + O(x^6).$$

Thus $\exp((\cos x)^{1/2})/(1 + x^2)^{1/2} = e(1 - x^2/4 + x^4/48)\cdot(1 - x^2/2 + 3x^4/8) + O(x^6)$

$$= e(1 - 3x^2/4) + 25ex^4/48 + O(x^6).$$

Estimate the error by the first neglected term

$$x = 0.1 \Rightarrow |R_T| \approx 25ex^4/48 \approx 1.4 \cdot 10^{-4}$$

$$x = 0.01 \Rightarrow |R_T| \approx 1.4 \cdot 10^{-8}.$$

3. (See also Example 3.1.11.)

(a) 2.704815; (b) 0.003334445.

4. $(1 + y)/(1 - y) = 1.2 \Rightarrow y = \frac{1}{11}$. *Answer:* $\ln(1.2) = 0.1823215$. Three terms of the series were used. Use the first neglected term $(y^7/7)$ as an error estimate.

If the series for $\ln(1 + y)$ is used, then nine terms are required.

5. *Answer:* 0.09975. (*Hint:* Expand the integrand in powers of $0.1 \sin t$; $\sin t$ can be replaced by t with sufficient accuracy.)

6. $y = x - x^3 + 3x^5 + O(x^7)$. $c_j = 0$ for j even, since $y(-x) = -y(x)$ for all x in a neighborhood of the origin.

7. Integrand $= x^{-3/2}(1 + x^{-2})^{-1/2} = x^{-1.5} - \frac{1}{2}\cdot x^{-3.5} + \frac{3}{8}\cdot x^{-5.5} - \ldots$

$$\text{Integral} = \frac{x^{-0.5}}{0.5} - \frac{x^{-2.5}}{5} + \frac{x^{-4.5}}{12} - \ldots \approx \frac{1.9980}{\sqrt{10}} = 0.6318.$$

8. The series is convergent according to Leibniz's criterion. Thus

$$R_n = (a_{n+1} + a_{n+2}) + (a_{n+3} + a_{n+4}) + \ldots.$$

Since $|a_{j+2}| \le |a_{j+1}|$, $\text{sign}(a_{j+1} + a_{j+2}) = \text{sign}\, a_{j+1}$. Use this for $j = n$, $n + 2$, $n + 4, \ldots$. But a_{n+1}, a_{n+3}, a_{n+5} all have the same sign. Thus all the terms in parentheses have the same sign as a_{n+1}, and hence sign R_n = sign a_{n+1}. This holds for all n. Replacing n by $n + 1$ we get

$$\text{sign}\, R_{n+1} = \text{sign}\, a_{n+2} = -\text{sign}\, R_n.$$

The theorem then follows from Theorem 3.1.3.

Section 3.2

1. (a) 0.693; (b) about 1,000.

2. (a)

```
real procedure AVESUM(a, k, n, eps);
value n, eps; real a, eps; integer k, n;
begin array S[0: n]; integer i, sgn; real slask, blask;
    k := 0; S[0] := a; k := 1; sgn := 1;
  L: slask := S[0]; sgn := -sgn; blask := S[0] := S[0] + sgn*a;
    for i := 1 step 1 until k - 1 do
    begin blask := (slask + blask)/2;
      slask := S[i];
      if abs(slask - blask) < eps then goto out;
      S[i] := blask
    end;
    S[k] := (slask + blask)/2; k := k + 1;
    if k ≤ n then goto L else
    OUTPUT(61, '/' POOR CONVERGENCE");
 OUT: AVESUM := blask
```

end $AVESUM$;

(b) $AVESUM(1/\ln(k+2), k, 50, {}_{10}-6)$;

3. (b) about 160,000;

(c) with the $f'(5)$ term, 2.612386.

4. Sum of three terms $= 0.65423 \pm 0.5 \cdot 10^{-5}$.

$$f(x) = x/(x^4+1) = x^{-3} - x^{-7} + x^{-11} - \dots.$$

$$\int_4^\infty f(x)\,dx = 0.03121, \quad |R| \le 10^{-5},$$

$$\tfrac{1}{2}f(4) = 0.00778, \quad |R| \le 0.3 \cdot 10^{-5},$$

$$-\tfrac{1}{12}f'(4) = 0.00098, \quad |R| \le 10^{-5}.$$

$0.03121 + 0.00778 + 0.00098 = 0.03997$, with $|R| < 2.3 \cdot 10^{-5}$. The first neglected term is $(-h^3/720)(f'''(\infty) - f'''(4))$. Its magnitude is less than $2.1 \cdot 10^{-5}$. Since $f^{\text{iv}}(x)$ has constant sign in $(4, \infty)$, we get sum $= S = 0.69420 \pm 5 \cdot 10^{-5}$.

5. 1.077.

6.

$$\frac{v_{n+1} - s}{v_n - s} \approx \frac{A(-k)^{n+1}}{A(-k)^n} = -k, \quad \text{where } 0 < k < 1.$$

Thus the conditions for Aitken extrapolation are satisfied. The Aitken-extrapolated sequence is obtained from

$$v'_{n+2} = v_{n+2} - \frac{(\Delta v_{n+1})^2}{\Delta^2 v_n}.$$

It is advisable to set up a difference scheme when performing Aitken extrapolation. In our case, we get

		Δ	Δ^2
v_0	659		
		-423	
v_1	236		650
		227	
v_2	463		-350
		-123	
v_3	340		189
		66	
v_4	406		

The extrapolated values become

$$v'_2 = v_2 - \frac{(\Delta v_1)^2}{\Delta^2 v_0} = 463 - \frac{227^2}{650} = 463 - 79.3 = 383.7,$$

$$v'_3 = v_3 - \frac{(\Delta v_2)^2}{\Delta^2 v_1} = 340 - \frac{123^2}{-350} = 340 + 43.2 = 383.2,$$

$$v'_4 = v_4 - \frac{(\Delta v_3)^2}{\Delta^2 v_2} = 406 - \frac{66^2}{189} = 406 - 23.0 = 383.0$$

Thus the new current s is about 383 units.

7. (a) The sequence is given by

$$y_{n+1} = (1 - \tfrac{1}{2})y_n + 1 = \tfrac{1}{2}y_n + 1, \quad \text{with} \quad y_0 = 1.$$

Difference scheme:

		Δ	Δ^2
y_0	1		
		0.5	
y_1	1.5		−0.25
		0.25	
y_2	1.75		−0.125
		0.125	
y_3	1.875		−0.0625
		0.0625	
y_4	1.9375		

$$y_2' = 1.75 - \frac{(0.25)^2}{-0.25} = 2, \quad y_3' = 2, \; y_4' = 2.$$

Thus the extrapolated sequence 2, 2, 2, . . . converges extremely fast to $\alpha = 1/N = 2$.

(b)
$$\frac{y_{n+1} - \alpha}{y_n - \alpha} = \frac{\frac{1}{2}y_n + 1 - 2}{y_n - 2} = \frac{1}{2} < 1.$$

The quotient between successive errors is thus exactly constant and therefore Aitken extrapolation gives the exact result.

8. Put $f(t) = g(t + h) - g(t)$, $f(t_j) = f_j$; then

$$f(t) = -kh + \sum_{n=1}^{\infty} a_n' e^{-\lambda_n t}, \quad \text{where } a_n' = a_n(1 - e^{-\lambda_n h}).$$

Aitken extrapolation is applicable to the sequence $\{f_j\}$, because $\lim_{j\to\infty} f_j = -kh$,

$$\frac{f_j - (-kh)}{f_{j-1} - (-kh)} = \frac{\sum a_n' e^{-\lambda_n t_j}}{\sum a_n' e^{-\lambda_n t_{j-1}}} \longrightarrow \lim \frac{a_1' e^{-\lambda_1 t_j}}{a_1' e^{-\lambda_1 t_{j-1}}} = e^{-\lambda_1 h}.$$

		$10^5\cdot\Delta f$	$10^5\cdot\Delta^2 f$
f_0	−0.32582		
		+10,138	
f_1	−0.22444		−6,777
		+3,361	
f_2	−0.19083		−2,174
		+1,187	
f_3	−0.17896		

$$f_3' = -0.17896 - \frac{(1{,}187)^2\cdot 10^{-10}}{(-2{,}174)\cdot 10^{-5}} = -0.17248 \Rightarrow k \approx \frac{0.17248}{h} = 1.7248.$$

$$f_2' = -0.19083 - \frac{(3{,}361)^2\cdot 10^{-10}}{(-6{,}777)\cdot 10^{-5}} = -0.17416 \Rightarrow k \approx \frac{0.17416}{h} = 1.7416.$$

f_3' gives a more reliable estimate than f_2'; even 1.7248 is probably a little too high. *Result:* $k \approx 1.72$.

9. (a) $a_n = (-1)^n(x/2)^{2n}/(n!)^2$, $x = 20$. The terms are alternating and monotone decreasing for $n > 10$. Thus we can estimate the remainder to be less than the magnitude of the first neglected term.

$$x = 20 \quad \text{yields} \quad |a_n| = 10^{2n}/(n!)^2.$$

To get six correct decimals, it is required that $|R_n| < 0.5 \cdot 10^{-6}$—i.e., that $10^{2(n+1)}/((n+1)!)^2 < 0.5 \cdot 10^{-6}$. Hence we must have $n > N = 31$.

(b) The largest terms are

$$|a_9| = |a_{10}| = 10^{18}/(3.63 \cdot 10^5)^2 = 7.6 \cdot 10^6.$$

(c) To get six correct decimals in the addition of thirty-one terms, we should compute with an error $< 0.5 \cdot 10^{-6}/31 = 1.6 \cdot 10^{-8}$—i.e., we should compute using eight decimals. Furthermore, seven digits to the left of the decimal point are required (see part (b)); thus fifteen digits are required. *Note:* $J_0(20) = 0.1670$. Thus very pronounced cancellation occurs in the summation of the series. Fortunately, there exist other formulas for computing Bessel functions of large arguments, see e.g. [27] or [37].

10. $$\int_x^\infty t^{-1}e^{-t}\,dt = [-t^{-1}e^{-t}]_x^\infty - \int_x^\infty -e^{-t}t^{-2}\,dt$$
$$= x^{-1}e^{-x} + \int_x^\infty e^{-t}t^{-2}\,dt.$$

Generally: $\displaystyle\int_x^\infty t^{-n}e^{-t}\,dt = x^{-n}e^{-x} - n\int_x^\infty e^{-t}t^{-n-1}\,dt.$

Hence $$f(x) = e^x \int_x^\infty e^{-t}t^{-1}\,dt$$
$$= x^{-1} - x^{-2} + 2x^{-3} - 3!x^{-4} + \ldots + (-1)^n n! e^x \int_x^\infty e^{-t}t^{-n-1}\,dt.$$

11. $x^{-1} - \frac{1}{2}x^{-3} + \frac{3}{4}x^{-5} - \ldots$ (Put $u = t^2$).

CHAPTER 4

Section 4.1

1. 1 and $\sqrt{2}/2 = 0.707$, respectively.

2. (a) $$\|f\| = \|f + g - g\| \le \|f - g\| + \|g\|.$$
$$\therefore \quad \|f - g\| \ge \|f\| - \|g\|.$$

(b) This is easy to prove by induction; for $n = 0$ the assertion reduces to axiom 2 of Sec. 4.1.3. Now put

$$v = \sum_0^n c_j f_j \quad \text{and suppose} \quad \left\|\sum_0^n c_j f_j\right\| \le \sum_0^n |c_j| \cdot \|f_j\|.$$

Then

$$\|v + c_{n+1}f_{n+1}\| \le \|v\| + \|c_{n+1}f_{n+1}\|$$

and $\|c_{n+1}f_{n+1}\| = |c_{n+1}| \cdot \|f_{n+1}\|$. So

$$\|v + c_{n+1}f_{n+1}\| \le \sum_0^{n+1} |c_j| \cdot \|f_j\|. \qquad \text{Q.E.D.}$$

Section 4.2

1. (a) $1.268 + 1.148x$; (b) $1.175 + 1.104x$;
(c) $1 + x$.
The errors at $x = 1$ are -0.302, 0.439, -0.718.

2. $c_0 = 2\pi^2/3$, $c_j = -4(-1)^j/j^2$ (independent of n) for $j > 0$.

3. The normal equations become

$$\begin{bmatrix} 6 & 0 & 0 \\ 0 & 3 & 0 \\ 0 & 0 & 3 \end{bmatrix}\begin{bmatrix} h_0 \\ a_1 \\ a_2 \end{bmatrix} = \begin{bmatrix} 5.6 \\ \sqrt{3} \\ 0.8 \end{bmatrix} \quad \text{with solution} \quad \begin{bmatrix} 0.933 \\ 0.557 \\ 0.267 \end{bmatrix}.$$

The normal equations are diagonal, since 1, $\sin x$, and $\cos x$ are orthogonal on the given point set.

4. (a) $(\varphi_j, \varphi_k) = \dfrac{1}{j+k+1}$, $0 \le j, k \le n$,

a so-called Hankel matrix, where the elements are constant along diagonals perpendicular to the main diagonal.

(b) $$(\varphi_j, \varphi_k) = \begin{cases} \dfrac{2}{j+k+1} & \text{if } j+k \text{ even,} \\ 0 & \text{if } j+k \text{ odd.} \end{cases}$$

The equations for the c_k^*, $k = 0, 2, 4, \ldots$ are separated from the equations for c_k^*, $k = 1, 3, 5, \ldots$. For example, $n = 3$:

$$\begin{bmatrix} 2 & 0 & \frac{2}{3} & 0 \\ 0 & \frac{2}{3} & 0 & \frac{2}{5} \\ \frac{2}{3} & 0 & \frac{2}{5} & 0 \\ 0 & \frac{2}{5} & 0 & \frac{2}{7} \end{bmatrix}\begin{bmatrix} c_0 \\ c_1 \\ c_2 \\ c_3 \end{bmatrix} = \begin{bmatrix} (f,\varphi_0) \\ (f,\varphi_1) \\ (f,\varphi_2) \\ (f,\varphi_3) \end{bmatrix} \Longleftrightarrow \begin{bmatrix} 2 & \frac{2}{3} & 0 & 0 \\ \frac{2}{3} & \frac{2}{5} & 0 & 0 \\ 0 & 0 & \frac{2}{3} & \frac{2}{5} \\ 0 & 0 & \frac{2}{5} & \frac{2}{7} \end{bmatrix}\begin{bmatrix} c_0 \\ c_2 \\ c_1 \\ c_3 \end{bmatrix} = \begin{bmatrix} (f,\varphi_0) \\ (f,\varphi_2) \\ (f,\varphi_1) \\ (f,\varphi_3) \end{bmatrix}.$$

Section 4.3

1. (a) $$\begin{bmatrix} H_0(x) \\ H_1(x) \\ H_2(x) \\ H_3(x) \\ H_4(x) \end{bmatrix} = \begin{bmatrix} 1 & & & & \\ 0 & 2 & & & \\ -2 & 0 & 4 & & \\ 0 & -12 & 0 & 8 & \\ 12 & 0 & -48 & 0 & 16 \end{bmatrix}\begin{bmatrix} 1 \\ x \\ x^2 \\ x^3 \\ x^4 \end{bmatrix}.$$

(b) $$\begin{bmatrix} 1 \\ x \\ x^2 \\ x^3 \\ x^4 \end{bmatrix} = \begin{bmatrix} 1 & & & & \\ 0 & \frac{1}{2} & & & \\ \frac{1}{2} & 0 & \frac{1}{4} & & \\ 0 & \frac{3}{4} & 0 & \frac{1}{8} & \\ \frac{3}{4} & 0 & \frac{3}{4} & 0 & \frac{1}{16} \end{bmatrix}\begin{bmatrix} H_0(x) \\ H_1(x) \\ H_2(x) \\ H_3(x) \\ H_4(x) \end{bmatrix}.$$

2. This recursion formula is clearly correct; it is analogous to Horner's rule (see Chap. 1) except that we multiply by $(\alpha - x_j)$ instead of by x at successive steps.

3, 4.

```
C     MAIN PROGRAM
      DOUBLE PRECISION C(20), D, F, P, RM, X(20), XARG, XI, Y(20)
      COMMON C, X, Y, M
C
      F(XARG)=DEXP(XARG)
C
      READ(5,900) M
      RM = 2.D0 * DFLOAT(M - 1)
      DO 5 I = 1,M
      X(I) = -1.D0 + 4.D0 * DFLOAT(I-1) / RM
      XI = X(I)
    5 Y(I) = F(XI)
      CALL INTERP(X,Y,C,M)
      WRITE(6,901)
      I2 = 4 * (M - 1) + 1
      DO 10 I = 1,I2
      XARG = -1.D0 + DFLOAT(I - 1)/RM
      D = P(XARG) - F(XARG)
   10 WRITE(6,902) XARG, D
      STOP
  900 FORMAT(I5)
  901 FORMAT(4X,'X',5X,'P(X)-F(X)')
  902 FORMAT(F8.3,D12.3)
      END
C
C     CALCULATE THE COEFFICIENTS C1, C2, ... , CM OF THE INTERPOLATION
C     POLYNOMIAL.
      SUBROUTINE INTERP(X,Y,C,N)
      DOUBLE PRECISION C(20), C1, PI, SIGMA, X(20), XI, Y(20)
      C(1) = Y(1)
      C1 = C(1)
      C(2) = (Y(2) - C1) / (X(2) - X(1))
      DO 100 I = 3,N
      PI = 1.D0
      SIGMA = 0.D0
      J2 = I - 1
      XI = X(I)
      DO 50 J = 2,J2
      PI = PI * (XI - X(J-1))
   50 SIGMA = SIGMA + C(J) * PI
  100 C(I) = (Y(I) - C1 - SIGMA) / (PI * (XI - X(I-I)))
      RETURN
      END
C
C     CALCULATE THE VALUE OF THE FUNCTION F FOR XARG USING THE
```

```
C      INTERPOLATION POLYNOMIAL.
       REAL FUNCTION P*8(ARG)
       DOUBLE PRECISION ARG, C(20), SIGMA, X(20), Y(20)
       COMMON C, X, Y, M
       SIGMA = C(M)
       I2 = M - 1
       DO 100 I = 1,I2
   100 SIGMA = (ARG - X(M-I)) * SIGMA + C(M-I)
       P = SIGMA
       RETURN
       END
```

With $M = 20$ the maximum error of the interpolation polynomial was estimated to be less than 10^{-12} for $f(x) = \exp(x)$ on $[-1, 1]$. The maximum error occurred near the end points of the interval. On the machine used, the round-off unit with double precision is $\approx 10^{-16}$.

Section 4.4

1. $(6 - 2x)/17$; 0.0294 (attained at $x = -1$).

2. $\varphi_0(x) = 1$, $\varphi_1(x) = x$, $\varphi_2(x) = x^2 - 2/5$, $\varphi_3(x) = x^3 - 9x/14$.

3. Write down the condition that the nth orthogonal polynomial with leading coefficient 1 should be orthogonal to x^k, $k = 0, 1, \ldots, n-1$, under the scalar product $(f, g) = \int_a^b fgw\,dx$—i.e.,

$$\int_a^b x^k\left(x^n + \sum_{j=0}^{n-1} c_j x^j\right)w(x)\,dx = 0, \quad k = 0, \ldots, n-1.$$

This just gives the system of equations shown in the formulation of the problem.

4. (a) $\varphi_n(x) = T_n(2x - 1)$; (b) $\varphi_n(x) = P_n(2x - 1)$.

5. (a) $U_0(x) = 1$, $U_1(x) = 2x$.

6. Recall that $\varphi_{-1} = 0$, $\varphi_0 = A_0$.

$$\varphi_{n+1}(x) = \alpha_n(x - \beta_n)\varphi_n(x) - \gamma_n\varphi_{n-1}(x).$$

Now, $f(x) = \sum_{k=0}^n c_k\varphi_k(x)$ by Eq. (4.4.18). Solve Eq. (4.4.19) for c_k; then

$$\begin{aligned} f(x) &= \sum_{k=0}^{n}(y_k - \alpha_k(x - \beta_k)y_{k+1} + \gamma_{k+1}y_{k+2})\varphi_k(x) \\ &= y_0\varphi_0 + y_1(\varphi_1(x) - \alpha_0(x - \beta_0)\varphi_0(x)) \\ &\quad + \sum_{k=2}^{n} y_k[\varphi_k(x) - \alpha_{k-1}(x - \beta_{k-1})\varphi_{k-1}(x) + \gamma_{k-1}\varphi_{k-2}(x)] \\ &= y_0A_0. \end{aligned}$$

7. (a)

```
      SUBROUTINE ORPOL(N,M,X,Y,B,C,S,P)
      DIMENSION X(1), Y(1), B(1), C(1), S(1), P(N,N)
      COMMON XJ(1)
```

```
      DO 10 I = 1,N
      DO 10 J = 1,N
   10 P(I,J) = 0.
      P(1,1) = 1.
      XJ(1) = 1.
      PP0 = FLOAT(M)
      PXP = 0.
      YP = 0.
      DO 20 I = 1,M
      PXP = PXP + X(I)
   20 YP = YP + Y(I)
      B(1) = - PXP / PP0
      C(1) = 0.
      S(1) = YP / PP0
      P(2,1) = 1.
      P(2,2) = B(1)
      DO 70 I = 3,N
      IM1 = I - 1
      IM2 = I - 2
      PP = 0.
      PXP = 0.
      YP = 0.
      DO 50 J = 1,M
      PJ = 0.
      XT = 1.
      DO 30 K = 2,IM1
      XT = XT * X(J)
   30 XJ(K) = XT
      DO 40 K = 1,IM1
   40 PJ = PJ + P(IM1,K) * XJ(IM1-K+1)
      PP = PP + PJ * PJ
      PXP = PXP + X(J) * PJ * PJ
   50 YP = YP + Y(J) * PJ
      B(IM1) = - PXP / PP
      C(IM1) = PP / PP0
      S(IM1) = YP / PP
      PP0 = PP
      P(I,1) = 1.
      P(I,2) = B(IM1) + P(IM1,2)
      IF (I .EQ. 3) GOTO 70
      DO 60 J = 3,IM1
   60 P(I,J) = P(IM1,J) + B(IM1) * P(IM1,J-1) - C(IM1) * P(IM2,J-2)
   70 P(I,I) = B(IM1) * P(IM1,IM1) - C(IM1) * P(IM2,IM2)
      PP = 0.
      YP = 0.
      DO 100 J = 1,M
      PJ = 0.
```

```
      XT = 1.
      DO 80 K = 2,N
      XT = XT * X(J)
   80 XJ(K) = XT
      DO 90 K = 1,N
   90 PJ = PJ + P(N,K) * XJ(N-K+1)
      PP = PP + PJ * PJ
  100 YP = YP + Y(J) * PJ
      S(N) = YP / PP
      RETURN
      END
```

(b) The following is a program which uses the above subroutine to find the best second-degree polynomial approximation to x^6 on the net $-2, -1, 1, 2$.

```
      DIMENSION X(10), Y(10), B(10), C(10), S(10), P(10,10)
      LOGICAL*1 T(80)
      COMMON Q(10)
      READ (5,800) T
      READ (5,801) N, M, (X(I),Y(I),I=1,M)
      WRITE (6,900) T, N, M, (X(I),I=1,M)
      N = N + 1
      CALL ORPOL(N,M,X,Y,B,C,S,P)
      CALL PPRINT(N,P)
      NM1 = N - 1
      WRITE (6,901) S(1), (I,S(I+1),I=1,NM1)
      CALL LPOL(N,S,P,Q)
      WRITE (6,902) Q(1), (I,Q(I+1),I=1,NM1)
      STOP
  800 FORMAT (80A1)
  801 FORMAT (2I5/8F10.0))
  900 FORMAT (1H0,'APPROXIMATE',80A1/'0WITH A POLYNOMIAL OF',I3,
     1        ' (=N) DEGREE'/'0ON THE FOLLOWING',I3,' (=M) POINTS:'/
     2       (1X,F7.2))
  901 FORMAT (1H0,'THE FOURIER-COEFFICIENTS:'/1X,'S( 0)=',F7.2/
     1       (1X,'S(',I2,')=',F7.2))
  902 FORMAT (1H0,'THE COEFFICIENTS IN THE POLYNOMIAL APPROXIMATION:'
     1       /1X,'P( 0)=',F7.2/(1X,'P(',I2,')=',F7.2))
      END

      SUBROUTINE LPOL(N,S,P,Q)
      DIMENSION S(1), P(N,N),Q(1)
      DO 10 I = 1,N
   10 Q(I) = 0.
      DO 20 I = 1,N
      K = N - I + 1
      DO 20 J = 1,I
```

```
      L = K + J - 1
   20 Q(K)  =  Q(K)  +  S(L)  *  P(L,J)
      RETURN
      END
      SUBROUTINE PPRINT(N,P)
      DIMENSION P(N,N)
      WRITE (6,900)
      DO 10 I = 1,N
   10 WRITE (6,901) (P(I,J),J=1,I)
      RETURN
  900 FORMAT (1H0,'THE COEFFICIENTS IN THE ORTHOGONAL POLYNOMIALS:')
  901 FORMAT (15F7.2)
      END
```

Result:

```
APPROXIMATE X ** 6
WITH A POLYNOMIAL OF 2. (=N) DEGREE
ON THE FOLLOWING  4 (=M) POINTS:
 -2.00
 -1.00
  1.00
  2.00
THE COEFFICIENTS IN THE ORTHOGONAL POLYNOMIALS:
  1.00
  1.00    0.0
  1.00    0.0    -2.50
THE FOURIER-COEFFICIENTS:
S( 0)=  32.50
S( 1)=    0.0
S( 2)=  21.00
THE COEFFICIENTS IN THE POLYNOMIAL APPROXIMATION:
P( 0)= -20.00
P( 1)=    0.0
P( 2)=  21.00
```

Thus for $x_i = -2, -1, 1, 2$ with $w = 1$, the first three orthogonal polynomials are computed to be

$$1, x, x^2 - 2.50,$$

and $x^6 \approx 32.5 + 0 \cdot x + 21(x^2 - 2.50)$, or $x^6 \approx 21(x^2) + 0 \cdot x - 20$.

8. (a) One shows that $\int_0^\infty x^k L_n(x) e^{-x}\,dx = 0$, $k = 0, 1, 2, \ldots, n-1$, by repeated partial integrations and using induction; see Sec. 4.4.4.

(b) The solution is analogous to (a).

9. Consider the expansion $p_n = \varphi_n/A_n + \sum_{j=0}^{n-1} c_j \varphi_j$ (p_n has leading coefficient 1). Then the problem is to find $c_0, \ldots, c_{n-1}$ such that $||p_n||^2$ is minimized, since any polynomial of degree n with leading coefficient 1 can be expanded in the above way. Now, using orthogonality,

$$\|p_n\|^2 = \frac{1}{A_n^2}\|\varphi_n\|^2 + \sum_{j=0}^{n-1} c_j^2 \|\varphi_j\|^2,$$

so the minimum occurs for $c_j = 0$, $j = 0, \ldots, n-1$—i.e., for $p_n = \varphi_n/A_n$.

Section 4.5

1. For $L(f) = \sum a_i f(x_i)$, take $f(x)$ to be any C^1 function for which $f(x_i) = \text{sign}(a_i)$, and $\|f\|_\infty = 1$.
   For $L(f) = 3f(0) - 2f(1)$, take $f(x) = 1 - 2x^2$.
   For $L(f) = \int_0^1 e^{-x} f(x)\,dx$, take $f(x) = 1$.
   For $L(f) = f(0) + f'(0)$, consider the sequence of functions $f_n = \sin(n\pi x)$, $f'_n = n\pi \cos(n\pi x)$.
   As $n \to \infty$, $L(f_n) \to \infty$, and since $\|f_n\|_\infty = 1$, we have $\|L\| = \infty$.
2. $\frac{15}{8}$.
3. (b) 0.222, 0.25. (Use the minimax property of the Tchebycheff polynomials!)
4. $8^{-1/2} - x/8$; $E_1(f) = 0.0214$ (see Problem 1 in Sec. 4.4 and Problem 5 in Sec. 4.5).
5. $(38 - 13x)/108$; 0.0278.
6. $0.000007 + 0.999479x + 0.249870x^2 + 0.057837x^3 + 0.010764x^4$.
7. $(x - 2)/(2x + 1)$.
8. By Theorem 4.5.1 and Stirling's formula,

$$E_n(f) \le \left(\frac{1}{2}\right)^{n+1} \cdot \frac{2e}{(n+1)!} \approx 2e[2\pi(n+1)]^{-1/2}\left[\frac{e}{2(n+1)}\right]^{n+1} \qquad (*)$$

Application of Bernstein's theorem: The major axis of the ellipse mentioned in the theorem is equal to $\frac{1}{2}(R + 1/R)$. Hence

$$M \le \exp\left(\tfrac{1}{2}(R + 1/R)\right),$$

$$E_n(f) \le 2M \cdot R^{-n} \cdot (R^2 - 1)^{-1/2} = 2\exp(R/2) \cdot R^{-(n+1)}(1 + O(R^{-1})). \qquad (**)$$

For given n, the last expression is minimized for $R = n + 1$. Hence

$$E_n(f) \le 2\left[\frac{e}{2(n+1)}\right]^{n+1} \cdot (1 + O(n^{-1})).$$

The bounds in Eqs. (*) and (**) are rather similar, the former being better by the factor $e \cdot [2\pi(n+1)]^{-1/2}$.

The optimal application of Jackson's theorem is with $k = n$. Then

$$E_n(f) \le e \cdot \frac{(\pi/2)^n}{(n+1)!}.$$

This result is inferior to the two others, approximately by the factor π^n.

For $n = 13$: Theorem 4.5.1 $\Rightarrow E_{13}(f) \lesssim 0.4 \cdot 10^{-14}$,
Bernstein's theorem $\Rightarrow E_{13}(f) \lesssim 1.3 \cdot 10^{-14}$,
Jackson's theorem $\Rightarrow E_{13}(f) \lesssim 10^{-8}$.

9. Put $L(f) = \sum_{k=0}^{n+1}{}' (-1)^k a_k f\{\cos[(k\pi)/(n+1)]\}$. It is evident that $\|L\| = \sum_{k=0}^{n+1}{}' |a_k| = n+1$. By Theorem 4.5.2 it suffices to show that $L(f) = 0$ for $f(x) = T_\nu(x)$, $\nu = 0, 1, \ldots, n$, because then $L(p) = 0$ for all nth degree polynomials. But

$$L(T_\nu) = \sum_{k=0}^{n+1}{}' (-1)^k a_k \cos\left(\frac{\nu k\pi}{n+1}\right).$$

By a calculation similar to the proof of Theorem 4.2.4 (discrete case), it follows that $L(T_\nu) = 0$, for $\nu = 0, 1, \ldots, n$.

Section 4.6

1. Equation (4.6.4) is easily verified by differentiation of Eq. (4.6.2) and substitution. A further differentiation gives (for convenience put $\alpha_{i-1} = k_{i-1} - d_i$, $\tilde{\alpha}_{i-1} = k_i - d_i$)

$$q_i''(x) = \frac{2}{h_i}\left[(3\alpha_{i-1} + 3\tilde{\alpha}_{i-1})t - 2\alpha_{i-1} - \tilde{\alpha}_{i-1}\right].$$

Then

For $t = 1$: $$q_i''(x_{i+1}) = \frac{2}{h_i}[\alpha_{i-1} + 2\tilde{\alpha}_{i-1}],$$

For $t = 0$: $$q_{i+1}''(x_{i+1}) = \frac{2}{h_{i+1}}[-2\alpha_i - \tilde{\alpha}_i].$$

The condition that the second derivative of the spline be continuous means that these last two expressions must be equal. Thus we get the conditions

$$\frac{2}{h_i}(k_{i-1} - d_i + 2k_i - 2d_i) = \frac{2}{h_{i+1}}(-2k_i + 2d_{i+1} - k_{i+1} + d_{i+1}).$$

Multiplying both sides by $\frac{1}{2}h_i h_{i+1}$ we get Eq. (4.6.3).

2. (a) $f(x) = x_+^3$ is a cubic polynomial in each of the intervals $(-\infty, 0)$ and $(0, \infty)$. Also f, f' and f'' are continuous everywhere. It follows that x_+^3 is a spline function for any set of nodes containing 0. We also have $f''(x) = 6$, $x \geq 0$ and $f''(x) = 0$, $x < 0$.

(b) Let $s(x)$ be a cubic spline which equals $s_i(x)$ on the interval $[x_{i-1}, x_i)$. Then we put $p(x) = s_1(x)$. Denote the jump in $s'''(x)$ at the node x_i by Δ_i,

$$\Delta_i = s_{i+1}''' - s_i''', \quad i = 1, 2, \ldots, m-1.$$

If we take $c_i = \frac{1}{6}\Delta_i$, then we have the wanted representation. By construction $p(x)$ and c_i are obviously uniquely determined.

3. Integral = trapezoidal sum + $(k_0 - k_m)h^2/12$.

4. Consider the spline with nodes $x_0, x_{1/2}, x_1, x_2, \ldots, x_{m-1}, x_{m-1/2}, x_m$. Differentiate the expression for q_i'' given in the answer to Problem 1.

$$q_i'''(x) = \frac{6}{h_i^2}(k_{i-1} + k_i - 2d_i).$$

Using this formula on the intervals $[x_0, x_{1/2}]$, $[x_{1/2}, x_1]$ both of length $h_1/2$ and requiring $q_{i/2}'''$ to be continuous gives

$$k_0 + k_{1/2} - 2\left(\frac{y_{1/2} - y_0}{h_1/2}\right) = k_{1/2} + k_1 - 2\left(\frac{y_1 - y_{1/2}}{h_1/2}\right)$$

or

$$k_1 - k_0 = \frac{4}{h_1}(y_1 - 2y_{1/2} + y_0).$$

Note that since $q'''_{1/2}$ is continuous this spline is represented by the same cubic polynomial in both $[x_0, x_{1/2})$ and $[x_{1/2}, x_1)$ and thus degenerates into a spline where the node $x_{1/2}$ is absent! The equation

$$k_m - k_{m-1} = \frac{4}{h_m}(y_{m-1} - 2y_{m-1/2} + y_m)$$

is obtained similarly, and the node $x_{m-1/2}$ can also be deleted.

5. The matrix is cyclic (circulant). All diagonal elements are equal to 4, all superdiagonal and subdiagonal elements as well as the corner elements a_{m1} and a_{1m} are equal to 1. The remaining elements are equal to zero. The elements of the right hand side are $d_i + d_{i+1}$, $i = 1, 2, \dots, m$, $(d_{m+1} = d_1)$.

6. (a) If we put

$$B_i(x) = \sum_{j=i-2}^{i+2} c_j(x - x_j)_+^3$$

then we immediately have

$$B_i'(x_{i-2}) = B_i''(x_{i-2}) = 0, \qquad B_i(x) = 0, \quad x \leq x_{i-2}.$$

By symmetry we should have $B'(x_i) = 0$, and so we get

$$B_i(x_i) = c_{i-2}(2h)^3 + c_{i-1}h^3 = 1,$$
$$B_i'(x_i) = 3c_{i-2}(2h)^2 + 3c_{i-1}h^2 = 0.$$

With $\gamma_j = h^3 c_j$ this gives $\gamma_{i-2} = \frac{1}{4}$, $\gamma_{i-1} = -1$. We find that $B_i(x_{i-1}) = \frac{1}{4}$ and

$$B_i'''(x) = 6h^{-3}(\gamma_{i-2} + \gamma_{i-1}) = 6h^{-3} \cdot \tfrac{3}{4}, \quad x \in [x_{i-1}, x_i).$$

Since B_i''' is antisymmetric at x_i we get $\gamma_i = \frac{3}{2}$. Also by symmetry it follows that $\gamma_{i+1} = -\gamma_{i-1} = 1$ and if we put $\gamma_{i+2} = -\gamma_{i-2} = -\frac{1}{4}$, then $B_i(x) = 0$, $x \geq x_{i+2}$.

To show $B_i(x) > 0$ for $x \in (x_{i-2}, x_{i+2})$ it suffices, by symmetry, to consider $x \in (x_{i-2}, x_i)$. Differentiate $B_i(x)$ and substitute the values for c_{i-1} and c_{i-2} computed above ($c_j = \gamma_j/h^3$). Then

$$\frac{4h^3}{3} \cdot B_i'(x) = \begin{cases} (x - x_{i-2})^2 & x \in (x_{i-2}, x_{i-1}] \\ (x - x_{i-2})^2 - 4(x - x_{i-1})^2 & x \in [x_{i-1}, x_i). \end{cases}$$

Hence $B_i'(x) > 0$ for $x \in (x_{i-2}, x_i)$, since

$$(x - x_{i-2})^2 - 4(x - x_{i-1})^2 = (3x - x_{i-2} - x_{i-1})(x_{i-1} - x + h) > 0$$
$$\text{for } x \in (x_{i-1}, x_i).$$

That $B_i(x) > 0$ for $x \in (x_{i-2}, x_i)$ follows, since $B_i(x_{i-2}) = 0$ and $B_i'(x) > 0$ on this interval.

(We remark that this construction can also be performed for nonequidistant nodes.)

(b) Any cubic spline $s(x)$ is uniquely determined on $[a, b]$ by the values $s_i = s(x_i)$ and $s'(x_0)$, $s'(x_m)$. (This follows from Theorem 4.6.1.) Since at the point x_i only $B_{i-1}(x)$, $B_i(x)$ and $B_{i+1}(x)$ are different from zero, we get the equations

$$\tfrac{1}{4}a_{i-1} + a_i + \tfrac{1}{4}a_{i+1} = s_i, \quad i = 0, 1, \ldots, m.$$

Also at x_i only $B_{i-1}(x)$ and $B_{i+1}(x)$ have a nonzero derivative, and thus

$$a_{i-1} + a_{i+1} = \tfrac{4}{3}hs'(x_i), \quad i = 0, m.$$

From these two equations a_{-1} and a_{m+1} can be eliminated; there remains a tridiagonal system of $m + 1$ linear equations for $m + 1$ unknowns. This system is strictly diagonally dominant and therefore nonsingular by Theorem 5.8.1. Thus all a_i are uniquely determined.

(c) We form base functions $B_i(x)$ and $B_j(y)$, $i = -1, 0, \ldots, m + 1$, $j = -1, 0, \ldots, n + 1$ as in (a) and use the representation

$$s(x, y) = \sum_{i=-1}^{m+1} \sum_{j=-1}^{n+1} c_{ij}\psi_{ij}(x, y), \quad \psi_{ij}(x, y) = B_i(x)B_j(y).$$

For each nodal point (x_r, y_s) only the nine base functions $\psi_{ij}(x, y)$ with $r - 1 \leq i \leq r + 1$, $s - 1 \leq j \leq s + 1$ are different from zero. For c_{ij} we therefore get a linear system of equations where each equation contains only nine unknowns. This can be solved, for example, with the methods for band matrices (Sec. 5.4.2).

CHAPTER 5

Section 5.2

1. (a) Characteristic polynomial: $\det(A - \lambda I) \equiv \prod_{i=1}^{n} (\lambda_i - \lambda)$. Put $\lambda = 0$.
   (b) The only eigenvalue is zero. $Ax = 0 \cdot x \Longleftrightarrow x_2 = 0$. Hence all eigenvectors are of the form $x^T = [x_1, 0]$—i.e., there is only one linearly independent eigenvector.
   (c) Yes; e.g., 0 or I or any Hermitian matrix with multiple eigenvalues.

2. (a) $Y = \begin{bmatrix} X & X \\ X & -X \end{bmatrix}$.
   (b) Show by induction that $A^k X = X\Lambda^k$, hence $P(A)X = XP(\Lambda)$ for any polynomial. Then

$$BY = Y\begin{bmatrix} P(\Lambda) + Q(\Lambda) & 0 \\ 0 & P(\Lambda) - Q(\Lambda) \end{bmatrix}.$$

   The eigenvalues of B are $P(\lambda_i) \pm Q(\lambda_i)$, where λ_i, $i = 1, 2, \ldots, n$ are the eigenvalues of A.

3. The premise implies that the columns of A are linearly independent; hence A is not singular.

4. (a) Set up Ax and λx and compare. There is only one eigenvector to each λ in the multiple-root case.
   (b) Characteristic equation $\lambda^{20} = -a_{20}$. For $a_{20} = 0$, $\lambda = 0$. For $a_{20} = -2^{-40}$, $\lambda = 0.25 \exp(2\pi ip/20)$, $p = 0, 1, 2, \ldots, 19$.

(c) It can be dangerous when an eigenvalue has a high multiplicity. Let the characteristic equation be $(\lambda + 1)^{20} = \epsilon$—i.e.,

$$a_j = \binom{20}{j}, \quad (j < 20), \quad a_{20} = 1 - \epsilon.$$

Compare $\epsilon = 0$ and $\epsilon = 2^{-40}$.

5. (a) *Hint:* Show first that the statements are valid for $A = \Sigma$, then use the relations $U^H U = I$, $V^H V = I$, $(ABC)^H = C^H B^H A^H$.

(b) Note that $A^I A = I$, $AA^I = A(A^H A)^{-1} A^H$; hence $(AA^I)^H = A(A^H A)^{-1} A^H$.

(c) In this case $\Sigma = \begin{bmatrix} D \\ 0 \end{bmatrix}$, $\Sigma^I = [D^{-1} \quad 0]$, $A^H A = VD^2 V^H$, $(A^H A)^{-1} A^H = VD^{-2}[D \quad 0]U^H = A^I$.

6. (a) By Eq. (5.2.4), $x^I = x^H/(x^H x)$.

(b) $B = A^H$. Apply Eqs. (5.2.5) and (5.2.4). Result: $B^I = B^H \cdot (BB^H)^{-1}$.

7. (a) Proceed as in Problem 5.

(b) The rank of a product cannot exceed the rank of any factor. $A^I A = I \Rightarrow$ rank $(A^I A) = n \Rightarrow r \geq n$. Since $r \leq \min(m, n)$, $r = n$. The converse was proved in Problem 5(b). The second relation is obtained by transposition of the first.

8. (a) By Eq. (5.2.4), $A^I = A^{-1}(A^H)^{-1} A^H = A^{-1}$.

(b) $A = [1, 1]^T$, $G = [1, 0]$, $GA = 1$, $AG = \begin{bmatrix} 1 & 0 \\ 1 & 0 \end{bmatrix}$, which is not Hermitian.

9. (a) $A = [1, 1]$, $B = [3, 4]^T$, $(A \cdot B)^I = 1/7$. By Problem 6(a) and (b), $A^I = [1, 1]^T/2$, $B^I = [3, 4]/25$, $B^I A^I = 7/50$.

(b) By Problem 7(b), $A^I A = BB^I = I$. Verify that Eq. (5.2.3) is valid for $A := AB$, $G := B^I A^I$, $ABB^I A^I AB = A \cdot I \cdot I \cdot B$; hence the first relation is true. The others are verified similarly—e.g., $AG = ABB^I A^I = AA^I$, which is Hermitian.

10. (a) See Eq. (5.2.3).

(b) $P^2 = A(A^I A A^I) = AA^I = P$.

(c) $\exists\, z \in R_n$ such that $x = Az$. $Px = AA^I Az = Az = x$.

(d) By (c), $x = Px$. $x^H(y - Py) = x^H P^H (I - P)y = x^H P(I - P)y = 0$.

Section 5.3

1. (a) $x = -(ab^{n-2}, ab^{n-3}, \ldots, ab, a, -1)^T$, $b = 1 - a$.

(b) Assume that $|x_i| \leq 2^{n-i-1}$. Since $|x_n| = 1$, $|x_{n-1}| \leq 1$, this is true for $i = n - 1$. Also,

$$|x_{i-1}| \leq \sum_{k=i}^{n} |u_{ik} x_k| \leq \sum_{k=i}^{n} |x_k| \leq 1 + 1 + 2 + \ldots + 2^{n-i-1} = 2^{n-i},$$

and the result follows by induction. The matrix $U_n(-1)$ from (a) shows that the bound can be attained.

2. (a) The equivalent triangular system becomes

$$\begin{pmatrix} 1 & 2 & 3 & 4 \\ 0 & 2 & 6 & 12 \\ 0 & 0 & 6 & 24 \\ 0 & 0 & 0 & 24 \end{pmatrix} \boldsymbol{x} = \begin{pmatrix} 2 \\ 8 \\ 18 \\ 24 \end{pmatrix},$$

and, by back substitution,

$$\boldsymbol{x} = (-1, 1, -1, 1)^T.$$

(b)

$$L = \begin{pmatrix} 1 & & & \\ 1 & 1 & & \\ 1 & 3 & 1 & \\ 1 & 7 & 6 & 1 \end{pmatrix}, \quad U = \begin{pmatrix} 1 & 2 & 3 & 4 \\ & 2 & 6 & 12 \\ & & 6 & 24 \\ & & & 24 \end{pmatrix},$$

and $\det(A) = 1 \cdot 2 \cdot 6 \cdot 24 = 288$.

3. (a)

```
Row   Algol

 1    for k := 1 step 1 until n − 1 do
 2    for i := k + 1 step 1 until n do
 3    begin m := a[i, k] := a[i, k]/a[k, k];
 4      for j := k + 1 step 1 until n do
 5      a[i, j] := a[i, j] − m × a[k, j]
 6    end:
 7    for i := 2 step 1 until n do
 8    for k := 1 step 1 until i − 1 do
 9    b[i] := b[i] − a[i, k] × b[k];
10    for i := n step − 1 until 1 do
11    begin for k := n step −1 until i + 1 do
12      b[i] := b[i] −a[i, k] × b[k];
13      b[i] := b[i]/a[i, i]
14    end;
```

```
Row        Fortran

 1          NM1 = N-1
 2          DO 1 K = 1,NM1
 3          KP1 = K+1
 4          DO 1 I = KP1,N
 5          T = A(I,K)/A(K,K)
 6          A(I,K) = T
 7          DO 1 J = KP1,N
 8      1   A(I,J) = A(I,J) - T*A(K,J)
 9          DO 2 I = 2,N
10          IM1 = I-1
11          DO 2 K = 1,IM1
12      2   B(I) = B(I) - A(I,K)*B(K)
13          B(N) = B(N)/A(N,N)
```

```
14        DO 4  IB = 2,N
15        I = N-IB+1
16        IP1 = I+1
17        DO 3 K = IP1,N
18     3  B(I) = B(I) - A(I,K)*B(K)
19     4  B(I) = B(I)/A(I,I)
```

Comments (Fortran row numbers in parentheses): The LU-decomposition of A is performed in rows 1–6 (1–8). Here the lower-triangular part of A is used to store the multipliers. The solution of $Ly = b$ is performed in rows 7–9 (9–12) and y is overwritten b. Note that the diagonal elements in L equal 1. Finally, $Ux = y$ is solved in rows 10–14 (13–19) and the solution x is stored in b.

(b) See Forsythe and Moler [67].

4.

$$\bar{L}\bar{L}^{\mathrm{T}} - \bar{A} = 10^{-6}\begin{pmatrix} 0 & 0 & 0 & 0 \\ 0 & 0.37 & 1.14 & -0.43 \\ 0 & 1.14 & -0.43 & 0.25 \\ 0 & -0.65 & 0.25 & -0.13 \end{pmatrix}.$$

The difference between A and $\bar{A}$ is much larger!

5.

$$A^{-1} = \frac{1}{2}\begin{pmatrix} -1 & 0 & 1 \\ -10 & 4 & 4 \\ 7 & -2 & -3 \end{pmatrix}.$$

6. (a) The inverse exists if $v^{\mathrm{T}}A^{-1}u \neq 1$.

(b) If the ith row is changed, then take $u = e_i$. Then

$$A^{-1}e_i = x_i = i\text{th column of } A^{-1}, \qquad z^{\mathrm{T}} = v^{\mathrm{T}}A^{-1} \quad (n^2 \text{ mult.}).$$

and we get $\alpha = 1/(1 - z_i)$, $\alpha A^{-1}uv^{\mathrm{T}}A^{-1} = \alpha x_i z^{\mathrm{T}}$ (n^2 mult.).

(c)

$$B^{-1} = \begin{pmatrix} 12.76 & 13.72 & 5.88 & 3.92 \\ 7.46 & -1.63 & 12.73 & 8.82 \\ 5.88 & 6.86 & 2.94 & 1.96 \\ 8.34 & 4.23 & 15.67 & 10.78 \end{pmatrix}.$$

(d) $A^{-1} + A^{-1}U(I - M)^{-1}V^{\mathrm{T}}A^{-1}$, $M = V^{\mathrm{T}}A^{-1}U$. Note that M is an $r \times r$ matrix.

7. (b) The computations can be arranged as follows:

1. A^{-1}, p^3 mult,; 2. $U = A^{-1}B$, $W = CA^{-1}$, $2p^2q$ mult.; 3. $(D - WB)^{-1}$, $pq^2 + q^3$ mult.; 4. $Y = -UV$, $Z = -VW$, $2pq^2$ mult.; 5. $X = A^{-1} - UZ$, p^3 mult. $\Rightarrow 2p^3 + 2p^2q + 3pq^2 + q^3$ mult.

Note that if $p = q = n$, then the total number is $8n^3 = (2n)^3$ mult.

Section 5.4

1. (a) By the symmetric version, Eq. (5.4.2), of Gaussian elimination we get successively the reduced matrices (only the upper-triangular part needs to be transformed!)

$$\begin{pmatrix} 10 & 7 & 8 & 7 \\ & 5 & 6 & 5 \\ & & 10 & 9 \\ & & & 10 \end{pmatrix} \begin{pmatrix} 0.1 & 0.4 & 0.1 \\ & 3.6 & 3.4 \\ & & 5.1 \end{pmatrix} \begin{pmatrix} 2 & 3 \\ & 5 \end{pmatrix} (0.5).$$

The pivotal elements 10, 0.1, 2, and 0.5 are all positive, which implies that A is positive-definite.

(b) We have $\boldsymbol{R} = \boldsymbol{D}^{-1/2}\boldsymbol{U}$, where, from (a),

$$\boldsymbol{R} = \begin{pmatrix} 10 & 7 & 8 & 7 \\ & 0.1 & 0.4 & 0.1 \\ & & 2 & 3 \\ & & & 0.5 \end{pmatrix}, \qquad \boldsymbol{D}^{-1/2} = \operatorname{diag}(10^{-1/2}, 10^{1/2}, 2^{-1/2}, 2^{1/2}).$$

2. The following rows in the programs should be changed:

Row

```
 3   begin m := a[k, i]/a[k, k];
 4      for j := i step 1 until n do
 9   b[i] := b[i] - a[k, i] × b[k]/a[k, k];
```

Row

```
5, 6      T = A(K,I)/A(K,K)
   7      DO 1  J = I,N
  12    2 B(I)=B(I) - A(K,I)*B(K)/A(K,K)
```

Comments (Fortran row numbers in parentheses): For computing the multipliers, row 3 (row 5, 6), the symmetry $a_{ij}^{(k)} = a_{ji}^{(k)}$ is used. The multipliers are not stored, and the lower half of A is never used. Only elements on or above the principal diagonal are transformed, row 4 (row 7). When solving the lower-triangular system we use the relation $m_{ik} = r_{ki}/r_{kk}$, row 9 (row 12).

3. (a) $\delta_k = \det(A_k) = \alpha_1 \cdot \alpha_2 \cdots \alpha_k$. If we put $\delta_0 = 1$, then from

$$\alpha_k = a_k - \beta_k c_{k-1} = a_k - b_k c_{k-1}/\alpha_{k-1}$$

it follows that

$$\delta_k = a_k \delta_{k-1} - b_k c_{k-1} \delta_{k-2}, \quad k = 2, 3, \ldots, n.$$

(b) From the recursion formula in (a) we get

$$\delta_0 = 1, \quad \delta_2 = 2, \quad \delta_k = 2\delta_{k-1} - (1.01)^2 \delta_{k-2}, \quad k \geq 2.$$

This difference equation has constant coefficients, and hence the solution can be written (see Sec. 8.5.1)

$$\delta_k = c_1 \cdot u_1^k + c_2 \cdot u_2^k,$$

where u_1 and u_2 are the roots of the equation

$$u^2 - 2u = -(1.01)^2, \quad u_{1,2} = 1 \pm i(0.0201)^{1/2}.$$

Determining c_1 and c_2 from δ_0 and δ_1 we get

$$\delta_k = (u_1^{k+1} - u_2^{k+1})/(u_1 - u_2) = 1.01^k \sin (k+1)\varphi/\sin \varphi,$$

where $\tan \varphi = (0.0201)^{1/2} = 0.14177$, and $\varphi = 8.069°$. Since $22\varphi < 180° < 23\varphi$, it follows that the largest n equals 21.

4.

$$(\beta_2, \ldots, \beta_5) = (1, 1, \tfrac{1}{2}, \tfrac{2}{7}),$$
$$(\alpha_1, \ldots, \alpha_5) = (1, 1, 2, \tfrac{7}{2}, \tfrac{33}{7}).$$

5. (b) We note that all matrices appearing are square and $m \times m$. If the inverses $\boldsymbol{\alpha}_k^{-1}$ are computed, we need $(n + 2(n-1))m^3 = (3n-2)m^3$ mult. for the decomposition, $((n-1) + 2(n-1) + 1)m^2 = (3n-2)m^2$ mult. for the forward and backward substitution. If instead the LU-decomposition of $\boldsymbol{\alpha}_k$ are used, we save $\frac{2}{3}nm^3$ multiplications.

6. Concerning Eq. (5.4.7), apart from column scaling factors, $\boldsymbol{L}$ is equal to the transpose of the upper-triangular matrix obtained by Gaussian elimination. During the elimination, no nonzero elements will be created in a column above the uppermost element of that column in the original matrix.

Equation (5.4.6) is fairly obvious.

Section 5.5

1. (a) $\boldsymbol{r}_1 = \begin{pmatrix} 0.001243 \\ 0.001572 \end{pmatrix}, \quad \boldsymbol{r}_2 = \begin{pmatrix} 0.000001 \\ 0 \end{pmatrix}.$

(b) $\boldsymbol{x} - \boldsymbol{x}_1 = \begin{pmatrix} 0.001 \\ 0.001 \end{pmatrix}, \quad \boldsymbol{x} - \boldsymbol{x}_2 = \begin{pmatrix} 0.659 \\ -0.913 \end{pmatrix}.$

2. (a) Put $s_i = \sum_j |a_{ij}|$, and suppose that s_i has a maximum when $i = I$.

$$\boldsymbol{y} = \boldsymbol{A}\boldsymbol{x} \Rightarrow y_i = \sum_j a_{ij}x_j \Rightarrow |y_i| \le s_i \cdot \|\boldsymbol{x}\|_\infty \Rightarrow \|\boldsymbol{A}\boldsymbol{x}\|_\infty \le s_I \cdot \|\boldsymbol{x}\|_\infty.$$

Hence $\|\boldsymbol{A}\|_\infty \le s_I$. On the other hand, choose the vector $\boldsymbol{x}$ such that $x_j = \operatorname{sgn}(a_{Ij})$. Then

$$\|\boldsymbol{A}\|_\infty \ge \|\boldsymbol{A}\boldsymbol{x}\| \ge |y_I| = \sum_j |a_{Ij}| = s_I.$$

Hence $\|\boldsymbol{A}\|_\infty = s_I$.

(b) $\|(\boldsymbol{A} + \boldsymbol{B})\boldsymbol{x}\| = \|\boldsymbol{A}\boldsymbol{x} + \boldsymbol{B}\boldsymbol{x}\| \le \|\boldsymbol{A}\boldsymbol{x}\| + \|\boldsymbol{B}\boldsymbol{x}\| \le (\|\boldsymbol{A}\| + \|\boldsymbol{B}\|) \cdot \|\boldsymbol{x}\|,$

$\|\boldsymbol{A} + \boldsymbol{B}\| = \sup \|(\boldsymbol{A} + \boldsymbol{B})\boldsymbol{x}\|/\|\boldsymbol{x}\| \le \|\boldsymbol{A}\| + \|\boldsymbol{B}\|,$

$\|\boldsymbol{A} \cdot \boldsymbol{B} \cdot \boldsymbol{x}\| \le \|\boldsymbol{A}\| \cdot \|\boldsymbol{B} \cdot \boldsymbol{x}\| \le \|\boldsymbol{A}\| \cdot \|\boldsymbol{B}\| \cdot \|\boldsymbol{x}\|,$

$\|\boldsymbol{A} \cdot \boldsymbol{B}\| = \sup \|\boldsymbol{A} \cdot \boldsymbol{B} \cdot \boldsymbol{x}\|/\|\boldsymbol{x}\| \le \|\boldsymbol{A}\| \cdot \|\boldsymbol{B}\|,$

3. (a) If $\boldsymbol{U}$ is unitary, then $\|\boldsymbol{U}\boldsymbol{y}\|_2 = \|\boldsymbol{y}\|_2$, since

$$\|\boldsymbol{U}\boldsymbol{y}\|_2^2 = \boldsymbol{y}^H\boldsymbol{U}^H\boldsymbol{U}\boldsymbol{y} = \boldsymbol{y}^H\boldsymbol{y} = \|\boldsymbol{y}\|_2^2.$$

Similarly, $\|\boldsymbol{V}\boldsymbol{x}\|_2 = \|\boldsymbol{x}\|_2$.

$$\|UAV\|_2 = \sup \|UAVx\|_2/\|x\|_2 = \sup \|AVx\|_2/\|Vx\|_2$$
$$= \sup \|Ay\|_2/\|y\|_2 = \|A\|_2.$$

(b) By Eq. (5.2.1), $\exists$ unitary U, V such that $UAV = D$ = diagonal matrix with nonnegative elements.

$$\|Dx\|_2^2 = \sum_{i=1}^{r} d_i^2 x_i^2 \leq \max(d_j^2) \cdot \sum_{i=1}^{n} x_i^2 = \max(d_j^2)\|x\|_2^2,$$

with equality if $x_i = 0$ when $d_i^2 < \max_j d_j^2$. Hence $\|D\|_2 = \max |d_j|$, and the result now follows from (a).

4. (a) $x \neq 0 \Longleftrightarrow Tx \neq 0 \Longrightarrow N(x) = \|Tx\| > 0,$

$x = 0 \Longrightarrow N(x) = 0,$

$N(\alpha x) = \|T(\alpha x)\| = |\alpha|\,\|Tx\| = |\alpha| \cdot N(x),$

$N(x + y) = \|Tx + Ty\| \leq \|Tx\| + \|Ty\| = N(x) + N(y).$

(b) $\sup_x \dfrac{N(Ax)}{N(x)} = \sup_x \dfrac{\|TAx\|}{\|Tx\|} = \sup_y \dfrac{\|TAT^{-1}y\|}{\|y\|} = \|TAT^{-1}\|.$

(c) Put $T = \operatorname{diag}(k_i)$. By Problem 4(b),

$$\sup N(Ax)/N(x) = \|TAT^{-1}\|_\infty = \max_i \sum_j |a_{ij}\, k_i/k_j|.$$

(d) It is possible to factorize G, $G = T^H T$—e.g., by Choleski's factorization. Hence $(x^H Gx)^{1/2} = \|Tx\|_2$. The result then follows from Problem 4(a).

(e) Since $x^T Gy$ is a real number,

$$x^T Gy = (x^T Gy)^T = y^T G^T x = y^T Gx.$$

Hence there is *commutativity*. The *linearity* condition is obvious, and the *positivity* condition follows from the positive-definiteness.

(f) Suppose that λ is an eigenvalue, $|\lambda| = \rho(A)$, and that v is a corresponding eigenvector.

$$Av = \lambda v \Longrightarrow \|Av\|/\|v\| = |\lambda| = \rho(A) \Longrightarrow \|A\| \geq \rho(A).$$

Suppose that $TAT^{-1} = D$ = diagonal. Then (compare Problem 3(b)) $\|D\|_2 = \rho(D) = \rho(A)$. Actually, for any p, $1 \leq p \leq \infty$, $\|D\|_p = \rho(D) = \rho(A)$. The result now follows, since, by Problem 4(b), $\|D\|_p$ is equal to a matrix-bound norm of A.

5. (a)

$$A^{-1} = \begin{pmatrix} 25 & -41 & 10 & -6 \\ -41 & 68 & -17 & 10 \\ 10 & -17 & 5 & -3 \\ -6 & 10 & -3 & 2 \end{pmatrix}, \qquad x = \begin{pmatrix} 1 \\ -1 \\ 1 \\ -1 \end{pmatrix}.$$

(b) $\delta x = A^{-1}\delta b$, $\|\delta x\|_\infty \leq \|A^{-1}\|_\infty \|\delta b\|_\infty = 136 \cdot 10^{-2} = 1.36.$

(c) $\kappa_\infty(A) = \|A\|_\infty \|A^{-1}\|_\infty = 33 \cdot 136 = 4{,}488.$

$\|\delta x\|_\infty/\|x\|_\infty = 1.36/1 = 1.36$, $\|\delta b\|_\infty/\|b\|_\infty = 0.01/4 = 0.0025,$

$1.36/0.0025 = 544.$

6. (a) If $(A + \delta A)$ is singular, then there is some $x \neq 0$ such that

$$(A + \delta A)x = 0 \Longrightarrow x = -A^{-1}\delta A x.$$

and thus

$$\|x\| \leq \|A^{-1}\| \|\delta A\| \|x\| \Longrightarrow \kappa(A) \geq \|A\|/\|\delta A\|.$$

(b) We have

$$(A + \delta A)x = AA^{-1}y - yx^Tx/x^Tx = 0,$$

so that $(A + \delta A)$ is singular. By definition,

$$\|\delta A\|_2 = \max_z \frac{\|yx^Tz\|_2}{x^Tx \cdot \|z\|_2} = \frac{\|y\|_2}{x^Tx} \max_z \frac{|x^Tz|}{\|z\|_2} = \frac{\|y\|_2}{\|x\|_2},$$

and from the property of y now follows

$$\|\delta A\|_2 = \frac{\|A^{-1}y\|_2}{\|x\|_2 \|A^{-1}\|_2} = \frac{1}{\|A^{-1}\|_2} = \frac{\|A\|_2}{\kappa(A)}.$$

(c) If we take

$$\delta A = \begin{pmatrix} 0 & 0 & 0 \\ 0 & -\epsilon & -\epsilon \\ 0 & -\epsilon & -\epsilon \end{pmatrix},$$

then $(A + \delta A)$ is singular, since $(A + \delta A)x = 0$ for $x = (0, 1, 1)^T$. Thus

$\kappa(A) \geq 3/(2|\epsilon|) = 1.5|\epsilon|^{-1}$ (true value $\kappa(A) = 2(1 + |\epsilon|^{-1})$).

7. (b) *Hint:* Use the inequality

$$\|A^{-1}\| \leq \|B^{-1}\| + \|A^{-1} - B^{-1}\| \leq \|B^{-1}\| + \delta\|A^{-1}\|.$$

(c) *Hint:* Deduce, from the equality in (a),

$$A^{-1}b - B^{-1}b = A^{-1}(B - A)B^{-1}b,$$

and the corresponding equality with A and B interchanged.

8. One obtains by Gaussian elimination the reduced matrices

$$\begin{pmatrix} 1 & 0 & 0 & 2 \\ -1 & 1 & 0 & 2 \\ -1 & -1 & 1 & 2 \\ -1 & -1 & -1 & 2 \end{pmatrix} \begin{pmatrix} 1 & 0 & 4 \\ -1 & 1 & 4 \\ -1 & -1 & 4 \end{pmatrix} \begin{pmatrix} 1 & 8 \\ -1 & 8 \end{pmatrix} (16).$$

9. (a) Without pivoting we obtain the triangular system

$$\begin{pmatrix} 10^{-5} & 10^{-5} & 1 \\ 0 & -2 \cdot 10^{-5} & 0 \\ 0 & 0 & -10^5 \end{pmatrix} \begin{pmatrix} x_1 \\ x_2 \\ x_3 \end{pmatrix} = \begin{pmatrix} 2 \cdot 10^{-5} \\ -4 \cdot 10^{-5} \\ -1 \end{pmatrix},$$

from which we get

$$x_3 = 10^{-5}, \quad x_2 = 2, \quad x_1 = -1,$$

which agrees with the exact solution to more than three digits.

(b) With complete pivoting we get, after one step in the elimination,

$$\begin{pmatrix} 2 & 1 & 1 \\ 0 & -\frac{1}{2} & -\frac{1}{2} \\ 0 & -\frac{1}{2} & -\frac{1}{2} \end{pmatrix} \begin{pmatrix} x_1 \\ x_2 \\ x_3 \end{pmatrix} = \begin{pmatrix} 2 \\ -2 \\ 1 \end{pmatrix},$$

which is a singular system. This shows that complete pivoting can fail if the unknowns are not suitably scaled.

(c) After scaling,

$$\begin{pmatrix} 1 & 1 & 1 \\ 1 & -1 & 1 \\ 1 & 1 & 2\cdot 10^{-5} \end{pmatrix} \begin{pmatrix} x_1 \\ x_2 \\ x_3 \end{pmatrix} = \begin{pmatrix} 2 \\ -2 \\ 1 \end{pmatrix}.$$

Complete pivoting will here give the same good result as in (a).

10. (a) $\|x^{(2)}\|_\infty = 36, \quad \|\delta x^{(1)}\|_\infty = 0.0329$, and hence

$$\kappa_\infty(A) \approx \frac{1}{3\cdot\frac{1}{2}\cdot 10^{-6}} \frac{0.0329}{36} \approx 0.61\cdot 10^3.$$

(b)

$$A^{-1} = \begin{pmatrix} 62 & -36 & -19 \\ -36 & 21 & 11 \\ -19 & 11 & 6 \end{pmatrix}, \qquad \|A\|_\infty = 20, \quad \|A^{-1}\|_\infty = 117, \quad \kappa_\infty(A) = 2{,}340.$$

Section 5.6

1. We get

$$\|B_J\|_\infty = 0.5, \qquad \|B_{GS}\| \le \max\left(\tfrac{1}{2}, \tfrac{1}{3}, \tfrac{1}{3}, 0\right) = 0.5.$$

Jacobi: $\|x^{(8)} - x\|_\infty \le \max(2.2, 1.9, 1.9, 2.2)\cdot 10^{-3} = 2.2\cdot 10^{-3}$.

Gauss-Seidel: $\|x^{(5)} - x\|_\infty \le \max(4.4, 2.2, 2.2, 1.1)\cdot 10^{-3} = 4.4\cdot 10^{-3}$.

2. Write the system as

$$x = 10^{-2} \begin{pmatrix} 5 & -7 & 0 & 0 & -5 & -1 \\ -7 & 5 & -7 & 0 & 0 & -4 \\ 0 & -7 & 5 & -6 & 0 & 0 \\ 0 & 0 & -6 & 5 & -6 & 0 \\ -5 & 0 & 0 & -6 & 5 & -6 \\ -1 & -4 & 0 & 0 & -6 & 5 \end{pmatrix} x + \begin{pmatrix} 0.5 \\ 0.5 \\ 0.5 \\ 0.5 \\ 0.5 \\ 0.5 \end{pmatrix},$$

and use Gauss-Seidel's method—i.e., the most recent approximation for each component of x. The solution is to four decimal places

$$x = (0.4658, 0.4378, 0.4644, 0.4691, 0.4422, 0.4750)^T.$$

3. (a) From the iteration formula follows that for $k \ge 0$,

$$I - AX_{k+1} = I - AX_k - AX_k(I - AX_k) = (I - AX_k)^2,$$

and thus

$$I - AX_k = (I - AX_0)^k.$$

Now, $\rho(I - AX_0) < 1$ is a necessary and sufficient condition for

$$\lim_{k\to\infty} (I - AX_0)^k x = 0$$

to hold for all vectors x, which is equivalent to

$$0 = \lim_{k\to\infty} (I - AX_0)^k = \lim_{k\to\infty} (I - AX_k) \quad \text{or} \quad \lim_{k\to\infty} X_k = A^{-1}.$$

(b) We get

$$X_1 = \begin{pmatrix} 1.99 & -0.99 \\ -0.99 & 0.99 \end{pmatrix} \qquad X_2 = \begin{pmatrix} 1.9999 & 0.9999 \\ -0.9999 & 0.9999 \end{pmatrix},$$

which is in good agreement with A^{-1}.

4. The sequence $z^{(k)}$ satisfies

$$z^{(k+1)} = (\alpha I + (1 - \alpha)B)z^{(k)} + (1 - \alpha)c,$$

so the iteration matrix is $B_\alpha = \alpha I + (1 - \alpha)B$, which has eigenvalues $\lambda(B_\alpha) = \alpha + (1 - \alpha)\lambda(B)$. Now $\lambda(B) \in [a, b]$, and thus

$$\max_{B\in\mathfrak{M}} \rho(B) = \max(|\alpha + (1 - \alpha)a|, |\alpha + (1 - \alpha)b|) = \rho_\alpha.$$

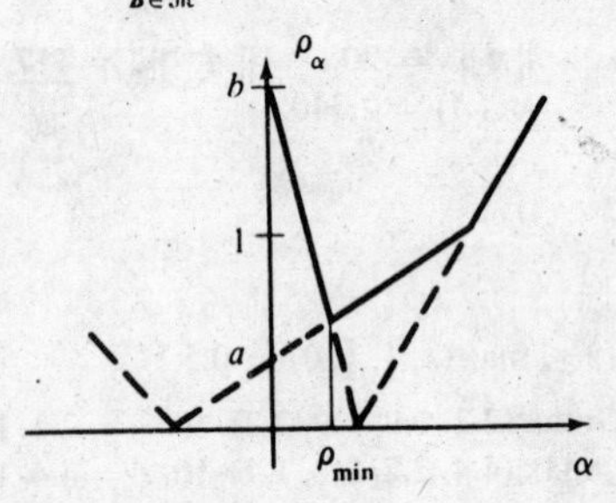

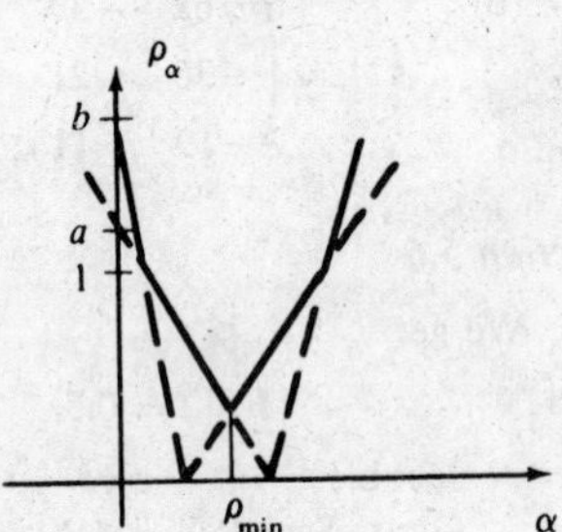

We now want min ρ_α. The figures above show the function ρ_α. (We get slightly different graphs depending on whether $1 \in [a, b]$ or not.) Obviously we have in the minimum that

$$\alpha + (1 - \alpha)a = \pm(\alpha + (1 - \alpha)b) = \rho_\alpha,$$

from which we get, if $\alpha \neq 1$,

$$2\alpha + (1 - \alpha)(a + b) = 0, \quad \alpha = -\frac{m}{1-m}, \quad m = \frac{a + b}{2}$$

(or $(1 - \alpha)(a - b) = 0$, which implies $\alpha = 1$ if $a \neq b$). Note that $\alpha = 1$ gives the trivial iteration $z^{(k+1)} = z^{(k)}$. If $\rho_{\min} = (a - b)/(2 - a - b) < 1$, then the minimum is $\rho_{\min}$; otherwise, we have the trivial minimum $\rho = 1$ for $\alpha = 1$.

5. The eigenvalues of the iteration matrix

$$B_\omega = \begin{pmatrix} 1 - \omega & \omega a \\ \omega a(1 - \omega) & \omega^2 a^2 + 1 - \omega \end{pmatrix}.$$

satisfy the quadratic equation

$$\lambda^2 - \lambda(2(1 - \omega) + \omega^2 a^2) + (1 - \omega)^2 = 0.$$

(a) For $\omega = 1$ we get the eigenvalues $\lambda_1 = 0$, $\lambda_2 = |a|^2$—i.e., $\rho(B_1) = |a|^2$. The method converges if $|a| < 1$.

(b) Solving the quadratic for the various values of ω we find that for $a = 0.5$, $\rho(B_\omega)$ is minimized by $\omega = 1.1$. The matrix A satisfies the conditions in Theorem 5.6.2, and thus $\rho(B_\omega)$ attains its absolute minimum for

$$\bar{\omega} = 2/(1 + (1 - 0.25)^{1/2}) = 4/(2 + \sqrt{3}) \approx 1.0718.$$

It can be shown that it is better to overestimate $\bar{\omega}$, since $\rho(B_\omega)$ has the following typical graph, with an infinite left derivative.

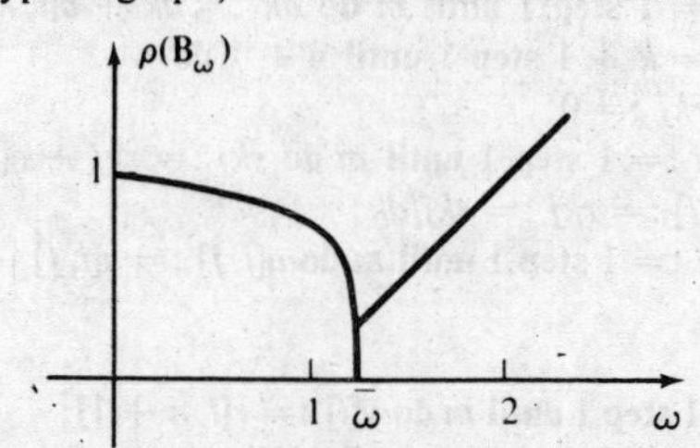

6. (a) We have $e^{(k+1)} = e^{(k)} + \gamma_k p^{(k)}$. The minimum of $\| e^{(k+1)} \|$ is attained when $e^{(k+1)}$ is orthogonal to $p^{(k)}$ with the scalar product $(x, y) = x^T A y$—i.e.,

$$(p^{(k)}, e^{(k)}) = -\gamma_k (p^{(k)}, p^{(k)}).$$

(b) One full iteration step in Gauss-Seidel's method corresponds to using the sequence of directions

$$p^{(0)} = e_1, \qquad p^{(1)} = e_2, \ldots, p^{(n-1)} = e_n,$$

where e_i is the ith unit vector.

Section 5.7

1. Take as unknowns $x_N - 14$ and $x_O - 16$. The normal equations are, then,

$$\begin{pmatrix} 18 & 29 \\ 29 & 56 \end{pmatrix} \begin{pmatrix} x_N - 14 \\ x_O - 16 \end{pmatrix} = \begin{pmatrix} 0.104 \\ 0.161 \end{pmatrix},$$

with the solution

$$x_N = 14.0069, \qquad x_O = 15.9993.$$

(From a table of atomic weights one gets 14.0067 and 15.9994.)

2. The relation is rewritten as

$$\frac{1}{p} - \frac{e}{p} \cos \varphi = \frac{1}{r},$$

which is linear in $1/p$ and e/p. The normal equations

$$\begin{pmatrix} 5 & -0.285 \\ -0.285 & 1.058 \end{pmatrix} \begin{pmatrix} 1/p \\ e/p \end{pmatrix} = \begin{pmatrix} 3.304 \\ 0.315 \end{pmatrix}$$

have the solution

$$1/p = 0.688, \qquad e/p = 0.482,$$

from which

$$p = 1.45 \quad \text{and} \quad e = 0.71 \quad \text{(i.e., an elliptic orbit).}$$

3.

Row	Algol
1	**for** i := 1 **step** 1 **until** m **do** a [i, n + 1] := b[i];
2	**for** k := 1 **step** 1 **until** n **do**
3	**begin** dk := 0;
4	**for** i := 1 **step** 1 **until** m **do** dk := dk + a[i, k] ↑ 2;
5	**for** j := k + 1 **step** 1 **until** n + 1 **do**
6	**begin** rkj := 0;
7	**for** i := 1 **step** 1 **until** m **do** rkj := rkj + a[i, k] × a[i, j];
8	r[k, j] := rkj := rkj/dk;
9	**for** i := 1 **step** 1 **until** m **do** a[i, j] := a[i, j] − rkj × a[i, k]
10	**end**
11	**end**;
12	**for** i := 1 **step** 1 **until** m **do** y[i] := r[i, n + 1];
13	**for** i := n − 1 **step** −1 **until** 1 **do**
14	**for** k := i + 1 **step** 1 **until** n **do** y[i] := y[i] − r[i, k] × y[k];

```
Row      Fortran
 1          DO 1  I = 1,M
 2      1   A(I,N+1) = B(I)
 3          DO 2  K = 1,N
 4          DK = 0
 5          DO 3  I = 1,M
 6      3   DK = DK+A(I,K)**2
 7          DO 2  JM1 = K,N
 8          J = JM1+1
 9          RKJ = 0
10          DO 4  I = 1,M
11      4   RKJ = RKJ+A(I,K)*A(I,J)
12          RKJ = RKJ/DK
13          R(K,J) = RKJ
14          DO 2  I = 1,M
15      2   A(I,J) = A(I,J)-RKJ*A(I,K)
16          DO 5  I = 1,M
17      5   Y(I) = R(I,N+1)
18          DO 6  IB = 2,N
19          I = N-IB+1
20          IP1 = I+1
21          DO 6  K = IP1,N
22      6   Y(I) = Y(I)-R(I,K)*Y(K)
```

Comments (Fortran row numbers in parentheses): The right-hand side is adjoined as (n + 1)st column to A in row 1 (rows 1–2), and therefore we get y

as the $(n+1)$st column in R, row 12 (rows 16–17). The computation of d_k, r_{kj}, and $a_j^{(k+1)}$ in Eq. (5.7.7) is performed in rows 3–9 (4–15). The triangular system $Rx = y$ is solved in rows 13–14 (18–22). Note that after the reduction the $(n+1)$st column of A contains the residual vector r.

4. We get

$$Q = \begin{pmatrix} 1 & 0 & 0 \\ \epsilon & -\epsilon & -\epsilon \\ 0 & \epsilon & 0 \\ 0 & 0 & \epsilon \end{pmatrix}, \qquad R = \begin{pmatrix} 1 & 1 & 1 \\ 0 & 1 & 0 \\ 0 & 0 & 1 \end{pmatrix}, \qquad D^{-1}Q^{\mathsf{T}}b = \begin{pmatrix} 1 \\ 0 \\ 0 \end{pmatrix},$$

and thus the computed solution is $x = (1, 0, 0)^{\mathsf{T}}$. (Bad!)

5. (a) By definition,

$$\|A\|_2 = \max_{x \neq 0} \|Ax\|_2/\|x\|_2 = \max_{\|x\|_2=1} \|Ax\|_2,$$

$$\|A^{-1}\|_2 = \max_{x \neq 0} \|A^{-1}x\|_2/\|x\|_2 = 1/\min_{\|y\|_2=1} \|Ay\|_2, \quad y = A^{-1}x.$$

(b) Since $A^{\mathsf{T}}A$ is symmetric, the eigenvectors x_i, $i = 1, 2, \ldots, n$ span the whole of R^n. Then, from $\|Ax_i\|_2^2 = x_i^{\mathsf{T}}(A^{\mathsf{T}}A)x_i = \lambda_i x_i^{\mathsf{T}} x_i$ it follows that for any vector x

$$\lambda_{\min} x^{\mathsf{T}}x \leq \|Ax\|_2^2 \leq \lambda_{\max} x^{\mathsf{T}}x,$$

where equality is attained for the eigenvectors corresponding to $\lambda_{\min}$ and $\lambda_{\max}$ respectively. In other words: the Euclidean condition number is equal to the ratio of the largest to the smallest singular value of A.

(c) We can write

$$A^{\mathsf{T}}A = \epsilon^2 I + ee^{\mathsf{T}}, \quad e^{\mathsf{T}} = (1, 1, \ldots, 1).$$

Thus $A^{\mathsf{T}}Ax = \epsilon^2 x + (e^{\mathsf{T}}x)e$, and it follows that

$$\lambda_{\min} = \epsilon^2 \quad (x \text{ orthogonal to } e), \qquad \lambda_{\max} = \epsilon^2 + n, \quad (x = e).$$

From this we get

$$\kappa_2(A) = (n/\epsilon^2 + 1)^{1/2} \approx \sqrt{n} \cdot \epsilon^{-1}.$$

CHAPTER 6

Section 6.2

1. (a) Graph $y = x - 1$ and $y = 4 \sin x$; roots: $\alpha_1 = -2.2$, $\alpha_2 = -0.3$, $\alpha_3 = 2.7$;
(b) $\alpha_1 = 0$, $\alpha_2 = 0.8$; (c) $\alpha = 0.6$; (d) $\alpha_1 = -1.0$, $\alpha_2 = 3.6$;
(e) no root; (f) $\alpha = 2.7$; (g) $\alpha = -0.3$.

2. (a) $1.325 < \alpha < 1.35$; (b) $0.35 < \alpha < 0.375$; (c) $0.775 < \alpha < 0.8$.

3. It is assumed that F(A)·F(B) < 0.

```
      FA = F(A)
      FB = F(B)
    5 XNEW = (A+B)/2
   10 IF (ABS(A-B) -  EPS) 90, 90, 20
   20 FNEW = F(XNEW)
      IF (FNEW*FA)  30, 90, 40
   30 B = XNEW
      FB = FNEW
      GO TO 5
   40 A = XNEW
      FA = FNEW
      GO TO 5
   90 ROOT = XNEW
```

Section 6.3

1. Newton-Raphson's method for solving $f(x) = x^2 - 1 = 0$ yields the iteration formula

$$x_{n+1} = x_n + h_n,$$

where $h_n = -f(x_n)/f'(x_n) = -\frac{1}{2}(x_n - 1/x_n)$. The formula can also be written (see Chap. 1 or Example 6.3.3) as

$$x_{n+1} = \tfrac{1}{2}(x_n + 1/x_n),$$

from which one can see that $x_i \geq 1$ for all i. Putting $x_0 = 2$ we get

n	x_n	h_n
0	2	-0.75
1	1.25	-0.225
2	1.025	-0.0247
3	1.0003	-0.000299955
4	$1 + 4.5 \cdot 10^{-8}$	$-4.5 \cdot 10^{-8} + 10 \cdot 10^{-16}$
5	$1 + 10 \cdot 10^{-16}$	

The convergence is very clearly seen to be quadratic.

2. 1.7963.

3. Write the equation as $f(x) = x - 1 + e^{-2x} = 0$. Then Newton-Raphson's method yields

$$x_{n+1} = x_n + h_n, \quad h_n = -f(x_n)/f'(x_n) = -(x_n - 1 + e^{-2x_n})/(1 - 2e^{-2x_n}).$$

With initial approximation $x_0 = 0.8$ we get $x_1 = 0.79682$, $h_1 \approx -10^{-5}$; so $\alpha = 0.7968$ to 4D.

4. 1.76322.

5.

n	z_n	h_n
0	$1 + i$	$-0.75 - i\cdot 0.25$
1	$0.25 + i\cdot 0.75$	$-0.325 + i\cdot 0.225$
2	$-0.075 + i\cdot 0.975$	$0.0767 + i\cdot 0.0218$
3	$0.0017 + i\cdot 0.9968$	$-0.001705 + i\cdot 0.003204$
4	$-0.000005 + i\cdot 1.000004$	

Correct value: $\alpha = i$. Note that a real initial estimate would not give convergence in this case.

Section 6.4

1. (a) Take $x_0 = 0.375$, $x_1 = 0.35$. $x_2 = 0.35174$, $x_3 = 0.351734$, $x_4 = 0.351734$, $\alpha = 0.35173$.

(b) Take $x_0 = -0.993$, $x_1 = -0.994$, $x_2 = -0.993549$, $x_3 = -0.993550$, $\alpha = -0.99355$.

2. Suppose that $f(x)$, a, and b are given and that $f(a)\cdot f(b) < 0$.

```
    x1 := a;   x2 := b;  y1 := f(x1);  y2 := f(x2);
L:  x3 := x2 - y2 * (x2 - x1)/(y2 - y1);
    y3 := f(x3);
    if  y3 * y2 < 0.0 then begin x1 := x2; y1 := y2 end
    else y1 := y1/2;
    x2 := x3; y2 := y3;
    if  abs(x2 - x1) > eps ∧ y2 ≠ 0 then goto L;
    root := x2;
```

4. (a) Newton's interpolation formula can be written:

$$f(x) = f_n + f[x_n, x_{n-1}](x - x_n) + f[x_n, x_{n-1}, x_{n-2}](x - x_{n-1})(x - x_n).$$

Differentiate:

$$f'(x) = f[x_n, x_{n-1}] + f[x_n, x_{n-1}, x_{n-2}](x - x_{n-1} + x - x_n).$$

Put $x = x_n$:

$$f'(x_n) = f[x_n, x_{n-1}] + f[x_n, x_{n-1}, x_{n-2}](x_n - x_{n-1}).$$

(b) Solve $f(x) = 0$ by the formula for the zeros of a quadratic polynomial.

$$x_{n+1} - x_n = \frac{-\omega \pm \sqrt{\omega^2 - 2f(x_n)f''(x_n)}}{f''(x_n)}$$

$$= \frac{\omega^2 - (\omega^2 - 2f(x_n)f''(x_n))}{f''(x_n)(-\omega \pm \sqrt{\omega^2 - 2f(x_n)f''(x_n)})}$$

$$x_{n+1} - x_n = -\frac{2f(x_n)}{(+\omega \pm \sqrt{\omega^2 - 2f(x_n)f''(x_n)})}. \quad \text{Q.E.D.}$$

5. (a) The equation corresponding to the cross-ratio theorem given in the problem can be written as:

$$\frac{x_{n+1} - x_n}{x_{n+1} - x_{n-2}} \cdot \frac{f_{n-2}}{f_n} = \frac{f[x_{n-1}, x_{n-2}]}{f[x_n, x_{n-1}]}.$$

Denote the right-hand side by $1/\beta$. Then we have

$$\beta(x_{n+1} - x_n)f_{n-2} = f_n(x_{n+1} - x_{n-2}). \qquad (*)$$

The "modified secant method,"

$$x_{n+1} = x_n - f_n\frac{(x_n - x_{n-2})}{f_n - \beta f_{n-2}},$$

yields

$$(f_n - f_{n-2}\beta)x_{n+1} = (f_n - f_{n-2}\beta)x_n - f_n(x_n - x_{n-2})$$
$$= f_n x_{n-2} - f_{n-2}\beta x_n,$$

which is equivalent to Eq. (*) above.

(b) Under the stated assumptions, the "modified secant method" just means that we compute x_{n+1} by a regula falsi ($\bar{f}_{n-2} \cdot f_n < 0$) iteration using the "points" (x_n, f_n) and $(x_{n-2}, \bar{f}_{n-2})$, so $x_{n+1} \in \text{int}\,(x_{n-2}, x_n)$.

(c) *Answer:* $x_4 = 1.93376,\ x_4 - \alpha = +0.00001.$

Section 6.5

1. (a), (b). For convergence, we must have $|\varphi'(\alpha)| \le m < 1$. The smaller m is, the faster the convergence.

I1: $x_{n+1} = -\ln x_n = \varphi(x_n) \Longrightarrow |\varphi'(\alpha)| = \left|\frac{1}{\alpha}\right| \approx 2.$

No convergence (unless $x_0 = \alpha$).

I2: $x_{n+1} = e^{-x_n} = \varphi(x_n) \Longrightarrow |\varphi'(\alpha)| = e^{-\alpha} \approx 0.6.$

This sequence converges and can be used with suitable x_0.

I3: $x_{n+1} = \dfrac{x_n + e^{-x_n}}{2} = \varphi(x_n) \Longrightarrow |\varphi'(\alpha)| = \dfrac{1}{2}|1 - e^{-\alpha}| \approx 0.20.$

This sequence converges faster than the one in I2 and should be used.

(c) We try

$$x_{n+1} = \frac{\beta x_n + e^{-x_n}}{\beta + 1}.$$

How should β be chosen to get the fastest convergence?

$$\varphi'(x) = (\beta - e^{-x})/(\beta + 1),$$
$$\varphi'(\alpha) = (\beta - \alpha)/(\beta + 1).$$

If we choose $\beta \approx \alpha$, we get rapid convergence. (Compare this choice with the Newton-Raphson formula applied to the equation $x = e^{-x}$.)

2. $x_0 < 2 \Longrightarrow$ convergence to $\alpha_1 = 1$,
$x_0 = 2 \Longrightarrow$ convergence to $\alpha_2 = 2$,
$x_0 > 2 \Longrightarrow$ divergence.

3. $x_{n+1} = \varphi(x_n), \quad \varphi(x) = x - \frac{1}{2}u(x) - \frac{1}{2}[u(x)/u'(x)],$

where

$$u = f/f', \qquad u' = 1 - ff''/(f')^2.$$

$$\varphi' = 1 - \tfrac{1}{2}u' - \tfrac{1}{2}(1 - uu''/(u')^2),$$

$$\varphi'' = -\tfrac{1}{2}u'' + \tfrac{1}{2}[(u')^2(u'u'' + uu''') - uu''\cdot 2u'u'']/(u')^4.$$

Hence, since $f(\xi) = 0$,

$$u(\xi) = 0, \qquad u'(\xi) = 1,$$

$$\varphi(\xi) = \xi, \qquad \varphi'(\xi) = 0, \qquad \varphi''(\xi) = -\tfrac{1}{2}u''(\xi) + \tfrac{1}{2}u''(\xi) = 0.$$

The last line shows that convergence is at least cubic.

4. $p = q = \frac{5}{9}, \qquad r = -\frac{1}{9}, \qquad x_{n+1} - a^{1/3} \approx \frac{5}{3}a^{-2/3}(x_n - a^{1/3})^3.$

Section 6.6

1. (a) Take $x_0 = 0.317$.

$$x_{n+1} = \frac{\cos x_n}{3} = \varphi(x_n)$$

yields $x_1 = 0.316725$, $x_2 = 0.316754$, $x_3 = 0.316750$. This is an alternating (about α) sequence, since $\varphi'(\alpha) \approx -0.10 < 0$. The root is contained between x_2 and x_3 except for possible rounding error in the computation of x_3. If we take account of this, we get

$$0.316749 < \alpha < 0.316754.$$

Thus we have attained the desired accuracy.

(b) $x_0 = 0.3$, $x_{n+1} = -(\tan x_n)/(3x_n)$ yields

$$x_4 = -0.347423, \qquad x_5 = -0.347426,$$

$$|x_5 - \alpha| \leq (0.08\cdot 3\cdot 10^{-6} + 0.5\cdot 10^{-6})/0.92 < 10^{-6},$$

$$\alpha = -0.347427 \pm 10^{-6}.$$

2. If we neglect rounding errors ($\delta = 0$), then the error Eq. (6.6.1) yields

$$|x_5 - \alpha| \leq [m/(1 - m)]\,|x_4 - x_5| = (0.4/0.6)25\cdot 10^{-5} < 17\cdot 10^{-5}.$$

We also have (see the proof of Theorem 6.5.1) that

$$|x_{5+k} - \alpha| \leq m^k\,|x_5 - \alpha|.$$

and thus we want the smallest k such that $(0.4)^k 17\cdot 10^{-5} < 5\cdot 10^{-5}$. $\therefore k = 2$ —i.e., two iterations will suffice.

3.

```
procedure IT(x0, F, x1, eps);
  real x0, F, x1, eps;
  begin real m, err, x2;
    x1 := x0; x2 := F;
  L: x0 := x1; x1 := x2; x2 := F;
```

```
        m := abs((x2 - x1)/(x1 - x0));
        err := m * abs(x2 - x1)/(1 - m);
        if err ≥ eps then goto L;
    x1 := x2
  end
```

The procedure presumes that the feature of Algol known as "Jensen's device", see e.g. [5], is used to compute values of F. A typical call of the procedure might be IT (0.5, exp $(-x1)$, $x1$, 0.01);—to solve $x = e^{-x}$.

Section 6.8

1. (a) 1. n add., n mult.; 2. $(3n - 2)$ add., $(4n - 2)$ mult.;
(b) $(2n + 1)$ add., $(2n + 2)$ mult., $n \geq 2$. Put $p = 2x_0$, $q = -(x_0^2 + y_0^2)$. Then $p(z_0) = b_{n-1}z_0 + b_n$, where

$$b_{-1} = 0, \qquad b_0 = a_0,$$
$$b_k = a_k + pb_{k-1} + qb_{k-2}, \quad k = 1, 2, \ldots, n-1,$$
$$b_n = a_n + qb_{n-2}.$$

Also, $p'(z_0) = 2iy_0(c_{n-3}z_0 + c_{n-2}) + b_{n-1}$, where

$$c_{-1} = 0, \qquad c_0 = b_0,$$
$$c_k = b_k + pc_{k-1} + qc_{k-2}, \qquad k = 1, 2, \ldots, n-3,$$
$$c_{n-2} = b_{n-2} + qc_{n-4}.$$

2. Use a graphical method to estimate the roots (all the roots are real in both (a) and (b)). Then use Newton-Raphson's method to compute them more accurately, using Horner's rule to evaluate the polynomials, and computing the roots in order of increasing magnitude.

(a) -3.273636, 3.437462, -10.663825; (b) 2.982281, -3.541918, -9.940363.

Section 6.9

2. (a) By Taylor's formula,

$$f(y) = f(x) - \lambda f'(x) \cdot f'(x)^{-1} f(x) + O(\lambda^2),$$
$$\|f(y)\| = (1 - \lambda) \cdot \|f(x)\| + O(\lambda^2) < \|f(x)\|,$$

if λ is small enough.

(b) The condition $\|f(x_{i+1})\|_2 < \|f(x_i)\|_2$ will be satisfied after a finite number of repetitions (if rounding errors are neglected).

3. Equation (6.9.7) can be written

$$J_i q = y, \qquad J_i p = J_{i-1} p, \qquad \forall\, p \perp q.$$

These n linearly independent conditions determine J_i *uniquely*. Direct insertion shows that the first equation in Eq. (6.9.8) implies Eq. (6.9.7). Then apply the Sherman-Morrison formula.

CHAPTER 7

Section 7.1

1. (a)

$$
\begin{array}{ccccc}
1 & & & & \\
 & 1 & & & \\
2 & & 1 & & \\
 & 2 & & 1 & \\
4 & & 2 & & 1 \\
 & 4 & & 2 & \\
8 & & 4 & & \\
 & 8 & & & \\
16 & & & &
\end{array}
$$

(b)

$$
\begin{array}{ccccc}
1 & & & & \\
 & -2 & & & \\
-1 & & 4 & & \\
 & 2 & & -8 & \\
1 & & -4 & & 16 \\
 & -2 & & 8 & \\
-1 & & 4 & & \\
 & 2 & & & \\
1 & & & &
\end{array}
$$

2.

	Δ	Δ^2	Δ^3	
0.852				
	1,399			
2,251		+1		
	1,400		−7	$+\epsilon$
3,651		−6		
	1,394		+25	-3ϵ
5,045		+19		
	1,413		−26	$+3\epsilon$
6,458		−7		
	1,406		+9	$+\epsilon$
7,864		+2		
	1,408			
9,272				

Comparison with the third differences in Example 7.1.1 shows that $\epsilon \approx -26/3 \approx -9$. Hence the element 5,045 should be 5,054.

3. The table was constructed so that the discontinuity $= 2{,}000$ at $t = 77.7398$. From the fourth differences one obtains that the discontinuity is $519 + 1{,}467 = 1{,}986$ at $t = 77 + 1{,}467/1{,}986 = 77.7387$.

4. Replace in Eq. (7.1.7) v_n by Δv_n. Replace 0 by 1 in the lower summation bound. Then

$$\sum_{n=1}^{N-1} u_n \Delta^2 v_{n-1} - u_N \Delta v_{N-1} - u_1 \Delta v_0 - \sum_{n=1}^{N-1} \Delta u_n \Delta v_n = -\sum_{n=1}^{N-1} \Delta u_n \Delta v_n.$$

Exchange u and v:

$$\sum_{n=1}^{N-1} v_n \Delta^2 u_{n-1} = -\sum_{n=1}^{N-1} \Delta v_n \Delta u_n.$$

5. (a) 0, 0; (b) $2^{-(n+4)}, 2^{-(n+2)}$.

Section 7.2

2. (a) $A_0 = \frac{12}{15},\ A_1 = \frac{16}{15},\ A_2 = \frac{2}{15}$; (b) $8h^3 f'''(0)/315$.

3. Letting $x = x_n + ht$, $h^3 \max|(t^3 - t)/6|$, $|t| \leq 1$, is obtained when $3t^2 - 1 = 0$; hence $\max = h^3/(243)^{1/2}$.

4. (a) a arbitrary, $b = 1 - 2a$, $c = a$; i.e., $g_n = f_n + a(f_{n+1} - 2f_n + f_{n-1})$;
(b) $a = 1/3$; i.e., $g_n = (f_{n+1} + f_n + f_{n-1})/3$;
(c) $a = 1/4$; i.e., $g_n = (f_{n+1} + 2f_n + f_{n-1})/4$.

5. Let $f(x) = e^x$. The following estimates for $f'(1)$ are obtained for $h = 1, 0.5, 0.25$.

$$\frac{7.3891 - 1.000}{2 \cdot 1} = 3.1946; \qquad \frac{4.4817 - 1.6987}{2 \cdot 0.5} = 2.8330;$$

$$\frac{3.4903 - 2.1170}{2 \cdot 0.25} = 2.7466.$$

Extrapolation scheme:

h		$\frac{\Delta}{3}$		$\frac{\Delta}{15}$	
1	3.1946				
0.5	2.8330	−1205	2.7125		
0.25	2.7466	−288	2.7178	4	2.7182

The final value is very good. The correct result is $e = 2.7183$.

6. See Theorem 7.1.4.

(a)

h	$f'(1.0)_h$		
0.4	−1.855556		
0.2	−1.872935	−1.878728	
0.1	−1.877320	−1.878782	−1.878786

(b)

h	$f''(1.0)_h$		
0.4	−0.82268		
0.2	−0.86430	−0.87817	
0.1	−0.87520	−0.87883	−0.87888

7. Richardson extrapolation is applicable!

P	X	$\frac{\Delta}{3}$		$\frac{\Delta}{7}$		
0.8	740					
0.4	487	−84.3	402.7			
0.2	475	−4.0	471.0	+9.8	480.8	
0.1	485	+3.3	488.3	+2.5	490.8	$\}\longrightarrow 491 = x(0)$
0.05	489	+1.3	490.3	+0.3	490.6	

8. (a) $c(h) = c_{1/h} = \dfrac{2}{h}\sin(\pi h) = 2\pi - \dfrac{\pi^3}{3}h^2 + \dfrac{\pi^5}{60}h^4 - \dots$.

(b) $c_2 = 4$, $c_3 = 3\sqrt{3} = 5.1962$, $c_6 = 6$. Repeated Richardson extrapolation gives $c(0) \approx 6.2823$, a better value than Archimedes got with his 96-sided polygon.

(c) $c_{2n} = 4n\sin\dfrac{\pi}{2n} = 4n\sqrt{\dfrac{1}{2}\left(1 - \cos\dfrac{\pi}{n}\right)} = \sqrt{8n^2 - 4n\sqrt{4n^2 - c_n^2}}$

(d) Cancellation. After changing the statement

$$x := \text{sqrt}(8*n\uparrow 2 - 4*n*\text{sqrt}(4*n\uparrow 2 - C[0]\uparrow 2));$$

to

$$x := 2*C[0]*\text{sqrt}(0.5/(1 + \text{sqrt}(1 - C[0]/N/2)\uparrow 2)));$$

much better results are obtained; see the table below.

6	6.000000000			
12	6.211657083	6.282209444		
24	6.265257227	6.283123942	6.283184908	
48	6.278700406	6.283181466	6.283185301	6.283185308
96	6.282063902	6.283185068	6.283185308	6.283185308

Section 7.3

1. (a) Choose $x_0 = 2$, $x_1 = 4$ (common points for the two alternatives).

$$f(x) = c_0 + c_1(x-2) + c_2(x-2)(x-4).$$

c_0 and c_1 are independent of the third point

$$x = 2 \Longrightarrow 2 = c_0 \Longrightarrow c_0 = 2,$$
$$x = 4 \Longrightarrow 12 = c_0 + c_1(4-2) \Longrightarrow c_1 = 5.$$

The value of c_2 depends on the choice of the third point

$$x = 1 \Longrightarrow 0 = 2 + 5(1-2) + c_2(1-2)(1-4) \Longrightarrow c_2 = 1,$$
$$f(3) = 2 + 5(3-2) + 1\cdot(3-2)(3-4) = 6,$$
$$x = 5 \Longrightarrow 21 = 2 + 5(5-2) + c_2(5-2)(5-4) \Longrightarrow c_2 = \tfrac{4}{3},$$
$$f(3) = 2 + 5(3-2) + (4/3)(3-2)(3-4) = 5.667.$$

From this we can estimate that $f(3) = 5.83 \pm 0.17$, approximately. The same results can, of course, be obtained by Newton's interpolation formula.

(b) Using a result from (a) we can put

$$f(x) = 2 + 5(x-2) + 1\cdot(x-2)(x-4) + c_3(x-2)(x-4)(x-1),$$
$$f(5) = 21 \Longrightarrow c_3 = \tfrac{1}{12} \Longrightarrow f(3) = 5.833.$$

2. (a) $f(0) = 67{,}882$.

(b) Upper bounds for the errors of the divided differences in f/ϵ are obtained by the following scheme:

x	f			
0.1	+0.5			
		−10		
0.2	−0.5		+50	
		+5		−89.3
0.4	+0.5		−12.5	
		−2.5		
0.8	−0.5			

$f(0) = f(0.1) - 0.1f[0.1, 0.2] + 0.02f[0.1, 0.2, 0.4] - 0.008f[0.1, 0.2, 0.4, 0.8]$.

A bound for the rounding error is given by:

$$(0.5 + 0.1 \cdot 10 + 0.02 \cdot 50 + 0.008 \cdot 89.3) < 3.5.$$

(c) The same result as in (a). (Why?)

3. (a) 1; (b) x. (Lagrange's interpolation formula is exact for polynomials of degree m or less.)

4. Let $Q(x)$ be the interpolation polynomial defined by Eq. (7.3.1). By Theorem 7.3.3 (Newton),

$$Q(x) = x^m \cdot f[x_0, x_1, \ldots, x_m] + (m-1)\text{th-degree polynomial.}$$

By Eq. (7.3.20),

$$\Phi(x)/(x - x_i) = x^m + (m-1)\text{th-degree polynomial.}$$

Hence, by Eqs. (7.3.17) and (7.3.21) (Lagrange),

$$Q(x) = x^m \sum_{i=0}^{m} f_i/\Phi'(x_i) + (m-1)\text{th-degree polynomial.}$$

The result follows by matching the two expressions for $Q(x)$.

Note: In order to prove Eq. (7.3.21) it is sufficient to verify that Eq. (7.3.18) is satisfied by the function defined by Eq. (7.3.21). Accordingly, for $j \neq i$:

$$\delta_i(x_j) = \frac{\Phi(x_j)}{(x_j - x_i)\Phi'(x_i)} = 0.$$

For $j = i$, apply L'Hospital's rule:

$$\delta_i(x_i) = \lim_{x \to x_i} \frac{\Phi(x)}{(x - x_i)\Phi'(x_i)} = \frac{\Phi'(x_i)}{1 \cdot \Phi'(x_i)} = 1.$$

5.

$$\begin{bmatrix} 1 & x_0 & \cdots & x_0^m \\ 1 & x_1 & \cdots & x_1^m \\ \cdots & & & \\ 1 & x_m & \cdots & x_m^m \end{bmatrix} \cdot \begin{bmatrix} c_{00} & \cdots & c_{0m} \\ c_{10} & \cdots & c_{1m} \\ \cdots & & \\ c_{m0} & \cdots & c_{mm} \end{bmatrix} = \{\sum_{k=0}^{m} c_{kj} x_i^k\} = \{\delta_j(x_i)\} = \boldsymbol{I}$$

by Eq. (7.3.18)

6. $x^2 - x^2(x-1) + \frac{1}{4}x^2(x-1)^2$.

7. Substituting in the formula for $\varphi(p)$ in Sec. 7.3.8, we get

$$\varphi(p) = \tfrac{1}{3}(2 - 1 - B_p''(2+2) - B_p'''(0)) = \tfrac{1}{3}(1 + p(1-p)),$$
$$p_0 = (2-1)/3 = \tfrac{1}{3},$$
$$p_1 = \varphi(p_0) = \tfrac{11}{27},$$
$$p_2 = \varphi(p_1) \approx 0.414, \qquad \therefore \quad \text{root} = x_0 + p \approx 1.414.$$

9. *Test case:* A: 1.104575, B: 1.105130, C and D: 1.105685.
(Exact value: 1.105171; linear interpolation: 1.110702.)

C and *D are mathematically equivalent*, because both formulas give that quadratic interpolation polynomial which is uniquely determined by the tabular values f_{-1}, f_0, f_1.

A gives the quadratic interpolation polynomial determined by the tabular values f_0, f_1, f_2, and hence *A is not equivalent to C and D*.

By the derivation in Sec. 7.3.4, *B* is the average of *A* and *C*, and hence *B is not equivalent* to any of the others. It requests f_{-1}, f_0, f_1, f_2 and does not define an interpolation polynomial in the proper sense of the word. (It gives the average of two interpolation polynomials.)

The equivalence of *C* and *D* can also be recognized by the following calculation, where RHS means "right-hand side":

$$\text{RHS of } C = f_0 + ph\frac{f_1 - f_0}{h} + p(p-1)h^2\cdot\left(\frac{f_1 - f_0}{h} - \frac{f_0 - f_{-1}}{h}\right)\cdot\frac{1}{2h}$$
$$= f_0 + p(f_1 - f_0) + (p^2 - p)\frac{f_1 - 2f_0 + f_{-1}}{2}$$
$$= f_0 + p\frac{f_1 - f_{-1}}{2} + p^2\frac{f_1 - 2f_0 + f_{-1}}{2} = \text{RHS of D}.$$

Section 7.4

1. 20,271. Only values at integer points need be used.

2. Proceed as in Sec. 7.4.2, but expand here $f(x)$ by Taylor's formula, around $x_{i-1/2}$, and integrate from x_{i-1} to x_i:

$$\int_{x_{i-1}}^{x_i} f(x)\,dx = hf(x_{i-1/2}) + 0 + \int_{x_{i-1}}^{x_i} \tfrac{1}{2}f''(\xi)\cdot(x - x_{i-1/2})^2\,dx,$$

Hence

$$\epsilon_i = -\tfrac{1}{2}f''(\xi_i)\cdot 2\cdot\tfrac{1}{3}h^3(\tfrac{1}{2})^3 = -f''(\xi_i)h^3/24.$$

Then Eq. (7.4.9) is obtained in the same way as Eq. (7.4.8).

3. 1.1276 and 1.128961, respectively. All the decimals are correct in the latter value.

4. (a) $F_1(h) + (F_1(h) - F_2(h))/(c_2/c_1 - 1)$;
(b) $R(h) + (T(h) - R(h))/3$ has an error which is $O(h^4)$. The formula is identical to Simpson's formula.

5. *Proposals:* (a) $\int_0^\infty = \int_0^{16} + \int_{16}^\infty$. Put $x = t^2$ in the first integral, and use Romberg's method. Treat the other integral in a way similar to Example 3.1.12.
(b) Put $sx = t$. Integrate using Romberg's method from $t = 0$ to $t = 12$.

The remainder is negligible. If one proceeds as in Example 3.2.6, then one can use a shorter interval of integration.

(c) Compute $s_n = \int_{\pi}^{n\pi}$ by Romberg's method for a few values of n, and apply repeated averaging according to Sec. 3.2.1.

6. (a) $\int_0^{2\pi} e^{ikt}\,dt = \begin{cases} 0, & k = 0, \\ 2\pi, & k \neq 0. \end{cases}$ The application of the trapezoidal rule to $\exp(ikt)$, k integer, $|k| \leq n$,

$$h(\tfrac{1}{2} + e^{ikh} + e^{2ikh} + \dots + e^{nikh} + \tfrac{1}{2}e^{(n+1)ikh}) = h \cdot \sum_{j=0}^{n} e^{jikh}$$

$$= \begin{cases} h(e^{(n+1)ikh} - 1)/(e^{ikh} - 1) = 0, & k \neq 0, \\ h(n+1) = 2\pi, & k = 0, \end{cases}$$

by the formula for the sum of a geometric series. Hence the trapezoidal rule is exact for the exponential functions considered, and hence also for all linear combinations of them.

Note: The trapezoidal rule and the rectangle rule are identical when a continuous periodic function is integrated over a full period.

(b) Let P be a trigonometric polynomial, $|f(t) - P(t)| < \epsilon$. Let I_f, I_p be the integrals of $f/2\pi$ and $P/2\pi$, respectively, and let T_f, T_p be the trapezoidal approximations. Then

$$|I_f - I_p| \leq \frac{1}{2\pi}\int_0^{2\pi} |f(t) - P(t)|\,dt < \frac{1}{2\pi}\int_0^{2\pi} \epsilon\,dt = \epsilon,$$

and similarly $|T_p - T_f| < \epsilon$. By the triangle inequality,

$$|I_f - T_f| \leq |I_f - I_p| + |I_p - T_p| + |T_p - T_f| < \epsilon + 0 + \epsilon = 2\epsilon.$$

(This type of error analysis has a wide applicability. See Theorem 4.5.2 and Example 4.5.3.)

(c) Use the hint,

$$E_7(g) \leq \frac{2.03}{8 \cdot 8!} 2^{-8} < 2.5 \cdot 10^{-8}.$$

Thus in Problem 3. $R_T < 5 \cdot 10^{-8}$.

7. The abscissae $(\sqrt{0.6}, 0, -\sqrt{0.6})$ are the roots of the Legendre polynomial $p_3(x) = x^3 - \frac{3}{5}x$. The weights are easily obtained by requiring that the integral be exact for $f = 1, x, x^2$ and using the method of undetermined coefficients. Since the desired Gaussian rule must also be exact for $f = 1, x, x^2$, the weights $(\frac{5}{9}, \frac{8}{9}, \frac{5}{9})$ obtained in this way must be correct, and by Theorem 7.4.3 this rule is in fact exact for polynomials of degree ≤ 5.

To compute $\int_0^1 [\sin x/(1 + x)]\,dx$, transform the interval in the same way as in Example 7.4.11. Result: 0.284249.

8. (a) $\frac{4}{3}[f(\sqrt{0.4}) + f(-\sqrt{0.4})]$; (b) $0.0108 f^{\text{iv}}(\xi)$.

9. Put $x = \cos t$. $f(\cos t) = F(t)$, F is a trigonometric polynomial of degree $(2n - 1)$, $F(2\pi - t) = F(t)$.

$$I = \int_{-1}^{1} f(x)(1 - x^2)^{-1/2}\, dx = \int_{0}^{\pi} F(t)\, dt = \tfrac{1}{2}\int_{0}^{2\pi} F(t)\, dt.$$

Substitute $2n - 1$ for n in Problem 6(a). By the trapezoidal rule *using the points* $2\pi(2j - 1)/(4n)$, $j = 0, 1, 2, \ldots, 2n$, we obtain the result, (since $F(t) = F(2\pi + t)$ and $F(2\pi - t) = F(t)$),

$$I = \frac{1}{2}\cdot\frac{2\pi}{2n}\sum_{j=1}^{2n} F\left(\frac{2j-1}{2n}\pi\right) = \frac{\pi}{n}\cdot\sum_{j=1}^{n} F\left(\frac{2j-1}{2n}\pi\right)$$
$$= \frac{\pi}{n}\sum_{j=1}^{n} f\left(\cos\frac{2j-1}{2n}\pi\right).$$

10. (a) Proceed as in Example 7.4.10, $f(x) = x^{-1}$, $h = 1$; (b) $\gamma = 0.5772157$; (c) $\gamma - F(M) = \sum c_i M^{-2i}$.

M^{-1}	$F(M)$	$\frac{\Delta}{3}$		$\frac{\Delta}{15}$	
1	.50000				
		1895			
0.5	.55685		57580		
		506		9	
0.25	.57204		57710		57719
		129		1	
0.125	.57592		57721		57722

$\gamma \approx 0.57722$ (actually correct to five decimal places).

Section 7.5

1. 1.1751 and 1.1752, respectively.

3. Put $a = (x_0 + x_{-1})/2$, $b = (x_0 + x_1)/2$. If f is quadratic, then

$$f'(a) = f[x_{-1}, x_0], \qquad f'(b) = f[x_0, x_1],$$

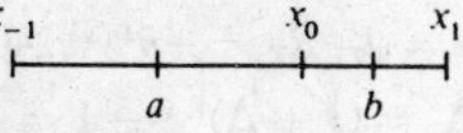

$$x_0 = (1 - \theta)\cdot a + \theta b, \quad \text{when} \quad \theta = \frac{x_0 - a}{b - a} = \frac{x_0 - x_{-1}}{x_1 - x_{-1}},$$

Hence the formula

$$f'(x_0) = (1 - \theta)\cdot f'(a) + \theta f'(b)$$

is exact when f' is linear (i.e., when f is quadratic).

Error analysis: For $f(x) = (x - x_0)^3$ the formula gives

$$f'(a) \approx \frac{(x_{-1} - x_0)^3 - 0}{x_{-1} - x_0} = (x_{-1} - x_0)^2, \quad f'(b) \approx (x_1 - x_0)^2,$$

$$f'(x_0) \approx \frac{(x_1 - x_0)(x_{-1} - x_0)^2}{x_1 - x_{-1}} + \frac{(x_0 - x_{-1})(x_1 - x_0)^2}{x_1 - x_{-1}}$$
$$= \frac{(x_0 - x_{-1})(x_1 - x_0)}{x_1 - x_{-1}}\cdot(x_0 - x_{-1} - x_1 - x_0)$$
$$= (x_0 - x_{-1})(x_1 - x_0).$$

Hence (see Example 7.2.1) an asymptotically correct error bound is

$$\frac{f'''(x_0)}{3!}(x_0 - x_{-1})(x_1 - x_0).$$

The same approximation can also be expressed in the form (see [80, p. 121]),

$$f'(x_0) \approx f[x_0, x_{-1}] + f[x_0, x_1] - f[x_{-1}, x_1],$$

the proof of which is left as a new exercise to the reader.

The variable θ, which goes from 0 to 1 as x goes from a to b, is a one-dimensional analog to the concept of barycentric coordinates, which is fundamental for polynomial approximation in several dimensions, see Sec. 7.7. See also Problem 3 of Sec. 10.5.

4. Put $\varphi_j(x) = (x - x_0)^j$, $P = \sum c_j\varphi_j$.

$$(\varphi_0, \varphi_0) = 5, \qquad (\varphi_0, \varphi_2) = (\varphi_1, \varphi_1) = 10h^2, \qquad (\varphi_2, \varphi_2) = 34h^4,$$

Otherwise, $(\varphi_i, \varphi_j) = 0$, because of antisymmetry.

Normal equations:

$$\begin{bmatrix} 5 & 0 & 10h^2 \\ 0 & 10h^2 & 0 \\ 10h^2 & 0 & 34h^4 \end{bmatrix} \begin{bmatrix} c_0 \\ c_1 \\ c_2 \end{bmatrix} = \begin{bmatrix} \sum f_i \\ \sum hif_i \\ \sum h^2i^2f_i \end{bmatrix},$$

$$\begin{aligned} 14h^2c_2 &= \sum_{-2}^{2} i^2f_i - 2\sum_{-2}^{2} f_i = 2(f_2 + f_{-2}) - (f_1 + f_{-1}) - 2f_0 \\ &= 7\delta^2 f_0 + 2\delta^4 f_0, \end{aligned}$$

$$P''(x_0) = \tfrac{1}{2}c_2, \qquad P(x_0) = c_0,$$

For $P'(x_0)$, see Examples 7.2.4 and 7.5.5.

Section 7.6

1. $a = \frac{1}{12}$, $R_T = -h^4y^{\text{vi}}/240$.

2. $\Delta_1 = (1 + \Delta)^{1/5} - 1 = \frac{1}{5}\Delta\cdot(1 - \frac{2}{5}\Delta + \frac{6}{25}\Delta^2 + \dots),$

$\Delta_1^2 = \frac{1}{25}\Delta^2\cdot(1 - \frac{4}{5}\Delta + \frac{16}{25}\Delta^2 + \dots),$

$\Delta_1^3 = \frac{1}{125}\Delta^3\cdot(1 - \frac{6}{5}\Delta + \frac{6}{5}\Delta^2 + \dots).$

3. (a)
$$\sum_{j=n}^{\infty} u_jz^j = \sum_{j=n}^{\infty} z^jE^ju_0 = \frac{z^nE^n}{1 - zE}u_0 = \frac{z^n}{1 - z - z\Delta}u_n$$
$$= \frac{z^n}{1 - z}\sum_{s=0}^{\infty}\left(\frac{z}{1 - z}\right)^s\Delta^s u_n.$$

(b) Faster convergence if

$$\left|\frac{a - 1}{2\sin(\varphi/2)}\right| < |a|,$$

i.e., if $[1 + 2\sin(\varphi/2)]^{-1} < a \le 1$. For alternating series, $\varphi = \pi$, $\frac{1}{3} < a \le 1$.

(c) $M_{n+1,k} - zM_{n,k}$

$$= \sum_{j=0}^{n} u_j z^j - \sum_{j=0}^{n-1} u_j z^{j+1} + \frac{z^{n+1}}{1-z} \sum_{s=0}^{k-1} \left(\frac{z}{1-z}\right)^s \Delta^s \cdot (u_{n+1} - u_n);$$

$$(1-z)M_{n,k+1} = \sum_{j=0}^{n-1} u_j z^j - \sum_{j=0}^{n-1} u_j z^{j+1} + z^n \sum_{t=0}^{k} \left(\frac{z}{1-z}\right)^t \Delta^t u_n.$$

Subtract right-hand sides:

$$u_n z^n + z^n \sum_{s=0}^{k-1} \left(\frac{z}{1-z}\right)^{s+1} \Delta^{s+1} u_n - z^n \left[u_n + \sum_{s=0}^{k-1} \left(\frac{z}{1-z}\right)^{s+1} \Delta^{s+1} u_n\right] = 0.$$

(d) Method of repeated averages; see Sec. 3.2.1.

(e) $M_{j+1,k} - M_{j,k} = (1-z)(M_{j,k+1} - M_{j,k}) = z^j \cdot \left(\frac{z}{1-z}\right)^k \Delta^k u_j.$

$$\therefore \quad M_{\nu,k} - M_{n,k} = \left(\frac{z}{1-z}\right)^k \sum_{j=n}^{\nu-1} z^j \Delta^k u_j, \quad \nu \geq n.$$

$$\lim_{\nu\to\infty} M_{\nu,k} = \sum_{j=0}^{\infty} u_j z^j + \frac{1}{1-z} \sum_{s=0}^{k-1} \left(\frac{z}{1-z}\right)^s \lim_{\nu\to\infty} (z^\nu \Delta^s u_\nu) = \sum_{j=0}^{\infty} u_j z^j$$

for $\lim |\Delta^s u_\nu| = 0, \quad |z| \leq 1.$

$$\therefore \quad \left|\sum_{j=0}^{\infty} u_j z^j - M_{n,k}\right| \leq \left|\frac{z}{1-z}\right|^k |z^n| \sum_{j=n}^{\infty} |\Delta^k u_j| = \left|\frac{z}{1-z}\right|^k |z^n \Delta^{k-1} u_n|$$

$$= |z| \cdot |M_{n,k} - M_{n,k-1}| \quad \text{for} \quad \sum_{j=n}^{\infty} |\Delta^k u_j| = \left|\sum_{j=n}^{\infty} \Delta^k u_j\right| = |\Delta^{k-1} u_n|,$$

because of the monotonicity assumption. Q.E.D.

4. (a) $E - 1 = hD(a_0 + a_1\nabla + a_2\nabla^2 + \dots)$. But $\nabla = 1 - E^{-1}$; hence $E = (1-\nabla)^{-1}$, $hD = \ln E = -\ln(1-\nabla)$.

Hence $(a_0 + a_1\nabla + a_2\nabla^2 + \dots) = \dfrac{(1-\nabla)^{-1} - 1}{-\ln(1-\nabla)}$,

$$(a_0 + a_1\nabla + a_2\nabla^2 + \dots)\cdot\left(\nabla + \frac{\nabla^2}{2} + \frac{\nabla^3}{3} + \dots\right)$$
$$\equiv \nabla + \nabla^2 + \nabla^3 + \dots.$$

Match coefficients: $a_0 = 1,$

$\frac{1}{2}a_0 + a_1 = 1 \Longrightarrow a_1 = \frac{1}{2},$

$\frac{1}{3}a_0 + \frac{1}{2}a_1 + a_2 = 1 \Longrightarrow a_2 = 1 - \frac{1}{4} - \frac{1}{3} = \frac{5}{12}.$

(b) $\nabla = hD(b_0 + b_1\nabla + b_2\nabla^2 + \dots),$

$$(b_0 + b_1\nabla + b^2\nabla^2 + \dots) = \frac{\nabla}{-\ln(1-\nabla)},$$

$$(b_0 + b_1\nabla + b_2\nabla^2 + \dots)\left(\nabla + \frac{\nabla^2}{2} + \frac{\nabla^3}{3} + \dots\right) \equiv \nabla.$$

Matching coefficients yields $b_0 = 1$, $b_1 = -\frac{1}{2}$, $b_2 = -\frac{1}{12}$.

5. (a) $y_i = (-\Delta)^i x_0$.

$$(-\Delta)^n y_0 = (1-E)^n y_0 = \sum_{i=0}^{n} \binom{n}{i}(-1)^i y_i = \sum_{i=0}^{n} \binom{n}{i} \Delta^i x_0 = (1+\Delta)^n x_0 = x_n.$$

(b) Put $y = (y_0, y_1, \ldots, y_N)$, $x = (x_0, x_1, \ldots, x_N)$,

$$y_i = (-\Delta)^i x_0 = \sum_{j=0}^{i} \binom{i}{j}(-1)^j x_j, \quad i = 0, 1, \ldots, N).$$

This relation can be expressed in the form $y = Ax$, $\forall x$. The result of (a) then shows that $x = Ay$—i.e., $y = A^{-1}x$, $\forall x$. Hence $A^{-1} = A$.

Section 7.7

1. Introduce polar coordinates (r, φ), $x = r\cos\varphi$, $y = r\sin\varphi$, $dx\,dy = r\,dr\,d\varphi$. After integration in the r-direction,

$$I = k^{-1}\int_0^{2\pi} [1 - \cos(k\sin\varphi)]\,d\varphi.$$

This integral is not expressible in finite terms of elementary functions. Its value is in fact $(1 - J_0(k))\cdot 2\pi/k$, where J_0 is a Bessel function; see, e.g., Abramowitz-Stegun [27] or Jahnke-Emde [36]. Note that the integrand is a periodic function of φ, whence the trapezoidal rule is very efficient. (See Sec. 7.4.5(c) and Problem 6 in Sec. 7.4.) This is a useful device for computer programs for Bessel functions and many other transcendental functions which have integral representations.

2. The functional determinant is equal to

$$a\cos t\cdot br\cos t - b\sin t(-ar\sin t) = abr.$$

Hence

$$I = ab\int_0^1 r\,dr\int_0^{2\pi} u(ar\cos t, br\sin t)\,dt.$$

The remark of the previous problem is valid here too.

3. Algorithm 198, *CACM* 6, no 8 (August 1963).

$a :=$ Integral (sin $(x) \uparrow 2*$Integral (2*sin $(y) \uparrow 2$/sqrt(1 + $x*x$ + $y*y$), y, 0, sqrt(1 − $x*x$), 5, $_{10}$ − 6), x, −1, sqrt(3)/2, 4, $_{10}$ − 5);

$a := a +$ Integral (sin $(x) \uparrow 2*$Integral (2* sin $(y) \uparrow 2$/sqrt(1 + $x*x$ + $y*y$), y, 0, 0.5, 5, $_{10}$ − 6), x, sqrt(3)/2, 3, 4, $_{10}$ − 5).

4. 2.125.

5. (a) $(u_{1,1} - u_{1,-1} - u_{-1,1} + u_{-1,-1})/(4hk)$;
(b) replace in Taylor's formula the derivatives by difference approximations valid for quadratic polynomials.

6. (b) Put $P = h\partial/\partial x$, $Q = h\partial/\partial y$. Then the left-hand side is

$$\sum_{i=1}^{3}(u_i - 2u_0 + u_{3+i})$$

$$= 2[\cosh P - 1] + 2\left[\cosh\left(\frac{P + Q\sqrt{3}}{2}\right) - 1\right] + 2\left[\cosh\left(\frac{P - Q\sqrt{3}}{2}\right) - 1\right]$$

$$= P^2 + \frac{2P^4}{24} + \dots + \left(\frac{P + Q\sqrt{3}}{2}\right)^2 + \frac{2}{24}\left(\frac{P + Q\sqrt{3}}{2}\right)^4 + \dots +$$

$$\left(\frac{P - Q\sqrt{3}}{2}\right)^2 + \frac{2}{24}\left(\frac{P - Q\sqrt{3}}{2}\right)^4$$

$$= \frac{3}{2}(P^2 + Q^2) + \frac{2}{24}\left(\frac{9P^4}{8} + \frac{18P^2Q^2}{8} + \frac{9Q^4}{8}\right) + O(h^6)$$

$$= \frac{3}{2}(P^2 + Q^2) + \frac{3}{32}(P^2 + Q^2)^2 + O(h^6).$$

7. Use operator technique, with the notations of the previous example, and show that the expansion of

$$h \cdot \frac{e^P - e^{-P}}{P} \cdot k \cdot \frac{e^Q - e^{-Q}}{Q} - \frac{4hk}{6}[2 \cosh P + 2 \cosh Q + 2]$$

only contains terms of the form P^iQ^j, where $i + j \geq 4$. Convince yourself that this is sufficient.

One can also split up the rectangle into two triangles.

Cubic polynomial = Quadratic polynomial + $\sum_{i=0}^{3} a_i(x - x_0)^i(y - y_0)^{3-i}$. The correctness for quadratic polynomials follows from Theorem 7.7.2, and the sum gives, by antisymmetry, zero contribution to the integral as well as to the approximation formula.

8. $\alpha\theta_i + (1 - \alpha)\theta'_i$.

9. Yes. In a quadratic polynomial,

$$a_1 + a_2x + a_3y + a_4x^2 + a_5xy + a_6y^2,$$

there are six coefficients. The matching to the six points mentioned gives six nonhomogeneous linear equations. Such a system has a unique solution, unless the matrix is singular. If the matrix were singular, there would be cases where the system had no solutions at all. However, Theorem 7.7.1 shows that there is *always* at least one solution. Hence the matrix is nonsingular, and the solution is unique.

10. See Simpson's rule. Take coordinate axes along SQ and QR, in general nonorthogonal.

CHAPTER 8

Section 8.1

1. $y'_1 = y_2$; $y'_2 = x^2 - y_2 - y_3^2$; $y'_3 = y_4$; $y'_4 = x + y_4 + y_1^3$.
   Initial conditions: $y_1(0) = 0$; $y_2(0) = 1$; $y_3(0) = 1$; $y_4(0) = 0$.

2. $y(1) = 0.3679$, $z(1) = 0.3606$, $y(1) - z(1) = 0.0073$. $L = -1$, $\epsilon_0 = 0.01$. The magnitude of the disturbance in the interval $(x, x + dx)$ is $|-0.01\, z \cdot dx| \leq 0.01 \cdot dx$. By Theorem 8.1.3 the total effect at $x = 1$ of disturbances is less than

$$0.01e^{-1} + \int_0^1 e^{-(1-x)}0.01\, dx = 0.01,$$

which is a slight overestimate of the actual difference, 0.0073. (A better estimate is obtained if one uses the approximation $z \cdot dx \approx e^{-x}\, dx$.)

3. (a) By Eq. (5.5.6), $\|I + \epsilon A\|_\infty = \max_i (|1 + \epsilon a_{ii}| + \sum_{j \neq i} |\epsilon a_{ij}|)$. Hence

$$\mu(A) = \lim_{\epsilon \downarrow 0} \max_i \left(\frac{|1 + \epsilon a_{ii}| - 1}{\epsilon} + \sum_{j, j \neq i} |a_{ij}| \right),$$

which yields the result.

(b) $\exp(\|A\|_\infty) = e^{11}$; $\exp(\mu_\infty[A]) = e^{-9}$.

Section 8.2

1. The extrapolation scheme becomes

$h = 0.5$	0.5000		
$h = 0.25$	0.5010	0.5020	
$h = 0.125$	0.5030	0.5050	0.5060

(Correct value: 0.5063)

2. 1.22861.

3.
```
for k := 0 step 1 until n do
begin y := y0; x := a; h := h0*2↑(−k);
  m := (b − a)/h + 0.5;
  for i := 1 step 1 until m do
  begin y := y + h*f(x, y);
    x := x + h
  end;
  T[k, 0] := y;
  for i := 0 step 1 until k − 1 do
  begin R := T[k, i] − T[k − 1, i];   result := T[k, i];
    if abs(R) < eps then goto out;
    T[k, i + 1] := T[k, i] + R/(2↑(i + 1) − 1)
  end
end; out:
```

4. (a) $(1 - h)^n$; (b) $0 \leq h \leq 2$; (c) $-\frac{1}{2} \cdot x \cdot e^{-x}$.

5. $h_{max} = 2/k$.

Section 8.3

1. Differentiation gives $y^{(i)}(0) = i!$, $i = 1, 2, \ldots$. The Taylor series is thus

$$y(x) = 1 + x + x^2 + x^3 + \ldots.$$

This series has radius of convergence 1; thus it is not usable for computing $y(1.2)$. The exact solution is

$$y = \frac{1}{1 - x}, \quad x < 1,$$

which has a vertical asymptote for $x = 1$; this explains why the series is no good for $x > 1$.

For $x = 0.2$ we get $y \doteq 1.25000$ using terms up to and including x^6; note that

$$|R_T| \leq \frac{(0.2)^7}{1 - 0.2} = 16 \cdot 10^{-6},$$
$$|R_C| = 16 \cdot 10^{-6},$$
$$|R| \leq (16 + 16) \cdot 10^{-6} < 0.5 \cdot 10^{-4}.$$

2. A suitable way to set up the computations is as follows:

$h = 0.2$	x	y	y'	$k = h \cdot y'$
	0	1	1	0.2
	0.1	1.1	1.2	0.24
	0.1	1.12	1.22	0.244
	0.2	1.244	1.444	0.2888

$y(0.2) = 1 + \frac{1}{6}(0.2 + 0.48 + 0.488 + 0.2888) = 1.242800$

$h = 0.1$	x	y	y'	k
	0	1	1	0.1
	0.05	1.05	1.1	0.11
	0.05	1.055	1.105	0.1105
	0.1	1.1105	1.2105	0.12105
	0.1	1.110342	1.210342	0.121034
	0.15	1.170859	1.320859	0.132086
	0.15	1.176385	1.326385	0.132638
	0.2	1.242980	1.442980	0.144298
	0.2	1.242805		

Extrapolation: $\quad y(0.2) = 1.242805 + \dfrac{5 \cdot 10^{-6}}{16 - 1} = 1.242805.$

The exact value is $2 \cdot e^{0.2} - 1.2 = 1.242806 \pm 1 \cdot 10^{-6}$.

3. (a) 0.84; (b) 0.84 without and 0.83672 with modification; (c) 0.83579; (d) 0.83600; (e) 0.83438. Correct result: 0.83578.

4. 0.83579.

5. By Eq. (8.3.2), $y_{N+1} = y_{N-1} + 2hf(y_N)$. Hence

$$\tfrac{1}{4}(y_{N+1} + 2y_N + y_{N-1}) = \tfrac{1}{4}(2y_N + 2y_{N-1} + 2hf(y_N))$$
$$= \tfrac{1}{2}(y_N + y_{N-1} + hf(y_N)).$$

which is identical to Eq. (8.3.3).

6. $y(x) = \displaystyle\int_0^\infty \frac{e^{-u^2}}{x + u}\, du,$

$$\frac{dy}{dx} = \int_0^\infty -\frac{e^{-u^2}}{(u + x)^2}\, du = \left[\frac{e^{-u^2}}{u + x}\right]_0^\infty + \int_0^\infty \frac{2ue^{-u^2}}{u + x}\, du$$
$$= -\frac{1}{x} - \int_0^\infty \frac{2x \cdot e^{-u^2}}{u + x}\, du + \int_0^\infty 2e^{-u^2}\, du = -\frac{1}{x} - 2xy + \sqrt{\pi}.$$

The Runge-Kutta method with step 0.2 then gives

$$y(1.2) = 0.5268, \qquad y(1.4) = 0.4692.$$

7. (a) $y_n = y_1^n$, where $y_1 = y_1(h) = 1 - h + h^2/2 - h^3/6 + h^4/24$;
(b) $0 \leq h \leq h_0$, where $h_0 \approx 2.7$ is a root of the equation $y_1(h) = 1$;
(c) $(y(x, h) - e^{-x})/h^4 \longrightarrow xe^{-x}/120$;
(d) $h_{\max} \approx 2.7/k$. Compare this with the experience in Examples 8.3.6 and 8.3.7!

8. (a) $A = 1$, $B = -\frac{1}{6}$; $R_T = \frac{7}{360}h^5y_0^{(5)}$;
(b) y_1 is computed with an error $= -R_T$;
(c) 0.1003567.

9. Order $= 2$.
(b) This is a worse case of the same trouble as was described in Example 8.3.8. When $h \ll 1$, very few significant digits of $f(x, y)$ affect the result of the $y_{n+2} = O(1) + h^3 f$ operator. A summed version of this formula can be devised, but we omit the details.

12. (a) Put $z(x) = y(-x)$; then $z(x)$ satisfies the same differential equation and initial conditions as $y(x)$. Since the initial-value problem has a unique solution, $y(-x) = y(x)$.
(b) Because of the symmetry, $y_{-1} = y_1$; hence $u_1 = u_{-1}$. By Eq. (8.3.23), u_1 can then be determined from

$$2u_1 - 2 = \tfrac{5}{6}h^2, \qquad y_n = \frac{u_n}{1 - (h^2/12)(1 - x_n^2)}.$$

Results:

$$y(0.4, h = 0.4) = 1.0787486,$$
$$y(0.4, h = 0.2) = 1.0788527,$$
$$y(0.4)_{\text{extrapol}} = 1.0788596.$$

13. (a) $y_3 = 1.4265$, $y_4 = 1.6596$; exact results: $y(x) = 1/(1 - x)$, $y_3 = \frac{10}{7} = 1.4286$, $y_4 = \frac{10}{6} = 1.6667$;
(b) $y_3 = 1.4287$, predicted $y_4 = 1.6630$, corrected $y_4 = 1.6670$, after one more iteration $y_4 = 1.6675$;
(c) the correction of y_4 is multiplied by approximately $\frac{5}{12}h(\partial f/\partial y) \approx 0.14$ in each iteration, if the effects of rounding are ignored.

Section 8.4

1. The secant method gives $y'(0) = -0.01833$.

2. (a) Put $x = s/L$, $y = u/c$, $z = (2 - x^2)y''$.
Then the system becomes

$$(2 - x^2)y'' = z,$$
$$z'' = -40y + 2 - x^2,$$

with boundary conditions $z(1) = z'(1) = 0$ and (because of symmetry) $y'(0) = z'(0) = 0$.

(b) $x_i = i/N$, $i = 0, 1, 2, \ldots, N$,

$$(2 - x_i^2)(y_{i-1} - 2y_i + y_{i+1}) = z_i/N^2,$$

$$z_{i-1} - 2z_i + z_{i+1} = (-40y_i + 2 - x_i^2)/N^2.$$

Insert the boundary conditions:

$$y_{-1} = y_1; \qquad z_{-1} = z_1; \qquad z_N = 0; \qquad z_{N+1} = z_{N-1}.$$

Then the number of unknowns becomes $(N + 2) + N$, which is equal to the number of equations.

(c) Using $N = 1$ and $N = 2$ we get $u(0) = 0.045527c$. Using $N = 2$ and $N = 5$ we get $u(0) = 0.045252c$.

(d) The numbering scheme $Y_0 = y_0$, $Y_1 = z_0$, $Y_2 = y_1$, $Y_3 = z_1$, $Y_4 = y_2$, etc. to $Y_{2N} = y_N$, $Y_{2N+1} = y_{N+1}$ yields a matrix where $a_{pq} = 0$ when $|p - q| > 3$.

Note: $N = 1$ gives a step size which is so coarse that the effect of the variation of $I(s)$ is not at all reproduced. Thus it is not surprising that the first extrapolation does not yield a better result.

3. (a) Using the difference approximation of Eq. (7.7.9) we get

$$\frac{1}{h^2}[(1 + (x_n + h/2)^2)(y_{n+1} - y_n) - (1 + (x_n - h/2)^2)(y_n - y_{n-1})] + \lambda y_n = 0.$$

The error in y_n is $O(h^2)$; thus the same holds for λ. With $h = \frac{2}{3}$, we get (using symmetry) $\lambda_1 = 3.25$. With $h = \frac{2}{5}$, $\lambda_1 \approx 3.496$. h^2-extrapolation yields $\lambda_1 \approx 3.64$.

(b) Represent the condition $y'(0) = 0$ by $y_{-1} = y_1$. $\lambda_1 = 3.50$ yields, after integration with step length $\frac{1}{5}$, $y_5 \approx 0.0244$; $\lambda_1 = 3.75$ yields $y_5 \approx -0.0282$. Linear inverse interpolation (to determine the zero of the function $y_5(\lambda)$) gives $\lambda_1 \approx 3.616$. This value of λ_1 then yields $y_5 \approx -2.6 \cdot 10^{-4}$. Then, using Richardson extrapolation ($h = \frac{2}{5}$, $\lambda_1 \approx 3.496$; $h = \frac{1}{5}$, $\lambda_1 \approx 3.616$), we get $\lambda_1 \approx 3.656$.

(c) The value $\lambda_1 = 3.496$ in part (a) came from the smallest root of the equation

$$(\lambda - \tfrac{29}{4}) \cdot (\lambda - \tfrac{70}{4}) - (\tfrac{29}{4})^2 = 0.$$

Roots: $$\frac{99 \pm \sqrt{5{,}045}}{8}.$$

As first approximation to λ_2, then, we can take $\lambda_2 \approx (99 + \sqrt{5045})/8 \approx 21.3$. Proceeding as in (b), we get (using $h = \frac{1}{5}$) finally $\lambda_2 = 26.7951$. Richardson extrapolation yields $\lambda_2 \approx 26.8 + (26.8 - 21.3)/3 \approx 28.6$.

4. After integration with the Runge-Kutta method, we have

$$y(1) = y_n + c_4 h^4 + c_5 h^5 + O(h^6).$$

The quotient between step lengths in the given data is not constant, so repeated extrapolation cannot be done directly. However, two separate extrapolations can be performed according to the tables below, and the results can be used to estimate the error.

$$\text{Denominator} = \left(\frac{h_1}{h_3}\right)^k - 1 = \left(\frac{20}{10}\right)^4 - 1 = 15.$$

h	$\lambda = 6.76$		$\lambda = 6.77$		$\lambda = 6.78$	
$\frac{1}{10}$	16,126		5,174		−5,752	
$\frac{1}{20}$	15,261	15,203	4,304	4,246	−6,627	−6,685

$$\text{Denominator} = \left(\frac{h_2}{h_3}\right)^k - 1 = \left(\frac{20}{15}\right)^4 - 1 = \frac{175}{81}.$$

h	$\lambda = 6.76$		$\lambda = 6.77$		$\lambda = 6.78$	
$\frac{1}{15}$	15,396		4,441		−6,490	
$\frac{1}{20}$	15,261	15,199	4,304	4,241	−6,627	−6,690

From these two tables we estimate the errors in the extrapolated values to be ± 5 units. Using the extrapolated values in the lower table, inverse interpolation yields $\lambda \approx 6.7739$.

5. (a) $\dfrac{2i}{2i+1}u_{i-1} - (2 - \lambda h^2)u_i + \dfrac{2i+2}{2i+1}u_{i+1} = 0, \qquad 0 \leq i \leq N-1.$
Boundary condition: $u_N = 0$ (note that the coefficient of u_{-1} is zero).
(b) $N = 1 \Rightarrow \lambda = 4.5$; $N = 2 \Rightarrow \lambda = 5.28$. Using extrapolation, one gets $\lambda = 5.72$. Note that $h_1/h_2 = \frac{5}{3}$. (The exact value is $\lambda = 5.784$.)
(c) If A is the matrix corresponding to the equations in (a) above, and if $D = \text{diag}\,[(2i+1)(2i+3)]^{1/2}$, then $D^{-1}AD$ is a symmetric matrix. This trick can generally be used for symmetrizing a tridiagonal matrix whose off-diagonal elements are positive.

Section 8.5

1. (a) $y_n = \frac{1}{4}(3^n - (-1)^n)$; (b) $y_n = ((2+i)^n - (2-i)^n)/i = 5^{n/2}\, 2 \sin(n \arctan \frac{1}{2})$; (c) $y_n = (3^n - (-1)^n)/(3^{10} - 1)$; (d) $y_n = (-1)^n(1-n)$.

2. (a) $y_n = 2^n - 1$; (b) $y_n = (3^n + (-1)^n - 2)/8$; (c) $y_n = \frac{1}{2}n(n-1)$; (d) $y_n = \frac{1}{4}n(n-1)2^n$.

3. (a) $y_n = cu^n \Rightarrow \Delta^i y_n = c(u-1)^i u^n$; (b) $y_n = 3^n - 2^n$; (c) $\sum b_i(1 - u^{-1})^i = 0$ or $\sum b_i u^{k-i}(u-1)^i = 0$.

4. (a) $u_1 = \frac{1}{2}(1 + \sqrt{5})$, $u_2 = \frac{1}{2}(1 - \sqrt{5})$, $y_n = (u_1^n - u_2^n)/(u_1 - u_2)$, $y_{n+1}/y_n \longrightarrow u_1$;
(b) Put $y_n = \log \epsilon_n$, $y_{n+1} - u_1 y_n \longrightarrow (u_1 - 1)\log C$. Hence $p = u_1$, $\epsilon_{n+1}/\epsilon_n^p \longrightarrow C^{p-1}$;
(c) $x_n = y_{n+1}/y_n \approx u_1 + (u_1 - u_2)(u_2/u_1)^n + O((u_2/u_1)^{2n})$. In the notation of Eq. (3.2.2), $x_6 = 1.625$, $x_6' = 1.6182$, $u_1 = 1.6180$.

5. (a) and (b): the same companion matrix (see Problem 4 of Sec. 5.2)

$$A = \begin{bmatrix} 0 & 1 & 0 & \cdots & 0 \\ 0 & 0 & 1 & \cdots & 0 \\ \cdots & & & \cdots & \\ 0 & 0 & 0 & \cdots & 1 \\ -a_k & -a_{k-1} & -a_{k-2} & & -a_1 \end{bmatrix}.$$

6. (a) By induction, $\|y_n\| \le z_n$, where $z_0 = \|y_0\|$, $z_{n+1} = az_n + b$; solve this equation;
(b) $\|y_n\| \le z_n$, $z_{n+1} = mz_n + b$, $z_n = b/(1-m) + (1 - b/(1-m))\|y_0\|$;
(c) put $y_n = Tx_n$, where $T^{-1}AT = D$ = diagonal matrix, $C_n = T^{-1}B_nT$,

$$x_{n+1} = (D + hC_n)x_n, \quad \|x_{n+1}\| \le (1 + hK_2)\|x_n\| \le e^{hK_2}\|x_n\|,$$

where $K_2 = K\|T^{-1}\|\cdot\|T\|$. $\|x_n\| \le \|x_0\| \exp(K_2 nh)$. The result follows with $K_1 = \|T^{-1}\|\cdot\|T\|$.
(d) $\|y_n\| \le (\|y_0\| + nb)\cdot a^n$. Note that this is bounded as $n \longrightarrow \infty$ if $a < 1$.

7. (a) $y_n = I_n + c(-5)^n$; (b) $y_n = I_n + \epsilon(-5)^n$; (c) $y_n = I_n - I_{10}(-5)^{n-10}$;
(d) by Problem 6 the effect in y_n of the initial rounding can be as large as $\frac{1}{2}\cdot 10^{-3}\cdot 5^n$, while the total effect of the subsequent rounding operations does not exceed $\frac{1}{2}\cdot 10^{-3}\cdot(5^n - 1)/(5-1) \le \frac{1}{8}\cdot 10^{-3}\cdot 5^n$.

8. (a) $u_1u_2 = 1$. Hence $|u_1| \le 1$, $|u_2| \le 1 \Longleftrightarrow u_1 = e^{i\varphi}$, $u_2 = e^{-i\varphi}$, φ real. Hence $2\lambda = e^{i\varphi} + e^{-i\varphi} = 2\cos\varphi$—i.e., $-1 \le \lambda \le 1$. The double-root cases $\lambda = \pm 1$ must be excluded.
(b) Diagonalize. The condition in (a) must hold for all eigenvalues of A.

9. (a) and (b): $\lambda = 2\cos(\pi k/10)$, $k = 1, 2, 3, \ldots, 9$. Then $y_n = \sin(\pi kn/10)$.

10. By the technique of Problem 9, $2 - \lambda_j(N)/N^2 = 2\cos(\pi j/N)$.

$$\lambda_j(N) = 2N^2(1 - \cos(\pi j/N)) = (\pi j)^2 - 2N^2\cdot\cos\xi\cdot(\pi j/N)^4/24,$$

by Table 3.1.1. The Taylor expansion of the cosine contains only even powers of $\pi j/N$.

11. (a) $|1 + qh| \le 1$, the interior of a circle; (b) $|1 - qh| \ge 1$, the exterior of a circle; (c) $p = 1$; (d) Euler is stable but not A-stable. Backward Euler is stable and A-stable.

12. $|1 + qh + (qh)^2/2 + (qh)^3/6 + (qh)^4/24| \le 1$. Real axis: $-2.7 \lesssim qh \le 0$, by Problem 7 of Sec. 8.3. Imaginary axis: put $qh = iw$, $|w| \le \sqrt{8} \approx 2.83$, because $(1 - \omega^2/2 + (\omega^4/24)^2 + (\omega - \omega^3/6)^2 \le 1 \Longleftrightarrow -\omega^6/72 + (\omega^4/24)^2 \le 0 \Longleftrightarrow -8 + \omega^2 \le 0$.

13. $0 < h < \sqrt{6}$.

14. Characteristic equation: $1 - 1/u + \frac{1}{2}(1 - 1/u)^2 = qh$—i.e., $(2 - 1/u)^2 = 1 + 2qh$.

$$0 \ge 2qh \ge -1 \Longrightarrow u \in [1/3, 1], \text{ simple root at } 1.$$

$$2qh < -1 \Longrightarrow 1/u = 2 - i\omega \Longrightarrow |u| \le 1/2.$$

15. (a) $\beta_1 = 4$, $\beta_2 = 2$, $c = 1/6$;

(b) roots of characteristic equation $u_1 \approx \exp(0.1) - 3\cdot 10^{-6} \approx 1.1$, $u_2 \approx -4.7$. The method is unstable. The total effect of the rounding errors is, however, less than $\frac{1}{2}\cdot 10^{-6}(|\zeta_2|^9 + |\zeta_2|^8 + \ldots + 1) \approx 0.7$. The larger part of the error comes from the fact that $|y_1 - \zeta_1| \approx |\exp(0.1) - \zeta_1| \approx 3\cdot 10^{-6}$. If rounding errors are neglected, $y_n = (1-a)\zeta_1^n + a\zeta_2^n$, where $a \approx -5\cdot 10^{-7}$, $\zeta_2^{10} \approx 5.3\cdot 10^6$—i.e., $a\zeta_2^{10} \approx -e$.

16. Oscillating component reduced from $|e^{-qx}\alpha^3/12|$ to $|e^{-qx}\alpha^5/48|$, i.e. from $2\cdot 10^{-4}$ to $5\cdot 10^{-7}$. Regular error component, $-e^{qx}\cdot qx\alpha^2/6$, increased by $e^{qx}\alpha^2/4$, i.e. from $6\cdot 10^{-4}$ to $15\cdot 10^{-4}$.

17. (a) $qh = \alpha$, $\rho(\zeta_j(\alpha)) - \alpha\sigma(\zeta_j(\alpha)) = 0$, $\zeta_j(0) = \zeta_j$, $\rho'(\zeta_j)\cdot\zeta_j'(0) - \sigma(\zeta_j) = 0$.
Hence $\zeta_j(\alpha) = \zeta_j(1 + \alpha\lambda_j + O(\alpha^2)) = \zeta_j \exp(\lambda_j qh) + O(h^2)$.
Particular solution: $y_n = [\zeta_j(\alpha)]^n$.
(b) Consistency $\Longleftrightarrow p \geq 1$. By Eq. (8.5.12),

$y(x) = 1, \quad \sum \alpha_i = 0 \Longleftrightarrow \rho(1) = 0.$

$y(x) = x, \quad -\sum \alpha_i ih - h\sum \beta_i = 0 \Longleftrightarrow -k\rho(1) + \rho'(1) - \sigma(1) = 0,$

Hence $\rho'(1) = \sigma(1)$—i.e., $\lambda_1 = 1$.
(c) $\rho(\zeta) = \zeta^2 - 1$, $\sigma(\zeta) = 2\zeta$, $\zeta_2 = -1$, $\lambda_2 = -1 < 0$, weak stability.
(d) $\rho(\zeta) = \zeta^2 - 1$, $\sigma(\zeta) = \frac{1}{3}(\zeta^2 + 4\zeta + 1)$ $\zeta_2 = -1$, $\lambda_2 = -1/3 < 0$, weak stability.

18. (a) $\sqrt{a} = (2/h)\sin h^{-1}\sqrt{\alpha h/2}$, $b = \beta a/\alpha$.
(b)
$$\begin{array}{ll} n=1: & 5.78\alpha + \beta = 32 \\ n=4: & 9.98\alpha + \beta = 71 \end{array}\Bigg\} \Longrightarrow \begin{cases}\alpha = 9.29 \\ \beta = -21.71\end{cases} \Longrightarrow \begin{cases} a = 9.21, \\ b = -21.53,\end{cases}$$
$n = 2$: $6.70\cdot 9.29 - 21.71 - 45 = -4.47$,
$n = 3$: $8.07\cdot 9.29 - 21.71 - 54 = -0.74$.

Better results may be obtained by the method of least squares. Note, however, that the errors in the four equations are not independent if the measurement of y_n is the primary error source.

Section 8.6

1. (a) $u_{i,j+1} = u_{i-1,j} - u_{ij} + u_{i+1,j}$; (b) -16;
(c) Useless (for ordinary recursive computation).

2. $u_{i,j+1} = u_{i,j} + \frac{1}{4}[(1 + x^2_{i+(1/2)})(u_{i+1,j} - u_{i,j}) - (1 + x^2_{i-(1/2)})(u_{i,j} - u_{i-1,j})]$.

$t \backslash x$	± 1	± 0.6	± 0.2
0	0	400	800
0.04	0	352	684
0.08	0	304	588
0.12	0	262	506

In the beginning, the singularity at the origin gives poor accuracy, but with time the singularity is smoothed out.

3. (a) Verify that $u(t)$ satisfies the differential equation and initial condition. Note that $\lambda_j v_j = A v_j$.
(b) See Problem 9 of Sec. 8.5.
(c) $d(u^T u)/dt = 2u^T\, du/dt = 2u^T A u \leq 0$, since A is negative-definite, by (b).
(d) $\|u_{n+1}\|^2 - \|u_n\|^2 = (u_{n+1} + u_n)^T(u_{n+1} - u_n) = (u_{n+1} + u_n)^T A (u_{n+1} + u_n)/2 < 0$, since A is negative-definite.
(e) See Example 5.4.3. The system is tridiagonal. $O(N)$ operations in both cases.
(f) The difference approximation of Eq. (7.7.9) yields a tridiagonal matrix A, where the diagonal elements are negative, and the row sums are zero, except at the boundaries. By Gerschgorin's theorem A is negative-definite or semidefinite. All conclusions in (c), (d), and (e) hold.

4. (a) In successive time steps the following values are obtained for $x = 0$: 1, 1, 1.5, 2.5, 4.375, 7.875, ...
(b) By Problem 3 one particular solution grows like

$$v_{N-1} \cdot \exp\,[(1 + \cos \pi/N) \cdot 2N^2 t] \approx v_{N-1} \exp\,(4N^2 t).$$

This indicates the possible speed of growth of disturbances. For $\partial u/\partial t = -\partial^2 u/\partial x^2$ there are solutions of arbitrarily fast growth.

5. $a = 106$, $c = 112$, $b = 108$. (The values should be computed in this order.)

6. We use the notation of Sec. 5.6:

$$\rho(B_{GS}) \sim 1 - 1/40, \qquad \bar{\omega} = 2/(1 + \sqrt{0.025}) \approx 1.72,$$
$$R_{GS} \approx 0.01, \qquad R_{\bar{\omega}} \approx 0.14,$$

R measures roughly the number of decimal digits gained per iteration.

7.
$$\frac{u_{i+1,j} + u_{i-1,j} + u_{i,j-1} - 4u_{i,j}}{h^2} + \lambda u_{i,j} + O(h^2) = 0.$$

$h = \frac{1}{4}$

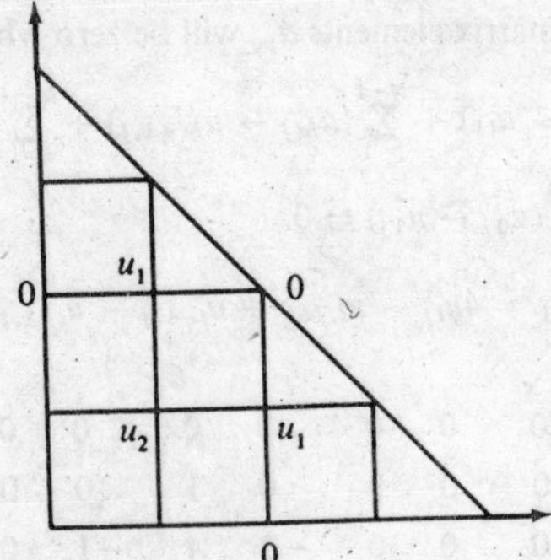

$$\begin{cases} u_2 - 4u_1 + \dfrac{\lambda}{16} \cdot u_1 = 0 \\ 2u_1 - 4u_2 + \dfrac{\lambda}{16} \cdot u_2 = 0 \end{cases} \Longrightarrow \begin{vmatrix} \dfrac{\lambda}{16} - 4 & 1 \\ 2 & \dfrac{\lambda}{16} - 4 \end{vmatrix} = 0 \Longrightarrow \lambda = 41.38.$$

$h = \frac{1}{5}$

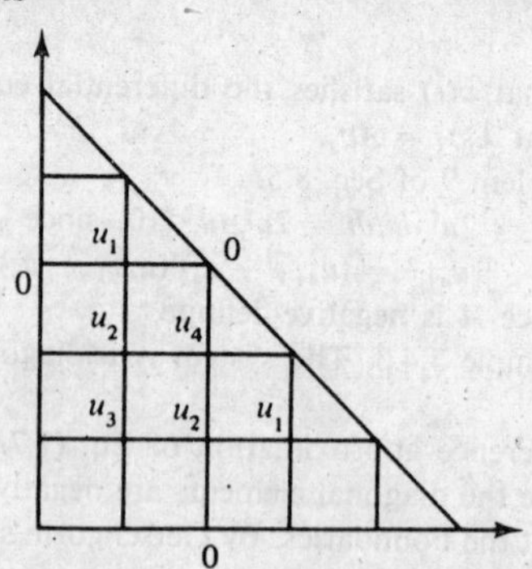

$$\begin{cases} u_2 - 4u_1 + \dfrac{\lambda}{25}u_1 = 0 \\ u_1 + u_3 + u_4 - 4u_2 + \dfrac{\lambda}{25}u_2 = 0 \\ \left.\begin{aligned} 2u_2 - 4u_3 + \dfrac{\lambda}{25}u_3 = 0 \\ 2u_2 - 4u_4 + \dfrac{\lambda}{25}u_4 = 0 \end{aligned}\right\} \Longrightarrow u_3 = u_4. \end{cases}$$

The determinant is zero for $\lambda \approx 44.07$.

Richardson: $\quad \lambda = 44.07 + \dfrac{44.07 - 41.38}{(5/4)^2 - 1} = 48.9.$

Exact: $\quad 5\pi^2 = 49.35.$

9. By introducing u as unknown in $(N + 1)(2N + 1) - 1$ points (i.e., also on the boundary but not in the corners) and v as unknown in $N \cdot 2N$ points (i.e., not on the boundary), one obtains $4N^2 + 3N$ unknowns. For a suitable ordering the matrix elements a_{pq} will be zero whenever $|p - q| > 2N + 2$.

10. $$\sum_{i=1}^{M-1} (u_{i0} - u_{i1}) + \sum_{j=1}^{N-1} (u_{Mj} - u_{M-1,j}) + \sum_{i=1}^{M-1} (u_{iN} - u_{i,N-1}) + \sum_{j=1}^{N-1} (u_{0j} - u_{1j}) = 0.$$

11. (a) $u_{i,j+1} = 4u_{ij} - u_{i,j-1} - u_{i-1,j} - u_{i+1,j}$.

$j\downarrow$								
0	0	0	0	0	0	0	0	0
0	0	0	0	1	0	0	0	0
0	0	0	−1	4	−1	0	0	0
0	0	1	−8	17	−8	1	0	0
0	−1	12	−49	80	−49	12	−1	0

Evidently an unstable computation.

(b) $2A = \begin{bmatrix} 4 & -1 & & \\ -1 & 4 & -1 & \\ & \cdots & & \end{bmatrix}$, $\lambda_j = 2 - \cos(\pi j/N) \notin (-1, 1)$,

unstable method.

12. $$2A = \begin{bmatrix} 2[1-(k/h)^2] & (k/h)^2 & 0 & 0 & \cdots \\ (k/h)^2 & 2[1-(k/h)^2] & (k/h)^2 & 0 & \cdots \\ 0 & \cdots & & & \end{bmatrix}.$$

$2\lambda_j = 2[1-(k/h)^2] + 2(k/h)^2 \cos(\pi j/N)$, (All $\lambda_j \in [-1, 1]$) $\Longleftrightarrow |k/h| \leq 1$.

13. (a) $v = u - x^4 - y^4 \Rightarrow \nabla^2 v = 1 - 12(x^2 + y^2)$ in D, $v = 0$ on ∂D.
(b) Take a particular solution of $\nabla^2 u = 1$—for example, $u_0 = \frac{1}{4}(x^2 + y^2)$. $v = u - u_0 \Rightarrow \nabla^2 v = 0$ in D, $v = (x^4 + y^4) - \frac{1}{4}(x^2 + y^2)$ on ∂D.
(c) If $u(x, y)$ satisfies the equation and boundary conditions, then so does $u(-x, y)$. Uniqueness of solution $\Rightarrow u(-x, y) = u(x, y)$. Similarly, $u(-x, -y) = u(x, y)$.

14. (b) The expansion has the form $u \approx \sum_{n=0}^{N/2} a_{2n} p_{2n}$.
(c) Use the recurrence relations

$$p_0(x, y) + iq_0(x, y) = 1,$$
$$p_{2n+2}(x, y) + iq_{2n+2}(x, y) = (x + iy)^2(p_{2n}(x, y) + iq_{2n}(x, y)),$$
$$a_{ij} = p_{2j-2}(x_i, y_i), \quad i = 1, 2, \ldots, M,$$
$$j = 1, 2, \ldots, N/2 + 1.$$

15. Yes, because $u = \sum c_j \psi_j$, $v = \sum d_i \psi_i \Rightarrow (v, Au) = \sum\sum d_i c_j (\psi_i, A\psi_j) = v_n^T A_n u_n$.

16. (a) $Au^* = f$.
$\langle u - u^*, u - u^* \rangle = \langle u, u \rangle - 2\langle u, u^* \rangle + \langle u^*, u^* \rangle = (u, Au) - 2(u, f) + \langle u^*, u^* \rangle$.
(b) Put $Au = -\nabla^2 u$. By Eq. (8.6.17) the integral is equal to $(u, Au) - 2(f, u)$, which by part (a) is minimized for $u = u^*$, where $-\nabla^2 u^* = f$.
(c) $-\nabla(p\nabla u) + g \cdot u - f = 0$.

17. (a) $$a_{ij} = (\psi_i', \psi_j') = \begin{cases} 2/h, & i = j, \\ -1/h, & |i - j| = 1, \end{cases}$$

$$(\psi_i, f) = \frac{h}{2}(f_{i-1/2} + f_{i+1/2}).$$

This yields the difference equation

$$(-u_{i+1} + 2u_i - u_{i-1})/h^2 = \tfrac{1}{2}(f_{i-1/2} + f_{i+1/2}).$$

The "explicit central difference method" and the Cowell method give the same left-hand side, but the right-hand side is f_i and $f_i + \frac{1}{12}\delta^2 f_i$, respectively. The latter gives $O(h^4)$-accuracy, while the finite element method and the "explicit central difference method" give $O(h^2)$-accuracy.

(b) $a_{ij} = (p\psi'_i, \psi'_j) + (g\psi_i, \psi_j)$. If p, g, f are treated as element-wise constants, one obtains the difference equation

$$(-p_{i-1/2} + \tfrac{1}{6}h^2 g_{i-1/2})u_{i-1} + [p_{i+1/2} + p_{i-1/2} + \tfrac{1}{3}h^2(g_{i-1/2} + g_{i+1/2})]u_i + (-p_{i+1/2} + \tfrac{1}{6}h^2 g_{i+1/2})u_{i+1} = \tfrac{1}{2}h^2(f_{i-1/2} + f_{i+1/2}).$$

19. (a) $\cos(x + y) = \cos x \cos y - \sin x \sin y$. Rank $= 2$.
$u(x) = c_1 \cos x + c_2 \sin x$,
$\hat{K}u(x) = \pi c_1 \cos x - \pi c_2 \sin x$; eigenvalues $= \pm\pi$.

(b) $u(x) = 1$.

(c) $\pi c_1 \cos x - \pi c_2 \cdot \sin x \equiv \cos x + ax$.
Solution exists iff $a = 0$. Then $c_1 = 1/\pi$, $c_2 = 0$.
General solution: $u(x) = (1/\pi)\cos x + v(x)$, where $v(x)$ is an arbitrary function which is orthogonal to $\cos x$ as well as to $\sin x$.

20. Yes, because $(\varphi_i, \varphi_j) = (\varphi_j, \varphi_i)$ and

$$(\varphi_i, \hat{K}\varphi_j) = \iint K(x, y)\varphi_i(x)\varphi_j(y)\,dx\,dy = \iint K(y, x)\varphi_i(x)\varphi_j(y)\,dx\,dy$$
$$= \iint K(x, y)\varphi_i(y)\varphi_j(x)\,dx\,dy = (\varphi_j, \hat{K}\varphi_i).$$

21. $x_i = h \cdot i$, $Nh = 1$, use the trapezoidal rule:
$du/dt = \boldsymbol{K}\boldsymbol{u} + \boldsymbol{f}(t)$,
where the elements of $\boldsymbol{K}$ are

$$K_{ij} = \begin{cases} \tfrac{1}{2}hK(x_i, x_j) & \text{if } j = 0 \text{ or } j = N, \\ hK(x_i, x_j) & \text{otherwise,} \end{cases}$$

$f^T(t) = [f(x_0, t), f(x_1, t), \ldots, f(x_N, t)]$.

Try different values of h, and use Richardson extrapolation. The differential equation can be treated by any of the methods in Sec. 8.3.

22. $x_i = a + (b - a)i/N$, $K_{ij}(u, v) = K(x_i, x_j, u, v)$, $f_i = f(x_i)$.

$u_i - \sum_j A_j K_{ij}(u_i, u_j) = f_i$.

The A_j are coefficients in a quadrature formula—e.g., $A_j = \frac{1}{2}(b - a)/N$ if $j = 0$ or $j = N$, $A_j = (b - a)/N$ otherwise.

One of the Newton-like methods of Sec. 6.9 can be applied. Imbedding, by the replacement of K by pK, where p grows from 0 to 1, may be efficient in order to obtain first approximations.

23. (a) $\sum_{j=1}^{n} (\varphi_i, (\hat{A} - \hat{K})\varphi_j)c_j = (\varphi_i, \hat{K}v)$, $i = 1, 2, \ldots, n$.

b) Put $u = \bar{u} + v$. Then $\hat{A}\bar{u} = \hat{K}\bar{u} + \hat{K}v$. Since $\{\hat{A}\varphi_j\}_1^n$ is a basis of the function space spanned by $\hat{K}$, this equation has a solution of the form $\bar{u} = \sum_1^n c_j\varphi_j$, unless the corresponding homogeneous problem has a nontrivial solution.

CHAPTER 9

Section 9.2

1. (a) By Eq. (9.2.6),

$$\pi b_j = \int_\pi^0 f(-x)\sin(-jx)d(-x) + \int_0^\pi f(x)\sin jx\,dx$$
$$= \int_0^\pi [f(x) - f(-x)]\sin jx\,dx.$$

Similarly, $\pi a_j = \int_0^\pi [f(x) + f(-x)]\cos jx\,dx.$

(b) f is an odd function, which satisfies the condition $f(\pi - x) = f(x)$.

2. (b) Use Eqs. (9.2.5) and (9.2.6). In the former case, integrate by parts.

Result: $f(x) = \dfrac{\pi}{2} + \dfrac{4}{\pi}\sum_{k=0}^{\infty}(2k+1)^{-2}\cos[(2k+1)x].$

(c) Put $x = 0$.

(d) $S - \frac{1}{4}S = 1 + 3^{-2} + 5^{-2} + \ldots = \pi^2/8$. Hence $S = \pi^2/6$.

(e) Put $g(x) = (\pi/4)(f(x) - \pi/2) = (\pi/4)(\pi/2 - |x|)$.

$c_j = \frac{1}{2}j^{-2}$, if j odd; $c_j = 0$, if j even (positive and negative!).

By Eq. (9.2.7),

$$2\cdot 2\pi \sum_0^\infty [\tfrac{1}{2}(2k+1)^{-2}]^2 = \int_{-\pi}^{\pi} g(x)^2\,dx = \pi^5/96.$$

Hence sum $= \pi^4/96$.

(f) The results are identical.

3. Put $t = x/2\pi$. Apply Eqs. (9.2.5) and (9.2.6) to $G_1(x/2\pi)$, but integrate from 0 to 2π, and use the symmetry.

Result: $a_j = 0, \quad b_j = (j\pi)^{-1}.$

Hence

$$G_1(t) = -\sum_{n=1}^{\infty}\frac{\sin 2n\pi t}{n\pi}.$$

$G_2(t)$ is obtained by termwise integration of this. The integration constant in the expansion for $G_2(t)$ will be zero because of the condition

$$\int_0^1 G_2(t)\,dt = 0.$$

Similarly, the expansion for $G_p(t)$ is obtained by repeated integration, and all integration constants will be zero. *Result:*

$$G_p(t) = \begin{cases} (-1)^{(1/2)p+1}\cdot 2\displaystyle\sum_{n=1}^{\infty}\frac{\cos 2n\pi t}{(2n\pi)^p}, & \text{if } p \text{ is even,}\\[2ex] (-1)^{(1/2)(p+1)}\cdot 2\displaystyle\sum_{n=1}^{\infty}\frac{\sin 2n\pi t}{(2n\pi)^p}, & \text{if } p \text{ is odd.}\end{cases}$$

By definition, $c_p = (-1)^p G_p(0)$. Hence $c_p = 0$ for odd p, and

$$c_p = (-1)^{(1/2)p+1} \cdot 2 \sum_{n=1}^{\infty} (2n\pi)^{-p}, \quad \text{if } p \text{ is even.}$$

i.e.,

$$\sum_{1}^{\infty} n^{-p} = \tfrac{1}{2}|c_p| \cdot (2\pi)^p, \quad \text{if } p \text{ is even.}$$

Suppose that p is even:

$$c_p - G_p(t) = (-1)^{(1/2)p+1} \cdot 2 \sum_{n=1}^{\infty} \frac{1 - \cos 2n\pi t}{(2n\pi)^p}.$$

Since $1 - \cos 2n\pi t \geq 0$, we have

$$\operatorname{sgn}(c_p - G_p(t)) = \operatorname{sgn}((-1)^{(1/2)p+1}) = \operatorname{sgn}(c_p).$$

Section 9.3

1. Representations of integers in the form of Eq. (9.3.3) are familiar in everyday life. The algorithm for the determination of the α_i is exactly what you do when you convert 720,000 seconds, say, into weeks, days, hours, minutes, and seconds. Equation (9.3.4) can be rewritten in the form:

$$\beta = k_1 N_0/N_0 + k_2 N_0/N_1 + \ldots = k_1 + k_2 r_1 + k_3 r_2 r_1 + \ldots + k_p r_{p-1} \ldots r_2 r_1.$$

Similarly ($\beta_0 = \beta$),

$$\beta_{i-1} = k_i + k_{i+1} r_i + k_{i+2} r_{i+1} r_i + \ldots + k_p r_{p-1} \ldots r_{i+1} r_i.$$

The remainder in the division β_{i-1}/r_i is evidently equal to k_i, while the quotient is equal to β_i.

3. Look in the index of algorithms (Chap. 13) under the heading "C6, Summation of Series. . .": Algorithm 339, *Comm. ACM* (Nov. 1968), 776, and (March 1969) 187.

Section 9.4

1. (a) By the formula for a sum of a geometric series,

$$A = \sum_{k=1}^{N} \exp(i\pi k/(N+1)) = \frac{-1 - \exp(i\pi/(N+1))}{\exp(i\pi/(N+1)) - 1}.$$

Divide numerator and denominator by $\exp[\frac{1}{2}i\pi/(N+1)]$:

$$A = \frac{-2\cos[\frac{1}{2}\pi/(N+1)]}{2i\sin[\frac{1}{2}\pi/(N+1)]} = i\cot[\tfrac{1}{2}\pi/(N+1)].$$

(b) By Eq. (9.4.3),

$$b_j = \frac{2}{N+1} \sum_{k=1}^{N} \sin[j\pi k/(N+1)].$$

By the procedure in (a), b_j is the imaginary part of

$$\frac{\exp(ji\pi) - \exp[ji\pi/(N+1)]}{\exp[ji\pi/(N+1)] - 1}, \quad \begin{cases} j \text{ even} \Rightarrow b_j = \operatorname{Im}(-1) = 0, \\ j \text{ odd} \Rightarrow b_j = \dfrac{2}{N+1}\cot[\frac{1}{2}j\pi/(N+1)]. \end{cases}$$

(c) For j odd, $\lim_{N\to\infty} b_j = 4/(j\pi)$ in agreement with Example 9.2.1.

2. (a) By Eq. (9.2.3),

$$2\pi c_j = \int_{-\pi}^{\pi} f(x)e^{-ijx}\,dx = [-f(x)/(ij)]_{-\pi}^{\pi} + (ij)^{-1}\int_{-\pi}^{\pi} f'(x)e^{-ijx}\,dx$$
$$= (ij)^{-1}\int_{-\pi}^{\pi} f'(x)e^{-ijx}\,dx.$$

Repeat this k times!

$$2\pi c_j = (ij)^{-(k+1)}\int_{-\pi}^{\pi} f^{(k+1)}(x)e^{-ijx}\,dx, \quad |c_j| \le j^{-(k+1)}\cdot\|f^{(k+1)}\|_\infty.$$

(b) $$|f(x) - \sum_{j=-n}^{n} c_j e^{ijx}| = |\sum_{|j|>n} c_j e^{ijx}| \le 2\sum_{j=n+1}^{\infty} |c_j|$$
$$\le 2\|f^{(k+1)}\|_\infty \cdot \sum_{j=n+1}^{\infty} j^{-(k+1)} < 2\|f^{(k+1)}\|_\infty \cdot \int_n^\infty t^{-(k+1)}\,dt$$
$$= 2\|f^{(k+1)}\|_\infty/(kn^k).$$

Section 9.5

2. (a) $g(t) = \int_{-\infty}^{\infty} \varphi(\xi)\exp(-2\pi i\xi t)\,d\xi,$

$$\int_{-\infty}^{\infty} \varphi(\xi+\alpha)\exp(-2\pi i\xi t)\,d\xi = \int_{-\infty}^{\infty} \varphi(x)\exp(-2\pi i(x-\alpha)t)\,dx$$
$$= \exp(2\pi i\alpha t)\int_{-\infty}^{\infty} \varphi(x)\exp(-2\pi ixt)\,dx$$
$$= \exp(2\pi i\alpha t)g(t).$$

(b) Integrate the first equation in Eq. (9.5.3) by parts:

$$g(t) = \left[\frac{e^{-2\pi i\xi t}}{-2\pi it}\varphi(\xi)\right]_{-\infty}^{\infty} + \frac{1}{2\pi it}\int_{-\infty}^{\infty} \varphi'(\xi)\exp(-2\pi i\xi t)\,d\xi.$$

Hence the Fourier transform of φ' is equal to $2\pi it\cdot g(t)$. The result follows after $k-1$ repetitions of this.

3. By Eq. (9.5.3),

$$\varphi(\xi) = \int_{-\infty}^{\infty} g(t)\exp(2\pi i\xi t)\,dt = \int_{-\infty}^{\infty}\int_{-\infty}^{\infty} g_1(x)g_2(t-x)\exp(2\pi i\xi t)\,dx\,dt.$$

Introduce new variables $u = x$, $v = t - x$. The Jacobian determinant is equal to 1, $t = u + v$

$$\varphi(\xi) = \int_{-\infty}^{\infty}\int_{-\infty}^{\infty} g_1(u)g_2(v)\exp(2\pi iu\xi)\exp(2\pi iv\xi)\,du\,dv$$
$$= \int_{-\infty}^{\infty} g_1(u)\exp(2\pi iu\xi)\,du\cdot\int_{-\infty}^{\infty} g_2(u)\exp(2\pi iv\xi)\,dv$$
$$= \varphi_1(\xi)\cdot\varphi_2(\xi).$$

4. Suppose that f is periodic, period $= L$.

$$f(t) = \sum_{j=-\infty}^{\infty} c_j\exp(2\pi ijt/L).$$

Denote complex conjugation by bars. Put

$$I = \int_{-L/2}^{L/2} |f(t)|^2\,dt = \int_{-L/2}^{L/2} \sum_j c_j \exp(2\pi ijt/L)\cdot \sum_k \bar{c}_k \exp(-2\pi ikt/L)\,dt$$
$$= \sum_j \sum_k c_j\bar{c}_k \int_{-L/2}^{L/2} \exp[2\pi i(j-k)t/L]\,dt.$$

The integrals are equal to L for $j = k$, and zero otherwise. Hence

$$I = L\cdot \sum_{j=-\infty}^{\infty} |c_j|^2.$$

Equation (9.2.7) is obtained for $L = 2\pi$. In order to derive Eq. (9.5.4) we proceed in the same "heuristic way" as in the derivation of Eq. (9.5.3). Then $c_j = L^{-1}g_L(j/L)$,

$$I = L^{-1} \sum_{j=-\infty}^{\infty} |g_L(j/L)|^2 \longrightarrow \int_{-\infty}^{\infty} |g(t)|^2\,dt, \quad \text{as } L \longrightarrow \infty.$$

CHAPTER 10

Section 10.2

1. (a) $x_1 = 0, x_2 = 1$, $\max f = 3$; (b) $x_1 = 0, x_2 = 1$, $\max f = 5$;
(c) $x_1 = x_2 = \frac{2}{3}$, $\max f = \frac{4}{3}$; (d) $x_1 = 0, x_2 = \frac{3}{2}$, $\max f = \frac{9}{2}$.
The equality sign holds in the same constraints as in (a) and (b).
(e) Among others, that the right-hand variables at an optimal feasible point can remain the same even when the values of the coefficients in the optimization problem are changed significantly. (However, this does not hold for all changes, as (c) shows.)

2. Note that x_5 and x_6 have to be zero in the optimal point, since they have positive coefficients in f and are constrained by the nonnegativity condition only. Hence x_5 and x_6 may be eliminated. Six slack variables are introduced. Then

$$c^T = [1,\ 2,\ 3,\ 4,\ 1,\ 0,\ 0,\ 0,\ 0,\ 0,\ 0]$$

$$A = \begin{bmatrix} 1 & 1 & 1 & 0 & 0 & 1 & 0 & 0 & 0 & 0 & 0 \\ 1 & 1 & 1 & 0 & 0 & 0 & -1 & 0 & 0 & 0 & 0 \\ 3 & 1 & 0 & 5 & 0 & 0 & 0 & 1 & 0 & 0 & 0 \\ 1 & 1 & 3 & 0 & 0 & 0 & 0 & 0 & -1 & 0 & 0 \\ 1 & -1 & 0 & 0 & 5 & 0 & 0 & 0 & 0 & 1 & 0 \\ 1 & -1 & 0 & 0 & 5 & 0 & 0 & 0 & 0 & 0 & -1 \end{bmatrix}, \quad b = \begin{bmatrix} 16 \\ -8 \\ 6 \\ 3 \\ -1 \\ -1 \end{bmatrix}.$$

3. $c'_{Li} = c_{Lj}/c_{ij}$, $c'_{LR} = c_{LR} - c_{Lj}c_{iR}/c_{ij}$, $(L \neq j, R \neq i)$,
$c'_{ji} = 1/c_{ij}$, $c'_{jR} = -c_{iR}/c_{ij}$, $(R \neq i)$.

Section 10.3

(a) Introduce two vectors y_+, y_-, which satisfy the relations $y = y_+ - y_-$, $y_+ \geq 0$, $y_- \geq 0$. Introduce a slack vector z, such that $y^T A - z^T = b$,

$z \geq 0$. Put $Y^{\mathsf{T}} = [y_+^{\mathsf{T}}, y_-^{\mathsf{T}}, z^{\mathsf{T}}]$. After some manipulation, we obtain the following result: Maximize $[-b^{\mathsf{T}}, b^{\mathsf{T}}, 0]\cdot Y$, subject to the constraints $[A^{\mathsf{T}}, -A^{\mathsf{T}}, -I]Y = c$, $Y \geq 0$.

(b) The dual of the dual problem reads: Minimize $-x^{\mathsf{T}}\cdot c$, subject to the constraints $-x^{\mathsf{T}}[A^{\mathsf{T}}, -A^{\mathsf{T}}, -I] \geq [-b^{\mathsf{T}}, b^{\mathsf{T}}, 0]$. This is equivalent to the maximization of $x^{\mathsf{T}}c$, subject to the constraints $-Ax \geq -b$, $Ax \geq b$, $x \geq 0$. Since the first two constraints are equivalent to the equation $Ax = b$, this is equivalent to the primal problem.

Section 10.5

1. (a) $g(x) = \varphi'(x)^{\mathsf{T}} = Gx - b$, Hessian $= G$. $\hat{x}$ exists iff the system $G\hat{x} = b$ is consistent. Taylor's formula yields Eq. (10.5.21).

(b) $x_1 = x_0 - G^{-1}(Gx_0 - b) = G^{-1}b = \hat{x}$. By Eq. (10.5.21), $\hat{x}$ is a minimum point if G is positive-definite.

2. Apply the Schwarz inequality to $\psi'(0) = -d^{\mathsf{T}}g(x_0)$.

3. If Q is quadratic, then Q' is linear. If λ' and θ are chosen according to Eq. (10.5.7), then

$$Q'(\lambda') = (1 - \theta)Q'(\tfrac{1}{2}(a + b)) + \theta Q'(\tfrac{1}{2}(b + c))$$
$$= (1 - \theta)\cdot\psi[a, b] + \theta\psi[b, c] = 0.$$

See also Problem 3 of Sec. 7.5.

5. (a) By Taylor's formula, $\psi(\lambda) = \psi(0) - \lambda g^{\mathsf{T}}d + \frac{1}{2}\lambda^2 d^{\mathsf{T}}Gd$. Put $\psi'(\lambda) = 0$.

(b) If $g^{\mathsf{T}}Gg > 0$ and $g^{\mathsf{T}}G^{-1}g > 0$, then Newton-Raphson's method gives a larger reduction of ψ than the steepest-descent method with optimal λ, if $g^{\mathsf{T}}G^{-1}g > (g^{\mathsf{T}}g)^2/g^{\mathsf{T}}Gg$. The conclusion holds if $\varphi(x_0 - \lambda d)$ can be approximated by a quadratic function of λ reasonably well in the relevant intervals.

6. Direct insertion and induction.

7. A symmetric rank-one matrix is always of the form uu^{T}.

$$H_v\gamma = H_{v-1}\gamma + uu^{\mathsf{T}}\gamma.$$

Hence

$$u = c(\delta - H_{v-1}\gamma), \quad c = 1/u^{\mathsf{T}}\gamma.$$

8. The Lagrange-multiplier method yields the equations

$$\varphi'(c_0) + \lambda^{\mathsf{T}}u'(c_0) = 0, \qquad u(c_0) = 0. \qquad (*)$$

The penalty function method gives

$$\varphi'(\hat{x}(k)) + 2k^{-1}u^{\mathsf{T}}(\hat{x}(k))\cdot u'(\hat{x}(k)) = 0.$$

Let $k \to 0$ and subtract Eq. $(*)$:

$$[\lim_{k\to 0} 2k^{-1}u^{\mathsf{T}}(\hat{x}(k)) - \lambda^{\mathsf{T}}]\cdot u'(c_0) = 0.$$

The result follows, since the rank of u' is equal to m.

CHAPTER 11

Section 11.2

1. (a) is correct; (b) is incorrect.
2. $P[\min\{R_1, R_2\} \leq x] = 1 - P[R_1 \geq x, R_2 \geq x] = 1 - (1-x)^2, 0 \leq x \leq 1.$ The frequency function is the derivative of this.
3. $P[R = i] = \alpha_i,$
$$f(x)\,dx = P[x \leq X \leq x + dx] = \sum_{i=1}^{k} P[R = i] \cdot f_i(x)\,dx.$$
4. Suppose we have k subintervals $[x_{i-1}, x_i]$, $i = 1, 2, \ldots, k$. For $x \in [x_{i-1}, x_1]$ put
$$x = x_{i-1} \cdot (1 - \theta) + x_i \cdot \theta, \quad 0 \leq \theta \leq 1, f(x) = a_i \cdot (1 - \theta) + b_i \cdot \theta.$$
Take four uniformly distributed random numbers R_1, R_2, R_3, R_4. i is chosen from R_1 as in Problem 3 with $\alpha_i = (x_i - x_{i-1})(a_i + b_i)/2$. Then
$$\theta = \begin{cases} R_3 & \text{if } R_2 \leq \dfrac{\min\{a_i, b_i\}}{(a_i + b_i)/2}, \\ \min(R_3, R_4) & \text{otherwise.} \end{cases}$$
If $a_i < b_i$, replace θ by $(1 - \theta)$.

Section 11.3

3. (a) 2,997 ± 991, 2,788 ± 204, 362 ± 151, 414 ± 53;
(b) $(991/53)^2 \approx 350$. (This judgment is based on the very uncertain estimate of the variances which one obtains with three observations.)

Section 11.4

1. (a) Statement 1, in conjunction with the previous statement, computes (I* 65539) mod P. (If underflow occurs in the product I*65539, then 2^{32} is added.) A further improvement in speed can be achieved by coding the three relevant statements "in-line" in the program where needed, instead of using a subroutine call.
(b) Simple Monte Carlo. (Exact value of integral = 1.71828.) (We just give the results of the program.)

N	Integral	|Error|	$\sqrt{N}\cdot$|Error|
25	1.76042	0.042	0.211
100	1.69529	0.023	0.230
225	1.69704	0.021	0.319
400	1.71364	0.005	0.093
625	1.69974	0.019	0.464

The error is (approximately) inversely proportional to $\sqrt{N}$.

(c) Control variate method.

N	Integral	$\lvert\text{Error}\rvert$	$\sqrt{N}\cdot\lvert\text{Error}\rvert$
25	1.72350	0.00522	0.026
100	1.71818	0.00010	0.001
225	1.71763	0.00065	0.010
400	1.71936	0.00108	0.022
625	1.71747	0.00081	0.020

3. (a) The covariance matrix of $\boldsymbol{y}$ is

$$V = E(\boldsymbol{y}\boldsymbol{y}^{\mathsf{T}}) = E(B\boldsymbol{x}\boldsymbol{x}^{\mathsf{T}}B^{\mathsf{T}}) = BE(\boldsymbol{x}\boldsymbol{x}^{\mathsf{T}})B^{\mathsf{T}} = B\sigma^2 IB^{\mathsf{T}}$$
$$= \sigma^2 BB^{\mathsf{T}} \qquad (E = \text{expected value}).$$

(b) Compute the Choleski decomposition, $V = LL^{\mathsf{T}}$, according to Sec. 5.3.5. If $\boldsymbol{x}$ has independent components, with mean zero and variance σ^2, then, by (a), the vector $\boldsymbol{y} = \sigma^{-1}L\boldsymbol{x}$ has the covariance matrix V.

13 BIBLIOGRAPHY AND PUBLISHED ALGORITHMS

13.1. INTRODUCTION

Since, for many readers, numerical analysis is an important practical subject, we shall here give a more complete overview of the literature than is usually given in an elementary textbook. The bibliography which we give is, however, by no means complete. It is, rather, a subjective selection which we feel should be a good guide to the reader who out of interest (or necessity!) wishes to deepen his knowledge. A rough division into various areas has been made in order to facilitate searching. A short commentary introduces each of these parts.

Numerical analysis is still in a dynamic stage of development, and it is important to keep track of the recent literature on the subject. (Notice that in the following bibliography more than one-third of the books were published after 1967.) Besides the reference periodicals (see Sec. 13.8), the periodical *Mathematics of Computation* publishes extensive reviews of most books of interest.

In order to make applications easier, some of the most modern books contain descriptions of algorithms in some programming language, usually Algol or Fortran. These books are marked with an asterisk in the bibliography

13.2. GENERAL LITERATURE IN NUMERICAL ANALYSIS

Some elementary textbooks which can be read as a complement to this book are Henrici [9], Noble [17], and Stiefel [23]. McCracken and Dorn [15]

and Prager [19] are introductory books in numerical analysis which are integrated with a course in programming. Ekman-Fröberg [5] is purely a programming text.

More advanced books include Fröberg [7], Isaacson and Keller [12], and Ralston [20]. Todd [24] and Walsh [25] contain monographs of varying degrees of difficulty. Both of these books are suitable for a reader who has a good mathematical background and who wants to orient himself in the methods and results of numerical analysis. For a more fundamental approach to computer arithmetic and algorithms for polynomials, the reader can consult Knuth [13], vol. 2. Ralston and Wilf [21] is a handbook which contains monographs on various subjects together with complete flow charts and programs for the algorithms. The principal questions in the production of software for mathematical computation are discussed in Rice (ed.) [22].

In scientific and technical applications, one often needs to supplement one's knowledge of mathematical analysis. Courant-Hilbert [3] is an excellent source, although it is very advanced in some sections. A aluable sourcebook to the literature before 1956 is Parke [18].

Short reviews of most of the books listed below can be found in: "Curriculum, 68: Recommendations for Academic Programs in Computer Science," *Communications of the ACM* 11 (March 1968), 184–85. This article also contains an excellent survey of literature and courses for the whole subject of computer science.

*1. CARNAHAN, B., H. A. LUTHER, and J. O. WILKES (1969), *Applied numerical methods.* John Wiley & Sons, New York.

2. COURANT, R. (1936), *Differential and integral calculus,* 2 vols. Blackie & Son Ltd. London and Glasgow, 616 pp.

3. COURANT, R., and D. HILBERT (1953 and 1962), *Methods of mathematical physics,* 2 vols. Interscience, New York, 561 + 830 pp.

4. DIEUDONNÉ, J. (1961), *Foundations of modern analysis.* Academic Press, New York, 361 pp.

*5. EKMAN, T., and C.-E. FRÖBERG (1967), *Introduction to Algol programming.* Oxford University Press, London.

6. FOX, L., and D. F. MAYERS (1968), *Computing methods for scientists and engineers.* Clarendon Press, Oxford, 255 pp.

7. FRÖBERG, C.-E. (1969), *Introduction to numerical analysis,* 2nd ed. Addison-Wesley Publishing Company, Reading, Mass, 434 pp.

8. HAMMING, R. W. (1962), *Numerical methods for scientists and engineers.* McGraw-Hill, New York, 411 pp.

9. HENRICI, P. (1964), *Elements of numerical analysis.* John Wiley & Sons, New York, 328 pp.

10. HILDEBRAND, F. B. (1956), *Introduction to numerical analysis.* McGraw-Hill, New York, 511 pp.

11. HOUSEHOLDER, A. S. (1953) *Principles of numerical analysis.* McGraw-Hill, New York, 274 pp.

12. ISAACSON, E., and H. B. KELLER (1966), *Analysis of numerical methods.* John Wiley & Sons, New York, 541 pp.

13. KNUTH, D. E. (1969), *The art of computer programming*, vol. 2: *Seminumerical algorithms.* Addison-Wesley Publishing Company, Reading, Mass.

14. LANCZOS, C. (1956), *Applied analysis.* Prentice-Hall, Englewood Cliffs, N.J., 539 pp.

*15. MCCRACKEN, D. D., and W. S. DORN (1964), *Numerical methods and FORTRAN programming, with applications in engineering and science.* John Wiley & Sons, New York, 457 pp.

16. NATIONAL PHYSICAL LABORATORY (1961), *Modern computing methods*, 2nd ed. Notes on applied science no. 16., H. M. Stationery Office, London, 170 pp.

17. NOBLE, B. (1964), *Numerical methods*, 2 Vols. Oliver and Boyd, New York, 372 pp.

18. PARKE, N. G. III (1958), *Guide to the literature of mathematics and physics*, 2nd rev. ed. Dover Publications, New York.

*19. PRAGER, W. (1965), *Introduction to basic FORTRAN programming and numerical methods.* Blaisdell, New York, 203 pp.

*20. RALSTON, A. (1965), *A first course in numerical analysis.* McGraw-Hill, New York, 541 pp.

*21. RALSTON, A., and H. S. WILF, eds. (1960 and 1967), *Mathematical methods for digital computers*, vols. 1 and 2. John Wiley & Sons, New York, 293 + 287 pp.

*22. RICE, J. R. (1971), *Mathematical software.* Academic Press, New York.

23. STIEFEL, E. L. (1963), *An introduction to numerical mathematics.* Academic Press, New York, 286 pp.

*23a. STOER, J. (1972), *Einführung in die Numerische Mathematik I.* Springer, Berlin, 250 pp.

24. TODD, J., ed. (1962), *A survey of numerical analysis.* McGraw-Hill, New York, 589 pp.

25. WALSH, J., ed. (1966), *Numerical analysis: An introduction.* Academic Press, New York, 212 pp.

26. WENDROFF, B. (1966), *Theoretical numerical analysis.* Academic Press, New York, 239 pp.

* Contains Algol or Fortran programs.

13.3. TABLES, COLLECTIONS OF FORMULAS, AND PROBLEMS

Mathematical tables are no longer as important for numerical calculations as they were previously. To let a computer perform a table look-up is seldom the best way to compute a mathematical function. However, tables can often be an important aid in checking a calculation or planning calculations on a computer. Detailed advice about the use and choice of tables is given in Todd [24, pp. 93ff]. Todd distinguishes among tables which should be immediately available, tables which should be accessible in a local library, and tables which one should just know exist, and how to get them. Only tables in the first category are included in the list below.

27. Abramowitz, M., and I. A. Stegun (1964), *Handbook of mathematical functions.* National Bureau of Standards, Applied Math. Series no. 55, Washington, D.C., 1,045 pp.

28. Chemical Rubber Company (1959), *C.R.C. Standard mathematical tables.* Chemical Rubber Publishing Company, Cleveland.

29. Comrie, L. J. (1941), *Barlow's tables of squares, cubes, . . .*, 4th ed. E. & F. N. Spoon, London, 258 pp.

30. ——— (1949), *Chambers' six-figure mathematical tables*, vol. 2. W. & R. Chambers, Edinburgh, 576 pp.

31. ——— (1955), *Chambers' shorter six-figure mathematical tables.* W. & R. Chambers, Edinburgh, 387 pp. (Referred to in this book as "Chambers.")

32. Fletcher, A., et al. (1962), *An index of mathematical tables*, vols. 1 and 2, 2nd ed. Addison-Wesley Publishing Company, Reading, Mass., 994 pp.

33. Fröberg, C.-E. (1952), *Hexadecimal Conversion Tables.* Gleerup, Lund, Sweden.

34. Greenwood, J. A., and H. O. Hartley (1962), *Guide to tables in mathematical statistics.* Princeton University Press, Princeton, N.J., 1,014 pp.

*35. Hagander, N., and Y. Sundblad (1969), *Lösta exempel i numeriska metoder*, 2 Vols. Gleerup, Lund, Sweden, 178 + 259 pp.

36. Jahnke, E., F. Emde, and F. Lösch (1960), *Tables of higher functions*, 6th ed. McGraw-Hill, New York, 318 pp.

37. James, G., and R. C. James (1959), *Mathematics dictionary*, multilingual ed. D. Van Nostrand Co., New York, 432 pp.

38. Lebedev, A. V., and R. M. Federova (1960), *A guide to mathematical tables.* Pergamon Press, Oxford, 586 pp.

*39. Naur, P. (1968), *Twelve exercises in algorithmic analysis.* Studentlitteratur, Lund, Sweden, 41 pp.

40. Scheid, F. (1967), *Numerical analysis, including 775 solved problems*. Schaum Publishing Company, New York, 442 pp.

* Contains Algol or Fortran programs.

13.4. ERROR ANALYSIS AND APPROXIMATION OF FUNCTIONS

The literature on approximation theory is very extensive. An excellent introduction to the mathematical theory can be found in Cheney [43]. Handscomb [48], which contains a series of lectures given at a summer course in approximation theory at Oxford in 1965, also gives practical aspects on approximation. Hart et al. [49] and Lyusternik et al. [54] are more in the nature of handbooks.

41. Achieser, N. I. (1956), *Theory of approximation*. Frederick Ungar Publishing Company, New York, 307 pp.

42. Ahlberg, J. H., E. N. Nilsson, and J. L. Walsh (1967), *The theory of splines and their applications*. Academic Press, New York, 284 pp.

43. Cheney, E. W. (1966), *Introduction to approximation theory*. McGraw-Hill, New York, 259 pp.

44. Fike, C. T. (1968), *Computer evaluation of mathematical functions*. Prentice-Hall, Englewood Cliffs, N.J., 227 pp.

45. Fox, L., and I. B. Parker (1968), *Chebyshev polynomials in numerical analysis*. Oxford University Press, London, 205 pp.

46. Garabedian, H. L., ed. (1965), *Approximation of functions*. Elsevier, Amsterdam, 220 pp.

47. Greville, T. N. E., ed. (1969), *Theory and applications of spline functions*. Academic Press, New York and London, 220 pp.

48. Handscomb, D. C., ed. (1966), *Methods of numerical approximation*. Pergamon Press, Oxford, 218 pp.

*49. Hart, J. F., et al. (1968), *Computer approximations*. John Wiley & Sons, New York, 343 pp.

50. Hayes, J. G., ed. (1970), *Numerical approximation to functions and data*. Athlone Press of the University of London, London, 177 pp.

51. Korovkin, P. P. (1960), *Linear operators and approximation theory*. Russian monographs and texts on advanced mathematics and physics, vol. 3. Hindustan Publishing Corp., Delhi, India, 222 pp.

52. Langer, R. E., ed. (1959), *On numerical approximation*. University of Wisconsin Press, Madison, 462 pp.

53. LUKE, Y. (1969), *The special functions and their approximations*, 2 vols. Academic Press, New York and London, 349 + 487 pp.

54. LYUSTERNIK, L. A., O. A. CHERVONENKIS, and A. R. YANPOLSKII (1965), *Handbook for computing elementary functions*. Pergamon Press, New York, 251 pp.

55. MEINARDUS, G. (1967), *Approximation of functions: Theory and numerical methods*. Springer, Berlin, 196 pp.

56. MOORE, R. E. (1966), *Interval analysis*. Prentice-Hall, Englewood Cliffs, N.J., 145 pp.

57. RICE, J. (1964), *The approximation of functions*, vol. 1: *Linear theory*. Addison-Wesley Publishing Company Reading, Mass., 200 pp.

58. ——— (1969), *The approximation of functions*, vol. 2: *Nonlinear and multivariate theory*. Addison-Wesley Publishing Company, Reading, Mass., 334 pp.

58a. SCHULTZ, M. H. (1973), *Spline analysis*, Prentice-Hall, Englewood Cliffs, N.J., 156 pp.

59. SNYDER, M. A. (1966), *Chebyshev methods in numerical approximation*. Prentice-Hall, Englewood Cliffs, N.J., 114 pp.

60. SZEGÖ, G. (1959), *Orthogonal polynomials*. American Mathematical Society, New York, 421 pp.

61. WILKINSON, J. H. (1963), *Rounding errors in algebraic processes*. H. M. Stationery Office, London, 161 pp.

* Contains Algol or Fortran programs.

13.5. LINEAR ALGEBRA AND NONLINEAR SYSTEMS OF EQUATIONS

Perhaps the most complete survey of classical matrix theory can be found in Faddeev and Faddeeva [65]. A concentrated, more modern presentation is given in Householder [71], but it is relatively difficult reading. Forsythe and Moler [67] and Fox [68] both give an elementary presentation of numerical methods in linear algebra. Lighter surveys are given by Forsythe [66] and Kahan [72]. The most complete discussion of numerical properties of methods in linear algebra can be found in a series of publications by Wilkinson [83]–[85], and Wilkinson-Reinsch [86]. The latter book contains Algol programs from the "Handbook" series of *Numerische Mathematik*.

Rabinowitz [76] is a good introduction to the solution of nonlinear systems of equations. Ortega-Rheinboldt [73] is a more complete work on this subject.

62. BJERHAMMAR, A. (1973), *Theory of errors and generalized matrix inverses.* Elsevier Scientific Publishing Company, Amsterdam.

63. BODEWIG, E. (1959), *Matrix calculus*, 2nd ed. North Holland Publishing Company, Amsterdam, 452 pp.

*64. DEKKER, T. J. and W. HOFFMAN (1968), *ALGOL 60 procedures in numerical algebra*, nos. 1 and 2. Math. Centre Tracts, Math. Centrum, Amsterdam, 110 pp.

65. FADDEEV, D. K., and V. N. FADDEEVA (1963), *Computational methods of linear algebra.* W. H. Freeman, San Francisco, 620 pp.

*66. FORSYTHE, G. E. (1967), "Today's computational methods of linear algebra." *SIAM Review* 9, no. 3, 489–515.

*67. FORSYTHE, G. E., and C. MOLER (1967), *Computer solutions of linear algebraic systems.* Prentice-Hall, Englewood Cliffs, N.J., 148 pp.

68. FOX, L. (1964), *Introduction to numerical linear algebra.* Clarendon Press, Oxford, 328 pp.

69. GANTMACHER, F. R. (1960), *The theory of matrices.* vols. 1 and 2. Chelsea Publishing Company, New York, 374 + 276 pp.

70. GREGORY, R. T., and L. K. KARNEY (1969), *A collection of matrices for testing computational algorithms.* John Wiley & Sons, New York.

71. HOUSEHOLDER, A. S. (1964), *The theory of matrices in numerical analysis.* Blaisdell, New York, 257 pp.

72. ——— (1970), *The numerical treatment of a single nonlinear equation.* McGraw-Hill, New York, 176 pp.

73. KAHAN, W. (1966), "Numerical linear algebra." *Canadian Math. Bulletin* 9, no. 6, 757–801.

73a. LAWSON, C. L. and HANSON, R. J. (1974), *Solving least squares problems.* Prentice-Hall, Englewood Cliffs, N.J.

74. ORTEGA, J. M., and W. C. RHEINBOLDT (1970), *Iterative solution of non-linear equations in several variables.* Academic Press, New York.

75. OSTROWSKI, A. M. (1966), *Solutions of equations and systems of equations*, 2nd ed. Academic Press, New York, 338 pp.

76. RABINOWITZ, P. (1970), *Numerical methods for non-linear algebraic equations.* Gordon and Breach Science Publishers, London, 199 pp.

77. RAO, C. R., and S. K. MITRA (1971), *Generalized inverse of matrices and its applications.* John Wiley & Sons, New York, 240 pp.

78. REID, J. K., ed. (1971), *Large sparse sets of linear equations.* Academic Press, London and New York, 284 pp.

79. ROSE, D. J., and R. A. WILLOUGHBY, eds. (1972), *Sparse matrices and their applications.* Plenum Press, New York and London, 215 pp.

79a. STEWART, G. W. (1973), *Introduction to matrix computations.* Academic Press, New York.

80. TRAUB, J. F. (1964), *Iterative methods for the solution of equations*. Prentice-Hall, Englewood Cliffs, N.J., 310 pp.

81. VARGA, R. S. (1962), *Matrix iterative analysis*. Prentice-Hall, Englewood Cliffs, N.J., 322 pp.

82. WESTLAKE, J. R. (1968), *A handbook of numerical matrix inversion and solution of linear equations*. John Wiley & Sons, New York, 171 pp.

83. WILKINSON, J. H. (1961), "Error analysis of direct methods of matrix inversion." *J. Assoc. Comput. Mach.* 8, 281–330.

84. ——— (1963), *Rounding errors in algebraic processes*. H. M. Stationery Office, London, 161 pp.

85. ——— (1965), *The algebraic eigenvalue problem*. Clarendon Press, Oxford, 662 pp.

*86. WILKINSON, J. H., and C. REINSCH (1971), *Handbook for automatic computation*, vol. 2: *Linear Algebra*. Springer, Berlin.

87. ZURMÜHL, R. (1961), *Matrizen und ihre technischen Anwendungen*. Springer, Berlin, 459 pp.

* Contains Algol or Fortran programs.

13.6. INTERPOLATION, NUMERICAL INTEGRATION, AND NUMERICAL TREATMENT OF DIFFERENTIAL EQUATIONS

The most complete presentations of the problem of numerical integrations of *ordinary* differential equations can be found in Gear [98] and Henrici [100]. Gear's book is the most up-to-date among *textbooks* available on the subject at the present time. Collatz [91] and Milne [106] are also very good, but somewhat older, references. Another modern presentation is given by Lapidus and Seinfeld [105], who also give much attention to stiff problems. Daniel and Moore [93] give a very readable survey of methods for ordinary differential equations; they emphasize combining analytical and numerical techniques.

For the treatment of boundary-value problems see Fox [96] and Keller [102].

A good introduction to numerical solution of *partial* differential equations is given in Smith [111]. A more complete treatment can be found in Mitchell [107]. Other more advanced treatments are Richtmyer and Morton [110] and Forsythe and Wasow [95].

88. BABUSKA, I., M. PRAGER, and E. VITÁSEK (1966), *Numerical processes in differential equations*. John Wiley & Sons, London and Prague, 351 pp.

89. BULIRSCH, R., and J. STOER (1965), "Numerical treatment of ordinary dif-

ferential equations by extrapolation methods." *Numerische Mathematik* 8, 1–13.

90. CODDINGTON, E. A., and N. LEVINSON (1955), *Theory of ordinary differential equations*. McGraw-Hill, New York.

91. COLLATZ, L. (1966), *The numerical treatment of differential equations*, 3rd ed. Springer, Berlin, 568 pp.

92. DAHLQUIST, G. (1959), *Stability and error bounds in the numerical integration of ordinary differential equations*. Transactions of the Royal Institute of Technology no. 130, Stockholm, 86 pp.

93. DANIEL, J. W., and R. MOORE (1970), *Computation and theory in ordinary differential equations*. W. H. Freeman, San Francisco, 172 pp.

*94. DAVIS, P. J., and P. RABINOWITZ (1967), *Numerical integration*. Blaisdell, London, 230 pp.

95. FORSYTHE, G. E., and W. R. WASOW (1960), *Finite difference methods for partial differential equations*. John Wiley & Sons, New York, 444 pp.

96. FOX, L. (1957), *Numerical solution of two-point boundary value problems*. Clarendon Press, Oxford, 371 pp.

97. ———, ed. (1962), *Numerical solution of ordinary and partial differential equations*. Addison-Wesley Publishing Company, Reading, Mass., 509 pp.

*98. GEAR, C. W. (1971), *Numerical initial value problems in ordinary differential equations*. Prentice-Hall, Englewood Cliffs, N. J.

*99. GEORGE, J. A. (1971), *Computer implementation of the finite element method*. Report STAN-CS-71-208, Stanford University, 222 pp.

100. HENRICI, P. (1962), *Discrete variable methods in ordinary differential equations*. John Wiley & Sons, New York, 407 pp.

101. ——— (1963), *Error propagation for difference methods*. John Wiley & Sons, New York, 73 pp.

102. KELLER, H. B. (1968), *Numerical methods for two-point boundary value problems*. Blaisdell, London, 184 pp.

103. KRYLOV, V. I. (1962), *Approximate calculation of integrals*. Macmillan, New York, 371 pp.

104. KUNTZMANN, J. (1959), *Méthodes numériques, interpolation—dérivées*. Dunod, Paris, 252 pp.

104a. LAMBERT, J. D. (1973), *Computational methods in ordinary differential equations*. John Wiley & Sons, New York, 278 pp.

105. LAPIDUS, L., and J. H. SEINFELD (1971), *Numerical solution of ordinary differential equations*. Academic Press, New York.

106. MILNE, W. E. (1953), *Numerical solution of differential equations*. John Wiley & Sons, New York, 275 pp.

107. MITCHELL, A. R. (1969), *Computation methods in partial differential equations*. John Wiley & Sons Ltd, London, 255 pp.

108. ODEN, J. T. (1972), *Finite elements of nonlinear continua*. McGraw-Hill, New York, 430 pp.

109. RALL, L. B. (1969), *Computational solution of nonlinear operator equations*. John Wiley & Sons, New York, 224 pp.

110. RICHTMYER, R. D., and K. W. MORTON (1967), *Difference methods for initial value problems*, 2nd ed. Interscience, New York, 405 pp.

111. SMITH, G. D. (1965), *Numerical solution of partial differential equations*. Oxford University Press, Oxford, 179 pp.

112. STEFFENSEN, J. F. (1950), *Interpolation*. Chelsea Publishing Company, New York, 248 pp.

113. STETTER, H. J. (1973), *Analysis of discretization methods for ordinary differential equations*. Springer, Berlin, 388 pp.

113a. STRANG, G. and FIX, G. (1973), *An analysis of the finite element method*. Prentice-Hall, Englewood Cliffs, N.J.

114. STROUD, A. H., and D. SECREST (1966), *Gaussian quadrature formulas*. Prentice-Hall, Englewood Cliffs, N. J., 374 pp.

115. ——— (1972), *Approximate calculation of multiple integrals*. Prentice-Hall, Englewood Cliffs, N. J.

116. YOUNG, D. M. (1972), *Iterative solution of large linear systems*. Academic Press, New York, 570 pp.

*117. ZONNEVELD, J. A. (1964), *Automatic numerical integration*. Math. Centre Tracts no. 8, Math. Centrum, Amsterdam, 110 pp.

* Contains Algol or Fortran programs.

13.7. OPTIMIZATION; SIMULATION

The best general reference in linear optimization is Dantzig [124], while a good presentation of the elementary theory can be found in Gass [130]. Künzi et al. [136] contains Algol and Fortran programs for the most important algorithms. For a survey of nonlinear optimization see Künzi [135] and Abadie [118].

Dixon [125] is a good general introduction to optimization. Daniel [123] gives an excellent treatment of the problem of minimizing functionals. Bellman and Kalaba [122] is perhaps the best introduction to the ideas behind dynamic programming.

Hammersley and Handscomb [133] is unsurpassed as a general reference on Monte Carlo methods.

118. ABADIE, J., ed. (1967), *Nonlinear programming*. North Holland Publishing Company, Amsterdam, 316 pp.

119. ——— (1970), *Integer and nonlinear programming*. North Holland Publishing Company, Amsterdam, 544 pp.

120. BAILEY, N. T. J. (1952), "A study of queues and appointment systems in hospital outpatient departments, with special reference to waiting times." *Journ. Roy. Stat. Soc.* 3, no. 14, 185.

121. BELLMAN, R. E., and S. E. DREYFUS (1962), *Applied Dynamic Programming*. Princeton University Press, Princeton.

122. BELLMAN, R. E., and R. KALABA (1971), *Dynamic programming and modern control theory*. Academic Press, New York and London, 112 pp.

122a. BRENT, R. P. (1973), *Algorithms for minimization without derivatives*. Prentice-Hall, Englewood Cliffs, N.J.

123. DANIEL, J. W. (1971), *The approximate minimization of functionals*. Prentice-Hall, Englewood Cliffs, N. J., 242 pp.

124. DANTZIG, G. B. (1965), *Linear programming and extensions*, 2nd ed. Princeton University, Press, Princeton, 623 pp.

125. DIXON, L. C. W. (1972), *Nonlinear optimization*. English Universities Press, London, 214 pp.

126. EVANS, G. W., G. F. WALLACE, and G. L. SUTHERLAND (1967), *Simulation using digital computers*. Prentice-Hall, Englewood Cliffs, N. J., 198 pp.

127. FELLER, W. (1950), *An introduction to probability theory and its applications*, vol. 1. John Wiley & Sons, New York.

128. FIACCO, A. V., and G. P. MCCORMICK (1968), *Sequential unconstrained minimization techniques*. John Wiley & Sons, New York, 210 pp.

129. FORD, L. R., and D. R. FULKERSON (1962), *Flows in networks*. Princeton University Press, Princeton, 194 pp.

130. GASS, S. I. (1969), *Linear programming, methods and applications*, 3rd ed. McGraw-Hill, New York, 384 pp.

131. GRAVES, L. G., and P. WOLFE (1963), *Recent advances in mathematical programming*. McGraw-Hill, New York, 346 pp.

132. HADLEY, G. (1962), *Linear programming*. McGraw-Hill, New York.

——— (1964), *Nonlinear and dynamic programming*. Addison-Wesley Publishing Company, Reading, Mass.

133. HAMMERSLEY J. M., and D. C. HANDSCOMB (1964), *Monte Carlo Methods*. Methuen & Co., London, 178 pp.

134. JANSSON, B. (1966), *Random number generators*. Victor Pettersons Bokindustri Aktiebolag, Stockholm.

135. KÜNZI, H. P., and W. KRELLE (1966), *Nonlinear programming*. Blaisdell, Waltham, Mass., 240 pp.

*136. KÜNZI, H. P., H. G. TZSCHACH, and C. A. ZEHNDER (1968), *Numerical methods of mathematical optimization*. Academic Press, New York, 171 pp.

137. MANGAZARIAN, O. L. (1969), *Computational methods in optimization*. McGraw-Hill, New York.

138. RAND CORPORATION (1955), *A million random digits with 100,000 normal deviates*. Free Press, Glencoe, Ill.

139. VAJDA, S. (1958), *Readings in linear programming*. Pitman and Sons, London.

140. ZANGWILL, W. I. (1969), *Nonlinear programming: A unified approach*. Prentice-Hall, Englewood Cliffs, N.J., 356 pp.

* Contains Algol or Fortran programs.

13.8. REVIEWS, ABSTRACTS, AND OTHER PERIODICALS

Computer Abstracts (1957–), Technical Information Company, New Jersey.

Computing Reviews (1960–), Association for Computing Machinery, New York.

Mathematical Reviews (1940–), American Mathematical Society, Providence, R. I.

**BIT* (*Nordisk Tidskrift for Informationsbehandling*) (1961–), Copenhagen.

**Communications of the Association for Computing Machinery* (1958–), New York.

**The Computer Journal* (1958–), British Computer Society, London.

**Computing. Archiv für Elektronisches Rechnen* (1965–), Springer, Vienna and New York.

Journal of Approximation Theory (1968–), Academic Press, New York.

**Journal of the Association for Computing Machinery* (1954–), New York.

Journal of Computational Physics (1966–), Academic Press, New York.

Journal of Mathematical Analysis and Applications (1945–), Academic Press, New York.

Journal of the Institute of Mathematics and Its Applications (1965–), Academic Press, New York.

Linear Algebra and Its Application (1968), Elsevier, New York.

**Mathematics of Computation* (1960–), previously called:

Mathematical Tables and Other Aids to Computation (1943–59), American Mathematical Society, Providence, R.I.

Mathematical Programming (1971–), Elsevier/North Holland, Amsterdam.

**Nordisk Tidskrift for Informationsbehandling*: see *BIT*.

Numerical Methods in Engineering (1969–), Wiley-Interscience, London.

**Numerische Mathematik* (1959–), Springer, Berlin.

SIAM Journal on Numerical Analysis (1964–), Society for Industrial and Applied Mathematics, Philadelphia.

SIAM Review (1958–), Society for Industrial and Applied Mathematics, Philadelphia.

U.S.S.R. Computational Mathematics and Mathematical Physics, Pergamon Press, Oxford (translated from Russian).

* Contains section with published algorithms.

13.9. SURVEY OF PUBLISHED ALGORITHMS

Many periodicals in numerical analysis now have a special section for the publication of algorithms in the form of programs, mostly in Algol (see Sec. 13.8). The publication *Communications of the Association for Computing Machinery* (*CACM*) published a list of such algorithms, the Collected Algorithms of the ACM, classified by subject, for 1960–68. (See *CACM* 11, no. 12 (Dec. 1968), 827–30.) The *ACM* publishes supplements to this list (in *CACM*) in the December issue of each year.

Collected Algorithms of the ACM is also published separately, and supplemented several times a year. The *complete list* of algorithms for 1960–1970 is reproduced (with the permission of the *ACM*) below. *Yearly updates* of this complete list are distributed to Collected Algorithms subscribers, and can also be purchased directly from the ACM.

It should be emphasized that the quality of these published algorithms varies considerably, and that printing errors can occur—at least in older published algorithms. In recent years, however, the quality has in general been very good. Occasionally, in the list of algorithms, *many references are given for the same program*. The later references contain those *corrections or other comments* of interest for the practical use of the program and *should thus be carefully studied before the program is used.*

The programs in the "Handbook" series of the periodical *Numerische Matematik* represent perhaps the best which can be attained at present. Since the development of an error-free computer program for a computational task is often quite a bit more work than one might at first believe, one should use these sources before attempting to solve the problem on one's own.

INDEX BY SUBJECT TO ALGORITHMS, 1960–1970

Key—1st column: A1, B1, B3, etc. is the key to the underlined Modified Share Classification heading each group of algorithms; 2d column: number of the algorithm in *CACM;* 3d column: title of algorithm; 4th column: month, year and page (in parens) in *CACM*, or reference elsewhere. This Index by Subject to Algorithms is cumulative to date (1960–1970) and replaces all previously published versions.

CLASSIFICATION SYSTEM (MODIFIED SHARE)

A1	REAL ARITHMETIC, NUMBER THEORY
A2	COMPLEX ARITHMETIC
B1	TRIG AND INVERSE TRIG FUNCTIONS
B2	HYPERBOLIC FUNCTIONS
B3	EXPONENTIAL AND LOGARITHMIC FUNCTIONS
B4	ROOTS AND POWERS
C1	OPERATIONS ON POLYNOMIALS AND POWER SERIES
C2	ZEROS OF POLYNOMIALS
C5	ZEROS OF ONE OR MORE TRANSCENDENTAL EQUATIONS
C6	SUMMATION OF SERIES, CONVERGENCE ACCELERATION
D1	QUADRATURE
D2	ORDINARY DIFFERENTIAL EQUATIONS
D3	PARTIAL DIFFERENTIAL EQUATIONS
D4	DIFFERENTIATION
D5	INTEGRAL EQUATIONS
E1	INTERPOLATION
E2	CURVE AND SURFACE FITTING
E3	SMOOTHING
E4	MINIMIZING OR MAXIMIZING A FUNCTION
F1	MATRIX OPERATIONS, INCLUDING INVERSION
F2	EIGENVALUES AND EIGENVECTORS OF MATRICES
F3	DETERMINANTS
F4	SIMULTANEOUS LINEAR EQUATIONS
F5	ORTHOGONALIZATION
G1	SIMPLE CALCULATIONS ON STATISTICAL DATA
G2	CORRELATION AND REGRESSION ANALYSIS
G5	RANDOM NUMBER GENERATORS
G6	PERMUTATIONS AND COMBINATIONS
G7	SUBSET GENERATORS
H	OPERATIONS RESEARCH, GRAPH STRUCTURES
I5	INPUT—COMPOSITE
J6	PLOTTING
K2	RELOCATION
L2	COMPILING
M1	SORTING
M2	DATA CONVERSION AND SCALING
O2	SIMULATION OF COMPUTING STRUCTURE
R2	SYMBOL MANIPULATION
S	APPROXIMATION OF SPECIAL FUNCTIONS...
S	FUNCTIONS ARE CLASSIFIED S01 TO S22, FOLLOWING
S	FLETCHER-MILLER-ROSENHEAD, INDEX OF MATH. TABLES
Z	ALL OTHERS

ABREVIATIONS FOR OTHER JOURNALS

BT	BIT
CB	COMPUTER BULLETIN
CH	CHIFFRES
CJ	COMPUTER JOURNAL
CP	COMPUTING
JC	JOURNAL OF THE ASSOCIATION FOR COMPUTING MACHINERY
IC	I.C.C. BULLETIN (INTERNATIONAL COMPUTATION CENTRE)
MC	MATHEMATICS OF COMPUTATION
NM	NUMERISCHE MATHEMATIK

REAL ARITHMETIC, NUMBER THEORY

A1	7	EUCLIDEAN ALGORITHM	4–60(240)
A1	35	SIEVE OF ERATOSTHENES	3–61(151), 4–62(209),
A1	35		8–62(438), 9–67(570)
A1	61	RANGE ARITHMETIC	7–61(319)
A1	68	AUGMENTATION	8–61(339), 11–61(498)
A1	72	COMPOSITIONS	11–61(498), 8–62(439)

A1	93	GENERALIZED ARITHMETIC	6-62(344), 62(514)
A1	95	PARTITIONS	6-62(344)
A1	99	JACOBI SYMBOL	6-62(345), 11-62(557)
A1	114	PARTITIONS	8-62(434)
A1	139	DIOPHANTINE EQUATION	11-62(556), 10-3-65(170)
A1	223	PRIME TWINS	4-64(243)
A1	237	GREATEST COMMON DIVISOR	8-64(481), 12-64(702)
A1	262	RESTRICTED PARTITIONS OF N	8-65(493)
A1	263	PARTITION GENERATOR	8-65(493)
A1	264	MAP OF PARTITIONS INTO INTEGERS	8-65(493)
A1	307	SYMMETRIC GROUP CHARACTERS	7-67(451), 1-68(14)
A1	310	PRIME NUMBER GENERATOR 1	9-67(569), 9-67(570),
A1	310		3-70(192)
A1	311	PRIME NUMBER GENERATOR 2	9-67(570), 9-67(570)
A1	313	MULTI-DIMENSION PARTITION GEN.	10-67(666)
A1	356	PRIME NUMBER GENERATOR	10-69(563)
A1	357	PRIME NUMBER GENERATOR	10-69(563)
A1	371	PARTITIONS IN NATURAL ORDER	1-70(52)
A1	372	COMPLEX PRIMES	1-70(52), 11-70(695)
A1	374	RESTRICTED PARTITIONS GEN.	2-70(120)
A1	375	DOUBLY RESTRICTED PARTITIONS	2-70(120)
A1	386	GCD OF N INTEGERS AND MLTPLRS	7-70(447)
A1	401	COMPLEX PRIMES	11-70(693), 11-70(695)
A1		PARTITION ECNS.(MODULO D)	BT, 9-69(83)
A1	25	OUTER PRODUCT OF SYMM. GROUP	BT, 10-70(106)
A1	18	SUM OF FACTORS OF N (SUMFAC)	CJ, 9-65(416)
A1	25	RATIONAL APPROX. OF REAL NO.	CJ, 11-68(347), 12-69(293)
A1		PARTITIONING INTEGERS, N DIMEN.	CJ, 13-70(282)

COMPLEX ARITHMETIC

A2	116	COMPLEX DIVIDE	8-62(435)
A2	186	COMPLEX ARITHMETIC	7-63(386)
A2	312	COMPLEX ABS, SQRT	10-67(665)
A2		COMPLEX ARITHMETIC	BT, 2-62(233)
A2	19	ADD, SUB, MULT, DIVD-COMPLEX	CJ, 10-66(112),10-66(208)
A2			V10(208)

TRIG AND INVERSE TRIG FUNCTIONS

B1	206	ARCCOSSIN	9-63(519), 2-65(104)
B1	229	ELEMENTARY FCNS. BY CONT. FRACT.	5-64(296), 12-69(692)
B1	241	ARCTAN(Z)	9-64(546)
B1		ARCSIN(Z)	BT, 2-62(236)
B1		ARCCOS(Z)	BT, 2-62(236)
B1		ARCTAN(Z)	BT, 2-62(236)
B1		SIN FCN. BY CHEBYSHEV EXPANSION	NM, 4-62(411), 7-65(194)
B1		COS FCN. BY CHEBYSHEV EXPANSION	NM, 4-62(411), 7-65(195)
B1		TAN FCN. BY CHEBYSHEV EXPANSION	NM, 4-62(412), 7-65(195)
B1		ARCSIN BY CHEBYSHEV EXPANSION	NM, 4-62(412)
B1		ARCTAN BY CHEBYSHEV EXPANSION	NM, 4-62(412)

HYPERBOLIC FUNCTIONS

B2		SINH(X)	BT, 2-62(235)
B2		COSH(X)	BT, 2-62(235)

EXPONENTIAL AND LOGARITHMIC FUNCTIONS

B3	46	EXP(Z), Z COMPLEX	4-61(178), 6-62(347)
B3	48	LOG(Z), Z COMPLEX	4-61(179), 6-62(347),
B3	48		7-62(391), 8-64(485)
B3	243	LOGARITHM OF COMPLEX NUMBER	11-64(660), 5-65(279)

B3		EXP FCN. BY CHEBYSHEV EXPANSION	NM, 4-62(410)
B3		LOG FCN. BY CHEBYSHEV EXPANSION	NM, 4-62(411)

ROOTS AND POWERS

B4	53	ROOTS OF COMPLEX NUMBERS	4-61(180), 7-61(322)
B4	106	POWERS OF COMPLEX NUMBER	7-62(388), 11-62(557)
B4	190	POWERS OF COMPLEX NUMBERS	7-63(388)

OPERATIONS ON POLYNOMIALS AND POWER SERIES

C1	92	POLYNOMIAL SHIFTER	11-60(604)
C1	131	DIVIDE POWER SERIES	11-62(551)
C1	134	EXPONENTIATE POWER SERIES	11-62(553), 7-63(390)
C1	158	EXPONENTIATE POWER SERIES	3-63(104), 7-63(390),
C1	158		9-63(522)
C1	193	REVERT POWER SERIES	7-63(388), 12-63(745)
C1	273	SOLN. OF EQNS. BY REVERSION	1-66(11)
C1	305	SYMMETRIC POLYNOMIALS	7-67(450), 4-68(272)
C1	337	POLY. AND DERIV. BY HORNER SCHEME	9-68(633), 1-69(39)
C1		EVALUATION OF CONTINUED FRACTNS.	CH, 9-XX(327)
C1	15	CALCULATION OF GRAM POLYS.	CJ, 9-65(323)

ZEROS OF POLYNOMIALS

C2	3	BAIRSTOW	2-60(74), 6-60(354),
C2	3		2-61(105), 3-61(153), 4-61(181)
C2	30	BAIRSTOW-NEWTON	12-60(643), 5-61(238),
C2	30		1-62(50), 5-67(293)
C2	59	RESULTANT METHOD	5-61(236)
C2	75	RATIONAL ROOTS-INTEGER COEFF.	1-62(48), 7-62(392),
C2	75		8-62(439)
C2	78	RATIONAL ROOTS-INTEGER COEFF.	2-62(97), 3-62(168),
C2	78		8-62(440)
C2	105	NEWTON-MAEHLY	7-62(387), 7-63(389)
C2	174	BOUNDS ON ZEROS	6-63(311)
C2	256	MODIFIED GRAEFFE METHOD	6-65(379), 9-66(687)
C2	283	REAL SIMPLE ROOTS	4-66(273)
C2	326	ROOTS OF LOW ORDER POLY EQNS.	4-68(269)
C2	340	RT-SQUARING AND RESULTANT METH.	11-68(779), 5-69(281)
C2	11	LEHMERS METHOD	BT, 4-64(255)
C2	21	ROTATING-CROSS METHOD	BT, 7-67(244)
C2	21	BAIRSTOW	CJ, 10-66(207)
C2		ZEROS OF POLYNOMIALS	CP, 5-70(109)
C2		PL/I PROG FOR LEHMERS METHOD	MC, 23-69(829)
C2		ZEROS IN THE RIGHT HALF PLANE	ZH. VYCH. MAT. MAT. FIZ.-1963(364)

ZEROS OF ONE OR MORE TRANSCENDENTAL EQUATIONS

C5	2	REGULA FALSI	2-60(74), 6-60(354),
C5	2		8-60(475), 3-61(153)
C5	4	BISECTION	3-60(174), 3-61(153)
C5	15	REGULA FALSI	8-60(475), 11-60(602),
C5	15		3-61(153)
C5	25	REAL ZEROS	11-60(602), 3-61(153),
C5	25		3-61(154)
C5	26	REGULA FALSI	11-60(603), 3-61(153)
C5	196	MULLERS METHOD	8-63(442), 1-68(12)
C5	314	N FUNCTIONAL EQNS. IN N UNKNOWNS	11-67(726), 1-69(38)
C5	315	DAMPEC TAYLR SERIES-NONL IN. SYS.	11-67(726), 9-69(513)
C5	316	NONL-LINEAR SYSTEM	11-67(728)
C5	365	COMPLEX ROOT FINDING	12-69(686)

C5	378	SOLVE NONLINEAR SYSTEM OF EQS.	4–70(259)
C5	6	ZEROS BY INTERP. OR BISECTION	BT, 3–63(205)
C5	44	SOLN. OF NONLINEAR EQUATIONS	CJ, 12–69(406), 13–70(219)
C5	12	SOLN. FIN. DIM. NONLINEAR SYSTEM	CP, 5–70(84)
C5		ROOTS OF FCNS. WITH ERROR BOUND	CP, 4–69(197)

SUMMATION OF SERIES, CONVERGENCE ACCELERATION

C6	8	EULER SUM	5–60(311), 11–63(663)
C6	128	FOURIER SERIES SUMMATION	10–62(513), 7–64(421)
C6	157	FOURIER SERIES APPROXIMATION	3–63(103), 9–63(521),
C6	157		10–63(618)
C6	215	EPSILON ALGORITHM	11–63(662), 5–64(297)
C6	255	FOURIER COEFFICIENTS	5–65(279), 11–69(636)
C6	277	CHEBYSHEV SERIES COEFFICIENTS	2–66(86)
C6	320	HARMONIC ANAL-SYM DISTR DATA	2–68(114)
C6	338	FAST FOURIER TRANSFORM	11–68(773)
C6	339	FAST FOURIER WITH ARB. FACTORS	11–68(776), 3–69(187)
C6	345	CONVOLUTION BASED ON FFT	3–69(179), 10–69(566)
C6	393	SUMMATION WITH ARB. PRECISION	9–70(570)
C6	1	FIND LIMIT OF SEQUENCE	BT, 1–61(64)
C6		EPSILON ALGORITHM	BT, 2–62(240)
C6		EPSILON ALGORITHM	NM, 6–64(22)
C6		EPSILON ALG.–CONTINUED FRACTNS	CH, 9–XX(327)
C6		SUM FOURIER SERIES	CJ, 6–63(248)
C6	31	COMPLEX FOURIER ANALYSIS	CJ, 10–67(414), 11–68(115)

QUADRATURE

D1	1	QUADRATURE	2–60(74)
D1	32	MULTIPLE INTEGRAL	2–61(106), 2–63(69),
D1	32		12–68(826)
D1	60	ROMBERG METHOD	6–61(255), 3–62(168),
D1	60		5–62(281), 7–64(420)
D1	84	SIMPSONS RULE	4–62(208), 7–62(392),
D1	84		8–62(440), 11–62(557)
D1	98	COMPLEX LINE INTEGRAL	6–62(345)
D1	103	SIMPSONS RULE	6–62(347)
D1	125	GAUSSIAN COEFFICIENTS	10–62(510)
D1	145	ADAPTIVE SIMPSON	12–62(604), 4–63(167),
D1	145		3–65(171)
D1	146	MULTIPLE INTEGRAL	12–62(604), 5–64(296)
D1	182	ADAPTIVE SIMPSON	6–63(315), 4–64(244)
D1	198	ADAPTIVE, MULTIPLE INTEGRAL	8–63(443)
D1	233	MULTIPLE INTEG.–SIMPSONS RULE	6–64(348), 8–70(512)
D1	257	HAVIE INTEGRATOR	6–65(381), 11–66(795),
D1	257		12–66(871)
D1	279	CHEBYSHEV QUADRATURE	4–66(270), 6–66(434),
D1	279		5–67(294), 10–67(666)
D1	280	GREGORY QUADRATURE COEFFICIENTS	4–66(271)
D1	281	ROMBERG QUADRATURE COEFFICENTS	4–66(271), 3–67(188)
D1	303	ADAPTIVE QUAD.–RANDOM PANEL SIZE	6–67(373)
D1	331	GAUSSIAN QUADRATURE FORMULAS	6–68(432), 5–69(280),
D1	331		8–70(512)
D1	351	MODIFIED ROMBERG QUADRATURE	6–69(324), 6–70(374),
D1	351		4–70(263), 7–70(449)
D1	353	FILON QUADRATURE	8–69(457), 4–70(263)
D1	379	ADAPTIVE SIMPSON QUADRATURE	4–70(260)
D1	400	MODIFIED HAVIE INTEGRATION	10–70(672)
D1	2	ADAPTIVE SIMPSONS RULE	BT, 1–61(290)
D1		MONTE CARLO QUADRATURE	CJ, 6–63(281)
D1		MSQ RESPONSE ANALYSIS INTEGRAL	CJ, 13–70(207)
D1		CALC. OF GAUSS QUAD. RULES	MC, 23–69(221)

D1	8	ROMBERG METHOD	BT, 4–64(58)
D1		ROMBERG METHOD	COLL. ANAL. NUM. (1961)
D1		ROMBERG METHOD	NM, 6–64(15)
D1		QUADRATURE BY EXTRAPOLATION	NM, 9–66(274)

ORDINARY DIFFERENTIAL EQUATIONS

D2	9	RUNGE-KUTTA	5–60(312), 4–66(273)
D2	194	ZEROS OF O.D.E. SYSTEM	8–63(441)
D2	218	KUTTA-MERSON	12–63(737), 10–64(585),
D2	218		4–66(273)
D2		EXTRAPOLATION METHOD	NM, 8–66(10)
D2		1ST ORDER, AUTO. STEP CHANGE	NM, 14–70(317)

PARTIAL DIFFERENTIAL EQUATIONS

D3	392	SYSTEMS OF HYPERBOLIC PDE	9–70(567)
D3		LINEAR ELLIPTIC BNDY. VAL. PROB.	AUTOMATISIERTE
D3		CONFORMAL MAP-ELLIPSE TO CIRCLE	BT, 2–62(243)
D3		KERNEL FCN. IN BNDY. VALUE PROBS.	NM, 3–61(209)
D3		BNDY. VALUE PROBS.-INTEGRAL OPRS	NM, 7–65(56)
D3		BEHANDLUNG ELLIPTISCHER RANDWER PROBLEME (PUB. 1962)	
D3		PDE SOLNS. BY INTEGRAL OPERATORS	STANFORD UNIV.–
D3		APPL. MATH.S TAT. REP. NONR 225(37) NO.24	

DIFFERENTIATION

D4	79	DIFFERENCE EXPRESSION COEFF.	2–62(97), 3–63(104)
D4		DIFFN. BY NEVILLES FORMULAS	NM, 8–66(462)

INTEGRAL EQUATIONS

D5	368	INVERSION OF LAPLACE TRANSFORM	1–70(47), 10–70(624)
D5		SYSTEM OF VOLTERRA EQNS.	ZH. VYCH. MAT. FIZ.–1965(933)

INTERPOLATION

E1	18	RATIONAL INTERP.-CONT. FRACT.	9–60(508), 8–62(437)
E1	70	AITKEN INTERPOLATION	11–61(497), 7–62(392)
E1	77	INTERPOLATION, DIFFN., INTEGRN.	2–62(96), 6–62(348),
E1	77		8–63(446), 11–63(663)
E1	167	CONFLUENT DIVIDED DIFFERENCES	4–63(164), 9–63(523)
E1	168	INTERPOLATION-DIVIDED DIFFCES.	4–63(165), 9–63(523)
E1	169	INTERPOLATION-DIVIDED DIFFCES.	4–63(165), 9–63(523)
E1	187	DIFFCES. AND DERIVS.-RECURSIVE	7–63(387)
E1	210	LAGRANGE INTERPOLATION	10–63(616)
E1	211	HERMITE INTERPOLATION	10–63(617), 10–63(619)
E1	264	INTERPOLATION IN A TABLE	10–65(602)
E1		SMOOTH CURVE INTERPOLATION	BT, 9–69(69)
E1	10	AITKEN INTERPOLATION	CJ, 9–66(211)
E1		NEVILLE INTERPOLATION	CJ, 9–66(212)
E1	40	SPLINE INTERPOLN OF DEGREE 3	CJ, 12–69(198), 12–69(409)
E1	42	QUINTIC SPLINES INTERPOLATION	CJ, 12–69(292), 13–70(115)
E1	10	TWO-DIMENSION. SMOOTH INTERPOL.	CP, 4–69(180)
E1		EXP. SPLINE INTERPOLATION	CP, 4–69(230)

CURVE AND SURFACE FITTING

E2	28	LEAST SQUARES BY ORTHOG. POLYN	11–60(604), 12–61(544),
E2	28		5–67(293)
E2	37	ECONOMIZATION	3–61(151), 8–62(438),
E2	37		8–63(445)

E2	38	ECONOMIZATION	3-61(151), 8-63(445)
E2	74	LEAST SQUARES WITH CONSTRAINTS	1-62(47), 6-63(316)
E2	91	CHEBYSHEV FIT	5-62(281), 4-63(167),
E2	91		5-64(290), 12-67(803)
E2	164	SURFACE FIT	4-63(162), 8-63(450)
E2	176	SURFACE FIT	6-63(313)
E2	177	LEAST SQUARES WITH CONSTRAINTS	6-63(313), 7-63(390)
E2	275	EXPONENTIAL CURVE FIT	2-66(85)
E2	276	CONSTRAINED EXPONENTIAL FIT	2-66(85)
E2	295	EXPONENTIAL CURVE FIT	2-67(87)
E2	296	LEAST SQ. FIT-ORTHOG. POLYS.	2-67(87), 6-67(377),
E2	296		11-69(636)
E2	318	CHEBYSHEV CURVE FIT (REVISED)	12-67(801)
E2	375	FIT DATA TO A*EXP (-B*X)	2-70(120)
E2	376	FIT DATA TO A*COS (B*X+C)	2-70(120)
E2		CONTINUED FRACTION EXPANSION	BT, 2-62(245)
E2	33	EXPONENTIAL FIT	CJ, 11-68(114)
E2	37	EXPONENTIALLY DAMPED LINEAR FIT	CJ, 12-69(100)
E2		RATIONAL CHEBYSHEV APPROX.	JC, 11-64(66)
E2		L1 APPROX. ON A DISCRETE SET	NM, 8-66(299)
E2		CHEBYSHEV APPROX.-DISCRETE SET	NM, 8-66(303)
E2		REMES ALGORITHM-GENERALIZED	NM, 10-67(203)
E2		RATIONAL CHEBYSHEV APPROX.	NM, 9-66(177)
E2		RATIONAL CHEBYSHEV APPROX.	NM, 10-67(291)
E2		RATIONAL CHEBYSHEV APPROX.	NM, 12-68(242)

SMOOTHING

E3	188	SMOOTHING	7-63(387)
E3	189	SMOOTHING	7-63(387)
E3	216	SMOOTHING	11-63(663)
E3		SMOOTHING BY SPLINE FCNS.	NM, 10-67(203)

MINIMIZING OR MAXIMIZING A FUNCTION

E4	129	MINIMIZE FUNCT. OF N VARIABLES	11-62(550), 9-63(521)
E4	178	DIRECT SEARCH	6-63(313), 9-66(684),
E4	178		7-68(498), 11-69(637), 11-69(638)
E4	203	MINIMIZE FUNCT. OF N VARIABLES	9-63(517), 10-64(585),
E4	203		3-65(171)
E4	204	MINIMIZE FUNCT. OF N VARIABLES	9-63(519)
E4	205	MINIMIZE FUNCT. OF N VARIABLES	9-63(519), 3-63(519), 3-65(171)
E4	251	FUNCTION MINIMIZATION	3-65(169), 9-66(686),
E4	251		9-69(512)
E4	315	MINIMIZING SUM OF SQUARES	11-67(726), 9-69(513)
E4	387	FCN. MINIMIZATION	8-70(509)
E4	2	FIBONACCI SEARCH	CB, 8-64(147), 9-65(105)
E4			CJ, 9-67(414), 9-67(416)
E4	7	MIN. OF UNIMODAL FCN. OF 1 VAR.	CB, 9-65(104), CJ, 9-67(414)
E4		MINIMIZING FCN.-CONJ. GRAD.	CJ, 7-64(151)
E4	16	MINIMIZING ARG. OF UNIMODAL FCN.	CJ, 9-66(415)
E4	17	MIN. OF A UNIMODAL FCN. OF 1 VAR.	CJ, 9-66(415)

MATRIX OPERATIONS, INCLUDING INVERSION

F1	42	INVERSION	4-61(176), 11-61(498),
F1	42		1-63(38), 8-63(445)
F1	50	INVERSE OF HILBERT MATRIX	4-61(179), 1-62(50),
F1	50		1-63(38)
F1	51	INVERSE OF PERTURBED MATRIX	4-61(180), 7-62(391)
F1	52	INVERSE OF TEST MATRIX	4-61(180), 8-61(339),
F1	52		8-62(438), 1-63(39), 8-63(446)
F1	58	INVERSION-GAUSSIAN ELIMINATION	5-61(236), 6-62(347),
F1	58		8-62(438), 12-62(606)

F1	66	INVERSION-SQRT METHOD	7-61(322), 1-62(50),
F1	66		6-62(348)
F1	67	CRAM MATRIX	7-61(322), 6-62(348)
F1	120	INVERSION-GAUSSIAN ELIMINATION	8-62(437), 1-63(40),
F1	120		8-63(445)
F1	140	INVERSION	11-62(556), 8-63(448)
F1	150	INVERSE OF SYMMETRIC MATRIX	2-63(67), 7-63(390),
F1	150		3-64(148)
F1	166	MONTE CARLO INVERSE	4-63(164), 9-63(523)
F1	197	MATRIX DIVISION	8-63(443), 3-64(148)
F1	230	MATRIX PERMUTATION	6-64(347)
F1	231	INVERSION-GAUSS. FILM.-COMP. PIV.	6-64(347), 4-65(220)
F1	274	HILBERT DERIVED TEST MATRIX	1-66(11), 7-69(407)
F1	287	INTEGER MATRIX TRIANGULATION	7-66(513)
F1	298	SQ. RT. OF A POS. DEFINITE MATRIX	3-67(182), 6-69(325)
F1	319	TRIANG FCTRS OF MODIFIED MATRIX	1-68(12)
F1	325	ADJUST INVERSE OF SYM MATRIX	2-68(118)
F1	348	MTRX SCALING BY INTGR PROGRAMING	4-69(212)
F1	358	SING. VAL. DECOMP.-COMPLEX MATRIX	10-69(564)
F1	380	TRANSPOSE RECTANGULAR MATX.	5-70(325), 5-70(327)
F1	20	SMITH NORMAL FORM	BT, 7-67(163)
F1	12	INVERSION-SYMM. POS. DEF. MTX	CJ, 9-66(321)
F1	20	PERMUTATIONS OF ROWS AND COLS.	CJ, 10-67(206)
F1		SYMM. DECOMP. OF POS. DEF. BAND MTX	CP, 1-66(77)
F1		ADJUST INVERSE OF SYM. MATRIX	CP, 3-68(76)
F1		EQUIVALENCE OF MATRICES	IC, 3-64(62)
F1		INVERSE-CONFL. VANDERMONDE MTR	NM, 5-63(429)
F1		SYMM. DECOMP. OF POS. DEF. BAND MTX	NM, 7-65(357)
F1		SYMM. DECOMP. OF POS. DEF. MTX	NM, 7-65(368)
F1		HOUSEHOLDER TRIDIAG OF SYM MTRX	NM, 11-68(184)
F1		TRIDIAG. OF SYM. BAND MATRIX	NM, 12-69(234)
F1		SIM. RED. GEN. MTX. HESSENBERG FORM	NM, 12-69(354)
F1		SINGULAR VALUE DEÇOMPOSITION	NM, 14-70(403)

EIGENVALUES AND EIGENVECTORS OF MATRICES

F2	85	JACOBI METHOD	4-62(208), 8-62(440),
F2	85		8-63(447)
F2	104	REDUCTION-BAND TO TRIDIAGONAL	7-62(387)
F2	122	GIVENS TRIDIAGONAL REDUCTION	9-62(482), 3-64(144)
F2	183	REDUCTION-BAND TO TRIDIAGONAL	6-63(315)
F2	253	SYMMETRIC QR-EIGENVALUES	4-65(217), 6-67(376)
F2	254	SYMMETRIC QR-EIVALUES,EIVECTORS	4-65(218), 6-67(376)
F2	270	EIGENVECTORS BY GAUSSIAN ELIM.	11-65(668)
F2	297	SYM.SYS. (A-LAM=B)X, EIVALS-VECS.	3-67(181)
F2	343	EIVALS-VECS. OF REAL GEN. MTRX	12-68(820), 2-70(122),
F2	343		11-70(694)
F2	384	EIVAL-EIVEC OF A SYMM. MATRIX	6-70(369), 12-70(750)
F2		SYMMETRIC-BISECTION, INV. ITN.	BT, 4-64(124)
F2		TRIDIAGONAL SIMIL. BY ELIM.	CJ, 4-61(175)
F2		EIGENVALUES BY QR-ALGORITHM	CJ, 4-61(344)
F2		SYMM. MAT.-LLT AND STURM SEQ.	CJ, 9-66(303)
F2	32	EIVALS-REAL SYM MTRX-DBL QR STP	CJ, 11-68(112)
F2		EIGENVALUES-LAGUERRES METHOD	MC, 18-64(474), 20-66(437)
F2		EIVECS FOR GENL COMPLEX MTRX	MC, 22-68(785)
F2		HOUSEHOLDERS METHOD	NM, 4-62(354)
F2		EIGENVALUES OF TRIDIAG. MATRIX	NM, 4-62(354)
F2		EIGENVECTORS OF TRIDIAG. MTRX	NM, 4-62(354)
F2		LR TRANSFORMATION METHOD	NM, 5-63(273)
F2		HOUSEHOLDER RED.-COMPLEX MAT.	NM, 8-66(79)
F2		JACOBI METHOD	NM, 9-66(3)
F2		EIGENVECTORS OF BAND MATRICES	NM, 9-67(285)
F2		EIVALUES OF SYMM. TRIDIAG. MATRIX	NM, 9-67(388)
F2		EIVALUES-EIVECTORS OF REAL MTRX	NM, 11-68(3)

F2		SYM EIPROBLEM A. X=LAM. B. X.	NM, 11–68(102)
F2		RATIONAL QR FOR SYM TRIDIAG	NM, 11–68(268)
F2		MOD. LR.–CMPLX HESSENBERG MATRIX	NM, 12–68(372)
F2		IMPLICIT QL ALGORITHM	NM, 12–68(379)
F2		BALANCING A MATRIX	NM, 13–69(293)
F2		QR FOR REAL HESSENBERG MTRX	NM, 13–70(219)
F2		HOUSEHOLDERS METHOD	STANFORD UNIV.–
F2		APPL. MATH. STAT.REP. NONR 225(37) NO.18	
F2		EIGENVALUES–LAGUERRES METHOD	STANFORD UNIV.–
F2		APPL. MATH. STAT. REP. NONR 225(37) NO.21	

DETERMINANTS

F3	41	DETERMINANT EVALUATION	4–61(176), 9–63(520),
F3	41		3–64(144), 9–66(686)
F3	159	DETERMINANT EVALUATION	3–63(104), 12–63(739)
F3	170	DETERMINANT–POLYNOMIAL ELEMENTS	4–63(165), 8–63(450),
F3	170		7–64(421)
F3	224	EVALUATION OF DETERMINANT	4–64(243), 12–64(702)
F3	269	DETERMINANT BY GAUSSIAN ELIM.	11–65(668), 9–66(686)

SIMULTANEOUS LINEAR EQUATIONS

F4	16	CROUT WITH PIVOTING	9–60(507), 10–60(540),
F4	16		3–61(154)
F4	17	SOLVE TRIDIAGONAL MATRIX	9–60(508)
F4	24	SOLVE TRIDIAGONAL MATRIX	11–60(602)
F4	43	CROUT WITH PIVOTING	4–61(176), 4–61(182),
F4	43		8–63(445)
F4	92	SIMULT. EQNS.–ITERATIVE SOLN.	5–62(286)
F4	107	GAUSSIAN ELIMINATION	7–62(388), 1–63(39),
F4	107		8–63(445)
F4	126	GAUSSIAN ELIMINATION	10–62(511)
F4	135	CROUT WITH EQUILIBRATION	11–62(553), 11–62(557),
F4	135		7–64(421), 2–65(104)
F4	195	BAND SOLVE	8–63(441)
F4	220	GAUSS–SEIDEL	12–63(739), 6–64(349)
F4	238	CONJUGATE GRADIENT METHOD	8–64(481)
F4	288	LINEAR DIOPHANTINE EQUATIONS	7–66(514)
F4	290	EXACT SOLUTION OF LINEAR EQNS.	9–66(683)
F4	328	CHEBY SOLN–OVERDET LINEAR SYS	6–68(428), 6–69(326)
F4	358	SING. VAL. DECOMP.–COMPLEX MATRIX	10–69(564)
F4	4	GAUSSIAN ELIMINATION	BT, 2–62(256), 3–63(60)
F4	7	LINEAR SYSTEM WITH BAND MATRIX	BT, 3–63(207)
F4	12	GAUSSIAN ELIMINATION	BT, 5–65(64)
F4	68	IT REFINEMENT–LEAST SQR SOLN	BT, 8–68(20)
F4		SOLUTION WITH REL ERR ESTIMATE	CJ, 11–68(92)
F4	51	GENL. 3–TERM LINEAR SYSTEMS	CJ, 13–70(324)
F4		CONJUGATE GRADIENT METHOD	NM, 5–63(195)
F4		LEAST SQUARES SOLUTION	NM, 7–65(271)
F4		ELIM. WITH WEIGHTED ROW COMB.	NM, 7–65(341)
F4		ITER. REFIN.–SOLN. OF POS. DEF. MTX	NM, 8–66(206)
F4		REAL AND COMPLEX LINEAR SYSTEM	NM, 8–66(222)
F4		SYMM. AND UNSYMM. BAND EQUATIONS	NM, 9–67(285)

ORTHOGONALIZATION

F5	127	ORTHONORMALIZATION	10–62(511), 2–70(122)
F5	358	SING. VAL. DECOMP.–COMPLEX MATRIX	10–69(564)
F5		SCHMIDT ORTHONORMALIZATION	CP, 1–66(159)

SIMPLE CALCULATIONS ON STATISTICAL DATA

G1	208	DISCRETE CONVOLUTION	10–63(615)
G1	212	DETERMINE DISTRIB. FCN. FROM DATA	10–63(617)

G1	289	CONFIDENCE INTERVAL FOR A RATIO	7–66(514)
G1	330	FACTORIAL ANALYSIS OF VARIANCE	6–68(431)
G1	359	FACTORIAL ANALYSIS OF VARIANCE	11–69(631), 7–70(449)
G1		TAIL AREA PROB. FOR 2×2 TABLE	CB, 9–65(56), CJ, 9–66(212), CJ, 9–67(416)

CORRELATION AND REGRESSION ANALYSIS

G2	39	CORRELATION COEFFICIENTS	3–61(152)
G2	142	TRIANGULAR REGRESSION	12–62(603)
G2	366	REGRESSION–DIR. PROD. OF MATRICES	12–69(687)
G2	367	ANALYSIS OF VAR.–DIRECT EXPMT.	12–69(688)

RANDOM NUMBER GENERATORS

G5	121	RANDOM NORMAL	9–62(482), 9–65(556)
G5	133	RANDOM FLAT	11–62(553), 12–62(606),
G5	133		3–63(105), 4–63(167)
G5	200	RANDOM NORMAL	8–63(444), 9–65(556)
G5	247	QUASI–RANDOM POINT SEQUENCE	12–64(701)
G5	266	PSEUDO–RANDOM NUMBERS	10–65(605), 9–66(687)
G5	267	RANDOM NORMAL DEVIATES	10–65(606)
G5	294	UNIFORM RANDOM	1–67(40)
G5	334	NORMAL RANDOM DEVIATES	7–68(498), 5–69(281)
G5	342	POISSON RANDOM NUMBERS	12–68(819)
G5	369	POISSON RANDOM NUMBERS	1–70(49)
G5	370	RANDOM NUMBER GENERATOR	1–70(49)
G5	381	RANDOM VECTORS IN SOLID ANGLE	5–70(326)
G5		RANDOM UNIFORM	CB, 9–65(105)
G5		RANDOM SAMPLES, VARIOUS DISTRIB.	CJ, 6–64(279)

PERMUTATIONS AND COMBINATIONS

G6	71	PERMUTATIONS	11–61(497), 4–62(209),
G6	71		8–62(439)
G6	86	PERMUTATIONS	4–62(208), 4–62(209),
G6	86		8–62(440)
G6	87	PERMUTATION GENERATOR	4–62(209), 8–62(440),
G6	87		10–62(514), 7–67(452)
G6	94	COMBINATIONS	6–62(344), 11–62(557),
G6	94		12–62(606)
G6	102	PERMUTATIONS IN LEXIC. ORDER	6–62(346), 10–62(514),
G6	102		7–67(452)
G6	115	PERMUTATIONS	8–62(434), 10–62(514),
G6	115		12–62(606)
G6	130	PERMUTE	11–62(551), 7–67(452)
G6	152	COMBINATIONS	2–63(68), 7–63(385)
G6	154	COMBINATION IN LEXIC. ORDER	3–63(103), 8–63(449)
G6	155	COMBINATION IN ANY ORDER	3–63(103), 8–63(449)
G6	156	ALGEBRA OF SETS	3–63(103), 8–63(450)
G6	160	COMB. OF M THINGS N AT A TIME	4–63(161), 8–63(450),
G6	160		10–63(618)
G6	161	COMBS. 1, 2, UP TO N AT A TIME	4–63(161), 8–63(450),
G6	161		10–63(619)
G6	202	PERMUTATIONS IN LEXIC. ORDER	9–63(517), 9–65(556),
G6	202		7–67(452)
G6	235	RANDOM PERMUTATION	7–64(420), 7–65(445)
G6	242	PERMUTATIONS WITH REPETITIONS	10–64(585)
G6	250	INVERSE PERMUTATION	2–65(104), 11–65(670)
G6	306	PERMUTATIONS WITH REPETITIONS	7–67(450)
G6	308	PERMUT. IN PSEUDOLEXIC. ORDER	7–67(452), 11–69(638)
G6	317	PERMUTATION	11–67(729)
G6	323	PERMUTATIONS IN LEXIC ORDER	2–68(117)
G6	329	DISTR OF INDISTINGUISHABLE OBJ	6–68(430), 3–69(187)
G6	361	PERMANENT FNC. OF SQUARE MATRIX	11–69(634), 6–70(376)

G6	362	RANDOM PERMUTATIONS	11-69(634)
G6	382	COMBINATIONS OF M OUT OF N	6-70(368), 6-70(376)
G6	383	PERMUTATIONS WITH REPETITIONS	6-70(368), 6-70(376)
G6		PARTITION FUNCTIONS-MOD(D)	BT, 9-69(83)
G6		ALL PERMUTATIONS OF N OBJECTS	CB, 9-65(104)
G6	28	PERMUTNS OF VECTOR-LEXIC ORDER	CJ, 10-67(311)
G6	29	PERMUTNS OF VECTOR	CJ, 10-67(311)
G6	30	FAST PERMUTN OF VECTOR	CJ, 10-67(311)

SUBSET GENERATORS

G7	81	SUBSEQUENCES	3-62(166)
G7	82	SUBSEQUENCES	3-62(167)
G7	47	A CLUSTERING ALGORITHM	CJ, 13-70(113), 13-70(326)
G7	49	INDEXING SUBARRAYS	CJ, 13-70(208), 13-70(219)
G7	52	HIERARCHIC CLUSTERING ALGORITHM	CJ, 13-70(325)

OPERATIONS RESEARCH, GRAPH STRUCTURES

H	27	ASSIGNMENT PROBLEM	11-60(603), 10-63(618),
H	27		12-63(739)
H	40	CRITICAL PATH SCHEDULING	3-61(152), 9-61(392),
H	40		10-62(513), 6-64(349)
H	69	CHAIN TRACING	9-61(392)
H	83	CLASSIFICATIONS	3-62(167)
H	96	ANCESTOR	6-62(344), 3-63(104)
H	97	SHORTEST PATH	6-62(345)
H	119	PERT NETWORK	8-62(436), 5-65(330)
H	141	FIND PATH	11-62(556)
H	153	INTEGER PROGRAMMING	2-63(68), 8-63(449)
H	217	MIN. EXCESS COST CURVE	12-63(737), 8-68(573)
H	219	TOPOLOGICAL ORDERING	12-63(738)
H	248	NETFLOW	2-65(103), 9-68(633),
H	248		9-68(633)
H	258	TRANSPORT	6-65(381), 7-65(445),
H	258		7-67(453)
H	263	INTEGER PROGRAMMING-GOMORY 1	10-65(601), 5-70(326)
H	285	MUTUAL PRIMAL-DUAL METHOD	5-66(326), 7-67(453)
H	286	EXAMINATION SCHEDULING	6-66(433), 11-66(795)
H	293	TRANSPORTATION PROBLEM	12-66(869), 7-67(453),
H	293		4-68(271)
H	324	MAXFLOW	2-68(117)
H	333	MINIT ALGORITHM FOR LIN PROG	6-68(437), 7-69(408)
H	336	NETFLOW	9-68(631), 3-70(192)
H	341	LINEAR PGMS. IN 0-1 VARIABLES	11-68(782), 12-69(692),
H	341		4-70(263)
H	350	SIMPLEX METHOD-LU DECOMPOSITION	5-69(275)
H	354	SPANNING TREE GENERATOR	9-69(511)
H	360	SHORTEST PATH FOREST-TOPOL. ORD.	11-69(632)
H	394	DECISION TABLE TRANSLATION	9-70(571)
H	397	INTEGER PROGRAMMING PROBLEM	10-70(620)
H	399	SPANNING TREE	10-70(621)
H	10	SIMPLEX	BT, 4-64(194), 6-66(82)
H		MINIMAL SPANNING TREE	CB, 8-64(67), 8-64(67), 8-64(109),
H			8-64(147), 9-67(18)
H	14	PROCESSING EVENT NETWORK	CJ, 9-66(323)
H	22	SHORTEST PATH-START TO END	CJ, 10-67(306)
H	23	SHORTEST PATH-START TO ANY	CJ, 10-67(307)
H	24	NODES ON SHORTEST PATH	CJ, 10-67(308)
H	8	BESTIMUNG ALLER MAXIMALEN	CP, 3-68(240), 4-69(75)
H	9	PROC. TO MOD. ALG. OF MINTY, MOORE	CP, 4-69(76)
H		LOSUNG DES OPTIMUM-MIX-PROBLEME	CP, 5-70(326)
H		MIN LENGTH FULL STEINER TREES	MC, 23-69(521)

INPUT—COMPOSITE

I5	239	FREE-FIELD READ	8–64(481)
I5	249	OUTREAL N	2–65(104)
I5	335	BASIC I/O PROCEDURES	8–68(567)
I5		OPTICAL SCANNING OF NUMBERS	ZH. VYCH. MAT. FIZ.–
I5			1962(236)

PLOTTING

J6	162	XY PLOTTER	4–63(161), 8–63(450),
J6	162		8–64(482)
J6	278	GRAPH PLOTTER	2–66(88)
J6	41	A CURVE PLOTTING PROCEDURE	CJ,12–69(291)

RELOCATION

K2	173	TRANSFER ARRAY VALUES	6–63(311), 10–63(619)
K2	284	INTERCHANGE 2 BLOCKS OF DATA	5–66(326)
K2	302	TRANSPOSE VECTOR STORED ARRAY	5–67(292), 6–69(326)

COMPILING

L2	265	FIND PRECEDENCE FUNCTIONS	10–65(604)
L2	13	EVALUATION OF FCNAL EXPRESSION	BT, 5–65(133)

SORTING

M1	23	SORT	11–60(610), 5–61(238)
M1	63	SORT	7–61(321), 8–62(439),
M1	63		8–63(446)
M1	64	SORT	7–61(321), 8–62(439),
M1	64		8–63(446)
M1	65	SORT	7–61(321), 8–62(439),
M1	65		8–63(446)
M1	76	SORT	1–62(48), 6–62(348)
M1	113	TREESORT	8–62(434)
M1	143	TREESORT	12–62(604)
M1	144	TREESORT	12–62(604)
M1	151	LOCATE IN A LIST	2–63(68)
M1	175	SHUTTLE SORT	6–63(312), 10–63(619),
M1	175		12–63(739), 5–64(296)
M1	201	SHELL SORT	8–63(445), 6–64(349),
M1	201		6–70(373)
M1	207	STRING SORT	10–63(615), 10–64(585)
M1	232	HEAPSORT	6–64(347)
M1	245	TREESORT 3	12–64(701), 7–65(445),
M1	245		6–70(371)
M1	271	QUICKERSORT	11–65(669), 5–66(354)
M1	347	SORT WITH MINIMAL STORAGE	3–69(185), 1–70(54),
M1	347		10–70(624)
M1	402	IMPROVED QUICKSORT	11–70(693)
M1		SORTING OF INTEGERS	CB, 9–67(63)
M1	25	SORT BY RANKING ELEMENTS	CJ, 10–67(308), 12–69(408),
M1	25		13–70(326)
M1	26	ORDER SUBSCRIPTS BY ELEMENT SIZE	CJ, 10–67(309), 12–69(408),
M1	26		13–70(326)
M1	27	SORT ON PERMUTN OF SUBSCRIPTS	CJ, 10–67(310)
M1	38	A SEARCHING ALGORITHM	CJ, 12–69(101)
M1	43	LISTED RADIX SORT	CJ, 12–69(406), 13–70(326)
M1	45	INTERNAL SORT BY TWO-WAY MERGE	CJ, 13–70(110)
M1	46	MODIFIED DAVIDON FOR FCN. MIN.	CJ, 13–70(111)
M1		SEARCH IN A LIST	JC, 9–62(23)
M1		INSERTION IN A LIST	JC, 9–62(23)

M1		DELETION FROM A LIST	JC, 9–62(24)
M1		SORTING WITH MINIMUM STORAGE	JC, 9–62(27)

DATA CONVERSION AND SCALING

M2		DATA PROCESSING–VECTORCARDIOGRM	2–62(121)

SIMULATION OF COMPUTING STRUCTURE

O2	100	PROCESSING OF CHAIN-LINKED LIST	6–62(346)
O2	101	PROCESSING OF CHAIN-LINKED LIST	6–62(346)
O2	137	NESTED FOR STATEMENT	11–62(555)
O2	138	NESTED FOR STATEMENT	11–62(555)
O2	13	EVALUATION OF FCNAL EXPRESSION	BT, 5–65(137)

SYMBOL MANIPULATION

R2	268	ALGOL 60 REF. LANG. EDITOR	11–65(667), 7–69(407)
R2	377	SYMB. EXPANSION OF EXPRESSIONS	3–70(191)
R2	17	BASIC LIST PROCESSING	BT, 6–66(166)
R2	18	SIMPLIFYING BOOLEAN EXPRESSIONS	BT, 6–66(260)
R2	24	FCN. VALUE FROM CHARAC. STRING	BT, 6–69(283)

APPROXIMATION OF SPECIAL FUNCTIONS . . . FUNCTIONS ARE CLASSIFIED S 01 TO S 22, FOLLOWING FLETCHER—MILLER—ROSENHEAD, INDEX OF MATH. TABLES

S03	19	BINOMIAL COEFFICIENTS	10–60(540), 6–62(347),
S03	19		8–62(438)
S03	33	FACTORIAL N	2–61(106)
S04		DIRICHLET L-SERIES	MC, 23–69(489)
S07		POLAR TRANSF. BY CHEBYSHEV EXP.	NM, 4–63(413)
S13	14	COMPLEX EXPONENTIAL INTEGRAL	7–60(406)
S13	20	REAL EXPONENTIAL INTEGRAL	10–60(540), 2–61(105),
S13	20		4–61(182)
S1f	108	EXPONENTIAL INTEGRAL	7–62(388), 7–62(393)
S13	109	EXPONENTIAL INTEGRAL	7–62(388), 7–62(393)
S13	385	EXPONENTIAL INTEGRAL EI(X)	7–70(446), 7–70(448)
S13	385		12–70(750)
S13		EXPONENTIAL INTEGRAL EXPANSION	CH, 6–XX(187)
S13		EI(X) BY CHEBYSHEV EXPANSION	NM, 4–63(413)
S13		SIN INTEGRAL SI(X)	NM, 9–67(381)
S13		COS INTEGRAL CI(X)	NM, 9–67(382)
S14	31	GAMMA FUNCTION	2–61(105), 12–62(605)
S14	34	GAMMA FUNCTION	2–61(106), 7–62(391),
S14	34		9–66(685)
S14	54	GAMMA FUNCTION	4–61(180), 9–66(685)
S14	80	GAMMA FUNCTION	3–62(166), 9–66(685)
S14	147	DERIVATIVE OF GAMMA FUNCTION	12–62(605), 4–63(168),
S14	147		12–69(691)
S14	179	BETA RATIO	6–63(314), 6–67(375)
S14	221	GAMMA FUNCTION	3–64(143), 10–64(586),
S14	221		9–66(685)
S14	222	INCOMPLETE BETA FCN. RATIOS	3–64(143), 4–64(244)
S14	225	GAMMA FCN WITH CONTROLLED ACCY.	5–64(295), 10–64(586)
S14	291	LOGARITHM OF GAMMA FCN.	9–66(684), 9–66(685),
S14	291		1–68(14)
S14	309	GAMMA FCN.-ARBITRARY PRECISION	8–67(511)
S14	321	T-TEST PROBABILITIES	2–68(115), 2–70(124)
S14	322	F-DISTRIBUTION	2–68(116), 1–69(39)
S14	344	STUDENT*S T-DISTRIBUTION	1–69(37), 2–70(124),
S14	344		7–70(449)

S21		INCOMPL. ELL. INT.–FIRST KIND(K)	NM, 5–63(297)
S21		INCOMPL. ELL. INT.–SECOND KIND(E)	NM, 5–63(298)
S21		INCOMPL. ELL. INT. (B)	NM, 5–63(299)
S21		JACOBIAN ELLIPTIC SIN FCN. (SN)	NM, 5–63(299)
S21		JACOBIAN ELLIPTIC COS FCN. (CN)	NM, 5–63(300)
S21		JACOBIAN ELLIPTIC FCN. (CN)	NM, 5–63(301)
S21		ELLIPTIC INTEGRALS–KIND 1, 2, 3	NM, 7–65(85), 7–65(353),
S21			13–69(305)
S21		JACOBIAN ELLIPTIC FUNCTIONS	NM, 7–65(89)
S22	10	CHEBYSHEV POLYNOMIAL	6–60(353), 4–61(181)
S22	12	LAGUERRE POLYNOMIAL	6–60(353)
S22	36	CHEBYSHEV POLYNOMIAL	3–61(151)
S22	110	PHYSICS INTEGRALS	7–62(389), 7–62(393)
S22	111	PHYSICS INTEGRALS	7–62(390)
S22	132	PHYSICS INTEGRALS	11–62(551)
S22	184	ERLANG PROBABILITY FUNCTION	7–63(386)
S22	191	HYPERGEOMETRIC FCN. (COMPLEX)	7–63(388), 4–64(244)
S22	192	CONFLUENT HYPERG. FCN. (COMPLEX)	7–63(388), 4–64(244)
S22	227	CHEBYSHEV POLYNOMIAL COEFF.	5–64(295)
S22	282	D/DX CF EXP(X)/X, COS(X)/X,	
S22	282	AND SIN(X)/X	4–69(272), 1–70(53)
S22	292	REGULAR COULOMB WAVE FCNS.	11–66(793), 5–69(278),
S22	292		5–69(280), 9–70(573)
S22	300	COULOMB WAVE FUNCTIONS	4–67(244), 5–69(279),
S22	300		12–69(692)
S22	327	DILOGARITHM	4–68(270)
S22	332	JACOBI POLYNOMIALS	6–68(436), 7–70(449)
S22	352	CHAR. VALS, SLNS OF MATHIEU*S DE.	7–69(399), 12–70(750)
S22	388	RADEMACHER FCN.	8–70(510)
S22	389	BINARY ORDERED WALSH FCNS.	8–70(511)
S22	390	SEQUENCY ORDERED WALSH FCNS.	8–70(511)
S22		CONFLUENT HYPERG. FCN (COMPLEX)	BT, 2–62(237)
S22	5	FERMI FUNCTION	BT, 3–63(141)
S22	13	RIEMANN ZETA FUNCTION	BT, 5–65(141)
S23	234	POISSON–CHARLIER POLYNOMIALS	7–64(420), 2–65(105)

ALL OTHERS

Z	45	INTEREST REFINEMENT	4–61(178), 9–63(520)
Z	112	POINT INSIDE POLYGON	8–62(434), 12–62(606)
Z	117	MAGIC SQUARE	8–62(435), 8–62(440),
Z	117		1–63(39), 3–63(105)
Z	118	MAGIC SQUARE	8–62(436), 8–62(440),
Z	118		12–62(606), 1–63(39), 3–63(105)
Z	136	ENLARGE A GROUP	11–62(555)
Z	148	MAGIC SQUARE	12–62(605), 4–63(168)
Z	199	CALENDAR CONVERSION	8–63(444), 11–64(661)
Z	240	COORDINATES ON AN ELLIPSOID	9–64(546)
Z	246	GRAYCODE	12–64(701), 6–65(382)
Z	252	VECTOR COUPLING COEFFICIENTS	4–65(217)
Z	260	6–J SYMBOLS	8–65(492)
Z	261	9–J SYMBOLS	8–65(492)
Z	355	GENERATE ISING CONFIGURATIONS	10–69(562)
Z	364	COLORING POLYGONAL REGIONS	12–69(685)
Z	391	UNITARY SYMMETRIC POLYNOMIALS	8–70(512)
Z	398	TABLELESS DATE CONVERSION	10–70(621)
Z		CALCULATION OF EASTER	4–62(209), 11–62(556)
Z		MANY–ELECTRON WAVE FUNCTIONS	4–66(278)
Z		GRADER PROGRAM	5–65(277)
Z		CALCULATION OF EASTER	CB, 9–65(18)
Z		SEASONAL ADJ–FORECASTING	CJ, 10–67(148), 11–68(25)
Z	50	PLAY LEGAL CHESS	CJ, 13–70(208)
Z		COMP. OF GALOIS GROUP ELEMENTS	MC, 23–69(425)

APPENDIX TABLES

Table I RANDOM DECIMAL DIGITS

	1	2	3	4	5	6	7	8	9	10
1	55693	02945	81723	43588	81350	76302	29361	88716	03196	84648
2	13940	47531	15339	18080	91414	14763	32771	60816	17027	18088
3	96613	78181	58246	30776	19676	69132	66252	88411	70567	76804
4	91972	57360	74189	54477	99037	72782	44173	97643	53216	96817
5	51557	75756	51723	34541	20391	67932	79315	48049	49960	75659
6	45761	26696	31236	10630	63127	74661	12320	29207	68326	12404
7	86335	13829	81805	60885	09315	59948	95651	28173	57617	63718
8	36076	42141	96852	67435	26279	99517	07930	61866	01108	14442
9	05338	61799	81240	43205	91375	11182	79736	51443	63711	85014
10	46890	40395	75917	91759	33656	47569	90537	64680	83667	46623

Table II NORMALLY DISTRIBUTED RANDOM NUMBERS ($m = 0, \sigma = 1$)

	1	2	3	4	5	6	7	8	9	10
1	−0.63	1.13	−1.35	0.28	0.93	0.79	−0.43	0.87	−0.04	−2.63
2	1.61	−0.49	0.48	−1.13	0.81	−0.87	1.46	−0.48	1.37	1.17
3	−1.38	−0.83	−0.21	0.34	0.48	1.88	−1.31	−0.96	−1.30	0.05
4	−0.25	0.04	−0.30	0.91	−1.09	−1.28	−1.24	−0.17	−0.42	−0.06
5	0.03	−0.62	−0.72	−1.02	−0.03	1.36	1.56	−1.33	−0.74	−0.55
6	−1.08	0.32	0.37	−0.99	0.53	0.28	1.39	0.41	0.06	−0.96
7	0.51	−1.41	−0.85	−1.19	0.19	−0.09	−0.87	−1.12	−0.16	−1.39
8	−0.41	−0.45	0.58	−0.27	1.25	0.56	0.00	1.08	−1.11	−0.59
9	0.88	−0.68	0.03	0.09	0.84	−0.30	−2.72	0.82	−1.25	0.35
10	−0.54	−0.40	1.21	−1.26	1.35	1.51	−0.18	−0.26	−0.53	0.80

One gets normally distributed random numbers with mean m and standard deviation σ by multiplying the values in the table by σ and then adding m.

Table III EXPONENTIALLY DISTRIBUTED RANDOM NUMBERS AND THEIR ANTITHETIC SERIES

MEAN = 100

(a) $X = \dfrac{|\log(1 - R)|}{0.01}$

211	20	104	37	170	84	77	168	73	86
3	18	51	156	171	19	336	164	229	165
53	32	189	38	70	130	73	217	278	229
159	51	185	107	118	43	29	55	167	150
24	3	129	43	72	123	20	106	273	4
35	39	62	44	0	47	215	159	169	79
54	53	16	35	8	28	92	19	5	43
39	159	21	99	76	73	6	117	99	7
44	217	35	73	25	0	149	74	484	140
13	19	83	89	163	39	341	67	65	277

(b) Antithetic series: $Y = \dfrac{|\log R|}{0.01}$

13	172	44	118	20	56	62	21	66	55
365	180	92	24	20	174	4	22	11	21
88	130	16	114	69	32	66	12	6	11
23	92	17	42	37	105	138	86	21	25
154	367	32	104	66	35	171	43	7	316
122	113	77	102	581	98	12	23	20	61
87	88	192	122	253	140	51	178	304	105
112	23	166	46	63	66	292	37	46	265
104	12	123	66	150	557	26	65	1	28
213	176	57	53	22	113	3	72	74	6

One obtains random numbers with frequency function $\lambda e^{-\lambda x}$ by dividing the values in the table by 100λ.

INDEX

References are made to *sections,* not page numbers.